Medicinal and Aromatic Plants

Traditional and Commercial Uses
Agrotechniques
Biodiversity Conservation

Medicinal and Aromatic Plants

Traditional and Commercial Uses
Agrotechniques
Biodiversity Conservation

R.K. Gupta

MSc, PhD (Poona), Dr és Sc (Toulouse-France), FNASc, FITE

Director
Center for Research on Ecology, Environmental
Applications, Training and Education (CREATE)
Dehradun-248 001, Uttarakhand, India

CBS

CBS Publishers & Distributors Pvt Ltd

New Delhi • Bangalore • Pune • Cochin • Chennai

Medicinal and Aromatic Plants

ISBN: 978-81-239-1814-3

First Edition: 2010

Published by Satish Kumar Jain and produced by Vinod K. Jain for
CBS Publishers & Distributors Pvt Ltd
4819/XI Prahlad Street, 24 Ansari Road, Daryaganj,
New Delhi 110 002, India. Website: www.cbspd.com
Ph: 23289259, 23266861, 23266867 Fax: 011-23243014 e-mail: delhi@cbspd.com

Branches

• Bangalore: Seema House 2975, 17th Cross, K.R. Road,
 Banasankari 2nd Stage, Bangalore 560 070, Karnataka
 Ph: 26771678/79 Fax: 080-26771680 e-mail: cbsbng@gmail.com

• Pune: Shaan Brahmha Complex, 631/632 Basement, Appa Balwant Chowk,
 Budhwar Peth, next to Ratan Talkies, Pune 411 002, Maharashtra
 Ph: 020-24464057/58 Fax: 020-24464059 e-mail: pune@cbspd.com

• Cochin: 36/14 Kalluvilakam, Lissie Hospital Road,
 Cochin 682 018, Kerala
 Ph: 0484-4059061-65 Fax: 0484-4059065 e-mail: cochin@cbspd.com

• Chennai: 20, West Park Road, Shenoy Nagar, Chennai 600 030, TN
 Ph: 044-26260666, 26202620 Fax: 044-45530020 email: chennai@cbspd.com

Printed at: Somya Printer, Delhi-110053

Disclaimer

Science and technology are constantly changing fields. New research and experience broaden the scope of information and knowledge. The author has tried his best in giving information available to him while preparing the material for this book. Although, all efforts have been made to ensure optimum accuracy of the material, yet it is quite possible some errors might have been left uncorrected. The publisher, printer and the author will not be held responsible for any inadvertent errors or inaccuracies.

to

my father

Vaid and Hakim Atma Ram Gupta (Late)

in reverence to have initiated me

on the subject

गरयस्ते पर्वता हिमवन्तोऽरण्यं ते पृथिवी स्योनमस्तु बभुं।
ष्णां रोहणीं विश्वरूपां ध्रुवां भूमिं पृथिवीमिन्द्रगुप्ताम्
अजीतोऽहतो अक्षतोऽध्यष्ठां पृथिवीमहम ११२२।।

अथर्ववेदीय पृथिवी सूक्त।

विशाल पहाड़ों और हरे-भरे जंगलों के विविध रंगों से सजी हमारी मातृभूमि महान है। यह
हमारी रक्षा करती है और हमें सुख देती है। बादलों से पोषित यह धरती हमें अनाज और
औषधियाँ उपलब्ध कराती है।

'सर्वे भवन्तु सुखीनः सर्वे सन्तु निरामयाः'

Preface

INDIA in general, and the hills in particular, always held a high place as the supplier of vegetable drugs and aromatic plants used for a variety of pharmaceutical purposes. It is because of the fact that climatic conditions varying from the torrid to the frigid zones exist in India, embraces vast tracts of tropical, subtropical plateaux, temperate hills and the valleys, dry alpine areas and irrigated soils, etc. The country has in fact been described as an epitome of climates, seasons and soils of the world. No wonder then that this country is a veritable emporium of medicinal and aromatic plants growing up to 75% of the vegetable drugs mentioned in the pharmacopoeias of different countries. In addition to the medicinal herbs, aromatic and essential oil-bearing plants constitute an important group of Indian flora and a large number of them are used in the indigenous systems of medicine, including some traditional drugs used by the rural, semi-rural population and the tribals. Out of the 17,500 species of the flowering plants, scientists know about 3000 species, but the tribals have known about 10,000 of them. Of these, they have recorded 8000 having medicinal use, 3500 edible use, 500 fibre use, 325 pesticidal use, and about 475 used as gum, resins and dyes.

Many crude vegetable drugs are still used in our country as such, or after limited purification in the indigenous systems of medicine. Even with the progress of synthetic organic chemistry, when the modern drugs and pharmaceutical industry is depending more and more upon chemical raw materials, vegetable drugs have by no means lost their importance and the industry has all been evincing keen interest in them. Inaugurating the 93rd Indian Science Congress on 2nd Jan 2006, the Prime Minister of India added 'utilization of herbal and other plants and the application of some to animal husbandry', to the five components of the second Green Revolution in the country.

In India, till recently, wild growing plant materials having therapeutic properties has been chiefly collected and marketed. As in the case of many other forest produce, the material is often collected by aborigines, forest dwellers, nomadic communities and villagers, and is thereafter sold to itinerant merchants. Even the auctions of minor forest produce of medicinal value are attended by middlemen who purchase the forest products and transship them to the sellers in big cities. It, thus, passes through a number of middlemen before it is collected in a sizeable quantity by the wholesale traders. As the material passes through a number of hands, it not only gets contaminated with extraneous matter and other organic materials, but is also subjected to willful mixing.

The result obviously is that sometimes substandard material reaches the consumer. The drug plant industry has thus earned a bad name, both in the country and outside, as and when the material is exported. It is, therefore, necessary that a process for standardization of the drug material is adopted with some mechanism involving the Indian Standards Organisation (ISO), as it is done for any other product.

The systematic cultivation of many vegetable drugs when undertaken on the plantation scale or on an industrial basis in other countries has proved very successful. Recently an effort has been made through the efforts of CIMAP and RRLs of the CSIR and the coordinated project of the ICAR, for the scientific cultivation of drug and aromatic plants. This is, indeed, a very healthy development which would not only improve the quality of such materials, marketed both within the country and abroad. It would also involve small and marginal farmers to accept cultivation of drug plants on a commercial scale to improve and supplement their returns from agriculture. However, the technology for the cultivation of medicinal and aromatic plants is different than for the simple cultivation of cereals and vegetables. Keeping this in view, a new section in this book has been added, providing *agrotechniques* for the profitable cultivation of these plants, which can be adopted by the farmers and others interested in improving their yields from such farms.

The forest department in some states, like Himachal, has also allowed cultivation of some drug plants in the forest areas which is a welcome change in the forest policy where people are involved in the forest floor management.

The quality of the drug plants refers to the intrinsic value of the drug, that is the amount of medicinal principle or active principles in the form of starches, mucilages, acids, bases, gums, fixed oils, volatile oils, resins, tanins, alkaloids, glycosides, hormones, vitamins, etc. It is well known that the quality and potency of the drug plants depend upon the conditions of its growth and habitat. Thus, varied soil and climatic conditions play a profound role on the yield as well as the constitution of the active principles. There is sufficient proof to indicate that there was a regular study and culture of botanical sciences in ancient India. The branch of science termed as *Vrikshayurveda*, dealt mainly with the cultivation of plants, prevention of plant diseases, etc. There is also an indication of using *Koopajala*, rich in chemical properties during the nursing of plant to get satisfactory medicinal properties. Such a science needs to be investigated in depth to improve the quality of drug plants. The current practices of applying inorganic manure to increase the total production without taking into account the quality of the drug produced needs to be replaced as per the standard fixed by the ISO. The potential for the export of drug plants can be achieved, which, at present, is one-tenth of the value earned by our neighbour China.

Today, the trade of medicinal plants is more or less in the hands of people not scientifically trained. They are not accustomed with the botanical names of various drug plants and as such local names are used, which are sometimes misleading. Different plants are known by more than one local name. The famous plant known as Brahmi is an example. Another common example is Safed Musli and so on. This confusion needs to be cleared once for all by using botanical names which are more valid and standardized and follow the International Code of Botanical Nomenclature all over the world.

While the estimation of refraction and extraneous matter and even that of chemical factors like ash, acid-insoluble ash are not likely to present much difficulty as standard procedures are available, the work of botanical identification does not appear to be so simple. The tests described in the pharmacopoeias aid in identification, but are not in all cases sufficient to establish proof of identity. Thus, there is great need to have easily available methods of botanical identification of crude drugs mentioned in the pharmacopoeias. Although, the problem of botanical identification has been all along engaging the attention, the progress achieved is far from satisfactory. Quite often well known botanists differ among themselves about the identity of the material submitted for their identification. Under the circumstances, it might be desirable to take the sample for botanical identification approved by the foreign buyers as the standard sample for botanical identification of the material, so that after the export of the material from the country, chances of complaints are reduced, if not altogether eliminated.

The standardization and control of crude drugs should also look to the aspects like gross morphology, micropial morphology, quantitative microscopy, polarizing microscopy, fluorescence testing, crude fiber determination, foreign organic matter, physical contents, quantitative chemical tests, paper chromatography and thin layer chromatography, etc. The adulteration control and purity of the drugs should be maintained at all costs. The drug dealers, who retail the same drug being very conversant, often mix them together, either intentionally or inadvertently. The nature of these types of adulteration may be due to one or the other of the following reasons: Sophistication, admixture, substitution, deterioration, spoilage and inferiority. Hence, standardization of crude drugs as per the ISO standards is an important aspect that needs priority and has been discussed in the book.

Newly created states like Uttarakhand, Jharkhand and Chhattisgarh have good potential for crude drugs export. In Uttaranchal (now Uttarakhand) business worth Rs 1.2 crores has been transacted (till mid-2004) against a target of Rs 4.9 crore. International export houses and herbal development organisations have been involved and entrusted with research and promotion of agrotechniques for the cultivation of such plants. It is estimated that 2% of landmass has 25% of biodiversity and 7000 species of herbs are available in different ecofloristic regions of India. Also 159 pharmaceutical companies are utilizing about 500 species of herbs in their formulations and the production has recently increased from Rs 100 crore in 1991 to Rs 400 crore. The resources have been declining rapidly as 95% of the plants used by industry in the country are collected from the wild, recklessly. The illegal export of the rare herbs still continues due to lack of surveillance by the responsible authorities though export of certain species has been banned. Even one-third of the plants included in the *Red Data Book of India* have medicinal value and adequate efforts have not been made for systematic collection of the germplasm of the plants.

In 2005 India's share in global trade was only worth ninety billion US dollar which is restricted to 0.9% of the total global exports worth $ 10,121 billion. This could be raised only if proper efforts are made keeping in view the agrotechniques suitable for plant species. In comparison with China the figure ranges from Rs 18,000 crore to Rs 22,000 crore.

There is still a wide scope for popularising the indigenous systems of medicine in both the developing and the developed countries, as not only the specialists but common man is concerned about the toxicity and adverse impact of inorganic compounds on the human system. Drugs derived from natural sources, like plants, are still safer and more effective in the long perspective.

It is hoped that this volume, apart from providing existing information on the subject, would inspire not only the scientists but also the drug farmers, export agencies and conservation managers to further explore the subject and work upon in the new millennium for people's benefit, good health and prosperity.

R.K. Gupta
MSc, PhD, Dr és Sc, FNASc, FITE

Director
Center for Research on Ecology,
Environmental Applications, Training
and Education (CREATE)
Dehradun-248 001
Uttarakhand, India

Contents

4. Identification and Nomenclature of Drug Plants 69

5. Traditional Medicinal Plants

A. Traditional Medicines and Medicinal Plants in Rural Folklore and Tribal Society 77

B. Medicinal Plants collected from Forest as a Product and as Weeds from Wild/Wastelands: Trees and Shrubs 93

C. Herbaceous Plants used as Drugs from Wastelands, Weeds of Croplands and Forest Understorey
187

 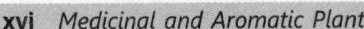

PART II

343

6. Potential Drug Plants, Quality Control Assessment, Commercial Exploitation and Biodiversity Conservation

345

7. Environmental Considerations: Farming of Aromatic and Medicinal Plants 387

8. Zonal Concept and Ecological Evaluation of the Habitat of Drug Plants 411

9. Cultivation and Multiplication of Drug and Aromatic Oil Plants 435

10. Agrotechniques for the Cultivation of Drug Plants 461

PART I

Introduction and History of Plant Explorations

History of medicine and exploration of medicinal plants can be traced to 4500 BC at the time of *Rig Veda* where sporadic mention of medicinal plants is made. Later in *Atharvaveda,* mention has been made of the plants used in charms and amulets for the cure of diseases. *Ayurveda*, written probably between 2500–900 BC, established real foundation of ancient medical science. Of the eight divisions of Ayurveda, *Sushruta Samhita* and *Charaka Samhita,* written about 1000 BC, are exclusively devoted to medical science; the former deals with surgery and the latter with medicines. The seventh chapter deals exclusively with *materia medica* of ancient Hindus. The method of administration describes even the injections, which were given in an orthodox way when immediate results were aimed. Anesthesia was not even overlooked in 'Sushruta' surgery before the operation was performed. In a treatise *Bhoja Prabandha* written about 980 AD mention is made of inhalation of medicament before surgical operation and an anesthetic called *Sammohini* is said to have been used in the times of Buddha.

In the Mesopotamian civilization the Sumerians (3000–1970 BC) and then Babylonians and Assyrians (1970–539 BC), plants were used in medicine and amulets. From Mesopotamia, especially from Babylonia, physicians of repute were called on to Egypt for treating members of the royal and noble families; Hamurabi who ruled Babylon (1728–1686 BC), issued 285 mandatory declarations, one of them referred to medicine. It declared that a man practising medicine must be a bonafide physician and unauthorized man is forbidden to practise medicine. Egyptians were equally advanced. Greeks carried the science of medicine from ancient Babylonia and Egyptians. Hippocrates (460–372 BC) regarded as a great medical man, was called by Hakims as *Abu-Altab*. He laid the foundation of medicine in Greece wherefrom it spread to Europe.

The study of drugs is perhaps also as old as the disease, though modern pharmacology is considered as one of the youngest of medical sciences. World's oldest therapeutic writings are perhaps from India and China. The great Herbal or Chinese Materia Medica *PanPsao* was probably written in 2735 BC and contained many vegetable and metallic preparations, including a few animal products such as todd's eyelids, elephants and tiger's bones, horns and fins, etc. Though the Greeks and later on the Arab's knowledge of Indian plants and medicines was brought to West, the teachings of Dioscorides and Avicennia owe a good deal to Indian sources. Although there are medical descriptions in *Rig Veda* (2500–3000 BC) it was Charaka, a renowned Indian physician and later Sushruta and Vagbhatta, who described various medical preparations, presently

included in Ayurveda. Initially these consisted of non-poisonous vegetable drugs. Thus, Charaka, described about 300 vegetable drugs and classified them according to their effects, mostly on symptoms, into five groups. The original *Ayurvedic Materia Medica* was later superseded to some extent by the alchemic or chemical substances at about the beginning of the Christian era. The earliest sources of western materia medica came from Egypt and the two kingdoms of Assyria and Babylonia. The *Papyri* were the first written account of medical experiences from Egypt and date back to 1900 BC. The papyrus discovered by Eber in 1872 was prepared in 1500 BC, and mentions about herbal remedies, including opium. A Babylonian clay tablet (700 BC) has been discovered which mentions about 300 drugs. Modern medicine is considered to date from Hipporates, a Greek physician who for the first time introduced the concept of disease as a pathologic process and tried to organise the science of medicine based on observation, analysis and deduction. Hippocratic practice, did not include extensive use of drugs, probably because he did not believe in shotgun or magical remedies, but instead recommended judicious use of simple and efficacious drugs.

Till the beginning of the 19th century, treatment of diseases was of such obnoxious remedies as flesh, excreta and blood of various animals along with a few metallic and plant preparations. James Gregory (1753–1821) was responsible for popularising heroic symptomatic treatment consisting of bloodletting, large doses of emetic and drastic purgative, often with disastrous results. Such treatment without any rational basis was called *allopathic* meaning other suffering, a term which is still wrongly used to denote the system of modern scientific medicine, as opposed to homeopathy.

The concept of homeopathy was first introduced in the early 19th century by Hanneman, who thought that *Like Cures Like* and that dilution potentiates the action of drugs. Homeopathy has outlived many remedies for various ailments, with drugs in very high dilutions.

Modern pharmacology, as a science, in fairly recent era started taking shape following introduction of experimental procedures in animals by Fransois Magendie (1783–1855) and Claude Bernard (1813–1878). Till then, treatment of diseases was purely empirical, based on a combination of guesswork and experience. Spectacular developments in physiology, biochemistry and organic chemistry, during recent years, greatly accelerated the advancement in pharmacology. For the action of drug, its effect is extremely important. Thus, pharmacology is the branch of biology to provide scientific data on both animals and humans. The word is derived from the Greek *Pharma Icon* (drug) and *logon* (a discourse). The word drug is derived from French word *Drogue*, a dry herb. According to WHO, "A drug is any substance or product that is used or intended to be used to modify or explore physiological systems or pathogen status for the benefit of the recepient." Materia medica is the older term for a branch of pharmacology concerned with sources, description and preparation of drugs. *Pharmacopoeia* is an official code containing a selective list of established drugs and medical preparations, with description of their physical properties and tests for the identity, purity and potency. Some well-known pharmacopoeias are the Indian Pharmacopoeia (IP), The British Pharmacopoeia (BP), The United States Pharmacopoeia (USP) and the European Pharmacopoeia (EP).

Various drug sources are mineral, animal, vegetable, synthetic and microorganism.

Pharmacologically active principles in various drugs are alkaloids, glycosides, oils-fixed and volatile, resins, oleoresins, gums, tannins and antibacterial substances.

Alkaloids are the basic substances containing cyclic nitrogen, which are insoluble in water, but combine with acids to form well-defined water soluble salts, e.g. morphine, atropine and emetine. *Glycosides* are either like combinations of sugars with other organic structures. A glycoside does not form salts with acids but when heated with mineral acids it is hydrolysed to sugar and a non-sugar component called aglycone or genin, e.g. digoxigenin. A glycoside which yields glucose on acid hydrolysis is called aglycoside, e.g. strophanthin. *Oils-fixed oils* are glycerides of oleic, palmatic and stearic acids. These are fats and may have food value. Many fixed oils are edible and arc employed for cooking and as solvent as peanut oil, coconut oil, and olive oil. Castor oil has certain pharmacological actions and it acts as a purgative. *Volatile oils* are volatilized by heat and possess aromas. They are also called essential or flavouring oils as aromas of plants and flowers reside in the volatile oil present in the plant. Chemically they are not fats and are without any calorific value. They contain the hydrocarbon terpene or some polymer of it, which serves as diluents or a solvent for a more active compound, e.g. menthol in peppermint oil. Volatile oils are used as carminatives for expulsion of gas from stomach, e.g. oil of eucalyptus, asfoeteda, ginger, antiseptics in mouthwash pastes. Counter irritants are, e.g. oil of winter green teurpentine oil. Flavouring agents, e.g. oil of peppermint and act as pain relieving agents, e.g. clove in toothache.

Resins are found in plants. They are formed by oxidation or polymerization of volatile oil and are insoluble in water but soluble in alcohol. Resins like *jalap* and *colocynth,* used formerly as purgatives, are now obsolete because they produce severe purgation. *Oleoresins* are mixtures of volatile oils and resins. Male fern extract used for tapeworm infestation contains an oleoresin.

Gums are secretory products of plants. These are dispersable in water and form thick mucilaginous colloids. Gums are pharmacologically inert and are mainly employed as suspending and emulsifying agents, e.g. gum acacia, gum tragacanth. *Agar* is used as bulk purgative.

Tannins are non-heterozygous plant constituents characterized by their astringent action upon mucous membrane. They precipitate proteins from cells of mucous membrane and thus have protective action. Substances, which release tannic acid in the small intestine such as tincture of catechu, were formerly employed in the treatment of diarrhoea.

Antibacterial substances are derived from moulds and fungi, e.g. penicillin and streptomycin.

Hindu period of herbal studies has been classified into four categories, viz. 1. vedic 2. original research. 3. period of compilation and 'Tantras' Sidhas and 4. period of decay and recompilation. During the second and the third period the progress was remarkable in Ayurveda. Towards the close of the period, ayurvedic medicines spread all over east and far beyond. In west, Greece and Rome took substantial material from Hindu and Arabic physicians to enrich medicine.

A large number of herbs were used by rural folklore. In villages where doctors and physicians are not present, these were used specially to cure children and old men. Many of these have now been tested. Eighty per-

cent of the pharmacopoeal drug records in BP, BPC, USD and IPC are successfully examined and no doubt many have been found useless also; snakebite being the main. Mhaskar and Cains, for a number of years working at Hafkin laboratory, refuted them as useless. Among some of the successful is *Rauvolfia serpentina*. Watt, records as the one which villagers think is chewed by a mangoose in the bush when bitten inadvertently by poisonous snake during fight.

Indian vegetation drug study started in the early part of the 19th century. Sir William Jones Memoir on *Botanical Observations on Select Plants* may be considered as the starting point. John Fleming's *Catalogue of Indian Medicinal Plants and Drugs* (1910) was followed by Ainslie's *Materia Medica of Hindustan* (1813), Roxburgh's *Flora Indica* (1924), Wallich, Royle and McNamara, later tried to resolve vast mass of unclassified botanical material on scientific lines. Later this was followed in 1841 by 'Shaughnessys' *The Bengal Dispensatory and Pharmacopoeia*. dealing with the properties and uses of medicinal herbs. In 1868 *Pharmacopoeia of India* was published under the able hands of Warming. The more important drugs are eventually incorporated in *British Pharmacopoeias*. Mohideen Sherif published a *Supplement to the Pharmacopoeia of India* in 1891. *Materia Medica of Madras* by Hopper is based on his work. Translation of Sanskrit "Materia Medica" by UC Dutta in 1877 brought to light the drugs used by Indian Physicians. Fluckiger and Handbury's *Pharmacographia India* published in 1879 is another good contribution to indigenous drugs. Dymock's *Vegetable Matria Indica* of West India (1883), Dymock, Warden and Hooper's *'Pharmacographia Indica'* (1890–1893) are the consolidated works in the field.

Most elaborate and significant work by Sir George Watt's *Economic Products of India* is a monumental contribution. The publication is in six volumes, covering six thousand pages is a mine of information (1889–1896). Of more recent publications are Day's *Indigenous drugs of India* (1896) and Kirtikar and Basu's *Indian Medicinal Plants* in 4 volumes (1923). Chopra's *Indigenous Drugs of India* published in 1958, must be considered as a significant landmark of the last century. Last but not the least *Wealth of India* by CSIR (1948–74) is the latest in the field. *Indian Materia Medica* by Nadkarni (1954) gives information from *Ayurvedic Materia Medica* and other references.

In India before1753, the date from Linnés publication of *Species Plantarum* and *Genera Plantarum* (6th ed.1754), Gracia de-Orta's as Coloquios, published in Goa in 1575 have a detailed account of the more striking Indian medicinal plants from firsthand knowledge, for it drew attention of the learned men of the west to the east's vast potentialities of the South Asiatic flora. This book became known in Europe mainly through the Latin summary of Clusius (Charles I Escluse) published in 1567, which in the course of half a century, went through several editions in the Latin European language or in the original Latin. Other important contributions of the pre-Linnaean era, which made possible the correct interpretation of the description and diagrams of the drug plants may be mentioned Hein Reich Van Rheede. *Hortus Malabaricus*, Linne's *Flora Zeylanica* (1746), J. Burmann's *Thesaurus Zeylanicus* (1737) and P. Hermann's, *Paradisus Bata-Cius* (1687), etc. The beginning of post-Linnaean botany in India may be said to date from the arrival of J.G. Koenig, a pupil of Linnaeus, in 1768 when he joined the Tranquebar Mission as a surgeon and naturalist and in the union of the United

Brethern such as Klein, Rottler, etc. made splendid collections in the neighbourhood of Madras. In Bengal the earlier works are associated with John Fleming, Sir William Jones, Roxburgh, Wallich, Buchanan-Hamilton, etc. In 1787, Robert Kyde founded the Botanical Garden of Calcutta and from the very beginning, even up to this date, it has occupied a prominent position in the development of studies on drug plants. In the region of Bombay (now Mumbai), the initial progress is associated with the names of Alexander Gibson, John S. law, J.E. Stocks and John Graham, while in the north and northwestern part the name of J.F. Royle stands out prominently. Other early explorers of medicinal and drug plants are W. S. Webb, William Moorcroft, Alexander and James S. Gerard, Henry Starchy, Richard Strachy, etc. and on these as well as on his own, Royle based his splendid *Illustrations.* Plantations of Cinchona were regularly organized with the appointment of Malmaduke Alexander Lawson, (1840–1896) in 1882. Dr. David Hopper's appointment as chemist in 1884 under Lawson, whose work got publicity through Dymock, Warden and Harpe's *Pharma-cographica Indica,* dealing with Indian drugs, while John Schroff wrote much. Edward John Waring's *Bazaar medicines* in 1860 is a book not to be overlooked.

With the publication of *Flora of British India* by Sir J.D. Hooker, there was a systematic progress in the identification of the drug and aromatic plants in India and from 1900 onwards, various regional floras and books on medicinal plants were published, which are referred to in the list of references.

Much useful knowledge relating to the habits of primitive man has been derived by the examination of ancient refuse dumps, which have been unearthed near the sites of human habitation of pre-historic man. What plants he collected from the forest to supplement his diet of fish. What plants he used as arrow poison to kill or stupefy his catch. Rejected seeds or tubers of edible plants thrown around his dwellings which germinated and produced their kind, must have given paleolithic man, the first idea of cultivating these plants, as plants that come up in soil manure by human refuse are found to be better than those cultivated from the wild. Many of the world's food crops like barley, rice, oats, maize and lentils and fruits like cucumber, melon and gourd, root crops like carrot, potato and yams have their origin as dunghill plants. The same may also be said to be the weed, *tobacco.*

Besides growing or collecting plants for food, primitive man was also interested in plants which he found useful for healing wounds and in rites and ceremonies connected with birth, death and marriage, all of which were associated in his mind with magic. The use of narcotics in connection with religion was common both in the new-world and the old-world and the reverence paid to such plants like *datura, neem,* the *bael* tree and the Tulsi plant in India may be relics of early man's appreciation of their medicinal properties. This aspect may be associated with an even earlier domestic worship of these plants whose medicinal or economic value was recognised by the elders of the tribe and kept as a secret from the common man. As a token, a plant or an animal cannot be killed or exterminated. Thus, we are very much indebted to the superstition of early man for the survival of much useful plants as the Turmeric, which has been cultivated from very early times in India. The detailed knowledge of the medicinal properties of our plants comes to us from the writings of scientifically minded sages who found the settlements in the vast forests of our country and studied the properties of surrounding plants around

their hermitage. Thus was found the indigenous system of medicine, the *Ayurveda,* which still remains a vast encyclopedia of the medicinal plants and for which we can still explore the value of our flora for the medical analysis. The case of *Rauvolfia,* which was in use for insanity is one such example. Texts like Bhaishejya, Ratnawali and Rastan TraSar are among the 84 books recognised as classical text under the Drugs and Cosmetic Act, 1940.

We know that until very recent times medicine relied almost entirely on plants, so that in common parlance the names Medical Doctor has nearly the same connotation as the name Botanist and *vice versa.* Even as late as the end of 19th century the key figures in the field in India's botany were medical men, often associated with the Army Medical Services. Through the Greeks and later on the Arabs, knowledge of Indian plants and medicine was brought to west, the teaching of Dioscorides and Avicennia, a good deal to Indian sources. Gradually, the West developed its own systems of medicine on the foundation received from the East, and thus came finally through the herbalists to the present very high standard of science. But western scientists were very alert to discover new sources of drugs for the alleviation of human misery, have once more turned the eyes to India, are now devoting much energy and inguinity to the study of the traditional Indian drugs. Numerous pharmaceutical firms both Indian and foreign, are making brave efforts to supply the external and internal markets with the valuable products; there are, however, many difficulties in the exploitation.

Medicinal plants are mercilessly removed from wherever they are found and often it is a calamity to point out the area where a valuable plant may be found. Pharmaceutical firms are not to be blamed for this state of affairs; botanists themselves are often the worst offenders in this line.

In 1952 Dr. H. Santapau found 100 clumps of *Rauwolfia serpentina* at the edge of a forest or in forest clearing and after striking results obtained in preparation of Resperine were published, there was a full scale clearing of this species when finally a ban on the export of raw material was imposed; there were practically no material. The position for this particular species has now improved due to large-scale cultivation.

Tylophora asthmatica (*T. indica*), some years ago, was a common climber over fences and hedges in Dharwar, and other places. The plant has high repute in indigenous system for control of asthma and allied troubles. Now, it is not that common and is available with difficulty.

Digitalis purpurea was similarly introduced in Kodaikanal from Europe as a garden plant in the second half of the last century. The plant succeeded well in the gardens and in times established as escape. A very attractive member of the local flora; both purple and white varieties, were until recently abundant all over the Kodaikanal hill. Now, it is very rare. A pharmaceutical firm has cleared the area with the hope that digitalis would come up again in future. From that time all our needs of digitalis are supplied from *D. lanata* from Kashmir.

For centuries India could carry on with the wild plants available in the country, without having to go into extensive cultivation of medicinal plants. Thus, the situation, however, has totally changed with the introduction of modern commercial methods in the exploitation of such plants. A modern big pharmaceutical laboratory will need hundreds or thousands of tons of any medicinal plant, before commercial experiments may be attempted. A single

laboratory will need more material of a single species for a day's work that have been needed in over a hundred years.

From all these points one is led to conclude that to survive we must cultivate such important plants in a large scale. We must do before very valuable plants completely disappear from the face of India and before foreign competitors have entirely captured the international market. In many cases, it is feared that it may be too late to obtain both aims, we must make efforts to save our best plants so that we may supply at least the internal needs of the country but better to stick to export such products to the world markets.

❑❑❑

Therapies used in Different Curing Systems

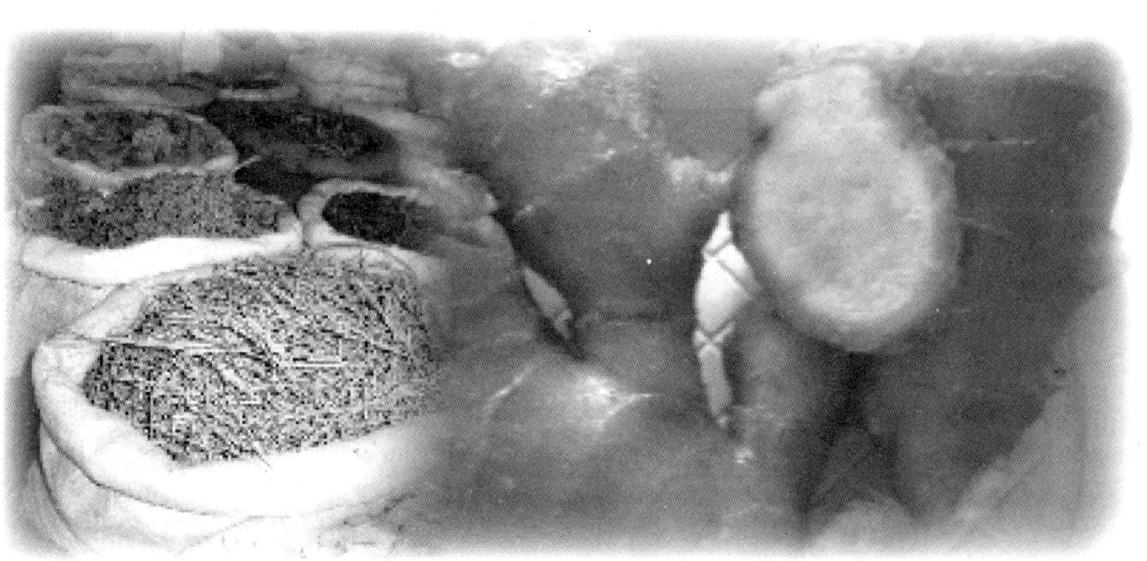

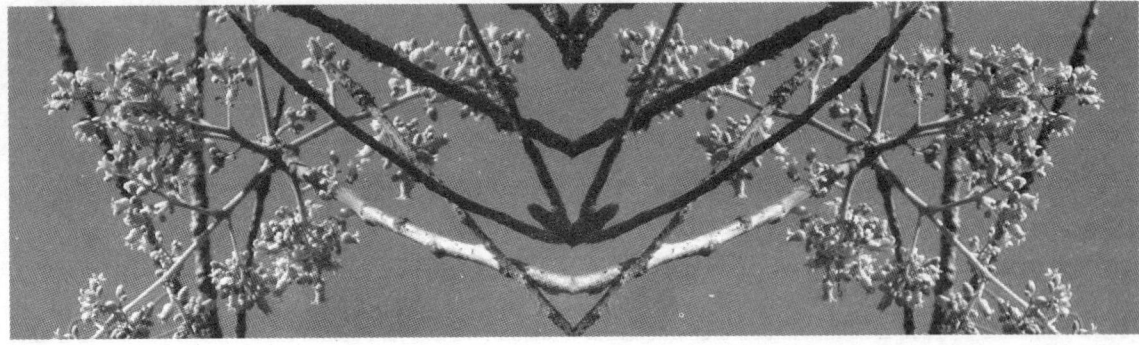

Traditional System of Treatment

Locally, drugs are utilized in the local Indian system of medicine (Ayurvedic and Yunani) in different forms. Ayurveda, the science of life is defined by *Charaka* as the science through the help of which one can obtain knowledge about useful and harmful types of life, happy and miserable types of life, things which are useful and harmful for such types of life, as well as the very nature of life. Ayurveda does not exclusively deal with treatment of human beings but also with diseases of animals and even plants. It is considered that total ingredients in the formula rather than the action of individual drugs that play role in the treatment.

In Ayurveda there are more than 8000 medicines—single drugs and compound preparations. Though minerals, including metals and animal products are used in these medicines but all of them are composed of drugs of vegetable origin. These can be broadly classified as *Kwath* (such as *Rasnadikwath*), *Avaleh* Chayanprash, *C. churna* (powder), *Bhashma* (oxides), *Arishth* or Asaw (fermented liquid) and *Vati* (pills). In the Unani system of medicine, the uses of the drugs may be broadly classified as *Sharbat* (syrup), *Joshanda* (decoction), *Deshanda arq* (distilled extract), *Safoof* (powder) and *Man-joon* (*chutneys*). Besides these, extract of many plants are being used in biochemic and homeopathic systems of medicine. Though synthetic drugs have assumed a great importance in the allopathic therapy, the use of extracts or tinctures from medicinal plants has not been ruled out and still a large number of them are being used.

Disease is believed to be caused by an accumulation of *Amm* (the contaminated mix of properly digested food and body fluids). The texts suggest two main therapies: *Shamna* through medicines and *Shodhna* (cleansing and rejuvenation) through *Panchkarma*. The *panchkarma* are *Vastie* (anema), *Vamna* (vomiting), *nasya* (the application of herbal preparations through the nostrils), *Virechna* (sweating) and *Rakta Mokshya* (the therapeutic release of toxic blood). *Pathya* (diet and other regimen) is also given due importance in Ayurveda. The *Shodhna* therapy has particularly developed in the Kerala state. Traditional texts provide clear instructions for day-to-day living in the *dincharya* and also suggest ways in which one should adapt to the various seasons in the *ritucharya*. If we adhere to the rules of healthy living, *Doshas* will remain balanced for a healthy and long life. The daily routines are advocated in Ayurveda classics and can be summarized as follows:

Early Rising

Individuals should get up before sunrise *(Brhammurta)* and follow the regime of (1) *cleanliness*, like brushing teeth scrapping the tongue, regular body massage, oil application on head, bathing, trimming of hair, nails, ablution of feet and excretory orifices, wearing clean clothes, etc. (2) *exercise*, depending on the age and *Prakriti (Nature)*. *Kaphaya* person should perform heavy exercises, *Paitika* dominant persons should do it in moderation avoiding mid-day and hot seasons, while *Vata* people should have regular exercise in moderation, preferably *Yoga* and not aerobic exercises.

It is also important as a part of daily regime, not to support natural physical urges such as urination, sneezing, defecation, hunger, yawning, vomiting, flatus, etc. since suppression of urges leads to various complications or diseases.

An equally important aspect of daily regimen concerns mental health for which a regimen of *sadavretta* (simple life) is prescribed. Strict mental discipline and adherence to moral values are considered vital to mental health. Key proponents of Ayurveda were long been aware, that abnormal codes of conduct produce stress and that errors of judgment are at the root of all stress. An improved code of conduct can free the body and mind from physical and mental disorders, preventing stress.

An Ayurvedic Lifestyle:
Diet and Conduct in the Rainy Season

The power of digestion during this period in the body, *Vata,* is also aggravated. It is advisable to be moderate in the diet. Astringent, bitter and pungent foods should not be eaten. Water should be boiled, cooled and mixed with honey, for maintaining normal digestion power. Wheat and rice along with vegetable soup are good. One should abstain from daytime sleep, moving in the sun, excessive physical exercise, etc. It is also advisable to massage the body by applying oil and take regular bath.

Diet and Conduct in the Fall

Food that is astringent or sweet in taste, preparation of milk with sugar or honey, rice, barley and wheat should be taken. In this season sweet, light, cold and bitter foods and drinks which alleviate *Pitta,* are to be taken in proper quantities *Ghee* prepared with bitter medicines, purgation and blood letting are also recommended in the fall, to remedy the aggravated *Pitta* of the previous season. Avoid excessive sun, bathing, fat, oil, yogurt and meat of animals living near water. One should sleep during the day and should not expose to frost or easterly winds. Exposure to autumnal flowers, sandalwood paste and rays of moon are all beneficial.

Diet and Conduct in the Winter

In this season, the *Kapha* that has already accumulated, is liquefied by the heat of the sun and as such disturbs the digestive capacity. One should avoid eating heavy, oily, sour and sweet food that may aggravate *Kapha.* One should take food of barley, wheat, rice, and soup made of bitter vegetables. At the beginning of spring, one should increase the amount of exercise, have more massages with oil and eat light foods. One should not sleep during day time. Elimination therapies like *Panchkarma* (emetics, purgation, decoction-oily enema and nasal at times) could be administered. It is wise to take measures that do not aggravate *Kapha.*

Generally, a person considers that he has some disease, when the symptoms appear. But it is well known that a lot of diseases remain as symptomatic for a long period of time. Hypertension is known to be a silent

killer. It may remain a symptomatic for a decade or more. According to the modern system of medicine, when the symptoms appear, the disease is usually at an advanced stage. The only way of detecting it at an early stage is annual check up but also usually it may not get detected in its formative stages.

Ayurveda, the ancient system of medicine, some 5000 years ago describes why some people are more prone to get a disease and how to prevent the same. Generally Ayurveda is discarded by the hardcore west-oriented modern medicine scientist without going into the depth of its wisdom and its logical scientific basis. According to Dr. Deepak Chopra of USA, the difference explained in his book *Perfect Health* is that the former talks about the disease which the patient is suffering from, on the contrary, an Ayurvedic physician asks who is my patient? Or what is the constitution of my patient's body? And therein lies the basis of Ayurveda.

The purpose of Ayurveda, according to Deepak, is to tell us how our lives can be influenced, shaped, extended and ultimately controlled without interference from sickness or old age. Ayurveda embraces the collected wisdom of sages, who began their tradition thousands of years ago. The exact origin of Ayurveda has been placed at around 6,000 BC; Charaka and Sushruta are its later exponents. The original works were revised, edited and supplemented by them and form the backbone of Ayurved Samhita (Compendia). The word philosophy refers to love of truth and the Ayurved truth is 'Being Pure Existence', the source of all life. Ayurved is a science of truth and it is expressed in life. The basis is the concept of close relationship between man and the universe and that is a cosmic energy that manifests in all living beings and non-living things.

Is Ayurveda a Mumbo-Jumbo?

Normally, a patient fight shy of Ayurveda vaids for dealing with health problems and is lured by the allopathic system. One reason could be that he wants quick relief. The allopathic doctor gives him a tablet or a prick that satisfies his mind so that the pain impulse does not reach it, not that the cause of the pain has been removed. A few days later, the pain revives and he seeks the same allopath's help, who repeats but does not warn the patient of the reaction the drug might cause to his nervous system, liver, kidney or any sensitive organs. The medical ettiquette, as it is called, puts a thick shroud over the mistake committed by another allopath. There is a misconception about the concept of *Tridosha* and questions, how a simplistic concept could deal with the disease? No amount of explaining works what he imagines but are the fundamental balancing mechanisms of the body, to keep it in good working condition, that in his system they are disturbed and they should be put in equilibrium.

Charaka explained Ayu as *sarirendriya-satvathmasam* yoga, that is a unified state-*sarira* (the physical body, *indriyas* the cog native organs), *manas* (the mind) and *atma* (the soul or the astral body), all of which combine to form the human system. This emphasises the fact that *sarira* has a distinctive place of its own, apart from the cognative organs, since any defects or abnormalities in the *sarira* could be easily noticed, whereas functional defects of the cognative organs are not easily detected except at an advanced stage of illness. The separate mention of the cognative organs also indicates their importance in dealing with the human system as a whole, an idea now being increasingly accepted by the western medical world.

Fig. 2.1: Sushruta, surgeon of old India

Surshruta (Fig. 2.1) has defined the healthy state (*swastha*) of an individual as: A person whose *doshas* (somatic and psychic) are in a state of equilibrium, i.e. the digestive system in good condition with normal functioning of the *dhatus* (tissues at best), an inadequate translation of all pervasive Sanskrit word and the *malas,* waste product, accompanied by the process of the *atma, indriyas* and *manas,* is said to be a healthy man.

The equilibrated state of the *doshas* in this definition of a healthy body indicates their importance in the maintenance of health: as also that any imbalance of the equilibrium vitiates any or all of the *doshas*, and this state is called disease. Since the mind plays a vital role in the maintenance of health, health should be visualised as psychological maintenance of the functions of a healthy living being and disease as a disturbance of this state. This concept is gaining increasing acceptance.

It is amazing that Ayurveda some millenniums ago propounded the theory that the cell was the fundamental unit of biological activity, apart from being the building block of the body. Despite the acceptance at the same time of the idea that cell divides and divided sub-cells retain many of the properties of the mother cell. It was understood that the cell was the crucial factor, that it has independent activity (movement, metabolism, sensitivity, reproduction and growth) and that it was also the transmitter of the flow of energy and the flow of the information. It was also understood that the cell maintained itself by utilising the energy, and that it had inbuilt mechanism to transform energy into work. In this process of chemical, somatic, electrical and mechanical work, motivated by the cell, the cell also regulated the output quantum of each activity to help the integrated working of human system. This supposes also infusion of energy by the cell for the bio-disintegration of macromolecules (complex building block) into microorganism needed to run the system.

As these, thousands of activities are conducted by the *tridoshas*. When the disequilibrium enters the arrangement, they are called *saririka* (somatic) doshas, both the mind and body being seats of disease. Two doshas could combine together and vitiate the third and cause mental (psychic) disorders.

Sushruta and Charaka have followed the *nyayavaiseshika* and *sankhya* systems of philosophy to identify gross matter as consisting of five *mahabhutas* (Agni, Prithvi, Vayu, Tejas and Akasha) in their atomic, molecular and cellular state. The human system is built of these *Panchbhutas*.

As the body, including the five cognative organs, is derived from five *bhutas,* the human body could be divided into *Parthiva Apya, Agneya, Vayaviya* and *Akashiya* aspects. Solid heavy, rough form or motionless parts of the body like bones, ligaments, muscles, nails, teeth, hairs even faces are

parthiva in nature: such parts as are liquids, spreading, soft like *Rasa, Pakna, Rakta* and even urine and sweat are *apya* in nature; the radiance of the body, the colour and sight (eyes) are *agneya*; inhalation, exhalation, movement, etc. are *vayaviya* and the visible and invisible pores and ducts of the body, the senses of hearing, etc. are *akashiya* in nature.

Since, the body (*Panchabhautika*) from birth (from the moment the ovum is fertilised by the sperm) is associated in the functions with the three doshas, the three somatic humors (with five divisions each) are also *Panchbhautikas. Akash* and *Vayu* are predominant in *Vata, Agni* in *Pitta* and *Akash* and *Prithvi* in *Kapha*. Further, the quality of *rasas* is seen in prominence in Vata, Satva and Rasas in Pitta and *Tamas* in Kaphas.

To elucidate further, though the *Panchbhutas* are the constituents of the body, (*Vata, Pitta* and *Kapha*) are the causative factors for their generation and functioning of the body. Any disturbance or misalignment in the fixed proportion of the *bhutas* constitute seat of disease. Since the combination of the *panchbhutas* in a human body varies infinitely with each creation, no human is exactly like other and so such seats of diseases are also myriad in number.

The ancients, therefore, came to the conclusion, in Ayurveda, that the body is composed of, and functions with, the three doshas and seven *dhatus* (*Rasa, Rakta, Mamsa, Meads, Ashier, Majja* and *Shukra*) and three *malas* (Purisha, Mutra, and Sweda).

VPK (Vata, Pitta and Kapha) are the causes of creation, preservation and destruction of the body. The evolution of the fertilised ovum to a full-grown child is helped by *nature,* in the form of Kapha, *Agni* in the form of Pitta, and *Vayu* in the form of Vata. In other words Vata helps in the fertilisation of ovum, *Pitta* to digest and assimilate the food and *Kapha* to give form to the body and sustain it.

VPK together control the body, but their special seats in the body in normal state is the lower, middle and upper regions respectively, further *Kapha* is predominant in childhood, *Pitta* in youth and Vata in old age.

Though to maintain health, the three doshas should be in equilibrium state, Vata with the predominance of *Akasha* and *Vayubhutas* becomes more important among them as the primary motive force like *lokavayu,* which *Charaka* explains is only deducible and it is invisible and cannot be seen. *Vata* is not only mobile but also capable of keeping *Pitta, Kapha* and *Dhatus* (which are incapable of mobility) in motion, like the wind which drives the clouds from the ocean to the land and carries soil from one place to another. In fact, it is said that so long as *Vata* lasts that long will life exist in the body.

Vata in association with the other two doshas activates them, when associated with *Pitta* it produces a feeling of hotness and when associated with *Kapha* a feeling of coldness. *Vata* is located in the region of pelvis and rectum. It is also supposed to be located in ears and the skin, which explain the phenomenon of feeling of these two sensory organs being affected by air currents. *Siras,* the thalamus, hypothalamus, lymphatic system and the lobes, also is supposed to have *Vata* as one of its location, as it is a somatic humor, controlling the motor and sensory functions, i.e. the nervous system. *Vata* effects and controls the muscles, motor functions and attainment of knowledge through the cognative organs. The concept of Vata is so obtrusive that even Hippocrates has to circumnavigate it in its humoral theory. The word *Vata* is derived from the Sanskrit *root* meaning: to move, to

enthuse, to make known, to become aware of and to enlighten.

Pitta conducts and controls the important function of digestive system and assimilation of food and conversion to nourish the *Sariram* or the body with the five sense organs.

The Six Rasa's

In Ayurveda food is as important as the medicine, and describes that it should include all the *rasas* (sweat, sour, bitter and astringent) and this equilibrium keeps the doshas also in an equilibrium state. Hence, in post *Panchkarma* the food should be carefully choosen as the medicine. Many diseases are known to have resulted with an imbalance of six rasas together with hygiene and cleanliness. It is claimed that all the disorders could be corrected by food alone, prescribed in a balanced way. This naturally presumes that all living matter is dependant upon food and that also body is a product of the food.

Sariram is so called because it gets degenerated and decays (natural characteristic of human body), to arrest with the intake of food is most important. *Pitta*, unlike *Vata*, is material and physical. *Kapha*, on the other hand being developed with a predominance of *AP* and *Prithvi bhutas*, is the basis for the structural integrity of the body. One of its seat is *moola*. It contributes to the welfare of the body by supplying it with watery components. Its psycho-chemical characteristic coincides with those of the cell and the protoplasm. Hence, *Kapha* contributes to the growth, bulk and weight of the body, its strength, plus its ability.

Though the functions of the *doshas, dhatus* and *malas* are in a sense compartmentalised in actual functioning, they are under an umbrella, complementary to each other and the living being is a complete, integrated whole.

For the identification of the ailments of the body, dictates have been elaborated which has gone into disequilibrium and set out in more detail the treatment for each. Medicinal properties of the herb around have been used to correct imbalances of the body. By the time of the *Rigveda* some coordinated knowledge of the body's functioning, malfunction and the therapy by using herbs, fruits, barks of trees, nuts and even products from the animals, which abounded the earth, has been evolved which seems to have brought together in a compendium in the *Atharvaveda*.

The Bench Rules

In the early century of the last two millennia Charaka, Sushruta, Vagbhata and others extended the vista and laid down more bench rules for preparing medicine and treatment. Three types of management of therapies have been laid down: spiritual; therapy based on reasoning and psychic. The therapies again are three: internal cleansing (*antah parimarjana*); external cleansing (*baahya parimarjana*); and surgical (*shastra parimarjana*.

Ayurveda is also called 'Ashtang Ayurveda' and 'Ashtanga'. Haridaya's eight branches include: *Kaya chakitsa* (internal medicine), toxicology: *Shalya chakitsa* (branch of medicine dealing with the diseases specific to the supra-clavicular region, i.e. eye, nose, ear, throat, mouth and head, etc. surgery; *bootha vidya* (psychology or the science of demon's seizures); pediatrics; *rasayana* (rejuvenation science) and the science of aphrodisiacs. *Kaya*, means *agni* (digestive fire) or the enzymes responsible for digestion and metabolism, and so *Kaya chakitsa* deals with the diseases caused by the impairment of digestion and metabolism.

At this stage the 'Panchkarma' theory was formulated as the curative line of treatment-cleansing and surgical—for all diseases which were not amenable to *shamand chakitsa* (palliative treatment). The cleansing therapy is by massage, fomentation, punction and kneading. Though this line is chiefly meant for the diseases of the body, it is efficaciously applied to cure mental disorders (epilepsy, insanity etc.)

According to Charka, the five *karmas* are (1) *Vamana* (emetic therapy) (2) *Virechna* (purgative therapy) (3) *Punctuous enema* therapy (4) decoction enema therapy and (5) *Naasya karma* (errhine therapy). Sushruta who has claimed 'rakta' as a fourth dosha (a much debated point), has clubbed the two (Vasti-enema) treatments together and marked *raktamokhana* (blood letting) as the fifth karma.

The *panchkarma* treatment is always preceded by *poorvikarmapre* operative, eliminative measures such as *snehana* (oleation) and *swedana* (sudation) therapies, the purpose being to bring the vitiated, morbid '*doshas*' from the *shakhaas* (including the *raktaadidhatus*) to the alimentary canal and from there to the elimination orifices of the body. The oleation canal and the sudation (sweating) processes require herbal preparations.

The foreign (*militia* as well cultural) invasion by the *Scythians* and the Huns in the first century AD, and much later, the Muslim and Christian, the burnt of which was borne by the Indo-Gangetic planes, seems to have dealt a death blow to such medical knowledge and therapy and the access to the herbs in question.

Malabar, far away from such foreign interferences has kept up with the oil therapy, a valuable vestige of the Panchkarma. Malabar, still the land of unpolluted green hills, abounding in herbs and culture, largely untouched by alien intrusions has bravely kept up its link with the ancient system. Across the centuries, several 'vaidyas' have been treating patients in their hamlets nestling among the lagoons and palm grooves with modest establishment and training a select group (barefoot doctors) keeping alive the tradition.

Ayurveda: The Science of Self-healing

In today's world allopathic treatment is being widely questioned, either because of over-drugging, due to dependence on antibiotics and steroids, or because of the numerous cases of spurious medicines appearing in the market, homeopathy has been making rapid inroads in long time treatment, more specially in chronic ail-ments; because the belief is that the innocuous white pills bear no side-effects. But a hundred percent diagnosis in homeopathy is still doubtful, and the patient feels that the trial and error method is being led recourse to.

The Ayurveda the ancient Indian System of Medicine, is endemic to India and that is the complete wisdom of *Rishis*, with surgeon *Sushruta* producing an Ayurvedic classic text on surgery more than 2000 years ago is a fact. An average Indian perhaps knows little about this science and whatever knowledge he has, remain on the periphery, conveyed by ill-equipped and 'unimpressive' *vaids*. Ayurveda regards the human body and its sensory experiences as manifestations of cosmic energy expressed in the fine basic elements.

According to Ayurvedas, 'tridoshic' concept the first requirement and other functions of the body is an understanding of the three *doshas* forming an equilibrium (for physiological functions). In the concept of Vat, Pitt and Cough (VPK), Vata is a principle of movement

and may be translated as the bodily air principle. Vata governs breathing, blinking, movements of muscle tissue, including function of nervous system, pulse and so on, signifies metabolism and heat production. Pitta signifies nutrition, meta-bolism and so on. The translation of Kapha, implies heat, homoeostasis of mucous, biological water, and this body principal is formed from the two elements, Earth and Water. This particular doasha maintains body resistance. The chest is the seat of Kapha. *Vata* people are generally physically under developed. Their complexion is brown, skin is cold, rough, dry and cracked. Besides they are either too tall or too short. Vata people tend to earn money quickly and also spend it quickly. Naturally they remain poor. *Pitta* people are of medium height, and their complexion may be yellowish or fair. The skin is soft and warm. They have a strong metabolism, good digestion and strong appetite. They are alert and intelligent and like to be leaders. The *Kapha* people have well developed bodies, with tendency to be overweight. Complexions are fair and bright. They are tolerant, forgiving and loving. They earn well, so they are generally wealthy. One dosha can overlap with another. Too many intruding traits, leading to contradictions make extremely difficult to zero-in on a particular dosha. As a result several doubts arise when enumerating disease progresses. "*Kapha* diseases" are repeated attacks of tonsillitis, sinusitis, bronchitis etc. The "pitta people" are prone to gallbladder, liver disorders, gastritis, skin disorders such as hives and rash. *Vata* natured people freque-ntly suffer from diseases, which have their origin in the large intestine.

Analysis of Categorisation of Body Structure

The key to health or disease is *Agni* and the root of ill-health are toxins called the *Ama*.

'Toxins can be created by emotional factors like repressed anger, which changes the flow of the gallbladder, bile duct, etc. If the tongue is coated with a white film, this symptom indicates that *ama* exists in the large intestine, small intestine, or stomach, depending on which, part of the tongue is coated.

The body produces three waste products, or *malas*. Faeces which is solid, Urine and Sweat, which are liquid.

The study of the pulse in *Ayurveda* and in *Unani* systems announces identification of the disease. It may be quick and slithery, like snake or jumpy like a frog, slow like a floating swan. The pulse rate is also checked. Other symptoms of diagnosis could be like vertical line between the brows, lower eyelids puffy, finger nails and the shape they are in. They express *Vatta-Pitta-Kapha* derangement, while spots on the nails speak of calcium or zinc deficiencies. *Stepped surface of nails* indicates malnutrition.

Much emphasis is also laid in *Panch-karma,* which is the physical elimination of body byproducts. The basic processes are vomiting, purgatives, or laxatives, enemas, etc. Thus, *Ayurveda* is a system which keeps the three *doshas* in an ecological equilibrium for a good health and is a way of life. Ayurveda believes that health and longevity result from the equilibrium of the three physiological functions, the *Tridosha*. Ayurveda also considers the balances working of the *dhatus* (lymph, blood, muscle, fat, etc.) and the *mala* waste is vital.

A scientific analysis of the categorisation of the body constitution reveals how useful Ayurveda is in modern diagnosis and treatment. It is not just an outdated form of medical enquiry, based on traditional belief. Everything that exists in the nature can be traced to its origin from five elements

consisting of both matter and energy in them. Them correspond to Air, Fire, Water, Earth and Space. Everybody has all the five elements in them but one must predominate over the other. The body consists of air, which is required for functioning of both respiration and digestive system. Air requires space to travel. Both together control the movement function of the body. The resultant motion allows one to breathe, circulate blood, pass food through the digestive tract and send nerve impulses to and from the brain.

The blood and the tissue fluids are mainly constituted by water. The body requires energy for living, which comes from the metabolism, digestion or what is called as fire. The water and the fire thus are responsible for metabolism of the body processes, air, food and water throughout the system.

The basic body structure is made of connective tissue, bones, minerals, etc. which represent 'earth'. The earthy components and water are required to hold the cells together and form muscles, fat, bone and nerves. Thus, all the above five elements are required to build the human body.

People with strong 'air' and 'space' elements are called 'Vata' type, with strong water and fire elements are called 'Pitta' type and with strong earth and 'water' elements are called 'Kapha' type personalities. Three basic types of personalities can thus be differentiated.

The Vata persons have tendency to 'airy' or Vayu, the Pitta to metabolism and the Kapha to structure disorders.

Being airy, typical vata persons, like air and light, are thin built. They can perform activity rather quickly but show variability in size, shape, mood and action. They are enthusiastic, imaginative, easily excitable, but have a tendency to worry, constipation

insomnia and irregular hunger and digestion. As they have tendency to overwork, they get tired easily. Typically, they are identified due to their thin and tall body, narrow hips and shoulders, may show physical irregularities, like large hands and feet, smaller or protruding teeth.

On the other hand 'Pitta' people are medium built with strong metabolism. They are tense, jealous, bold, argumentative, ambitious, and sharp-witted, outspoken and bold, basically they are warm, ardent to their emotions, loving and content. They generally live by their watch, feel restless if dinner is half an hour late. Typically, they are medium built in size, well-proportioned, with warm and moist-skin and a tendency to premature graying of hair.

Lastly, the *Kapha* personalities can be identified due to their heavy-built and great physical strength. They are basically strong, heavy and watery in everything, Heavy prolonged sleep, strong in emotions, slow digestion, affectionate, tolerant, forgiving, slow to grasp new information; but with good retentive memory, tranquility of mind and relaxed personality type are the main features.

You can be one personality type with predominant Vata, Pitta or Kapha or two personality type Vata-Pitta Pitta-Vata, Kapha-Pitta, Vata-Kapha, Kapha-Vata or three personality type V-P-K in equal distribution.

In a situation like when the cricket match is at the climax a large number of people are watching the match on TV and suddenly the light goes off, one can find that different people react differently. A typical 'Vata' will become emotional, restless, he may start smoking or start moving up and down. He may even get easily excitable and start asking every 2–3 minutes when will power be restored. The typical Pitta will become angry, may start abusing the electricity department, starts ringing telephone and firing them. The typical Kapha type can be seen to remain calm, indifferent and cool, waiting for the light to come.

A disease in Ayurveda is the 'dosha imbalance' and typical dosha disorder will be nothing but excess of that *Dosha* and the basic treatment lies in adopting 'Dosha pacifying lifestyle'.

Being airy the Vatas are cold, dry, thin, moving, quick and rough. They are prone to get cold hands and feet, dislike for cold climate, irregular heart rhythm, high blood pressure, intestinal gas, constipation, bowel irregularities, nervousness, frightening dreams, dry and rough skin and coarse textured hair. *Pitta* person has strong metabolism. Basically they have hot-skin, hot sensations in the skin, urine and anus. They are of sharp mind and have sharp speech. They are moist which show up as profuse sweating, hot sweaty palms. They have a sour smell leading to bad breath, sour body odour, bad smelling urine and faeces. They are more prone to get acidity, ulcer, acne, boils over skin, heat stroke, kidney stones, jaundice, gallbladder stones etc. 'Kapha' typically are heavy, swet steady, soft and slow. They are prone to get obesity, heavy digestion, oppressive type of depression, weight gain and diabetes. They basically are soft spoken with soft skin and hair and a soft look in the eyes.

A simple way of differentiating them is, that Vata are thin built, tall, dry and cold. Pitta are medium built, hot and moist and Kapha are strong built, cold and moist.

In cold weather Vata and Kapha people may feel uncomfortable but Pitta will enjoy it and *vice-versa* in hot season. Similarly, a Pitta may enjoy or tolerate in cold, while Vata and Kapha may feel uncomfortable.

The treatment of any disorder is aimed at bringing back the 'dosha' into normal. A Vata imbalanced person therefore should avoid anything which is cold, dry, light and prefer anything which is hot chily, heavy, sweet, sour and salty. They can not tolerate dry or cold weather, dry air, ice cold drinks, food, popcorn, cabbage, coriander, beans, etc. They will get better with hot beverage, well-cooked food, etc. Pitta imbalance will not tolerate anything which is hot, light, oily, pungent, sour and salty. They will prefer anything that is cold, light, dry, sweet, bitter and astringent. They should avoid taking bath with hot water, hot weather, hot beverages, hot food, hot pepper, etc. Kapha type will not tolerate anything which is heavy, oily, cold, sweet, sour and salty and will prefer light, dry hot, pungent, bitter and astringent. Likewise, they should take warm dry food, cooked without much water with minimum of oil, butter and sugar.

Vata people have qualities of poise, briskness, agility, co-ordination and inner cheerfulness and hence they always perform better in games like yoga, aerobic dances, light bicycling walking. *Pitta* type of persons have more drive than endurance. They always like a challenge and hence do better in exercises like sking, hiking, mountaineering, swimming. Kapha type people being strong and with steady energy, perform better in exercises like weight lifting, running, aerobic, rowing and dancing. When it comes to endurance sports, Kapha always excel and have a natural built for a long run and persistence rowing. The combination of Pitta and Kapha provides determination and endurance. This is a common type seen in professional basketball and football players.

Each dosha type reports different to five senses. Vata has preference for hearing and touch, Pitta for sight and Kapha for taste. Vatas are extremely sensitive to loud noise and are very sensitive to slightest touch. Thus, usually responding to light music. Pittas are very responsive to visual beauty and cannot tolerate bright light. They often require a semi-dim lighting. The Kaphas are always sensitive to natural taste and smell.

Vata people always get balanced with smell of orange, dove, basil and rose. Pitta with sandalwood, rose, mint, cinnamon and jasmine and Kapha with eucalyptus, camphor and clove. It is always better for a body type to have a smelling oil, steam inhalation or scent of material of type, which is soothing to their respective body type.

Massage also is a good way of balancing your doshas. The Pittas will need a cold oil and Vata and Kapha will need a warm oil massage (called *Abhyanga*) on the body. The usual oil used is til oil (seasmum oil).

Meditation helps all types of disorders but is specially used for Vata disorder. Similarly, Panchkarma can also balance all the dosha disorders.

Some people grow older for their age. Their biological age is more than their chronological age. Ageing is nothing but typical Vata imbalance and the same can be controlled or regressed with Vata pacifying lifestyle. Addictions similarly are seen in people with Vata imbalance. The main treatment of addictions thus will be Vata pacifying lifestyle.

The concept is simple and logical. In modern medicine there are many unans-wered questions. We do not know why some alcoholics get liver disease and others cardiomyopathy but not both and why some smokers gets lung disease and others a heart attack.

The Ayurvedic way one can understand it better. An alcoholic Pitta will get a liver disease and a *Kapha* a cardiomyopathy. A smoker *Kapha* will be prone to bronchitis, a smoker *Pitta* to a heart attack, and a smoker

Vata to high blood pressure. A *Vata* patient with rheumatic fever, will get joint pains and *Kapha* will get joint pain with swelling. An age related arthritis called Osteoarthritis will also behave differently according to body type. A Vata will have arthritis without swelling with typical cracking noise in the joints, while a Kapha will get arthritis with swelling and fluid collection. Similarly, a *Vata* getting a viral infection of the respiratory tract will get dry nose, dryness of the sinus and dry cough, while on the other hand a *Kapha* will be prone to get watery discharge from the nose, sinus and productive cough.

Quite often we see patients with non-specific chest pains or anxiety. This is nothing but *Vata* imbalance, often there will be a history of exposure to cold, dry air or the person might have eaten *Vata*-precipitating diet with dry items, mushroom, coriander, beans or cabbage. Instead of giving them pain killers they only need to live a Vata pacifying lifestyle. The same is true for anxiety. The antioxytics prescribed in tons can thus be avoided.

We come across patients, and if we combine the Ayurveda way of body type classification, the particular *dosha* pacifying lifestyles diagnosing and monitoring the disease, the allopathic way will make perfect sense. Giving allopathic medicine for specific disease is important when required. When it comes to acute emergencies Ayurveda usually finds no place but Ayurvedic principles underlying the various body types and therapy with specific pacifying diet and lifestyle will definitely improve the status of currently available therapies in modern medicine, especially in prevention of disease or when adopted in its formative stage or early stages when symptoms appear.

In Ayurveda, diagnosis is achieved from determination of the root cause of the

disease and the vitiated doshas (body types), accumulation of toxic metabolic products, one has to find out the constitution, habits and temperament and most importantly the state of host defense. These factors determine the therapeutic modalities.

The goal of therapy in Ayurveda is to achieve complete health which is achieved through four essential pillars of therapy, the physician, his various approaches for treatment (pacifying the dosha) aroma therapy, *mantra* therapy, music therapy, purification therapy (by various steps of *Panchkarma* and Meditation), the herbs (which help in bringing the body's defenses in alignment), the attendants and the important is patient himself.

CHARAKA (Su 9:6) *states Thorough knowledge of science, keen observation and practice and experiences in various spheres, skill and integrity are four qualities of the physician.*

CHARAKA (Su 9:4) "Disease is the disequilibrium of the 3 Dhatus."

VAGBHVNTA (Su I:2) "Disease is the disequilibrium of the Doshas."

CHARAKA (Su I:53). Objective of any therapeutic measure is primarily to reactivate a state of equilibrium. The act of achieving the equilibrium of the doshas is the objective of this science.

Of all the therapeutic approaches an important one of the *Doshas* has been discussed above. It is however, heartening to note that *Ayurveda* is being accepted by modern medical doctors in various parts of the world and it is high time that we drop our prejudice against Ayurvedic system of treatment and it is made a part of training in medical colleges as is accepted by allopathetic doctors in India—the land of origin of Ayurveda. It is to be understood in terms of contemporary medicine modified, if anywhere necessary, and practised according to its lofty principles in totality.

Indian Influence on Arab Medicine and Evolution of Unani System

The Arabs had close commercial and cultural relations with the Indians in the pre-Islamic days. Being active traders and adventurous navigators, they traded in Indian commodities with the west and had built permanent settlements on the western coast of India. India was far ahead of the world in science, technology and literature, which is also borne out by the fact that Indian exports were highly prized, both in west and the east. As a result of the age-old relations of Indo-Arab relations, the Arabs were acquainted with some facts of the rich intellectual heritage of India; long before the advent of Islam in this country. It is written in "Majma jmaul-Mussannifia" (vol: 11, 60) that the Indians are people of noble ideas and sound brains, who are created with having made significant researches in mathematics, engineering, medicinal, astronomy, science and philosophy.

After the rise of Islam, when the Arab's developed a keen sense of intellectual curiosity and veracious aptitude of learning, they were, but naturally, attracted towards the intellectual legacy of India and drew large benefits from its different branches of knowledge and learning; while India's contribution to Arab medicine remains to be introduced to the academic world for a better appreciation of the richness of the ancient Indian legacy.

Prophet Mohammad on Indian Drugs

In pre-Islamic days medicine among the Arabs was practised on primitive and superstitious lines. Legitimate remedies mingled with medical charms and talismans against the evil eye. But when the prophet came he gave so much importance to the maintenance of health as to give its primacy over religion itself. He not only laid down fundamental principles of preventive medicine, but also exhorted his followers to undertake extensive research in the field of medicine. In this connection, he himself used a number of Indian drugs and recommended their use to his followers. Of them *Qust Indica* (a kind of Indian incense), *Zanzabi* and *Zarorah* are well known. He has described the medicinal properties of *Qust* in the following words: Treat with the Indian incense for it has cure for seven diseases: it is to be sniffed by one having throat-trouble, and to be put into one side of the mouth of one suffering from pleurisy (inflammation of the lungs).

Indian Physicians in Baghdad

Later on, in the Abbasid period, several Indian physicians were invited to the court of Baghdad, the capital of Abbasid Empire and employed in the state hospitals or in translation works. Some of them were *Manakh, Iba Dahn, Salich-Bin-Bahiah, Kamkah Sinjhal* and *Shanak*.

Manakh is mentioned as a distinguished Indian physician by Ibn Usaybiat in his famous book *Uyun-ub anba fi Tabaquat* il-*Attibah* (sources of information on the classes of physicians). This is an elaborate collection of some 400 biographs of Arab, Indian and Greek medical men. It is related that once Harun-al-Rashid, the famous Abbasid Caliph (986–'09 AD) fell seriously ill and his court physician failed to cure him, *Manakh* was invited from India to treat the *Caliph*, when he cured him he was given a high rank and status. He stayed in Baghdad and contributed a great deal to the development of Arab medicine.

When moving in a market in Baghdad, Manakh is said to have come across a juggler who was displaying his drugs and explaining their medicinal properties. During this he described a certain medicine prepared by him as a panacea for all the diseases of man. Hearing this, the critical Manakh said if the juggler was imposter, the Caliph should kill him as the law permits the massacre of such persons. For, if he was not killed, he will kill several persons daily. This shows how bold, confident and conscientious he was as a physician.

Ibn Dahan was also invited for the treatment of the above Caliph and was appointed as Chief Medical Officer at the prestigious Barmaked Hospital in Baghdad and assisted in translation of several Sanskrit works into Arabic.

Translation into Arabic of Sanskrit Books on Medicine

The *Charaka Samhita* and *Sushruta Samhita* were the two major Sanskrit books on medicine that were translated into Arabic. The book *Sasru,* containing 10 chapters on diseases, was translated by Manakh at the instance of Yahya Bin Khalis-al Barmaki. He is also the translator of another book on Indian drugs and herbs, a book on female diseases written by an Indian lady doctor named Rawsa, a book on pregnant women; a treatise on a hundred disease and their treatment as well as a book on the suspicion of disease and their effects by Naukishnal.

The Unani system of medicine, such as Grecko-Arab medicine, Ionian medicine, Islamic medicine, oriental medicine, etc. in different parts of the world is known by different names. The system is practised in India in a quite different form, from that of its original Greek one. In the country of its origin it is at a low profile. But now they are getting anxious about it, as Italy has been importing many Unani medicines from

Indian companies. In Arabic *Unani* means the Greek and *Tibb* means medicine. Therefore, the complete name is "Tibb-é Unani" It was Hippocrates in Greece in 460 who introduced this concept and it was stabilised by Galen. In reality it was the Arab community that actually cashed on it by researching more about its efficacious use in their countries. In 750 AD when the Asia Minor came to be ruled by Abbasid, Arabs got the books of the Unani system translated into Arabic. Abu Sena (Avicenna) was the most revered man practising the profession. His book 'Al-Quanan' is considered to be the bible of Unani system. After the Mongols attacked Persian and Central Asian cities, like Shiraz, Gilan, Ifshan, etc. the scholars and physicians fled to the safety, to India, and were patronised by the Delhi Sultans like the Khiljis and Tuglaks and later by the Mughals.

Infact, the Unani system of medicine got enriched by imbibing what was the best in contemporary systems of traditional medicine IW vogue in various parts of Central Asia. It was introduced by Arabs. As mentioned above, after Hippocrates and Galen the system was amalgamated into full-fledged science by Arab apothecaries like Rhazes (850–920AD) and Avicennia or Abu Sina (980–1037 AD). After that there was a big lapse and the moment this art of medicine was about to die, Masih-ul Mulk Hakim Ajmal khan (1864–1927) came on the scene in India and the system was rejuvenated.

Hakim Ajmal khan, who lived in Sharif-Manzil, 'The Dawakhana' is embibed in a rich historical past, when it covered new ground in the school of Unani and to some extent by Ayurvedic medicines by Sharif Khani family. Hakim Ajmal Khan, experimented with various herbs and plants and today's most Unani-medicines in the market are the result

of extensive research done by Hakim Ajmal Khan. According to the will his prescriptions were only to be secretly used by Tibbia college and Hindustani Dawakhana at Delhi. Hakim Abu Razaq in his book "Unani System of Medicine in India" a profile, describes the "History of Development in India".

According to the Unani system of medicine, a healthy status is retained on account of the balance of four humors or constituent substances—*Akhlat*. These humors are blood, phlegm, yellow bile and black bile (Mizaj) or temperament is of permanent importance. Concepts like sanguine, phlegmatic, choleric or melancholic according to the respective preponderance of humors depict the temperament of the patient. Take for example, if a person has a history of fevers, he will have hot and dry yellow bile. Most of the cure in Unani medicine actually do not depend on the medicine but on the regulation of certain kind of diets as per quality and quantity. Hakim Masood Ahmad Baqui, is of the view that foods are capable of pharmacological action too, for instance, many foods have laxative, diuretic and diaphoretic properties. Some of the wonder medicines of Hindustani Dawakhana include Sharbat-Sadar (for chronic sinusitis and cough), Dawaush Shifa (for mental disorders), Seeko (for paralysis), Hubb-é-Mussafi (for purifying blood), Tiryaq-e-Yarqun (for jaundice), Safoof Ziabatis (for headache), Kusta-e-Aquiq (an aphrodisiac), Majoon-Suhag-Soonth (for post delivery control), Namak Margang (for stomach ailments).

Hakim Abdul Hameed had been experimenting, lately, with various natural medicines for the best results. In fact the medically advanced west sought answers to what appear to be incurable diseases, it found them in indigenous pharmacology. *Hamdard* began in 1906 after Hakim Abdul

Hameed's father Hakim Late Abdul Majeed opened his clinic after leaving his job from Hakeem Ajmal Khan's Hindustani Dawakhana. Abdul Hameed's *Khamira Abresham Arshadwala* was discovered against heart ailments. Lately Hamdard has attained a reputation for its various medicine brand names.

The Tibetan System of Medicine

This system has a separate way of differentiating diseases which makes the diagnosis easier, as the process is based on three-phase-observation, and interrogation. As Tibet is surrounded by mountains, rich in minerals and medicinal plants, the physician draws herbs from the mountains for treatment. A Tibetan doctor, though is familiar with different kinds of medicinal plants. The poisonous part of the plant is eliminated by the process of precipitation and then finally burnt to ashes with other compositions. It is, however, important to know about the tastes and potency of each individual herb for the preparation of medicines.

The Tibetan scholars followed the Indian practice of describing authorship of their works to their gods or preceptors, e.g. in basic treatises on Tibetan medicine, *Budha* is described as the source of all knowledge. The works on medicine are based on religious tradition which are composed by Khyung-Sprul 'Jigs-medanam-nikha-irdoge during 1937–1950 have recently been published by Tibetan Bon-Po monastic center.

Unlike the Indian Ayurvedic system which offers only herbal remedies, Tibetan system uses all kinds of metals, stones, animals' horns, bones, etc. as medicinal trees, medicinal oils, decoction from medicinal plants, fruits and flowers. The medicines are prepared in the following forms:

Decoction: extraction of medicine by boiling (Thonsman)

Powder, Pills, Syrups, Oils Ashes (Thalsman)

Concentrated medicine, (Khanda and medicinal wine).

Nagarjuna and Buston, are the two well-known authors of the work Yagasalaka (Tibetan slyor-ba-Brigyapa). Nagarjuna was a philosopher and his contributions to alchemy and Tantric studies are well-known.

The Tibetan System of Therapy

The four works on medicine based on 'Bon'religious tradition which are composed by khyung-sprul Jigs-medanam-nikha-i-dorje during 1937–50 have recently been published by Tibetan Bon po Monastric Center. Propagation of Ayurveda in Tibet was done by *Budha-Bikshu* of India along with sermon of Lord Budha, and many plants were described which were growing in the Himalayan region. During the regime of *Nyatri Tsenpo*, the first king of Tibet, Viajay and Garhjay went to Tibet and imparted the knowledge of healing. In Ayurveda classics, prior to Budha, many plants now available were described. The treatment is based on the belief that imbalances of the four elements namely air, water, fire and earth, cause ailments and herbal medicines are used to cure them. The constituent of the medicines are fully nurtured by the four elements and are made up of wild plants and flowers. The Tibetan system was kept a secret for over a thousand years by physicians, who practised it as their hereditary profession and was subsequently handed over by the teacher to a single disciple.

The most powerful ruler of Tibet, Emperor Srugtsen Gonpo (617–650 AD) contributed to the development of medicine. He invited three famous physicians from India,

who jointly compiled a volume. It was during the great king, Trisong Dentsen's regime (742–798 AD), the medicines received royal patronage and inspiration, when an international conference of medicine attended by physicians from all over the world was convened.

It is a unique natural system of medical science with a practical way of therapeutic approach, using natural medicines. Besides, Desi Sangyae Gyatso, who was the 5th Dalai Lama and Telhasa Institute of Astro-Medical Science, founded by Khenrab Norbu (1830–1962), contributed greatly to the advancement of herbal medicines the world over.

The treatment attempts to treat both mind and body with holistic approach, believing that one's mental disposition influences and determines to a vast degree, the bodily functions. The primary aim of this system is to relieve human sufferings and to restore the equilibrium in imbalance in the normal functioning of the wind, bile and phlegm elements in the body.

The following plants are commonly used:

Fever: Decoction of *Tinospora cordifolia, Cyperus rotundus: Fagonia cretica.*

Chronic Fever: Solanum xanthocarpum, Tinospora cordifolia, Piper longum.

Fever accompanied by burning in abdomen and bleeding from different parts: Decoction of *Fagonia cretica, Fumaria parviflora, Terminalia chebula, Adhatoda vasica,; Picrorhiza kurroa, Swertia chirata, Callicarpa macrophylla.*

Diabetes: Decoction with honey: *Berberis aristata, Terminalia chebula, Emblica officinalis, Citrulus colocynthis.*

Antifilarial: Argyreia speciosa (Roots and leaves powder given internally with milk), *Calotropis procera, Breynia patens, Strebelus asper*-seeds) and *Derris indica.*

Chronic diarrhoea blood diarrhoea and acute pain in abdomen: Decoction of *Coleus vettiveroides, Aconitum hetrophyllum, Cyperus rotundus, Holarrhena antidysenterica.*

Skin diseases accompanied by anemia, oedema, salivation and heaviness in stomach: Plants mixed with cow's urine-*Boerhavia verticillata, Berberis aristata, Terminalia chebula, Tinospora cordifolia, Commiphora mukul.*

Worms: Decoction of *Cyperus rotundus, Berberis aristata, Embeli ribes, Piper longum,* fruits of *Balliospermum montanum. Punica granatum, Artemisia siversiana*-fluid extract of bark, (Nadkarni 1976).

Gout: Decoction with castor oil: *Adhatoda vasica, Tinospora cordifolia* and *Cassia fistula.*

Tumor, heaviness of stomach, chronic fever and chronic diarrhoea: Vida, a type of salt with *Ferrula asfoetida, Acorus calamus, Zigiber officinalis, Cuminum cyminum, Terminalia chebula, Inula racemosa, Saussurea lappa.*

Abdominal disorders, anemia: Powder from burnt leaves mixed with molasses and lavana (rock salt) *Pongamia pinnata, Citrullus colocynthis, Plumbago zeylanica, Piper longum* and *Gingiber officinalis.*

Fever, cold and cough: Oil extract of the plant with salt, six times the buttermilk and powder of fried paddy. *Zingiber officinalis, Saussurea lappa, Senseviera roxburghii, Curcuma longa, Rubia cordifolia.*

Chronic fever, coughing, phantom-tumor, asthma: Decoction of *Tinospora cordifolia, Solanum xanthocarpum* and *Adhatoda vasica.*

Diseases of the tongue, mouth and body: Tinospora cordifolia, Berberis aristata, Jasminum grandiflorum, Vitis vinifera,

Holarrhena antidysenterica, Emblica officinalis (fruits) and *Terminalia chebula*.

General health and black hair: Plant powder with butter and honey: *Tribulus terrestris, Emblica officinalis, Tinospora cordifolia*.

Various drug sources in this system are mineral, animal, vegetable, synthetic and micro-organisms. Pharmacologically active principle in indigenous drugs are alkaloids, glycosides, oils-fixed and volatile resins, oleoresins, gums, tannings and antibacterial substances.

The drug plants given above may broadly be classified as follows:

Plants recognised on pharmacopoeias: Though a large number of drug plants grow in India and some are cultivated too, yet all are not taken into the pharmacopoeias. The active principles in most cases come up to the standards laid down in the pharmacopoeias.

Substitutes for pharmacopoeial drugs: A large number of drugs possess properties similar to the imports and other plants mentioned in pharmacopoeias, examples of such plants are *Colchicum luteum* from northwest Himalaya for *C. autumnale, Scilla indica* for *S. maritima, Picrasma quassioides* and *Gentiana kurroa* for *P. excelsa* and *Gentiana lutea, Ferula anthrax* for *F. asfeotida, Rheum emodi* for *R. palmatum, Aconitum chasmanthum* for *A. nepallus*. A number of species of the male fern *Dryopteris* are as effective as *D. filix-max* of British Pharmacopoeia. *Physo-chlaina praealata* from Ladakh and a number of *Artemisia* species give a similar or even higher yield of alkaloids and valuable essential oils.

Homeopathy

The credit of Homeopathy therapy's discovery, as a healing science, goes to Dr. Samual Hahneman, a German physician in the early 19th century. Homeopathy has outlined many remedies for various ailments with drugs in very high dilutions. The founder has tried to find a form of medication which cures rather than suppresses. He observed, that, wherever the natives had fever accompanied by chill it was cured by drinking a decoction made by boiling the bark of Chincona tree. Hahneman felt that if fever was cured by drinking the decoction than fever should also be caused if a healthy person drank it. Thus, he came upon the *Law of Similars*.

On July 4, 1850, the then President of USA, Zachary Taylor, had a meal of fresh Cherries; iced milk and fresh vegetables. Soon after he developed gastroenteritis and acute diarrhoea. Five days later he was dead. Could fruit or raw vegetables, washed down with cold milk, really kill a man? Taylor's symptoms as vomiting, cramps, diarrhoea and progressive weakening, these could have resulted only from arsenic poisoning. Of course, he was not poisoned by arsenic, yet he died from similar symptoms. Alb 30 on his tongue at half hour interval could have saved the President's life: But then lack of proper propagation of the science of homeopathy not only killed the President, in the 19th century, but is doing so even today. Innumerable people are dying mainly because the drugs administered suppress the symptoms and not affect as cure. Nature did not intend us to be dependent on indiscriminate medication and pill popping! Is not it surprising therefore, that whenever the body temperature rises due to an infection, you take medicine for fever and not the infection? The fever becomes our primary concern and one can end up suppressing the body's natural fighting mechanism. Similar is the case with headaches, joint pains and cough. Interestingly, each one of us has a vital force

in one's body, that works as a natural healing power which we invariably suppress by indiscriminate use of medicines.

It is surprising that though homeopathy is 200 years old, its knowledge is restricted, and people still refer to it as faith healing, when actually it is based on definite scientific principles. Each and every remedy is derived from nature, whether it is metal like, iron, gold, silver, aluminum, tin, platinum, etc. or poison like arsenic, conium, (hemlock) snake poison like lochesis, or acid like nitric, phosphoric or picric acid. Medicines are extracted even from bees like *Afis mel,* salt Natrum Mur, even from flint stone like silica! Each one of these remedies has been tested by physicians, either on themselves or by provers and symptoms written down in minutest detail for every remedy.

Constant research in the field of homeopathy led to discover that any disease can be cured by a minimum dose of the indicated remedy in a potent form. For example, if there are symptoms of iron deficiency—potentised form of iron would be the cure and not the iron in material form, i.e. in capsules. Homeopathy thus is based on this law.

People often ask what can the small sugar pill or globule that make a homeopathy dose, do for a sick body? It must be very clearly understood that though these globules are helpless in fighting the disease themselves, they strengthen the vital force of the body and energises it to fight the disease.

It is amazing that in modern times, people keep talking about diseases and related diagnostic tests, but there is no remedy for the patients whose tests are all clear and yet the patient says, "I am not well".

This is where homeopathy scores over other medical sciences. It does not cure individual diseases but takes care of the patient in "totality", which is why the homeopath asks seemingly weird questions like whether the patient feels cold or hot, likes to have a bath or not, the type of food he relishes, whether he likes to stay outdoors or indoors. This enables him to decide on one single remedy or complimentary follow up remedies to cure the individual.

A homeopath does not have any specific remedies for headaches, or rheumatism or indigestion, he treats every individual as separate and that is why he may give the same medicine to four people suffering from four different problems, or even four different remedies to people suffering from a single problem, e.g. headache. A homeopath thus, abides by the principle—never generalize, always individualise.

A patient once asked Dr. Hahneman, what was the name of medicine he was giving him. To which he replied, "The name of your disease is no concern of mine and the name of the remedy I give is no concern of yours".

Homeopathy is thus considered to be the only science which takes care of the mind, since most of physical problems accrue from the mental state of the individual. Very often there may be no evident physical problem yet the patient does not feel happy and well. It is as Bernard Shaw said, "A healthy body is the product of a healthy mind."

(A part of the text extracted from an article by Vijay Lakshmi, published in The *Hindustan Times—Be with Nature.*

Medicinal Plants having Therapeutic Value in Homeopathy

1. *A. arabica–Babul*
2. *A. italicum*
3. *A. maculatum*

4. *A. quinquefolia*
5. *A. syriaca*
6. *A. triphyllum*
7. *A. tuberosa*
8. *A. vincetoxicum*
9. *Abelomoschus spesculentus*
10. *Abies canadensis*
11. *Abies nigra*
12. *Abrus precatorius*
13. *Acacia arabica*
14. *Acacia catechu willd*
15. *Acalypha indica*
16. *Acanthus vulgaris*
17. *Accicia avcbeca*
18. *Aconitum cammarum*
19. *Aconitum cammorum*
20. *Aconitum ferox*
21. *Aconitum fisheri*
22. *Aconitum lycotonum*
23. *Aconitum mycotonum*
24. *Aconitum nepallus*
25. *Aconitum radi*
26. *Aconitum tartaricum*
27. *Acorus calamus*
28. *Actaea racemosa*
29. *Actaea spicata*
30. *Adhatoda vasica nees (Justicea adhatoda)*
31. *Adonis vernalis*
32. *Aegle marmalos* (Báel)
33. *Aesculus glabra*
34. *Aesculus hippocastanus*
35. *Agave americana*
36. *Agnus castaus*
37. *Agrimonia eupatorium*
38. *Agropyrum repens*
39. *Agrostema githago*
40. *Agrostis* sp
41. *Ailanthus glandulosa*
42. *Alangium pamarekii*
43. *Alectris farinosa*
44. *Alfalfa*
45. *Alisma plantago*
46. *Allium ceppa*
47. *Alnus rubra*
48. *Aloe socotrina*
49. *Alstonia constricta*
50. *Alstonia scholaris*
51. *Althaea officinalis*
52. *Ampelopsuis quinquefolia*
53. *Amygdalus amara*
54. *Amygdalus persica*
55. *Anacardium occidentale*
56. *Anagallis arvensis*
57. *Angelica archangelica*
58. *Anisum stellatum*
59. *Anthemis cotula*
60. *Anthemis nobilis*
61. *Antirrhinium linearum*
62. *Apium graveolens*
63. *Apocynum cannabis*
64. *Aquilegia canadensis*
65. *Aquilegia vulgaris*
66. *Aralaea scinencia*
67. *Aralia hispida*
68. *Aralia nudicaulis*
69. *Aralia racemosa*
70. *Aralia spinosa*
71. *Aranaea diadema*
72. *Arctium lappa*
73. *Argemone mexicana*
74. *Aristolochia serpentaria*
75. *Arnica montana*
76. *Artemisia vulgaris*
77. *Arthante elongata*
78. *Arum draconium*
79. *Arundo mauritanica*
80. *Asclepias curassavica*
81. *Asimina triloba*
82. *Asparagus officinalis*
83. *Asperula odorata*
84. *Asplenium scolopend*
85. *Astacus fluviatalis*
86. *Astacus flviatalis*
87. *Asteras rubens*
88. *Asterias rubens*
89. *Astragalus menzieai*
90. *Astragalus* sp

91. *Avena sativa*
92. *B. vulgaris*
93. *Balsamum peruvianum*
94. *Barbus fluviatalis*
95. *Barosma crenata*
96. *Berberis aquifolia*
97. *Betonica officinalis*
98. *Blatta orientalis*
99. *Bothropus lanceolata*
100. *Brassica janea* (Vern Rai)
101. *Bryonia alba*
102. *Buxus sempervirens*
103. *C. duratinus*
104. *C. florida*
105. *C. horridus*
106. *C. sericea*
107. *Cactus grandiflora*
108. *Caladium seguinum*
109. *Calamus aromnaticus*
110. *Calotropis gigantea*
111. *Canna angustifolia*
112. *Canna glauca*
113. *Capparis decidua* (Vern Karil)
114. *Capsicum annum*
115. *Capsicum baccatus*
116. *Carduus marianus*
117. *Cassytha filiformis* L
118. *Cicer acietinum* (Vern Chana)
119. *Citrus decumana*
120. *Citrus limonum*
121. *Clematis erecta*
122. *Cocculus indicus*
123. *Cochlearia armoracia*
124. *Colchicum acutumnalis*
125. *Collinsonia canadensis*
126. *Colocynthis coccinia*
127. *Coniumma culatum*
128. *Convolvulus arvensis*
129. *Copaiva officinalls*
130. *Coptus trifolia*
131. *Corallium rubrum*
132. *Cornus alternifolia*
133. *Cornus circinata*
134. *Corydalis formosa*

135. *Covalaria majalis*
136. *Crataegus oxycantha*
137. *Crocus sativa*
138. *Crotalus cascavella*
139. *Croton tiglium*
140. *Cubeba officinalis*
141. *Cucurbita citrullus*
142. *Cuphea viscosissima*
143. *Cupressus lawsonia*
144. *Cupressus sempervirens*
145. *Cuscuta americana*
146. *Cyclamen europaeum*
147. *Cydonia vulgaris*
148. *Cypripedium pubescens*
149. *Cysticus scoparius*
150. *Cystisus laburnum*
151. *D. arvensis*
152. *D. batula*
153. *Daphne indica*
154. *Datura arborea*
155. *Delphinium consolida*
156. *Dicentra canadensis*
157. *Dictamnus albus*
158. *Digitalis purpurea*
159. *Dioscorea villosa*
160. *Diosma crenata*
161. *Dipsacus sylvestris*
162. *Diptres odorata*
163. *Dirca palustris*
164. *Dolicus pruriens*
165. *Draconiumfoetidum*
166. *Drosera rotundifolia*
167. *Dulcamara* sp
168. *E. europaeus*
169. *E. glutinosum*
170. *E. hypericifolia*
171. *E. lathyris*
172. *E. marginata*
173. *E. martimum*
174. *E. pilulifera*
175. *E. purpureum*
176. *E. resinifera*
177. *E. rostrata*
178. *Echinacea angustifolia*

179. *Echinops echinalus* (Veru Utkantra)
180. *Elaps corallinus*
181. *Ephedra vulgaris*
182. *Epigea repens*
183. *Epiphehus virginiana*
184. *Equisteum arvense*
185. *Erechthites haeracifolia*
186. *Erigeron canadense*
187. *Eriodictcon canadense*
188. *Eryngium aquaticum*
189. *Erythronium americana*
190. *Escoba amargo*
191. *Eucalyptus globulus*
192. *Eugenia jambos*
193. *Euopnymus atropurpureum*
194. *Eupatorium aromaticum*
195. *Euphorbia corolina*
196. *Euphrasia* sp.
197. *Exogonium purga*
198. *Fabiana imbricata*
199. *Fagopyrum esculentum*
200. *Fasfoetida* L (Heeng)
201. *Fbenglensis* L (Vat Vriksh)
202. *Ferula sumbul*
203. *Ficus vesiculosus*
204. *Fraxinus americana*
205. *G. quinqueflora*
206. *G. robertianum*
207. *G. squarrosa*
208. *G. ulginosa*
209. *Galipea cusparia*
210. *Galium aparine*
211. *Garcinia morella*
212. *Gelsmium sempervirens*
213. *Genista tinctoria*
214. *Gentiana cruciata*
215. *Geranium maculatum*
216. *Geum urbanum*
217. *Ginseng*
218. *Glechoma hederacea*
219. *Gnaphalium polycepa*
220. *Gossypium herbaceum*
221. *Granatum punica* (*Punica granatum*)
222. *Gratiola officinalis*
223. *Grindelia robusta*
224. *Guarea trichillioides*
225. *Gueraria maritima*
226. *Gutta gum*
227. *Gymnocladus canadensis*
228. *Gynandris pentaphylla* (Vern Hulhul)
229. *H. niger*
230. *Hammamelis verginica*
231. *Hedeoma pilegioides*
232. *Hedeoma pulegioides*
233. *Hedera helix*
234. *Hedysarum ildefonsi*
235. *Hegenia abyssinica*
236. *Helianthus annuus*
237. *Heliotropium peruvianum*
238. *Heloderma horridus*
239. *Helonia dioica*
240. *Hepatica triloba*
241. *Heracleum spondylium*
242. *Hippomane mancinella*
243. *Humulus lupulus*
244. *Hura brasiliensis*
245. *Hydrangea arborescens*
246. *Hydrastis canadensis*
247. *Hyoscyamus niger*
248. *Hypericum perforatum*
249. *I. foetidissima*
250. *I. germanica*
251. *Iberis amara*
252. *Ictodes foetidus*
253. *Ignatia amara*
254. *Ilex opaca*
255. *Illicum anisatum*
256. *Impatiens fulva*
257. *Inula helenium*
258. *Ipomoea* sp
259. *Iris florentina*
260. *J. gualandai*
261. *J. pilosus*
262. *J. viginiana*
263. *Jacaranda caroba*
264. *Jambosa vulgaris*
265. *Jatropha curcas*
266. *Juglans regia*

267. *Juncus effuses*
268. *Juniperus communis*
269. *Kalmia latifolia*
270. *L. cicera*
271. *L. coerulea*
272. *L. crinus*
273. *L. inflata*
274. *L. syphilitica*
275. *L. tigrinum*
276. *L. usita*
277. *L. viginicus*
278. *L. virosa*
279. *Lactuca sativa*
280. *Lamium album*
281. *Lapathum acutum*
282. *Lathyrus sativus*
283. *Ledum palustre*
284. *Lemna minor*
285. *Leonurus cardiaca*
286. *Lepidium bonariense*
287. *Leptandra virginica*
288. *Leptilon canadense*
289. *Lespedeza capitata*
290. *Liatris spicata*
291. *Lignum vitae*
292. *Lilium superbum*
293. *Limulus cyclops*
294. *Linaria vulgaris*
295. *Lippia mexicana*
296. *Liriodendron tulipifera*
297. *Lobelia cardinalis*
298. *Lolium temulentum*
299. *Lonicera japonica*
300. *Lufa amara* (Tori-Bitter)
301. *Lycopodium clavatum*
302. *Lycopus europaeus*
303. *Lycospersicum esculentum*
304. *M. officinalis*
305. *M. pudica* (Vern Lajwanti)
306. *M. pulegium*
307. *Magnolia grandiflora*
308. *Mangifera indica*
309. *Medicago sativa*
310. *Melastoma* sp
311. *Meliolotus alba*
312. *Menispermum candida*
313. *Mentha piperita*
314. *Menyanthes trifoliata*
315. *Micromeria* sp
316. *Mimosa humilis*
317. *Momordica balsamica*
318. *Monotropa uniflora*
319. *Mucuna pruriens*
320. *Musa sapientum*
321. *Myosotis palustris*
322. *Myrica cerifera*
323. *Myrtus communis*
324. *N. sativa*
325. *Najas tripudiens*
326. *Narcissus poeticus*
327. *Nasturtium officinalis*
328. *Negundium americana*
329. *Nepeta cataria*
330. *Nerium odorum*
331. *Nigella damascana*
332. Nux-vomica (*Strychnos nux-vomica*)
333. *Nymphaea odorata*
334. *O. chenopodium*
335. *O. eroeata*
336. *O. gaultheria*
337. *O. majorana*
338. *O. muricata*
339. *O. umbellatum*
340. *O. vulgare*
341. *O. wittnebianum*
342. *Ocimum canum*
343. *Oenanthe cordata*
344. *Oleum chaulmoogra*
345. *Opopanax gummi*
346. *Opuntia vulgaris*
347. *Oreodaphne californica*
348. *Origanum creticum*
349. *Ornithogalum maritima*
350. *Orobanche virginiana*
351. *Oxalis acetosella*
352. *Oxydendron arboreum*
353. *Oxytropis lamberti*
354. *P. anisum* (Methi Jira or Saunf)

355. *P. aviculare*
356. *P. capitis*
357. *P. coronaria*
358. *P. hydropiper*
359. *P. major*
360. *P. malus*
361. *P. nana*
362. *P. nigrum*
363. *P. pinicola*
364. *P. punctatum*
365. *P. rovburghii* (Urad or mash)
366. *P. rubra*
367. *P. sagittatum*
368. *P. sorbilis*
369. *P. spinosa*
370. *P. sylvestris*
371. *P. tremuloide*
372. *P. virginica*
373. *P. vulgaris*
374. *P. vulgaris*
375. *Paeonia officinalis*
376. *Panacea arvensis*
377. Papaya (Carica papaya)
378. *Papver somniferum*
379. *Pariera brava*
380. *Paris quadrifolia*
381. *Parthenium hystrophorus*
382. *Passiflora incarnata*
383. *Paullinia pinnata*
384. *Pavia ohioensis*
385. *Pedicularis canadensis*
386. *Penthorum sedoides*
387. *Petasites vulgaris*
388. *Petiveria tetrandra*
389. *Petroselinum sativum*
390. *Phaseolus alba*
391. *Phelum pratense*
392. *Phoradendron flavescens*
393. *Physalis alkekengi*
394. *Phytolacca decandra*
395. *Picraena excelsa*
396. *Pilicarpus microphyllus*
397. *Pillocarpus pinnatifolius*
398. *Pimeta officinalis*
399. *Pimpinella saxifraga*
400. *Pinus palustris*
401. *Piper methysticum*
402. *Piscidia eruthrina*
403. *Plantago lancifolia*
404. *Platanus occidentalis*
405. *Plectranthus fruticosus*
406. *Plumbago littorale*
407. *Ptychotis ajowan*
408. *Podophyllum peltatum*
409. *Polyanthes tuberosa*
410. *Polygonum acre*
411. *Polymnia uvedalia*
412. *Polyporis officinalis*
413. *Polytrichum juniperina*
414. *Populus candicans*
415. *Potentilla norvegica*
416. *Pothos foetida*
417. *Primula obconica*
418. *Prunella vulgaris*
419. *Prunus padus*
420. *Prunus virginiana*
421. *Psoralea corylifolia* (Bavchi—vakuchi Beng)
422. *Psoralea corylifolia* (Bavxhi)
423. *Ptelea trifoliata*
424. *Pulsatilla nigricans*
425. *Pyrus americana*
426. *Q. glandium*
427. *Quecus alba*
428. *Quillaya saponaria*
429. *R. bulbosus*
430. *R. californica*
431. *R. catharticus*
432. *R. diversiloba*
433. *R. flammula*
434. *R. frangula*
435. *R. glabra*
436. *R. purshiana*
437. *R. radicans*
438. *R. repens*
439. *R. scleratus*
440. *R. toxicodendron*
441. *R. veneta*

442. *Ranunculus acris*
443. *Raphanus sativa*
444. *Reseda luteola*
445. *Rhamnus californica*
446. *Rheum sp*
447. *Rhododendron*
448. *Rhubarb sp.*
449. *Rhus aromatica*
450. *Ricinus communis* (Vern Aurad)
451. *Robinia pseud-acaci*
452. *Rosa centifolia*
453. *Rosa damascana*
454. *Rosmarinus officinalis*
455. *Rottlera tinctoria*
456. *Rubia tinctoria*
457. *Rubus villosus*
458. *Rudbeckia hirta*
459. *Rumex acetosella*
460. *Ruta graveolens*
461. *S. domesticus*
462. *S. ebulis*
463. *S. ebulis*
464. *S. lycospersicum*
465. *S. mammosum*
466. *S. marilandica*
467. *S. melogena* (Begun)
468. *S. nigra*
469. *S. nigrum* (Mako)
470. *S. oleraceum*
471. *S. pseudocapsicum*
472. *S. purpurea*
473. *S. tuberosum*
474. *S. vesicarium*
475. *Sabal serrulata*
476. *Sabbatia angularis*
477. *Saccharum lactis*
478. *Sagittaria sagittifolia*
479. *Salix alba*
480. *Salvia officinalis*
481. *Sambucus canadensis*
482. *Sanguinaria canadensis*
483. *Sanguinaria canadensis*
484. *Sanicula marilandica*
485. *Santalum alba*
486. *Saponaria officinalis*
487. *Sedum telephium*
488. *Selnium sp*
489. *Sempervivum tectorum*
490. *Senecio aureus*
491. *Senna*
492. *Solanum carolinense*
493. *Solidago odora*
494. *Solidago virgaurea*
495. *Spartium scoparium*
496. *Spigella anthelmintica*
497. *Spiraea ulmaria*
498. *Spiranthes*
499. *Spongia tosta*
500. *Squilla hispanica*
501. *Stachys betonica*
502. *Stellaria media*
503. *Stercularia acuminata*
504. *Sticta pulmonaria*
505. *Stigmata maydis*
506. *Stillingia sylvatica*
507. *Strychnos nux-vomica*
508. *Symphytum officinalis*
509. *Syzygium jambolana* (cumini)
510. *T. chebila* (Vern Hariki, Harad)
511. *T. dioicum*
512. *T. fibrinum*
513. *T. hispana*
514. *T. pendulum*
515. *T. petasites*
516. *T. pratense*
517. *T. repens*
518. *T. selescea* (Bahera)
519. *Tamus communis*
520. *Tanacetum vulgare*
521. *Tarantua cubensis*
522. *Taraxacum officinale*
523. *Taxaus baccata*
524. *Terminalia arjuna* (Arjun)
525. *Teucrium maritima*
526. *Thalaspi bursa-pastoris*
527. *Thalictrum cornuti*
528. *Thaspium aureum*

529. *Thea sinensis*
530. *Thuja occidentalis*
531. *Thymus serpyllum*
532. *Tiglium officinale*
533. *Tilia europaea*
534. *Tillandsia usneoides*
535. *Tradescantia diuretica*
536. *Trifolium arvensis*
537. *Trillium cernum*
538. *Triosetum perfoliatum*
539. *Triticum repens*
540. *Tussilago farfara*
541. *Typha latifolia*
542. *U. urens*
543. *Ulmus fulva*
544. *Urtica dioica* (Bichu ghass)
545. *Usnea barbata*
546. *V. officinalis*
547. *V. officinelis*
548. *V. opulus*
549. *V. prunifolium*
550. *V. tricolor*
551. *V. urticifolia*
552. *Vaccinium myrtillus*
553. *Valeriana officinalis*
554. *Veratrum viride*
555. *Verbascum thapsus*
556. *Verbena hastata*
557. *Veronica beccabunga*
558. *Vesicaria officinalis*
559. *Viburnum acerifolium*
560. *Vinca minor*
561. *Vinca toxicum*
562. *Viola odorata*
563. *Viscum album*
564. *Withania somnifera* (Aswagandha or winter cherry)
565. *Xanthium spinosum*
566. *Xanthophyllum fraxinum*
567. *Xerophyllum tenax*
568. *Xiphosura amaricanus*
569. *Yucca*
570. *Zingiber officinalis*

Mother Tincture of Some Indian Drugs Used:

Abroma angustifolia
Acalypha indica
Achyranthus aspera
Aegle foliosa (Marmelos)
Ammora rohituka
Andrographis purpurea
Atista indica, A. radix
Azadirachta indica
Blatta orientalis
Blumea odorata
Boerhaavia diffusa
B. repens
Brahmi *(Centella asiatica)* * also called Mandukparni
Bryophyllum sp
Caesalpinia bunduc
Calotropis gigantea
Carica papaya
Cassia sophera
Cephalandra indica
Clerodendron indica
Cynodon dactylon
Desmodium gigantea
Embelia ribes
Ficus indica, F. religiosa, F. venosa
Gentiana chirata (Swertia chirata)
Gymnema sylvestris
Hemidesmus indicus
Holarrhena antidysentrica
Hydrocotyl asiatica (Centella asiatica)
Hygrophilla sp
Adhatoda vasica
Leucas aspera
Luffa amara
Momordica charantia
Nyctanthes arbor-tristis
Ocimum radix (Tulsi mool)
O. sanctum (Sweet tulsi),
O. dulal. (Tulsi, Ram tulsi)
Oldenlandia herbacea
Rauwolfia serpentina

Psoralia (Babchi)
Saussurea lappa (Kuth)
Saraca indica (Asoke)
Solanum xanthocarpum
Syzygium cumini
Terminalia arjuna
Tinospra cordifolia
Tribulus terrestris
Trichosanthes indica
Vernonia anthelmentica
Vitex nigundo
Withania somnifera

   * According to CDRI, Brahmi is *Bacopa moneri.*

Therapeutic Indications of Some Important Tinctures

1. *Abroma angustifolia* (A. radix): Most effective uterine tonic, regulates irregularities of menstruation and relieves menstrual pains.
2. *Acalypha indica* (Muktajhuru): Remedy for severe cough associated with bleeding from lungs, haemoptysis and in incipient phthisis.
3. *Aegle marmelos* (Bilwa-patra): Cure of dropsy, bilious fever and other biliary complaints. Mild purgative in dropsy during pregnancy.
4. *Adonis vernalis*: Remedy for dropsy, where heart is much involved. Heart beats increase in force and urine rapidly reduced.
5. *Alfalfa* : Wonderful tonic for all run-down conditions, tones up appetite, improves mental and physical vigour. Milk-producing agency for feeding mothers.
6. *Alstonia scholaris*: Tonic, all wasting fevers, especially in malaria associated with diarrhoea, dysentery and impaired digestion.
7. *Amygdalus persica*: Vomiting in pregnancy.
8. *Apocynum cannabinum* (decoction): specific for dropsy, also in doughty skin and oedema.
9. *Saraca indica*: For irregular menstruation with severe pain and leucorrhoea.
10. *Terminalia arjuna*: Most effective and wonderful heart tonic.
11. *Withania somnifera*: Powerful nerve tonic and tones up shettered nerves, sharpens dull memory, removes depilated condition of old age and invaluable in impotency.
12. *Shorea indica* (ash sheora): Most powerful antimalarial tonic and can be used with full confidence in cases of biliary colic, worm colic, diarrhoea and dysentery.
13. *Avena sativa:* Brain and nerve tonic, effective in case of spermatorrhoea and impotency, palpitation and sleeplessness morphea habit and its bad effects are checked with it.
14. *Azadirachta indica*: Powerful antimalarial tonic, afternoon fever and rheumatic pains.
15. *Berberis aquifolium:* In small doses, a drop or so of tincture drive away pimples from the face, also a remedy for "bilious headache".
16. *Blatta orientalis*: A marvellous remedy for asthma including worst cases, pronounced "worth in weight by gold".
17. *Centilla asiatica* (Brahmi): Brain tonic, sharpens dull memory. Group and catarrhal complainers are checked.
18. *Cactus grandiflorus* (Akda): A night blooming Cereus, useful in heart therapy.
19. *Calendula succus*: Pure aqueous juice with alcohol added to prevent fermentation; an excellent remedy for ordinary cuts and bruises and also as surgical dressing.

20. *Calotropis gigantea* (Akda): In elephantiasis, lupus, tuberculous, leprosy, fever, mercurial cachexia, syphilic ulceration, pain in the hand and feet and chronic rheumatism. Useful in scorpion bites and other kind of poisonous bites. Also useful in gangrenous ulcers and in ulcers of leprosy. Good medicine in night blindness.

21. *Carica papaya*: Helps, digestion, checks acidity, regulate liver function, improves appetite.

22. *Ceanothus americans*: Specific remedy for spleen, and malarial fever with faulty spleen and liver; also useful for anemic patient.

23. *Chimaphila umbellate*: Good remedy for chronic bladder troubles where pus and blood are in evidence in the urine.

24. *Crataegus oxycantha*: In cases of heart disease with special indications as pain in heart, rapid pulse, angina pectoris, palpitataion and dyspnoea.

25. *Cuphea viscosissiana*: As indicated in cholera infantum with vomiting of undigested food, acid stool and passing of food right through, or dysenteric stools, small, frequent and bloody with pain, restlessness and fever.

26. *Echinacea angustifolia*: Used in malignant form of disease such as dangerous diarrhoea of typhoid, malignant diphtheria, foul ulcers, bad effects of vaccination, vitiation of blood.

27. *Fraxinus americana*: In hypertrophy of the uterus.

28. *Ginseng*: Specific in stiffness of joints, lumbago, sciatica and chronic rheumatism.

29. *Gossypium herbaceum*: It aids the flow of menstruation and used as emmenagogue in dysmenorrhoea and as abortifacient.

30. *Helonias asiatica*: Uterine tonic, acts well in the stomach in chlorosis of anemia and beneficial in Bright's disease.

31. *Hydrocotyle asiatica*: Medicine in leprosy, skin and female generative organs are chief centres of attack. It has powerful influence on liver, the nerves and mucous membrane; usefull in gout and rheumatism. In white or bloody dysentery action is unique and supreme. Great tonic to keep up strength and memory.

32. *Adhatoda vasica*: Practically specific to cough and highly efficacious in colds, coryza, cough, bronchitis, pneumonia, spitting of blood, fever, jaundice and vomiting

33. *Andrographis (Kalmegh)*: Domestic medicine for flatulence and diarrhoea in children, and also used for worm symptom and is anthelmintic, also used in torpidity of liver, general debility and convalescence after fever.

34. *Lemna minor*: It acts in cases where nose is never clear, contains polypi, sense of smell poor, ozena and catarrh.

35. *Negundium americana* (box elder) painful engorgements of rectum and pile sacs, leading to haemorrhoids.

36. *Nyctanthes arbor-tristis* : Remittent and bilious fever due to irregularities of Liver function and malarial fevers associated with constipation.

37. *Nymphaea colorata*: Used as suppository in uterine cancer, prolapsusuteri, vaginitis, ulceration, inflamed condition.

38. *Oenanthe crocata*: It is quite poisonous and should not be given, not stronger than 5 drops of mother tincture for curing epilepsy.

39. *Origanum major ana*: Effective against sexual passion, in mansturbation and in excessively aroused sexual impulses.

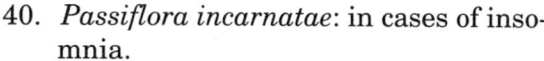

40. *Passiflora incarnatae*: in cases of insomnia.
41. *Oxydendron arboreum*: In cases of dropsy, where there is great accumulation of water.
42. *Passiflora incarnata*: Extraordinary in cases of insomnia.
43. *Phaseolus nana*: In cancer, have strong action on heart when indicated gives magical relief.
44. *Phytolacca decandra*: Large oedematous, cases destroyed in a few weeks and nothing left but a faint scar.
45. *Rauwolfia serpentina*: Effective in blood pressure and insanity.
46. *Rhus aromatica*: Cures diabetes and enuresis.
47. *Solidago virgaurea*: In kidney troubles, where urine cannot void without catheter.
48. *Stellaria media*: A remedy for darting, and shifting rheumatism.
49. *Symphoricarpus racemosa*: Remedy for morning sickness of pregnancy.
50. *Syzygium cumini*: Jamun in diabetes, reduces specific gravity of urine, lessens sugar and effective in chronic dyspepsia along with diabetic symptoms.
51. *Thlaspi bursa-pastoris*: Clears kidneys in cases where there is uric acid, sand or 'brick dust' in urine.
52. *Tribulus terrestris*: In debilitated conditions, sexual weakness, tendency towards impotency and other urinary troubles.

Alternate Healing Therapies
Reiki

In this system, all you need is a pair of hands and a good teacher to plug you into the cosmic energy. The Chinese *ki* or *chi* that surrounds us. The teacher opens up the students channels into bingo. Reiki can be summoned at will to pass through the hands into the affected person's body. This system can be traced back to one Dr. Usui of Japan who delved into the mysteries of cosmic energy with the help of Tibetan masters and Indian Tantriks. Disease is an imbalance of energy caused by stress (the lack of ease) and Reiki facilitates the sense of ease by balancing body's energy. At the basic level, Reiki relaxes the body and allows its immune system to do the work it was meant to be, but was unable to do because of the impediments caused by an unhealthy lifestyle (reading, smoking, eating, drinking, overeating, etc.).

Pranic Healing

This method does not even require direct contact with the body. Disaese is set to manifest itself first in the 'etheric body' (the bioplasmic or energy body, the other non-visible body that every living being has before it percopates down into the visible material body. Pranic healing believes in the nipping it in the bud and catches the disease in the energy body. A pranic healer typically run his hands a few inches away from the physical body. The healer senses the bumps and dips in the energy body, becomes aware of the source of pain or disease, cleans out the negative energies and imbalances, and restores the energy to its original equilibrium.

Colour Therapy

At different points of the day, various parts of the body absorb from the light, the colours that require for their well-being. Orange, for example, is good for respiratory disorders. Purple facilitates sedation and sleep. Red, yellow and lemon are the hot and stimulating colours. Tourquoise, blue, indigo and violet

are the sedative colours. Feeding purple to the head, neck and lungs and heart, for instance, would treat migraine.

Mantra Therapy

There are six *Chakras* or energy centres in the body along the spine. Modern medicine has found six glands corresponding to these "Chakras". According to mantra healers, the right mantra or sound triggers off the gland which secretes the neuro-chemical lacking in the body, which itself is the cause of the disease. The *Vishudh Chakra*, for example, corresponds to the thyroid, which is in conssonance with the sound of 'Aa', and so a mantra based on the 'Aa' sound can correct a thyroid imbalance.

For cure a couplet from Ramayana might be of interest to quote.

as: दैहिक दैविक भौतिक तापा। रामराज नहि काहुहि वयापा।।

Acupuncture and Acupressure

All the parts of the body have their correspondence points on the hands and the feet which are supposed to be miniature versions of the body. They are like 'switch boards' that can be used to operate parts of the anatomy. Pressures of a few needles at the right place, can clear the paths for energy flow and good health. For instance, pressure applied at the 'Su-Sam-Li' (meaning 3 more miles) point below the knee, can provide relief from fatigue and give the walker three more miles.

Chart of various important reflex points is provided in Fig. 2.2. Similar points exist both in hands and feet.

Aroma Therapy and the Aromatic Oils

Aroma therapy is the use of plant-based essential oils to promote well-being. There are two ways for the aroma therapy to work on the body. They are through inhalations or applications on the body or in the form of massages or wraps. In case of inhalations, it could be dry or wet which means that the oil is applied to a cloth or diluted in water. In this case, the tiny terminators in the nose trap the oils and take them to the brain, while in case of applications, it is the process of osmosis that makes aroma therapy effective. To use the essential oils just four drops of the oils must be diluted in 10 ml of the base oil. Sweat almond oil is the most effective base oil. Each essential oil has three notes. One is the top note that evaporates, the other is the medium note, while the third is the base note or lingering note, which is the calming note.

Aroma therapy oils have a shelf life of 3–9 months but are sensitive to sunlight.

There are several aroma therapy brands, available in the market. There is Blossom Kochhar's Aroma Magic, which is divided into bath and body oils, cosmetic and hair oils. In the bath and body oil section there are oils for skin, peace of mind, happiness and as well even as dreams. In the cosmetic sections Blossom has cleansers, toners, under eye gels, dream packs and for hair care there are shampoos and hair oils. Blossom Kochhar is one of the well-known personalities of hair and beauty care products.

Saloni by Sushma Hora is another popular brand, which has oils for all problems including the home peaceful. RK's aroma has oils for depressing fatigue hangover, headaches, insomnia, stress, high blood pressure, colds, cellular stretch marks, muscle aches and stimulating romance. The aroma garden by Jyoti Devjani and Deepika Nangia has a large selection of oils that can help in uplifting, soothing and relaxing the mind and the body. "Breathe" by Deepika Bhatia has stimulating massage oil's bath, and skin oils and oils spirit. Some of the common products

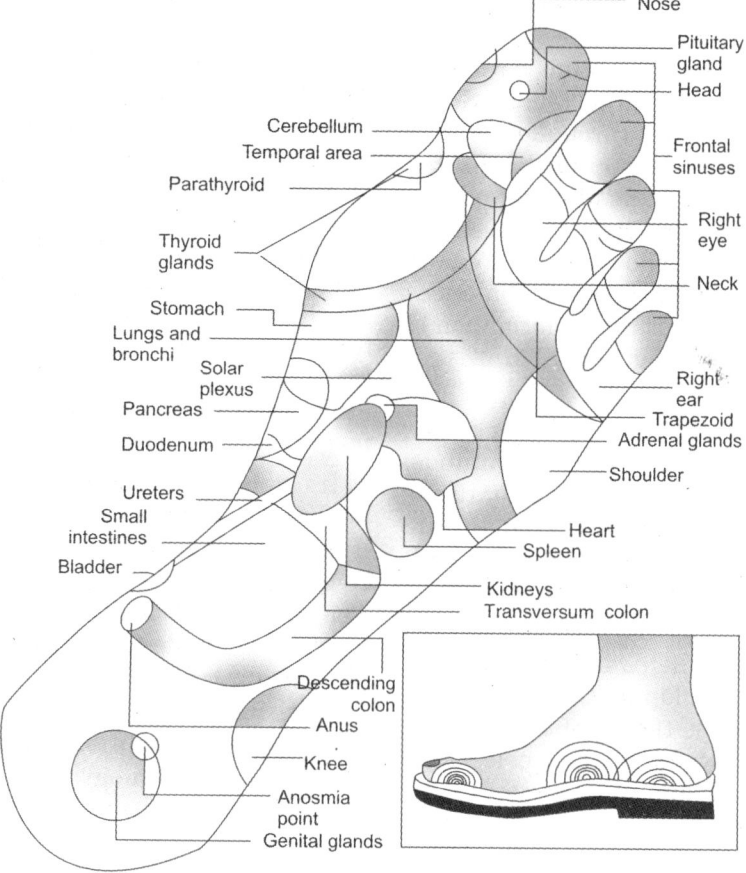

Fig. 2.2: Chart of various important reflex points. Similar points exist both in hands and feet

used for falling hair are Jojoba, Rosemarie, Lavender and for dandruff it is Tea tree. For cellulite it is Fennel, Juniper and Lemon as application and for headaches Lavender, Sandal and Basil can be used to inhale.

Essential oils have healing properties that even improve the energy system of the body and stimulate the immune responses keeping the body healthy. All essential oils are antiseptic, so they help the whole body.

Bergamot, derived from the peel of the fruit is clean. Citrus scent has both balancing and uplifting qualities. Diffuse scent throughout a room by heating a few drops on an oil burner or an electric diffuser.

Chamomile, aids digestion, relaxes and calms, lessens the severity of headache and acts as antiinflammatory. Soak away worries and tension by adding 8 drops to your bathwater, just before stepping into the tub, but make sure to mix well.

Champa oil cools, soothes and softens skin.

Clary Sage (a complex scent) is often used to treat premenstrual symptoms and calm emotional crisis quickly; 4–5 drops are diluted in one tablespoon of light unscented vegetable oil (such as Jojoba). Massage into skin and breathing deeply are effective. It has a slight narcotic effect that may be exaggerated if you are drinking alcohol.

Frank incense reduces inflammation and encourages growth of skin cells.

Lavender is a soft scent that has a relaxing effect that can coax you to sleep. Some hospitals diffuse lavender throughout rooms to help ease patients' insomnia. It absorbs quickly into the bloodstream. Massage oil into palms and soles of feet, to help in sleeplessness and add a drop to your pillow case (it does not stain fabric) for sweetdreams. It balances sebum, soothes and heals, helps formation of scar tissue and encourages growth of skin cells.

Rose oil extracted from rose flower petals is effective for easening emotional traumas. Put a few drops on the handkerchief or tissue and tuck it into your pocket to inhale throughout the day. Rosewood oil is cell stimulant and tissue regenerating.

Rose otto is used for dry and dehydrated skin. Sandalwood oil helps in formulation of scar tissue, reduces inflammation and soothes and softens skin. In terms of therapeutic properties sandalwood is analgesic, anti-depressant and anti-inflammatory, antiseptic (urinary and pulmonary) anti-spasmodic, aphrodisiac, astringent, carminative, cicatrizant, diuretic, expectorant, sedative and tonic. For the mind and spirit it has calming effect and useful for stresses of busy life as it helps reduce tension, nervous depression, anxiety and stress. Being a heavy oil it is more sedative than uplifting and useful for insomnia caused by worry, stills mind and allows creativity and higher consciousness to flower. An important meditation aid, as it calms the mind, stills 'mental chatter' and aids in opening 'third eye'. It opens us up emotionally and helps us to accept others. It is widely known as aphrodisiac although it can also be used to help transmute sexual energy for those who are practising celibacy.

Physically too it has a pronounced effect on genitourinary tract and useful in treating urinary tract infections including cystitis and gonorrhoea. A good pulmonary anti-septic, effective for coughs and for chronic bronchitis and sore throat. If can also be used for treating diarrhoea, irritable bowel syndrome and colic and for skin care. It is the most useful oil and is excellent for dehydrated skin and relieves itching, inflammation and burning sensation. Can be used as astringent for oily skin conditions, extensively used in perfumeries as an excellent base and fixative for other perfumes. By itself it has a deep but mild, long-lasting sweet aroma and can blend with other perfumes and does not impart its fragrance when used as base.

It can be used to prepare for and facilitate healing and visualisation. The *yogis* describe it as a 'fragrance for the subtle body, the centre of highest insight and enlightment. The oil has affinity with the *Base chakra* and works at the level of *Crown Chakra* in facilitating spiritual development.

Gerarium oil helps formation of lymphatic tone, balances sebum and regenerates tissues.

Jasmine and mogra oil aleviate inflammation and are used for oily and scared skins.

Palmarosa oil is a cell-stimulant. It regulates sebum, calms and aids regeneration.

Melodius Cures

Normally considered as the hype of over enthusiastic music lovers, music is finding increasing acceptance worldwide, as a tool in therapy, while music in general works in psychiatric hospitals, rehabilitation facilities, outpatient clinics, day-care treatment centres for the mentally disabled, mental health care centres, drug and alcohol programmes, old age homes and juvenile delinquent homes. Clinical trials with music therapy have revealed a reduction in heart rate, blood pressure, breathing rate, insomnia, depression and anxiety. Cardiologists use it in cardiac cases and insomnia disorders. Psychiatrists claim a calming effect on nervous system and is effective on patients with neurosis or psychosis. Music is also reported useful in reducing the intensity, duration and frequency of migraine head-aches. A fifty-years study in France concluded that people with defective speech or inability to communicate easily and clearly, improved their expression power; when they listened to Mosari's music every day for one hour, over a period of 6–7 months. Music therapy with antiemetic drugs (drugs that relieve nausea and vomiting) in patients receiving high dose chemotherapy can help cases with physical symptom of nausea and vomiting. A number of trials have shown that music with pain-relieving drugs decrease the overall intensity of the pain and even help reduce use of pain-killers.

There is a theory that our muscles including the heart muscle, learn to synchro-nise to the best of the music. Another theory proposes that music and sound distract the mind from focusing on pain and anxiety, perhaps due to its origin in divinity and hence endowed with great powers.

The music therapy department of Apollo Hospital at Chennai says that it is the use of one more opportunity to heal their patients better and faster. This department testifies its utility in paediatric, gynae-cological and neurology wards. Passive form of music therapy is believed to reduce hyperactivity and problematic behaviour of children, while active form enhances speech, reading fluency, concentration power and improving classroom performances.

The hospital claims that premature babies can gain weight faster, exposure to soothing music during labour makes it an easier experience and is effective in curing neurological disorder patients who cannot or won't talk but can sometimes hum, even dance to music.

It is, however, admitted that music cannot cure all ailments and diseases though music can cure help healing, eases pain, hastens recovery and can contribute to a visible reduction of intensity of symptoms in many illnesses. However, the use of word *cure* is carefully avoided in the western approach making it clear that music is a therapeutic "tool". Though, in India, even resurrection powers are attributed to music. These as such mentioned in Veda (*Samveda*), while great music composer Thyagaraja is said to have resurrected a dead man with his rendering *Kriti* (musical composition), Naa Jee-Vadhara and Jayadeva with his "Sanjeenai". Muthuswamy Diskshitar is said to have cured his disciple's chronic severe stoma-chahe with his Navagraha Kirtis. Late Emani Shankari said "the powers of music to calm, soothe and resusat is limitless, one only has to approach with respect and faith".

Healing with Stone Therapy

The use of stones on and around the body as a healing tool has been popular through-out the history of our planet. Stones have been used in treatments dating back as far as the South America, the Egyptians and the Chinese. In India too, with rural households, heated stones are still being used to relieve body pains during preg-nancy and deliveries after childbirth. The stone is the modern way of working. This age-old practice by Mary D. Hannigam (1993), has been adapted for use by many different workers. One such therapy is conducted at Ananda and goes under the name LITHOs. Traditionally, the therapy uses a combination of massage, thermo-therapy (alternate application of hot and cold temperatures), energy work and healing. Stone can also be used in other treatments to make these treatments much more effective, they are aromatic massage, deep-tissue work, reflexology, facials and

waxing, manicures and pedicures, remedial massage and relaxation therapy.

Stone therapy brings to the body a profound sense to balance and peace. If one's system is sluggish, this type of treatment assists in opening and clearing the alternative pathways to stimulate the system and increase the energy levels. If someone is under stress, unsettled and cold, then the outcome of this system would be to feel much more grounded and calm, safe and warm. It not only balances to the physical body but settles to the mind, emotions and the spirit.

Use of basalt stones (grey, black, dark green) assists in staying hottest and longest. Black stones have always been used in healing for this reason. The stone therapy is good for sprains, strains, bruises, chronic tension, stress, muscular pain, headache, insomnia and constipation.

The shape and colour of the stone have a bearing on the use of certain stone. The more round the stone and the lighter the colour, the more feminine or *Yih* is its nature. The longer or more oblong stone along with darker colour offers more *yang* or masculine qualities to the body. Use of hot stones over an isolated area causes passive dilation of the blood vessels of the related organ. It also pulls blood from the organ to the surrounding tissue, resulting in warm flushed skin. Hot stones are used to expand or dilate blood vessels, increase circulation, metabolism, cell metabolism, lymph function, pulse rate and sedate the muscles and the nervous system. Cold stone therapy is used in an isolated area and causes an immediate numbing effect, followed by dilation of blood vessels in related organs. They are also used to rehabilitate injuries and chronic muscle spasms, speeding up the recovery, where motion increases and muscles and nervous system are stimulated.

In fresh injuries and burns cold stones are used to constrict the blood vessels, push blood away from the area, stimulate and improve circulation, produce an analgesic effect, limit further damage to injured muscles, reduce inflammation and stimulate the nervous system.

A natural response to cold is that the body will bring itself back to a homeostasis, thus the body heats itself. This treatment is done as part of a massage treatment, mainly because it feels fantastic and is relaxing and invigorating, depending on what an individual requires. The body will find its own individual balance.

References

Acharya Vaidya, JT. 1980. *Sushruta samhita of Sushruta*. (Rept.) pp. 16. 824. (Sanskrit)

Dahanukar S. 1989. *Ayurveda Revisited*-Ayurveda in the Light of contemporary Medicine. 11. 189.

Dey AC. 1998. *Indian Medicinal Plants used in Ayurvedic Preparations*. pp. 202.

Eiri *Handbook of Ayurvedic* and *Herbal Medicines with formulaes*. 18, pp. 274.

Ghai CM 2004. *Health Rejuvination and Longevity through Ayurveda: Holistic and Preventive Herbs for Better Health*. 7. pp. 272.

Godagama S. *The Handbook of Ayurveda: India's Medical Wisdom Explained*.

Govil JN, *et al. Medicinal Plants: New Vistas of Researches*. Pt. 1 and 2.

Gupta KRL 1997. *Madhava Nidana: Ayurvedic System of Pathology*. 9–11. pp. 270.

Karnick CR. 1996. *Pharmacology of Ayurvedic Medicinal Plants*. 67 pp.

Kulkarni PH. 2000. *Ayurvedic Herbs*. pp. 147.

Kulkarni PH. 2004. *The Ayurvedic Plants*. 16. pp 334.

Nadkarni AK Dr. KM Nadkarni's *Indian Materia Medica with Ayurvedic-Unani-Tibbi, Sidha, Allopathic, Homeopathic, Naturopathic and Home Remedies*.

Nadkarni, AK 1976. *Indian Materia Medica (3rd ed)*. Popular Prakasan. Bombay.

Nagar S. 2000. *Bontanical and Medicinal Plants: As Depicted in Ancient Texts Art and Archaeology from Dawn to Civilization.* 14. pp. 488.

Narayan Rao UB. 2003. *Indigenous Medicinal Specialities.* 32. pp 304.

Pandey G. 1994. *Uncommon Plants-Drugs of Ayurveda.*

Pandey G. 1996. *Some Lesser Known Herbal Drugs in Ayurveda.*

Nair CKN. 1998. *Medicinal Plants of India: with special Reference to Ayurveda.*

Paranjpe P. 2001. *Indian Medicinal Plants-Forgotten Healers– A Guide to Ayurveda Herbal Medicine.* 32. pp 316, colour plates 135.

Sensarma P. 1998. *Ethnobiological information in Kautiliya Arthshastra.* pp. 184.

Sharma R. 2003. *Dictionary of Ayurveda* 7. pp. 412. appendices. Bibl.

Singh R. 1998. *Vedic Medicine.* 12. pp 306.

Singh S. 2000. *Traditional knowledge of the medicinal plants of Ayurveda.* 6. pp. 392.

Sircar NN. *Vriksayurveda of Parasara: A treatise on plant Science.* 42. pp. 166.

Thatte DG and Tiwari GP. *Current trends in the study of Sarira.* 20. pp.80.

Sandhu DB, 1987. *Indian Therapeutics: Dealing with the Comparative Diagnosis. Ayurvedic Materia Medica. Dietetics prescriptions.*

Salini MS. 1998. *Vedic Leguminous Plants; Medical and Microbioloigcal Study.*

Sarin YK. 1996. *Illustrated Manual of Herbal Drugs used in Ayurveda.*39 pp. 422.

Vaidya Shrikar D. 1999. *Instant and Fast Acting Ayurvedic Treatment: Drugs Formulas and Therapies Asukari Chikitsa in Ayurvedic* Jl-lukar. 10. pp 206.

Unani System

Hakim MS. 1990, 1997. *Hamdard Pharmacopoea of Eastern Medicine.* 16. pp 544. with planter and line drawings.

Hamdard ? *Diseases and Treatments.* Hamdard wakf. Delhi. pp. 1–32.

Tibetan

Dewa Narbu *Introduction to Tibetan Medicine.*

Dash Vaidya B. 1994. *Encyclopaedia of Tibetan Medicine.* 7 vols 22. pp. 694. figs. Vol. 2

Finchke E. *Foundations of Tibetan Medicine.* Vol 1.

Tsarong 1994. *Tibetan Medicinal Plants.*

Das Vaidya B. *Tibetan Medicine with special Reference to Yoga.* Salaka Lama S. and Santra SC. 1979. Development of Tibetan Plant medicine. *Sci and Cult.* **45(7)**: 262–265.

Chinese

Huang KC. 1994. *The Pharmacology of Chinese Herbs.* 21. pp 512.

Stuart GA. 1911. *Chinese Materia Medica. Vegetable Kingdom.* 6. pp. 558.

Homeopathy

Gopi KS. *Encyclopaedia of Medicinal Plants used in Homeopathy.* Vol 1.

Hamilton E. 2001. *The Flora Homeopathica-Illustrations and Descriptions of medicinal plants used in Homeopathic remedies.* 34. pp 523.

❏❏❏

Aromatic Plants
and
Essential Oils

Out of a total of about 17,500 species reported to occur in the Flora of earlier British India by Hooker (1904), about 6700 are reported to be endemic to India of which Himalaya alone accounts for 3,169 dicotyledons and about 1000 monocotyledons. In addition, 1300 species are reported to contain odoriferous substances which made this land famous for its scented roots, leaves, flowers and essential oils, attracting large number of travelers and traders from all over the world. The position of essential oils production is not any better, of late. The export of sandalwood oil showed a slow decline from 1000 tons to 200 tons with Guatemala, pushing its produce in the world. We are losing our hold on Palma Rosa oil market, because of our inability to meet the increased demand; Indonesia and now Brazil have come up with its regular plantations. The fusarium rot of geranium and nematode attack on Patchculi oil do not allow these crops to expand to large areas. Indian Celery seed could reach a modest value of a little more than Rs 2,000. Delays in export and some marketing malpractices enabled Israel to enter the markets in Germany and Italy. Increase of local consumption of aromatic oils also reduced the export values. *Citronella javanica* cultivation in the meantime extended to Assam and Karnataka states, while remarkable progress was achieved in the indigenous production of mint oil in the Tarai area of U.P. now Uttaranchal, supporting research in agro-techniques, weed control, and distillation techniques has enabled the country to augment export capabilities at par with many other countries like China, Japan and Taiwan. (Ref. the Chapter Agro-techniques for details).

Perfume and perfumery material in India are associated with aristocracy, while in other societies of Europe and America these are the common articles of use in affluent societies. Actually more than 280 million US dollars of natural essential oils and allied perfumery materials are purchased by developed countries of the world in which India's share is just not more than 4–5 per cent. One could push export of the perfumery material and essential oils for higher income. Export of Davana (*Artemisia pallens*), Panari (*Pogostemon cablin*), Costus (*Saussurea lappa*), French basil (*Ocimum basilicum*), Khus (*Vetiveria zizanioides*), floral perfumes like jasmine, Champaca, Nag-kesar, Keora, Acacia and many other similar oriental flowers could be pushed with their commercial cultivation, aggressive marketing and conscious quality control. However, also there is dearth of trained personals for manning these plantations. Essential oil-bearing plants like anse, celery, java cironella, geranium, mints, palma rosa grass oil, patchouli and vetiveria

have received special attention under co-ordi-nated programme of the ICAR and CSIR institutes; crop research is a continuous process which involves release of improved plant stock and cultivation devices to cultivators.

In the field of essential oil, development of a tetraploid strain in Japanese mint containing 5 per cent essential oil (zero moisture) and screening trials at CIMPO farm had two findings positive impact in the 'Sin bar' herbicide effective in control of weeds in Japanese mint and peppermint crops. Another significant work is the isolation of high yielding cultivars of ill with large-sized fruiting umbel. Lemon Grass Research Station in Oodakaly (Kerala) developed a strain of lemon grass called as OD 19 which gave larger herbage and oil yield than the existing plant type. However, details regarding the agrotechniques including improved varieties and measures to control pests insect pests, have been discussed in detail for some species the author section of this work the information is based on various experiments and trials progress in various parts of the country by organisations like CIMPO, CIMAP, regional research laboratories of CSIR and the coordinated project of the ICAR.

Nature still remains the best maker of flavours, for its preparation is beyond the reach of men. Spices, herbs and seeds are the nature's self-made aromatic botanic products, some have a delightful flavour, while some are strong and pungent but invariably they are indispensable in the preparation of delectable dishes. These are in a large way, alternative to alcohols, modern stimulant drugs and are certainly cheaper, less habit forming and due to their adjuvant and alleviative qualities are more likely to promote natural and healthy body functions. Apart from their culinary and medicinal uses, many spices possess bactericidal and fungicidal properties and quite a few of them also are powerful germicides. Aromatic plant species have become the inevitable concomitant of urban living the world over, where everything has to be had readily, in a handy form. Without spices it would be impossible for the processed and tinned food industry of today to present their natural taste.

A large number of aromatic plants contain essential oils which are important from their utilitarian point of view. Natural perfume is one of the most remarkable phenomena of plant metabolism and the history of aromatic plants has been associated with India, since times immemorial. During the Greek and Roman periods, the trade flourished between India and the West.

The spices and aromatic plants were at one time considered as weapons of mystic cult and magic and earthly symbols of supernatural powers. Basil, for instance, as claimed could, when kept hidden under a vessel or cut and mingled, transformed into a scorpion. It could draw out poison from the bite of numerous creatures. It was sacred to the Indians as the protecting spirit of the family. Leaves and flowers of borage dipped in drinks could bring about absolute forgetfullness. Portions of caraway on the other hand could force lovers to be faithful, help detect thieves and tame pegions. cardamoms were grains of paradise, with matchless aphrodisiac powers. coriander, it is said was one of the first and most ancient spices to be used by mankind. It had magical powers and romantic virtues, it could bestow immortality, conjure up spirits and cunjugate those in love. Garlic served as supper Hecate, the Greek underworld goddess of charms and enchantments. Persians used cloves in 'Clove pithers'; it was an ingredient to several traditional medicines and sacred offerings in religious ceremonies. The quality and efficacy of spices were directly attributed to the Gods, the Sun, the Moon and the Stars.

Aromatic plants and spices were used to work for good as well as evil. Doctors applied spice balms to heal the wounds of soldiers and administered doses of aromatics to ease sufferings of the sick. They were basic to ancient perfumery and cosmetics. The subtle green and floral fresh notes of sweet smelling spices were essential to total naturalness. Fragrant fumes of Cinnamon perfumed the ladies of pleasure in Biblical times and the bodies of Egyptian women. Essences refined from cinnamon, saffron and ginger added distinction to beauty; spice recipes and sweetness made of saffron, sandalwood and myrrh gave personal freshness to the body; and aromatic dyes tinted the men and women of India in delicate hues and decorative ways. Cumin paste applied to the face gave the Roman an exclusive look. Spice perfumes breathed a new air into households and spices entered into cooking for flavour, but to take away the ill-effects of putrefaction of foods.

The essential oils occurring in aromatic plants form an important group. These oils are present in small quantity, e.g. in Indian cloves, it is as much as 16–18%, while in Indian rose flowers as little as 0.02 per cent and in Indian jasmine one tenth of this quantity. Chemically they are a combination of such substances such as terpenes, sesquieterpenes, phenols, alcohols, acids, esters, aldehydes, ketones, nitrogen and sulphur compounds. While vegetable oils are combination of glycerine and various fatty acids . Their functions are not well understood but they are said to help in the cross pollination, and their presence in the bark to have protective value against insect attacks. They are also said to regulate the rate of transpiration in plants. These oils are used variously in everyday life, such as in the manufacture of soaps, cosmetics, pharmaceutical preparations, confectionery, aerated waters, tobacco manufacture, disinfectants, detergents, and incenses, etc. According to Sadagopal, 1000 different aromatic plants out of a total of 1500 are used in perfumery throughout the world are found in India. The production of the essential oils is in an unorganised way. Since there is a great demand of certain oils, like eucalyptus oil, rose oil, sandalwood oil, lemongrass oil, turpentine oil, palmers oil, cinnamon oil, citrus oil, etc. particular attention need to be paid to the cultivation of these plants on a more scientific line (Table 3.1).

Table 3.1: Common spices, condiments and herbs used in Indian dietary

	Local name	*Parts used*
1.	Anise	Seeds
2.	Asfoeida	Rhizome, oleoresin
3.	Bishop's weed	Seeds, leaves
4.	Carawy	Seeds
5.	Cardamom	Capsule, seeds
6.	Cassia	Bark/Bud
7.	Cayense	Dried pods or fruits
8.	Celery	Seeds
9.	Chilly	Fruit
10.	Cinnamon	Bark and bud
11.	Clove	Unopened flower buds
12.	Coriander	Seeds, leaves
13.	Cumin	Seeds
14.	Cumin black	Seeds
15.	Curry leaf	Leaves
16.	Dill	Seeds
17.	Fennel	Seeds
18.	Fenngre	Seeds
19.	Garlic	Bulb
20.	Ginger	Rhizome
21.	Mace	Arilus

Table 3.1: Common spices, condiments and herbs used in Indian dietary (*Contd..*)

	Local name	*Parts used*
22.	Mango	Fruit
23.	Mint	Leaves
24.	Mustard	Seeds
25.	Nutmeg	Seeds
26.	Onion	Bulb
27.	Paprika	Dried pods
28.	Parsley	Seeds
29.	Pepper	Berry
30.	Pomgranate	Seeds
31.	Poppy	Seeds
32.	Safforn	Stigma
33.	Tamarind	Fruits
34.	Turmeric	Rhizome
35.	Vanilla	Bean

Essential Oils and Perfumes

A very exhaustive treatise has been published by Guenther on essential oils. In spite of the huge volume of work included, information on many oriental species is lacking. Older works of Parry, though brief, yet give more information on the oriental herbs used in perfumery. The reports of Schimmel and Co. and bulletins of Roure Bertrand Fils, American Perfumer, and others also include much information in the analytical aspects of perfumes and essential oils.

Essential oils are the byproducts of plant metabolism and are important item of our economy, it is not only the flower that bears perfume but other parts also are odorous. The actual mechanism and precise manner by which essential oils develop is still a matter of much theoretical hypothesis and the entire field remains to be explored in spite of recent advances. How they are linked with other products in the functioning of plants is an aspect, the knowledge about which is still far from adequate.

Unlike Lady Macbeth, we may not need perfumes to remove the smell of blood. We, however, certainly need them for a better, cleaner, happier and more sophisticated living. We need them for use in our cosmetics, soap and detergents, foods and beverages, sweets and confectionary, chewing and smoking tobacco, joss-sticks and *aggarbattis*, disinfectants and pharmaceutical preparations and a host of other articles of everyday use. Most modern technologies are engaged in scientifically unraveling and elusive mysteries of exciting new odours. They are using their imagination and ingenuity for skillful blends, subtle nuances and new tonalities in perfume.

Man has not become fond of perfumes only in modern times. From times immemorial, his aesthetics have led him in search of pleasant exotic, exhilarating, soothing odours. Right from the dawn of civilisation perfumes have been used for religious ceremonies at the alters of God, for thanks offering, for anointing and embalming and for scenting the air in places of worship and courts of kings.

If it were possible to delve into the past, it would probably be found that the romance of perfumes has its beginning with the Allanmans, who flourished thousands of years before the Christian era. Earliest records of perfuming have been found in Egypt. The opening of tomb of Tutankhaman, who ruled about 1350 BC has brought to light many excellent specimens of the early perfumers art, the unguent vases exquisitely executed in alabaster contained quantities of aromatics which are still elusively fragrant. Like Egyptians, Greeks also held the perfumes in high esteem. Homer, frequently refers to perfumes in his "Iliad" and "Odyssey". Theophrastus (370 BC) was probably the first to write on perfumes from plants. He also speaks of the compounded perfumes as distinct from the flower perfume. Romans during their early history showed very little interest in perfumes. Nero, became the empress of Rome in 54 AD and by this time both perfumes and cosmetics had assumed an important role at her court. In the centuries that followed, the Arabs delved into the scientific side of perfumery more than any other race. In 10th century, *Avicenna* made efforts to extract perfumes from flower by distillation. He was, however, lucky to isolate some perfumes in the form of oil and to produce supplies of rose water.

Famous travelers of the past, like Fa-Hien, described India as the land of aromatic plants, fruits, flowers, woods, roots, resin and grasses, of these best known was Sandalwood and its oil. There happens to be a regular barter trade from ancient days for sandalwood, when caravans carried precious wood to countries like Egypt, Greece and Rome. Moghul emperors were great patrons of perfumes and references in 'Ain-e Akrabary' are testimony to it.

The essential oils are extracted from the roots, stem, bark, leaves, flowers, seeds and peels and are different from vegetable oils in being volatile and composed of chemical constituents of a different nature. Vegetable oils are combinations of glycerin and various fatty acids.

In India, major part of essential oils are still produced by water distillation on an open fire with old-fashioned stills and condensers. Modern steam operated stills with arrangement for stripping columns have come into operation in case of oils of linaloe, palmarosa, sandalwood, vetiver, chenopodium, peppermint, etc.

During the last few decades, the export of essential oils increased many fold. Since the olfactory spectrum of the natural products is relatively narrow, the marketing needs demand a constant creation of new perfumes and flavors. With the country production of synthetic perfumes is very recent, e.g. benzyl and phenyl series of products are based on coaltar derivatives, amylesters from fusil oil derivatives, amyl cinnamic derived from castor oil cracking, yara-yara from beta nephthol. Similarly, manufacture of ionones and methyl ionone from citral (extract oil of lemongrass) as well as camphor and allied products is based on indigenous oil of turpentine.

Annotated list of plants providing aromatic and other essential oils is as follows.

1. *Amaryllidaceae*

a. *Hemerocallis fulva* Linn. Commonly cultivated in the gardens. Natural perfume is not exploited but synthetic "yellow tuberose" is made by using turpineol, hydroxy-citronellal amyl-salycilate, etc. which does not fully match with the natural perfume.

b. *Narcissus poeticus, N. tazetta, N. jonquilla.* Absolutes and concretes of Narcissus are obtained from flowers. Steam distillation gives poor oil quality. Used in perfumery.

2. *Anacardiaceae*

a. *Schinus mollel:* Indigenous to America but commonly grown along avenues and other shady places. The berries are armoatic and possess sharp spicy smell. The leaves also contain essential oil. Used for flavours.

b. *Pistacia mutica* Fisch. et Mey: Occurs in Baluchistan, Afghanistan. It along with *P. integrrima* yields an oleoresin known as gummastic. The gum is used in medicines, flavours and lacquer.

3. *Araceae*

Acorus calamus: In semi-aquatic conditions, found all over the world. In warmer climates, it develops more essential oil than in cold. Rhizome is commonly used for aromatic vinegar. Also in powder form used for preparation of toilet and sachet powder; flavouring of gin and beer. It is also used as insecticide.

4. *Annonaceae*

Cananga odorata: Found throughout humid tropics of Asia (Indonesia, Philippines), Madagascar and Burma (Myanmar). Cananga oil is obtained from the flowers which appear from June to December. It is also called Ylang-Ylang oil, while oil from *C. odorata forma macrophylla* is called Canagaa oil.

5. *Bignoniaceae*

Bignonia suaveolens: Flowers are used for extracting Begnonia oil. In ancient literature it has been described as quiver of *Kamdeva* (Indian cupid), Kamduti (Massenger of love) but not in common use now.

6. *Burseraceae*

a. *Commiphora wightii* (Gugul)*:* distributed in arid Rajasthan,
   (= *Balsamodendron roxburghi*)*:* Pakistan, Sind, Baluchistan
   (= *B. pubescens*) Baluchistan and Sind: The balsam gum obtained by giving incision on the stem is used in the manufacture of heavy oriental perfumes. It is often adulterated with turpentine, Canada balsam and Rosin. *C. myrrha* var. *molmol*, source of oil of myrrh is not found indigenously, but may be tried.
b. *Boswellia serrata* (var *foliosa, ovata, lineari,* and *lanceolata:* The essential oil (olibanum) obtained from the gum-resin dissolves varnish resin. Indian olibanum is obtained from these species. The varnish dries rapidly. Tincture of olibanum is used as a fixative in perfumery.

7. *Cannabinaceae*

Humulus lupulus: The essential oil is obtained from the flowers, which is principally used for flavouring and given a special note to perfume.

8. *Caprifoliaceae*

a. *Lonicera webbiana, L. quinquelocularis, L.sempervirens, L. japonica, L. confusa.* The first two species are in the western Himalaya while others in cultivated gardens. The essential oil extracted from species are in the gardens. The essential oil extracted from the flowers is obtained by treating them with petroleum ether and alcohol. Very little is known about the chemistry of the oil.
b. *Viburnum grandiflorum Wall. V. nervosum, V. foetens* and *V. cotinifolium:* Distributed in temperate regions of W. Himalaya. The flowers have a pleasant odour and can safely find use in perfume industry.

9. *Caryophyllaceae*

Dianthus caryophyllus: Cultivated in gardens and many varieties have been evolved. Oil is extracted from the flowers having intense odour. It is used in perfumery.

10. *Chenopodiaceae*

Chenopodium ambrosoides: Oil is obtained from the fresh plant in flowering stage. It is pale liquid with heavy odour. It is used primarily in medicine but with suitable dilution can be used in soap industry.

11. *Compositae* or *Asteraceae*

a. *Achillea millefolium:* The plant is found wild. The oil is obtained by steam distillation. It was used as an adulterant of chamomile. It is used in flavours and perfumery.

b. *Anthemis nobilis:* Cultivated sometimes as an ornamental plant. The flowers yield chamomile oil and used as like *Matricaria chamomilla*. German chamomile is cultivated in more than 40 countries like Argentina, Egypt, Italy and Switzerland. Blue oil is extracted from flower and used in scented cream and shampoo.

c. *Artemisia absinthium:* The oil of wormwood or oil of absinthe is obtained from the leaves and buds of various species. The characteristics of oil vary with locality and conditions of growth. It is used to a considerable extent in flavouring and soap industry and to a small extent in perfumery.

d. *Artemisia dracunculus:* Distillation of the flowering herb yields 'Estragon oil', which is yellowish green and similar in odour to aniseed. It is employed for the manufacture of wines and in aromatic and toilet vinegars and as flavouring agent.

e. *Artemisia maritima, A. cine:* Both these species yield Levant wormseed oil from the flower heads in bud condition. It yields 2–3 per cent volatile oil as a byproduct of santonin extraction. The oil is a source of cineole.

f. *Artemisia pallens* (Davana): Oil is extracted through vapour distillation of flowers. Mysore and Bangalore produce oil as industry. Oil used as medicinal.

g. *Artemisia vulgaris:* All parts of this plant contain oil called Yomugi oil. Not used commercially.

h. *Blumea balsamifera* (Kakronda), *B.malcomii* (Panjirut) *B. aromatica* (Panjrut): Yield Blumea oil, which is obtained from the fresh material (leaves and stalks) during the production of Ngai camphor. It is an aromatic liquid used for the manufacture of Ngai camphor, Chinese ink, incense manufacture and medicine. The plant is common in Malaysia, Indonesia and other parts of tropical Asia. *A. vistata* provides oil through steam distillation of flower and is used in cosmetic and aroma industry. First in 1962 oil was extracted by Handa and Vashist.

i. *Conyza stricta:* Hilly shrub available in Mahabaleshwar and Chota Nagpur. A new diterpene resin acid, known as Conzic acid, has been isolated in good yield by Dutta in 1975.

j. *Erigeron canadensis:* A plant of the western Himalaya. Flebane oil is extracted from the fresh herb collected at the flowering time. It is used in cheap toilet preparations.

k. *Inula graveolens:* A plant of temperate W. Himalaya. Inula oil has an odour of lemons. May be used as a substitute flavour for lemons.

l. *Matricaria chamomilla:* Commonly cultivated as an ornamental. The flower yields a deep blue essential oil used in medicine, as flavouring agent and perfume.

m. *Saussurea lappa:* Distributed in Himalaya. The roots give costus oil having strong antiseptic and disinfectant properties. It is used in perfumery and in medicine.

n. *Tagetlis minuta:* Yields an essential oil which is used in pharmaceutical and perfumery industry. Maximum yield of both the herb 324.59/h and oil 50.637 kg/ha is obtained with application of $N_{120} P_{120}$ kg/ha qtls. having an increase to the extent

of 73.6 and 55.7% as compared under control. Seeds also yield Pyrethrins.

12. Coniferae

a. *Abies pindrow:* A tree of temperate West Himalaya. Essential oil obtained from the leaves is used in cheap perfumery and soap industry.

b. *Cedrus deodara:* A tree of the temperate West Himalaya. The wood yields Cedar oil having a balsamic pleasant odour. It is used in cheap perfumery, where heavy smell is needed.

c. *Cupressus torulosa:* A tree of temperate W. Himalaya. Essential oil obtained from the leaves can be used in perfumery industry.

d. *Juniperus communis:* The Juniper oil extracted from the berries. Oxidises rapidly on keeping. It is used as a flavouring agent in liquors, and to a limited extent in perfumery.

e. *Juniperus macropoda:* A plant of the western Himalaya. Oil is extracted from the roots. It is not much in use except in flavouring gins and liquors.

f. *Pinus roxburghii:* An important conifer of the Himalaya. The wood yields turpentine oil from rosin, used intensively in perfume industry as a raw material for turpineol, camphor and as an adulterant in essential oils. It is also used in the medicines. Use of pine-needle oil in perfumery is restricted due to compositional variation from place to place.

g. *Pinus excelsa:* In temperate regions of W. Himalaya. The essential oil is obtained from the wood and needles is used in medicines, varnishes and cheap type of perfumery.

13. Convolvulaceae

a. *Rivea hypocrateriformis:* The flowers yield a perfume having a limited use in the perfumery industry.

b. *Ipomoea grandiflora (I. bonanox var.grandiflora L):* It is a *clove scented Ipomoea,* but not much exploited.

Cyperaceae

a. *Cyperus rotundus, C. longus* and *C. niveus :* The tuberous roots are powdered and used as a perfume. *C. longus* has a most sweet odour.

b. *C. scariosis* (a weed)*:* Provides oil through steam distillation of the rhizome. Used as an alternative to patcholic oil that is extracted in UP, Bengal and Punjab.

14. Dipterocarpaceae

Dipterocarpus turbinatus: Gurjan balsam is the exudation obtained by giving a blaze or cut in the wood and is collected the like the rosin. There are two balsams-kanyin oil from *D. turbinatus,* and also an oil from *D. tuberculatus.* Consistency and colour of the oil depends upon the method of extraction. It is used as a fixative perfume and also for scenting soaps.

15. Ericaeeae

a. *Rhododendron arboruem, R. lepidotum* and *R. campanulatum:* Plants of high altitudes of W. Himalaya. Flowers yield rhododendron oil, having a pungent aromatic odour. Not used extensively in the perfumery industry.

b. *Gaultheria procumbens, G. fragantissima:* All the plant parts are used to extract 'sweet birch oil' or 'oil of wintergreens'. The former is inferior. It is used in the perfume manufacture in industry.

Erythroxylaceae

Erythroxylum monogynum Roxb. (Bustardsandalwood) is aromatic and used as an adulterant in the industry.

16. Euphorbiaceae

a. *Exocaeria agalloha* (Agar)*:* Plant of the tidal forests. Heartwood of the young

trees is odorous and on distillation gives an ester with pleasant heavy smell. Wood is used in *agarbatti* trade and in medicine.

b. *Ricinus communis:* The plant is cultivated in many parts of India. Castor oil obtained from the seeds is nearly a colourless pale liquid and is an ingredient of collodium and as lubricant for stomach to ease way for waste products of the intestines.

17. Geraniaceae

Pelargonium: It is commonly cultivated. Horticultural term "geranium" is used for this genus. *P. graveolens* and *P. roseum* are some of the species yielding geranium oils (Rose geranium). Geranium oil is a colourless to yellowish green liquid and employed in the manufacture or toilet praparations. The truc geranium oil should not be confused with Indian or Turkish Geranium oil which is obtained from *Cymbopogon martinii*. The oil is obtained from leaves picked at the beginning of flowering season. The oil from various origins is put to different uses. It is used for scenting soaps, powders, creams and perfumes. Cultivated mostly in Sevroy hills of South India and in the Nilgiris. About 14,000 ha or more are under cultivation producing about more than 20 tonnes of oil in a year. Oil used in cosmetics, perfumery and soap industry.

Gramineae (Poaceae)

a. *Anthoxanthum odoratum:* A gass of hilly districts which is sweet-scented perennial. Native of Europe and North Asia.

b. *Cymbopogon species:* There are many species of this genus which have been described at length in detail in other chapters of this book. The oils from different species are used for various purposes like extraction of citral, as flavours and prepartion of β-ionone.

c. *C. winterianus* (Java grass)*:* It is being cultivated in India, oil used as cosmetics and pharmaceuticals. About 7000–8000 ha is under cultivation producing 5000–6000 tonnes of oil. *C. martinii* leaves provide oil and the natural source of Geraniol and Geraniol acetate used in cosmetics. Cultivated in Maha-rashtra, Andhra, M.P. *Cymbopogon flucuosus* oil from leaves, provide 80.90% citrol used in manufacture of α and β-Ionon, the latter is used for vitamin A production in Kerala and in UP.

d. *Vetiveria zizanioides* Stapf: Oil obtained from the roots is a popular perfume. The essential oil is thick brownish liquid with a persistent strong odour, compounded in oriental perfumes. It is used by perfumers and in different flavours. It is mentioned in Sanskrit as *Usira* and *Bala*. Bharatpur and Kanally are the centres for oil production, more than 12 tonnes of oil annually is produced.

18. Iridaceae

Iris germanica: Irris root oil is extracted from the roots that are dried slowly. It is much in demand in the perfumery for powders or in the form of tincture.

19. Juglandaceae

Juglans regia: Fresh leaves yield an oil having an odour resembling tea and amber. It is used also as a flavour.

20. Labiatae (Lamiaceae)

a. *Calamintha umbrosa:* A plant of temperate regions in Himalaya. Oil is obtained from the leaves and stem and is rather a strong and complex oil, used in perfumery. *C. nepeta* is another species from Sicily.

b. *Hyssopus officinalis:* A plant of North west Himalaya. Hyssop oil is obtained from flowers and is a pleasant aromatic

liquid with sweetish odour. It is used as a flavouring agent in Sicily (Italy).

c. *Lavendula officinalis:* A plant of the Mediterranean region, but introduced in India at Kashmir and other states and in Pakistan at many places. The stalk and flower yields lavender oil which is extensively employed in the perfumery, cosmetic soap and toilet preparations.

d. *Mentha piperita* (developed from Nil-giris hills, the oil is exported), *M. viridis, M. sylvestris* and *M. arvensis:* (Japanese Mint) Menthol Mint. Mentha oil is mostly distilled from the dried herb. The oil is used in flavouring of the medicine and perfumery. Menthol is now exported. 1995–96 cultivation was in 60,000 ha provided, 6000–10,000 tonnes of oil that contains 70–80% menthol. *M. spicata* was introduced and about 30 years back. Spear mint oil is distilled and is used in food products. Carvin is more than 60% *M. citrata* introduced in 1959, about 100 tonnes of the oil is extracted, and used in medicine.

e. *Nepeta cataria (N. calaminthoides):* There are many species in the Himalaya except this species. The Catnip oil is obtained from the plant which is a native of Kurram valley, extending to the West Europe. It is used as a lure for trapping bobcats, lynx and other predatory felines.

f. *Ocimum basilicum, O.canum, O. grantis-simum, O. viridis, O. sanctum* and *O. min-imm:* Various species of ocimum are distributed throughout India. The entire plant is used for the extraction of Basil oil, it has fragrant smell. The oil is used in the manufacture of perfumes of the types of violet, mignonette and jonquil. *Ocimum kilimandscharicum* is introduced at many places for camphor extraction.

g. *Origanum vulgare, O. marjorana:* The former species is found in the W. Him-alaya. The oil of marjorana is obtained principally from the later species. Oil from *O. vulgare* contains thymol, car-vacrol, sesquiterpenes and has a spicy, aromatic basil like odour. It is used in perfumery and in medicine.

h. *Perilla frutescens Brit:* The plant yields perilla oil which is used as a flavour and in China known as pesow.

i. *Pogostemon patchouli Pellet:* The plant is cultivated at many places but is a native of Sylhet and Malaysia. The specific epithet is a Bengali name for the plant. More than a century ago Indian shawls and other fabrics are reported to have been scented wih the perfume of this plant. The leaves powdered or as such are used to ward off insects and moths. The oil is obtained by distillation with steam and is used in perfumery, soap and incense. It is being extensively cultivated in Eastern Himalayas and needs encourgement. First recorded in MP in 1941.

j. *Salvia officinalis:* A native of *Dalmatia* but several varieties are now cultivated. The leaves are highly aromatic and used for seasoning in meat. The Sage oil is obtained by steam distillation. It is also used as convulsant like wormwood oil though it is less powerful.

k. *S. selerea:* Oil is distilled from the leaves. Mainly cultivated in Bulgaria, Russia and France. Now introduced in Kashmir. The oil is used as a medicine.

l. *Thymus serpyllum:* The oil of wild thyme is mainly used in pharmaceutical industry.

m. *Ziziphora clinopodioides:* A plant of Kurram and Baluchistan, having an odour of peppermint, *Z. tenior*, is a native of Persia and is a sweet scented annual having the odour like that of the peppermint.

21. *Lauraceae*

a. *Cinnamomum camphora:* A native of China, but cultivated at many countries. Camphor oil obtained from the twigs and branches are used in perfumery and the manufacture of Heliotropin.

b. *Cinnamomum cecicodaphne:* Wood oil and oil from the leaves is used as a flavouring agent.

c. *C. tamala, C. obtusifolium:* The leaves give an essential oil. Both of the species are used a spice and for dressing. The former species on distillation is recorded to give 2% oil, major constituent being eugenol (80%); other components being α-pinene, β-pinene, myrcene, limonene, cineole and b cymene.

d. *C. zeylanicum:* It is cultivated in countries India, Pakistan, Myanmar and Sri Lanka. Oil from the bark is extracted which is used in perfumery. Oil from the leaves consisting of primarily Eugenol (70–87%) is used in the manufacture of Vanillin.

e. *Melaleuca leucadendron, M. minor:* The plants are distributed at Malaysia. The leaves and twigs yield cajuput oil and used in native medicine as a stomachic, in intestinal troubles, skin diseases and as an insecticide. The oil can be replaced by Eucalyptus oil.

22. *Leguminosae*

a. *Acacias:* The two perfumes *mimosa* and *cassie* are obtained from the flowers (*A. farnesiana* and *A. decurrens* var *dealbata*). The perfume is used in the preparation of Pomades. Either as Cassie, pomade concrete or absolute, also used in the preparation of *Violet boquets*. It is also used in toilet preparations. *Mimosa* perfume merits much wider application in fine perfumery.

b. *Lathyrus odoratus:* It is cultivated in the gardens. The perfume from the flowers is obtained, though mostly synthetic perfume is largely used.

c. *Dalbergia sissoo:* An essential oil derived from the wood is a pale yellow, aromatic oil.

d. *Spartium Junceum:* It is a native of Mediterranean countries and Canary island, but cultivated in many parts of the Asia. The Jenet concrete or 'absolute' is obtained from the flowers. Concrete is a solid mass with heavy honey like odour. The *absolute* blends well and produces remarkable effects in heavier types of perfumes.

23. *Liliaceae*

a. *Hyacinthus orientalis, H. glaucus* and *H. ciliata:* The first one is native of W. Asian countries which is now cultivated in many parts of Europe. Other two species are indigenous to Indo-Pakistan subcontinent. The flowers yield an oil having sweet heavy odour and is available as concrete. It is used in perfumery industry.

b. *Notholirion thomsonianum:* Native of West Himalaya in nature but sometimes also cultivated in the gardens. It emits a sweet fragrance.

24. *Linaceae*

Linum usitatissimum: Cultivated as a crop. Linseed oil obtained from the seeds, with a characteristic odour. It is used in paint industry more than in the perfume manufacture.

Magnoliaceae

a. *Magnolia grandiflora, M. pterocarpa:* Both these species are cultivated in the gardens and produce large fragrant flowers which are sometimes sold in the

market. Oil is obtained both from the leaves and flowers which is fit for the production of heavy perfumes.

b. *Michelia champaca:* Cultivated mostly in the gardens. Oil from the flowers is a volatile liquid with a pronounced but pleasant aroma. The seeds which are yellow as compared to red in the Phillipine variety also yield a fatty oil having a strong flavour. The *champa oil* is used in perfumery, the odour is heavy and persistent. Yellow pigment from flowers is used for dyeing.

Malvaceae

a. *Hibiscus abelomoschus (Abelomoschus moschatus Medic):* Music mellow, Musk-dana. Distributed throughout the hotter parts of India. Seeds are used for the extraction of ambrette oil having an odour of amber. Seed coat contains 0.2–0.6% of the essential oil having a musky-odour. (Krishna and Badhwar, 1947). The oil is used in perfumery industry and preparation of farnesol. Marsh mallow (*Althaea officinalis.* L) leaves contain an essential oil (0.22%) with palmatic glycerides. In flowers the essential oil is only 0.024%.

b. *Pavonia odorata:* Wild flowers and roots have aromatic constituents.

c. *Urena lobata:* Distributed in hotter parts India. The leaves are used to adulterate *patchouli* leaves. Fresh plants of *Abutilon hirtum* (Lam.) Sweet (*A. graveolens* and *A. theophrasstii* Medic) emit a strong unpleasant odour, reminding of ammonia and amines. (Krishna and Badhwar, 1947).

Meliaceae

Toona ciliata, Cedrela microcarpa: Both of these species have scented wood. *C. odorata* has a West Indies wood of the same type and smell. Essential oil is extracted from the wood in West Indies and the same can be better done from the native species.

Myrtaceae

a. *Decaspermum paniculatum:* Nuanna oil is yellowish green with amber like odour is obtained by steam distillation of the leaves. It is used in perfumery.

b. *Eucalyptus globulus, E. citriodora, E. tereticornis:* These species are native of Australia. Oil is obtained by steam distillation of leaves. It is used in scenting soaps, useful in catarrh and colds. These grow in South India (Karnataka, T. Nadu and Kerala) and contain 85–90% Citronlol, that is used in medicine.

Oleaceae

a. *Jasminum officinale, J. grandiflorum, J. sambac* and *J. humile:* These species are both either wild or cultivated. Jasmine oil is obtained from the flowers. It is used in perfumery and application is hair oils.

b. *Nyctanthes arbor-tristis:* Distributed in the sub-Himalayan tracts but widely cultivated in gardens for its flowers. Flowers yield an oil which is used in perfumery.

c. *Syringa vulgaris, S. emodi: S. vulgaris,* is the native *liliac of Iran*, while *S. emodi* is found in West Himalaya. The perfume has not been successfully isolated as yet but synthetic perfume is in market.

Piperaceae

a. *Piper betel:* It is cultivated in many parts of the country. The betel oil from the fresh leaves is used as a flavouring agent in the masticatories.

b. *Piper longum:* A slender creeper. The essential oil obtained from its fruits is

a light viscous liquid. Nothing is known about its chemical composition. It is used as a flavour and in medicine.

25. Resedaceae

Reseda odorata: Cultivated commercially in Southern France and commonly grown as a winter annual in the gardens. The whitish flowers with brick red anthers, on steam distillation or petroleum-ether extraction, gives an essential oil which exhibits characteristic sweet odour on strong dilution. The oil is used in perfumery industry.

26. Rosaceae

a. *Prunus amygdalus* var. *amara:* It is cultivated in semi-temperate regions of Mediterranean climate and also in Kashmir. The kernels are used for the extraction of almond oil. The oil consists of almost Benzaldehyde with several per cent of hydrocyanic acid. Pure bitter almond oil free from Prussic acid is oxidised rapidly into benzoic acid, if not stored carefully. The benzaldehyde is obtained and is used in perfumery.

b. *Rosa damascana, R. alba (R. damascuns* var *alba), R. vorviana, R. centifolia, R. moschata:* These species of rose are cultivated for oil, while the last is found as wild in Western Himalaya. Flowers yield an essential oil. Rose perfume manufacture is an important and ancient industry. The percentage of free alcohol varies and is estimated as the geraniol. *R. damascana* is also important and the oil is preferred because of scent. Now more than 35,000 ha are under cultivation in India. Total production in India is more than 300 kg of oil and some part of it is exported. Rose water, rose oil, gulkand and dry leaves are also used apart from its use in the decoration and in God worship.

c. *Spiraea* sp: There are many species of this genus wild in the Himalaya, besides one which is cultivated, that is the Chinese species *S. purnifolia.* The oil of *Spiraea* is an essential oil which contains Salicyladehyde and other compounds, e.g. methyl salicylate, pipetonal, vanillin.

27. Rubiaceae

a. *Anthocephalus cadamba:* A large tree, indigenous in the sub-Himalayan tracts. The flowers yield an essential oil (Gupta 1954).

b. *Gardenia florida:* Native of tropical Africa but cultivated in the gardens. Gardenia oil is obtained from the flowers by maceration with hydrocarbon oil and its subsequent distillation, it is used in perfumery. In China the flowers are used for flavouring tea.

c. *Randia dumetorum, R. ulginosa:* A plant of the sub-Himalayan regions. The flowers yield an essential oil, like Gardenia oil. The perfume is not extracted commercially.

d. *Rondeletia exserta, R. odorata, R. tinctoria:* R. odorata is a native of W. Indies, Mexico, while other species, are native. The flowers yield an essential oil under the name *Rondeletia* and was popular as handkerchief perfume. It is used in the perfumery industry.

28. Rutaceae

a. *Citrus aurantium,* ssp. *amara:* It is mostly cultivated. The *Bergamot oil* has a soft sweet odour and is obtained from the fruit rind. The *neroli* and *bigarade oils* are extracted from the flowers. Petit grain bigarade oil is obtained from leaves. The sweet orange yields lesser esteemed oil. Flowers and water are together distilled to obtain

flower water. Absolute is also prepared. The oils are extensively used in the manufacture of flavours and perfumes, *Eau-de Cologne*, Lavender waters and other fancy perfumes. The *neroli oil* is highly valued by the perfumers and is an essential component of *Eau de Cologne* and other purposes.

b. *Citrus aurantiun* ssp. *bergamia:* Origin of the subspecies is rather obscure. *Bergamot oil* is obtained from the peel and the main constituent is linylacetate. *Petit grain* in bergamot oil is obtained from the leaves which contains citral, d-limonene, linalool, terpenes and other compounds.The Bergamot oil is one of the classic perfumes and is extensively used.

c. *Citrus aurantifolia* (*C. medica* var. *acida):* It is indigenous having two varieties, the *acid lime* and *sweet lime:* the latter is insignificant as far as the production of the oil is concerned. The acid variety has two forms, viz. small-fruited, key or west Indian lime the and large-fruited, *Persian* or *Tahitian Seedless Lime*. The latter may have arisen through hybridisation of Lime and Lemon. The peel is primarily the source of oil. The juice containing peel oil is steam distilled. The oil distills in a number of fractions it flavours. Orange flower water is also prepared.

d. *Citrus aurantifolia (C. limetta):* It is cultivated in many parts of the country and is considered to be a hybrid between the Mexican type and the Sweet lemon. The peel oil gives the sweet lime oil which is a flavouring agent. It is not produced on a commercial lines.

e. *Citrus decumana* (*C. gandis, C.maxima):* It is mainly cultivated for the fruits. The *Shaddock oil* is obtained from the peel of the fruit which has a lemon orange odour.

It can be used in flavours and perfumes but it is not expressed for marketing.

f. *Citrus limon:* It is extensively cultivated, the origin of the plant is obscure and is regarded by *Swingle* to be a hybrid between citron (*C. medica*) and lime (*C. aurantifolila*). Oil of citrus lemon is derived from peel. The oil extracted from the twigs and leaves and immature fruits is called *petit grain oil*, it contains more or less the same constituents as *lemon oil* but in different proportions. This has a considerable demand in perfumery and flavour.

g. *Citrus medica:* It is cultivated and gives *citron oil* which is golden yellow liquid obtained from the fruits rind and leaves. It has a sharp but characteristic odour. The hand pressed oil is considered much better in quality. The chief component is citral. The petit grain oil does exist but is never extracted. The oil is largely used as a flavouring agent and in Eau-de Cologne but is seldom commercially produced. The peel is made into candy and produced commercially.

h. *Citrus paradisica:* It is cultivated in many parts of India. The grapefruit oil is obtained from the peel by expression. Another oil from leaves and twigs (petit grain grapefruit) resembles *nirol bigarade* with more of rose and honey smell.

i. *Citrus reticulata* var. *mandarin* (*C. nobilis* var. *deliciosa):* A native of Cochin but cultivated in many parts of India. *Mandarin oil* is obtained from the peel, it is used primarily as a modifier of sweet orange oil. In addition, it finds use as both flavour and perfume, Petit grain oil of Mandarin is not used in perfume industry.

j. *Citrus reticulat* var. *tangerine:* Cultivated in many parts of India for the fruits and the oil. The peel yields an oil, which is distinct from Mandarin oil

-in flavour and chemical composition. It contains limonene actyle aldehyde, dectyl aldehyde, lilnalool and other compounds. On standing tangeretin separates out at low temperatures and it is used as a flavouring agent.

k. *Citrus sinensis:* A native of south-east Asia but cultivated in many parts of the world. Orange oil is obtained from peel and is yellowish brown in colour. Leaves petiole and twiglets are steam distilled to give oil of petit grain sweet orange. Expressed oil is used (in perfumery) in the manufacture of some types of *Eau de Cologne* and for flavouring purposes.

l. *Murraya koenigii, M. paniculata:* The fruits and flowers yield *Murraya oil* which is a yellow coloured liquid having strong odour like neroli oil. It is used in perfumery industry industry.

29. Salcicaceae

Populus nigra, P. balsamifera: P. nigra is indigenous to Europe and cultivated in the Himalaya, while the latter is a native species. The sticky buds yield Poplar oil of pale yellow colour having smell of chamomiles. It contains small quantity of esters of free alcohols. It is not very common in the perfumery industry.

30. Santalaceae (Vern chandan)

Santalum album L. Indigenous in peninsular India. The hard wood is distilled; it is used in the perfumery and as medicine in local ayurvedic and Unani systems. Karnataka and T. Nadu are the main centres now, also done in Assam, U.P. and M.P.

31. Sapotaceae

Mimusops elengi: A medium-sized tree, often cultivated in the gardens. The flowers yield an essential oil which is used for perfumery and for preparation of hair oils.

32. Solanaceae

Cestrum nocturnum, C. parqui and *C. aurantiacum:* Commoly cultivated in the gardens also seen as a weed. The plant bears greenish yellow flower in panicles which emits very strong fragrance in the evening that continues at night. There is a strong possibility of its use in perfumery.

33. Nicotiana tabaccum L.

It is cultivated commercially in many parts of the country. The fermented leaves contain an aromatic, viscid oil which is dark in colour and recalls that of chamomile. It contains small quantities of the phenol.

34. Sterculiaceae pterosp ermum

Fragrant flowers give an essential oil that have not been studied in detail.

35. Ternstroemiaceae

Thea chinensis: Cultivated in many parts of the country. The oil is obtained from the dried leaves and contains an alcohol and Methyl salicylate. It is used as a flavouring agent.

36. Umbelliferae (Apiaceeae)

a. *Anethum graveolens:* Cultivated in many parts of India, Pakistan and other countries of the world. Immature seeds and mature seeds both give *dill oil* having an odour, somewhat similar to that of Caraway oil. It is used as a flavouring agent and in medicines.

b. *Archangelica officinalis (A. archangelica, Angelica glauca):* Distributed in temperate and subalpine regions of W. Himalaya. *A. anomala* is grown in Japan. *Angelica oil* is obtained from fresh roots. *Japanese oil* has a slight odour of *musk*. The root contains d.α-phellendrene lactone of 15 hydroxypenta-de-

canoic acid and other compounds. Oil from the seeds contains β-phellandrene and many other compounds but is not so important as root oil. The root oil is used for flavouring liquors and to some extent in the perfumery.

c. *Apium graveolens:* It is native of Eurasia and cultivated in many parts of the world. Celery oil is obtained from the fruits and it owes its odour to the presence of lactone sedanolide. Besides, this oil contains 60% d-Linonene, 10–15% Selinene and other compounds. Oil when distilled, from cultivated varieties, is of considerable value to the perfumer as a fixative and as an addition to floral perfumes such as sweet pea. It is a valuable Flavouring agent.

d. *Carum carvi:* Distributed in temperate regions of W. Himalaya and also in Baluchistan, West and Northern Asia and Europe. Caraway oil is obtained from the fruits. The chief constituent is *Carvone*. Carvone is sometimes employed in place of the oil. Light oil is used in cheap soaps and perfumery. It is used in flavours and as a mild stomachic and carminative.

e. *Trachyspermum copticum, Carum copticum, Ptychotis ajowan, Carum ajowan:* Cultivated in many parts of India, Pakistan and middle east countries. Seeds are used for the extraction of oil and thymol. The oil has a pungent aromatic odour. Oil is used for the preparation of thymol and carvacrol.

f. *Coriandrum sativum:* Cultivated in many countries. Coriander oil is obtained from ripe fruits and contains a number of aromatic compounds. Coriander herb oil is also obtained. It flavours and is used in manufacture of *Eau de Cologne* as well as in indigenous medicine.

g. *Cuminum cyminum (Zira-saféd):* Grows wild but often cultivated in many parts of India. The fruit is used for the extraction of oil which contains cumin aldehyde, cuminyl alcohol, perillaldehyde, etc. It is used as a flavour.

h. *Daucus carota:* Cultivated in many parts of the country. Daucus oil is obtained from the seeds and contains terpenes, divalent sesquiperpene alcohol, carotol, bisabolene, duacol, α-pinene, asarone. The oil has a definite perfume value. It blends well with many type of scents.

i. *Ferula foetida, F. jaeschkeana, F. narthex:* Distributed in dry regions of W. Himalaya with Mediterranean influence. Gum is also imported from Iran and Afghanistan. Gum resin exudes from incised cortex of stem and roots of various species and is the article of commerce. Principal constituents are ferulic ester of asaresino tannol and allyl sulphides. The resinous gum has a strong offensive odour and is used in medicine, perfumery and as a flavouring agent.

j. *Foeniculum vulgare* sp. *capillaceum* var. *dulce* (sweet fennel), var. *vulgare* (bitter fennel) var. *panmorium:* Cultivated in plains and hills ascending to 2000 m. var. *panmorium* is mainly cultivated in the Indo-Pakistan subcontinent. Two types of oils are distinguished, depending upon the variety, viz. bitter and sweet oil. The oil is used in flavouring and medicine.

k. *Ferula jaeschkeaina:* Oil extracted through steam distillation of rhizome. Oil used as medicine.

l. *Osmorhiza longistylis* D.C. (*O.clay-tonia* C.B.Clarke) Occurs in North-West Himalaya from 1600–2600 m. The roots yield an oil whose principal com-

ponent is anethol and has a marked odour of aniseed and fennel. It is used as a flavouring agent.

m. *Pimpinella anisum* L: Cultivated for the seeds widely in Europe, Egypt and USA. Aniseed oil is obtained from seeds and is composed of anethol (80–90%), Methyl chavicol, anise ketone, etc. It is used in medicine and as flavouring agent.

n. *Heracleum candicans* Wall. Kaindal (Kashmir) The herb is used as cure against some diseases of sheep and goats. Fruits yield a Furanocoumarin 'Begapten' and two other crystalline compounds of Lactone nature. A fixed oil, about 9 per cent yield, has also been reported after distillation from the roots. An essential oil in 0.1 per cent yield has been distilled from the roots, characterised by sweet fragrance.

o. *Prangos pabularia indl. (Kurungas-Kash, Komal-Hindi):* Fruits used as carminative, stimulant and diuretic. Flower tops and young leaves are employed as insect repellent in paddy godowns locally in Kashmir. Essential oil contains 0.65 and 1.02 per cent of oil of the 1.125 sp. gr. (at 15°C) in fruits and roots respectively. The oil contains Myrcene, α-pinene, camphene, borneol and dihydrocuminol. From roots coumarin 'osthol' has been isolated from roots which causes rise in BP and well marked stimulation of respiration along with analeptic effect.

p. *Seseli sibiricum* Benth. (Bhootkeshi-Kash): The roots are indigenously used in Kashmir to blend beverages and a yield of 0.6% essential oil with 0.867% sp.gr. at 25° C emitting a tenaciously clinging musky odour.

q. *Selenium vaginatum* C.B. Clarke (Bhooykeshi-Kashmir): The roots are

used as incense and on distillation yield about 8% of essential oil having sweet odour, It consists of α-penene, p-pinene, limonene, phellandrene and fenchone and fenchyl alcohol. The oil possess hypotensive, sedative and analgesic properties.

37. Valerianaceae

Valeriana wallichii, V. officinalis: Distributed in W. Himalaya. Besides these two species there are many other species. The essential oil known as "kesso oil" is obtained from the roots. It is pale yellow oil with strong odour which on suitable dilution is favoured as perfume. It is used in perfumery, hair washes, ointments, soaps and medicines.

38. Verbenaceae

Lantana camara, L. odorata: L. camara is found as escape in many parts of the country while *L. odorata* is native of W. Indies. The leaves yield an oil having a strong, pleasant, *sage* like smell. The uses have not been determined. It contains 1-α phellendrene (16%) and 80% sesquiterpene and other compounds.

Vitex agnus-costus: The plant is cultivated. The leaves yield a brown essential oil which has odour of Hyssops. The seeds yield oil with more spicy odour. It is not much used in perfumery. Used in medicine and for baths.

Vitex trifolia: Distributed in India and Pakistan. The leaves yield an oil of spicy odour and contains Leavopinene, camphene, terpinyl acetate and diterpene, alcohol. The leaves are used for bath.

39. Violaceae

Viola odorata, V. serpens, V. canescens: V. odorata is cultivated and there are many varieties of this species, while other species

are natural in temperate regions of W. Himalaya. The flowers yield violet perfume. The essential oil has faintly green colour and fragrant smell. The yield is very low. The absolute of flowers contains Benzyl alcohol, 2,6 monadien 1-al (violet leaf aldehyde), parmone, eugenol, etc. The perfume is in extensive demand but synthetics are generally in the market.

40. *Vitaceae*

Vitis vinifera: Cultivated in many parts of the country and some species also grow wild. The Cognac oil is obtained from the spent up residue of the fermented liquid juice by distillation from the cells remaining after the removal of wine. It contains ethyl ether of oenanthylic acid (pelargonic acid) and a mixture of esters and amyl alcohols and capric and caprylic acids. It is used in flavouring and to a limited extent in perfumery.

41. *Zingiberaceae*

a. *Alpinia galanga, A. malaccensis:* These are cultivated in Malaysia, Malacca and many parts of the country. *A. officinarum* is a native of China and cultivated from its roots. The fresh roots yield *Alpinia oil* which contains methyl cinnamate, cineol, d-pinene and camphor. The leaves contain pinene and methyl cinnamate. It is used in perfumery for the preparation of methyl cinnamate and cineole. It is not produced on a commercial scale.

b. *Ammomum aromaticum:* Many other species besides this like *A. dealbatum, A. linguiforme* and *A. costatum* are distributed in evergreen forests. The fruit gives camphraceous odour which contains cineole.

c. *Curcuma aromatica:* It is wild and found in many parts of the country. A greenish brown oil is obtained from rhizomes by steam distillation. It has camphraceous odour and contains curcumene (65%), sesquiterpene alcohols, d-camphor and other compounds. It is used in medicine and to a limited extent in flavours.

d. *Curcuma longa:* Cultivated in many parts. The rhizome yields an orange oil which is used in flavours and in heavy perfumes. *Curcuma zedoaria* dry rhizome on steam distillation yields oil of Zedoary which is a viscous greenish red liquid recalling that of ginger, cineole and camphor. It is used in medicine and as flavours.

e. *Elletaria cardamomum* var. *minuscule:* Mostly cultivated in tropics in India and Sri Lanka. Fruit is used for the extraction of cardamon oil. The fruits are used as flavouring agent and in medicine. Oil is used in perfumery to a very limited extent.

f. *Hedychium spicatum, H. coronarium:* Distributed in W. Himalaya. Guenther considers var. *flavum* of *H. coronarium* as distinct species. *Kapur-kachri* oil recalls the odour of Hyacinths and is obtained from roots. The *H. coronarium* has sweet fragrant white flowers. Chemistry of the root oil is not known. Similarly, perfume from flowers has not been worked out. The variety *flavum* produces longoze flower oil of dark brown colour possibly containing indole and methyl anthranilate. It is used in preparatin of perfume and incense.

g. *Kampferia galanga:* The underground rhizome yields a heavy essential oil which has not been established and hence cannot be used as a flavouring agent.

h. *Zingiber officinale:* Cultivated in tropical and subtropical regions. Other plants are also cultivated as an adulterant for true ginger. The rhizomes are used for oil distillation which has highly aromatic odour. The oil is free from pungency resient in the resinous content of the rhizome. It contains zingiberene, zingberol, d-camphene, borneol, citral, cineol, zingeron, phellandrene, etc. The oil has highly fixative value and can be advantageously used in perfumery as well as flavouring agent. It is also used in medicine.

Zygophyllaeeae

a. *Peganum harmala:* Seeds contain a soft resin with a deep carmine like colour used in houses for smell specially at the childbirth.

b. *Tribulus terrestris:* Contains a small quality of an essential oil. The essential oils isoealed from this family has yielded Guaiol, an alcohol and azulene, a sesquiterpene.

References

Anonymous 1910, 1912. Perfumery and Essential Oils Records.

Anonymous 1920. Perfumery and Essential Oil Records.

Anonymous 1960. List of medicinal Plants deposited in various herbaria of the Botanical Survey of India. *Bull. bot. Surv. India.* 2(1–2): 190–273.

Anonymous 1975. National Symposium on recent advances and development, production and utilisation of medicinal and aromastic plants. February 24–26. CIMPO, Lucknow.

Ahmad CD and Thind AS. 1948. Cultivation of *Rosha* in the Punjab. *Indian Farming.* 9(5): 184.

American Pharmaceutical Association. 1939. Volatile oils of Anise, Caraway, Celery fruit, Coriander, Cubebs and Fennel. *Rept. Bull. Nat.Farm Comm.* 7.

Arnaud R. 1949. Culture des Plantes á perfumes in France. *Ind. Perfume.* 4: 368

Bhadwar RL, Rao PS and Sethi HC, 1964.Some useful aromatic plants. FRI Dehradun. Manager of Publications. Delhi.pp.1–63.

Bhattacharjee SK. 2000. Handbook of Aromatic Plants, Pointer Publishers Jaipur. 28 pp. 544 (Reptel 2004) 30 pp 490.

Chauhan NS. 1999. Medicinal and Aromatic Plants of Himachal Pradesh.

Cheverger JF. 1939. Volatile oils of Anise, Caraway, Celery fruit, Coriander and Fennel. *Natl. Forum Comm. Bul.* 7. American, Pharmaceutical Association.

Crooks and Stevers. 1941. Condiment Plants, *ISDA Bull. Plant Ind.* July 17.

Clayton DA. 1937. A taxonomic study of the genus Lavendula. *J. Linn.* Soc. 51.

Deogun PN. 1950. Camphor, its possible source and production. *Indian Forester*, 76(4): 139.

EIRI. 2004. Modern Technology of Essential oils. pp 22 465.

Frach L.1939. *Acta Univ. Asiae Med.*(Tashkent) 34: 13.

Gaponen-Kar, TK 1935. *Journ. Appl. Chemistry.* (USSR) 8: 1050–57.

Ghosh TP and Verma BS.1942. Medicinal Products of *Pinus longifolia* Tar. *Leaflet* No U2 (Chemistry) *Forest Recoedsch* Institute, Dehradun.

Grove DD. 1941. Assay of Peppermint, Rosmary and Sandalwood oils. *Journ. Assn. Agri. Chemists.* 24.

Guenther E. 1942. Japan's vast Camphor industry. *Amer. Perfume.* March, April, May.

Guenther E. 1949–52. *The Essential oils.* Vols I-IV. Van Nostrand and Co oils. Perfume and Essential Oil Records. 45(3).

Gupta GN. 1954. Chemical examination of some new Indian essential oils. *Perfume and Essential Oil Records.* 45(3).

Gurgen 1948. Essential oils in Turkey. *Journ. Ankara Higher Inst. Agriculture*, Ankara. Vol. 92 No 18.pp. 332–360.

Gupta R. 1975. *Kuth* has many uses. *Indian Farm.*

Gupta R. 1972. Grow scented leaved Geranium. *Indian Farm.*

Gupta R. 1972. Vital drugs and essential oil bearing plants as future cash crops in India. *Indian farming.*

Hussain, A. 1994 status reports on Aromatic and Essential oil bearing plants in NAM Countries. pp 186.

Iglen G. 1936. Essential oils of Carrot. *Perfume, France*. 14.

Janaki-Ammal EK and Gupta Balkrishna 1975. The aromatic grasses of India –and appraisal. *Proc. Nat. symp.* (Abstr) p 2.

Jame-Verghe, Gulati IC and Joshi ML. 1949. Production of Thymol from ajwain seeds. *Curr. Sci.* 18 (I).

Kapoor LD. 1977. Advances in Essential Oil Industry. pp 20. 249.

Karim A, Azam MA and Walliullah M. 1941. A study on the development of Gurjan oil in Bengal. Pt. I. Govt. *Bengal Bull*. 910.

Khan AH. 1951. Perfume Grasses of Pakistan. *Cooperation and Marketing review* 5 (3).

Khan AH. 1952. Essential oil bearing exotics suitable for Pakistan *Agriculture Pakistan*. 3 (1&2): 48–53.

Krause K. 1916. Die Duftstoffplanzen- kleinasins. *Deutsch Perfumerie*-Zeitung. 1916. Vol. 3, 273, 299, 314.

Krause K. 1976. Rosenolindustrie and Duftpflanzen in Sudwestliche, Kleinasien Berichte van Schmnel. Leipzing. 1976. pp 1–4.

Krishna S and Bhadwar RL. 1947, 1948–49 Aromatic Plants of India . *J. Sci and Industr. Res*. Series of paper.

Kathpalia YP and Dutt S. 1951. Chemical examination of essential oil derived from *Dalbergia sissoo* wood. *Indian Soap Journ*. 17.

LA Face 1923. La Perfumerie Moderene.

Lutkdv AN. 1962. Polyploidy and its significance in essential oil crops.*Tr. Moskova ISYPT Periody. Otd Bio.* 5: 260.

Misra DN, 2000. Cultivation of Aromatic Plants in India. 9 pp. 330.

Md. Miaruddin 1955. Studies on Bazna oil. *Pakistan J. Sci. Res* **7 (2)**

Naves et Mazuyer 1924. Les Perfumes de France. 1939. Les perfumie Naturelle. Paris.

Parry EJ. 1925. *The Cyclopaedia of Perfumery*. JA Churchile Ltd. London.

Parry EJ. 1925. *The chemistry of essential oils and artificial perfumes*. Scott Greenwood and Sons. London.

Patel IS and Guha PC. 1950. Indian Rosin oil from *Pinus excelsa.Curr. Sci.* 19(4): 128.

Quazilbaksh MA. 1956. Assay of *Artemisia*. *Pakistan J. Sci. Res*. 8 (2).

Rikshit JN. 1939. Development of essential oil Industry. *Sc. And Cult*. 1954. *Sci and Cult* 19 (7).

Redgrove HS. 1933. Spices and Condiments.

Sarin J and Baru ML. 1939. North Indian Essential Oils. *American Perfum*. 38 (2).

Simonsen JL. 1922. Constituents of some Indian essential oils. Pt. 8 Essential oil from the gum rosin of *Boswellia*. *Indian Forest Rec*. 9(6).

Sarin YK and Kapoor LD. 1963. Some additions to the essential oil bearing plants from NW Himalaya. *Perfumerie and essential oil record*. July: 1–6.

Simonson JL and Gopalrao M. 1922. Constituents of some Indian essential Oils. *Indian For. Rec.* 9(4). Pt. II 10 (10).

Simonson JL. 1922. Essential oil from the leaves of *Abies Pindrow*.

Singh DP 1950. *Kuth* cultivation in Lahul and its future. *Indian for* 77: 71.

Sharma ML, Sharma O_5 and Singh A. 1975. Exploitation of Indigenous Aromatic Plants for the Production of Perfumery Chemicals. Proc. *National Symposium* (Abst) pp.8

Singh P. 1916. Constituents of Indian Geranium oil. *Indian For Rec*. 5: 415.

Singh P. 1916. Eucalyptus oil industry in Nilgiris. *Indian For. Rec*.5(8).

Shiva MP, 2002. Alok Lahiri and Shiva, A. Aromatic and Medicinal and Plants Yielding Essential oil for Pharmaceutical, Perfumery and Cosmetic Industry and Trade. pp 8. 341. 20. col. pI. (ISBN: **81**–7089 287–2.

Singh P. 1912. Distillation and composition of Turpentine oil from the Chir resin. *Indian For. Rec.* **4(1).**

Singh P. Perfumery and Essential oils. Indian For. Rec. 8 pp.

Singh PB. 1999. Illustrated Field Guide to Commercially Important Medicinal and Aromatic Plants of Himachal Pradesh (with special reference of Mandi District. pp. 117. Col Pl. 32. map 1. (Soc. for Herbal Medicine and Himalaya Products).

Sobti SN, Bodhraj and Singh P. 1923. Perfumery and Essential Oils. *Indian For. Rec.* **14.**

Sarin J. and Bari ML. 1939. Northern India Essential Oils. *American Perfumer.* **38(2).**

Sinha G.K. 1975. Study of Essential oils from odoriferous plants of Kumaon. *Proe. Nat. symp.* (Abstr).

Verma SN. 1969. Cultivation of aromatic and Medicinal Plants in India. *Perfume, Essential oil. Rec.* 1969. **60 (1)** 19–22.

Verma BS. 1951. Isolation of Costus oil from costus root. *Indian For.* **77**:513.

Vgolew G. 1936. Essential oils of Carrot. *Perfume,* France. 14.

Trotter 1944. Manual of Indian Forest Utilisation. FRI Dehradun.

Willis JC and Bamber. MK. 1902. Camphor. *Indian For.* 28(4:156.

Identification and Nomenclature of Drug Plants

References to Indian medicinal plants are to be found in Sanskrit classics. The Treatise on Botany, like the *Ayurveda, Charaka Samhita* and *Sushruta Samhita*, deal with plants mainly in relation to medicine, agriculture and horticulture. The two common names were generally given for a plant, one for the use of the ordinary man and the other for a physician, e.g. the castor plant was popularly known as *Chitravija* (plant with painted seeds) and *Vatahari* (enemy of rheumatism) by the physicians. Furthermore, the plants were classified on their external morphological characters, medicinal properties and environmental associations. In 'Sanskrit' nomenclature, the genus is indicated by a name, while the species is described by a prefix denoting its distinctive character, viz. *Bala* is the modern genus *Sida* and the *Mahabala* is *Sida rhomboidea*, *Atibala*, the *S. rhombifolia* and *Nagtala* the *S. spinosa*. Thus, plant nomenclature arose because at that time this was the only best method of communication.

Botanical Identification

While the estimation of refraction and extraneous matter, and even that of chemical factors like ash, are not likely to present much difficulty as standard procedures are available, the work of botanical identification does not appear to be so simple. The tests prescribed in the pharmacopoeias aid in the identification, but are not in all cases sufficient to establish proof of identity. Thus, there is a great need to have easily available methods of botanical identification of crude drugs mentioned in the pharmacopoeia. Although the problem of botanical identification has been all along engaging the attention of the drug administration, the progress achieved is rather slow. Quite often, well-known botanists differ among themselves about the identity of plant material submitted for their identification. Under the circumstances it might be desirable to take the sample approved by the foreign buyer as the standard sample for botanical identification so that after the export and standardisation the chances of complaint and confusion are eliminated.

Today, the trend in the crude drug's trade is still more or less in the hands of local people who are not scientifically trained. They are not accustomed with the botanical names of the plants needed and as such local names are used which are sometimes misleading. Since the modern botany in India originated in Europe in the 16th century AD, various languages could not furnish precise names for all plants and the botanists were compelled to invent vocabulary of their own, which is based on the Binomial System of Nomenclature consistently used by Linnaeus since 1753.

The vernacular names present difficulty as similar local names are used for different plants and *vice versa* (Table 4.1). Some plants with different local names as given in Table 4.2 are a few examples.

For an effective chemo-taxonomical correlation in the study of economic plants, it is imperative that the correct identity and nomenclature of the plant(s) concerned should be first determined. The actual plant users may employ botanists to work out the problems and furnish the correct names for the plants that concern them. In addition, a voucher specimen should be deposited in a herbarium of repute for verification, if necessary, as to what plant was studied. The majority of the recent name changes in Indian plants are due to the strict application of the *International Code of Botanical Nomenclature*, while others are either due to a better understanding of the identity of plants or even to the proper judgment of the taxonomic status of a taxon. These changes are annoying to ecologists, foresters, economic botanists and other plant users, who feel that the name should be stabilised. Stabilisation is not fixation; stabilisation should be achieved through the International Code of Botanical Nomenclature.

Table 4.1: Similar local names used for different plants

Plant names used	Vern. name used	Uses
Centella asiatica (L.) Urb.	Brahmi	Alterative, used in skin diseases, leprosy.
Bacopa monnieri Wettst.	do	Asthma, epilepsy and for memory plus.
Hydocolyle rotundifolia	do	
Terminalia arjuna W & A.	Arjuna	Tonic, aspringent, used in heart diseases as cardiac tonic, scorpion sting and earache, etc.
Lagerstroemia speciosa Lin.	do	Narcotic, purgative, astringent, febrifuge, etc.
Hibiscus tiliaceous Lin.	Bela	Febrifuge employed in preparation of embrocations.
Sida acuta Burm.	Bala	Astringent, cooling, tonic, useful in nervous and urinary diseases.
Cissampelos pariera Lin.	Patha	Antiperiodic in dropsy, cough and urinary troubles like custitis.
Cinnamomum camphora Nees and Ebern	Kapur	Sedative, anodyne, antiseptic, stimulant and anthelmintic.
Limnophila gratissima Blume	Kapur	Juice of plant antiseptic used as cooling medicine in fever.
Symplocos racemosa Roxb.	Lodh	Cooling, astringent, useful in menorrhoea and eye diseases.
** *Litsea polyantha* Juss.	do	Astringent, used in diarrhoea and stomachic etc.

* Also called 'Mandukparni'.

**though a different local name is given in the book, the plants are commonly known as Lodh. In the market of Calcutta the bark of Lodh procured from different plants were identified pharmacognostically to be *L. Polyantha*.

Table 4.2: Same plant with different local names

Botanical name	Family	Local names
1. *Rauwolfia serpentina* Benth.	Apocynaceae	Chandra, Choto-chand, Sarpagandha, Bon-potol
2. *Gloriosa superba* Lin.	Liliaceae	Olotchndal, Bil-languli, Langalica, Bihalanguli.
3. *Trichosanthes cucumerina* Lin.	Cucurbitaceae	Bon-chinchinga
4. *Cucumis melo* Lin.	Cucurbitaceae	Kankur, Phuti, Kharbuja.
5. *Momordica charantia* Lin.	Cucurbitaceae	Karéla, Uchchey.

Pharmacognostic Identification of Drug Plants

When the dispute about the local vernacular name is settled, the importance arises in the identification of drugs by means of pharmacognosy. If the pharmacognostic data are wanting for the plants, the results will necessarily be indifferent. The description of plants used in the indigenous system of medicine given in the literature is meagre and vague which has resulted in considerable confusion. Again, with the passage of time, the plants themselves have masked their identity. Descriptions are alone not sufficient for settling disputed questions and recourse to actual specimens is absolutely necessary. Ayurvedic classics were written some three thousand years back and some have gone extinct. The plant *Soma*, which is mentioned even in *Vedas*, is described by *Sushruta* to have 15 leaves in total. On the new-moon day, the plant becomes leafless and everyday thereafter one leaf is added to it, till it has fifteen leaves on the full-moon day. Its root is said to contain an exhilerating sweet juice in large quantity. The plant of this descri-ption and some such other plants are not seen now a days.

For standardisation the following general aspects are usually considered important:

1. *Gross morphology*: It gives a definite indication about the plants authenticity in most cases. For instance in bark drugs, the outer and inner colour of the bark, shape and other physical characters such as cracks, fissures, formation of wrinkles, etc. are some of the important identifying characters. In case of Cinchona, different species can be identified by their morphological characters alone, while the official cinnamon bark can be identified and distinguished from cassia and other species of cinnamon from the morphological characters of the outer surface. Genuine 'Buchu' grass (local name) leaf can be identified from the other species of Bichu as well as *Myrtus* leaves from the size, shape, nature of apex and character of the margin. Similarly, leaves of *Cassia angustifolia* and *C. acutifolia* can be distinguished from their common adulterant such as leaves of *Tephrosia, Colutea* and other species of Cassia by the critical examination of the morphological characters.

2. *Microscopical morphology*: This is to control the identity of ground and underground crude drugs. Trichomes, Calcium oxalate crystals, nature of stomata, epidermal cells, palisade cells, stone cells, fibres, tracheids and trachea are some of the important microscopical structures for the identity and quality control of drugs. The microscopical morphology of the common adulterants has also been

closely studied vis-à-vis genuine crude drugs. For instance, *senna* leaf powder can be distinguished from Palthe-senna, its common adulterant, from the size and shape of the trichomes. *Digitalis purpurea,* on the other hand, can be distinguished from the *D. lanata, D. lutea* and *D. thapsi*, from the nature of glandular hairs. Different species of Rauwolfia can be distinguished from the official *R. serpentina* from the nature of cork cells and by the presence or absence of stone cells in the cortical region. *Kurchi* bark can be identified from *Wrightia* bark by the absence of fibers in the fomer, while Ashok bark can be identified from *Trema* bark by the presence of prismatic crystals of calcium oxalate and crystal fibres. Different species of *Aloe* can be distinguished by the presence or absence of aloin crystals.

3. *Quantitative microscopy:* Powdered crude drugs are more often adulterated. In such cases where physico-chemical methods cannot be applied, Lycopodium spore method of Wallis, has been successfully used. The average number of starch grains per mg has been estimated by this method. The good quality of ginger should be 2,84,000 and Speacuncha stem should contain about 33 stone cells per mg. These facts form the basis of quantitative determination of these two important groups of drugs. The quality of Pytethrum flower is mostly assessed by the pyrethrin content, but quantitative microscopy is still used to detect what type of flower should contain 1000 to 2000 pollen grains per mg and sample with less than 500/mg of pollen are considered to be of inferior quality.

Data observing islet numbers, veinlet termination numbers pallisade ratio, stomatal number, stomatal index, measurements of total area of epidermal tissue, sclerenchyma tissue, the area of fibre per gm in cinnamon and areas of weaker cells per sq mm of the testa such as in case of Cardamon seeds are some of the other useful quantitative microscopic determinations which permit evaluation of powdered drugs.

4. *Polarising microscopy*: Polarising microscopy is now being widely used for the quality control of many crude drugs. The purity of *Balsam tolu* is achieved by identifying cinnamic acid under polarising microscope. Similarly, Siam and Sumatra gum benzoins can be differentiated by this technique.

5. *Flourescence tests*: Many alkaloids in solid form show distinct colour under ultravoilet light. Different kinds of oils, fats and waxes also give different distinct fluorescence which can be employed for their identity and quality control. Drugs like Indian and Chinese Rhubarb, in different species of derris, similarly are tested for their purity by this method. Likewise different species of hydrastis, viburnum, *Chirata*, wild Cherry bark, Cinchona bark, henbane, etc. can be evaluated under UV light. When aloe is suspected to be adulterated with black catechu, this can easily be detected by the flourescence test.

6. *Crude fibre determination*: The determination of crude fibre in samples of drug plants enables the analyst to detect certain types of adultration and to determine the proportion of added materials. Quantitative data for a few drugs are now available and can be used as basis for their evaluation, such as crude fibre from cinnamon (Sri Lanka) is 26–36 per cent, whereas that of *Cassia* is 18–24 per cent only. The American *Podophyllum* contains 7.5 per cent crude fibre,

whereas in Indian type this value is about 10.5 per cent. A good quality of clove should not yield more than 10 per cent of crude fibre.

7. *Foreign organic matter*: In pharmacopoea and other standard texts, limit for foreign matter in crude drugs have been laid down. Foreign organic matter are generally the parts of the organisms from which the drug is derived but other than those named and described in the definition and description of the drug itself. This impurity in some cases is determined easily by inspection with unaided eye or with a lense. However, in case of powdered drugs *Lycopodium spore method* is applied as mentioned earlier. Many crude drugs, viz. rhubarb, ginger, malefern, starches, ergot and yeast powder are easily infested with insects. All pharmacopoeas specify that vegetable drugs are required to be free from insect, animal matter and animal excreta. Methods are now available for the quantitative determination of insect parts and animal excreta in vegetable drugs which are of considerable assistance in ascertaining the quality.

8. *Physical constants*: Physical constants such as specific gravity, optical rotation, viscosity, refractive index, etc. are some of the important characteristics for the quality control of certain categories of crude drugs, such as fats, oils, waxes, resins, gums, oleoresins, oleo gum-resins and balsams. In pharmacopoeas, limits and standard values have been given in such cases and good quality drugs are expected to comply with the same.

9. *Qualitative chemical tests*: These tests are generally specific colour reactions for certain drugs and are useful for their identification and/or checking sophistication. Ruthenium red-test and Iodine-test immediately distinguish Persian and Smyrna *Tragacanth*, *Sterculiagum* and Carol gum. Anthraquinone test distinguishes official *Senna* leaf from other species of Cassia having no medicinal properties. Copper acetate test will detect the admixture of colophony in Balsam tolu. Benzidine test will distinguish gum Acacia from certain other allied gums, while tannic acid test will detect the presence of gelatine in agar powder. Apart from these typical examples, there is considerable scope for newer and more specific tests based upon chemical structure of the active principle involved.

Chemical analysis: Total ash, acid and water insoluble ash, water and alcohol insoluble matter, are some of the valuable chemical estimations which help to detect low grade, exhausted or adulterated drugs. Total ash content of some crude drugs is due to variable amount of calcium oxalate crystals, present in them. Total ash content when compared with acid or water insoluble ash, gives indication of the presence of earthy matter in many drugs, viz. leaves of digitalis, henbane, some gums and resins. Alcohol insoluble matter in balsam tolu and gum benzoin give an important index of the quality of these drugs.

In case of those vegetable drugs where the active constituents are well defined and can be estimated either chemically or biologically, standards and test methods have been laid down in official Compendia. For example drugs, strychnine in nux-vomica seeds, phenolic and non-phenolic alkaloids in Ipecacuanha, resperine in Rauwolfia, glycosides in Digitalis, Strophanthus and Scilla have been evaluated biologically. As illustrated below requirements of some crude drugs are shown in the Table 4.3.

Table 4.3: Requirements of some crude drugs

	Botanical name	Ash% max	Acid (insol) ash%	Foreign matter%	organic matter%	Active principle
1.	*Swertia chirata*	–	1.0	2.0	1.3	Bitter and tonic
2.	*Cinchona calisya* wood	4.0	–	–	–	Antimalarial
3.	*Digitalis purpurea*	–	5.0	2.0	–	Cardiotonic
4.	*Glycirrhiza glabra* (peeled)	6.0	–	–	0.04	Cardiovascular depressant
	(unpeeled)	10.0	–			
5.	*Cephaelis acuminata*	5.0	2.0	1.0	2.0	Emetic and expectorant
	(Ipecacuanha roots)					
6.	*Saussurea lappa*	–	–	2.0	–	Expectorant
7.	*Strychnos nux-vomica*	3.0	–	1.0	1.02	Bitter
	(dried ripe seeds)					
8.	*Rauvolfia serpentina* (roots)	8.0	2.0	2.0	0.15	Hypotension
9.	*Cassia angustifolia*	–	2.0	1.0	–	Laxative
	(leaves and pods)					
10.	*Valeriana wallichii* (roots)	12.0	–	2.0	–	Sedative
11.	*Urgina indica*	6.0	–	2.0	–	Cardiotonic and expectorant

Paper chromatography: It is an inexpen-sive tool for the analysis and evaluation of many crude fibre drugs, particularly those which are difficult to evaluate otherwise. For instance, fruits of *Ammi visnaga* can be distinguished from its very similar adulterant, viz. the inactive fruits of *Ammi majus* only by paper chromatography. Similarly, in different species of Rhubarb, Frangula bark, *Digitalis* leaves, *Aloes*, etc. paper chromatography is an important method of evaluation.

Thin layer chromatography, the most up-to-date analytical tool, for its simplicity, rapidity of execution, specificity and sensi-tivity is playing an indispensable role in the evaluation of vegetable drugs in recent years. In Umbelliferous drugs, their substitutes and adulterants can be easily identified by thin layer chromatography. This method is particularly useful in distinguishing *Anethum sowa* from *A. graveolens*. *Podophyllum emodi* and its resins can be differentiated from *P. peltatum* in TLC plates by multiple development technique. In *Pimpinella* roots, the roots of *Heracleum* can be easily detected by this technique. By the use of TLC, it has also been possible to establish the fact that the bitter principles of Jamaica and Brazilian *Quassia* are not the same. It has also been possible to distinguish natural Canada Balsam from synthetic one, Sumatra Benzoin from Siam Benzoin, natural Asfoetida from locally prepared one by the use of TLC. In these cases, usual pharmacognostical and/or chemical methods are not satisfactory.

References

Gupta AK. 2003. *Quality Standard of Indian Medicinal Plants*. vol. 1. pp 16. 262.

ISI—Indian Standards Instituion 1967, 11th Indian standards convention Chandigarh. 24th sept-2nd Oct. 1967. Doc. S-1/1 to S-1/26. Mimeo.

MC. Guffin M. *American Herbal Products Association's Botanical Safety Handbook*.

❑❑❑

Traditional Medicinal Plants

A. Traditional Medicines and Medicinal Plants in Rural Folklore and Tribal Society

B. Medicinal Plants collected from Forest as a Product and as Weeds from Wild/Wastelands: Trees and Shrubs

C. Herbaceous Plants used as Drugs from Wastelands, Weeds of Croplands and Forest Understorey

5

A. Traditional Medicines and Medicinal Plants in Rural Folklore and Tribal Society

India has a rich and diverse folk-tradition including age old agricultural practices, home health remedies and advice about do's and don'ts of everyday life. Medical aids to the rural areas have always been provided from the earliest times and had been individualistic in nature. *Hippocrates,* the father of medicine, used to wander about in villages, seek out patients and treat them in their homes. *Charaka* has given necessary directions with regard to the health of the rural population and areas. *Avicennia* has laid down the criteria for choosing a place for human habitations. *Kautiliya* and *Ibn Khaldun* have discussed the definition of a village and the habits and characteristics of the villagers. In the pre-Independence era, during the period of monarchy and feudalism, along with the establishment of hospitals in the cities, the physicians were deputed to carry medicaments to the villages. Particularly during the epidemics, physicians used to carry medicines and provide medical aids. Charitable institutions vied with one another to provide medical aid and had a share in such works. Local *hakims* and *vaids* also did not spare themselves to the difficulties of rural life; though all these services were individualistic and of limited nature. Folk health practitioners in some villages still take care of serious conditions such as bone injury and poisoning.

With the advancement in medical sciences, great strides were made in surgery and in medicines. After Independence, though problems of rural health and medical aid received attention, the phenomenal increase in population could not solve the problem in its totality. Still there are places where adequate medical support is not available in far-flung rural areas. The medical education has hardly equipped the graduates to provide cheap and effective medical care to the poor and the deprived. Another important aspect is that the countries which are considered to have solved their problems are either not very populous or are bracted with high income groups. Even in China, where the population is tremendous, the concept of 'Bare Foot Doctors' was mooted as early as in the seventies. The influence of modernisation and change in lifestyles, however, have resulted in a decline in traditional health and agricultural practices. Many modern practitioners show indifference or negative attitude towards traditional practices.

In India, the first organised attempt was suggested in a *Report on the Health Survey and Development Committee*, where several recommendations were made, which subsequently formed the basis of the programmes of health and medical aid of the *National Extension Services* and the

Community Development Blocks (CDB) with occasional modifications. Each community Development Block includes the establishment of a *Primary Health Centre*. One CD block normally covers about 100 villages, having a population of about 66,000 and an estimated area of about 300 sq km. Attempt is made to provide three centres in a CD Block in each of which a midwife is employed, apart from one medical officer, one com-pounder, one lady health visitor, one sanitary inspector and two general atten-dants. However, the same standards are applicable to hilly areas, like in the Himalayas, where the topographical conditions are entirely different than those from the plain areas.

Another mode is that of enabling rural population to bear the responsibility of the solution of all its economic, social, health and medical aid system in the villages to become capable of self-help. Traditional aid for medical systems in the villages has been in vogue since centuries but is still used effectively in many parts of the country. The drugs available within the villages, which are and could be used, by which 90% of the ailments are safely treated in the villages.

It is time now that such medicines which are effective and so nicely used by the villagers, are recognised officially and studied further to expand their scope and put to use on scientific lines. Though some beginning has been made but much remains to be done so that the treatment is cheap, effective and help all sections of the society. Information provided in the following pages, it is hoped, would be useful and effective to meet the goals of *Health of All*.

Since 1980, a movement 'Patriotic and People-oriented Movement' aimed to explore various aspects of Traditional Indian Science;

Centre for Indian Knowledge Systems (CIKS) established against the backdrop of Green Revolution. Since then, this organisation has been working for searching texts. In 1994 various experiments were scanned and prescriptions of *Vrakshayurveda* under field conditions (Balsubramanian *et al* 2000) for curing plant diseases with solutions of plant and animal products.

Many wild plants are sources of expectorants, sedatives, tonics, carminatives, aphrodisiacs and abortifacients. Ayurveda acquired from the aboriginals, who possibly learnt it by closely observing the remedial measures taken by the ailing wild animals.

Many tribal communities control their population by using the plants with anti-fertility agents; *Annona squamosa* is one such plant. Medicines for snakebite and high blood pressure from *Rauwolfia* species (*serpentina*), for sexual disorders from *Hmidesmus indicus* (Anantmul), for dysentery from *Halorrhena antidyseneterica*, for epilepsy from *Borassus flabellifer* (tari) and for numerous other disorders were tried.

India has one of the largest collections of ancient manuscripts in the world. While there has never been a precise count, estimates suggest there may be as many as 300 million texts. These old Indian scripts pay considerable attention to philosophy, religion, health care, livestock, agriculture (rains and harvests). These include prayers, hymns, *mantras* and ancient prescriptions. The classical Indian system (Ayurveda) and plant science (Vrakshayurveda) are highly advanced; scripts are in many local languages including Sanskrit, Tamil and Telgu.

Three types of 'Vrakshayurveda' literature can be distinguished. The first category consists of general texts with only specific sections devoted to traditional plant

science. The second category involves more general texts and Vrakshayurveda is an essential part of the content. These texts provide basic theoretical framework allowing us to understand 'Vrakshayurveda' in literature. Thirdly, there are those manuscripts that are devoted entirely to plant science and are of great interest.

The subject of 'Vrakshayurveda' is vast, including subjects like seed collection, germination, cultivation, planting nursery techniques, soil manuring, cultivation under favourable climatic conditions, pest and disease management, as well as traditional names of plants. Some of the prescriptions of Vrakshayurveda are of general nature, other prescriptions define a particular species precisely suited. Quite often prescriptions list a set of indigenous species without specifying the proportions to be used.

Many wild plants and their various parts such as seeds and fruits provide nourishment and energy to the body and are also essential for the growth of human body. In villages fruits like mango, melons, papaya, plums, banana, cucumber, amla, guava, lemon, prunes, plums, figs are grown in abundance. Sugarcane, along with certain vegetables like carrot, reddish, tomato, etc. are taken raw and are good sources of expectorants, sedatives, tonics, carminatives and abortifacients. Ayurvedic and Unani systems are mainly based on the knowledge acquired from the aboriginals, who possibly learnt it by closely observing the remedial measures taken by the ailing wild animals. For example, in Uttaranchal the indigenous medical system recommends the following cures based on indigenous resource system (after Singh, 1981).

Piles: **Kalmishora** (KNO_3) and **nisoth** (*Ipomoea turpethum*) taken orally or applied locally in the morning and evening, in the ketchup form, removes irritation and chronic piles.

Bleeding to Haemorrhoids: *Nagkesar* or *Sahinjana* (*Moringa oleifera*) taken along with *misri* (sugar) and butter, orally for the six months at morning and evening stops bleeding.

Irritation of piles: Application of butter or suran (*Euphorbia royleana*) made to pulp with curd relieves irritation of piles.

Asthma and bronchitis: Ashtvarg, rishivak, kakoli, chirkakoli are taken orally.

Diarrhoea and dysentery: Bél (*Aegle marmelos*) pulp is given with water.

Enorrhoea and colic: *celtis eroriocarpa* fruit is taken.

Demulcent and stringent: (*Lisora*) *Cordia vestita* fruits are taken.

Tonic: Fruits of *Corylus colurna* nuts are eaten like those of almond.

Astringent and diuretic. Fruits of *Fragaria vesca* (local strawberry) are eaten.

How Humans Started Using Plant Drug as Cure

Monkeys eat nuts and other plant species to keep them fit and allay their hunger. A case has been reported in December 1994, when a monkey, electrocuted, did not die even his left arm got charred, when mostly they die instantly. The juvenile after lying on the ground for 10 minutes (overcoming the shock) dragged himself near the *neem* (*Azadirachta indica*) and spent one week on this tree, eating only its leaves and fruits. The next one

---

Singh AP. 1981. Indigenous medical system in Uttarakhand. *Himalaya Man and Nature*. 5(6): 10-12.

week he spent on this tree in total seclusion and when he came down his lower arm had fallen off but the skin and the point of electrocution was still raw. For the next month this juvenile was seen applying his own saliva on the wound and eating the tender leaves, branches and flowers of 'kethri' (*Sonchus arvensis*), at least 15 times a day, and then he recovered fully.

The instance triggers the idea for a new line of research and as the author states he wanted to know all about the other plants which monkeys eat when they are sick. Some of his observations are reported in the Hindustan Times, dated December, 1994 (Iqabal Malik).

1. The mating season (from October to Jan.) is an interesting season in the field. There is a severe competition amongst males for mating opportu-nities. During this time, the tubers of "barbunda" (*Oxystelma sycamore*) are eaten by males. During this season there is increased inter-individual aggression among the males, and very often it leads to cuts and bleedings. All such males were observed by the author to be eating the buds of akh (*Calotropis procera*) and they also apply the milky secretion on the bleeding parts. This leads to blood coagulation, thus stopping the bleeding.

2. By the end of the mating season, 80 per cent of the females between the ages of three to fifteen years are pregnant. A major part of their natural diet comes from *imli* (*Tamarindus indica*) tree. The leaves, bark and the fruits are consumed in large quantity. Birth season starts in the month of May and for the first three months infants only depend on mother's milk and the sucking continues till they are almost a year old. During the yearly lactation period the tender brownish pink leaves of

Peepul (*Ficus religiosa*) are the sought-after food by the new mothers.

Between the food eaten by human beings like *paratha* (oiled chapattis) and the bananas, monkeys prefer banana, rich in potassium, excellent for vitality and stamina, so highly recommended for athletes. The low rank members of the group, if perforce end up eating chocolate and parathas often lead to vomiting and diarrhoea in them. After vomiting these monkeys have been observed eating Doob grass (*Cynodon dactylon*) tender grassblades and tender white shoots, normally wrapped by the blades.

Gokhru (*Pedalium murex*) fruits are consumed during diarrhoea. The monkeys with blisters in their mouth eat the gum secreted by *Acacia nilotica* ssp. *indica* (kikkar). If they get feverish, the pods of *Cleome gynandra* (hulhull) are eaten. Coughing monkeys eat leaves of *Adhatoda vasica* (bansa or *vasaca*).

Thus, it is estimated that over 150 plant species provide food for the 'rheas' and in the forest monkeys have an understanding of the flora of their habitats. Their diet is fat free, free of proverised food and artificial sweeteners. They eat what their bodies demand in a given situation.

Tribal Medicines and Treatment

The tribals have a quite simple remedy for each disease, for snakebite a simple *Kurvo* fruit has to be taken, rub it on the stone and apply it to the bite and to the eyes. For dysentery, the roots of another plant would be crushed and soaked in a pot of water and taken twice a day or you might have to make a sacrifice to appease the Gods or exercise evil spirits. Tribal remedies can be found in the forest on the abundance of plants that grow there. These herbal medicines are described by the tribal medicine-men who is also a **witchdoctor**.

Tribals ascribe most diseases, especially epidemics, to evil-spirits. They believe that a breach of tabo, spirit, intrusion or loss of soul have let the spirits in and to free the patient from the wrath of the supernatural, they worship spirits and benevolent deities and calamities. For common illness, like cough and cold or the quite common snakebite they usually the herbal medicines, along with magic. The most tribes follow a similar system of medicine, there are small differences in the kind of plants used, depending on what is available and exact where it is applied.

The knowledge of these medicines is passed down the Generations. A father takes his son into the forest and shows him the various plants and roots, used to cure diseases and the method of preparation. Tribals even have closely guarded secrets regarding cures.

Among the Garcia tribals, plants such as *Lunia* are used to cure tiger bite. *Aval*, for syphilis, *Akd* for a scorpion bite, *Char-nus-vela* for gonorrhoea, *Muli-ne-vel*, *Kala sadad*, *Kid muli* or *Gorhai* for cough, Siluti for fever, *Velva* for sore eyes, *sitri* for headache, *munjal* for sprains. In each case the plant is rubbed on a stone, powdered and soaked in water and made into a paste.

The people, who live in eastern Orissa use turmeric and water for eye trouble, a mixture of rice, ginger for yellow fever, and ground *valia* seeds for scabies. For a snakebite they use turmeric water and ground root of *nageshwar* plant. Often a *Chamak* stone is fixed on the wound to absorb the poison.

For roundworms a banana and a plant called *rundel* are consumed and for tonsillitis, a silver neck ornament is used for the treatment.

Another tribe has a way of discerning whether a bite is poisonous or not. This is done by giving the patient chili, salt and *neem* leaves to eat. If the patient finds it sweet it mean he has been poisoned.

For the treatment a small bag with a concoction of the *barha* tree plant and water is held under the patients nose. As he breathes in the smell, the poison is said to flow out of his nose and mouth.

Some tribals even use a tuber of ginger for curing heart trouble. They also have a herb for infertility and diabetes.

The most common diseases among the tribals concentrated in the hills and forests of Central India are Yaws, tuberculosis, leprosy, malaria and venereal diseases. The *Santhals*, call these the "Sir" and spraining, dislocation or twisting of any part of the SIR is supposed to cause physical complaints.

Iskir, a kind of massage is considered by them as the best remedy for this kind of trouble. If it does not work, the cause is *teja* or worms, rabies, epilepsy and scabies are in this category.

Vertika *et al* (2004) from *Kols* tribe reports that many plants are used as medicine against various diseases and has reported medicinal use of about 60 plants; as an example Kols use Sadaphuli (*Catharanthus roseus*) or periwinkle (*Vinea rosea*) and Jaurl (*Lagerstroemia speciosa*) for curing diabetes. Other plants like anjan (*Ailanthus excelsa*) and *Nux-vomica* (*Strychnos nuxvomica*) are also used in combination with Jarul and Sadaphuli for treating debility, caused from diabetes. *Jarul* leaves are crushed to make a paste and taken (20 ml) with milk. *Nux-vomica* and anjan leaves are taken before meal. Thus, the type of multiple medicine dose like in allopathy or in Unani and ayurvedic systems is used by the tribals also.

Eastern part of the Himalayas like Nagaland (*Shepoumaramth*), Mikir hills

(Karbi-Anglong), Lepchas (Sikkim) are given as examples only and it is hoped that a detailed scientific study on medicines used by tribals shall provide interesting results which could be scientifically confirmed and applied in the larger interest of the poors of the country.

Vitex peduncularis (vern. *Charai-godra*) is used in ethno-medicinal lores of *Paharias* of Rajmahal hills and Adivasi. *Aushad Horopathy* (a term coined for ethno-medicine of Austrie linguistic group **Munda** tribe of Chhotanagpur) by is also used as herbal cure for Aids.

Under medico-religious beliefs *Achyranthes aspera* root is used as hairdo to expedite childbirth by tribals. Flower of *Bambusa arundinacea* var. *orientalis* is considered as an evil symbol. *Calotropis procera* (white flowered), if planted in a house on any day except Thursday, after wearing a new piece of cloth, keeps epidemic diseases like pox and cholera away. Similarly, roots of *Jatropha gossypifolia* when worn as amulet on right hand of the male and left hand of the female works as antidote to pox. Similarly, for cholera, roots of *Tragia involucrata* are worn to stop spreading of disease. *Mimosa pudica* roots kept under the pillow induce sleep in children but should be thrown after 2–3 hours. Pin drop acrid juice of *Semercarpus anacardium* nut put on the head of newly born baby is considered as a safeguard against evil eyes. *Loranthes*, a parasite on *Strychnos nux-vomica*, cut into pieces is used as amulet by women to stop childbirth. This portion is called locally as *Athkura*. Similarly, *Loranthus* on *Vitex negundo* is also used. Perhaps the fact that fruit production in plants is reduced due to parasitic infestation. Aerial roots of *Tinospora cordifolia*, worn as amulet on Saturday. Sunday or Tuesday, by pregnant women to protect from supernatural powers like evil spirits. Before wearing it is touched on Ocimum bed (*Tulsibeds*). Leaf paste of *Antidesma diandrum* is prescribed to cure blood dysentery.

Plants used by Tribals of Karbi-Anglong (Mikir) in Assam

1. *Aristolochia saccata* Wall. (Mik Rikang atelong)
   Used in stomachache. Fresh underground part is washed and made into paste. It is mixed with water (50 gm paste in 100 cc water for each dose).
2. *Baccaurea sapinda* (Roxb.) Muell-Arg. (Vern. *Tampaink*)
   Bark is useful in constipation. Adults chew it as such. In case of children 100 cc of sap is given early morning.
3. *Bidens biternata* (Lour.) Merr. and Sherff. (Vern Bap-nak-he)
   Bruised leaves are applied on forehead in headache.
4. *Blumea balsamifera* (L.) DC. (Vern. Jania Nagamese)
   Disorder of bowl due to defective mother's milk. Spoonful of sap of young leaf mixed with water given once a day. Sap is also applied to the body as protection from evil spirit.
5. *Blumea lanceolaria* (Roxb.) Druce (Vern. Hanmoichu)
   Leaves of *B. sessilifera, B. balsamifera, Mikania cordata* and this plant in equal parts boiled in, as decoction; water used for bathing, to cure body pain.
6. *Brucea mollies* Wall.ex Kurz. (Vern. Koinine) Fam. Simaroubaceae.

Borthakur SK. Less known medicinal uses of plants among the tribes of Karbi-Anglong (Mikir).

Useful in malaria. Powder of dry seeds (50 g) mixed with water (50 cc) given thrice a day. Acts as stomachic.

7. *Buettneria pilosa* Roxb. (Vern. Champhat) Fam. Sterculiaceae.
Paste of leaves applied on sores of cattle.

8. *Cajanus cajan* (L.) Millsp. (Vern. Rahban. Thaka Arhar (Hindi)
Paste of leaf or flowers applied on sores of mouth or tongue.

9. *Callcarpa arborea* (Wall.) Roxb. ex C.B Clarke (Vern. Arhi)
Paste of bark or leaf applied on sting of scorpion.

10. *Clausenia excavata* Burm.f. (Vern. Thonkuk)
Sap of leaf rubbed on all kinds of muscular pains.

11. *Clerodendrum colebrookianum* Walp. (Vern. Hunching)
Young leaves used as vegetable, useful as anthelmintic

12. *Coffea bengalensis* Hyne ex R. and S. (vern. Mihirai)
Bath with leaf infusion is useful when taken for 2–3 weeks for patients suffering from fever. 100 cc of infusion of leaves mixed with about 5 lit. of warm water. Bath given for 3 consecutive days.

13. *Dalhousiea petiolata* Grahm ex Benth. (Vern. Longyogthu)
Leaf paste applied to cuts.

14. *Dracaena petiolata* Hook.f. (Vern. Longla)
Useful in stomach pain and vomiting. Two spoonful of leaf sap given thrice daily.

15. *Eranthemum platiferum* Nees (Vern. Long-lamak-Mikir)
Paste of roots and leaves of *Narvelia zeylanica*, equal parts by weight, applied on bone fracture. Treatment renewed each day till bone heals.

16. *Elatostema lineolatum* Wight (6 Vern. NaglanAimbu)
Paste of leaf applied to cuts caused by rocks, stone and iron pieces.

17. *E. platylhyllumWild.* (Vern. Tangnep-Mikir)
Paste of leaves or bark applied on cuts caused by rocks.

18. *Euphorbia anticorum* (Vern. Hi juarong)
Latex applied on burns.

19. *Forestia hookeri Hassk*. (ern. Chahalubor, Chagukaduakhuan)
Paste of roots applied to cuts.

20. *Hedyotis scandens* Roxb. ex D.Don (Vern. Haniktu)
Paste of leaves (warm) applied to boils in early stage.

21. *Hyptianthera stricta* Wight and Arn. (Vern. Mirhirai)
Infusion or dry crushed leaves mixed with hot water given once a day to expectant mother.

22. *Homalomena aromatica* Schott. (Vern. Okhihalachang)
Useful in influenza. Aroma of rhizomes inhaled. The rhizome has itching effect.

23. *Houttuynia cordata* Thumb. (Vern. Nag Hakongpi)
Given in muscular pain as 200 cc of leaf sap, thrice a day.

24. *Hoya globulosa* Hook.f. (Vern. Mithnadai)
Leaf ash applied to dog bite repeatedly.

25. *Ixora acuminata* Roxb. (Longlapranpitheka)
Underground parts made into paste and applied on wounds. Root decoction is glactagogue. 1–2 spoonful is prescribed twice a day.

26. *Justicia gendarussa* Burmf. (Vern. Trachamai)
Leaf-paste applied on bone fracture in cattle; also used for relief in human beings in pains caused by displacement of bones.

27. Merremia umbellate (L.) Hall.f. (Vern. Torlongchok)
Leaves mixed with equal parts of leaves of *Mikania cordata*, made into a paste and applied on cuts.

28. *Mikania cordata* (Burm.f.) RL Robinson (Ranusinga)
Useful in stomach pain and dysentery. 50 cc of leaf sap is given thrice daily.

29. *Milletia caudate* Baker (Vern. Longtanap)
Leaves with *Ixora acuminata* and soots of *Stauranthera grandiflora* mixed in equal parts by weight and applied as paste on snakebite.

30. *Morinda angustifolia* Roxb. (Vern. Konthu)
Useful in sores of feet, which are dipped in hot leaf decoction, until it cools. It is repeated 3–4 times.

31. *Musa velutiana* Wendle and Rude (Vern. Khoyancham)
Sap of the pseudostem (100 cc) is given thrice a day to cure dysentery; also useful in stopping bleeding.

32. *Natsiatum herpericum* Buch.-Ham. ex R.Br. (Vern. Hanpalu)
Paste of leaf applied to cuts.

33. *Olax acuminata* Wall.ex Benth. (Vern. Han Kangyang)
Young leaves used as vegetable and acts as cathartic.

34. *Ophiorhiza ochroleuca* Hook.f. (Vern. Longlamihik)
In headache, 5 cc of leaf sap is given three times a day.

35. *Piper attenuatum* Buch. –Ham. ex Miq. (Vern. Mikaibithi)
To cure cold, 25 cc of root sap is given three times a day.

36. *Piper diffusum* Vahl. (Vern. Okang)
To cure indigestion half spoonful of root sap given two times.

37. *Piper griffithii* DC. (Vern. Cachapan-Nagaland)
20 cc of leaf sap given twice, relieves constipation.

38. *Pogostemon villosus* Benth. (Vern. Hanbila)
In stomach pain, 20 cc of leaf decoction is given three times a day. The decoction is mixed with mustard oil and rubbed on the body.

39. *Saprosma ternatum* (Wall.ex Roxb.) Hook. f. Vern. Thabai, Thabaibanghi)
In indigestion; 25 cc of bark-sap mixed with common salt and *Piper* given three times.

40. *Solanum myriacanthum* Dun. (Vern. Charha)
Ripe and unripe fruits are boiled and cut into two halves and kept over decayed teeth for half an hour. Alternately fruits are crushed, kept and boiled in a pot covered with banana leaves. A small hole is made in the leaf and through this hole a hollow bamboo is inserted and the vapours are allowed to foment the teeth and the mouth.

41. *Stereospermum personatum* (Hassk.) Chatt. (Vern. Inhet)
Useful in stomachache with vomiting but without stool. 2–3 spoonful of raw sap of bark is given thrice.

42. *Thunbergia grandiflora* Roxb. (Vern. Nagthahghadur)
Paste of leaves applied to cuts.

43. *Turpinia Nepalese's* (Wall.) Wight. (Vern. Thebangi)
Leaf paste and seeds applied to relieve muscular pain.

Plants used as Traditional Folk Medicine by Nagas

1. *Hemiphragma heterophyllum* Wall. (Shyrama Kipro-Mao-Nagaland)

Purohit VK, Nandi SK, Palno LM, Bag N. and Rawat DS. 2004. *Science letters*, NAS, **27**(5–6).

Used in cattle fever also of dogs and human. The plant is washed, boiled and the soup is drunk.

2. *Sarcopyranis nepalensis* L. (Vern. Oramikipra-Mao).
   Used in malarial fever; entire plant is washed, boiled and the soup used.

3. *Thalictrum foliolosum* DC. (Vern. Eveakorie, Mao)
   Roots are crushed after washing and taken raw or boiled and the soup drunk in fever, stomachache, hypertension, dysentery, emetic, also in headache, indigestion and abdominal flatulence.

4. *Ranunculus scleratus* L. (Vern. Othukoshir)
   The root and leaves are washed, crushed and applied on infectious part of the body, in boils and snakebite.

5. *Hydrocotyl sibthropioides* Lam. (Vern. Changobou)
   The leaves are washed, boiled and soup drunk in bodyache, dry cough, fever, gastric troubles and high blood pressure.

6. *Oxalis corniculata* (Vern. Oziambi)
   The leaves are folded in banana leaves and warmed in the ash of fire. The squeezed juice is taken as cure to diarrhoea, dysentery, lip and mouth infection.

7. *Rhus javanica* L. (Vern. Emosiibou)
   The fruits, leaves and flowers are boiled with sugar and the soup taken orally, in fever, dysentery, emetic, due the intake of contaminated oily food.

8. *Ocimum sanctum* L. (Vern. Napeou)
   The leaves and inflorescence are cooked along with rice or in some cases taken raw in fever, headache, also in high blood pressure, tonsil and gastric trouble.

9. *Spilanthes paniculata* (Vern. Chirli)
   The plants are washed, boiled and tak-
en in diarrhoea, high blood pressure abdominal flatulence, etc. The raw leaves are taken along with cooked rice and egg yolk to cure jaundice.

10. *Solanum nigrum* L. (Vern. Ohukostin)
    Leaves are washed, boiled and soup taken in kidney disorder and trouble in pancreas. Crushed fruits are mixed with cooked rice and given to hen in fever.

11. *Geranium nepalense* Sw. (Vern. Heniatopro. Likhodaphroshi) (and *Fragaria nigherensis* Sch-Rosaceae)
    The plants are taken in equal quantity, washed, boiled and the soup is taken to cure stone in urinary bladder, body swelling or oedema, dysuria/strangury. Extract of boiled leaves of Fragaria is given to cure diabetes.

12. *Zanthoxylum rhetsa* DC. (Vern. Motiibou, Nagothe)
    Leaf juice is squeezed and applied to nose in sinus, raw fruits taken in abdominal flatulence.

13. *Potentilla anserina* L. (Vern. Ojiipan)
    Washed roots are chewed while the juice cures diarrhoea and stomachache.

14. *Cajanus cajan* (L.) Mill (Vern. Kolemalepro)
    Paste of the leaves is applied to boils and swelling chicks.

15. *Melothrea madaraspatena* (L.) Cag. (Katamai, Kanganti. Maram)
    Fruits are washed, boiled and the soup drunk to cure jaundice.

16. *Catharanthus roseus* (L.) Don (Vern. Adam Evepa-Maram)
    Washed leaves are boiled and the soup taken to cure recurrent fever.

17. *Cynodon dactylon* (L.) Pers. (Vern. Tamsii-Maram)
    Leaves and tender stem applied externally after washing and crushing for cut or injury.

18. *Chenopodium ambrosioides* L. (Vern. Napouroi-Maram)

Leaves are boiled by adding salt and applied to dislocation, cut and broken bone. The roots are washed and boiled; soup taken in fever and headache.

19. *Clerodendrum viscos*um (Vern. Pfidigam)

Leaves after wash are boiled and taken as foodstuff in hypertension.

20. *Oroxylon indicum* (L.) Vern. (Kabi-maram, Kathemei, Dokre-maram)

The stem/root bark is crushed, boiled and drunk or taken raw to cure disorders, jaundice, blood purification and indigestion; also used as fish poison and for treatment of blood cancer, etc.

21. *Taxus baccata* L. (Vern. Khonghosii-Mao)

Leaf decoction is given as remedy for asthma, bronchitis, hiccough, and epilepsy and for indigestion. It is also used as fish poison and for treatment of blood cancer/leukemia.

22. *Pilea umbrosa* Wedd. (Vern. Ehrou-Mao)

Leaves are warmed on fire and pasted to the injury of body, as it is believed to prevent further infection caused by germs, bacteria and other pathogens.

23. *Impatiens balsamiana* L. (Vern. Eshou-Mao)

The leaves and tender stems/branches are boiled and the soup is given in diarrhoea and gastric disorders.

24. *Betula alnoides* Buch.-Ham. ex D. Don (Vern. Raisii-Paomei)

Root/stem bark is crushed and boiled. The soup is drunk to cure diarrhoea, dysentery and stomache.

25. *Quercus serrata* Thunb. (Vern. Chissi-Poumel)

The stems/branches and the juice is collected. One teaspoonful is mixed in 1 lit. of water and boiled by adding 4–5 teaspoonful. of salt and applied to acute anal infection. The juice without mixing water is taken internally to cure dysentery, diarrhoea and stomachache.

26. *Mimosa pudica* L. (Vern. Rheohii-Maram. Ahriipro-Mao).

The root is crushed and mixed with water and drunk for curing fever. This mixture is also used for curing dental caries.

27. *Melia azdairach* Willd. (Vern. Chetousii)

The leaves along with peach leaves are boiled and the soup is drunk in malarial fever. The fleshy soft tissue of the fruit is also taken raw to cure fever (malarial) and stomachache.

28. *Hedychium spicatum* Buch.-Ham. (Vern. Hraikama-Mao)

Rootstock is boiled after a wash by adding salt and the soup is taken to cure stomachache as carminative, and stimulant.

29. *Nasturtium indica* L. (Lerriovu-Mao)

Leaves of this plant are boiled after washing by adding salt and the soup is taken to cure bodyache and malarial fever.

30. *Ocimum basilicum* L. (Vern. Nepeou-Mao)

Leaves of this herb are boiled by adding chilly and other ingredients and taken in headache, fever and high blood pressure.

31. *Artemisia vulgaris* L. (Vern. Shipraikro-Mao)

The leaves are washed and squeezed and the juice is taken to cure bleeding during dysentery, diarrhoea, etc.

32. *Emblica officinalis* Gaertn. (Vern. Chiihroshibou-Mao)

Fleshy inner bark of the stem is crushed and boiled to cure diarrhoea and acute stomachache.

33. *Drymaria cordata* Willd. (Vern. Pfiipfii-pro-Mao)

    The crushed leaves after washing emit a flavour, which is inhaled in sinus. Leaf paste is used against boils and snakebite.

34. *Gynura cusimba* (D. Don) Moore (Vern. Mozhatobo-Mao)

    Extract of the leaves and flowers used against blocking bleeding during cut, injuries. The leaves are also cooked along foodstuff of pig against swine fever.

35. *Lantana camara* L. (Vern. Lingmen-Khollw)

    Juice extracted from the fresh and raw leaves and leaf paste is applied against bleeding during cut/injury/tumours. Also mixed with honey is given in fever.

36. *Meriandra bengalensis* Benth.(Vern. Kafur-pat)

    Extract of the juice from fresh raw leaves is used against dysentery, diarrhoea, stomachache, etc. Leaf paste applied on forehead in fever.

37. *Mussaenda roxburghii* Roxb. (Vern. Saleniapasii-Mao)

    Leaves and roots from the shrub are mixed and crushed into pieces, the juice is squeezed out and applied to snakebite. Fresh leaves when warmed on fire, juice squeezed and drunk to cure jaundice. Flowers are prescribed as remedy as diuretic and used in dropsy, asthma and recurring fever.

38. *Centella Asiatic* (L.) Urban (Vern. Korieu-Mao) The plant is washed, boiled and taken as foodstuff. Also the extracted juice from fresh plant is a remedy for gastric troubles, stomach ulcers urinary troubles, digestive complaints and dysentery, etc.

39. *Achyranthes aspera*. L. (Vern. Chaghapi-khra-Mao)

    Crushed roots are applied to infected tooth/teeth and cure dental caries pyorrhoea and gum complaints. Extracted juice is applied as remedy for irregular menstruation and piles. Seeds used as remedy for hydrophobia.

40. *Solanum xanthocarpum* Schard. (Vern. Prepribishou-Mao)

    Smoke of the seeds with dry fruits of nutmeg burnt and the smoke injected into infected tooth, to cure dental caries. Crushed fruits with water poured into nostrils of cow, buffaloe to kill and remove leaches from the nose of the animals.

41. *Eryngium foetidum* (Vern. Burmdiena-Mao)

    Leaf juice applied for cut/injury by rusting nail or broken glass. The leaves are boiled and the soup taken as remedy for prevention of titanus.

42. *Carica papaya* L. (Vern. Siiphrosii-Mao)

    Fruit juice is applied to anus for cure of pile/haemorrhoid.

43. *Punica granatum* L. (Vern. Shikriishi-Mao)

    Roots are crushed and mixed with water and boiled; the soup is taken in bleeding during dysentery. (cf. Photo)

44. *Nicotiana tabaecum* L. (Vern.Siidabou-Mao)

    The warm leaves are squeezed and the juice is applied to infected part (toes) caused by germs/bacteria present in contaminated water.

45. *Parkia roxburghii* L. (Vern. Yongchak-Mao)

    Outer skin of the pod is dried and boiled; the soup is taken in gastric troubles, bleeding piles. The bark extract is also used to cure diarrhoea and dysentery.

46. *Cinnamomum tamala* L. (Vern. Dalchinisi-*Mao*)

The leaves and bark washed, boiled and the soup is drunk in rheumatism, cold disease, diarrhoea and snakebites. Fresh bark is taken as raw. Leaves used as spice.

47. *Bombax ceiba* L. (Vern. Oekrisiibou-*Mao*)

Stem bark is crushed and placed in mouth for curing toothache in dental caries.

48. *Michelia champaca* L. (Vern. Zhoviesi-*Mao*)

Flower decoction taken against stomachache and is carminative, used in dyspepsia, nausea, fever, etc. Dried root and bark decoction is prescribed as purgative, stimulant and in diarrhoea.

49. *Erythrina stricta* Rox. (Vern. Letousii-*Mao*)

Stem bark is crushed into pieces along with that of *Kashii* and kept in a sac and shaken in river water. The foam thus produced either kils or stinks the fishes.

50. *Sehima wallichii* L. (Vern. Kashin-*Maram*)

Stem bark is crushed with of *Kashiin* small pieces and shaken in river. The foam produced kills fishes or stinks them.

51. *Alnus nepalensis* D. Don (Vern. Episii-bou-*Mao*)

The stem bark is crushed and boiled in water and taken orally to cure diarrhoea, dysentery and stomachache.

52. *Zanthoxyllum aromaticum* (Vern. Ora-momoshi-*Mao*)

Decoction of the stem bark taken orally to cure stomachache and possess anthelmintic properties. Fruits and seeds are used as aromatic, tonic and given during fever and dysentery. Fruit extract is effective against piles/heamorrhoids.

53. *Rubus ellipticus Sm*. (Vern. Shin-gnushi-*Mao*)

After washing, roots are crushed and boiled and the soup is taken in stomachache and bleeding during dysentery and diarrhoea.

54. *Solanum surattense* Burn. (Vern. Hrozippro-*Mao*)

Leaves boiled or raw leaf extract is applied direct to anus against piles/heamorrhoids.

55. *Bergenia ligulata* Wall. Engl. (Vern. Esiipro-*Mao*)

Crushed leaves are taken in prolonged and acute stomachache.

56. *Saussurea deltoidea* L. (Vern. Pfusa-ropro-*Mao*)

Fresh leaves washed and extracted juice is mixed with small amount of water and drunk against gastric trouble, stomachache caused by witchcraft and food poison, i.e. antidote.

57. *Myrica esculenta* Buch-Ham. Vern. Pashibou/Pzhea/shibou)

Stem bark/root is crushed and juice squeezed out and water is mixed and drunk in jaundice and gallstone. Also 2–3 drops are applied to ear against infection. (cf. photo)

58. *Jatropha curcas* L. (Vern. Monkeyfac-es-*Maram*)

Stem/branch juice is applied to body to cure injury caused by fire burns; it needs careful handling since it is poisonous.

The plant has assumed great significance recently as a source of motor-diesel and is promoted extensively all over the country as an economic alternative for utilisation of waste-

lands. It can easily be propagated by air layering.

59. *Pouzolzia benneteiana* Wight. (Vern. Ohunakary-*Poumei*).

    Roots are washed, crushed and applied in cuts/injury for blocking bleeding.

60. *Swertia pulchella* C.B. Clarke (Vern. Rasopro-*Poumei*)

    Soup of leaves given to cure typhoid fever associated with malarial fever accompanied by chill and sweating.

61. *Crawfurdia* affinis Wall. (Vern. Letihorie-*Mao*)

    Leaves or flowers are boiled and taken to cure stomachache, fever, headache and high blood pressure.

62. *Desmodium triquetrum* (Vern. Mukunamalyii-*Maran*)

    Leaves are washed and boiled and the soup is given against diarrhoea, dysentery and is nervine tonic.

Plants used by Lepchas of Sikkim

Lepchas presently are concentrated in Darjeeling district of W. Bengal including Kalimpong Division, east and north of Sikkim and maintain age-old tradition. Though various authors contributed to the ethnobotanical studies of the region, Sinha and Chauhan, 2000 listed the drug plants used by Lepchas of Sikkim. Though cardamom is the cash crop of the region, various medicinal plants occur which could also be cultivated as a cash crop. Few seasonal vegetables are grown traditionally along with cabbage, peas, beans and squash various wild plants like ferns (Angiopters) bamboo shoots, tender canes are used both as feed and cheap bear prepared from *Angiopteris ovata*, *Manihot esculenta* and millets. Table 5.1 lists plants and provides information regarding the species used by tribesmen like jhankris, Bonglinghs, etc. The list is not complete but given only as an example.

Table 5.1: Plants in tribal usage in Sikkim

	Plant names	Local names	Uses
1.	*Aconitum bisma*	Bikhma	Malaria
2.	*Mahonia nepalensis*	Termoking	Dhoop in rituals
3.	*Drymeria cordata*	Abhyal	Jaundice
4.	*Hibiscus rosa-sinensis*	Javakusum	Menstrual disorders
5.	*Prunus cerasioides*	Kamki	Fodder
6.	*Rubus ellipticus*	Kesimpt	Cough and Stomachache
7.	*R. foliosus*	Chillum	Edible fruit
8.	*R. nepaulensis*	Salum	Edible fruit
9.	*Astilbe rivularis*	Burokhat	Menstrual dysfunction, back acne
10.	*Bergenia ciliata*	Pakahnbed	Food poisoning, labour pains
11.	*Dichoma febrifuge*	Gesbokhnak	Dysentery
12.	*Trichosanthes tricuspidata*	Indrani	Food and pickle
13.	*Nardostachys grandiflora*	Panpu	Epilepsy, Dhoop

Singh DN. 1995. Use of medicinal plants of Sikkim in Ayurvedic medicine. HIMVIKAS. Occasional Publication 7 pp c5–68 = 68. GB Pant Institute of Himalayan Enviroment and Development.

Table 5.1: Plants in tribal usage in Sikkim (*Contd...*)

	Plant names	Local names	Uses
14.	*Saussurea gossyphora*	Yakchephabo	Burns
15.	*Lyonia ovalifolia*	Taksoinak	Skin diseases
16.	*Diploknema butyracea*	Nybe	Edible fruit
17.	*Budleja paniculata*	Pandamkung	Fermenting agent
18.	*Datura metel*	Kujuphimyungaman	Dog-bite
19.	*Picrorrhiza scrophularia*	Humle	Headache and fever
20.	*Phytolaca acinosa*	Jaringo	Food poisoning
21.	*Aconogonum sp.*	Jaringo	Edible as chutney
22.	*Rheum austeali*	Chhucha	As tea
23.	*R. nobile*	Chuka	Vegetable and pickle
24.	*Houttunia cordata*	Gandhejhar	Dysentery
25.	*Daphne bholua*	Nambongkantuh	Food poisoning
26.	*Edgeworthia gardneri*	Kuntkung	Rope and paper
27.	*Litsea cubeba*	Siltimur	Skin itch
28.	*Viscum articulatum*	Singleut	Cast for bone setting
29.	*Elatostemma platyphyllum*	Kanchelbhi	Deworming
30.	*Girardinia divesifolia*	Kujusrung	Cloth for traditional garment
31.	*Ficus glabrrima*	Ringjiking	Vegetable and pickle
32.	*Juglans regia*	Kalking	Leech repellant
33.	*Dactylorhiza hatigra*	Smbulkapa	Heal fractures
34.	*Amomum subulatum*	Ainilaichi	Gastric troubles
35.	*Costus speciosus*	Kafer	Infertility
36.	*Dioscorea glabra*	Kiewe	Edible
37.	*Arisaema griffthi*	Tangchit	Burning urine

Plants used in Tribal Areas of Himachal Pradesh and Uttaranchal (high altitudes)

Table 5.2 lists plants used in the tribal areas of Himachal Pradesh and Uttaranchal.

Table 5.2: Plants used in the high altitude tribal areas of Himachal Pradesh and Uttaranchal

	Plant names	Local names	Uses
1.	*Annselia optera*	Sathjalari	Digestion
2.	*Aconitum heterophyllum*	Mithi Patish	Cold, cough and fever
3.	*Acorus calamus*	Bach	Cardiovascular, cough, cold
4.	*Angelica glauca*	Chora	Arthritis, cold, cough
5.	*Atropa belladonna*	Belladona	Antimalarial

Table 5.2: Plants used in the high altitude tribal areas of Himachal Pradesh and Uttaranchal (*Contd...*)

	Plant names	Local names	Uses
6.	*Artemisia brevifolia*	Seski	Antimalarial, arthritis
7.	*Berberis aristata*	Kashmal	Opthalmia
8.	*Datura stramonium*	Dhatura	Mydritic
9.	*Dioscorea deltoidea*	Shingly-Mingly	Arthritis, fertility, cardiovascular and tonic
10.	*Hypericum patulum*	Basant	Antidepressent, weightloss
11.	*Jurinea macrocephala*	Dhoop	Incense
12.	*Picorrhiza kurroa*	Karoo	Antihepatotoxic, antiperiodic
13.	*Potentilla nepalensis*	Dori, Vajradanti	Antipyretic
14.	*Podophyllum hexandrum*	Bankakri	Antineoplastic
15.	*Morchella esculentus*	Guchhi	Cardiovascular
16.	*Pistacia integerrima*	Kakrasinghi	Asthma, chest congestion
17.	*Rauwolfia serpentina*	Sarpgandha	Tranquiliser, antihypertensic
18.	*Salvia moorcroftiana*		Arthritis, tonic in liver disorders
19.	*Swertia chirata*	Chrayta	Antimalaria, bronchial asthma
20.	*Rheum emodi*	Chuchi	Purgative, astringent, tonic
21.	*Taxus baccata*	Rakhal	Anticancerous
22.	*Valeriana wallichii*	Muskbala	Antispasmodic, hysteria
23.	*V. hardwickii*	Nehani	Perfumery
24.	*Viola odorata*	Banefsha	Cold and cough
25.	*Ipomoea turpetlhum*	Nisoth	Chronic piles (applied on piles orally mixed with KNO_3 (Kalmisora)
26.	*Moringa oleifera*	Nagkesar/Satingana	Stop bleeding haemorrhoids, taken along with *Misri* (sugar candy) and butter
27.	*Euphorbia royleana*	Suran	Irritating piles. Applied with curd the pulp made to a putty
28.	*Fragaria vesca* (fruits)		Diuretic, astringent
29.	*Aegle marmelos* (dry pulp)	Bel	Astringent in diarrhoea and dysentery
30.	*Cordia vestita*	Drupes	Drupes are demulcent and astringent
31.	*Corylus colurna*	Nuts	Nuts used as tonic and substitute for almonds
32.	*Seasmum oil*	Til ka tel	Oil with butter on irritating piles

B. Medicinal Plants collected from Forest as a Product and as Weeds from Wild/Wastelands: Trees and Shrubs

1. *Acacia catechu* Willd.

Family: Mimosoideae/Mimosaceae.

Local names: Khádirà (Sanskrit); Khair-babul Khair (Hindi); Khadéri (Mumbai); Kuth (Bihar).

Description: A moderatesized tree with thorny branches; stipular spines in pairs, short and recurved, shining brown or nearly black Bark rough, dark grey or ash coloured, exfoliating in long narrow rectangular flakes. Leaves 10–17 cm long; rachis prickly and with 4–5 glands; pinnae 20–60; leaflets 60–100 on each pinna, ligulate. Flowers pale-yellow or cream coloured, in lax-axillary cylindrical spikes. Corolla two to three times longer than calyx. Stamens numerous, much excerted. Pods flat, dark brown, 5–6 seeded. Seeds orbicular, dark brown.

Distribution: Generally distributed throughout the country. In the Himalaya it ascends to 1700 m up to Sikkim, generally in the beds of the streams, either pure or mixed with *Dalbergia sissoo*; occasionally also in mixed-scrub forests and on hillsides away from the streambeds. In Gujarat and Deccan ascends to 1000 m in Khandesh, Akrani and South Maharashtra state, mixed with other thorny species in the dry open thorn forests. In north Kanara and the Konkan it occurs nearly pure in larger or small patches on the low level; physiologically dry laterite soils, near seacoast. It is sporadic in the most monsoon forests but always occurs under dry conditions of soil.

Prain *Journ. As Soc. Bengal* **66** : 508.1898 distinguished three varieties. 1. *catechu* (distributed in Garhwal, Kumaon, Bihar, Ganjam and in Irrwady valley; also in north Kanara and Konkan) 2. *catechuoides* (in Sikkim Tarai and Assam, upper Myanmar, Mysore and Nilgiris) 3. *sundra* in Indian peninsula).

All these varieties have been raised to the rank of species by some authors. *A. catechu* proper can be distinguished by its calyx, petals and rachis covered with spreading hairs, while the other two species have calyx and glabrous petals.

Uses: The parts used medicinally are bark, root bark, flowering top and catechu.

The *bark* is bitter and acrid. It is astringent and useful in diarrhoea and ailments of gums, mouth and throat. Paste of the bark is useful in conjunctivitis.

Flower-tops mixed with other substances is used in gonorrhoea. From the wood three articles of commerce are obtained: *cutch*, *kwath* and *kheersal*. The first two are obtained by boiling the softer parts of the wood; cutch is dark but Kwath is pale as most of the tannic acid has been removed from it.

Kwath has extensive medicinal use in the treatment of diarrhoea, ailments of throat, mouth and gums. It is used widely with betel-leaf in India. In bleeding piles, uterine haemorr-hages, leucorrhoea, gleet, dyspepsia, chr-onic bronchitis it is used in various forms. A lozenge from catechu with gum Arabic and sugar is very useful in hoarseness., loss of voice, etc. The cavity of an aching tooth is plugged with catechu to relieve pain. The tincture of catechu is particularly useful for mammary glands and bedsores.

Khersal is a remedy for diseases of the chest.

Katha, turmeric and sugar candy powder relieve dry cough. With arecanut katha is ground and applied to teeth, if these are bleeding.

Other uses: The wood of catechu tree is very durable. The heartwood is an important source of tannin; it contains the highly active catechin, 1-epicatechin. The gum from the tree is considered as the substitute for Gum arabic. The tree is a host for lac insect. Finely grounded 'katha' applied to inflammation of throat organs, uvula of throat if given to drink, it also neutralises arsenic poisoning.

Extent of utilization: Katha is extracted in a fairly large scale and forms one of the important cottage industries in the *Tarai* region and is exported to various parts of the country.

Catechin, catechu tannic acid and tannin are the active principle in the heartwood, in addition to a, b and g catchin; l-epicat-echin.

Other species

a. *Acacia leueophloea* Willd. (Reonja-Rajasthan)

Tree with gnarled and crooked trunk is often on the sandy tracts. Gum is used for the same purpose as of *A. senegal*. Locally called *Salai-ka-gond*. Used externally to cover inflamed surfaces like burn and sore nipples, etc.

b. *Acacia jacquemontii* Benth. (Vern. Bawli, Bhu, Bambul-Rajasthan)

A shrub, usually found in the desert region on piedmont plains near the sand dunes and the gum from the stem is used in medicine.

2. *Acacia nilotica (Linn.)* Del. ssp. indica (Benth.) Bren. (Acacia arabica (Lamk.) Willd.

Family: Mimosaceae.

Local names: Kikar (H); Babbula (Sanskrit); Babla, Babul (Mumbai); Babhula-Kikar (Maharashtra).

Description: A moderate-sized thorny tree; with a large spreading crown; bark dark brown, almost black, much fissured; spines straight, white, sharply pointed. Leaves 2–5 cm. long; rachis and pinnae downy; pinnae 3–6 pairs with several glands at their insertions. Leaflets 10–20 pairs, linear, glabrous. Flowers yellow, fragrant, in axillary globose heads; peduncles in fascicles of 3–5, short, slender, grey downy, with a whorl of bracts above the middle. Corolla twice the length of calyx. Pod generally solitary, 8–12 seeded, stalked, deeply indented between the seeds, densely grey downy, tardily dehiscent. Important varieties are *Telia* with shady crown, *Kandia* with smaller trunk and spreading branches and *Ram Kanta*, branches arising at acute angles like in *Cupressus*.

Distribution: The plant is planted widely and self-sown throughout the dry and hot regions of India. It is absent from the humid regions of west beyond Jhelum river. It is common in the lower part of the river Ganga and throughout Deccan, Carnatic and Circars in dry localities, either gregarious in patches of forest, especially on old tank-beds and black cotton soil or in groups or single trees among the fields. It is often cultivated near field-bunds.

Uses: The leaves, flowers, bark, gum and pods are medicinally used. The tender *leaves* are made into pulp and used in diarrhoea and diabetes. The leaves are chewed as antiscorbutic. Poultice of the leaves is used for sore eyes and ulcers. In decoction the leaves are useful when sprouting, gargle is done in sore throat, spongy gum and wash for bledding ulcers. Leaves when dried in shade powdered and sieved when taken empty stomach, used in spermatorrhoea premature ejaculation and tenuity of semen, prevents abortion. Twigs used as toothbrush and prevents gum bleeding. Growing tops and leaves used as douche in cases of gonorrhoea, dropsy and leucorrhoea. Leaf decoction used as gargle for sore throat and, spongy gums and wash for bleeding ulcers.

Gum from the bark contains tannin and gallic acid and is a powerful astringent. A decoction of the bark is administered in diarrhoea, dysentery and diabetes, as an astringent lotion for ulcers, cancerous and syphilitic affections and diseases of eye and throat. The gum is the base of many medici-nal pastes and as a mucilage is useful in pulmonary and catarrhal dis-orders. Bark powder useful in diarrhoea. Fresh bark chewed to cure weak gums and in decoction cures prolapsus of utertus and leukorrhoea.

Infusion of the *pods* is used in diseases of mucous membrane, sore mouth and for healing syphilitic ulcers. Powder of the pods is used for stopping bleeding from sting of leeches. The pods are expectorant and used in coughs and as astringent for diseases of mucous membrane such as diarrhoea, dysentery, leucorrhoea discharges, sores of mouth, premature ejaulation, etc.

Extent of utilization: The bark and gum are collected on a small scale and exported to plains from the hills. The leaves contain tannin (32%) and fruits contains 41.7% tannin. Fruit contains a volatile and bitter principle 'balsa mine' resembling odour of 'Balsam of Paris'.

Acacia senegal Wild, (Kumat-Rajasthan). A thorny deciduous tree. It gives a gum called 'katira-kágond'. Occurs on rocky hills, sandy plains and stabilised dunes. Tree gum is odourless used as demulcent and as an emulsifying agent.

Gum exudation in *Acacia senegal* and *Commiphora mukul*

Acacia senegal, yields 'gum Arabic' occurs on rocky areas and no systematic tapping has been done on scientific lines. Since the gum exudation is very low, investigations to study the effect of periodicity and length of freshening and chemical stimulants have been initiated at Central Arid Zone Research Institute (CAZRI), Jodhpur. Chemical stimulants like Sulphuric acid of 50% concentration, Hydrochloric acid of 50% concentration and two length of blaze, viz. 10 and 20 cms and two periods of freshening, viz. one week and two weeks were used. The studies indicated that 50% concentration of sulp-huric acid yielded on an average 427 g of gum per tree in 30 days, whereas in the control, where no stimulants were used and the length of freshening was 20 cm, average yield per tree was 20 g only. Further studies are needed on this aspect. Similarly, in *Commiphora mukul* which occurs almost on the same environment, as the Acacia the tapping method is very crude yielding low quantities of gum of

inferior quality, often resulting in death of the tapped trees. The study conducted at the CAZRI with the same treatments as given above showed that 50% concentration of sulphuric acid yielded 100 gram per tree during a period of 30 days, as compared to control yielding about 30 gram during the same period.

However, the seeds from other areas where *Acacia senegal* grows need to be sown and result evaluated with improved material.

The *Acacia senegal* is reported to give gum in any quantity in zones where the average rainfall is below 500 mm. It has been reported that on lands where moisture is conserved, the Acacias exudes no gum or insignificant amount only, this is an indisputable fact. While certain authors have attributed the phenomenon to invasion of plant by bacteria or fungi, however, it has been consistently maintained that it was difficult to explain it otherwise than by biological imperatives of which the chief is the conservation of water in the actual structure of the plant. Today, it is generally accepted that water retention by hydrophilic colloids is one of the most effective xerophytic mechanism.

Gum-Arabic has emollient properties and emulsifying power of which much use is still to be made in pharmaceutics. It continues to be employed extensively in the manufacture of confectionary glues, finishes, etc. and it remains a valuable article of trade. It is one of the main cash crops of the desert region and its market value depends, above all, on the care taken in its collection, cleaning and packaging.

Babool Gum

The babool or *kikargum* exudes from the wounds in bark during March-May. Though some tree yield a maximum of one kg of gum in a year, average yield is only a few grams. The gum occurs in the form of rounded or ovoid tears, pale yellow, brown or almost black in colour. The darker samples contain tannin. The gum of *Acacia nilotica* is not the true *gum Arabic* which is obtained from *Acacia senegal*. Babool gum is generally considered inferior to the true *gum Arabic,* while in indigenous medicine it is credited with many virtues. It is given in diseases like diarrhoea and diabetes mellitus. The bark is considered astringent and demulcent. In Rajasthan it is used locally against asthma. The gum is also given in diarrhoea. The gum and sugar cubes are kept in mouth to cure mouth boils (*Chhal'e*). The gum is also used in many food preparations or as tonic, especially for ladies after childbirth.

3. *Achillea millefolium* L.

Family: Compositae or Asteraceae.

Local names: Gandana (H), Momadruchopandiga (Kash.), Birranjasif (Trade) Rajmari (Bombay); Bloodwort, Carpenter's weed, Devils Nettle, Mil-foil (English).

Description: An erect, pubescent herb with leafy stem, grooved. Leaves alternate, oblong-lanceolate, 3–pinnatisect, segments linear, acute. Radical leaves stalked, upper sessile. Heads radiate, small, crowded in compound corymbs. Involucral bracts few, erect; outer ones shorter; receptacle flat, covered with thin oblong scales, nearly as long as the flowers. Flowers white or pale pink; pappus none; ligules rounded, reflexed. Corolla of discflowers 5-lobed. Achenes shining, oblong, flattened.

Distribution: In Western Himalaya from Kashmir to Kumaon, at altitudes between 2000–3000 m. In Kashmir recorded from Gilgit, Murree, also in Lahul, Kunawar, Chur, Shimla, Chamba (in HP) and in Garhwal and Kumaon hills.

Uses: The herb is diaphoretic, tonic, asteringent and stimulant, it is used in

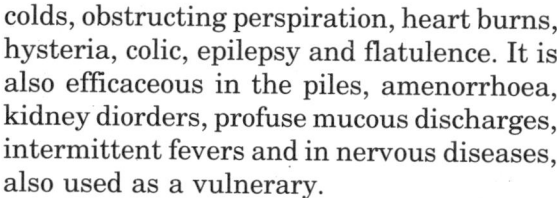

colds, obstructing perspiration, heart burns, hysteria, colic, epilepsy and flatulence. It is also efficaceous in the piles, amenorrhoea, kidney diorders, profuse mucous discharges, intermittent fevers and in nervous diseases, also used as a vulnerary.

The leaves are used as a tonic; decoction of leaves is injected in bleeding piles and vaginal haemorrhages. Decoction is also used for sore throat as gargle, hair wash and lotion for sore nipples. In other countries leaf decoction is also regarded as a specific remedy against colds, haemorrhages and nervous affections. The flowers are stimulant, aromatic and vulnerary. It yields an essential oil, useful as appetizer and is given in disorders of urinary system and of female reproductive organs.

Extent of utilization: The plant yields an essential oil having *azulene* as the main constituent, as tested on rabbit. Plants also possess weak antipyretic action.

4. *Aegle marmelos* Correa.

Family: Rutaceae.

Local names: Bilvá, Shriphalá (S); Bél (Hindi, Mumbai, in other parts of Maharashtra).

Description: A moderate-sized tree armed with axillary, straight spines, about 2–3 cm long; thick, grey coloured. Leaves aromatic, trifoliate; leaflet membranous, long. Flowers in panicles, greenish white, fragrant. Fruit smooth, greenish when unripe and gets yellow after ripening, pulp of fruit yellow when ripe. The fruit is pulp with peculiar smell (Colour Plate 5.1).

Distribution: Throughout dry regions of India in the sub-Himalayan tract, central and south India ascending to 1300 m; frequently planted all over India, near the temples of Lord Shiva.

Punjab plains, Dehradun, Assam, Gonda, Sambalpur, Koncan, Coorg, Cudpah, Ganjam, Godavari, also in UP, Uttaranchal at low elevations, Madhya Pradesh.

Uses: The drug called *Bel* comprises fresh, ripe or half ripe fruits.

Fresh juice of the leaves is antibilous, laxative and febrifuge, given in cases of fever with disorder of the liver, dropsy, constipation and jaundice. In decoction it is much valued in asthmatic complaints. Hot poultice of leaves is used externally in ophthalmia and is applied on the inflamed parts of the body, to the head in delirium of fevers and to the chest in acute bronchitis.

Juice of the bark is used in seminal debility and is useful in controlling palpitation of the heart. Root bark is used in hypochondriasis, melancholia and in the intermittent fevers. Mixed with sugar and fried, rice decoction of root bark is given to cure diarrhoea and gastric irritability of infants.

The fruit is considered best for its medicinal utility when it just begins to ripe. The ripe fruit is sweet, aromatic, astringent, cooling and laxative. It is chiefly valued for its mucilage and pectin and is one of the most popular remedies for curing irregularity of the bowls. It is best given in cases of chronic diarrhoea and dysentery, particularly for patients having diarrhoea alternating with spells of constipitation. It is also used as an adjunct after treatment of bacillary dysentery. Sun-dried slices of ripe fruit are astringent, digestive, antiscorbutic and stomachic, useful in chronic mucous diarrhoea and dysentery. The ripe fruit is described as mild aperient and juicy pulp when mixed with water and sweetened is used as a refrigerant drink in fever and inflammatory affections attended with the thirst. Preserved juice of the ripe fruit (*sherbet*) is beneficial in habitual costiveness. The rind of the fruit on distillation gives Marmelle oil.

Locally half-ripe fruit is placed on fire and when the clay is baked pulp is taken out and eaten empty stomach for controlling diarrhoea and dysentery. Pulp of unripe fruit with white cumin and cardamom ground and strained is given to children for relief from diarrhoea. Fruit syrup is sold in the market and is a simple cure for dyspepsia, used for drinking in summer season. Fruits administered as (1) extract of dry fruit (2) liquid extract from sun-dried unriped slices and as powdered dry syrup.

Extent of utilization: The fruit contains tannic acid, volatile principle and a bitter principle, resembling in odour Balsam of Peru. Pulp of the unripe fruit is dried and sold commercially as Bélgiri. Herbal granules are marketed as ICBEL by Baidyanath as a cure for intestinal disorders, amoebiosis dyspepsia and dysentery, etc.

Other uses: The wood is suitable for making charcoal for producer gas plants. The gummy substance, in which the seeds remain embedded is used as an adhesive, in varnished cementing mixtures. The twigs and leaves are used as fodder. Twigs are also used as toothbrushs. The wood is suitable for making small agricultural implements. The pulp of the fruits is used as substitute for soap.

5. *Adina cordifolia* (Roxb.) Benth. et Hook. f. ex Brandis

Family: *Rubiaceae.*

Local names: Dharákadambá (S); Haldu (H); Kéli-Kadam (Mumbai).

Description: A large, deciduous tree, often buttressed; bark grey, young parts pubescent. Leaves cordate-orbicular, pubescent beneath, leathery, 10–30 cm long; leafstalk thick, pubescent. Flower heads stalked, yellow, about 2 cm in diameter, 1–3 from one leaf axils; flowers densely pubescent. Fruit head is a collection of numerous small capsules. The tree can be easily recognised from a distance by its dark green umbrageous crown. The tree usually overtops all other trees.

Distribution: In the deciduous forests throughout the moister regions of India and in the sub-Himalayan tract up to 1000 m, from Yamuna eastwards. It is familiar in sal forests of Uttar Pradesh and common in mixed/deciduous and *sal* forests of Chhotanagpur and scattered throughout the state of Madhya Pradesh in erstwhile Bombay presidency and peninsular India, where it is gregarious in suitable localities but generally a scattered tree.

Uses: Bark of the plant is febrifuge and antiseptic.

Juice of the leaves is used to kill worms in sores.

Other uses: The freshly cut wood is yellowish which later turns reddish brown and is moderately strong; (40–50 lbs per cu ft). It seasons well and works easily, used for construction work, furniture and agricultural implements.

Extent of utilization: Commercially utilized to a limited extent.

The plant remains leafless for a short time in May. It flowers during the months of June and July. The seeds ripen in the cold season. *Haldwani,* a city name in Uttaranchal, is based on this plant which was abundant.

6. *Aesculus indica* Colebr ex Cambess.

Family: Hippocastanaceae.

Local names: Bankhor (Hindi), Hanudun (Kash.); Pángar (Garh., Kum.), Kandur (Jaunsar).

Description: A large tree with scaly buds. Old bark peeling off upwards in long thick bands. Leaves opposite, digitate, exstipulate, deciduous; leaflets 5–9, the centre one largest, oblanceolate or oblong, acuminate, sharply serrate, glabrous; lateral nerves 15–22 pairs, arcuate; base acute, bud scales oblong, membranous, caducous. Flowers white, horizontal, in large thyroid cyme bearing terminal panicles. Calyx tubular, with 5, short rounded lobes, often split longitudinally in open flowers. Petals 4, the place of 5th usually vacant, white and yellow, clawed, unequal in breadth. Stamens 7, filiform, curved upwards, longer than the petals, anthers versatile. Disk one-sided. Ovary sessile, 3-celled; style simple, slender. Fruit a 1–3 celled capsule, ovoid, rough outside. Seed dark-brown, smooth, shining.

Distribution: In the temperate regions of north-west Himalaya from the Indus to Nepal, between 1300–3300 m, on hill tops. At Chitra, Kashitwar, Pahalgam, Chamba, Shimla, Murree, Lambatch, Kedarkanta, Munsiari, Chaur, Uttaranchal Nainital. Cultivated at Kodaikanal.

It is a close vicariant of *A. hippocastanum*, which is common in Europe, while in the eastern Himalaya it is replaced by *A. assamica* Griff. *A. indica* was considered to have been introduced into Europe till it was sown by Heldrich in 1879, it forms a very limited area in Balkans. The out-standing difference between the two (*A. hippocastanum* and *A. indica*) is in the flowering date which is much later in *A. hippocastanum*.

Uses: The fruit is used for horses in colic, hence called *horse chestnut*. Paste of the fruit is applied in rheumatism and for this purpose an oil is extracted from the seeds, and used for the same purpose as above. In Kumaon roots are used to cure leukorrhoea. Crushed seeds are given to cattle to increase supply and quality of milk.

Extent of utilization: The plant is utilized to a very limited extent by the local villagers in the region and no commercial use is made.

Other uses: The wood of the plant is of a cream colour, soft and close-grained with very fine and many medullay veins; weight about 35 lbs per cu ft. It is turned into cups, dishes and platters which are used to hold milk, *ghee*, etc. The embryo of the seed is eaten by the hill people ground and mixed with flour.

Other species: Extract from the leaves of *A. hippocastanum* is used in whooping cough. The plant is cultivated as an ornamental tree.

7. *Ailanthus excelsa* Roxb.

Family: Simarubaceae.

Local names: Mahanim, Maharukh (Hindi); Tree of Heaven (Eng.).

Description: A lofty, beautiful tree, having large branches and rough, light-grey bark. Leaves alternate, pinnate, up to 1 m long, covered with fine hairs, when young, glabrous when old, falcate-lanceolate when young. Flowers small, white or yellowish, unisexual and bisexual on the same plant, in axillary panicles. Fruit winged, red coloured, twisted at the base. Seed only one per fruit.

Distribution: Indigenous in central and south India, Bihar and western peninsula. Cultivated on the roadsides in *Shekhawati* region of Rajasthan.

Uses: Bark and the gum-resin, exuded from the trunk, are used medicinally.

The leaves and the bark prescribed as tonic, in debility after childbirth; particularly useful in dyspepsia. Leaf juice of fresh bark is given mixed with rice porridge (*Khir*) or coconut milk or honey to stop pains after childbirth.

Bark is used as an astringent in the dysentery and bloody stool, also prescribed in dyspepsia, general debility, and as powerful febrifuge, stomachic and tonic. The bark is also expectorant and antispasmodic, prescribed in bronchitis, asthma and as an emetic when given in high doses.

Gum is anthelmintic and parasiticide.

Ailanthus malabarica DC. (Local name-Guggaldhupa, Mahanimbu) is a lofty tree, shedding its leaves annually at a time. Bark grey, rough, distinguished from *A. excelsa* in having leaflets entire Petals erect. Stamens longer than the ovate-cordate anthers. Fruit winged, not twisted.

The tree occurs on westernghats from north Canara to the erstwhile Travancore state in the south, ascending to 1000 m in Kerala state.

Bark and the gum-resin exuded from the trunk are used as medicine. The bark contains ailanthic acid and is a valuable tonic and febrifuge, used in dyspepsia. Juice of fresh bark mixed with milk or curd is used to cure dysentery and diarrhoea.

The roots are used as antidysenteric.

Leaves administerd in headache and gastralgia.

Gum-resin from the tree called "Muttipal" is used as a remedy for dysentery and bronchitis; a god stimulant in bronchitic disorders.

8. *Annona squamosa* L.

Family: Annonaceae.

Local names: Sharifá, Sitáphal (Hindi), Custard apple, Sugar apple (Eng).

Description: A low straggling, glabrous tree with grey, thin bark. Leaves glabrous above, downy beneath, pubescent when young, oblanceolate, pellucid dotted, scented when crushed. Flowers solitary or 2–4 together, drooping, short stalked, greenish. Sepals minute, triangular. Petals outer 3, about 2–3 cm long, valvate, fleshy. Fruit irregularly globose, tubercled, green. Seeds covered with a white, sweet pulp, oblong, deep brownish black.

Distribution: A native of south America, naturalised in many parts of India and gone wild in western and central parts of the country.

Uses: Leaves and roots are used. Leaves are insecticidal, used externally. Infusion is used in prolapse of the anus of children. Leaf poultice, mixed with salt, is applied to boils, maggot infested ulcers and malignant tumours for suppuration. Crushed leaves are applied to the nostrils of people having hysterical or fainting fits; also juice of the leaves is poured into the nostrils. Leaves are used in the form of paste and poultice on maggot infested sores, kills lice of the hairs and works as an aid in the extraction of guinea worms; also given in dysentery. Bark is astringent, used in diarrhoea, contains the alkaloid *anonaine*. Roots classed as drastic purgative and prescribed in spinal diseases, melancholia and acute disentery.

Fruit is edible and regarded as food after acute illness. Ripe fruit with salt is used as maturant for malignant ulcers while unripe fruit is astringent and is prescribed in diarrhoea and dyspepsia, mixed with ginger it is given in vertigo. Fruit powder is vermicidal.

The seeds are antiparasitic, powdered seeds used to kill maggots in cattle wounds, extracting guinea worms. Also used as hairwash, when used as powder mixed with gram flour it kills the lice.

9. *Alhagi camalorum Fisch.* (Alhalagi maurorum Medic.)

Family: Papilionaceae.

Local names: Jawása, Duralába, Taranjabin, Yavasa (Hindi); Arabian Manna Camel's thorn, Hebrew Manna Plant (Eng.).

Description: A small shrub, armed with many hard prickles. Leaves simple, drooping, oblong, leathery, smooth with rounded apex. Flowers 1–6, borne on a short stalk. Corolla reddish, much longer than the calyx. Pods up to Koncan area in Maharashtra.

Uses: All parts of the plant are used in medicine. The plant is laxative, diuretic and expectorant. Fresh juice of the plant is used in combination with aromatics for relieving suppression of urine. Poultice from the plant is used in piles. The expressed juice is applied for opacities of the cornea; and as snuff to cure migraine.

The extract of the plant by evaporating the decoction of the plant (called Yavasarkar) is used as demulcent and sedative in cases of cough in children. It is also cholagogue and acts as an aphrodisiac and restorative when taken with milk.

Mana or the sweet exudation on the leaves and branches (called taranjabin) when taken with milk is restorative and aphrodisiac: also used as an aperient, cholagogue, diuretic and blood purifier.

Oil prepared from the leaves is used in rheumatism. The flowers are used for removing piles.

10. *Albizia amara Boiv.* (syn. Albizia wightii Grah.; Mimosa amara Roxb.)

Family: Mimosaceae.

Local names: Lallei, Siris (Hindi).

Description: A tree, found in the dry mixed-deciduous forest of the Deccan, Karnataka and Kerala, up to 1000 m; generally planted along the roadsides. The pods are about 15 cm long.

Uses: The bark of the tree is used as mouthwash in case of pain or toothache.

Acne is cured by continuous application of the bark ground with water. Bark decoction is given to drink in cases of dropsy and oedema.

The seeds are used in collyrium for eyes and relieve itching, night blindness, macula, nebula and haziness.

The snuff prepared from the powdered seeds produces sneezing, in case of congestion due to cold and catarrh and makes it run; hemicrania is also relieved by smelling this powder. Seeds are taken along with milk to relieve complaints of spermatorrhoea, premature ejaculation and tenacity of sperms also act as aphrodisiac.

The seeds are also useful in cases of scrofula. These are powdered, sifted, mixed twice their weight of honey and preserved for 40 days.

Dried leaves are used as a substitute for soap and as green manure.

Other species

a. *Albizia chinensis* (Osbeck) Merr. (-A.*stipulata* Boiv.)

   The leaves and twigs of this plant are used as fodder; writing and printing paper is made of the wood pulp. The tree is found in the sub-Himalayan tract, the western peninsula and the Nicobar islands.

b. *Albizia Pebbeck* (L.) Benth (Siris-Hindi)

   The bark is used in dropsy, tonic and blood purifier. It is boiled in water with *Commiphora mukul* and given to camels when sick. Leaves used in night blindness and eyepain. Seeds used in leukoderma, hysteria and piles. Bark and seeds are astringent and diuretic.

c. *Albizia odoratissima* Benth is shown in Fig. 5.1.

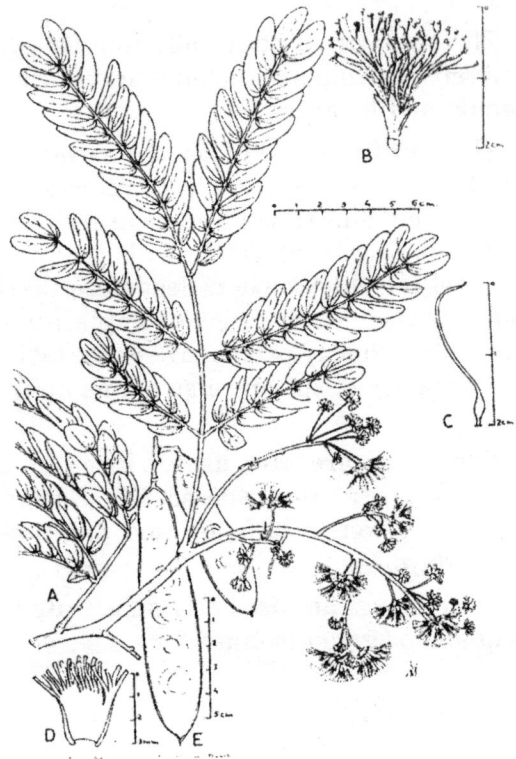

Fig. 5.1: *Albizia odoratissima* Benth. A. flowering twig; B. flower; C. gynoecium; D. staminal tube split open; E. pods

11. *Alstonia scholaris* R. Br.

Family: Apocynaceae.

Local names: Chétwan, Chatium, Lationj, Satiana, Pálà, Palágarduda, Rukatan.

Distribution: Moist regions of India, Westcoast forests.

Uses: It affords the bark of commerce which is used in medicine as an astringent tonic, anthelmintic, alterative and antiperiodic. It is a remedy in cases of chronic diarrhoea and dysentery, also of catarrhal fever and stomachic debility. Externally the milky juice is applied to foul ulcers and is also used with oil in earache. *Ditain* is obtained from the bark, and also tincture of Alstonia, appear to be useful in cases where quinine produces secondary symptoms. The effectiveness of the drug in diarrhoea and dysentery is beyond doubt but its effect as febriguge is not lasting.

The timber is not durable. But easily worked. Used for boxes, furniture, scaoboards and also for blackboards, hence the name scholaris. Young wood is used as a substitute for cork in bottles.

12. *Alangium salvifolium* (L.f.) Wang. VK = Thw (Alanguim lamarchii.

Family: Alangiaceae.

Local names: (Sage-leaved alangium (Eng); Dhera Akola (Hindi).

A small tree with edible fruits. Throughout the drier parts of India; especially in forests of South India.

Uses: The inner heartwood is ground in water and used as antidote for opium poison and in infantile tuberculosis.

13. *Anacarduim occidentale* L.

Family: Anacardiaceae.

Local names: Cashewnut (English), Kaju (Hindi).

Destribution: A small tree native of Brazil but extensively cultivated in Kerala, Karnataka and Andhra Pradesh.

Uses: The bark yields a gum which is obnoxious to the insects juice from incisions made in bark give ink of indelible marking. Cashew apple is edible as dry fruit. Oil from fresh kernels is used against white ants and is rubifacient and vasicant. Spirit distilled from fresh fruits juice is antiscorbutic.

14. *Andrachne cordifolia* Muell. Arg.

Family: Euphorbiaceae.

Local names: Gurguli (P); Bhatula (Garh); Bharti, Bhartoi (Jaun).

Description: A small shrub with slender branches. Leaves alternate, ovate-oblong, obtuse, entire, thin, rounded at the base, glabrous above, pale and thinly pubescent beneath, especially on the nerves; petiole slender. Flowers green, monoecious, axillary, solitary or several together; pedicels filliform, variable in length. Male flowers; calyx 5 partite, hairy outside, segments obovate; petals 5, spathulate, shorter than the calyx segments; disk 10-lobed, conspicuous; stamens 5, free, surrounding a small rudimentary pistil. Female flowers; calyx as in the male, accrescent in fruit; petals reduced to small glands; disk a fleshy ring surrounding the base of the globose ovary; ovary 3-celled; ovules 2 in each cell; styles 3, deeply divided into two long branches. Capsule globose, supported by the enlarged calyx, 6-valved.

Distribution: Temperate regions of the Himalaya westwards Muree, 1300–2500 m, Kashmir, Jaunsar; Chachpur peak, Lambatch, Kathiyan; Chamba; Shimla, Dehradun.

Uses: The plant is poisonous to cattle. Twigs and leaves are said to kill cattle when browsed in early morning on an empty stomach.

15. **Azadirachta indica** A Jussieu (-Melia azadirachta L.) (Fam. Meliaceae)

Local names: Neem Tree, Limba, kadunimba Nimba, Margosa tree.

Distribution: A native of Burma (Myanmar), is grown all over the country.

Uses: The plant properties were recently patented by USA, which has, however, been vacated by the court. Azadirectin is an effective ingredient extracted from the seeds. The powdered kernel when mixed with wheat, protects it against rive weevil (*Sitophilus oryzae*), lesser grain borer (*Rhizopentha dominica*) and grubs of khapra beetle (*Trogoderma granarium*), for about 9–13 months respectively. The extract of seed acts as gustatory repellent against rice weevil and flour wheat. Crushed seeds (1.2%) when mixed with the seeds of green gram, Bengal gram, cowpea and peas protect them against the bruchid (*Callosobruchus inaculatus*) for 8–11 months. Well known for its antiseptic and insecticidal properties.

For preventive hair loss, conditioning of hair and in treatment of dandruff, preparations are made from the kernels of the neem seeds, along with almonds and seeds of cucurbits (magzkaddu). Equal amount of these components are finly ground and mixed with water to form a paste which is spread over the scalp, left for 2–3 hours and washed off. For effective treatment, regular application of the preparation is required for several months.

Neembark gum acts as stimulant, demulcent and tonic, useful in catarrhal and other affections. Along with other herbs, it is used for the treatment of jaundice. It is also beneficial in malarial fever and useful in cutaneous diseases.

The two water soluble polysaccharides isolated from the bark show tumour activity against *Sarcoma* mice. The stem and root-bark yield *nimbin* and *nimbidin*, show anti-allergic activities. Nimbin inhibits stimulation produced by Histamine in guineapigs ileum at a dose of 1 mg/ml. At doses of 2 mg/ml nimbdin blocks the stimulant action of acetychlonie in frog rectus abdominus muscles. Nimbidin is effective against eosinophilia with marked symptomatic relief.

Preparations containing extract of the wood bark of neem tree are useful for preventing and curing gingivitis and periodotitis. Two parts by weight of finely made bark is refluxed with 8 parts of 90% ethanol for one hour, its alcohol extract after addition of a suitable aromatic combination gives the mouthwash. Nimbidin gargle and dentifrices are effective in bleeding gums and pyorrhoea.

Nimbidin, nimbin, nimbidol and neem oil are very effective against various detrimental fungi.

Stem bark is antiseptic, astringent, bitter tonic and applied on boils pimples, eczema, etc. Bark gum is demulcent and tonic.

Neem fruit called *nimboli* is also used as tonic, antiperiodontitic, purgative, emollient and as an anthelmintic. It is beneficial in urinary diseases and in the treatment of piles. The dry fruits are bruised in water and employed to treat cutaneous diseases. The pulp water when sprayed on crops protects them from locusts. In the rural areas children are given a bath with water boiled with *neem* leaves after the attack of chicken/smallpox. The branch with leaves is also kept in the house as an antidote. Unripe fruits are greenish and bitter while they turn yellow when ripe and also sweetish.

Neem leaves are useful during dry periods when no other fodders are available (Fig. 5.2.) In Gujarat, during the drought and scarcity period, 15–20 kg of neem leaves are fed daily to the cattle. A minimum of 3–5 kg of green neem per day is sufficient to provide an adequate supply of vitamin A. The leaves are also useful in combating worm infestation in cattle. The leaves contain adequate quantities of trace minerals except zinc, but alleviate copper deficiency in most straw and dry fodders. Neem leaves which are known for bitter-

Fig. 5.2: Branch of *Azadirachta indica*

ness also contain nimbin, nimbinene, nimbandiol, nimbolide, de acetylnibinene and quercetun, which are useful for ayurvedic medicines, pastes and diseases control. Leaf decoction is used in fevers.

Bhil tribe of Jhabua destrict (MP) believe that brushing teeth with neem, the body becomes resistant against snakebite.

Baigas of Mandla (MP) take orally 1½ leaf daily and apply leaf-ash with oil to cure skin diseases. The leaves are ground with pepper in water, strained and taken in cases of scabies, ringworm, abscesses, pimples and even in leprosy. Crushed leaves are also used externally to wash body and wounds.

The collyrium of flowers is effective in cases of itching eyes.

Kernel of the fruit stone is useful for piles (even bloody piles).

Ulcers, kill leaf-lice and effective in killing human sperms, thus making an efficient agent for family planning.

Sap of neem purifies blood and acts as tonic.

Gum is demulcent and tonic.

16. *Balanitis aegyptiaca* Planch.
(Balanitis roxburghii Planch.)

Family: Simaroubaceae.

Local names: Hingan, Hingu, Ingudi (Hindi).

Description: A small thorny tree or shrub with yellow or cinerous bark, spines long, sharp, often leaf and flower bearing; young parts pubescent or hairy tomentose. Leaves short-stalked, 2-foliate; leaflets elliptic or obovate, downy, entire, leathery. Flowers scented, small, whitish or greenish white, crowded in clusters, axillary, in 4–10 fld. cymes. Sepals and petals white, pubescent. Fruit a drupe, ovoid, woody, large, 5-grooved, rind light-grey, pulp bitter with an offensive greasy smell; stone hard, tubercled.

Distribution: In plains on drier parts of India, alluvial plains of Rajasthan, usually dominant on heavy soils.

Uses: The seeds, fruits, bark and leaves are anthelmintic and purgative.

Fruit (outer rind) pulp is used in cough mixtures, leukoderma and other skin diseases and boils.

A fixed oil as zachun is extracted from the seeds. Pulp of the fruit is used for washing silk as it contains saponin. Oil is also used to treat cuts and wounds.

Seeds are expectorant and given in colic.

In Uganda in S. Africa it is used as a remedy for sleeping sickness and as a purgative.

17. *Bauhinia variegata* L. (B. candida Roxb.)

Family: Caesalpiniaceae.

Local names: Mountain Ebony, Variegated Bauhinia (English) Kachnár (Hindi).

Description: A large tree with white flowers, having red and yellow stripes. The leaves and flowers are eaten as vegetable and the bark used for tanning and dyeing.

Distribution: It is found in Punjab, the western peninsula and Assam.

Uses: The bark, flowers and root with rice water are used as a cataplasm. Root decoction is given in dyspepsia. Flowers are taken with sugar as laxative. The bark is regarded as tonic and anthelmintic.

Extent of utilization: Used on a limited scale by rural housefold.

Other species

a. *Bauhinia acuminata* (Saféd-kachna). A small ornamental tree.

b. *Bauhinia corymbosa* Roxb. (Saféd-Kachnár-Hindi). A scandent shrub, grown as an ornamental.

c. *Bauhinia galpinii* N.E.Br. (Galpin Bauhinia). A straggling, prostrate shrub, native of south and tropical Africa. In garden for its bright scarlet flowers.

d. *Bauhinia malabarica* Roxb. (Malabar Mountain Ebony-English). A common tree of south India and Assam-Bengal region. The leaves are used for flavouring food stuffs and the bark as a tan. The tender seeds and leaftips are edible.

e. *Bauhinia monandra* Kurz. (Butterfly flower, Jerusalem date). Grown as a hedge plant.

f. *Bauhinia purpurea* L. (Camel's foot, Pink Bauhinia-English, Lal-Kachnár). The bark is used as fibre, also for dyeing and tanning. Flower buds are vegetable, leaves as fodder. Bark used for tannin extraction.

g. *Bauhinia racemosa* Lamk. (Kachnár, Guiral-Hindi). Small bushy tree, found in outer Himalaya on calcium rich soils. The bark yields fibre and the flower buds and fruits are edible.

h. *Bauhinia retusa* Roxb, (Vern. Kandla –Hindi). Yield gum, used for sizing cloth and paper. Found in Bihar, MP and Punjab.

i. *Bauhinia tomentosa* L. (Yellow Bauhinia, St.Thomas tree-Eng.). The bark used as cordage. Common in S. India, Assam and Bihar.

j. *Bauhinia vahlii* Wight and Arn. (Camel's foot climber Eng, Maljhan –Hindi). A climber used for ropes. Leaves sold in market for cups, the bark gives tannin.

18. *Betula jacquemontii* Spach. (B. utilis D. Don) B. bhojpatra Wall.

Family: Betulaceae.

Local names: Bhurja (S), Bhoja (Garhwal), Bhojpatra Tree (English).

Description: A large tree; outer bark white, inner layers pink, peeling off in thin sheets. Leaves stalked, ovate, long pointed, sharpely and irregularly toothed. Male catkins drooping; flowers bracteate in groups of usually three; perianth 4-parted; stamens 2, filaments minutely forked. Female cones usually drooping, single; flowers bracteate; in groups of three, perianth none. Ovary 2-celled. Nuts minute, flattened, winged on both sides; bracts 3-lobed, lobes of fruiting bract nearly equal.

Distribution: In temperate regions of the Himalaya from Kashmir, 2300–4000 m to Sikkim, 3000–4500 m. In Shimla and Garhwal the plant is not found under 3300 m. West of Sutlej, Swat, north of Jangla, Chitral, Burzil valley Zanskar, Kagan, Banihál, Kolahi, Lahul, Shimla, Kunawar, Chamba Chansil, Deota, Harkidun, on way to Gangotri.

The name *B. utilis* and *B. bhojpatra* are synonyms and based on the same species collected at Gosainthan in the Himalayan region. Being earlier, *B. utilis* has the priority and extends up to Kumaon. All other sheets from Kumaon and westwards to Kashmir and Hazara are named as *B. jacquemontii*.

Uses: The Bhujhera (a fungus formation of *B. utilis*) is used for alimentary disorders of the animals.

Extent of utilization and trade: A small quantity is colleted and consumed locally. However, *Bhojpatri* as sold commercially in the market though some old manuscripts have been written on this bark and are still preserved in museums; the *bhojpatra* is not used as paper now.

19. *Bombax ceiba* L. (Bombax malabaricum DC., Salmalia malabarica Schott and Endl)

Family: Bombacaceae.

Local names: Sémul (H); Salmili (S); Roktosémul (Bo), Silk cotton tree (Eng).

Description: A large deciduous tree, stem more or less buttressed at the base when old, covered with large conical prickles when young, branches whorled, horizontally spreading. Leaves digitate; leaflets 5–7, lanceolate or obovate, acuminate, entire, more or less coriaceous; stipules small, caducous. Flowers 10–12 cm. across, fleshy. Petals crimson, white tomentose outside. Stamens usually 5-adelphous, filaments numerous, the inner most forked. Ovary 5-celled. Capsule oblong-ovoid; seeds surrounded by long silky hairs (Colour Plates 5.2 and 5.3).

Distribution: Throughout hotter parts of India, in the hills ascending to 1300 m in the outer ranges and in hot valleys generally affecting water-logged places and dry beds of river. Roots, bark and gum used as drug apart from flower buds, which are used as vegetable.

Uses: Roots of the young tree, 1–2 years old (semul musli), are stimulant, tonic, demulcent, emetic and alterative and is useful in dysentery and phthisis. These are also used as an aphrodisiac and in gonorrhoea and dysentery, effective in cases of seminal debility spermatorrhoea and tenuity of semen.

Bark of the stem is mucilaginous. Its infusion is given as demulcent and aphrodisiac in seminal weakness. Externally it is valuable as a styptic in abnormal uterine bleeding and its paste is used over inflammations and eruptions of the skin. Stem bark is given in blood dysentery and loose motions by 'Bhils of Jhabua'.

Gum from the bark called Mochras or Supari-ka-phool" contains gallic and tannic acids and is an astringent, tonic, alterative, styptic, demulcent and aphrodisiac. It is used with beneficial results in diarrhoea, dysentery, haemoptysis of pulmonary tuberculosis and influenza, haematemesis and menorrhagia. It is also useful for checking menses after delivery. Bark chips are tied in small bundles and add to *Salphi* (Caryota urens) wine quality to increase its intoxication.

Dried flowers with Poppy seeds are used in the treatment of piles and a paste of flowers is applied over boils, sores and itch.

Young fruits are useful in calculous affections, chronic inflammation and ulceration of the bladder and kidneys. It is also useful in weakness of genital organs.

The seeds are used in gonorrhoea, gleet, chronic cystitis consumption and catarrhal disease. Silk rounded seeds are externally used for dressing burnt and inflamed surface. It is used extensively and is available on a commercial scale.

20. *Berberis aristata* Roxb. ex DC. Var. asiatica Ahrendt.

Family: Berberidaceae.

Local names: Kingorá (Garhwal), Kashmoi, Kingor (Hindi), Daruhaldi (S). The Indian Barberry.

Description: A large thorny shrub with grey bark; young branches red. Leaves ovate, sharp-toothed and pale green beneath. The stamens possess a curious irritability, if touched by an insect they spring forward and jerk out the pollen. The small orange or red spot seen on the leaves is the aecidial stage of *Puccinia*.

Distribution: In temperate zone of the oak and pine forests between 1500–2700 m and on dry aspects of the hill slopes. It is generally found in association with *Rubus ellipticus*, *Rosa moschata*, *Berberis chitria* and *Cotonester* sp.

Uses: The plant has got properties similar to those of turmeric. The wood and rootbark gives an extract called "Rasot" which is considered alterative, deobstruent, astringent, antiperiodic and diaphoretic, used in skin diseases, menorrhagia, diarrhoea, jaundice, bilious complaints and diseases of the eye. It is also used as a purgative for children, blood purifier, alterative, tonic and febrifuge. With rose water it is dropped into the eyes in conjunctivitis and ophthalmia. It is also used for gastric and duodenal indolent ulcers.

Rasot (prepared from Berberis) along with *Neem* (Azadirachta indica) leaves is made into a paste and applied locally on piles. With camphor and butter, it is applied locally to subside swelling and to check blood coagulation.

The root decoction is used internally in menorrhagia. The root bark is useful as tonic, antiperiodic and diaphoretic. It is as valuable as quinine because its decoction is very efficacious in malaria. It is particularly useful in relieving pyrexia, checking return of paroxynis of intermittent fevers. The decoction is used as a wash for unhealthy ulcers.

The bark is of great value in intermittent and remittent fevers and in debility after fever. It has some uses in general debility, diarrhoea and dyspepsia. The decoction of the stem is efficacious in painful urination and jaundice. The wood furnishes a yellow dye used for colouring leather. It is rich in alkaloid content (Sci. and Cult. 1941)

Ripe berries along with *Cinnamomum tamala* (stem bark) and honey are given internally in leukorrhoea. Decoction the fruits are used as mouth wash, in gum swelling and toothache.

Time of collection: The roots are collected during April to June and contain berberina, an alkaloid.

The drug is collected during August and September. Collectors sometimes prepare Rasaut in the forest and bring it for disposal to the market. The current rate for Rasaut is more than Rs 180/kg and for the stem at more than Rs 50/kg. In the indigenous system dariadikath, darvidiavlehya and darviaditila are prepared from this plant. Extraction of roots on steep slopes in Kumaon hills has been baned for fear of soil erosion from steep slopes.

The roots contain an alkaloid Berberine.

a. *Berberis asiatica* Roxb. ex DC.

Family: Berberidaceae.

Local names: Dáruhaldi, Kingorá, Kasmal (Chamba).

Description: A much branched spreading shrub with pale, furrowed, corky bark. The leaves are hard, elliptic, with distant, spinous teeth, strongly coriaceous, grey on undersurface with pale reticulation, exstipulate. Flowers pale yellow. Fruit a blue black berry (Fig. 5.3).

Distribution: The plant is abundant between 600–3000 m in hills of Kumaon and Garhwal.

Fig. 5.3: *Berberis asiatica* Roxb. ex DC.

Uses: The root is bitter sharp in taste, used for healing ulcers, urethral discharges, leukorrhoea, fever, diseases of the skin, eye and ear. It is also useful in ophthalmia, jaundice and diseases of the mouth. It is good in eye sores, toothache, asthma and skin pigmentation as fomentation removes inflammation and swelling.

The extract of the bark called *Rasot* is prepared and used as described earlier and is said analogous in properties to tumeric. Total alkaloid content of root is 40%, of stem 1.5% of which Berperine is 2.09 and 1.29% (Indian Journ. Med. Res. 1929)

Extent of utilization and trade: A fair quantity of the plant is exported and sold in the form of Rasot.

b. *Berberis lycium* Royle. var. lycium

Family: Berberidaceae.

Local names: Kingorá (Garhwal); Kashmal (H).

Description: Shrub with pale-grey bark, branches angular. Leaves sessile, tough, narrow lanceolate, acute, entire or with few

teeth, upper surface bright green, lower pale. Racemes shortly stalked, simple, barely longer than the leaves. Flowers pale yellow. Style short but distinct. Berry ovoid, violet, covered with bloom.

Distribution: In temperate regions of W. Himalaya in Garhwal and Kumaon hills, from 1000–3000 m.

Uses: Root is bitter with unpleasant taste and used in spleen troubles; tonic and a good febrifuge, internal astringent, good for cough, chest and throat troubles, eye sores and itching of the eyes; piles and menorrhagia. It is useful in chronic diarrhoea, allays thirst, as a gargle strengthens gums and a good application to boils. Root extract gives *Rasot* which is used as from other species.

The leaves are used as a cure for jaundice.

Extent of utilization and trade: A fair quantity of the plant is exported. It has been reported that it contains less of *Rasot* than in *B. aristata*.

The active principle is Umbellatine; alkaloid in roots.

21. *Boswellia serrata* Roxb.

Family: Burseraceae.

Local names: Dhup, Gugal, Saleh, Salai (Hindi), Incense Tree, Indian Frank Incence Tree, Indian Oilbanum Tree. (Eng.).

Description: A deciduous tree of moderate size, bark greenish, peeling off in smooth, thick flakes; young parts and leaves hairy leaves crowded near the end of the branches, odd-pinnate, about 30 cm long; leaflets 8–15 pairs, opposite, lanceolate or ovate, deeply broadly crenate, unequally sided. Flowers bisexual. Fruit a 3-valved drupe, 3-seeded.

Distribution: Common in the moist forests of Central India, Maharashtra south, and Orissa, Karnataka states. Also occurs in Rajasthan, north Gujarat, Uttar Pradesh, Punjab and forests foothills of the Himalaya.

Uses: The tree is best known for the gum-*oleoresin,* known as *Salaiguggal,* that exudes from the stem and the branches. It consists of an essential oil, used as a substitute for the American pine turpentine, employed for making varnishes, etc.

The gum resin is locally sold as 'loban' which is burnt as an incense.

Internally, the gum is used as a stimulant; expectorant, diuretic, diaphoretic, astringent and stomachic. It is prescribed in chronic lung diseases such as chronic laryngitis, bronchorrhoea; as stimulant for liver in some cases of jaundice, chronic cases of diarrhoea, dysentery, dyspepsia, pulmonary affections, amenorrhoea, dysmenorrhoea, gonorrhoea, placenta previa and piles. With aromatics, the gum is given in rheumatic and nervous diseases, scrofulous affections and urinary disorders. The gum is also used for reducing obesity.

Oil from the gum is used in gonorrhoea mixed in demulcent drinks.

Externally, the gum is used with coconut oil as *lép* applied to sores, syphilitic ulcers, indolent swellings, sprains, bruises, dislocations, carbuncles, ringworm, nipple sores, etc.

Gum fumigation is used in fevers, laryngitis and bronchorrhoea.

Fragrant resin, sold as *loban,* (Vern. *Kundur* and *Ral*) is diaphoretic, astringent, used in osteo-arthritis, nervous and skin diseases and has been recently marketed by Dabur on a commercial scale.

Ral is commercially sold in the market and used to improve efficiency of fire used to burn human bodies. Loban is also sold commercially in the market as an incense.

22. *Butea monosperma* (Lamk.) Taubert. (-Butea frondosa Roxb.)

Family: Papilionaceae.

Local names: Dhák, Palás, Tésu (Hindi), Flame of the forest, Bengal-kino (Eng.).

Description: A small tree, serving as host for lac insects. The flowers appear in February and March and are the source of yellow dye used during the Holi festival. The leaves are also used mainly as plates in plain areas of UP. The ripe fruits are edible (Fig. 5.4).

Distribution: Found throughout the plains of north India in the forests and is called the *Flame of the Forest*, because of the colour of flowers during the season.

Uses: The bark, flowers, gum and the seeds are used in medicine.

The leafshoots, gum and bark of the tree are useful in spermatorrhoea, tenuity of semen, premature ejaculation and leukorrhoea. The unopened leafshoots when dried in shade, powdered, sifted and mixed with equal weight of sugar, are taken every morning with milk or water for a fortnight to get relief from the above complaints. A kg of roots when boiled in 5 ltr. of water, till 2 liters is left, then strained. *Santhi rice* (seeds of *Eleusine indica*) steeped into. When the water has dried up, the rice is ground to a meal and fried into a pudding with sugar. The product is eaten to cure spermatorrhoea and leukorrhea.

Gum from the bark, locally called *Chunia gond* contains large mucilage, and is a stringent, also cures many forms of chronic diarrhoea, round worms, leukorrhoea and excessive menstrual flow. It is given to women after childbirth.

The flowers with poppy seeds when boiled with water are used to get a relief from testicle pains. These are astringent diuretic, and aphrodisiac. Seeds useful in quartan and malarial fevers after removing the red coating, and ground to a fine powder with equal quantity of Karanj (*Pomgamia pinnata*) kernels.

Paste of the seeds made in water and applied to cure ringworm disease. Freshly powdered seeds given against Ascaris and when pounded with lemon juice act as powerful rubifacient.

Extent of utilization: The flowers, seeds and the gum is available in the market by the name of Mochras and used extensively.

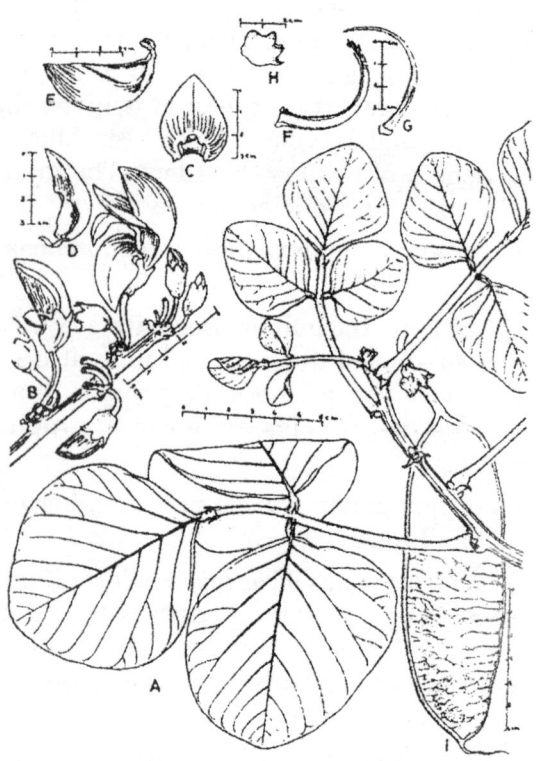

Fig. 5.4: *Butea monosperma* Taub. A. leafy twig; B. flowering twig; C. standard; D. wing; E. keel; F. androecium; G. gynoecium; H. calyx; I. pod.

23. *Caesalpinia bonducella* Flem.

Family: Caesalpiniaceae.

Local names: Fevernut tree (English), Karanjwa, Nata, Sagargota, Karanja, Karbat (Hindi).

Description: A thorny scandent bush, often gregarious, with fruits like castor, divided into compartments, having bluishgrey stone which is very hard. The kernel is obtained on breaking the stone and tastes bitter but used in medicine

Distribution: It is often gregarious in the hills up to 1000 m; also near seashore in the tropics. The name 'bonducella' is given, because of bullet like seeds. It does not lose viability, even if carried by sea.

Uses: The seeds are highly esteemed for the treatment of intermittent fevers, especially, is associated with skin diseases. It is used to cure malarial fever. It purifies the blood, used in cases of hydroceole, cases of scabies and flatulence of stomach.

Kernel of 'Karanjwa' and *Palas* Papara (*Butea monosperma*), leafshoot of Babool (*Acacia arabica*) in equal weight are finely powdered and made into pills with little water (of 2 g each) and taken thrice daily to cure malarial fever, even if it is quartan.

Externally, it is applied as an ointment with castor oil in hydrocoele.

Leaves are ground in water, strained and given to drink for killing intestinal worms. The roots and root bark possess properties similar to seeds but weaker in intensity. The kernel is ground and made into paste and applied warm in hydrocoele, the water is absorbed and the testicles resume their proper size, if applied patiently for 2–3 weeks. Three stones of 'Karanj' when buried in hot ashes, till the shell is almost burnt and the kernel is taken out and given during asthma attack. With dry ginger, kernel is ground and the powder is administered to allay pain due to flatulence. '*Karanj oil*' cures itch, when applied locally to the affected parts.

Extent of utilization: Seeds are available in the market.

24. *Calligonum polygonoides* L.

Family: Polygonaceae.

Local names: Phog (Hindi), Phogáro (Rajasthan).

Description: An evergreen, xerophytic shrub, almost leafless; stipules membranous, cup shaped. Branches flexuous, terete with slender branchlets. Leaves minute, bristles at distant nodes. Flowers pinkish, fascicled in the axil of ochrea. Nut oblong, hard, densely clothed with many series of branching intricate, rigid, redbrown, flexuous bristles.

Distribution: A genus of Mediterranean regions, while the species is restricted to Rajasthan, some parts of Haryana and Punjab extending westwards to Pakistan. Common on sand dunes and sandy plains.

Uses: The roots bruised and boiled in combination with *catechu* are used for gargle in soregums.

Flowers are rich in protein; flowerbuds are effective in treating sun-stroke. Flower buds locally called *lasun* are also used as food during famine.

25. *Cadaba fruticosa* (L.) Druce (Syn. C. farinosa Forssk. ; C. indica Lamk.)

Family: Capparidaceae.

Local names: Kodháb, Dábi (Hindi).

Description: A shrub or small tree, unarmed, straggling, much branched; leaves simple, ovate-elliptic. Flowers greenish white, in terminal raceme. Bracts small,

5

subulate. Petals 4, limb oblong. Stamens 4. Fruit cylindric, dehiscent; pulp orange coloured.

Distribution: In tropical regions throughout India, on old walls, dry river-beds, hillocks and waste places; most abundant. Jafri (Pakistan J. for. 7:204, 1958) regards *C. farinosa* to be a distinct from *C. fruticosa* (L.) Druce, and the two taxa can be separated based on the a number of stamens (5–8 vs 4–5 in *farinosa* and *fruticosa*), kinds of hairs and 2 kinds of filaments (discoid vs filamentous). Bhandari (*Ann. Arid zone* 6: 201, 1967) examined a number of sheets in the herbarium and found that Rajasthan plants possess two kinds of hairs. However, collections put under *C. indica* in Blatter's herbarium at Bombay, have mixture of two Taxas. *C. farinosa* is confined to Arabia and Africa, while *C. fruticosa* is in the Indo- Pakistan subcontinent.

Uses: The roots and leaves are medicinal, anthelmintic and deobstruent; often prescribed as decoction in urine obstruction.

The leaves are used as poultice, also as purgative, anthelmintic and antisyphilitic.

26. *Calophyllum inophyllum* L.

Family: Guttiferae.

Local names: Indian Laurel, Laurelwood (Eng.), Surpan, Sultan-champa. Poonang (Hindi).

Description: An evergreen tree of moderate-size, glabrous, bark smooth grey or blackish brown. Leaves opposite, leathery, broadly-elliptic, shining on both sides. Flower buds with minute rusty hairs, flowers stalked, white, fragrant, in lax, few flowered,

axillary racemes; unisexual and bisexual on the same tree. Sepals and petals 4 each. Stamens many in 4 bundles. Ovary usually purple. Fruit a roundish drupe, green with scanty pulp.

Distribution: Indigenous in Malabar area, T.Nadu. Pondicherry, Karnataka; throughout the coasts of south India including Orissa.

Uses: Stem when wounded gives a gum that is bright green, pleasantly scented; which is neither collected nor made use of.

The seeds yield an oil (60% by weight), having disagreeable odour and flavour but valued as an excellent application in rheumatic affections, and in elephantiasis. Mixed with 'chaumagra oil' it is employed for exanthematous affections. It is believed to be efficacious in curing scabies in Pondicherry.

The leaves contain saponin and hydrocyanic acid and used as fish poison.

A volatile oil extracted from the gum 'tacamahca gum' through distillation is used as perfume in Tahiti.

Extent of utilization: The chief centres of production of the oil are Mumbai, Goa, Thiruvananthapuram, Tanjore and Puri in Orissa state.

27. *Callicarpa macrophylla* Vahl.

Family: Verbenaceae.

Local names: Dáyá (Hindi), Priyamágu (Sans), Dahiya (Tehri Garhwal), Daia (DDN).

Description: An erect shrub, branches, leafstalk and inflorescence covered with wool like tomentum. Leaves stalked shortly, lanceolate, crenate, upper surface wrinkled,

Madhu Agarwal. 1991. Phytochemical and Pharmacological studies of Indian Medicinal Plants. M. Phil. Thesis AMU.

stellately pubescent, lower tomentose. Flowers pink. Fruit succulent, globose, white, containing 4-one seeded butlets.

Distribution: In sub-Himalayan tracts from Hazara (in Pakistan) eastwards to Assam the variety *griffithii* occurs in Koraput, Sikkim, Singbhum, Palamau, Purnea, Jalpaiguri and east Himalayan regions. Also on swampy locations in Doon and along the Himalaya.

Uses: In Kumaon the woodpaste, obtained by rubbing against a stone, is used against mouth and tongue sores, locally known as *khap*.

Oil from the root is used as a remedy for disorders of the stomach.

Leaves on warming are applied to rheumatic joints. *Callicarpa arborea* Roxb. forma farinosa at Sikkim is known for its bark, which is carminative, aromatic, bitter and in decoction applied to cutaneous diseases. It is also considered as tonic and carminative. However, the plant is a tree with thick trunk found in lower hills from Kumaon eastwards.

Extent of use and utilization: The plant is used locally.

28. *Calotropis gigantea* (L.) Dryand ex W. Aiton.

Family: Asclepiadaceae.

Local names: Aák, Madár (Hindi).

Description: A large shrub, covered with woolly tomentum. Leaves opposite, ovate, tomentose, without stalk. Flowers in lateral, umbelate cymes, purplish white. Corolla lobes hairy, spreading or reflexed, appendages of corona longer than broad. Follicles boat shaped, green, up to 10 cm.

Distribution: Common on abandoned fields and fallow lands. A lactiferous shrub of arid and semi-arid regions: Milk from plant, flowers and leaves used in medicine.

Uses: The powder of the rootbark is emetic, expectorant, diaphoretic and purgative. It is used in chronic rheumatism, jaundice, enlargement of viscera, intestinal worms and cough. Externally paste of root bark is applied to elephantiasis to legs and other cutaneous affections such as leprosy. Dried bark is an excellent substitute for *Ipeca cunada* and is reported to promote gastric secretion and acts as mild stimulant, which may be given with carminations in dyspepsia. It is also given as febrifuge. Powder of the root in goat milk is used in ear trouble and boils in Kumaon.

The leaves in powder form are dusted upon wounds to destroy excessive granulation and promote healthy actions. Tincture from the leaf is tried as an antiperiodic in cases of intermittent fevers as effectively as quinine. Oil in which the leaves have been boiled are applied to paralysed parts. Yellow ripe leaf warmed before fire, crushed and juice dropped into ear to relieve earache. Oil prepared by burning leaves in seasmum oil relieves pain of patients suffering from rheumatism, lumbago and sciatica.

In Kumaon hills the latex is used to cure leprosy and ringworm. Economically, a sort of 'guttaparche' is obtained from the latex. Dried latex is nervine tonic, antispasmodic and anodyne. A snuf is made from the milk, causes congested catarrah to run, thus relieving headache. The flowers are considered digestive, stomachic, tonic and useful in asthma, catarrah and loss of appetite. They are said to be digestive and expectrorant and are prescribed locally in cough, and catarrh.

Time of collection: The plant flowers throughout the year.

Extent of utilization: The active principle is Calotropine. A proteolytic enzyme, somewhat similar to 'papain' has been found in the milky juice. The *Kondh, Saora* and *Bhumij* tribes of Orissa use roots for curing leprosy. Root is pounded in water mixed with liquor prepared from *Mahua* flowers and mixture applied externally, and in Kumaon a mixture of bark and pepper (*Piper nigrum*) given orally in epilepsy by 'Santals' and 'Paharia' tribes. Heated leaves applied on swellings.

29. *Calotropis procera* R. Br.

It is also a common shrub in the plains but differs from the above species in its coronal scales, being fleshy and smooth. Its distribution is somewhat restricted to the plains of northern India (Fig. 5.5).

The rootbark known as 'Madar Bark' is reputed as a remedy for the piles, cough, dropsy and skin diseases. The latex is purgative and tincture of the leaves is used in intermittent fevers. It has also been reported to have anticoagulant properties. The acrid juice is poisonous to human beings but the leaves are taken by goats with impunity. It has 5 principles from which an indentical 'genin' can be extracted by hydrolysis this is called *Caqatropagenin. P* root powder is a reputed remedy for piles, cough, dropsy and skin diseases.

Latex is a poison to human being but leaves are taken by goats

Active principle is *Capotropine*, a proteolytic enzyme, somewhat similar to Papain.

Other uses: A valuable fibre is obtained from the stem which can be spun into a finest thread.

30. *Capparis aphylla* Roth. (Capparis decidua (Forsk.) Pax.

Family: Capparidaceae.

Local names: Karél, Kér, Karirá (Hindi).

Description: A large twiggy shrub or small tree up to 3.5 m sometimes leafless when mature, having grey bark which is thick, corky, irregularly fissured; branches zig-zag. Leaves linear, small, thick, sometimes red, having a pair of straight thorns at the base. Flowers reddish-brown, in many flowered corymbs, stalks slender and tomentose. Sepals 4 in 2 whorls. Petals 4, unequal, scarlet, filaments red, many. Fruit globose, red when ripe, fleshy, many seeded, Fls May-June (Colour Plate 5.4).

Distribution: A common plant of the rocky as well as consolidated sandy plains in Kutch, Rajasthan, Punjab, Haryana, upper Gangetic plains and Central India.

Fig. 5.5: *Calotropis procera,* a common plant of the ruderal habitat throughout North India on calcium rich soils. Flee leaves are being collected for use against guinea worms.

Uses: The leaves and branches are used as plaster for boils and when chewed relieve toothache.

The wood is bright yellow, shining, hard and bitter and resistant to white ant attacks, favoured for making knees of boats. Bark is useful in rheumatism. Charcoal powder with honey used in cases of phlegmatic cough, rheumatism and lumbago.

The rootbark is pungent and given in cases of intermittent fevers, asthma, inflammations and rheumatism.

The bark is acrid, laxative, diaphoretic, and anthelmintic, useful in cough, asthma and inflammations, in dropsy ground, rootbark applied externally to ribs in case of pleurisy.

The fruits locally called *tent* are astringent, are useful in cardiac troubles. The young flowerbud and fruits are pickled. Fruit is eaten both as green or ripe. Useful in facial paralysis and solves problem of enlarged spleen, kills intestinal worms.

Other species

a. *Capparis grandis* L.F. (Vern. Antéra) produces an oil used in medicine. Rootbark is burnt to ash, mixed with honey and milk to cure cholera, skin diseases and blood purification. Infusion of the bark used internally for swellings and eruption.

b. *Capparis spinosa* L. (Vern. Kábrà-Hindi) provides flower buds known as *capers*, used as a *condiment*, yields a volatile oil having properties of *garlic oil*, and employed in affections of liver and spleen, also in amenor-rhoea. All the parts are regarded as stimulating and astringent, when externally applied. Young flowers and fruits are edible and pickled when green.

Extent of utilization: It is used commercially and sold in the market.

31. *Carissa carandus* L. sensu Hook. f.

Family: Apocynaceae.

Local names: Garinga, Karaundá (Hindi); Karamchá (Bihar); Kanachuka (Sansk.). Bengal Currents (Eng.).

Description: A large erect, evergreen, spinous shrub with rigid, glabrous branches having spines that are long, horizontal, sharp and straight. Leaves leathery, broadly ovate, dark green, shining above and paler beneath. Flowers white, often tinted pink, fragrant, in terminal, stalked, 10–12 fld cymes. Lower part of the corolla tube cylindrical, upper swollen and pubescent. Fruit a globose berry, deep purple or black when ripe, filled with sticky, milky juice. Seeds flat, usually 4.

Distribution: Wild throughout India, cultivated for the fruits.

Uses: The fruit is edible, used for making preserves. The ripe fruit is antiscorbutic, while the unripe one is astringent.

The roots are used as fly repellent, when made into a paste is reported as bitter, stomachic and anthelmintic.

An alkaloid and salicylic acid have been reported from the fruit.

Other species

1. *Carissa spinarum* L. (Vern. Name Karaunda), is a small ever-green shrub, in dry and rocky situations often forming extensive undergrowth in forests of the Shiwalik and teak forests. Flowers April-June and the fruits ripen during cold season. The berries are like those of the above species and eaten raw or as cooked preserve.

The roots are ground and put in worm infested sores of animals.

2. *Carissa opaca* Stapf: The details are depicted is Fig. 5.6.

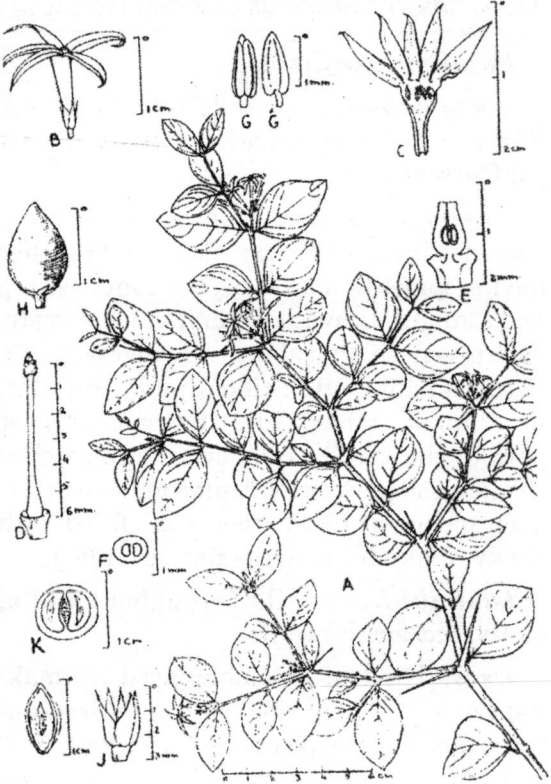

Fig. 5.6: *Carissa opaca* Stapf. A. flowering twig; B. flower; C. corolla opened to show androecium; D. gynoecium; E. longitudinal section of gynoecium; F. transverse section of gynoecium; G and Ǵ outer and inner views of a stamen; H. fruit; I. seed; J. calyx; K. transverse section of fruit.

32. *Cassia fistula* L.

Family: Caesalpiniaceae.

Local names: Gurmálá, Amaltás (Hindi). Purging Cassia (Eng.).

Description: A deciduous tree with greenish grey bark. Leaves compound. Leaflets each 5–12 cm long. 4–8 pairs, ovate. Flowers large, scented, yellow, in drooping racemes. Fruit a cylindric pod, up to 50 cm long, pendulous. Seeds many, in black, sweet pulp separated by transverse partitions. The long pods which are green, when unripe, turn black on ripening after flowers shed (Colour Plate 5.5).

Distribution: In deciduous and mixed-monsoon forests throughout greater parts of India, ascending to 1300 m in outer Himalaya. In Maharashtra, it occurs as a scattered tree throughout the Deccan and Konkan.

Uses: The root is prescribed as a tonic, febrifuge and strong purgative. Extract of the rootbark with alcohol can be used for backwart fever.

The leaves are laxative and used externally as emollient, a poultice is used for chilblains, in rheumatism and facial paralysis. Juice of the leaves is useful as dressing for ringworm, relieving irritation and relief of dropsical swelling.

The pulp of the fruit around the seeds is a mild purgative when given with other medicines, but when given alone it is apt to cause nausea, griping and flatulence. It is also used in biliousness and in diabetes. Externally, it is useful for evacuation in flatulent colic, as dressing for gouty or rheumatic joints, for removal of obstructions of the abdominal viscera. The pulp contains sugar, tannic matter, albuminous starch, oxalate of calcium and other unimportant constituents. The pith is particularly useful if there is swelling in stomach, liver or intestine, in cases of diphtheria it contains sugar, tannins, albuminous starch, oxalate of calcium and other unimportant constituents.

The seeds are emetic and have cathartic properties. Leaves and flowers are both purgative like the pulp. They contain anthraquinone, tannin, oxyanthra quinone, rhein, volatile oil. Ashes from burnt pod

mixed with little salt is used with honey taken 3–4 times a day to relieve cough.

Extent of utilization: Fair quantity of this plant is exploited commercially.

Other species

(Out of the 40 species existing only some are used as medicinal).

a. *Cassia angustifolia* Vahl.

Local names: Indian Senna (Eng), Senna (Hindi).

It is extensively cultivated in south India on red, or black clay loam soils in Tinevelley, Maudurai and Trichy districts. It is exported and often adultrated with many species. The land is liberally ploughed and manured, the sowing is done in May. Weeding has been attended to but irrigation is hardly necessary. The season for leaf collection is June-December and the yield expected is 500 kg for an acre. *C. acutifolia* is next in importance but little used. *C. obovata,* has never achieved commercial success, though included in French pharmacopoea.

b. *Cassia absus* Linn

Local names: Cháksu (Trade name-Hindi)

The plant is found in south India. The seeds are used in the treatment of ophthalmia and are cathartic, used also in skin diseases (Fig. 5.7).

c. *Cassia auriculata* Linn

Local names: The Tanner's cassia, (Eng.) Tarwar (Hindi), Avaram (Tam.).

A shrub of heavy soils. The seeds of the plant are like those of *C. absus* and valued as a local application in ophthalmia. The infusion of leaves is esteemed as a cooling medicine and as substitute for tea. The leaves are also eaten as green vegetable in

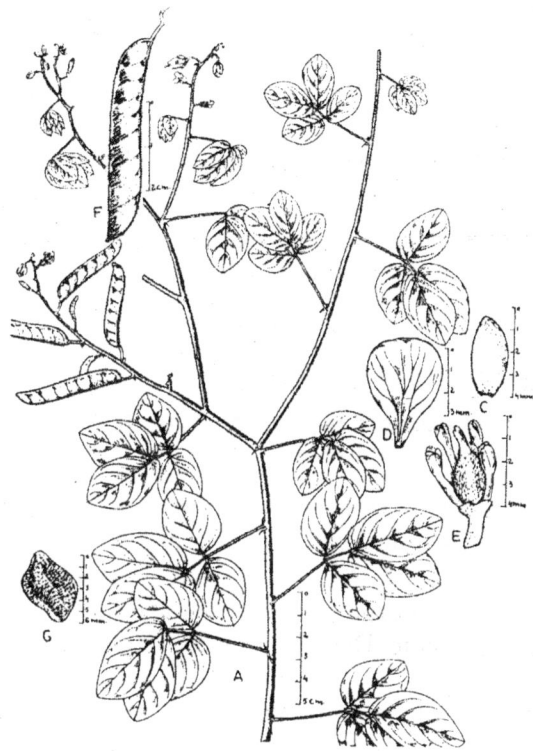

Fig. 5.7: *Cassia absus* L. A. flowering and fruiting branch; B. flower; C. sepal; D. petal; E. androecium and gynoecium; F. pod; G. seed.

times of famine. The shoots are largely used as native toothbrush and root is spoken of as great value in tempering iron metal and used in skin diseases. The bark is largely used in tanning. Bark is astringent. Leaves and fruits are anthelmintic.

d. *Cassia alata* L.

Local names: Dadmaran, Dadmurbam (H); Ringworm Cassia (Eng.)

The leaves when rubbed into a thin paste and mixed with vaseline constitute an effectual remedy for ringworm.

e. *Cassia laevigata* Willd.

Local names: Shyansundri (H).

A weed of cultivation in the subtropical zone during the rainy season. The cooked pods are used internally in skin diseases.

f. *Cassia lanceolata* (Vern. Sonamukhi, Svarnpatri). ·

Leaves used to relieve constipation, also in cases of rheumatism, lumbago, sciatica, asthma and malarial fevers.

g. *Cassia occidentalis* Linn.

Local names: Káli-Herwan (Jammu), also known as Talawphali.

A weed of rainy season, abundant on roadsides and similar localities, particularly on the rubbish heaps. The leaves, roots and seeds are purgative. The root is diuretic and antiperiodic. Decoction of the herb is highly prized in hysteria and to relieve spasm. Dried seeds used as substitute for coffee and as an antibilious: The variety with bluish black branches is considered more efficacious.

h. *Cassia sophora* Linn.

Local names: Sophera senna, Kasaundi-hindi.

A cosmopolitan weed sprouting after the rainy season in the tropics. The bark, leaves and seeds are cathartic and juice of the leaves is specific for ringworm.

i. *Cassia tora* Linn. (C. obtusifolia Linn.)

Local names: Banarah (Kum.), Herwan (Jammu), Panwár (Trade name), Chakaunda (Hindi).

A plant is abundant after the rains as a weed. The powder of the roots and seeds mixed with lemon juice or whey is applied externally for ringworms in Kumaon. The seeds boiled with tea are also used in cold.

In Jammu region. The leaves and seeds constitute a valuable remedy for skin diseases. Both these parts contain a glucoside resembling chrysophanic acid and certain Oxytoxic properties. Leaf decoction is laxative.

33. *Careya arborea* Roxb.

Family: Lecythidaceae.

Local names: Gadáva, Gaválda, Márà, Kambà, Kumbi, Vakamba (Hindi), Slow Match Tree, Wild Guava (Eng.).

Description: A large, stout, deciduous tree with dark grey bark, peeling off in narrow flakes, reddish and fibrous within. Leaves usually without stalks, obovate, large, crenate and glabrous. Flowers pinkish-white, large but without stalk, clustered at the ends of branches, having unpleasant smell. Petals 4, stamens pink or purple, numerous. Fruit green, globose, fleshy, crowned with persistent calyx and style. Seeds many.

Distribution: In Bengal, Maharashtra, MP, in most regions of the sub-Himalayan tract from Yamuna eastwards; also in the peninsula to some extent.

Uses: The bark yields a gum and broadband of fibres used for cordage. The wood is light-red to clear red or reddish brown, durable and can stand water; hence used for making vessels and agri-implements, furniture, etc.

The bark is astringent, when moistened it gives much mucilage is utilized in preparation of emollient embrocations. Flowers are given as tonic after childbirth and the dried calyces of the flower are sold as demulcent in cough and cold. The seeds are poisonous, while some record that the seeds are edible. Leaves are also fed to *Tassar* silkworms, also for making *bidis*.

Extent of utilization: The leaves are used extensively, while the wood is used as fuel and for furniture.

34. *Cedrus deodara* (Roxb.) Loud.

Family: Coniferae.

Local names: Deodar, Déwar, Kilar (Hindi); Himalayan Cedar (Eng.).

Description: A tall, evergreen tree with leading shoots and extremities of branchlets droping. Foliage dark green or bluish green. Leaves 3-sided, single on elongated shoot, but in dense clusters on arrested branchlets. Catkins cylindric, single, at the end of arrested branches. Cones erect, 10–12 cm long and 7–10 cm in diameter. Scales thin, rounded. Seeds up to 2 cm long with triangular wings.

Distribution: In Western Himalaya, extends up to some western parts of Nepal, 1300–3300 m alt. also in Afghanistan.

Uses: The wood gives an oleo-resin called Kél-ka-tél, resembling crude turpentine oil, which is used in ulcers, skin diseases, in veterinary practices, and sore feet of calf. The wood powder is carminative, diaphoretic and diuretic, useful in fever, costiveness, pulmonary complaints and urinary disorders. The wood is pounded with water in a stone and the paste is applied to the temples to relieve the pain.

The oil called 'deodar-tar-oil' is obtained by destructive distillation of the wood, also used as perfumery, fixative in soap and cosmetic industry, skin diseases and for ulcers. The sesquiterpenes present in alcoholic extract of stem gets to papaverine-like spasmolytic activity.

Extent of utilization: The oil is used in trade and available in the market. The wood is used extensively and supposed to be the strongest of all conifers.

35. *Celastrus paniculata* Willd.

Family: Celastraceae.

Local names: Málkangni (Hindi).

Description: A large woody climber with corky bark; young shoots marked with lenticels. Leaves variable in size and shape obovate-orbicular elliptic or oblong lanceolate, short acuminate, crenate, lateral nerves 4–6 pairs, parallel to margins. Flowers in terminal panicles, drooping, branching into compound cymes. Flowers pale green, calyx-lobes rounded. Petals oblong. Anthers large, about the size of petals. Capsule globose, bright yellow when ripe. Seeds black, enclosed in a red aril (Fig. 5.8).

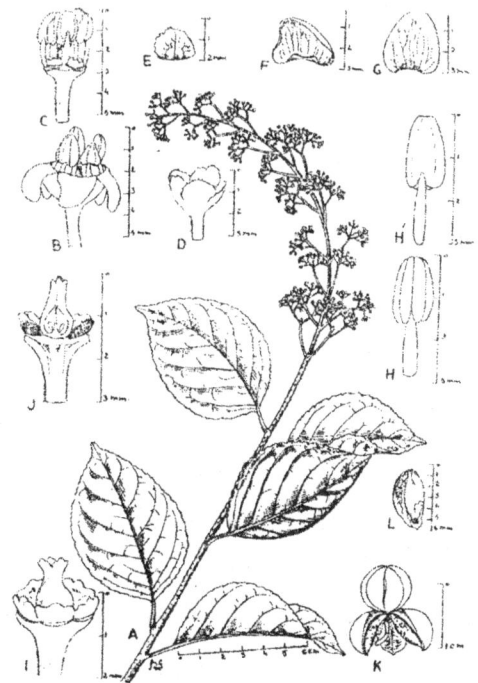

Fig. 5.8: *Celastrus paniculata* Willd. A. flowering twig; B. flower; C. same with calyx and corolla removed; D. calyx; E. sepal; F. petal (outer view); G. same (inner view); H and H̄ stamen, (inner and outer view); I. calyx, corolla and androecium removed to show disc and gynoecium; J. longitudinal section of gynoecium; K. dehiscing capsule; L. seed.

Distribution: Common in low forests, near Dehradun, Jaunsar and Saharanpur forests, having striking appearance when in fruit, which hang like bunches of yellow beads.

Uses: The leaves of this woody and deciduous climber are emmenagogue and the juice is given as antidote to neutralise overdose of opium.

The bark is abortifacient, used to purify blood and the oil from bark to cure sores.

The seeds are bitter appetizer, laxative, emetie and powerful brain tonic. As an external and internal remedy, seeds are used in rheumatism, gout, paralysis and leprosy; also used as stimulant and diaphoretic in rheumatism and gout. Destructive distillation of the oil with benzoin, cloves, nutmeg and mace is obtained and employed in treatment of beri-beri.

The black oil from seeds which is deep reddish, is stimulant and diaphoretic. Seed oil if taken in small quantity also cures abdominal complaints and used to increase intelligence and memory.

The seeds are supposed to have the property of stimulating the intellect and sharpening the memory. The oil is obtaining by distillation.

Extent of utilization and trade: The seeds are traded in the market at Dehradun. *C. stylosa*, shorter racemes and style being 3-lobed also occurs under the same ecology and forests.

Bastar tribals eat the flowers as vegetable wheras the 'Marias' eat young fruits. The 'Bhil tribe' of 'Jhabua' apply seed oil in eczema. Sambalpur tribals make root into a paste with Talmuli (*Curcilago orchoides*), satavar (*Asparagus racemosus*) and lilmutri (*Smilax sp*) and take twice daily for 4–5 days in spermatorrhoea.

36. *Ceriops tagal* (Perr.) C.B. Robins (Ceriops candolleana Arn.)

Family: Rhizophoraceae.

Local names: Kirari, Kiri, Goran. The vernacular name also applied to both (*C. tagal* and *C. roxburghii* Arn).

Description: A small tree of the mangrove vegetation of coastal areas in tidal forests. The tree is used for fuel by the locals.

Distribution: In tidal forests of Sundarban and Andaman islands, a small tree of muddy shores and tidal creeks, while *C. roxurghii* is a large shrub.

Uses: The plant is astringent and bark decoction is applied to stop haemorrhage. On Agrican coast the young shoots are applied as substitute for quinine.

The bark is used as a dye to colour brownish red, but especially a gold black and purple in conjunction with indigo.

The leaves are used for tanning.

37. *Cinnamomum tamala* Nees.

Family: Lauraceae.

Local names: Dálchini, Kikrà. Téjpát, Tamálà (Hindi), Gurándra (Jaunsar).

Description: A moderate-sized tree, evergreen, aromatic with wrinkled, thin and brown bark. Leaves shortly stalked, in opposite pairs, ovate-oblong or lanceolate, leathery, shining above, 7–14 cm long. Flowers white, numerous, in pubescent panicles, perianth 6-parted, small. Fruit succulent, ellipsoid, black when ripe.

Uses: The bark of commerce, locally called *Tejpata* gives a fragrant oil which is used in soap manufacture.

The leaves are carminative and stimulant, used in colic pain of diarrhoea, rheumatism, tonic for brain, good for liver and

spleen. Also used in the manufacture of vinegar and for flavouring food ('*Pulao*') and tea. In decoction for lochia after childbirth. Used as substitute for betel leaves.

***Extent of utilization*:** Leaves and bark are collected in April-may. Uniyal and Issar, (1967) reported a collection of 2 qtls of leaves and 10 qtls of bark from Kanatal forest only. The leaves given oil, similar to cinnamon leaf oil containing d-phellandren and Eugenol which is pale yellow containing 70–85% cinnamic aldehyde. In eastern Himalaya *C. cecicodaphne* (Meissn). has been reported. *C. zeylanicum* (Dalchini) is described in section agrotechniques for cultivation.

38. *Citrus limon* (L.) Burm.f.

(Citrus medica var. *limonum, pseudolimon, C. medica* L. var. *psueudolimon).*

***Family*:** Rutaceae.

***Local names*:** Bará-Nimbu, Gorá-nimbu, Pahári-nimbu (Hindi), Mahá-Nimbu (Sansk.) Commonly grown locally, called 'Galgal'.

***Discription*:** A small tree or shrub with strong axillary spines, bark greenish-grey, smooth. Leaves one –foliate, petiole often winged. Flowers often unisexual, white, sweet scented, in axillary cymes or solitary. Calyx cup-shaped. Petals 4–8, fleshy with glandular dots. Stamens 20–40 around a disc. Fruits longish with a thick rind, yellow when ripe, pulp juicy, usually acrid.

***Distribution*:** Cultivated in home gardens for local use.

***Uses*:** Juice of the ripe fruit is antiscorbutic, refrigerant and useful in scurvy, rheumatism, dysentery and diarrhoea.

Rind of the ripe fruit is also used as carminative and in stomach problems, to infants locally called jhoka in Kumaon.

***Other species*:** There are a number of Citrus species that are used as medicinal, a few are as below:

Citrus aurantifolia (Christ.)Swingle (*C.medica* var *acida*) Vern. Name: Kagzi Nimbu (Hindi).

The fruit is refrigerant, appetizer, antiseptic and anti-scorbutic in bilous vomiting. It contains citric acid, volatile oil containing *citral, limonene, linalool, linayle acetate, terpineol* and *cymene*. The oil is largely a by-product of lime juice industry.

Citrus aurantifolia (Christm.) var. *limetta* (*C. limattoides* Tanaka, *C. medica* var. *limetta* Wight and Arn). Vern. Mitha- Nimbu. The shrub is cultivated for edible fruits. Also used as rootstock for sweet oranges and mandarins. Leaves yield an essential oil called 'petit grain oil' used in cosmetics, etc.

Citrus aurantium L. Vern. Sour-orange, Seville orange (Eng.) Karna-Khatta (Hindi).

A small tree or shrub. The fruits are edible, used for preparing Marmalade. Leaves, a source of essential oil, used in confectionary, cosmetics and perfumery. Also used as rootstock for sweet orange. Oil from young leaves used for skin creams. Fruit is laxative and stomachic, rich in vitamin A and B. Var. *bergemia,* rarely met in India.

Citrus decumana L (*Aurantium maximum* Burm.) Vern. Pomelo, Chakotra (Hindi).

A large shrub or tree grown in Dehradun, UP, Punjab and south Indian states. The fruits are large and edible, a glucoside *Naringin* from fruits is used for flavouring beverages. Fruits are cardiotonic. Leaves useful in epilepsy and convulsive cough.

Citrus limon (L.) Burm. f. (-*C. medica* var. *limon, C. medica* var. *limonum* L.) Vern. Bará-Nimbu. Pahári-nimbu (Hindi) Small thorny tree, commonly grown variety is *galgal*. Rind of the fruit used as stomachic and carminative, the Juice is antiscorbutic, refrigerant in scurvy, in rheumatism, diarrhoea and dysentery.

Other varieties of citrus in common use are *C. jambhiri* (*C. limonis.* Lush) Hindi Khatta; *C. japonica* Thunb. Karna; *C. margarita* Lour. *C. medica* L, *C. Paradisi* Macf, *C. reticulatis* and *C. sinensis.*

39. *Croton tiglium* L.

Family: Euphorbiaceae.

Local names: Jaipal, Jamálgota (Hindi), The purging croton (Eng.).

Description: A small evergreen tree; young shoots sparsely hairy. Leaves ovate-elliptc or oblong, narrowly pointed, 3–5 nerved, thinly membranous, toothed, varying in colour; from metallic green to bronze and orange, yellowish coloured when dry. Flowers in raceme, 5–7 cm long, male flower stalk having stellate hairs, while female flowers without petals. Fruits a capsule, up to 3 cm long, white, ovoid.

Distribution: Naturalised and widely cultivated in Assam and Bengal, moist forests on the ghats to the south of peninsula. Met within under cultivation in greater parts of India. It grows on the poorest soils, such as wastelands, from sea level up to 1000 m. Under cultivation no special care is needed. It fruits in the second year of plantation and has been grown as a plant for coffee.

Uses: All plant parts are medicinal; specially a drastic purgative.

The wood, known as *lignum pavavae*, used in small doses as diuretic, mild emetic and a powerful diaphoretic. Pulverised root a drastic purgative given in dropsy also as abortifacient.

The nut yields a large amount of oil which is medicinal and administered as a violent purgative; in large doses even may cause death. Seeds are used after removing the seed coast and the embryo. Very small doses such as one or two seeds should be used with honey as it is a poison in overdoses. The seeds are used in epilepsy, insanity, dropsy, enlargement of visera, ascites, convulsions, obstinate constipation, intestinal worms, tympanites, gout and calculous affection.

The oil is used by fraudulent manufacturers an an adulterant of tincture of iodine. The oil is of great medicinal value, a powerful hydragogue purgative, prescribed in spasmodic cholera, epilepsy, lockjaw. In case the oil causes ill effect such as gripping, vomiting, excessive purging, a drink of lime juice is an excellent antidote.

Oil can be used externally as a stimulant and rubifacient to skin, a liniment and skin stimulant and embrocation in infantile bronchitis, asthma, paralysis, gout, chronic rheumatism, arthritis, indolent tumours, laryngitis, neuralgia, sciatica and diseases of the joints. Croton oil mixed with mustard oils, coconut oil, olive oil in 99 parts makes a liniment. With oils in 99 parts it makes a good hair growth promoting oil.

Extent of utilization and trade: Croton oil is used by fraudulent manufacturers as an adulterant of tincture of Iodine. The seeds are used largly by natives.

40. *Cinchona ledgeriana* Moens ex Trimen.

Family: Rubiaceae.

Local names: Cincona-Hindi, Quinine-plant (Eng.).

Description: A tree, grown in Bengal, Khasi hills and south India.

Distribution: The plant is grown for quinine in various parts of the country; however it is not indigenous to India.

Uses: The utilization of many synthetic anti-malarial drugs led to the neglect of this plant for a considerable period. The quinine (the active antimalarial compound) has bet-

ter drug value as compared to its synthetic replicas, the use of which is now being discouraged for the reason of its toxicity and side effects. It is reported that quinidine, present in chincona, is the best heart regulator. There is also possibility of conversion of quinine to quinidine by chemical methods.

Other species of cinchona are *C. calisaya* Wedd. A tree, native of S. America, but found in the Nilgiris and Sikkim. Bark is the source of quinine.

Extent of utilization and trade: *Cinchona succirubra* Pavon ex Klotzsch., is grown in hilly regions of south, Sikkim and used in treating malarial fever.

41. *Clerodendrum multiflorum* (Burm.f.) O. Kuntze (non G. Don 1824). (Clerodendron phlomoides L.f., Volkameria multiflora Burm. f.)

Family: Verbenaceae.

Local names: Arni, Yerná (Rajasthan).

Description: A tall shrub, shoots pubescent. Leaves ovate-sinuate. Flowers fragrant in dichotomous cymes, collectively forming terminal panicle. Calyx puberulous, in fruit slightly enlarged. Corolla puberulous, white. Fruit a drupe, succulent.

Distribution: Throughout India, westwards to Pakistan. Frequent on freshly deposited sand, also a hedge plant of sand dunes. Flowers scented during evening.

Uses: The roots are astringent and the decoction is used as demulcent in gonorrhoea and in convalescence after measles. The leaves are applied locally against Guineaworms. Leafjuice is alterative and given in syphilis; also given to cattle as a cure for diarrhoea and worms.

Other species: Species of this genus like *Clorodendrum fragrans* Vent, *C. indicum* (L.) Kuntze, *C. siphonanthus R.Br. C. squamatum* Vahl. are used as ornamental.

42. *Commiphora wightii* (Arnott) Bhandari (-C. mukul Engl., C. roxburghii (Stocks) Engl. Balsamodendron wightii Arnott, B. roxburghii Stocks

Family: Burserace.

Local names: Guggal, Indian Bdellium tree (Eng).

Description: A small thorny tree or shrub, 1–2 m high with knotty trunk, with outer bark greenish yellow, peeling off in rough flakes; branches ending in sharp spine. Leaves alternate, simple or 3- foliate; lateral leaflets smaller than the end one, smooth, shining. Flowers uni or bisexual, solitary or in clusters. Calyx tubular, lobes as long as the tube, glandular. Petals 4. Fruit a roundish fleshy drupe.

Uses: The gum is used for a variety of diseases, as a substitute for myrrh. A bitter stomachic, carminative, demulcent, astrin-gent, aperient and antiseptic; also used as diphoretic, diuretic and expectorant. Internally used as urine stimulant and emmenagogue; regulates menstrual function.

Resin is used as lotion for indolent ulcers.

Inhalation of guggal fumes is recommended in acute bronchitis and in hay fever.

Guggal is an effective remedy for arthritis and is prescribed in many preparations like *Kaishor Guggal, Kanchnár Guggal, Triodashang Guggal* and many others. Also administered in neuralgia, skin diseases and diseases of genito-urinary organs.

43. *Colebrookea oppsitifolia* Sm.

Family: Labiatae or Lamiaceae.

Local names: Bindá, Pansrá (Kumaon, Garhwal); Dasari (Mumbai); Shakardáná (Himachal); Dosul (Nepal).

Description: An erect tomentose shrub, 2–3 m high. Leaves opposite or in threes, lanceolate, stalked shortly, crenate, long pointed, upper surface pubescent, wrinkled, lower grey-tomentose. Flowers white, minute; male and female often on different plants; also bisexual., Calyx deeply 5-lobed, lobes linear, hairy, feathery in fruit stage when tips often turn purple. Corolla pubescent, tube as long as the calyx, limb spreading, 4-lobed. Stamens 4, equal, protruding in male fls. included in female; style protruding in female fls. Nutlet 1, with hairy tip (Fig. 5.9).

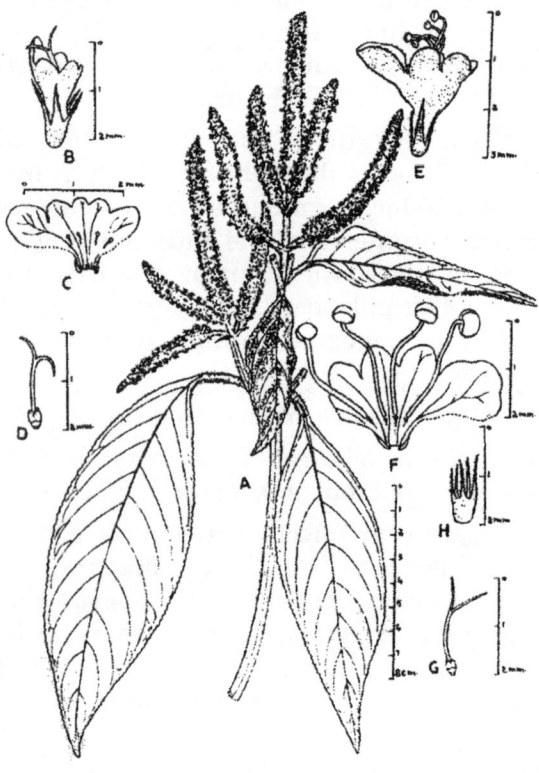

Fig. 5.9: *Colebrookea oppositifolia* Sm. A. flowering twig; B. functionally female flower; C. corolla of B opened; D. gynoecium from B; E. functionally male flower; F. corolla of E opened to show androecium; G. gynoecium from E; H. calyx from E.

Distribution: A very common shrub on stony wastelands in the valley region throughout Himalaya, ascending 1300 m; often considered as weed.

Uses: Not much is known to locals in Garhwal region about its medicinal uses. Preparation from the root is used in epilepsy, while the leaves are locally applied to wounds and bruises.

44. *Cordia dichotoma* Forst. f.

Family: Boraginaceae.

Local names: Barágund, Bhokar, Buhal, Lasurá (Hindi), Sebastán, Sapistán (Urdu). Clammy Cherry, Indian Cherry (Eng.).

Description: A middle-sized, low spreading tree, with crooked stem and glabrous branchlets. Leaves alternate, stalked, entire or slightly dentate, rough when fully grown, variable in shape from elliptic-lanceolate to broad-ovate. Flowers small, unisexual and bisexual on the same plant. Whitish, fragrant in small open panicles. Berries yellowish brown, nearly black when ripe, shining, minutely rugose, ovoid, about 1 cm long, stone wrinkled, 1–4 seeded.

Distribution: Throughout India, ascending up to 1300 m in the Himalayan region.

Uses: The dry fruits are used in cases of rattling of throat and windpipe, thickness of the phlegmatic fluid, of catarrh and makes it easy to expel. It acts as laxative. Fresh ripe fruits remove defects of consistency of seminal fluid and spermatorrhoea, dry cough and dysentery. The dry fruits are an ingredient of *Joshanda*.

Extent of utilization: A *chutney*, locally called '*lawook Sapistan*' is available in the market, which is used to cure whooping cough.

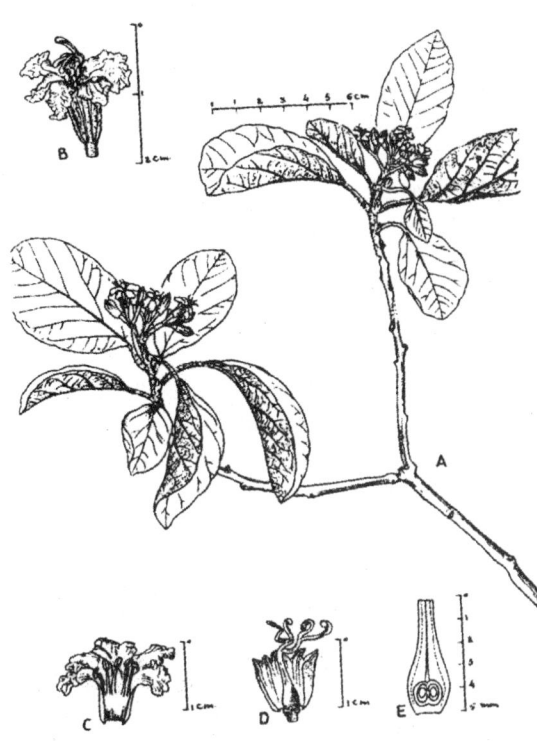

Fig. 5.10: *Cordia vestita* HK. F. and T. A. flowering twig; B. flower; C. corolla opened to show androecium; D. calyx opened to show gynoecium; E. longitudinal section of gynoecium.

Details of *Cordia vestita* HK. F and T are given in Fig. 5.10.

45. *Coriaria nepalensis* Wall.

Family: Coriariaceae.

Local names: Mansuri, Masuri, Masroi, Gangéru, Gangárá (Garhwal, Jaunsàr).

Description: A large shrub with long spreading branches; bark reddish brown; branchlets quadrangular. Leaves ovate-oblong cordate, acuminate, 3–7 nerved at the base, sometimes puberulous beneath. Flowers clustered in racemes, greenish-yellow. Fruit dark brown or black.

Distribution: Fairly common among shrubby vegetation on hillsides and ravines. The summer resort Mussoorie derives its name from this plant.

Uses: The leaves are used as tan, also used as adulterant to 'Senna' leaves and act as poison in large doses. The fruit produces symptoms like that of tetanus.

46. *Cotoneaster microphylla* Wall.var. buxifolia.

Family: Rosaceae.

Local names: Bhédda.

Description: A small, much-branched, evergreen shrub with almost black bark. Leaves ovate or elliptic, coriaceous, dark-green and glossy above, pubescent beneath. Flowers white axillary solitary. Fruits globose, scarlet when ripe.

Distribution: Trailing on rocks or spreading on hillsides. 2000–3300 m.

Uses: The stolons are used as astringent.

Branches used to make baskets.

47. *Cryptolepis buchanani* Roem. and Schult.

Family: Asclepediaceae.

Local names: Méda-singhi (DDN) Karánta (Hindi).

Description: A large glabrous, twining shrub with terete, whitish branches. Leaves coriaceous, shortly stalked, elliptic or oblong ovate with a rounded or apiculate apex, usually acute at the base, dark green above, glaucous beneath. Main lateral nerves many, slender, horizonal and uniting within the margin. Flowers pale greenish yellow, in short axillary panicled cymes. Bracts ovate-lanceolate with scarious margins. Calyx ovate, acute. Corolla linear

or linear-lanceolate. Corona of 5-clavate scales. Follicles 5–10 cm long, stout, straight, terete, tapering, seeds black, compressed, ovate-oblong, coma 3 cm long (Fig 5.11).

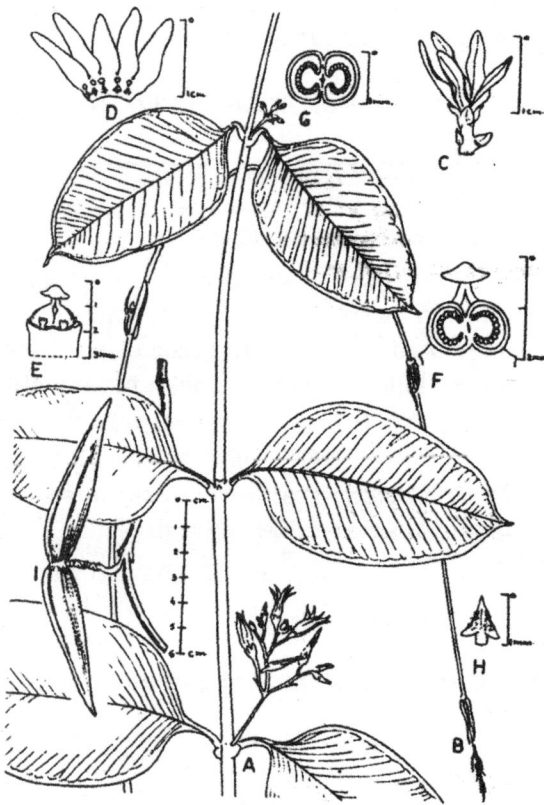

Fig. 5.11: *Cryptolepis buchanani* Roem. and Schult. A. flowering twig; B. upper part of A; C. flower; D. corolla opened to show coronal scales and stamens: E. calyx lobes and corolla removed to show gynoecium and the scales inside the calyx; F. longitudinal section of gynoecium; G. transverse section of gynoecium; H. stamen, outer view.

Distribution: Throughout India in deciduous forests in the sub-Himalayan tracts in hedges, also in hotter parts ascending 1300 m in the hills. On Siwalik hills up to 1300 m.

Uses: Preparation from this plant given to rickety children, also combined with *Euphorbia microphyla*, given to women when milk supply fails. The latex coagulates as rubber.

48. *Dalbergia sissoo* Roxb.

Family: Papilionaceae.

Local names: Shisham, Sissu; Táli (Punjab); The Sissoo tree (Eng.).

Description: A fairly large tree, bark exfoliating in narrow strips, young parts grey downy. Leaflets 3–5, alternate, rhomboid, lateral nerves very slender. Flowers pale-white in racemes, arranged in short axillary panicles. Calyx downy, standard with long claw. Stamens 9, united in a sheath which is slit on top. Pods strap shaped, pale brown, 1-seeded; seeds kidneyshaped, flat.

Uses: It is mostly known as timber tree. The sawdust purifies the blood and used after steeping in water and the decanted water mixed with syrup of *jejubs*. It also cures boils and pimples. A paste of leaves heals pinching of shoe. Also useful in cases of spermatorrhoea, when taken internally. About 20 gms of leaves steeped in water overnight and water decanted and the viscid fluid is strained. It is taken mixed with sugar. Also effective in curing leukorrhoea.

Helps in stopping bleeding from the nose.

49. *Desmodium tilaefolium* G. Don

Family: Papilionaceae.

Local names: Mártoi, Matoi (Jaunsar); Chamlài, Chamyát. Shámru Samber (Kum.)

Description: Diffuse shrub, up to 3 m in height. Leaves 3-foliate, pubescent; leaflets rhomboid, mucronate at apex, silky pubescent beneath, lateral nerves about 6-pairs. Flowers pale-liliac, in terminal panicles and axillary racemes. Calyx downy, teeth shorter than the tube. Pods falcate, silky hairy.

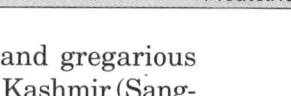

***Distribution*:** Common and gregarious between 1300–3000 m. From Kashmir (Sangpo valley) 3 forms, viz. *rhabdoclada*, form *typical* and form *calva* have been reported.

***Uses*:** Roots diuretic and used in bilious complaints. Bitter in taste and tonic to chest and brain, reduce oedema and cure bad-smell of ozoena. It improves appetite and enrich blood.

1. *D. laburnifolium* in south India is prescribed as cure for scorpion and snake bite (*Indian for.* 1968).
2. *D. gangeticum* (L.) DC. Roots and seeds are used as febrifuge and as an anticathartic medicine.

50. *Dendrophthoe falcata* (L.f.) Ettingsh (-Loranthus falcatum L.f L. longiflorus).

***Family*:** Loranthaceae.

***Local names*:** Banda (Hindi); Baramanda (Bihar); Vanda (Sanskrit.); Pilluri (Tam.); Badanika (Telgu).

***Description*:** A woody branched parasite with glabrous branches and broad usually, but variously shaped, opposite or alternate leaves, sessile or petioled, thickly coriaceous, glabrous leaves, about 7–9 cm long. Flowers showy scarlet orange, sometimes also pink. In axillary and extra-axillary secund racemes, 2–10 cm long. Hypanthium with caliculus which is distinct, broadly tubular and faintly toothed. Perianth slender tubular, and faintly toothed at the back and curved slightly with 5-linear-oblong, often green lobes. Fruit oblong, crowned with caliculus. Flowers and fruits Nov-March (Fig. 5.12).

***Uses* :** The bark is astringent and used for the wounds and menstrual problems in women. Used also as a remedy for asthma, mania and consumption. Also a substitute for betel nuts (vern *supari*).

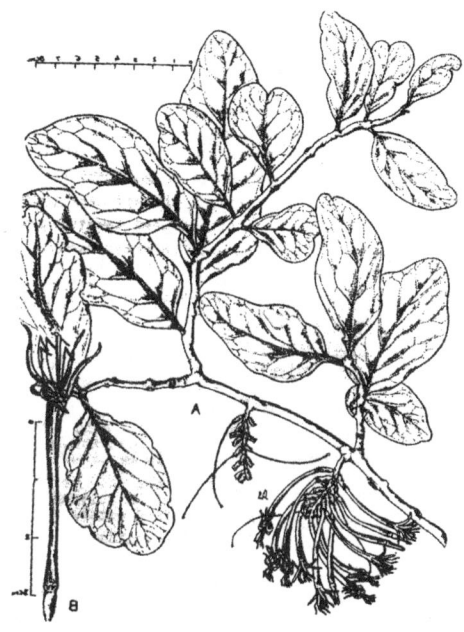

Fig. 5.12: *Dendrophthoe falcata* Ettings. A. flowering twig; B. flower.

***Distribution*:** Throughout India, a common loranthus is a parasite on a number of trees in the Sal forests.

Other species

a. *Dendrophthoe elastica* (Descr.) Danser. This parasite occurs mostly in Southern India in states of Tami Nadu, Kerala and Andhra. However, the leaves are used locally to prevent abortion and also to resolve stone in kidney and bladder.

b. *Dendrophthoe pentandra* (L.) Miq. The leaves are used as poultice for ulcers and sore. Twigs are reported to contain quercitrin and a wax.

51. *Dodonaea viscosa* L.

***Family* :** Sapindaceae.

***Local names*:** Sonátta, Wilayati-Mahéndi (Hindi).

Description: An evergreen shrub with glabrous red branchlets. Leaves alternate, simple, oblanceolate to linear, cuneate, shining above, more or less viscid with a yellow resin, lateral nerves numerous but not prominent. A short panicle with racemose branches. Flowers polygamous, with greenish yellow sepals. Petals nil. In male flowers, the disc is obsolete, small in female flowers. Fruit a compressed capsule, septicidally 2–4 valved, each valve with an oblong membranous wing (Fig. 5.13).

Distribution: In the lower Himalaya, up to 1500 m. Used as hedge at many places in gardens, particularly in arid areas.

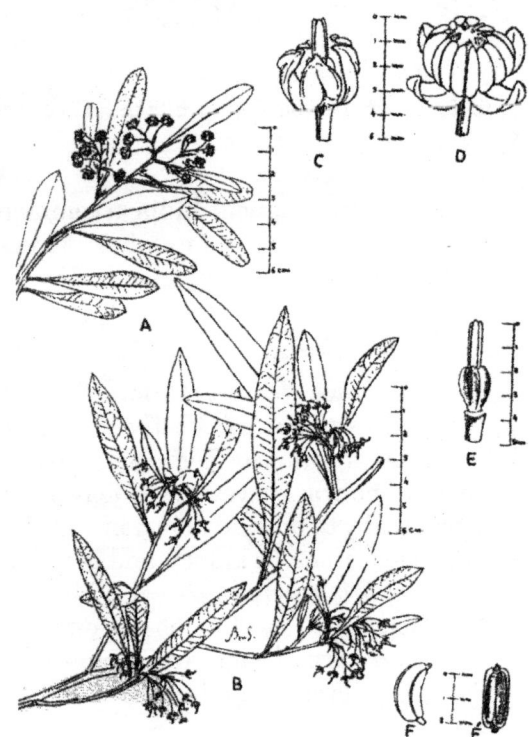

Fig. 5.13: *Dodonaea viscosa* Jacq. A. flowering twig of a male plant; B. flowering twig of a female plant; C. female flower; D. male flower; E. gynoecium; F and F.′ stamens.

Uses: The plant is reputed as febrifuge. The leaves are esteemed as sudorific in gout and rheumatism. A poultice from the leaves is made which is supposed to retain the heat like a Linseed meal poultice.

Bark is used as an astringent for bath and fomentation, but used only as a home remedy.

52. *Elaeagnus umbellata* Thunb. Var. parvifolia (Royle) Schneider. (E. umbellata sensu Hook.f. in Flora of British India—FBI)

Family: Elegnaceae.

Local names: Ginroi (Jaunsar), in distt. Dehradun near Chakrata.

Description: A thorny deciduous shrub with many branches, often forms dense bush. Leaves oblong or lanceolate, acute at both ends, clothed below with bright silvery scales, pubescent above. Flowers dull-white, silvery outside, appearing with leaves, in small axillary clusters. Perianth triangular, ovate. Fruit succulent, oblong-ovoid, ribbed outside.

Distribution: In Himalaya on outer hill slopes at 1000–2300 m, Kashmir to Nepal.

Uses: The seeds are said to be stimulant in cough, expressed oil is used in pulmonary affections. Flowers are cardiac stimulant and astringent. Fruits are edible.

E. angustifolia L. (syn *E. hortensis* Bieb.) Shiulik (Hindi.) *E. latifolia* L. (Bastard oleaster-Eng.; Fiwain-Hindi) also have edible fruits.

53. *Elaeagnus umbellata* Thumb. Var. parvifolia (Royle) Schneider (E. umbellata sensu Hook. f. in FBI 5: 201, 1886.)

Family: Elaeagnaceae.
Local names: Ginroi (H).

Description: A thorny, deciduous shrub with numerous branches, often forming a dense bush. Leaves oblong or lanceolate, acute at both ends, clothed below with bright silvery scales; sparsely scaly or pubescent above, nerves indistinct. Flowers dull white, silvery outside. Appearing with the leaves in small axillary clusters. Perianth lobes triangular, ovate. Fruit oblong-ovoid, succulent, ribbed outside.

Distribution: On the outer slopes of the Himalaya from Kashmir to Nepal, 1000–2300 m. Common on dry slopes. Indigenous to swampy parts of the sub-Himalayan tracts from Yamuna eastwards.

Uses: The fruits are edible, stimulant in cough. Seeds are used in pulmonary affections.

Flowers are astringent and cardiac.

The expressed oil from the seeds is also useful in pulmonary affections.

Extent of utilization: Utilized on a limited scale in the area.

54. *Emblica officinalis* Gaertn.
(Phyllanthus emblica L.)

Family: Euphorbiaceae.

Local names: Emblic (Eng.); Amlá, Áonla (Hindi).

Description: A moderate-sized deciduous tree. Bark exfoliating in irregular patches, red inside, branchlets finely pubescent. Leaves linear-oblong, acute or mucronate, distichously close-set on deciduous branchlets, together having the appearance of pinnate leaves. Flowers apetalous, monoecious, green-yellow, in axillary clusters. Male flowers with 3 stamens in a short column, disc of distinct glands, alternating with calyx-segments, rarely 0 sepals in male flowers, in the female flowers ovary 3-celled with 2 ovules in each cell; styles 3, connate at the base, twice bifid.

Fruit of 3 to 2-valved cocci, 6-lobed, globose, fleshy, pale-yellow, dehiscent, only when dry (Colour Plate 5.6).

Distribution: Throughout India, a common tree with edible fruits.

Uses: The root-bark is effective in apthous stomatitis. A fermented preparation from root is used to cure jaundice, dyspepsia and cough.

Tender shoots of the plant are used as remedy in indigestion and diarrhoea.

The leaf infusion is used as a bitter tonic and a remedy for chronic dysentery. Leaf decoction is useful as mouthwash in apthae and eye wash for sore eyes. Young leaves are dried in shed, powdered, sieved, and mixed with sugar and milk to cure spermatorrhoea. Juice of the fresh bark with honey and turmeric is used to cure gonorrhoea.

Fruit is the most useful part and the richest source of vitamin C. *Banarsi amla* is well known for its size and made into preserve and pickle. Fresh fruit is refrigerant, tonic, antiscorbutic, diuretic and laxative; used in summers to quench thirst. A syrup from the fruit is antibilous, diuretic and cooling; used in fevers, vomiting, indigestion and habitual constipation along with other digestive system complaints. Fresh fruit is vermifuge and when dried is an excellent intestinal astringent, cooling, stomachic, antiscorbutic, blood purifier and given by vaids in diarrhoea, jaundice and dyspepsia. Rich in vitamin C, even higher than citrus fruits. The liquid that exudes from the cut end of fresh fruit, while it is on the tree, is used as eye wash and as an external application on the inflamed eyes. Dried fruit is an important ingredient of famous *Triphala*, comprising *Hard*, *Baheda* and *Amla*.

Dry *amla* fruits are used in anxiety, palpitation of the heart and melancholia. It invigorates stomach and intestines, also

used to allay thirst, fever of sanguine and bilious humors and stops haematuria or stools. The dried fruits are made into small pills and after grounding, mixed with little salt and taken to cure diarrhoea.

Amla is used for washing hairs and made into hair oil. An important medicine, *Chayvanprash* contains a fair quantity of the fruit juice, and is taken with milk for general tonic during the winter season. Commercially, it has a market of several billion Rupees, both in India and abroad.

A paste of the fried powder of fruits is applied on pubes in cases of irritability of bladder and retention of urine. Decoction of the dried fruit is injected with benefit in gonorrhoea. Infusion prepared by steeping dried fruit overnight in water in a new earthen vessel is effective in ophthalmia and in hair diseases.

Seed-decoction is used as gargle, for loss of taste after fever and an ointment of burnt part of the body. Seeds are useful for itching.

Extent of utilization and trade: The fruits are extensively utilized and grown commercially. Many varieties have been tried at Pratapgarh in UP. Fruits traded both as green and dry. Amla fruits are extensively collected and traded in different forms, (both as green and dry). Its use in hair oils, *Chayvanprash* and *Murabbas* (canned fruit) makes it an important commodity. The tree has been grown widely on a large scale at Pratapgarh in UP and various improved varieties are existing.

55. *Euonymus tingens* Wall.

Family: Celastraceae.

Local names: Bhambéli, Konkon, Roini (Hindi).

Description: Evergreen shrub with dark ash-coloured bark, bright yellow inside. Leaves elliptic-oblong, dark green above, pale beneath, stipules brown. Flowers in dichotomous cymes, usually 5-merous, with dull white petals marked with brown or purple veins. Stamens shorter than petals. Fruit a capsule, 3–5 angled, not winged. Seeds in an orange coloured arillode.

Distribution: Common in the Himalaya from 2000–3300 m.

Uses: Bark used in eye diseases and chronic constipations.

Extent of utilization: Utilized on a limited scale as a home remedy.

56. *Echinops echinatus* DC.

Family: Compositae or Asteraceae (nom alt).

Local names: Gokhru (Garhwal, local in Hindi). The name is also applied to other plant (*Tribulus terrestris*).

Description: A thistle-like herb with white felt. Leaves alternate, pinnatifid, spinous, thickly white-felted beneath. Heads one flowered, crowded in a globose white-ball, 5–7 cm across. Fruit an obconic, cypsela.

Distribution: Found on fallow lands and as an undergrowth in *chirpine* forests.

Uses: The flowers are boiled in water and the decoction is mixed with sugar and ghee (butter). It is injected through nostrils as a cure for jaundice. Yellow fluid comes out from the nose. The practice is followed in villages of Pauri Garhwal.

57. *Eucalyptus globulus* Labill.

Family: Myrtaceae.

Local names: Blue Gum (E); Karpura-máram (Tamil).

Description: A large tree, growing to a height of 90 m or more with straight clean bole, when grown under forest conditions, but tending to branch freely when growing

in open. Seedling leaves are opposite and ovate, while in adult trees and older branches leaves are alternate and lanceolate about 15 to 30 × 2.5 to 5 cm.

Distribution: The tree is indigenous to Australia but has been cultivated in south Indian hills, particularly the Nilgiris and Palnis, also in Shimla, Darjeeling, Kumaon and Shillong above 1200 m. It is a fast growing species and there are extensive plantations of Eucalyptus in various parts of the country.

Uses: The leaves yield well-known 'eucalyptus oil' which is used for medicinal purposes and also for removing incrustations in boilers and locomotives and also as insect repellant on account of its antiseptic and deodorant properties. In medicine it is used to relieve cough in chronic bronchitis and asthma when inhaled with steam. Pastilles containing eucalyptus oil, often with menthol are used in the symptomatic relief of the common cold. It is also used in treatment of burns, as a vermifuge, and as cloth cleaner, spot and stain remover.

Extent of utilization and trade: Distillation of leaves and terminal branchlets of *E. globulus* is a sizeable cottage industry in Palni hills and Nilgiris where oil is obtained by water and steam distillation. The leaves are dried in shade for 3 days before they are distilled. The yield varies from 0.75 to 1.25 per cent and oil valued about 1.65 lakhs is reported to have been distilled as far back as in 1949–50 in south India. The oil is colourless or pale yellow and has a spicy and pungent odour.

The tree requires a moist cool, equitable climate with a deep fertile soil which is not calcareous or saline. It should be cultivated below 1300 m where there is no snowfall, since the plant is liable to break. It is propagated from seeds sown in nursery beds; young seedlings require protection from frost. It is fast growing species and coppices well.

a. Eucalyptus polybracteate R.T. Baker.

Local names: Blue Mallee (Eng.); Blue green tree.

It is a shrubby species about 3–6 m high. Mature leaves are narrowly lanceolate. A native of Australia, occurring in low rainfall regions of New south Wales and Victoria. The yield of the oil is much higher than in *E. globulus;* the oil has a much higher 'cineole' content; it is desirable to raise plantations of this species. The tree flourishes in continental climate with high summer temperatures, frequently exceeding 38° C and occasional winter frosts. Annual rainfall of 250–500 mm is sufficient for the plant growth. It grows on poor sandy soils and also on dry-rocky ridges. It is propagated by seeds, sown in nursery beds which are taken in pots before planting. The oil content from leaves and terminal branches is about 1.5 to 2.5 per cent., which is extracted on steam distillation.

Other species like *E. cinearifolia* DC., which have higher *cineole* content need to be planted in areas with scanty rainfall which may prove to be more suitable for better oil yield.

b. Eucalyptus citriodora Hook.

Sowing of seeds is done on raised seed beds with fine leafmould and sand, in germination boxes. The seedings are pricked out when 5–10 cm in height and kept in shade for 2–3 days. Second pricking may be done when seedlings are 15 cm in height. Planting done in rainy season, of 30–45 cm long seedlings. It does not tolerate water logging.

58. *Euphorbia royleana* L.

Family: Euphorbiaceae.

5

Local names: Thor, Thuru, Senhur (Hindi); Suru (Jaunsar).

Description: A large milky shrub with thick fleshy, 5–7 gonous branches, ridges wavy with a pair of stipular prickles at the crest of each wave. Leaves sessile, alternate, 1–15 cm long, spathulate, shortly mucronate, lateral nerves 6–8 pairs, quite indistinct until dry. Involucres hemispheric, greenish yellow, 3–4 together in axillary, subsessile clusters. Styles combined up to middle. Capsule 3-gonous.

Distribution: In outer Himalaya and valleys, on dry aspects up to 1300 m; leafless during cold and hot season. Grows readily from cuttings, associated with *Carissa* and *Rhus*; planted as hedge.

Uses: The latex is acrid, rubifacient, purgative, expectorant and is liable to cause dermatitis. Mixed with *Calotropis* procera latex, it is used as purgative. It also relieves earache.

Other species

a. *Euphorbia antiquorum* L. (Vern. Mingut, Narashi, Sayord, Thor, Tridhari-shend) Spurge, Cactus (Eng.).
A variable fleshy small tree, branches jointed, stout, soft, 3–5 angled, thorny with undulating ridges. Leaves small, falling early. Flowers minute, yellow, in clusters, borne near the end of the branches. Fruit compressed.

Throughout hotter parts in dry places, ascending 600 m. The milky juice is a drastic purgative, powerful emetic and a deobstruent, given in visceral obstruction, dropsical affections and relief of pain in the loins. Juice from stem is used in nervine diseases, deafness and total blindness. Pills made with a mixture of juice and gramflour, roasted together are given for gonorrhoea; juice is a household remedy for warts and other skin diseases, rheumatism and painful joints.

Powder of fried stem is dusted over old ulcers to promote healing. Decoction of the stem is given in gout.

Rootbark is used as purgative. A paste made of the roots with asfoetida, is applied to stomach of children, suffering from worms. Root juice also used to clean maggot-infested wounds.

b. *Euphorbia neriifolia* L. (Mingut, Nivrang, Patrasanuk, Tonkiend, Snuhi, Thear—Thor). A small tree or shrub, stem obscurely 5-angled, smooth, hairless, fleshy, cylindric; bark reticulated; branches bearing small nodes, arranged vertically or spirally, a pair of spines present at each node. Leaves few, fleshy, without stalks, bore near the end of small branches. Flowers in yellowish or reddish clusters, near the end of the branches; each cluster containing 3-lobes, each lobe containing a seed. It is found throughout the western peninsula, Gujarat and some parts of Western Rajasthan. Also used as hedge, and as a tree guard, because of its sharp thorn. Leaves when plucked produce latex, which is purgative.
Juice of the plant is used medicinally. Leaf juice is purgative and deobstruent in visceral obstructions and dropsical disorders, and long-continued intermittent fevers. The leaves are diuretic. The latex is acidic, used both externally and internally. Internally it is a drastic purgative, expectorant, cathartic and deobstruent in visceral obstructions and drospical affections in combination with other drugs such as *Terminalia chebula* (Harad), and *Piper longum*

(Pipli). For loosening bowel. The latex is given after it is boiled with butter milk.

Externally the juice is applied to glandular swellings and prevent suppuration; mixed with butter, used as ointment for ulcers and scabies. An ointment with turmeric powder it is applied to the piles. Mixed with oil, the juice is applied to the limbs suffering from rheumatism, dermatitis and cutaneous eruptions. An ointment is also prepared from the pith of this plant. Half kg of the pith is charred, in equal weight of oil, bees-wax is melted in it and blue citrol is also mixed and made into an ointment by triturating it thoroughly. It is also used in dropsy, syphilis, phlegmatic cough. Oil from this plant is prepared and used in paralysis and rheumatism. Latex also cures toothache, if locally applied, also to ringworm or psoriasis when an ointment is applied.

The latex is acidic and in Kumaon region used for a disease called 'Khor' which has symptoms like falling of hairs and the eyebrows. It is purgative, rubifacient and liable to cause dermatitis, in asthma it gives relief in mixture with other compounds. Latex gives *Guttaparcha*. When fresh, does not blister fingers but injurious to the eyes and strongly flavors anything. The plant is supposed to keep away the evil spirits and work as lightening conductor and hence kept on the rooftops.

c. *Euphorbia antiquorum*. L. Tridhari-(Hindi); Sehund, Vajrakanta (Sansk.).
The plant has 3–5 angled branches and grows throughout India on warmer regions.
The juice is used in dropsy and as a nervine tonic.

d. *Euphorbia barnhartii* Croizat. (*E. trigona Roxb.*).
The plant occurs on dry rocky hills of the Deccan region and on Andaman islands. The powdered leaves are used as a poultice on boils.

e. *Euphorbia nivulia* Buch.-Ham. -Vern. Kathuar (Hindi); Patrasnuhi (Sansk.).
The branches are cylindrical. It occurs almost throughout the country and is often grown as hedges.
The juice is used by the tribals of Bastar for treating wounds in cattle and is believed to be diuretic.

f. *Euphorbia tirucalli* L. (Sihund-Hindi; Vajradruma-Sansk.)
A shrub or small tree. Branches cylindric, very much branched. An African species, planted for hedges and on the roadsides, in drier parts of the country; chiefly in Deccan and Eastern regions.
The milky juice of the plant is acrid and well known as purgative and counter irritant. It is also painful, when applied to wounds or to the eyes and that is why the cattle do not attempt to break the hedge of this plant.
The juice is also considered useful in local applications for curing rheumatic pains, toothache, etc. and when mixed with mud it is used to construct the flat roof of houses in north Arcot district of Tamil Nadu state.
Latex of this plant on boiling becomes brittle, though whilst warm; it is ductile and elastic. Nitric acid causes separation of rubber. In small doses it is given internally.
Local belief is, that where this plant is kept, it repels mosquitoes and so kept in the house, particularly in Maharashtra state.

59. *Ficus benghalensis* L.

Family: Moraceae.

Local names: Banyan Tree, (Eng.), Bargad (Hindi); Vat-Vriksh (Sansk.).

Description: A large, evergreen, spreading tree, extending laterally by sending down lateral roots from branches. Bark greyish white, smooth, exfoliating in irregular flakes. Leaves alternate, ovate or elliptic, glabrescent above, with 3–5 basal nerves and 4–6 pairs of lateral nerves. Male flowers. crowded near the mouth of the receptacle, female in the same receptacle, which is red coloured when ripe.

Distribution: Fairly common in *Sal* forests. It is sacred and worshipped particularly on Amavasya of the months May/June depending on Hindu Calender.

The *Great Banyan Tree* of the Indian Botanic Garden looks like a miniature forest and is over 250 yrs old with 2800 prop roots covering an area of 1.5 hac.

Uses: The latex from the plant is used externally for relieving pains, and on bruises as an anodyne in rheumatism. Infusion of the bark is regarded as a powerful tonic in the treatment of diabetes.

The warmed leaves are used as a poultice.

The fruits are eaten in times of famine; the twigs and leaves are grazed by animals.

Extent of utilization: The plant as drug is utilized on a limited scale by Vaids and Hakims and no regular exploitation of the plant is done.

Other Species

a. *Ficus carica* (The edible fig tree of Europe; Anjir-Hindi)

   The tree is cultivated in many parts of the country; especially in Jammu and Kashmir state. There are several cultivated varieties of this plant.

A rich and mouldy soil is required with a considerable quantity of lime combined with thorough drainage.

The trees are propagated by cuttings of one year old wood, planted in shadybeds at a distance of 3–5 m apart. Fertilizers @ 25 kg of FYM is required for each tree, after the crop is gathered. The plant begins to bear fruits in the 2nd or 3rd years, after the transplantation and continues for 12–15 years. It fruits twice in a year. The first season commences in June-July, but the fruit is not allowed to ripen, lest it should injure the crop which commences in January and is by far the most valuable period.

Figs for drying should be cut from the tree and carefully placed in trays and boxes. To improve the colour and soften the skin the figs before drying are sometimes exposed to the fumes of burning sulphur or are dipped in a hot solution of salt and Saltpeter, but the former practice gives the fruits a very unpleasant taste and is injurious to the health of the consumer.

The drying ground should be a clean space outside the orchard, where figs may be exposed to the full rays of the sun. The figs should be turned twice a day at first and once a day in the later stages. Drying for 6–7 days, yield the best quality of figs. So far India is concerned the best variety recommended is *Khed-Shivpur*.

The dried fruits are demulcent, emollient, nutritive and laxative. The fruits are also sometimes added to *Joshanda*, if the patient suffers from constipation.

b. *Ficus palmata* Forssk. (Bhédu, Bery-Kumaon, Garhwal)

   It is a shrub and the fruits are boiled and mixed with curd and other condiments (like coriander, cumin, black

mustard and common salt) and taken to cure dysentery.

The plant is very so much integrated in the social life of the inhabitants that a famous song has been written and liked very much by the people and others (*Bhédu pako bárá mas, kaiphal páko Chaita*).

c. *Ficus religiosa* L. (Vern. Pipal, Ashwat-Hindi; The Pipal Tree-Eng.)

A well-known tree; both cultivated and growing wild, having typical leaves. It is considered to be sacred and grown near the temples, all over the country. The plant also gets established naturally on the walls, rooftops and other areas during the months of rainy season.

The leaves, bark and ripe fruits are medicinal.

Burnt bark, quenched in water, is given to patients suffering from malaria, cholera and also to quench the thirst.

Ripe berries, dried in shade and powdered, are useful in cases of sperma torrhoea, nocturnal emissions, premature ejaculation and leukorrhoea when given with milk.

It is laxative and also useful in inducing pregnancy, if taken a week after menstruation, provided there is no uterine defect.

60. *Fraxinus floribunda* Wall.

Family: Oleaceae.

Local names: Shir-Khist (Kashmir); Ust-Khadus (Urdu).

Description: A large deciduous tree, bark ash-grey, smooth on young poles, corky and deeply furrowed on mature trees. Leaves imparipinnate; leaflets 3–7, opposite, petioled, ovate-oblong, long pointed. Flowers small, in terminal panicles. Corolla white, longer than the calyx. Stamens longer than

the petals, attached to their base. Fruits winged, one-seeded nut.

Distribution: In temperate regions from Kashmir to Bhutan, 1500–3000 m. Occasionally cultivated in the Himalayan region at suitable elevations. In wild state by no means it is dominant; its distribution being somewhat local and confined to rich and shady situations. It has been reported from Nainital district in Kumaon region in the neighbourhood of limestone rocks and reproduced fairly well on loose soils free from the weeds.

Uses: 'Mana' a sort of gum, is obtained by incision from the stem which is used for its sweetening and laxative properties.

The leaves are purgative and the bark is bitter and astringent.

Extent of utilization: The plant is not exploited largely on a commercial scale.

61. *Garcinia morella* Desr.

Family: Clusiaceae.

Local name: Tamal (Hindi); Ceylon Gamboge (Eng.)

Description: A small evergreen tree. Leaves leathery, 10–15 cm long, elliptic-obovate to ovate-lanceolate. Flowers unisexual and bi-sexual; male about 8 cm across in the axil of fallen leaves; flower stalks up to 75 mm long, stamens nume-rous. Female flowers larger than the male, solitary axillary, stalk, if present, very short. Fruit the size of cherry, roundish, slightly 4-lobed, 4 seeded.

Distribution: In forests of Bangladesh, Khasi hills and the western Ghats from south Canara to Mysore, Kerala and Eastern peninsula.

Uses: The true Gamboge of commerce is the gum-resin that exudes from this tree. It is hydragogue and drastic cathartic and anthelmintic.

From the seeds a semi-solid oil or fat is obtained, used as substitute for oil.

Gamboge of commerce is utilized for dyeing silk fabrics by Budhist priests and also for putting 'Tilak' on the forehead by Hindus.

The gum resin with other drugs is used for treatment of obstinate constipation, liver disorders, etc. It is used with cathartics like *aloes* and aromatics like 'cinnamon'.

A common ingredient of many remedies; used for expelling tapeworms from stomach.

Externally applied as a paste to sprains, bruises and to the swollen hands and feet.

Extent of utilization and trade: It is used in cooking, confectionary and candle making. The yellow resin is used for preparing water colours and gold-coloured varnishes for metals and for dyeing silk fabrics.

The *Gamboge* of European commerce comes from southeast Asia and is obtained from *G. hanburyi* Hook.f.

G. gambogia yields an adhesive gum and is insoluble in water. Oil is used in medicine.

G. cowa, It produces a yellow gum and dye.

G.undica seeds give an oil which is considered nutritive, demulcent, astringent and emollient.

62. *Gardenia gummifera* L.f.

Family: Rubiaceae.

Local name: Dhikámali (Hindi)

Description: A small, unarmed, nearly glabrous shrub with resinous buds. Leaves 3–6 cm long, obovate, shining, base obtuse, acute or cordate. Flowers 1–3 together. Calyx pubescent. Corolla white, turning to yellow, tube 3–5 cm, limb 3–7 cm across, lobes 5. Fruit ellipsoid, smooth.

Distribution: Southwards from Chota Nagpur region.

Uses: The gum is obtained from the plant, called *Dhikámali*, which is used for wounds. In a preparation made with 'Gugal' (gum from *commiphora mukul*, soap, black mustard, white bee wax and sesamum oil). The gum is used in cases of infection with guinea worms and administered orally with water in a dose of about 2 g at a time. Dry cough is relieved, when the gum is boiled in water with 'bansa' or arusa (*Adhatoda vasica*) leaves. Intestinal worms are killed, when the gum dissolved in water is given to drink.

63. *Gmelina arborea* L.

Family: Verbenaceae.

Local names: Gumbhári (Sansk); Gumári (Hindi) : Khambári, Gamári (Bihar) Gumhar (Himachal); Kattanám (Tam.) Gummádi (Telgu), Candahar tree, Coomb Teak, Kashmir tree (Eng).

Description: A moderate-sized or large tree with straight trunk, bark smooth, grayish yellow or whitish, young parts tomentose. Leaves opposite, broadly ovate or cordate, glabrous above when mature, yellow tomentose beneath. Leaf stalk 5–8 cm long, glandular at the top. Flowers in terminal yellowish tomentose panicle. Calyx yellow-ish tomentose. Corolla brownish yellow, pubescent outside, lobes 5, 2-lipped. Drupes yellow when ripe, ovoid or pyriform, 2–3 cm long (Fig. 5.14).

Distribution: In India throughout the country; scattered over large parts in tropical and subtropical regions, ascending 1700 m in deciduous forests.

Uses: The wood is yellowish brown and of excellent quality, used for various purposes. Bark and the root are of medicinal value and used by Santhal and Gond *Bheels*. The fruit juice is demulcent and used in curing gonorrhoea, and cough, also to remove foetid discharges and worms from the ulcers.

Fig. 5.14: *Gmelina arborea* Roxb. A. flowering and fruiting twig; B. corolla opened to show androecium; C. calyx; D. gynoecium; E. longitudinal section of gynoecium; F. transverse section of gynoecium; G and H. upper portion of stamen in two views.

64. *Hedychium spicatum* Buch-Ham. ex Sm. Var. acuminatum

Family: Scitamineae.

Local names: Karchurá (H); Shéldu (Tehri G), Kapur-kachri (Trade Name)

Description: A robust perennial shrub. Leaves sessile, broad, lanceolate, ending in a tail like tip. Bract 1-flowered, green, oblong, obtuse, inner short, membranous. Flowers fragrant. Calyx membranous, slit on one side, 3-toothed. Corolla-tube longer than calyx, linear, spreading, pale yellow. Lateral staminodes linear-spathulate, spreading, white except the orange base, lower spreading, deeply divided in two ovate, pointed lobes and narrowed downwards in an orangered claw. Filaments red, curved, margin unrolled over the style.

Distribution: In subtropical and temperate regions, between 1200–2700 m in moist shady forest. Shimla, Dalhousiae, Dehradun, Bageshar, Nainital, Darjeeling and Sikkim.

Uses: The rhizome is bitter, acrid, pungent and useful to cure inflammation, asthma pains, bronchitis and as tonic to brain. These are also useful as stomachic, carminative, stimulant emmenagogue and expectorant. The antimicrobial activity of essential oil of *H. coronarium* is better than that of *H. spicatuim.*

65. *Hippophae rhamnoides* Servattaz ssp. salicifolia (D.Don) Servettaz. (*H. salicifolia* D.Don)

Family: Elaeagnaceae.

Local name: Amil (Tehri Garhwal).

Description: An erect, thorny shrub. Leaves linear-lanceolate, glabrous or stellately pubescent on upper surface, softly white tomentose on lower side, except the rusty midrib, edges recurved. Male and female flowers on different plants. Male flowers sessile, clustered in the axils of fallen leaves, perianth of 2, opposite, concave, rounded, leaflike segments. Female flowers stalked, axillary, solitary or clustered, perianth tubular, stigma protruding. Fruit ovoid, orange or scarlet when ripe.

Dixit VK and Verma KC 1975. Antimicrobial activity of essential oil obtainmed from rhizomes of Hedychium coronarium and H. spicatum. Pro. Nat. Symp. (Abst) p.45

Time of collection: The plant flowers during the months from April to May and fruits from October to March.

Distribution: In temperate and subalpine regions from Kunawar to Kumaon, 1500–3500 m. Fairly common on riverbanks and damp places. Most of the lateral twigs seem to be deciduous, sometimes drying back to form thorns on young plants. Kunawar, Bashahr, Tons valley, Bhagirathi valley at Gangotri.

Uses: The fruits are used as a sauce and the bark for cuts and wounds. It is also used in cases of lung diseases. Oil from seeds is used in China for heart diseases.

Extent of utilization: Large quantity of the drug is collected for sale in the market. The oil from seeds is produced commercially in China and traded as medicine for heart diseases. Dried twigs are used for fuel in Ladakh (Leh) region.

66. *Holarrhena antidysenterica* (L.) Wall. ex G. Don.

Family: Apocynaceae.

Local names: Indrajau, Korai, Kurchi (Hindi), Kéwar (Punjabi), Kutaja, Kalinga (Sans.), Keora (Nainital); Kuda (Gujarat).

The trade name 'Kurchi' is based on local Indian name of the plant; the specific epithet *antidysenterica* indicates its chief usages. The Sanskrit name denoted that it grows on the peaks.

Description: A small, deciduous tree, sometimes up to 10 m tall. Leaves 10–30 cm long, ovate, thin, nerves on the leaves conspicuous. Leaf stalk very small. Flowers white, fragrant 1.0–1.5 cm in diameter, in large terminal bunches. Fruits slender, cylindric, 20–45 cm long, blackish with white specks all over. Seeds about 1 cm long, having a tuft of long (2–2.5 cm) brown hairs

at the top. All parts of the plant on incision give out milky juice (Fig. 5.15).

Distribution: Throughout India, in the Himalaya ascending to 1200 m and to a similar height in south India. An associate of 'Sal' forests in Northern and Central India. Common on forest fringes.

Uses: The plant leaves resembles with that of *Adhatoda vasica*. The seeds and the bark are used as astringent and styptic, for stopping bleeding from the intestine or haemorrhoids. Owing to its great merit in the treatment in dysentery, it has been called also as *Herba malabarica* in Malabar region. This plant and *Picorrhiza kurroo*, have been considered to have similar properties.

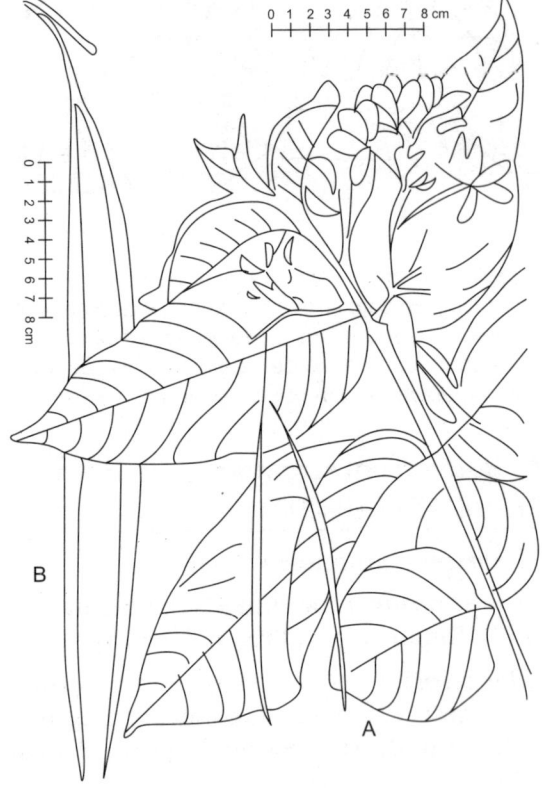

Fig. 5.15: *Holarrhena antidysenterica* Wall. A. flowering and fruiting twig; B. follicles.

Bark of an allied plant, *Wrightia tinctoria*, has often been confused and substituted for true *Kurchi* and thus led to the later having fallen in disrepute. The bark may be distinguished from *Wrightia* bark by its darker colour and by its not exfoliating in patches. An ayurvedic medicine 'Kutja Aristh', is prepared from the bark and is sold by the dispensaries.

The seeds yield a fixed-oil and also possess alkaloids, effective in dysentery. Certain medicinal properties are also ascribed to the leaves. The bark contains *Kurchincine* or 'conessine', 'holarrhenine' and 'Kurchine'. It can be safely given to expectant mothers and has no action on uterus.

67. *Jasminum grandiflorum* L.

Family: Oleaceae.

Local names: Cheméli, Jati (H) Catalonian Jasmine, Spanish Jasmine (Eng).

Description: A climbing shrub. Leaves unequally branched, opposite. Flowers fragrant, in terminal panicles. Calyx lobes long linear. Corolla white, reddish beneath, lobes spreading.

Distribution: Common on the Western ghats, the Nilgiris, Pulney hills, often cultivated for the flowers used by ladies for putting in hairs.

Uses: Leaves are astringent and chewed for sore-gums, toothache, stomatitis and ulceration of mucous membrane of the mouth and also for aphthous conditions of the mouth.

Fresh juice is used to remove soft corns between toes. Oil prepared from the leaves is poured in ears for curing otorrhoea.

The flowers are used in skin diseases, weak eyes and headache, and for the preparation of scented oil.

Other species

a. *Jasminum sambac* (Moghrà, Balphul) is a climber found throughout the country; very often also cultivated. Poultice of dried leaves is used for indolent ulcers. Flowers are lactifuge. Flowers without being moistened are applied to breast of nursing mothers to check secretion of milk in cases of threatened abscess. Oil from flowers is used as deodorant in case of foul smelling discharges from the nose and ears.

b. *Jasminum officinale* L. Flowers are used in heart disease, diabetes, burning sensation, thirst, and diseases of the skin.

68. *Jatropha curcas* L.

Family: Euphorbiaceae.

Local name: Saféd Arand, Jamalgota, Jungli-arand, Saféd Arand (Hindi), Physic nut (Eng.).

Description: A soft wooded shrub or small evergreen tree with smooth bark, greenish, young shoots often pubescent. Leaves long-stalked, angular, globose or cordate. Flowers unisexual, in stalked cymes, greenish yellow; male and female on the same plant. Sepals and petals 5 each. Capsules ovoid, 3–4 cm long.

Distribution: Native of Armenia; grown often as a hedge-plant to keep animals away, now cultivated on wastelands, on a large scale promoted as a source for biodiesel (biofuel) from the fruits. It has assumed importance as a potential source for biofuel in the coming years and being proposed to cover wastelands of the country.

Uses: The albuminous seeds, being rich in oil, it is of commercial importance, particularly after its value as biofuel has been realized. It contains croton resin used as a stimulant for the growth of hairs; it burns

without smoke and had been in use as illuminant, for the manufacture of hard soaps and candles, as lubricant and making varnish in conjuction with iron-oxide.

From the seeds 3 kinds of poison are obtained. The shell of the seed contains; the albuminol "curcin" of the group of ricin and crotin, present in the albumen of the seed; poultice of the seed or their oil contains a third poison which acts as irritant on the skin.

When dried in sun, the juice forms a reddish-brown brittle substance like shellac or Kino and also stated to dye-linen as black.

The oil percentage in seeds is around 30%, known as croton oil which is purgative and emetic, used for application in cutaneous diseases. It has also been recommended as a substitute for olive oil, in dressing of wollen clothes and as a good drying oil.

Numerous cases of poisoning are reported through eating the entire seed, though used by natives as purgative.

Extent of utilization and trade: Because of its importance as a source of biofuels, it is planned to cultivate this plant particularly on wastelands in thousands of hac as a prospective source of biofuels in the coming years.

Jatropha glandulifera, J. hastata, J. podagrica are cutivated as ornamentals. *J. glandulifera* is a small shrub in hotter and damp tract of the country, often grown as hedge. It is reputed to produce green dye, the seeds produce light oil, yellow which has for long been used for external application in rheumatism and paralytic affections, but seldom administered internally being a purgative.

69. *Juglans regia* Linn.

Family: Juglandaceae.

Local names: The Walnut tree (Eng.); Akhrot, Krot (H); Askkhor, Akhor (Kum) Kabshing, (Kash) Kol, Akhrot (Dun).

Descriptions: A large tree with alternate, pinnate leaves; leaflets 7–9. Flowers green, male and female on the same plant. Male flowers numerous, inpendulous catkins, with narrow perianth, irregularly 5-lobed. Female flowers 1–3 clustered; bracts combined in a pubescent ovoid involucre, adnate to the ovary. Ovary 1-celled, style arms 2. Fruit a drupe.

Distribution: In temperate regions from Kashmir eastwards, 1000–3000 m. Cultivated in villages both on the hills and as agroforestry.

Kashmir, Kagan valley, Hazara, Shimla, Chakrata, Murree, Yamuna valley, Konain, Ramgarh, Naitwar. (Himachal Pradesh and Uttaranchal hills).

Uses: The bark is used as an anthelmintic, detergent and lactifuge. In is used for eczema and scrofula, etc. Decoction of the bark is useful for stopping mammary secretions, as a gargle for sore throat.

The leaves are astringent, alterative, tonic and detergent; useful in rickets, scrofula, eczema, syphilis and as anthelmintic. Externally, the decoction is useful as a wash for malignant sores, skin eruptions, ulcers, etc. Dried leaves, in Kumaon, are used as insecticide and are kept in food stores of houses.

The fruit is alterative and used in rheumatism. Unripe fruits is given to children as anthelmintic. Vinegar pickled fruits are used in sore throat. Unripe shell of the nut is useful in syphilis. The spirit distilled from the fruits and leaves acts as antispasmodic, ripe fruit and meat of the nut is laxative, cholagogue and vermifuge for tapeworms. Externally, oil is used in dressing for skin diseases of leprous type. The grind-rind of the fruit was in great demand for syphilis, but it has now diminished from use. In Kumaon, the paste of

fruit rind is used in toe-sores called 'Katya'. Ash of the nuts is used as tooth powder.

In Malaya, Walnut Kernels are said to produce plumpness of body and to strengthen and lubricate the muscles. They are recommended in heart burn, colic and dysentery.

Extent of utilization and trade: The utility of this tree is from very remote times. Its chief value lies in timber for gunstock, carving, turnery and fancy work. Huge warts or burs on the stems are also valued. Bark is employed as a dye and is medicinal. It is exported to plains since the ladies use it for dyeing their lips and is sold in the market as "Dasunda".

The fruit is an important article of food. The kernel yields an oil. Rind is used for tanning and dyeing. Twigs and leaves are used as fodder.

70. *Juniperus communis* L.

Family: Cupressaceae.

Local name: Dhoop plant (trade name).

Description: A shrub or small tree with reddish brown bark, scalling off in papery ridges. Young shoots slender, triangular, with projecting ridges. Buds with acuminate scales. Leaves awn-shaped, persisting for 3-years, sessile, upper surface concave with a broad white band of stomata; lower surface bluntly keeled. Male and female flowers on different plants. Male solitary, stamens yellow in 5–6 whorls. Female solitary, green. Fruit ripens in second or third year, green when young, bluish or black when ripe with 3 minute points at the tip, Seeds 2–3, elongated, ovoid, 3-cornered with depressions between.

Distribution: In drier inner ranges, 3000–4000 m exposed open positions in company with other members of the genus.

Uses: The wood is resinous and used as incense. Fruit is aromatic, volatile oil is used to flavour spirit in Europe. Also used in scanty urination, chronic cough and pectoral affections.

Powder of the berries is rubbed on rheumatic swellings and also as substitute for coffee.

Other species

a. *Juniperus macropoda* Boiss. *(J. polycarpos)* (Vern. name Shugpa-Lahul; Dhup-Hindi). A moderate-sized tree. Leaves pungent, upper scale like, closely appressed, ovate, with a dorsal gland. Dense formations are seen in some regions. The female young *berries* yield an oil which is a substitute for imported *Juniper oil* which is used in industries Berries collected from Lahul in June yielded 1.25% oil. Fruits used in manufacture of Dhoop, in place of *J. communis* Hauber. Berries collected in October, when essential oil content is maximum.

b. *Juniperus squamata* Buch.-Ham. Also provides oblong berries, dark purple when ripe, but small quantities are available only for local use. Wood powder is also aromatic.

71. *Lawsonia inermis* L. (*L.alba* Lamk.)

Family : Lythraceae.

Local names : Méhndi, Hénna (Hindi); Egyptian privet, Mignonette tree (Eng).

Description : A large shrub, deciduous or small, straggling tree, bark rusty brown, fairly smooth; branchlets quadrangular, often spiniscent. Leaves elliptic, opposite,

glabrous. Flowers small, in dense terminal panicles, fragrant, white or rose coloured. Sepals and petals 4, roundish, margins undulate. Capsules green, shining when young, later turning to red; drying ultimately with a brittle outer mass, tipped with persistent style.

Distribution : Native to Arabia and Persia, cultivated in many parts like Rajasthan, Gujarat, Haryana and MP.

Uses : The plant is known for its cosmetic dye obtained from its leaves called *Henna'*; widely used both by men and women for colouring hands, hairs and feet. Leaves are also used in the manufacture of ottos and perfumed oil. The flowers on distillation yield an aromatic essential oil, used in perfumery and embalming. Oil can also be extracted from the seeds.

The bark is given in jaundice and enlargement of the spleen, in calculus affections and is applied in skin diseases like leprosy. Externally, the leaves are applied for relief in headache and rubbed over the soles in burning sensation of feet.

Leaf decoction is used as an astringent and for relaxation when used as gargle in sore throat. Leaf juice, mixed with water and sugar, is given as a remedy in spermatorrhoea.

Infusion of the flowers is administered as a cure for headache. The flowers are also reported to be sporific and as such are used for stuffing pillows. Seeds with honey and tragacanth are prescribed as cure for headache.

72. *Helecteres ixora* L.in

Family: Sterculiaceae.

Local names: Mriga-shingá (Sansk); Marorphali (Hindi), Atmora (Mumbai). The name 'Marorphali' is derived from the shape of the fruit that are spirally twisted.

Description: A shrub with thin spreading branches and grey bark, young parts covered with stellate hairs. Leaves, bifarous, broadly-ovate or orbicular, often lobed, obliquely cordate or rounded at the base, short-acuminate, irregularly toothed, scabrous above, pubescent beneath. Flowers, axillary, usually 2–4 together. Calyx tubular, indistinct 2-lipped, brown floccose outside. Petals 5, scarlet, clawed twice the length of calyx, claws winged. Ovary 5-celled, on a gonophore which is elongated in fruit, styles 5, more or less connate. Fruit cylindrical, pubescent composed of 5-spirally twisted carpels.

Distribution: Common shrub of coppice areas in outer hills of Dehradun and Saharanpur, also in central and western India from Bihar as far west as Jammu and Western peninsula.

Uses: The fruits and leaves are used in Indian medicine. Fruits are demulcent, astringent, used in gripping of the bowels and flatulence of children. The fruits after grinding are soaked in water (hot), filtered and the filtrate is allowed to settle. Drops of decanted infusion are poured in the ear (1–2 drops), twice daily to relieve headache by the tribals of Ganjam district. In Sambalpur the fruit is powdered and put in the ear to relieve earache. In Mandla district of Madhya Pradesh, the powdered fruits are given to cure fever, while in Jhabua district the aqueous extract of the seed in small quantity is given to children in dysentery.

Root juice is used in diabetes, stomach affections. Both root and bark are expectorant, astringent to bowels, antigalactagogue, lessen gripping and is a cure for scabies when applied topically.

73. *Lyonia ovalifolia* (Wall.) Drude (-Pieris ovalifolia D.Don)

Family: Ericaceae.

Local names: Aniyár (TG); Anyar, Ayár (Jaunsar), Angyár, Ayanar (Kumaon).

Description: A small deciduous tree, bark brown, peeling off in narrow strips, often deeply furrowed with spiral cleft. Leaves elliptic-ovate, entire, rounded at the base, coriaceous, glabrous, often pilose beneath when young. Inflorescence, a simple raceme, rarely falsely fascicled; bracts lanceolate or linear, deciduous. Flowers white. Calyx lobes white, triangular, lanceolate, connate at the base. Corolla elongate-ovoid, pubescent, lobes-5, short, recurved. Stamens 10, hypogynous, filaments subulate, anthers open by terminal pores. Ovary 5-celled with many ovules in each cell. Capsules globose, 5-valved. Seeds minute, many.

Distribution: In temperate regions from Indus to Bhutan, 1500–2500 m. Osmaston, reports that at high elevations it becomes shrubby, while some have separated it as *P. villosa,* but the distinctive features are not constant. It is associated with 'banjoak' and 'burans', however, it is predominant because it is not touched by the animals.

Uses: The leaves if eaten are poisonous enough to kill any animal.

Plant decoction is used externally for killing worms and in cutaneous diseases. In Kumaon, people keep the cuttings of the tree in paddy fields to protect the crop from worms and insects. Good quantity is collected locally.

74. *Manilkara hexandra* (Roxb.) Dub. (-Mimusops hexandra Roxb.)

Family: Sapotaceae.

Local names: Khirni (Hindi), Milk tree (Eng.).

Description: A tree often cultivated at Delhi, MP, Gujarat, Maharashtra, and Andhra Pradesh.

Uses: It is used as a rootstock for *Sapota.* The fruits are edible. Timber is useful.

In Bastar the ripe fruit is eaten by the tribals, while the 'Bhils' of Jhabua district boil the stembark in water and take the bath with this water to cure bodyache.

75. *Madhuca indica* J.F. Gmel (-Bassia latifolia Roxb., Madhuca latifolia (Roxb) Macb)

Family: Sapotaceae.

Local names: Mahuá tree, Mohua, Moha, Modhuka (Hindi), Illipe butter (Eng.).

Description: Tree with mango look but leaves larger in size. The flowers are white, having sickly sweethish smell. The fruits contain a stone which gives oil called the 'mahua oil'. The flowers are edible and a liquor is made from them.

Distribution: A large tree, cultivated mainly in Uttar Pradesh and Bihar. The flowers are used as vegetable and are also the source of alcohol. The seed oil is used both for cooking and soap making. It is planted by almost every family near the houses in eastern part of Uttar Pradesh for its economic uses.

Uses: The fruit contains a stone which gives the oil. The flowers are edible and a liquor is made from them. They also act as aphrodisiac and used in chronic cases of cough, when taken after cleaning their pollens. It is cooked in milk and taken at bed and no other food is taken.

Boiling *Mahua* leaves in water when applied relieves swelling of the testicles, by exposing them to vapours and fomenting with it. All kinds of pains, due to phlegm or flatulence, are relieved by rubbing of 'mahua oil' and so is the case with rheumatism, lumbago, pleurisy and pain in the chest gets relief by rubbing the oil and bandaging the part with warm cotton wool.

The oil is also used for cooking purposes; the bark is used as tan.

Extent of utilization and trade: The oil is extensively used and traded for soap making.

Other species: *Diplokenma butyracea* Roxb (Bassia butyracea Roxb.) provides a butter-like substance and used locally as substitute for oil in the Himalayan region.

76. *Maesa indica* (Roxb.) Wall.

Family: Myrsinaceae.

Local names: Kramighnaphal, Nagpadhéra (Hindi).

Description: Large shrub with long straggling branches; twigs with numerous small lenticels. Leaves lanceolate or oblong lanceolate, acuminate, distantly serrate-dentate, membranous, glabrous above, pale beneath, lateral nerves 5–10 pairs. Flowers white, in simple or compound axillary racemes; one bract at the base of each pedicel. Corolla tube twice as long as the calyx. Berry globose, pinkish-white, almost covered with persistent calyx and usually tipped with the style.

Distribution: Common in the ravines of *chirpine* forest but often in open oak forests in the northern aspects. It is often gregarious in patches. Found in sub-Himalayan tracts, ascending to 2000 m from Ravi eastwards.

Uses: The roots are used in syphilis, while the fruit is anthelmintic, but edible.

Extent of utilization and trade: Utilized on a commercial scale in a very limited quantity; the fruits are available in the market.

Other species

a. *Maesa chisia* D.Don. Distributed in Nepal and Assam. Roots and leaves possess insecticidal properties.

b. *Maesa ramentacea* Wall. A small tree, found in Nepal, Bhutan, NE Himalayan region, Andaman islands and Vizag in Andhra. Leaves used for skin diseases and pounded for application to the skin. Wood used for *tree nails* used for boat building.

77. *Mahonia nepaulensis* DC. (Berberis nepalensis Spreng.)

Family: Lardizabalaceae.

Local names: Gurm, Haldia (Garhwal).

Description: An erect shrub with coriaceous leaves that are shining, imparipinnate, crowded towards the end of the branches bearing 2, small, subulate, spiniscent stipules on its broad sheathing base; leaflets ovate-lanceolate, sessile, with 3–8 spinous teeth. Bracts persistent, clothing at the end of branches. Flowers yellow. Petals in two series. Berries purple, elliptic.

Distribution: In central and inner ranges of Garhwal and Kumaon, Sikkim, North Bengal and Nagar hills, also in Nepal. Fairly common in moist, shady *oak forests* as undergrowth, between 2000–2500 m.

Uses: The berries are considered diuretic and demulcent in dysentery.

Extent of utilization and trade: Drug used locally, not commercially exploited.

78. *Mallotus philippensis* Muell and Arg.

Family: Euphorbiaceae.

Local names: Raini (Dehradun), Kambel (Jaun), Rohni, Roini (Kash,), Kaméla. Trade name-Kaméla Senduri (H); Kapila (Guj); Shendri (Mah), Gasarba (Bihar, Singhbhum), Kalupatti (Tamil.).

Description: A small tree with buttressed trunk. Leaves alternate; usually ovate-lanceolate, 1–17 cm long with numerous crimson glands. Flowers unisexual; male and female on separate individuals in terminal spike, brick red. Male yellow, usually 3 together in slender, drooping racemes, 7–15 cm long, anthers indefinite. Female flowers solitary in stiff spikes, 5–7 long. Ovary glandular, scarlet. Fruit a capsule, 3 valved, covered with bright red powder consisting of resin mixed with hairs (Fig. 5.16).

Distribution: Distributed throughout the tropical regions of India from 1500 m; from the Himalaya to Kerala.

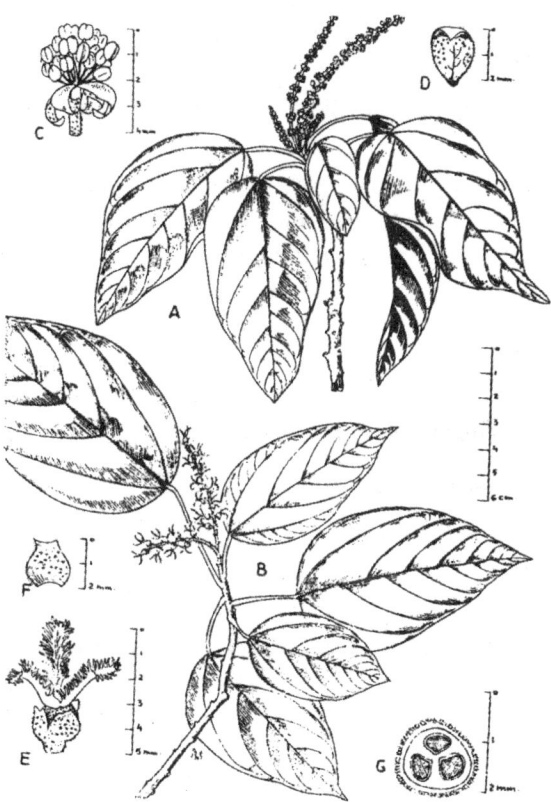

Fig. 5.16: *Mallotus philippensis* Muell. Arg. A. flowering twig from male plant; B. flowering twig from female plant; C. male flower; D. perianth lobe from C; E. female flower; F. perianth lobe from E; G. transverse section of the ovary.

Uses: The roots are particularly useful for the treatment of skin diseases and for removing frickles and pustules.

The powder from the female flowers constitute the drug locally called *Kamela* and is used as astringent and anthelmintic (taenicide). It is also cathartic and purgative.

The powder is taken with milk or curd and if in one dose it does not expel the worm another dose is repeated. Sometimes a dose of Castor oil is also necessary to expel dead worms. Externally, the powder is used in treatment of skin diseases, like ringworm and scabies.

The hairs on the fruits have been tried on animals as oral contraceptives and have been found to reduce fertility in female rats and guinea pigs.

Extent of utilization and trade: The seeds yield 5.83% of a bland oil which is used in paints and varnishes and due to its excellent drying properties it is in much demand for painting work. The plant is extensively utilized for the powder. Puri Division in Orissa, U.P., Kumaon and sub-Himalayan tracts in Dehradun, Punjab, Kangra, Belgaum and Ganjam districts are the important centres of production.

The ripe fruits are placed in a cloth or sack and beaten until the glandular pubescence is removed. In some places the fruit is simply rubbed between the palms or is kneeded with the feet on the ground. The powder thus obtained is then sifted to free it from the fruit and broken pieces and in this condition it is ready for market.

Other uses: The bark of the plant is reported to be used for tanning leather, the most important use is in dyeing, for imparting to silk a bright orange or flame colour. The wood of the tree is used for minor domestic articles and fuel, it is suitable for matchboxes. The leaves provide good fodder. The red powder is used by ladies as vermilion on their forehead.

79. *Mangifera indica* L.

Family: Anacardiaceae.

Local names: Mango tree (Eng.); Am, Amrá (Hindi); Mángà (South).

Description: A large, evergreen tree, glabrous, except the inflorescence; bark thick, rough, brown or blackish. Leaves alternate, leathery, oblong-lanceolate or oblong, entire, margins often undulate, dark green. Flowers unisexual and bisexual, very small, in large terminal, erect, pubescent panicles. Fruit 5–20 cm long, smooth variable in shape.

Distribution: Indigenous in eastern part like Sikkim, Assam, Khasi hills, Satpura range and the western ghats; but widely cultivated for the fruits throughout the country. Various improved varieties have been created different taste and shapes, in Maharashtra, Andhra, Tamil Nadu and UP (Dushheri, Langra, Banarsi, etc.)

Uses: Plant leaves are strung in the houses in religious ceremonies and considered to be auspicious by the Hindus. The wood is used in *Havan* by Hindus. Wood also used for furniture, boats, (lasting in water) and making shoe heels.

The unripe fruit, where the stone has not formed, called Kéri is largely used to ward off the effects of hot winds. The ripe fruit is sucked, while a grafted mango is sliced and eaten. It is considered nourishing, relieves constipation and contains both vitamin A and C. Drinking milk after sucking mango juice, is considered very nutritious.

Kernal of the mango seed, cures diarrhoea, stops profuse bleeding from piles, excessive menstruation, leukorrhoea and excessive urination in diabetes.

Inner bark of the mango tree, stops diarrhoea and haemorrhage. The bark is boiled in water overnight in about 30 cc. of water, strained in the morning and taken for a week, is reported to cure gonorrhoea.

The flowers of mango called 'baur' or 'maur' are useful in spermatorrhoea and leukorrhoea. These are dried in shade, powdered, sifted and mixed with equal weight of sugar (brown) and taken along with milk or water. In case of bleeding from the intestine or the piles or urethra, leafshoots of mango are ground in water, strained and drunk to stop diarrhoea.

Powdered mango, locally called *Amchur* is ground in water and applied to skin on which a spider has been cruhed.

Mango pickle oil, if applied to head cures alopecia.

Mango kernel may be used for preparing flour, starch and cattlefeed. The leaves contain a dye in the form of yellow crystalline substance (euxanthic acid), known as *piury*. Dry unripe fruit has also been used as mordant.

80. *Marsdenia roylei* Wight

Family: Asclepiadaceae.

Local names: Murkula, Pathor, Tar, Veri (Hindi).

Description: A softly tomentose, twining shrub, juice milky. Leaves cordate. Flowers in umbellate cymes. Calyx half as long as the corolla; coronal scales 5, attached at the base of the staminal tube. Follicles hairy, deeply wrinkled, shortly pointed.

Distribution: In temperate regions from Kashmir to Sikkim, 1000–2300 m. Fairly common in the scrub and oak forest, often on rocks. Shimla, Bashahr, Kashmir, Sarju valley, Mussoorie, Jaunsar, Bodyar, Garhwal, Sikkim.

Uses: The unripe fruit is powdered and given as a cooling medicine. A decoction is used as a remedy in gonorrhoea.

Extent of utilization and trade: The drug is not used as a trade commodity but locally it is used.

Other uses: The stem yields a fibre from which fishing nets, lines and ropes are made. In *M. tenacissima* bark of the stem also yields a valuable fibre which was said by Royle to be the second best in India.

81. *Melia azaderach* Linn.

Family: Meliaceae.

Local names: Bakain, Bakáyan, Mahánim (H); Dek (Dun), Deknoi (Jaun).

Description: A moderate-sized tree, fast growing, with smooth dark grey bark. Leaves bipinnate, sometimes tripinnate, pinnae usually opposite; leaflets 7–14 cm, longer than Neem tree leaves, each pinna ovate-lanceolate, serrate, acuminate, the lateral more or less oblique. Flowers liliac blue, in numerous axillary cymes bearing panicles. Calyx deeply 5-lobed. Petals linear-oblanceolate. Staminal tube purple, cylindrical with 20–30 linear teeth; anthers at the mouth of the tube. Ovary 5-celled. Fruit drupe globose, generally 5-celled, more or less dry, yellow when ripe, at first smooth, afterwards wrinkled; remaining on the tree long after drying.

Distribution: It is commonly cultivated in the hot valleys and planted throughout India. Doubtfully indigenous. Wild in the north-west Himalaya up to 1500 m though commonly met with up to 2000 m. The tree is fast growing and coppices extremely well.

Uses: Leaves used for purifying blood in ailments like scabies, ringworm, leprosy and leukoderma. Externally used for fomentation to resolve swelling or allay pain. Leaf extract applied as collyrium to eyes to cure cataract.

The bark is bitter and employed as anthelmintic. Kernel of fruit used as remedy for piles along with *rasault* (extract from Berberis root). Leaves and pulp of the fruit and seeds are used extensively.

Inner soft bark of this plant, *Kasni* (*Cichoriunmintybus*) seeds steeped in water, strained and drunk to relieve old fever. Burnt bark and catechu in equal weight used to cure mouthsores.

Extent of utilization and trade: The drug is used extensively in the Indian medicine and exploited commercially. The fruit yields an oil.

Phenology and time of collection: The plant is leafless during December to March. Flowers during March-May and fruits during the cold season.

Other uses: Like *neem*, it also yields a brown adhesive gum; oil from the seed is not so important. It has long been used in medicine by Arabs and Persians but Indians have neglected it in favour of *neem*, stone from fruit is employed as bead.

82. *Meriandra strobilifera* Benth.

Family: Labiatae.

Local name: Kafur-ka-pát (Hindi).

Description: An erect, aromatic shrub. Leaves thick, shortly stalked, oblong or lanceolate, petioles and undersurface clothed with thick, stellate, white-woolly tomentum, base rounded or cordate; lateral nerves 8–12 pairs. Flowers small, white or pale-red in whorls, crowded in erect, tomentose, 4-sided, often paniculate spikes. Floral leaves small, bract like, ovate, overlapping. Calyx tubular, ovoid, 2-lipped; upperlip concave, entire, lower 2-toothed. Corolla tube as long as the calyx; limb spreading, 4-lobed.

Stamens 2, anthers protruding. Nutlets oblong, compressed, tip produced into a wing.

Distribution: Throughout western Himalaya, 1500–2000 m, in open situations and stony ground. Shimla, Bashahr, Punjab, Sirmur, Jaunsar, Dungri, Konain, Garhwal.

Uses : Decoction of the plant is a good lotion for ulcers and heal new abrasions of skin. It is reported to decrease the supply of breast milk.

Extent of utilization and trade: The drug is not exploited on a commercial scale but used locally on a limited scale.

83. *Leptadenia reticulata* (Retz.) Wight and Arn. (-Gymnema aurantiacum Wall. ex Hook.f.)

Family: Asclepiadaceae.

Local names: Jivánti (S); Dori (Hindi).

Description: Much-branched twining leafy shrub. Bark of older stems corky; branchlets more or less minutely pubescent. Leaves thin, sub-coroeaceous, ovate-lanceolate, acute or acuminate, rounded or acute at the base, glabrous above or minutely pubescent beneath. Flowers pale brown, in many fld. umbelliform cymes. Calyx silky outside, divided halfway down into ovate-oblong, subacute segments. Corolla pubescent outside, tube very short. Follicles usually solitary. Seeds narrowly ovate, flat and margined.

Distribution: Punjab southwards to Western peninsula, found in the forests near thorny trees, seen in Gujarat, Khasi hills, Konkan, Nilgiris and in south India. Mention of this plant is made in *Atharva Veda* (4500–1600 BC) as life and strength giver and propagator of milk.

Uses: The plant is stimulant, galactagogue, oestrogenic, eye tonic, astringent and increases milk after parturition, prolapse of uterus, vagina controlling habitual abortion, maintain pregnancy, induce heat, soothen hard milkers, induce milk letting and useful in skin affections and wounds.

The leaf paste and roots taken orally with water to cure gangrene by the *Bhils* of S. Rajasthan.

Alcoholic extract of roots and leaves show antibacterial activity against gram (+) and gram (–) bacteria, inducing *Micrococcus pyogenes* var. aureus, albus and citreus. *Bacillus megatherium, Escherchia colii, Salmonella typhi, Proteus vulgaris* and *Trychophyton rubrum*. Sterols such as stigmasterol in major amounts and i-setosterol in minor quantities are also present. A fructosan of the inulin type has been separated from the rubers. Important ingredient of the Chayvanprash.

Unscientific and unsustainable harvest of the produce led to a crisis in the market. When propagated under field condition, the FYM treated plants followed by a dose of NPK plus Hexameal mixed with NPK and full dose after 4 months recorded maximum collar diam. in NPK: half dose plus Hexameal combination, followed by NPK full dose plus Hexameal and the minimum under control.

Above ground and below ground biomass ranged between 1.09 to 6.15 and 0.32–0.83 grams per plant (dry weight) respectively.

Leptadenia pyrotechnica (Forsk.) Decne. (-Cynanchum pyrotechnicum Wight,) *L. spartium* Wight. (Vern. Khimp, Kip Raj) An erect shrub with slender stem , leafless when leaves present, are acuminate, leathery, shortly petioled. Flowers yellow, in small umbellate cymes, few fld. hoary. Follicles with slender long beak, terete.

Throughout northwest India, extending eastwards to Yamuna. One of the most common leafless shrubs of arid-zone.

84. *Mimosa hamata* Willd.

Family: Mimosaceae.

Local names: Jinjanio, Jinjani, Aráti, Arkar (Hindi).

Description: A prickly shrub. Stem and rachis copiously prickly. Leaves bipinnate, pinnae 3–4 pairs, nearly sessile. Leaflets 6–10 pairs, minute, oblique, ligulate oblong. Flowers in small peduncled heads. Pods 4–6 seeded.

Distribution: On calcium rich soils, where clay is extracted for brick making or where there is *Kankar* (stony) zone below. Through-out India.

Uses: The seeds are boiled in buffaloe's milk and taken as tonic and stimulant also against weakness. It should not be taken in excess.

85. *Murraya Koenigii* Spreng.

Family: Rutaceae.

Local names: Katneem (Hindi), Surabhinimba (Sansk); Karrinim (Mumbai). Gamdaneem (Punjab); Karuveppilai (Tamil), Karvipaku (Telgu.).

Description: A small pubescent tree. Leaves gland dotted, odd pinnate, 30–35 cm, often crowded towards the end of branches, stalk not winged; leaflets 11–21, alternate, shortly stalked, lanceolate, 3–5 cm, entire or with small teeth near the tip. Flowers white, bisexual, in terminal panicles. Calyx 5-parted, persistent. Petals 5, nearly erect, much longer than the calyx, dotted with green glands. Stamens 10, free, five long and five shorter. Ovary ovoid, 2-celled; style terminal, stigma capitate. Berry ovoid, black, wrinkled, seeds one or two (Fig. 5.17).

Distribution: Throughout India, ascending to 1700 m in the hills in sub-Himalayan region, western ghats up to Kerala, Tamil Nadu and from Kumaon to Sikkim.

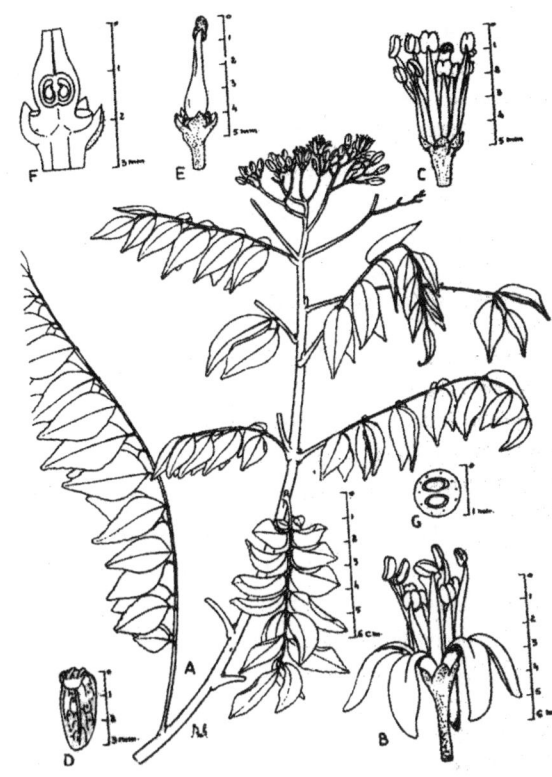

Fig. 5.17: *Murraya koenigii* Spreng. A. flowering twig, some of the leaves cut away; B. flower; C. petal removed to show androecium and gynoecium; D. petal; E. gynoecium; F. longitudinal section of gynoecium; G. transverse section of gynoecium.

Uses: Leaves, bark and roots are medicinal.

The green leaves are eaten raw to cure dysentery; when bruised these are applied externally to cure eruption and in decoction as febrifuge.

The bark and the roots are stimulant and used externally to cure eruptions.

Leaves contain an essential oil, glucoside *Koeinigin*

Other species: *Murraya paniculata* (L.) Jack. (*M. exotica* L.).

(Vern Name: Márchula, Kamini, Juti, Atal-Hindi; Pandry-Kanara, Simalikkonji-

Tam). Chinese-Box, Chinese-Myrtl, Satin wood. (Eng.).

Evergreen shrub or small tree, bark grey, pubescent. Leaves dark, shining, odd-pinnate; leaflets alternate, ovate, glabrous. Flowers fragrant, in terminal cymes. Sepals glandular. Petals white, gland dotted. Berries globose, red when ripe, usually 2-seeded, seated on a persistent style.

Indigenous in the sub-Himalayan ranges, Garhwal to Assam, ascending 1200 m and south to Bihar, Orissa, and Central India.

Leaves and bark are medicinal. Leaf-powder used as application to fresh cuts, while in decoction it is drunk in dropsy.

Ground bark of root is eaten in bodyache, bark of stem is antidiarrhoeal.

Leaves administered in diarrhoea and dysentery.

Flowers contain a glucoside *murrayin*.

86. *Myrica nagi* Thunb. (Myrica esculenta Buch. Ham ex D.Don var sapida)

Family: Myricaceae.

Local names: Kaiphal, Katphalá (S); Kaphal (TG); Kaphaw (Kum.).

Description: A small tree. Leaves alternate, crowded towards the end of branches, 7–12 cm long, lance shaped, gland dotted. Flowers minute, unisexual, the two sexes being on different trees. Male flowers in small catkins, solitary on drooping stalks; perianth absent. Female in small, erect spikes; perianth absent. Fruit a red drupe, fleshy without stalk (Fig. 5.18).

Distribution: Temperate regions from Ravi eastwards in ravines, 1200–2100 m. It is found growing between *Pinus roxbutrhii* and *Quercus incana* zones in association with *Rhodendron arboreum* and *Lyonia*

Fig. 5.18: *Myrica esculenta.*

ovalifolia. It is especially common in 'Banj' forests. The leaves of young plants or of coppice shoots are often serrate.

Uses: The bark of the plant is aromatic, antiseptic, expectorant, astringent, stimulant, carminative, tonic and resolvent. It is a valuable remedy in fevers, catarrh of mucous membrane, affections of chest, asthma, diarrhoea, typhoid, dysentery, chronic gonorrhoea, gleet, chronic bronchitis cough, etc. Externally the powder of the bark is used as a rubefacient and a stimulant application to limbs during collapse stage of cholera. With vinegar it is applied to strengthen gums and for relief in toothache. Decoction of bark is used for cleaning scrofulous and apthous affections and putrid sore. Powdered bark is also used in menorrhagia and as snuff in cattarh with headache and for inducing sneezing and as a dust for cleaning putrid sore. With honey the powdered bark is given for alimentary disorders. It is applied in the form of plaster in rheumatism. Oil from the bark is used in earache. The pressure of the bark acts on uterus and promotes menses in ladies. In Kumaon decoction of the stem bark is used during fevers.

The fruits are edible and a syrup is made which is commercially available, said to cure heart problems.

Extent of utilization and trade: In Ayurvedic system of medicine Katphali-kwath, Kathphaladi nasya, and Katphala-di churna are prepare from the bark. The bark is collected in March and April; the reddish portion being used only. Uniyal and Issar (1967) recorded about 5–6 quintals bark collected from Kantal forest in Tehri Garhwal and sold at a rate of Rs. 125/per quintal. The fruit is edible.

Phenology and time of collection: The tree flowers during the months of Aug-Oct. and fruits from May to June.

87. *Moghania fruticulosa* (Wall.) Mukerjee (Moghania strobilifera R. Br., Flemingia fruticulosa Wall.)

Family: Papilionaceae.

Local names: Kushrunt (Hindi), Kursunt (Hindi), Nundar (Bihar).

Description: An erect shrub, branches slender, procumbent, terete spreading from the base. Leaves simple, cordate-ovate or orbicular, upper surface nearly glabrous, prominent veins on the lower surface. Flowers pink or white, in small clusters, enclosed by folded membranes, orbicular bracts. Arranged in two rows in short racemes, keel slightly incurved, style tip glabrous. Pods oblique, 2-seeded.

Distribution: In temperate and subtropical regions from Indus eastwards to Kumaon. 1300–3000 m, on grassy slopes in *chirpine* and oak forests. New leaves often have coppery brown colours.

Uses: The roots are used in epilepsy and hysteria; in powdered form it is taken to have sound sleep.

Other species: *Moghania bracteata* (Roxb.) HL Li (*Flerningea bracteata Wight; F. stroibilifera* var. bracteata Baker). Grows under *Shorea robusta* forests in the sub-Himalayan tracts and the Siwaliks. *Moghania chaapar* or *Feminingia chaapar* Ham., have properties similar to *M. strobilifera*.

Other species of *Moghania* like *M. macrophylla* vern. Hara-salpab (H), roots are used for external application to ulcers and swellings, mainly of the neck, species *nana* Roxb. (*F. congesta* var. *nana*) are also used for ulcers and swellings.

88. *Moringa oleifera* Lam. (Moringa pterigosperma Gaertn.)

Family: Moringaceae.

Local names: Sainjná (H), Saguna (H); Biskandrá (T. Garhwal), Moringkai (Tamil). The Horse raddish tree (E)

Description: Small unarmed tree with corky thick bark, longitudinally cracked leaves 3-pinnate, 30–75 cm long, main axis and its branch jointed, glandular at joints; leaflets glabrous, entire. Flowers white, scented in large axillary downy panicles, pods pendulous, ribbed, seeds 3-angled.

Distribution: The tree is wild in the sub-Himalayan tracts from Chenab to Oudh but very commonly cultivated near houses in Bengal, Assam and peninsular India. It is a prolific coppicer.

Uses: The tree produces pods which are made into curries, pickles, etc. An oil, almost colourless, is obtained from the seeds, commonly called as "Benoil" and is highly valued as lubricant by watch makers. Owing to its power of absorbing or retaining odours, it is fairly largely employed by perfumers.

Flowers are stimulant and aphrodisiac. Seeds used in venereal affections.

The plant yields a gum which is white when it exudes but turns to a mahogany colour gradually. It is used in dental caries with seasmum oil used in otalgea.

Fruit used in diseases of liver and spleen, and in articular the pain. Roots are used in paralytic affections, epilepsy, chronic rheumatism, and also as stomachic, cardiac and circulatory tonic.

Extent of utilization and trade: The oil is colourless called 'Ben Oil' is seldom made in India and does not form an article of export. Leaves used as *Curry patta* and sold in market (Fig. 5.19).

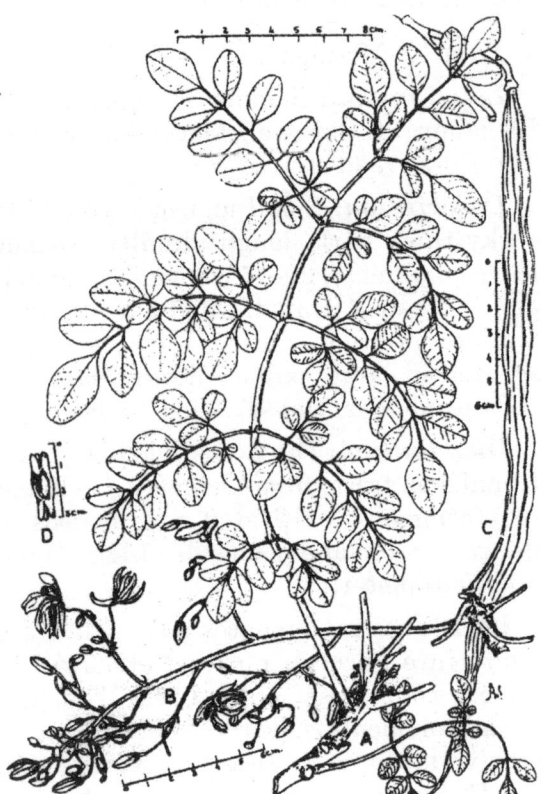

Fig. 5.19: *Moringa oleifera* Lamk. A. twig bearing a young and a fully developed leaf, other leaves cut away from a little above the base; B. inflorescence; C. capsule; D. seed.

89. *Mucuna prurita* Hook. (-M. pruriens Bak non DC.)

Family: Papilionaceae.

Local names: Kawanch, Kaunchá (Hindi); Atmagupta (Sansk.). Alkusa, Atkir, Etká (Bihar), Kuhili (Mumbai); Dulagondi (Tel.) Punaikkali (Tamil).

Description: A slender climber, leaflets appressed, hairy beneath, lateral exceeding the terminal, semi-cordate ovate, 7–15 cm long with 6–7 rather strong lateral nerves, terminal rhomboid. Flowers purple, in dense, drooping, short peduncled racemes, 15–30 cm. Pod turgid linear with the ends curved in opposite directions, densely clothed with brown or grey, intensely irritating bristles. Seeds about 6, in a papery endocarp, black and shining. Flowers Sept.-Nov. Fruits Jan-April (Fig. 5.20).

Distribution: In dry forests from the base of Himalaya to SriLanka and Myanmar; along the riverain forests.

Uses: Various medicinal virtues are ascribed to the plant some probably imaginary.

Roots are purgative, prescribed as a remedy for delirium in fever; powdered and made into a paste applied to the body in dropsy.

Pods are anthelmintic. The hairs were at one time included in the British Pharmacopoeia.

Both the roots and the seeds are medicinal, the seeds are aphrodisiac and nervine tonic. An intoxicant liquor, "Khasuna" is prepared from the plant in Bihar (Palamau). The young pods are also eaten as a vegetable after the removal of hairs.

The cultivated varieties of this plant are:

Var. *utilis* Wall. (Vern. Mal Pah-Kursar) and is cultivated in Bihar.

Var. *nivea* DC. (Vern. Khamach-Bengal)

over the affected parts. Pod bristles are considered to be poisonous.

Mucuna monosperma DC. A large climber with glabrescent branches; leaflets elliptic, dark green with red nerves and scanty ferruginous hairs, shortly and suddenly cuspidate. Fls. large, purple. Pods subglose when young, single seeded. Seeds used in cough and asthma. Externally applied as sedative.

90. *Myrica asculenta Myrsine africana* Linn.

Family: Myrsinaceae.

Local names: Báibarang, Bebrang (Hindi).

Description: A small, erect, pubescent evergreen shrub. Leaves subsessile, lanceolate, sharply toothed; petiole ferruginous, pubescent, midrib prominent. Flowers small, nearly sessile in axillary clusters of 4–6. Anthers exceeding the corolla. Style short, stigma flat. Berry usually solitary, red, dark purple or black when ripe fully.

Distribution: In temperate regions from Trans-Indus to Nepal, 600–3000 m, very common and gregarious in shady oak forests, usually on the northern aspects.

Muree, Kishtwar, Hazara, Swat, Chamba, Shimla, Mussoorie, Lambatch, Ranikhet, Jogeshwar, Nepal, Mundali.

Uses: Decoction of the leaf is used as a blood purifier.

Gum from the plant is a remedy for dysmenorrhoea.

The fruit is anthelmintic (for tapeworm) and used often as substitute for that of *Embelia ribes*. It is also laxative in dropsy and colic.

After the childbirth, the mother is bathed in the water boiled with *Baibarang* seeds.

Extent of utilization and trade: The fruits are collected and sold in the market. The fruits are used as an adulterant for black pepper.

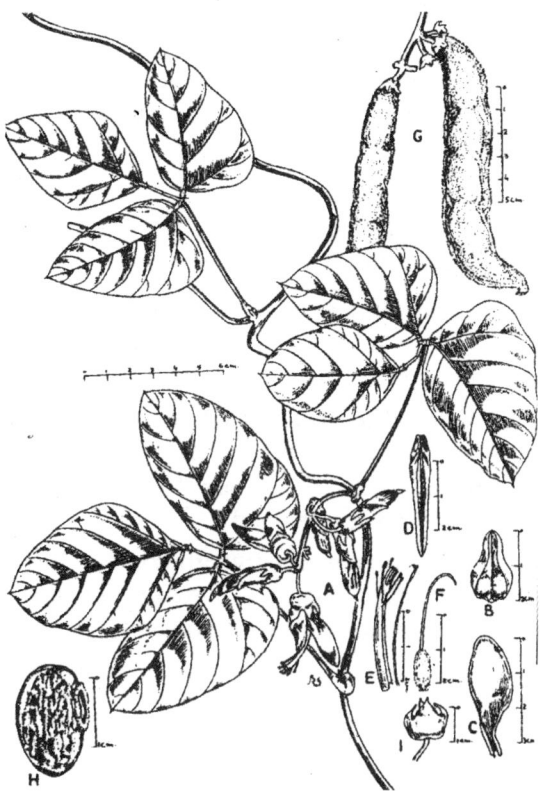

Fig. 5.20: *Mucuna pruriens* DC. A. flowering twig; B. standard; C. wing; D. keel; E. androecium; F. gynoecium; G. pods; H. seed; I. calyx.

Leaflets glabrescent beneath and flowers white. Cultivated in Jharkhand.

Var. capitata (-Mucuna capitata Wight and Arn.) Branches glabrescent, leaflets smaller than in *M. purrita,* thinly appressed hairy beneath, petiole often shorter than the leaflet. Racemes are few fld. short peduncled. Pods purple,12–15 cm long, bristles dense subsequently few and deep brown or blackish, when dried marked with faint, oblique grooves.

Other species: *Mucuna gigantea* DC. A littoral species on the costal areas in south in TamilNadu, Kerala and Andhra. The bark is used in rheumatic complaints. It is pulverized and mixed with dried ginger, rubbed

5

91. *Nerium indicum* Mill. (-Nerium odorum Solander)

Family: Apocynaceae.

Local names: Kanér (Hindi), Oleander (Eng.).

Description: A large shrub, much branched. Leaves in threes, entire, lanceolate, 12–15 cm long, leathery, dark green above, pale beneath. Flowers rose-coloured or white, odorous or inodorous, in large terminal corymbs. Calyx deeply divided. Corolla tube cone-shaped or tubuler, lobes spreading, often divided. Fruit with elongated follicles, linear, cylindric, 15–23 cm long; seeds numerous, covered with fine hairs (Fig. 5.21).

Fig. 5.21: *Nerium indicum* Mill. A. flowering twig; B. follicles.

Distribution: In western Himlayan region, central India and plains of Punjab, but cultivated throughout the country.

Uses: The Indian oleander does not grow to the height of a tree. Usually grown as an ornamental.

It is useful in all kinds of itching and cures ringworm.

The rootbark is ground in water and applied to skin which has become dark, hardened and thick.

Powdered leaves and flowers are used as snuff to cure headache, due to congestion. It can also be used as a rodenticide, for which bark and wood are used.

The plant contains two glucosides: 'nerin' and 'oleandrin'. Death has resulted from the use of the wood as meat skewers.

Cutaneous vermins are destroyed by application of leaf decoction.

A decoction of the leaves is used for reducing swelling. Fresh juice of young leaves is dropped in the eyes for inducing lacrimation in ophthalmia.

Roots contain two bitter principles: 'neridorin' and 'neriodorein', used as resolvent and attenuant. A paste made with water is applied locally to piles, ulcers on the penis, ringworm and other diseases. In decoction used as wash for reducing inflammatory swellings. Roots also used to procure abortion.

Oil from rootbark used locally in skin diseases, such as eczema, impetigo and leprosy.

92. *Nyctanthes arbor-tristis* L.

Family: Oleaceae (Some recent authors have placed it in Oleaceae, refer vide pp 80 Empire Forestry Review. 1954.)

Local names: Hársingar, Parijáttaka, Singhár (Hindi), Sorrowful Tree. (Eng.).

Description: A large deciduous shrub or small tree, roughly hairy all over; branchlets quadrangular; bark thick, pale brown. Leaves dark green, ovate, opposite and leathery, rough above with bulbous white hairs, paler and tomentose beneath. Flowers without stalks, sweet scented in clusters of 3–5, forming terminal cymes. Corolla white, tube orange. Capsules flat, roundish, 2-seeded (Fig. 5.22).

Distribution: In forests of MP and sub-Himalayan tracts and the Tarai regions.

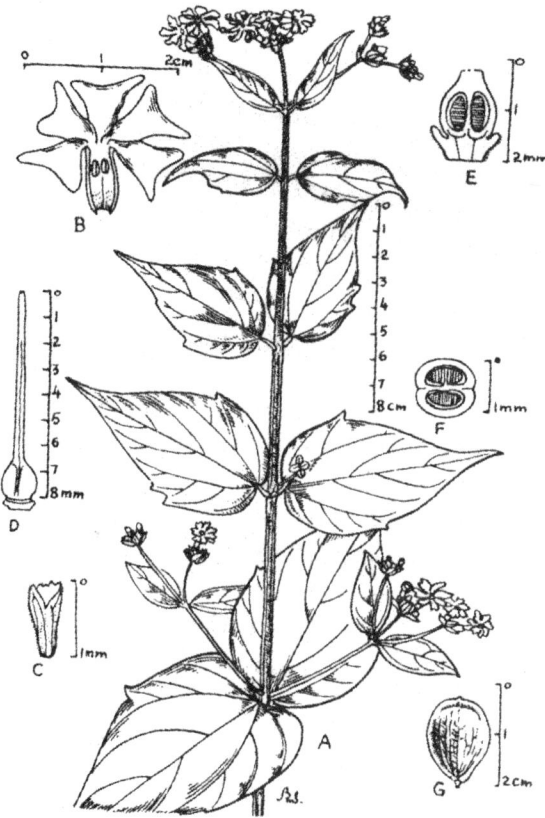

Fig. 5.22: *Nyctanthes arbor-tristis* L. A. flowering twig; B. corolla opened to show androecium; C. calyx; D. gynoecium; E. longitudinal section of gynoecium; F. transverse section of gynoecium; G. fruit.

Often grown in garden for flowers, gregarious on hot southern aspects.

Uses: The flowers open during night and shed early morning. Orange-coloured tube is rich in water soluble colour which is orange or golden yellow, often used as dye and often used to colour liquors.

Leaves are useful in fever and rheumatism. Fresh leaf juice with honey and ginger is used in chronic fevers. Expressed use of the juice acts as chloragogue, laxative and tonic and given to children to expel intestinal worms. These are also reputed to be effective in malaria, juice with honey and common salt is a useful vermifuge for expelling roundworms and thread worms. Leaf decoction prepared over slow fire is reported as a specific remedy for obstinate sciatica. The powdered seeds are used to cure severe affections of the scalp; it is antibilious and expectorant, useful in bilious fevers.

The bark is chloragogue and laxative. Bark is chewed with betel leaves (vern. *Pán*) to promote expectoration, also used for tanning. Leaves often used to polish wood in place of sand paper. *Patanjali* refers to cloth dyed of this flowers as *Sephalika*.

Extent of utilization and trade: Seeds are a source of commerce and available in market. Used medicinally on a limited scale. The flowers are sacred and also used as dye.

93. *Olea cuspidata* Wall.(O. ferruginea Royle)

Family: Oleaceae.

Local names: Káu (Jaun); Kahu (H).

Description: A middle-sized tree, bark thin, smooth when young, exfoliating in narrow strips when old. Leaves oblong lanceolate, cuspidate, entire, very coriaceous, dark green and shining above, thickly clothes beneath with a dense film of minute red scales, margins slightly incurved, midrib prominent. Flowers bisexual, whitish, in axillary trichotomous cymes. Calyx nearly truncate or with 4 short teeth. Corolla deeply divided, lobed, elliptic, obtuse or acute with a ridge along the middle, induplicate-valvate in bud. Anthers ovate, dehiscing laterally. Style short, stigma bifid. Drupe ovoid, black when ripe, supported by the remains of the calyx.

Distribution: Fairly common in the outer Himalaya, between 1000–2000 m.

Swat, Hazara, Kagan, Chamba, Kumaon, Jaunsar, Kashmir and central salt ranges of Pakistan.

Uses: The leaves and bark are bitter and astringent and used as an antiperiodic in fever and debility. Ashes of the root are useful in rheumatism and in brain diseases.

The fruit is tonic, appetizer, useful in lever complaints, thirst, burning of eyes, caries of teeth and toothache.

Oil from the fruit is rubefacient and has a bad taste. It is purgative, useful in gripping, liver troubles, pain in joints, rheumatism, lumbago, old wounds. From the green fruits the oil is astringent.

Extent of utilization and trade: The seeds are exploited commercially and sold in the market. The oil is also a market commodity.

Other species: *Olea glandulifera* Wall. (Gair, Gaild-H) is a moderate-sized tree, fairly common along the outer Himalayan tracts between 1000–2000 m from Kashmir to Nepal, common along the river banks and in shady ravines. The bark and the leaves are astringent and used as antiperiodic in fever.

94. *Osyris wightiana* Wall. ex. Wight (= O. arborea Wall. ex DC.)

Family: Santalaceae.

Local names: Popli (Mumbai), Tamparal (Kan.), Bakraja (Kumaon hills).

Description: An erect shrub, young shoots 3-angled. Leaves nearly sessile, oblong-ovate, tip acute. Flowers minute, yellow green, shortly stalked. Male flowers small, stalked, axillary clusters often forming short panicles. Female or 2-sexual flowers, solitary on axillary stalks. Style short, stigma recurved. Fruit globose, red when ripe.

Distribution: In sub-tropical regions of Himalaya, from Kangra eastwards to Bhutan, 300–2000 m, also in south Indian hills. Fairly common in open type of forests but especially in secondary miscellaneous forests. Sometimes also parasitic on the roots of other plants.

Uses: Infusion of the leaves is a powerful emetic.

Extent of utilization and trade: The plant is not exploited on a commercial basis.

95. *Periploca aphylla* Decne.

Family: Periplocaceae (Asclepiadaceae).

Local names: Buraye (Mumbai), Barri (Punj).

Description: An erect shrub, branched, commonly leafless. Bark greenish. Leaves when present thick, nerveless. Flowers fragrant, purple, many in cymes. Coronal lobes very long, glabrous. Follicles woody, on short peduncles, widely divergent.

Distribution: Throughout northwest India, extending westwards to Pakistan. (Punjab, Baluchistan and Sindh)

Uses: The milky juice is used for cough. The latex is combined with *Euphorbia nerifolia* juice and is a drastic purgative. The milky juice is also used externally for tumours.

Flowers in small doses are useful in the treatment of cold and asthma.

The rootbark is diaphoretic, expectorant and useful in dysentery and in the form of paste is applied in elephantiasis.

96. *Pergularia daemia* (Forsk.) Chiov. (P. extensa N, E. Br; Daemia extensa R.Br. ; Asclepias daemia Forsk).

Family: Asclepiadaceae.

Local names: Utran (Hindi); Gadaria-ki-Bél (Rajasthan).

Description: A twiner, juice milky with foetid smell. Leaves broadly ovate or suborbicular, acuminate, velvety pubescent beneath. Flowers greenish yellow or white, in lateral cymes, pedicels capillary. Sepals small, ovate. Petals pale-yellowish green and red. Column large, tips of inner coronal process often twisted together. Follicles reflexed lanceolate, beaked with soft spines.

Distribution: On rocky situations and on field hedges.

Uses: The plant is emetic and diuretic, given internally in asthma to cure rheumatic swelling.

Leaf juice is used as an expectorant in catarrhal affections, in infantile diarrhoea.

Rootbark is mixed in cow's milk and used as purgative in rheumatic diseases.

The plant contains glucoside and three steroids. It compares favourably with *Pitutrin* in action on uterus and produces the same intensity of contraction. It exerts a stimulating influence on smooth muscles of intestine and gastric sections are stimulated in increasing total acidity of the gastric juice. A generalised stimulating effect on the involuntary muscle, both plain and striated, produces a pronounced effect of circulatory system raising arterial blood pressure. The stimulation action is partly due to direct stimulating of the involuntary muscles and partly to stimulation of postganglionic exonergic nerves.

97. *Picrasma quassioides* Bennett.

Family: Simaroubaceae.

Local names: Báringi, Bháringi, Káshshing (H); Kárui, Tithai (Jaunar).

Description: A large deciduous shrub. Young shoots clothed with rusty pubescence. Twigs and branches smooth with numerous circular lenticels. Flowers green, unisexual, or polygamous, pale green in axillary panicles. Calyx 4–5 lobed. Petals 4–5, enlarged and coriaceous in fruit. Stamens 4–5, filaments with thick hairy base, anthers versatile. Fruit membranous drupe, black when ripe, each containing a seed.

Distribution: In the outer Himalaya, in shady ravines, 1500–2500 m. On the banks of the rivers, not very abundant. Rajpur near Dehradun, Jaunsar and Tehri Garhwal.

Uses: The bark of the stem and root is used as a febrifuge.

The leaves are applied to itch.

Extent of utilization and trade: The bark is used in the Indian system of medicine and exploited commercially on a very limited scale.

Other uses: The wood is of cream colour and close-grained. The bark has the same property as the quassiawood.

98. *Pimenta dioica* (L.) Merr. (P. officinalis Lindl.)

Family: Myrtaceae.

Local names: Pimento tree, All spice tree, Jamaica Pepper tree (Eng).

Discription: An evergreen bushy tree, 6–9 m high. Bark smooth, greyish, leaves

opposite, oblong-lanceolate, polished green, gland dotted (on the under surface). Flowers white or greenish white, fragrant. Fruit black or purple, 2-seeded berry, about the size of pea. The dried unripe fruit constitutes the pepper of Jamaica. The name originates from the resemblence of the spice in perfume and flavour to a combination of cinnamon, clove and nutmeg.

Distribution: Native of West Indies and tropical America. Also cultivated in Mexico for export of fruit, but the Jamaican product is more valued. In India the cultivation of this tree is only in the gardens, especially in west Bengal, Bihar and Orissa.

The tree flourishes best in hot dry climate, on poor, calcareous soils; heavy or clayey soils are not suitable. Propagation is by seeds. The tree begins to bear when it is about 7 years old but does not come into full bearing until 15–20 years of age. The yield from a full-grown tree is about 45 kg of dried fruits annually. It continues to bear well for a long time. However, some trees are sterile and do not bear fruit at all. The berries are picked when mature but still green and if allowed to ripen fully they lose some of their aroma. Harvested berries are carefully dried in sun for 4–10 days.

The dried berries have minute glands having volatile oil in the shells. The seeds are not so aromatic and have slightly nutty flavour.

Uses: The fuits are extensively used as condiment and as flavouring agent for meat, soups, ketchups, pickles, preserves, etc. and is an valuable ingredient of whole mixed pickling spice. It is also used in medicine as aromatic stimulant and carminative, resembling clove in action.

Oil from the fruits is used for flavouring all kinds of food products and in medicine as carminative also an adjuvant to aperient medicines. It is also used in perfumery.

Leaf oil is used as substitute for the berry oil.

Extent of utilization and trade: The extent of volatile oil present in the berry ranges from 3 to 4.5 percent. The odour and flavour of leaf oil is not comparable to that of berry oil and the yield are also low, 0.5–1.25 percent. Chief constiutent of the oil is *eugenol* (75–80%) which is also rich in the leaves (up to 97%). *P. racemosa* J.W. Mooro. (*P. acris Kostel*) powdered fruit used in dyspepsia, diarrhoea and in flatulence in W. Indies.

99. *Pinus roxburghii* Sarg. (P. longifolia Roxb.)

Family: Pinaceae.

Local names: Chir (H); Sháu (Kum.), Sarol, Sirli, Kulhain (Garh); Tellia (Ass); Saralagach (Beng); Chir (Punj), Simaidevadari (Tam), Long-leaved pine (E).

Description: A large, light loving tree with a clear straight bole. Bark thick outer corky and in thin crisp plates, reddish brown, inner brick red, vertically and spirally furrowed; branches rough, soon corky. Leaves in bundle of 3, obscured triquetrous, light green, sheath greyish brown. Male catkins ovoid cylindric. Cones solitary or 3–5 together, ovoid on short stalks, beak 4–6 gonous, reflexed, much thickened, generally with a sharp black tip. Seeds obliquely oblanceolate, compressed with a membranous wing, which is rather longer than the seed.

Distribution: It forms extensive forests which are remarkable for the absence of other tree species, though *Quercus incana* (oak) is frequently associated at higher levels. It is distributed throughout the subtropical zone on exposed hill slopes between 900–1800 m, sometimes also cultivated for ornament in the garden in the plains of North India, even at Delhi.

Uses: The drug oil of turpentine is obtained by purification from turpentine, an oleoresin obtained from this plant. The oil has a local irritant action. In controlled small doses, it acts as stimulant, expectorant and is useful in chronic bronchitis. It cures flatulent colic. It has limited use in typhoid, minor haemorrhages (such as from gums, nose, etc). Given as an enema, it cures constipation. Its commonest use is as a liniment in rheumatic pains. Inhaling the vapour of turpentine is useful in bronchitis. The resin, externally, is applied as a plaster to buboes and abscesses for suppuration. Internally, it acts chiefly on the mucous membrane the of genito-urinary organs.

In kumaon the resin is used in urine troubles and as a plaster in swelling, sprains, boils and on bone fractures. Pollens mixed with red-loam soil are used as a plaster in bone fractures. The wood is considered stimulant, diaphoretic and useful in cough, ulcers and cures insect stings.

Other uses: The timber is largely used for various purposes, e.g. house building, furniture, for chests, match industry, sports goods, musical instrument, etc.

1. *Pinus excelsa* Mc Clel. (*P. wallichiana* A.B. Jackson)Vern Kail. Five-leaved Pine. A large tree with leaves in bundles of 5, margin sparingly toothed, triquetrous, drooping, the foliage of a blue-green colour. Cones 2–3 together, long when ripe, cylinrical, erect before fertilisation, afterwards pendulous. The properties are the same as for *P. roxburghii*. The oil is used for external application in sores of animals. The plant is wild along the main Himalayan range, between 1500–4000 m. It grows generally as an associate of Deodar, for which species, if properly guarded against, it is an excellent nurse in early life. It is seldom gregarious and sometimes as low as 1300 m, e.g. in Giri valley. It is abundant in the inner dry ranges where it often occurs pure as a belt immediately below that of *Betula utilis*. The cone remains on branches, long after seed-shed.

2. *Pinus gerardiana*: A moderate-sized tree of inner dry and arid regions in north-west Himalaya, west of Bashahar, 2000–3300 m. The seeds of the plant are locally called *Chilgoza* and are sold in the market. Pinus khasya gives the most valuable resin.

100. *Pistacia integerrima* Stewart ex Brandis.

Family: Anacardiaceae

Local names: Kakrá-Singi (H); Kakrro (Tehri Garhwal), Kakkar (Punj), Kakra (Kum.)

Description: A middle-sized, deciduous tree with rough grey bark. Leaves alternate, pari or imparipinnate; leaflets 4–6 pairs, subopposite, lanceolate, long-acuminate, entire, hard, coriaceous, glabrous, main lateral nerves about 20 pairs, slender, base oblique. Inflorescence, a lateral panicle. Flowers small, dioecious, apetalous. Male flowers: panicles compact, pubescent, calyx gamosepalous, 3–5 fid; stamens 5–7 on a small disc, anthers large, red. Female flowers: panicles lax, sepals 4, free, linear, deciduous, ovary sessile, 1-celled, styles 3, cohering only near the base. Drupe oblique, broader than long, glabrous and rugose.

Distribution: On dry slopes and river valleys in the sub-tropical zone from salt ranges in Pakistan to Kumaon, in Uttaranchal ascending to 2000 m.

Uses: The galls which are in the shape of a horn and produced by a hemipterous insect are the drug.

These are tonic and expectorant. To cure dysentery these are given after frying with *ghee* and adding a little sugar. The drug has

a good reputation as a tonic, expectorant and useful in cough, phthisis, asthma, fever, want of appetite, irritability of stomach and diseases of respiratoy tract because of the presence of a fair amount of essential oil. It may be classed with terbinthinate astringents. In association with other drugs like *atis, pipal, nagar motha,* when boiled in water the extract is given in acute dysentery and in acute cases when even water could not be digested and retained in the body (tested on the author).

***Extent of utilization and trade*:** The galls are used extensively in Indian medicine and are collected by forest contractors.

***Phenology and time of collection*:** New leaves and flowers appear in March-May, while the fruits from June-Oct. The hollow galls are formed during the month of October.

***Other uses*:** The galls are used to a small extent in dyeing and tannin. Wood is used for furniture.

101. *Premna mucronata* Roxb.
(-P. latifolia var. mucronata Clarke)

***Family*:** Verbenaceae

***Local names*:** Bákar, Bosotá (Hindi)

***Description*:** Medium-sized or small, deciduous tree, branchces and young leaves pubescent or velvety. Leaves drying black, membranous, broadly ovate, sharply acuminate, usually quite entire, base cuneate, upper surface glabrous. When mature, with lower hairy especially on the midrib. Flowers in corymbs, usually terminating in short leafy branchlets, rusty pubescent. Calyx clothed with spreading hairs. Corolla green white, pubescent within. Fruit a globose drupe (Fig. 5.23).

***Distribution*:** In sub-Himalayan tracts, often on the outskirts of Sal (*Shorea robusta*)

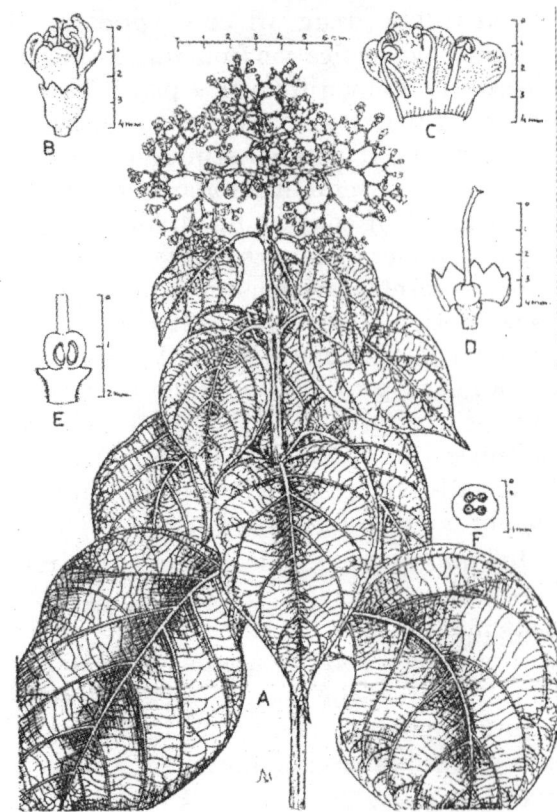

Fig. 5.23: *Premna latifolia* var. *mucronata* Cl. A. flowering twig; B. flower; C. corolla opened to show androecium; D. calyx opened to show gynoecium; E. longitudinal section of gynoecium; F. transverse section of gynoecium.

forests, in Siwalik ranges and in Gangetic plains. In some areas leaves remain green on drying.

***Uses*:** Juice of the bark is medicinal, usually given to cattle in colic.

The leaves are diuretic, given internally in dropsy. Milk of the bark is applied to boils.

Other species

a. *Premna tomentosa* Wild. (Vern. Name: Chambara Maharashtra, Kattutekha (Malayalam); Náguru Podanganari Narve, Ije.(Telugu).

A small to medium-sized tree, deciduous, bark grayish brown, leaves, inflorescence and branches densely clothed with soft tomentum. Leaves ovate, acuminate, entire, 12–20 cm long, Fls. yellow with 2-lipped calyx, which is leathery and hairy in throat.
Oil from the root is aromatic and used for stomach disorders.

b. *P. esculenta* Roxb. Grows in Assam, has leaves that are diuretic and externally applied in dropsy.

c. *P. herbacea Roxb*. Vern. Bhumjámbu (Sansk); Bharangi (Hindi) Bamanhati Bihar). The root preparations are given internally for rheumatism.

d. *P. integerifolia* L. Vern. Ganákashika (Sansk); Arni (Hindi) Root decoction given in liver complaints; plant used in neuralgia and in rheumatic complaints. Stem bark contains alkaloid *Premnine* which decreases force of action of heart and produces dilation of the pupil. Another alkaloid present is *ganiarine*. Leaf decoction is given for flatulence and in the form of soup used as carminative.

102. *Prinsepia utilis* Royle

Family: Rosaceae

Local names: L. Bhékal, Bhékor (Gar), Bhékoi, Bhék (Jaun); Jhataloo (Kum).

Description: A dark green spinous shrub, spines often leafbearing. Leaves alternate, simple, narrow, lanceolate, minutely toothed, long pointed. Flowers white, in short axillary racemes. Calyx cup-shaped, persistent. Petals orbicular, short clawed. Stamens numerous in many series; filaments short, anther cells separated by a broad connective. Carpel one, inserted at the base of calyx tube. Fruit a drupe, purple, obliquely oblong; cotyledons oily.

Distribution: In temperate regions from Hazara (in Pakistan) eastwards, 1300–2500 m in secondary scrub forest, especially in sunny open situations.

Uses: The fruits yield an oil, which is used as rubefacient and as an application in rheumatism and pains from over-fatigue. It is as relief also used for burning sensation.

Other uses: Branches of the shrub are supposed to be efficacious in doing away the spirits "chal". A toy-gun is made of hollow stem and unripe fruits are used as bullets by the local children.

Extent of utilization and trade: The fruits are collected locally and pressed for oil, it is not expoited on a commercial basis.

103. *Prunus cerasioides* D.Don

Family: Rosaceae.

Local names: Padmakha, Padmakshta (S); Padam (H), Phája (Jaun); Páyan (Kum) Panya (TG).

Description : A moderate-sized tree, bark brownish grey, peeling off in thin shining horizontal strips. Leaves ovate-lanceolate, long acuminate, closely doubly serrate, glabrous, shining, conduplicate in bud; petiole with 2–4 glands at the base. Flowers rose-coloured fading to white; peduncles in umbellate fascicles. Calyx glabrous, campanulate. Petals obovate or oblong. Stigma 3-lobed. Fruit a drupe, ovoid, obtuse at both ends, yellow or red. Stone rugose.

Distribution: In temperate regions from Indus eastwards to Bhutan, 1000–2000 m; mostly in the outskirts of the villages, apparently cultivated or run wild.

Uses : It is considered as a sacred-plant. Locally the oil from the wood is used as tonic. The bark ground into paste is applied as a sedative. The bark is used in the preparation of hair oil for massage. The paste of

the bark in Kumaon, is applied over the fore-head for hemicrania and is also used as a plaster for fractures and dislocations. Ash of the bark mixed with mustard oil is used externally for cuts, wounds and burns.

Phenology and time of collection: New leaves appear in May-June, Flower in April-May and sometime also in autumn. Fruits appear two months after the flowers. Stem and bark are collected during March-April. Stem is cut into pieces of convenient sizes and dried sun.

Other species: *Prunus cerasus* Linn. A moderate-sized tree, often cultivated in the Himalaya for its fruits. The oil from the seeds is used in rheumatic plains and also for burning.

Prunus persica Batsch. (Aru-H) A mode-rate-sized tree is cultivated. The seeds are pounded and an oil is extracted which is used to cure eczema.

Extent of utilization and trade (P. cerasioides): In Ayurvedic preparations Dasansausterker churna and Raktapi-tanashak churna are the main products in which it is used as an ingredient. A total of about 250 kg of bark reported to have been collected from the Kanatal Forests (near Mussoori) annually (Uniyal 1967).

104. *Punica granatum* Linn.

Family: Punicaceae

Local names: Dárim (Kumaon, Garhw-al, Jaunsar) Anár (H), Pomengranate (Eng).

Description: A small tree with pale brown bark. Leaves opposite, sessile, lan-ceolate. Flowers terminal, red, solitary in twos or threes. Sepals red. Petals 5 or 7. Fruit globular with persistent calyx. Seeds many (Colour Plate 5.7).

Distribution: The plant runs wild in val-leys below 2000 m in Jaunsar and Tehri Garhwal. Cultivated in many places.

Uses: The rootbark is specific in cases of tapeworm as astringent, it is often used as a decoction. This is also effective as febrifuge and for treatment of leukorrhoea, passive haemorrhages and wasting disease of children. It is also useful as mouthwash, in relaxed sore throat and as dressing for ulcer, of uterus or rectum.

Externally, the juice of the leaves and flowers is used as styptic for checking bleeding from nose. Paste of the green leaves is also used as a cure of conjunctivitis, decoction used locally as an eyewash and mouthwash.

The flowerbuds called *Gulé nar* are astringent and tonic for treatment of leuko-rrhoea and chronic diarrhoea. Strained with sugar the flowerbuds when ground in water are given to prevent abortion, in complaints of habitual miscarriage. With juice of "Dub" leaves (*Cyanodon dactylon*), it is used for stopping bleeding from the nose. Powder of the flower buds is used as vermifuge and as snuff for checking nasal haemorrhage. Infusion is useful in dysentery and inflam-mation of the throat.

The fruit is nutritious containing protein, sugar, iron and phosphorus valued as stomachic and refrigerant. Its juice with saffron is given in dyspepsia and fever to allay thirst. *Sherbat é anar*, Syrup of the fruit is given in typhus, gastric and asthmatic fever, inflammation of urinary tract and haemorrhages. Rind of the fruit is astringent and used as a remedy for diarrhoea and chronic dysentery for children, mucous discharges, prolapsus of rectum and uterus. In fevers, due to sanguine and bilious nature it quenches thirst, reduces fever and maintains energy. Rind of the fruit, wood bark and seeds are used in cough. It is also prescribed with considerable benefit in profuse perspirations accompanying the last stages of phthisis, pulmonatis and in

diarrhoea which is so distressing a symptom towards the close of the disease. It cures diarrhoea and dysentery (when rind is ground with water). Decoction is used as mouthwash, strengthens loose gums the gargle cures uvula.

The seeds are cooling and given in fever. Paste made with milk is used on patients suffering form stone or gravel in kidney during pregnancy. In Kumaon crushed seeds are employed on pimples, by the natives. A cooling *Sherbat* is made in Unani system and sold in market. Dried seeds are used in *Churan* (digestive powders) and in *chutneys*, also as an ingredient of locally used spices.

Extent of utilization and trade: The rootbark contains *punicin*, which is highly toxic to tapeworm. The quality of the plant varies in different localities. In Bengal it is considered to be inferior than in north-west India. The plant is used extensively on a commercial scale. Many varieties are cultivated, the wild variety called *Darim* occurs in West Himalaya.

Other uses: The seeds of the plant are dried and sold as *Anardana* for making chutneys. The extract of the seeds is made into a syrup and preserved as a refrigerating drink. In places where the tree is abundant, the fruit rind is employed for tanning leather.

105. *Pyrus pashia* Buch-Ham.

Family: Malaceae (Rosaceae).

Local names: Méhal, Mélu (H), Garhmehau (Kum); Kaint, Kainth (Jaun), Mohalm Mehal, Meyel, Mahol (Garh).

Description: A moderate-sized deciduous tree, bark on old stems almost black, both longitudinally and transversely split into small, thick, rectangular scales, on branches occur small white dots. Leaves simple, long acuminate, lobed and often woolly beneath, lateral nerves 8–10 pairs; stipules setaceous, early deciduous. Inflorescence, a corymb or umbel. Flowers white, appearing shortly before the leaves. Fruit globose, dark brown, covered with raised white dots.

Distribution: In sub-tropical and temperate regions up to 2500 m, also found in outer Himalayan ranges.

Uses: The fruit is given externally in corneal ulcers of cattle. In Kumaon the juice of the leaves is used in eye diseases.

Extent of utilization and trade: The drug is of local significance and not used on a large scale.

Phenology and time of collection: New leaves and flowers March-April. Fruit Sept Dec.

Other uses: The wood is used for walking sticks, combs, tobacco pipes and such other purposes. The leaves and twigs are lopped for fodder. The fruit is eaten when half rotten.

106. *Pterocarpus marsupium* Roxb.

Family: Papilionaceae.

Local names: Piásel, Bijásál (H); Biblá (Guj); Honné (Kan); Kino (Mal); Honi (Marh); Mahakutaj (S); The Indian Kino tree (E). The trade name is based on the gum *Kino* exuded from the bark of the tree.

Description: A large handsome tree. Leaves compound, having 5–7 leaflets; leaflets 8–13 cm long, oblong or elliptic. Flowers about 1.5 cm long, yellow, in very large, dense bunches. Fruit 2.5 cm long, roundish, winged, with one seed.

Distribution: In mixed-deciduous forest of Central and Peninsular India, ascending to about 1000 m in Gujarat and Madhya Pradesh.

Uses: Leaves, flowers and gum of the tree are medicinal. The water in which a block of the wood is soaked overnight is believed to be useful for diabetic patient. In a clinical experiment about 7% of the diabetic patients, treated with this drug, showed improvement.

The bruised leaves are applied on boils, sores and other skin diseases.

The gum called *Kino*, exudes from incisions in the bark, is astringent and useful in diarrhoea, also used for toothache. The gum is also considered to be useful in fevers and urinary discharges.

Extent of utilization and trade: The sum *kino* is a commodity of commerce and exploited on a commercial scale. However, the plant is much utilized for its wood which is valued for high class furniture and construction work. The containers from the wood are commercially available as *Magic glass* used by diabetics.

Other species: *Pterocarpus santalinus* L.f (Lálchandan or Red Sander's wood) is a tree having a restricted distribution. The heart wood is used externally as cooling for headache and inflammation. The wood is also used in decoction mixed with other drugs. The water with wood dipped overnight is used to lower the blood pressure. The wood is valuable for making musical instruments and largely exported.

107. *Populus ciliata* Wall. ex Royle

Family: Salicaceae

Local names: Pahárai-pipal (H); Biaon, Sharphara, Tilaunju, Kapási (Jaun).

Description: A loft tree, bark grey smooth on young parts with vertical wrinkles or fissures on old stem, leaf buds sticky. Leaves broad-ovate, acuminate, finely dentate, margins gland ciliate, base usually cordate, 3–5 nerved, lateral nerves 4–6 pairs above the basal, compressed above. Flowers appear before the leaves, in lateral catkins. Male catkins somewhat interrupted, female lax in fruit. Capsule ovoid, opening by 3–4 valves.

Distribution: In temperate regions from trans-Indus to Bhutan, 2000–3300 m. In Jaunsar and Tehri-Garhwal it is between 1300–3300 m, sometimes cultivated. Affects sunny blanks in the forest with a light soil. Male trees are very scarce. In Kumaon it is fairly common on the inner ranges usually bordering streams and occasionally forming small gregarious patches on the banks of larger water courses. The plant is often found on the shady hill sides and sometimes associated with bluepine and silverfir.

Uses: The bark of the plant is used as a tonic, stimulant and blood purifier.

Extent of utilization and trade: The plant is not exploited very much commercially.

Phenology and time of collection: Leafless during Oct-February and flowers from March-April. The male flowers earlier than the female flowers. It fruits in May-June.

108. *Ricinus communis* L.

Family: Euphorbiaceae

Local names: Erandá (Sansk), Bheréndá (Bihar); Eréndi (Mumbai); Eri (Asam); Arandi (Hindi); Mandá (Kan), Erándam (Mal.) Erándamu (Tam); Arandá (Beng); Diveli (Guj); Amidamu (Tél.) The Castor oil plant (Eng).

Description: A well-known tree having large leaves divided into 5-parts, like a big palm of the hand; about 30 cm in diam. Flowers unisexual, male and female on the same plant, in terminal racemes; female flowers above the male; petals absent. Male calyx 3–5 alved, stamens numerous. Female

calyx spathaceous; styles red or yellow. Capsules 3-seeded, 1–2 cm long, globosely oblong, generally softly spiny, mottled grey with brown purple streaks (Fig. 5.24).

Fig. 5.24: *Ricinus communis* L. A. twig; B. inflorescence; C. branching stamens; D. female flower; E. transverse section of ovary; F. longitudinal section of ovary; G. infructescence.

Distribution: Northwest Himalaya from Jhelum eastwards, Khasi hills (600–2700 m); on waste lands; sometimes self-sown on lands left out in habitations.

Uses: The leaves, kernel of the seeds and castor oil are used in medicine.

The leaves resolve swellings and allay pains; leaves are smeared with sesame oil or cooked like a vegetable and applied warm to swollen parts in cases of swelling of neck or rheumatism. The leaves have also antidotal properties against the poison of opium, aconite, etc. The leaves are crushed, juice expressed and given to drink. Vomiting and diarrhoea will follow and the poison is expelled. The leaves also relieve hiccup when dry leaves are smoked in place of tobacco (smoke should not be swallowed).

Kernel of the castor seeds is a purgative and much more efficaceous than castor oil. Ingestion of 4–5 kernels purges and cures diseases like paralysis, rheumatism, facial paralysis, cough and asthma.

The kernels made as *Halwa* (pudding) used in cases of paralysis, rheumatism and asthma. The kernals (about 10) soaked in water overnight, grinded and boiled in milk if taken for a month, cause purging but cured rheumatism (the prescription has been tried on the author). However, one must take care that the patient does not suffer from heart ailment.

Castor oil, a well-known purgative, is administered equally to children, adults or old people. In case of scybalous dysentery, the oil expels the scybala. If lime were to enter the eyes, one or two drops of castor oil introduced into the eye give immediate relief. This oil also relieves headache. Alkaloid ricinine, toxalbumin ricinin have been reported from the oil which is 45–50% of fixed oil of the beans.

109. *Rhododendron arboreum* Smith

Family: Ericaceae.

Local names: Buraûs (Garh.and Kum.) The Rhododendron tree (E).

Description: A large tree; bark reddish brown, peeling off in small flakes. Leaves oblong-lanceolate, crowded at the end of branches, acute at both ends, coriaceous, glabrous above, rusty tomentose or covered with small silvery scales beneath, nerves and midrib prominent beneath, depressed

above; buds viscous. Flowers red or pink, in corymbose fascicles, at the end of branches. Capsule cylindrical, curved, longitudinally ribbed (Colour Plate 5.8).

Distribution: In temperate regions from Indus eastwards, 2000–3000 m almost always associated with *Lyonia ovalifolia* (Pieris ovalifolia) and *Quercus incana.* (*Q. leueotrichophora*).

Uses: The flowers are used with common salt, as sauce. They are said to be used in the preparation of some expectorant and astringent and are believed to be good for dysentery ; applied as poultice in headaches. Locally in Kumaon, they are used in dysentery and are taken particularly when the fish cartilage stucks in the throat.

Extent of utilization and trade: The flowers are collected and marketed to Bombay during the month of April. About 5 quintals are reported to have been collected only from Kantal forests in Tehri Garhwal during 1963–64 (Uniyal & Issar 1967). Burans syrup is being prepared as a cottage industry product and is good as expectorant, astringent and cardiac tonic.

Phenology and time of collection: Fls. March-May, occasionally again in July and August, if the spring flowering has been checked by drought hailstorm, etc. it fruits, in autumn and cold season.

Other species: *Rhododendron campanulatum* D.Don (Simris TG) is an evergreen shrub with elliptic-oblong leaves, rounded at both ends, crowded at the end of branches. Flowers whitish pink purple or liliac in lax terminal corymbs. Distributed above 3000 m in the Himalaya, above the timber line generally. The leaves are poisonous to goats and when mixed with tobacco these are made into a medicinal snuff useful in cold and hemicrania. These are also used in chronic rheumatism, syphilis and sciatica.

Dried twigs and wood is used in phthisis and chronic fevers.

Large seale collection of Burans flowers might threaten the regeneration of this plant in nature.

110. *Quercus leucotrichophora*
Bahadur (Quercus incana Roxb.)

Family: Cupuliferae (Fagaceae).

Local names: Ban (Jaun), Banj (Garh), The Himalayan Oak (Eng.).

Description: A large evergreen tree, bark dark grey, rough with cracks and fissures. Leaves stalked, ovate-lanceolate, spinous-toothed towards the tip, often entire on old plants, coriaceous, glabrous and dark green above, generally rusty tomentose below, lateral nerves 6–12 pairs. Male spike crowded and congested in seed years, interrupted in other years, perianth segments obtuse. Stamens indefinite, anthers apiculate. Female spike short, at the tip of current years shoot. Style long, recurved. Acorn single or in pairs at first, covering the whole nut but when mature it covers half.

Distribution: In temperate regions from Trans-Indus to Nepal, 13000–2500 m. Common and usually gregarious, often found on hot aspects and more conspicuous on the outer than on inner ranges. Often found growing with *chirpine*, as an undergrowth towards the upperlimit of chirpine, less often it is found with bluepine and rarely with Deodar; chief associates are *Lyonia ovalifolia* and *Rhododendron arboreum*. The regeneration is usually bad, especially when oak forms a large portion of the crop.

Uses: The acorns are used medicinally and are diuretic in gonorrhoea and also an astringent in indigestion, diarrhoea, of especially children. It is also used in asthma. Before being administered they are usually

buried in the earth to remove their bitter principle, then washed and ground.

Extent of utilization and trade: The acorns are widely collected and sold in the market. Acorns are widely collected and sold in the market.

Phenology and time of collection: New leaves and flower appear from May-June. The female flower appears with the male flowers and becomes fertilized by their pollen but the acorns begin to grow only in the next spring, often before the male flowers of the season begin to open. By the first week of June, the young acorns are about the size of pea, while the flowers (which produce acorns next year) are then just visible. The fruits are formed just 15 months after flowering.

Other species: The excrescences on *Querucs lusitanica* (a plant found in Greece, Asia minor, Syria extending to Persia) caused by an insect are used medicinally either in the form of a powder or as ointment.

Other uses: *Q. dilatata, Q. semercapifolia* and *Q. incana* provide wood which is used for charcoal and for agricultural implements. The bark of some oaks is also used for tanning. The leaves provide fodder and are invariably lopped for it, thus creating blanks in the forests, where grasslands form the stable communities.

The oaks are getting over-exploited, thus affecting the under and overground water availability. These have been exploited at the expense of oaks. Conservation of oak forests for water availability needs to be emphasized and propagated for better moisture conservation of the Himalayan region.

111. *Randia dumetorum* Lam.
(-Gardenia spinosa L.f., R. spinosa Poir.)

Family: Rubiaceae

Local names: Mainphal, Méndphal, Maindal (Hindi), Madana (Sansk), Geláphal (Mumbai); Mangára (Telgu).

Description: A large rigid shrub or small tree with long opposite axillary spreading spines. Leaves usually fascicled obovate, obtuse or subacute, 3–5 cm narrowed into a short marginate petiole, pubescent or hispid, rarely glabrous, deciduous; stipules ovate acuminate. Flowers at the end of short leaf-bearing branchlets, solitary or in pairs, greenish yellow or almost white, fragrant. Calyx tube campanulate, strigose. Corolla as long as the calyx lobes, hairy outside, lobes spreading. Berry yellow when ripe often ribbed. Seeds many, compressed, embedded in pulp.

Distribution: Sub-Himalayan tracts, Siwalik hills and forests of Dehradun, sub-tropical Himalaya from Jammu to Sikkim up to 1300 m.

Uses: The fruit is eaten when ripe, either raw or cooked; pulp of unripe fruit is fish poison and is irritating and emetic. Fruit pulp used in dysentery, as anthelmintic, abortifacient. In powder form it is applied to the tongue and palate for fevers and incidental ailments of children during the teething period.

The bark is given internally and applied externally as anodyne in rheumatism when bones ache during fevers.

The aqueous extract of rootbark is insecticidal. Fruits are also reported to contain neutral and acid saponin as the active constituent and lead in seeds.

Other species

a. *Randia densiflora* Benth. A plant of eastern region the Assam, Nagahills, Kerala and Andamans. The bark is bitter and used in fever, also called forest fever in China.

b. *Randia ulginosa* DC. (Vern Pindálu-Hindi; Gangali-Sansk., Pendari (Mumbai). A shrub of eastern, western, central and south Indian forests. The roots are given in dysentery and diarrhoea, while the roasted fruits prescribed in dysentery and diarrhoea. The seeds as astringent and usually rejected with the fruit.

c. *Randia spinosa* Poir is shown in Fig. 5.25.

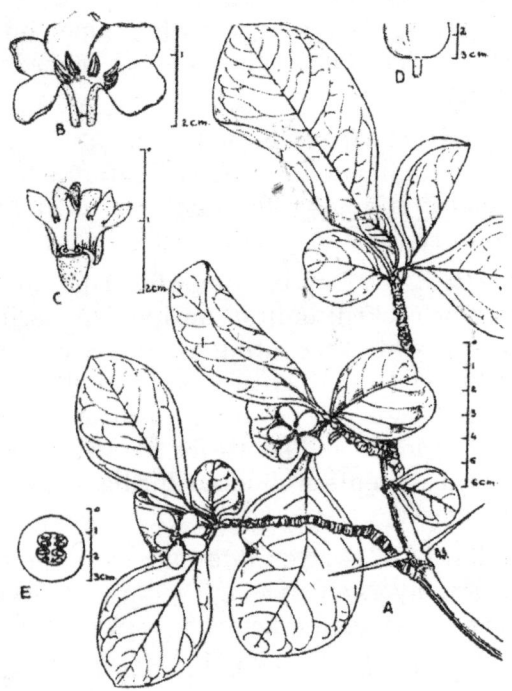

Fig. 5.25: *Randia spinosa* Poir. A. flowering twig; B. corolla opened to show androecium; C. corolla removed and calyx opened to show gynoecium; D. fruit; E. transverse section of fruit.

112. *Rhamnus purpureus* Edgew.

Family: Rhamnaceae.

Local names: Lhith, Luhish (H).

Description: A large deciduous shrub with spreading branches. Twigs with numerous pale-lenticels, emiting foetid smell when bruised. Leaves alternate, elliptic, serrate with long petiole. Flowers greenish purple, 5-merous in axillary clusters. Fruit pink, turning to black when fully ripe.

Distribution: In temperate regions from Murree (in Pakistan) to Nepal, 2000–3000 m, fairly common in moist ravines and in shady oak forests on open and shady aspects.

Uses: The fruit is used locally as purgative.

Extent of utilization and trade: The plant is not exploited as an article of commerce.

Phenology and time of collection: Fl: April-May Fr. July-oct. The flowers appear with the new leaves.

Other species: *Rhamnus virgatus* Roxb. (*Rhamnus dahuricus* sensu Lawson) is a deciduous, spinous shrub, spines at the fork of the branches. Leaves are opposite, sometimes in clusters, near the end of branches. Flowers 4-merous. Drupe globose. In temperate regions from Indus eastwards. The fruit which is bitter when ripe has emetic and purgative properties and is used in affections of spleen. Often the fruits and flowers are present throughout the year, but generally Flowers in: March-June Fr. June-Oct.

Rhamnus triquetra Wall. ex Roxb. is shown in Fig. 5.26.

113. *Rhus wallichii* Hook. f. (syn. R. vernicifera DC. in part)

Family: Anacardiaceae.

Local names: Akoria, Arkhar, Arkhoi, Arkol, Ambal (Hindi).

Description: A small or middle-sized tree, deciduous, with smooth grey bark, rusty tomentose, exuding a black acrid

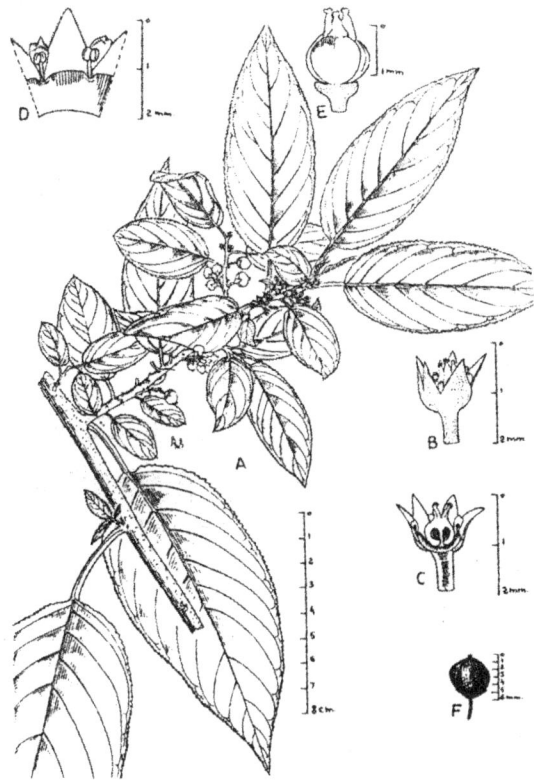

Fig. 5.26: *Rhamnus triquetra* Wall. ex. Roxb. A. flowering twig; B. flower; C. same in longitudinal section; D. portion of the flower opened to show two of the petals and stamens and part of the calyx; E. gynoecium; F. fruit.

varnish; young parts covered with rusty brown tomentum. Leaves almost near the end of branches, 30–45 cm long, odd-pinnate; leaflets 7–11 pairs. Flowers almost without stalks, in axillary panicles which are compact from the axils of the dark veins, much larger than the sepals. Drupes in compact, pyramidal panicles, hairy when young, glabrous when ripe. Rind dry, papery. Seeds smooth, hard and embedded in the mass of pulp.

Distribution: West Himalaya, from Kashmir to Nepal, 1300–2300 m in hot and dry localities.

Uses: The juice of the leaves is corossive and causes blisters. People avoid touching it or even going near it.

Other species: *Rhus parviflora* Roxb. (vern Dansria-Rajasthan).

A cooling agent in fever and useful in vomiting. Fruits mixed with salt act as tamarind.

Rhus chinensis Mill. (Tatri-Hindi). The galls are astringent and expectorant; employed in preparation of ointment, used in swellings and wounds. Fruits are used in treatment of colic and dysentery.

Rhus punjabensis J.L. Stew. ex Brand (Vern. Titri). Fruits are edible and used in preparation of syrups (sherbets).

Rhus succedanea L. (Japanese wax-Eng); Kakrásingi-Hindi. A shrub or tree. Mesocarp of the fruit is edible. Galls are astringent and expectorant.

114. *Salvadora oleoides* Decne.

Family: Salvadoraceae.

Local names: Pilu, Barápilu, Jhál (Hindi).

Description: A large, much-branched, evergreen shrub or small tree, usually twisted or the trunk is bent. Branches stiff, whitish. Leaves narrowly lanceolate, acute. Flowers greenish white, in erect, axillary panicled spike. Drupes clustered subsessile, often touching each other.

Distribution: A common plant of arid tracts of Punjab, Haryana and Rajasthan, in the trans-Indus region it ascends to 1000 m while in Salt range up to 800 m; also westwards to Sindh, Baluchistan and to Aden.

(Gupta RK and Saxena SK 1968. Trop. Ecology **9(2)** 140–152)

Uses: Leaves are used as purgative and as a cure to relieve cough. The fruit is aphrodisiac and sweet in taste. Oil from the seeds is used as stimulant for application in painful rheumatic affections and after childbirth. Fruits are edible and taste like currents.

A good source for non-edible oil for which it has some potential for exploitaion. However, that the scattered trees and sporadic fruiting which do not mature simultaneously make seed collection a bit costly.

Other species

Salvadora persica L. (Vern. Pilu-Chhota, Jhal, Kharjap (Hindi), Mustard tree (Eng.).

A much-branched evergreen tree or a shrub; bark whitish with yellowish wood, branches drooping. Leaves somewhat fleshy, ovate or oblong. Flowers greenish yellow, pedicelled in lax panicles. Calyx lobes ovate. Petals almost 5-partite. Fruits scattered, drupe.

Fruits and leaves are eaten, pungent like that of Mustard and Cress. Leaves used in rheumatism externally, and internally given in scurvy. Decoction of root bark used as vesicant. Fruits are carminative, diuretic and deobstruent.

Decoction of the root is given against gonorrhoea and vesical catarrh.

Leaf decoction is used in asthma and cough, poultice is applied to painful tumours and piles.

The fruit is carminative, diuretic and deobstruent.

The twigs consist of cellulose fibre, 10% of salts, mainly sodium and potassium chlorides and sulphates in addition to calcium oxalate and traces of ethereal oil with small quantity of aromatic resin.

Oil from the fruit is of commercial value and used in soap manufacture.

Stembark is given as decoction in fever and as stimulant and tonic in amenorrhoea.

115. *Santalum album* Linn.

Family: Santalaceae.

Local names: Chandan, Sandal (H); Sukhad (Guj, Beng), Malayaja, Bhogivallabha (S); Agarugandha (Kan); Chandan (Tam, Tel); Sandalwood tree (Eng.).

The trade name Sandalwood and *Chandan* is based on local names.

Description: A moderate-sized evergreen tree. Branchces almost drooping; bark dark, rough with vertical cracks. Leaves opposite, shining on upper surface, 4–7 cm long. Flowers small, dull purplish in small bunches. Fruit roundish, purple black, succulent.

Distribution: In Deccan peninsula, in forests of T.Nadu, Karnataka, Andhra and Kerala, much of the forests are over-exploited either by the official agencies or by intruders.

Uses: The wood is ground up with water into a paste and applied on local inflammations, on forehead in fever and skin diseases.

Oil from the heartwood is used in treatment of dysuria (to promote and facilitate urination, cystitis, inflammation of bladder, gonorrhoea and cough). The drug is useful in tuberculosis of gallbadder. Oil from the seeds is used in skin diseases.

Exent of utilization and trade: The wood is largely utilized on an extensive scale and the oil is extracted particularly in Karnataka and Kerala States.

Other uses: The wood is used for making small domestic items, on account of the perfume which lasts for a long time. Wood powder is used for making *Agarbatti* and an incense powder. Oil is used for perfuming soaps and other toilet preparations. Paste

of the wood is used by 'Pundit' on the forehead.

Sandalwood is used for the pyre of selected people.

116. *Sapindus trifoliatus* Linn.

Family: Sapindaceae.

Local names: Areetha, Bará-ritha, Ritha (H); The Soapnut tree (Eng.).

Description: A large deciduous tree. Leaves alternate; leaflets 2–3 pairs, elliptic. Flowers in terminal, rusty pubescent panicles, unisexual and bisexual flowers on the same plant, white, hairy. Fruit fleshy, rusty tomentoes saponaceous, outer skin wrinkled and darkish yellow with black-stone containing white kernel inside.

Distribution: Common in evergreen forests of Konkan and Kanara, along the western ghats.

Uses: The fruits are useful in hysteria and melancholia, when given in the form of fumigation. Kernel of the seeds stimulate uterus in childbirth. In small dose, pulp is given by mouth for various conditions and has no injurious effect on stomach. In large doses it is a powerful gastro-intestinal irritant. It is useful in colic. Pulp of the fruit in small doses is anthelmintic.

The root is said to be expectorant and is administered internally. The rind of fruit is ground, stored and used as antidote for snake poison. Powder mixed with water and drunk to induce vomiting and diarrhoea. Also applied externally to wounds.

The rind of soapnut ground in water is applied to face for removing blemishes and improving complexion. Hemicrania is relieved when nut ribbed in water is dropped into the nose.

In case of facial paralysis rind of soapnut is powdered along with jaggery and made into pills of the size of a berry and taken twice daily after eating *Halwa* (pudding). Stone of soapnut is aphrodisiac, finely grounded powder is mixed with sugar and taken along with milk. The kernel also cures infantile tuberculosis when kernels from stones are triturated with black goat milk.

Extent of utilization and trade: The fruits are exploited on a large scale and sold in the market.

117. *Skimmia arborescens* T. Anders. ex Gamble (-S. laureola Hook.f in part non Sieb.non Sieb. and Zucc.

Family: Rutaceae.

Local names: Guralpattá, kédarpattá, Kathurchurá (Hindi).

Description: An aromatic, small, evergreen shrub, with white bark. Leaves alternate, simple, oblong-oblanceolate or obovate, alternate, coriaceous, rather succulent, gland dotted and crowded at the end of branches. Midrib stout but the nerves indistinct. Flowers polygamous, yellow, in compact terminal panicles. Calyx 4–5 lobed. Petals 4–5, oblong, valvate. Stamens 4–5, filaments stout. Ovary 2–5-celled; style 1. Fruit fleshy drupe, red when ripe, with 2–3 one-seeded stones.

Distribution: In shady locations: in dense gregarious patches, 2000–3300 m. The odour of the musk-dear, *Kastura,* is popularly supposed to be derived from it (Smythies A).

Uses: The entire plant except the roots is used as an insect repellent.

The leaves are regarded as sacred and kept as souvenir on the head, while returning from the pilgrimage of Kedarnath-dham.

The smoke of the burning leaves purifies the air and is used as incense.

During March-April, aerial parts of the tree are collected for its essential oil, which

is obtained from steam distillation of the leaves.

It contains mainly linayl acetate, a Linolol, unidentified hydrocarbons and a complex mixture of sesquiterpens, alcohol and esters.

Leaf availability of the tree can be judged from the fact that in 1997, Uniyal and Issar reported about 10 quintals of leaves to have been collected from the Kanatal forests and Mussorie-Chamba forests in Uttaranchal state.

118. *Stephania glabra* (Roxb.) Miers.

Family: Menispermaceae.

Local names: Purha (D. Dn), Gindaru (Garh.), Nimilahara (Nepal).

Description: A climbing shrub. Roots bulbous. Twigs ribbed. Leaves broad-ovate or orbicular, often sinuate, thin, pale beneath, peltate. Umbels axillary. Flowers yellowish green. Petals shorter and broader than the sepals. Drupe red, pisiform.

Distribution: In temperate regions from Sutlej eastwards, ascending to 2000 m.

Uses: In Kumaon the root sap is used for massage in headache and bodyache. Ash of the root is also used in eye troubles.

Extent of utilization and trade: The drug is not exploited on a commercial scale.

119. *Sterculia urens* Roxb.

Family: Sterculiaceae.

Local names: Kulu, Gulu, Karái (H); Kandol, Karáyo (Guj); Kulu (Punj); Bhutáli (Kanara); Tonti (Malyalam); Grindola (Or); Vellaipputtali (Tam); Ponaku (Tel), Telhhech- (Santal Pargans).

The trade name *Karáyà* is based on the local name.

Description: A middle-sized deciduous tree; bark white or greenish white, shining, papery smooth. Leaves are crowded at the end of branches, shallowly 5-lobed, very densely hairy on undersurface. Flowers yellowish brown, small, in large, very densely hairy, erect bunches. Fruit of 4–5, thick, red, carpels, densely covered with stinging hairs.

Distribution: Tropical region of the Himalaya, central and Southern India, usually in dry or mixed deciduous forests.

Uses: The gum from the stem has medicinal properties and is called *Karáyá* gum or *Kadirá* gum. It develops an odour of vinager when kept in closed bottles. It is laxative and is used as a substitute for *Tragacanth* gum in throat affections and in dental fixture powders, lozenges and pastes. The mucilage when applied to skin, has a softer action than even 'Tragacanth gum'.

Leaves and tender shoots ,steeped in water, yield a mucilaginous extract, used in pleuropneumonia of cattle. *Tribals of Bastar* are reported to pound the bark of the tree and give it to women before childbirth; this is believed to facilitate delivery.

Santhals of Netrahat plateau in Palaman district use it for throat affection.

Other uses: The wood is used for miscellaneous domestic purposes such as doors, boats, carrings, packing cases, etc. The bark yields a very strong fibre used for ropes, cloth, etc. The seeds are edible and so is the gum in curries. The bark gives fibre. Fruits are pounded, seeds separated and cooked as vegetable by *Baster tribals*.

Other species: *Sterculia villosa* Roxb. (Udál-H) occurring almost throughout India. *Sterculia foetida* L. (Jangli-Bádám), a large tree of westcoast occurring in peninsular India, yields products of medicinal value. Flowers are remarkable for disagreeable odour. Gum resembles to that of Tragacanth

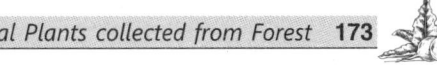

extracted from the seeds when boiled. Oil extracted from the seeds of *S. foetida* is laxative.

120. *Spermadictyon suaveolens* Roxb. (Hamiltonia suaveolens)

Family: Rubiaceae.

Local names: Padéra (Garhwal).

Description: An erect shrub with divaricate branches. Leaves opposite, elliptic-oblong or ovate, smelling foetid when crushed, roughly pubescent; stipules intra-petiolar, short, persistent flowers bluish purple, in small bracteate-head like clusters at the end of short forking branches, forming terminal pubescent panicles. Capsule ellipsoid, 1-celled, 5-seeded (Fig. 5.27).

Distribution: In the sub-Himalayan tract and outer ranges from Indus eastwards, ascending to 2000 m. Very common in dry miscellaneous forests, often on hot-southern aspects and rocky ground, frequently associated with *chirpine*. Young leaves have a foetid smell which disappears almost or quite when the leaves mature.

Uses: Infusion of the roots is useful to cure curvature of bone.

Extent of utilization and trade: Not exploited on a commercial scale.

121. *Strychnos nux-vomica* L.

Family: Loganiaceae.

Local names: Kuchlá (H), Visha-mushti (S), Kajra (Mumbai), Kanjira (Kan), Etti (Tam), Mushti (Tel).

Description: Smaller medium-sized tree with smooth greyish thin bark, branch codilated at nodes, often converted to spines. Leaves broadly elliptic, thin, 5–12 cm long. Flowers greenish white, in short terminal, downy cymes. Fruit globose 2–5 cm in diam, yellow to orange red when ripe.

Fig. 5.27: *Spermadictyon suaveolens* Roxb. A and A'. upper and middle portions of a flowering twig; B. flower, C. corolla opened to show androecium; D. corolla removed to show style and stigmas; E. transverse section of ovary; F and F'. front and side view of a stamen.

Distribution: Common in the moist forests of western ghats, Bengal, Circcar, Orissa and Deccan carnatic; also in Gorakhpur forests of north India.

Uses: The seeds from an important source of the alkaloid *strychnine* and *brucine*. By the hill tribes of Nilgiris the seeds are used as fish poison. The bark and the wood also contain 'brucine' and employed medicinally.

After treatment the seeds are used in *Ayurvedic* medicines, as all the parts except fruit pulp are poisonous.

Extent of utilization and trade: The fruits are collected and the seeds washed out and dried in sun or the seeds are simply gathered from the ground but in the later case have little market value.

Dry deciduous forests of Kerala, Ganjam, Nellore and Godavari districts of Andhra are the important centres of production from where considerable quantities are exported.

An important source of alkaloids like *strychinine* and *brucine*, obtained from the seeds. The bark contains *Brucine*. In addition to alkaloid, seeds give a dye and oil though seeds are considered poisonous and are purified before selling as a medicine in Ayurveda (SudhKapilu). The product is used in the treatment of rheumatism along with other components.

Strychnos potatorum L. f. (Nirmali-H; clearning nut tree (Eng.). A small tree, seeds are nonpoisonous and used as emetic. Have properties of clearing muddy water in wells.

122. *Syzygium cumini* (L) Skeels (-Eugenia jambolana Lamk., Myrtus cumini L.)

Family: Myrtaceae.

Local names: Java Plum, Jambolana, Black Plum, Indian Blackberry, (Eng); Jámun, Jambu, Jambavá, kala-Jámb (Hindi).

Description: A large tree, evergreen, with thick, smooth, light grey bark, exfoliating in thick, irregular scales. Leaves opposite, elliptic, smooth, leathery, aromatic and shining, Flowers without stalks, whitish or pale green, scented, crowded in small heads at the end of lax panicled cymes. Calyx tube yellow within, rough outside. Petals 4, rounded. Stamens numerous. Berries juicy, pink when ripening, black when ripe (Fig. 5.28).

Fig. 5.28: A stand of *Syzygium cumini* in Yercaud, Salem district

Distribution: Throughout India, but indigenous to western ghats: however not present in arid regions of the country.

Uses: It is well known for its fruit. In can grow in standing water, hence preferred on water stagnated area for plantation of sites.

The kernels of the stone is a well known medicine for curing diabetes.

Jamun fruit acts as tonic for stomach and in fever. An ayurvedic medicine (Asava) is prepared from the fruits which is commercially sold.

Kernels of the jamun stone, kernels of mango and small myrobalan, in equal quantities, are powdered and a teaspoonful of this powder is taken with whey to stop diarrhoea and in cases of diabetes.

Bark of *Jamun* tree is astringent and a decoction is used as mouthwash to strengthen gums and stop bleeding from gums.

Sugared extract, syrup or vineager is also prepared from *Jamun* fruit which is used in cases of stomachache, spleen and liver problems; particularly vineager is used in cases of enlarged spleen.

Blossoms are a source of honey.

Alcoholic extract of the seeds has been reported to reduce level of blood sugar in diabetic patients.

Other species

a. *Syzygium aqueum* (Burm. f.) Alston Watery roseapple (E) LalJumrul (Hindi). A small tree in Assam, Sikkim, Meghalaya. Fruits are edible. Tanin obtained from stem bark.

b. *S. arnottianum* (Wight) Walp. Distributed in western ghats, Nilgiris and Palney hills. Fruits edible. Tannin from stem bark.

c. *S. aromaticum* (L.) Merr. and Perry (Eng. Clove Lavang-Hindi). Cultivated in south. Discussed in agro-techniques section.

d. *S. claviflorum* (Roxb.) Wall. ex Cowan and Cowan. Tree of Andamans. Fruits acidic and edible.

e. *S. fruticosum* (Roxb.) DC. Wild Jamun (Eng.Jangli-Jamun) Grown as an avenue tree.

f. *S. zeylanicum* (L.) DC. A shrub of western and southern parts. Fruits aromatic and edible. Leaves are vermifuge.

123. *Terminalia belerica* (Gaertn.) Roxb. ex Fleming

Family: Combretaceae.

Local names: Bahéra (H); Télaphala (S); Bahéra (Guj, Punj, Marh); Akkam, Tanri (Tam), Tadi, Tandra (Tel); Boara gota (Chittagong), Lopon (Santal); The Belleric Myrobalan (E). The trade name *Bahera* is based on local name and English name.

Description: A large tree, deciduous, with bluish or ash grey bark, uneven with longitudinal furrows. Leaves alternate, crowded towards the ends of branches, obovate-elliptic, coriaceous, pale beneath, sub-acute or acuminate; lateral nerves 5–8 on either side of the thick midrib; petiole usually glandular. Flowers small, pale green, bad smelling, in simple spikes, the upper often male. Fruit ovoid, brownish, densely covered with hairs, nut thick-walled and hard.

Distribution: Throughout India, ascending to 1000 m, except in the dry regions of western India; common associate in mixed deciduous forests and a companion of *Shorea robusta*.

Uses: The dried fruits constitute the drug, as astringent, tonic and laxative, used in piles, diarrhoea and dropsy.

Bark of the plant is used to increase diuresis.

Dried ripe fruit is astringent, tonic and laxative, used in disorders of the stomach, piles, diarrhoea and dropsy. It is given in the form of infusion or decoction; useful in cough and hoarseness. It is also used in fever, sore throat and dyspepsia. It is also given as brain-tonic and is applied on eyes as a soothing lotion. Fruit strengthens the intestine, stops excessive salivation stops euphoria (watering of eyes). Half roasted rind powder used to cure cough, asthma. The pulp of the ripe fruit is laxative and is a constituent of the *Triphala* which is used in a great variety of diseases (*Triphala* is a combination of three —*Harad, Bahera* and *Anwala*). It is given either in the form of infusion or as decoction. With honey the pulp is also used in ophthalmia. The half ripe fruit is considered to be purgative but the ripe and dried fruit has the opposite property. Oil expressed from the seeds is hair tonic and used as a liniment in rheumatism and as a salve in ophthalmia. The kernel is astringent and applied to inflamed parts either in the form of plaster or as poultice.

Extent of utilization and trade: The plant is extensively exploited for its fruits which are sold in the market. The main centres of production are Madhya Pradesh and outer Siwalik ranges in the Himalaya.

Phenology and time of collection: Fls. April-June after the new leaves Fr. Dec-Feb.

Other uses: The timber is used for boats and for miscellaneous agricultural tools, since it keeps well under water. The fruit yields a tannin which is used for dyeing cloth and leather; also for making inks.

Other species: *Terminalia chebula* Retz. (Chébulic Myrobalab, Har, Harad. Harrá, Haritaki-H; Harrakoj-Tam). A large deciduous tree with dark brown bark, young parts covered with rusty hairs. Leaves mostly sub-opposite, lateral nerves arcuate, 6–12 on either side of the midrib, petiole hardly exceeding 2 cm often with 2 or more glands. The leaves are not crowded at the end of branches but in almost opposite pairs. Flowers dull-white in spikes at the ends of branches. Fruit obovid, distinctly ribbed, with a rough grooved surface (Fig. 5.29). Fruits are of 3 types:

1. As long as stone has not formed, it drops from tree assumes black colour when dry. Fruit called *choti-harad*.
2. When it is half ripe and yellow, stone has formed called large harad or 'yellow harad'.
3. Fully ripe called 'kabuli' or *Julafa harad*.

The distribution is similar to *T. bellirica*. A companion of *Shorea robusta* but less common; fairly common in certain limited areas. In some early literature it has been called *abyatha*, which denotes, it takes away pain (byatha). Fruits form an important ingredient of *Triphala*, along with *Bahera* and *Amla* fruits.

The bark is a useful cardiac tonic, as it raises the blood pressure and is an effective diuretic. The fruit is an astringent and used as mild safe-laxative in stomachic and as antibilious and alterative. The unripe fruit is used in dysentery and diarrhoea, ripe fruits are safe purgative, it does not cause any gripping, nausea, etc. The pulp stimulates liver and is used in piles, dysentery,

Fig. 5.29: *Terminalia chebula* Retz. A. flowering twig; B. flower and subtending bract; C. perianth opened to show arrangement of stamens and gynoecium cut longitudinally.

costiveness, flatulence, asthma, urinary disorders, vomiting, hiccup, intestinal worms, enlarged spleen and liver. The seeds are generally rejected. Externally, the pulp is applied as an astringent. Powder is used as a dust over chronic ulcers and wounds, as a dentrifice in carious teeth, in bleeding and ulceration of gums; a gargle of powder is useful in stomatitis. Decoction of the fruit is an efficacious astringent and often used as a wash for treatment of bleeding piles and some vaginal discharges. An ointment of the fruit is used in piles. Powdered fruit with salt, is commonly used at bed-time as mild laxative. The fruits are also preserved with sugar known as *Harr-ká-Murabba*. It strenghthens brain and eyes.

The fruit is an ingredient of the well known preparation *Triphala*. Large harad is used for stopping stools in children, white small *Harad* soaked in salt water is used in cases of piles. Small *Harad* fried in a little castor oil, powdered acts as an effective cure to relieve from constipation.

***Extent of utilization and trade*:** The tree yields a gum which is largely collected in Berar. The dried fruits constitute the 'Chebulic' and used as a dye. The rind of the fruit being powdered and steeped in water with *alum* gives permanent yellow colour. It gives tan which is of commercial value. The chief exporting centres are Madhya Pradesh, Mumbai, Rajasthan, Bengal, Tamilnadu and chief importing towns are Kolkata, Chennai, UP and Bengal. Foreign trade is large and important.

Terminalia arjuna Wight and Arnott (Arjun-H; Indradrum-S) A tall tree, 25–30 m high. Throughout India, chiefly along water channels and in moist places. It has a greyish white trunk 3–6 m in girth, bark is astringent and is used in fevers and in fractures and contusions; it is also taken as a cardiac tonic, as a cure for angina. In early Sanskrit works it has been named as *Nadisarjja*, which denotes that the tree grows on river banks. The fruit is 5–6 sided, resembling Chinese gooseberry (Vern. Kamrakh) as seen in Fig. 5.30.

The bark is boiled in slow fire with milk and till half the liquid is left, strained and drunk with sugar. Bark powder mixed with butter (ghee) applied to wounds and injuries, it relieves diarrhoea and dysentery, when taken with goat's milk, relieves spermatorrhoea and increases sexual power. Gargle of bark decoction is helpful in swelling of gums.

Fig. 5.30: *Terminalia arjuna* on the banks of Cauvery in Salem district. The bank of this plant when dipped in water ones night is considered effective in heart disease.

124. *Symplocos racemosa* Roxb.

***Family*:** Symplocaceae.

***Local names*:** Lodh (H); Lodhrá (S); Lodh (Beng); Lodrá (Guj); Ludam-daru (Sighbhum); Lapong-dang, Dieng Lamaki Assam).

The Sanskrit name denotes that the plant stops ocular discharges. The trade name is based on the Hindi name.

***Description*:** A small tree, about 6 m high. Leaves dark green, 8–20 cm long, leathery, usually pointed at tip, margins entire or toothed; leafstalk small, about 8–20 mm long. Flowers small, pale yellow or white, in small axillary clusters. Fruit purplish black, 1 to 1.5 cm long.

***Distribution*:** In the plains and lower hills of Eastern and Central India.

***Uses*:** Fresh air dried bark of the plant constitutes the drug. The drug is useful in digestive disorders, eye diseases and ulcers. Decoction of the bark is useful in bleeding gums. It is used in plasters and applied on wounds for promoting maturation of wounds. It is astringent and used in excessive bleeding during menstruation. The astringent properties are utilized for curing loose

motions. It is considered useful in elephantiasis and in troubles of urine.

Extent of utilization and trade: The drug is utilized on a commercial scale in *Ayurvedic* system of medicine but exploited on a limited scale.

Other species

a. *Symplocos chinensis* (Lour.) Druce (=*S. crantaegoides* Buch.-Ham. ex D. Don) is a small tree with white flowers, fragrant in terminal panicle. The plant goes under the same local names as above. Locally, the bark is used as an astringent and used in ophthalmia. It is supposed to promote maturation and resolution of stagnant ulcers. The tree is common in temperate regions from Indus eastwards to Bhutan, ascending to 2500 m.

b. *Symplocos paniculata* Wall: Bark has same uses as *S. racemosa* and considered as tonic.

Other uses: The wood is durable and the bark yields a yellow dye but generally employed as a mordant with other dyes (as *Morinda tinctoria*), *Butea monosperma*, *Caesalpinia sappan*, etc.

125. *Tamarindus indica* Linn.

Family: Caesalpiniaceae.

Local names: Imli (H); Ámbli (Beng, Guj), Chinch (Kan); Amlam (Mal); Ambli, Chinch (Marh); Tentuli (or); Amlika (S), Puli maram (Tam) Amlika (Tel).

The trade name tamarind is based on the English name.

Description: A large tree. Leaves compound; leaflets 10–20 pairs about 1 cm long. Flowers yellowish with reddish streaks, in small erect clusters among the leaves. Fruits 8–20 cm long, 2–3 cm broad, fleshy, pendulous, brown in colour. Seeds 3–12, dark brown, embedded in the fleshy fibrous mass which is the well known acrid pulp, tastes sweet-sour enclosing hard red seeds.

Distribution: Central and southern regions of India, planted throughout India on roadsides and in garden. Also self sown in waste and forest and. It is said to be indigenous to Africa.

Uses: The pulp of the fruit is medicinal, and used as *chutney*.

Tamarind pulp has laxative properties; its infusion in water is a refreshing drink and is useful in fevers. As a laxative, it is taken singly or with in mixture with other purgative drugs. Used as protectant against summer heat. When taken as syrup, also brings down bilious fevers, allays thirst, relieves nausea and vomiting; it is cooling, allays preponderance of sanguine and bilious humors expelling it along with stools.

It acts as appetizer, helps to digest food. The kernel is useful in spermatorrhoea, nocturnal emission in men and leukorrhoea in women. Seeds are roasted and powdered, sifted and mixed with equal weight of sugar and taken with water. Swelling produced by *marking nut* (Bhilawa) is resolved by drinking 10 gm of leaves of tamarind of ground in water and strained.

Other uses: The timber of the tree is resistant to insects attack and largely used for agricultural tools and domestic articles. It makes very good charcoal and is used for gunpowder. The leaves yield a yellow dye. Pulp is used to clean silver and brassware and other utensils. Seeds are used in jam and jelly. Seed kernel is a source of starch. It yields a dirty gum of no value, old trees are sometimes seen to have a liquid exudation consisting of calcium oxalate. Infusion of leaves is believed to yield a red dye and to impart a yellow shade to cloth previously

dyed with Indigo. Oil of an amber colour is prepared from the seeds by expression and has been valued as an antiscorbutic.

126. *Tamarix gallica* L. (-Tamarix dioica Roxb.)

Family: Tamaricaceae.

Local names: Jháu, Jharewk (Hindi); French Tamarisk (Eng.)

Description: A bushy shrub, growing abundantly on river banks and bears all malformed fruits called *Bari mayeen.* The branches are slender, articulated. Leaves minute, covered with a fine bloom, white margined, not sheathing the branch on which they are borne, smooth, scale like. Flowers bisexual, minute, white or pink, crowded on slender-panicled racemes at the end of branches. Capsules minute, surrounded by withered perianth.

Distribution: Throughout India from Western Himalaya, southwards of western peninsula, on riverbeds and on sea coasts.

Uses: The leaves and wood are used to restore enlarged spleen when taken as decoction or drinking water from a cup made of *Jhau* wood.

Wounds, especially haemorrhoids, dry up by allowing the smoke from the burning leaves to play on wounds.

Prolapsus of the anus is prevented by washing the part with decoction of *Jhau* leaves.

The fruit is astringent and stops bleeding. It is also used in diseases like spermatorrhoea, tenous semen, premature ejaculation and leukorrhoea. Decoction of the fruit is used as mouthwash to strengthen gums and as gargle which restores enlarged uvula and heal throat sores.

A *mana,* known as *gazanbeem* or *guzunjabin* is formed on the plant as a result of insect puncture, which contains dextrine and organic acids.

Other species

a. *Tamarix aphlla* (L) Karst. (*Tamarix articulata* Vahl.) (Vern. name: Asréli, Faràsh, Laljhau). A middle-sized tree with minute, sheathing leaves which sometimes get to triangular tooth. Indigenous to Punjab region, extending to Yamuna. The tree is planted as a wind-break. Galls are formed. The bark is bitter astringent, contains tannin used for tanning leather. The powdered bark in combination with kamlea (*Mallotus phillipenses*) powder is used as aphrodisiac and also applied on eczema.

b. *Tamarix dioica* Roxb. (Vern Jháu, jau, Pilchi) is a moderate-sized shrub or a large tree with dark-coloured bark. Leaves minute with a broad white margin. Flowers unisexual; male and female found on different plants; purple or light rose coloured. It extends from west Rajasthan to Assam and the western peninsula. The galls contain 50% tannin, while leaves and bark contain 8% tannin. Branches are used for making baskets. Leaves and wood used in cases of swelling of spleen. Smoke from burning leaves is allowed to play upon sores of smallpox, haemorrhoids, to make them dry up.

127. *Taxus wallichiana* Zucc.Taxus baccata L.

Family: Coniferae.

Local names: The yew (Eng.), Thunér, Thuniara-Uttaranchal; Talispatra (Hindi), Sthuneyaka (Sanskrit).

Description: An evergreen tree, generally of middle size but sometimes very large,

having spreading crown. Bark reddish grey, peeling off in longitudinal shreds. Leaves are distichous, linear, often cuspidate-acuminate, dark green and shining above, brownish yellow and somewhat mealy beneath, one nerved, narrowed into a short petiole. Flowers usually dioecious, axillary. Male fls. a pedicelled whorl of 3–8 anther cells on peltate scales. Female fls. a single erect ovule, surrounded by a disc. Fruit a berry, the disc developing into a fleshy covering which overtops and nearly conceals the compressed, wingless, olive-green seed (Colour Plate 5.9).

Distribution: In shady ravines in the Himalayan region, up to 3300 m, not abundant, but being over-exploited. Hence efforts have been made to propagate the plant by tissue culture method in the laboratory.

Uses: The leaves and the bark are used for their emmenagogue, expectorant, aphrodisiac, sedative and antispasmodic activities. It is also employed in cancerous complaints.

The leaves and young fruit act as narcotic and are poisonous to horses and cows and fatal cases of poisoning have been reported from swallowing the fruit.

The plant has recently attained importance for the active principle; Taxol, which has been used recently with efficacy.

The tincture prepared from young shoots is used for headache, giddiness, diarrhoea and biliousness.

In ayurvedic medicines Talisadichurna, *Talismodak, Marigankbatika, Yks jamantakarlahe* are prepared where this plant is an ingredient. The leaves are collected during March and dried, in partial shade. The availability is however limited. Recent reports noted that about 20 qtls of leaves could only be collected from Kantal forests in Garhwal Himalaya. However, *Rhododendron*

anthopogon has been found as an adulterant to the genuine leaves.

128. *Thymus serpyllum* L.

Family: Labiatae or Lamiaceae nom alt.

Local names: Ban-ajwain (Hindi); Ban-Yabani (T.Garhwal); Màsho (Himachal).

Description: An aromatic shrub, often tufted. Leaves nearly sessile; glands dotted, oblong-ovate, entire, obtuse. Flowers small, purple, sometimes 1-sexual, in whorls, crowded in short terminal spikes. Calyx hairy, gland dotted, 2-lipped. Stamens 4, all nearly equal.

Distribution: In temperate regions of western Himalaya, from Kashmir to Kumaon, 1700–3000 m. Common on the downs.

Uses: The whole plant is powdered and mixed with jaggery (*gur*) and used as vermicide. It is also used in weak vision, complaints of the stomach and liver and suppression of urine.

The leaves are laxative, good for kidney and purify blood.

Seeds are used as vermifuge.

Oil from the leaves is sometimes applied as a remedy in toothache. It contains 0.5% essential oil, having *phenols*, beta *cymenemterpenes* and *terpine alcohols*.

Extent of utilization and trade: Very small quantity of this plant is collected; however in the Himlayan region it is used locally.

129. *Toona ciliata* M. Roem. (= Cedrela toona Roxb.ex Rottl.et Willd.)

Family: Meliaceae.

Local names: Tun (H); Red Cedar, Sandal Nim, Moulmeik Cedar, Indian Mahagany (Eng.).

Description: A large tree with dense spreading crown and thin dark grey bark

which is smooth up to the middle age. Leaves paripinnate, generally glabrous; leaflets 8–30, usually opposite, ovate-lanceolate, acuminate, sometimes pubescent beneath; margins entire, usually wavy, base cream coloured, scented like honey, in drooping panicles. Capsule septifragally dehiscent, dark brown. Seeds reddish brown, light with a membranous wing at either end.

Distribution: In subtropical regions and valleys in the hills up to 1300 m generally along the banks of streams or in marshy locatlities; scarce with *Shorea robusta*. Largely planted in avenues, along canals and in gardens.

Uses: The bark is used in infantile dysentery and as local astringent application in various forms of ulcers. Finely powdered wood yields an oil.

Extent of utilization and trade: The plant is principally exploited for the wood and used medicinally on a limited scale.

130. *Ulmus wallichiana* Planch

Family: Ulmaceae.

Local names: Haimar, Keetamára (TG); Embroi, Imroi (Jaun).

Description: A large deciduous tree, bark rough, grey, exfoliating in diamond-shaped scales; young branches pubescent. Leaves alternate, elliptic-ovate, long acuminate, doubly serrate, pubescent beneath, nerves 15–20 pairs; stipules subulate. Flowers bisexual, in dense lateral racemes, appearing before the leaves from the leafscars of last years shoots. Fruit a flat stipitate Samara.

Distribution: In temperate regions from Indus to Nepal, 1500–3000 m. It is common in mixed-deciduous forests of *Aesculus, Juglans* and in *Silverfir*.

Uses: Locally, the bark is used as a plaster in fracture or dislocation of bones. It is prepared by powdering the bark with 'Saral' (*Pinus roxburghii gum*) along with *Loban* (trade name). The powder is mixed with clay and applied as plaster on the affected part.

Extent of utilization and trade: Small quantities of the plant are collected for local consumption.

Phenology and time of collection: Fls. March-April when leafless. Fr. June.

131. *Viburnum cotinifolium* D.Don

Family: Caprifoliaceae.

Local names: Bhatyánu, Guyanya (TG); Bhutnoi (Jaunsar); Gyána (Kumaon).

Description: A large, deciduous shrub with greyish bark. Leaves ovate-elliptic or orbicular, woolly tomentose beneath, lateral nerves 5–6 pairs. Flowers generally in dense terminal corymbs. Drupe oblong, red, compressed.

Distribution: Temperate regions from Kashmir to Bhutan, 2000–3000 m in forests of oak; especially gregarious.

Uses: The bark of the plant is said to be used as a prescription which is for menorrhagia and metrorrhagia also used in decoction, for menstrual disorders.

Extent of utilization and trade: The plant is not extensively and intensively exploited. About 4 qtls. are reported to have been collected from Kanátal forests (district of Tehri Garhwal) in 1963–64 (Uniyal and Issar, 1967).

Phenology and time of collection: Flowers, April-June Fruits, August-Oct. The leaves turn purplish before flowering.

Other species: *Viburnum nervosum* D. Don (Tileen-TG). A shrub with stout grey or brown branches. Leaves elliptic-oblong, serrate, glabrous above, hairy on the nerves beneath, lateral nerves 8–10 pairs. Panicles compact or lax, silky; branches short. In temperate and subalpine regions above 3000 m. the bark is febrifuge and laxative.

Viburnum foetidum D. Don. Vern Bhutnoi (Jaunsa). The plant is astringent and emmenagogue; juice of leaves given internally in menorrhagia and in postpartum haemorrhage.

132. *Viscum album* Linn.

Family: Loranthaceae.

Local names: Chulu-kà-Banda (Hindi, Jaun. and Garh.).

Description: A large parasitic shrub, green all over; branches dichotomous or whorled. Leaves sessile, cuneate-oblong or oblanceolate, with 3–5 longitudinal basal nerves. Flowers dioecious, sessile, in clusters of 3–5, supported by concave bracts. Perianth segments, 3–4, triangular, dioecious. Fruit ellipsoid, white, smooth, almost transparent.

Distribution: Chiefly on rosaceous shrubs and on *elm, walnut, willow* and *oaks*, ascending up to 2000 m from Kashmir to Nepal.

Uses: The berry is sweet and given in enlargement of spleen in cases of wound, tumour and diseases of ear.

The plant is used occasionally as an antispasmodic and diaphoretic, it is also given in epilepsy.

Extent of utilization and trade: The drug is not exploited on a commercial scale for medicinal purposes.

Phenology and time of collection: Flowers, March-May, Fruits, Sept-Nov.

133. *Vitex negundo* Linn.

Family: Verbenaceae.

Local names: Shimàlu, Sumàlu, Chhatimal, Shivlingi Nirgundi (H); Nishinda (Dun); Soni (Jaun); Malla (SRE); Sewain, Shiwai (Kum).

Description: An erect shrub, 2–2.5 m high, leafstalk and inflorescence densely grey pubescent. Leaves stalked, digitately compound; leaflets 3–5, lanceolate, unequal, upper surface glabrous, lower densely pubescent. Flowers blue purple, crowded in short cymes, forming erect, narrow tapering terminal panicles. Fruit succulent, black, ovoid (Fig. 5.31).

Distribution: Throughout India, ascending to 1500 m. In sub-Himalayan tracts from Indus eastwards, often gregarious in

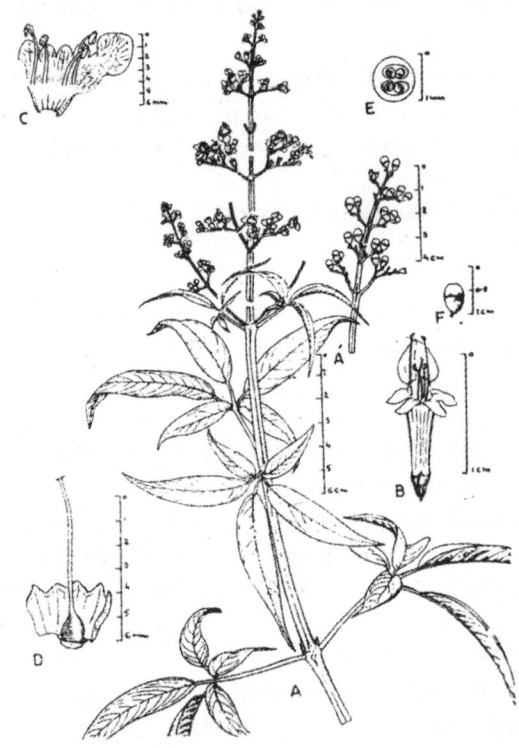

Fig. 5.31: *Vitex negundo* L. A. flowering twig; A'. fruiting twig; B. flower; C. corolla opened to show androecium; D. calyx opened to show gynoecium; E. transverse section of gynoecium; F. fruit.

small patches on the banks of streams and similar places. Much used as a live hedge, for soil conservation.

Uses: The plant is a powerful anthelmintic, expels worms when leaf-juice is taken mixed with curd for 3–4 days. The seeds form the main drug, seed oil is applied to sinuses and scrofulous sores. Vapour bath of the plant is useful in febrile, catarrhal and rheumatic diseases.

The root is considered tonic, febrifuge and expectorant. It is given in a wide variety of affections, in dyspepsia, colic, rheumatism, worms, boils and leprosy. Decoction of the root is given in intermittent and typhus fever. Tincture of the root bark is given in case of irritable bladder and rheumatism. Powdered root is used as anthelmintic and as demulcent in dysentery and piles.

Juice with heated mustard oil, 150 gm each, till all water evaporates mixed with 10 gm bees wax produces an ointment that heals worst wounds. Leaves of this plant with *Cuscuta reflexa* (Akashbél), when boiled in water, is used to relieve arthritic pains.

The leaves are aromatic, discutient and anodyne, cures colic, rheumatic swelling, etc. Externally, it is used in the form of fomentation and bath in sprains, rheumatism, swelled testicles, contusions, etc. The juice of leaves is said to remove foetid discharges and worm from ulcers. Decoction of the leaves, with other drugs, is given in catarrhal fevers, rheumatism and enlargement of spleen. In headache and catarrah the dried leaves are smoked. Decoction of the leaves is used as a bath for women after childbirth and for those suffering from beriberi. A pillow stuffed with dried leaves is placed under the head for relief in headache. An oil from the leaves is applied to sinuses and scrofulous sores. The leaves are antiparasitic and are placed between leaves of

the books to preserve them from insects. Leaves heated in earthen pots are used as a fomentation in rheumatism and swelling of the body. The decoction mixed with pepper is used in cold.

The flowers are astringent and used in diarrhoea, cholera fever, discharge of blood from stomach and bowels, it is recommended as a cardiac tonic.

Dried fruits are vermifuge and used as nervine tonic and cephalic. The seeds are reputed to be cooling in skin diseases and in leprosy.

Seeds with other drugs like *Ashwagandha* (*Withania somnnifera*), *Nirgundi* (*Irtex negundo*), *Punarnuva* roots (*Boerhaavia diffusa*) and *Nagar motha* (*Cyperus scariosus*) powder are prescribed to be taken with Dashmool kwath, as a cure for osteoarthritis.

Extent of utilization and trade: The plant is extensively used in Indian medicine and exploited commercially. The seeds are sold in the market.

Phenology and time of collection: Fls. May-June.

134. *Wattakaka volubilis* (L.) Staff. (Marsdenia volubilis T.Cook, Dregea volubilis Benth ex Hook.f.)

Family: Asclepiadaceae.

Local names: Madhumálti (Sansk); Nakchikni (Hindi); Dodhi (Bihar) Akad-Bél (Rajasthan).

Description: A perennial climbing shrub with watery sap. Stem softly greyish, cylindrical, hoary when young, warty. Lenticelled when old. Leaves broadly ovate, shortly acuminate, rounded at base, numerous glands at the base of midrib on upper surface, strongly nerved with 4–5 pairs of lateral nerves. Flowers yellowish green, in many-fld., drooping, umbellate cymes;

bracts caducous. Calyx pubescent outside. Corolla rotate; corona staminal of 5-fleshy lobes. Staminal column arising from the base of corolla (Fig. 5.32).

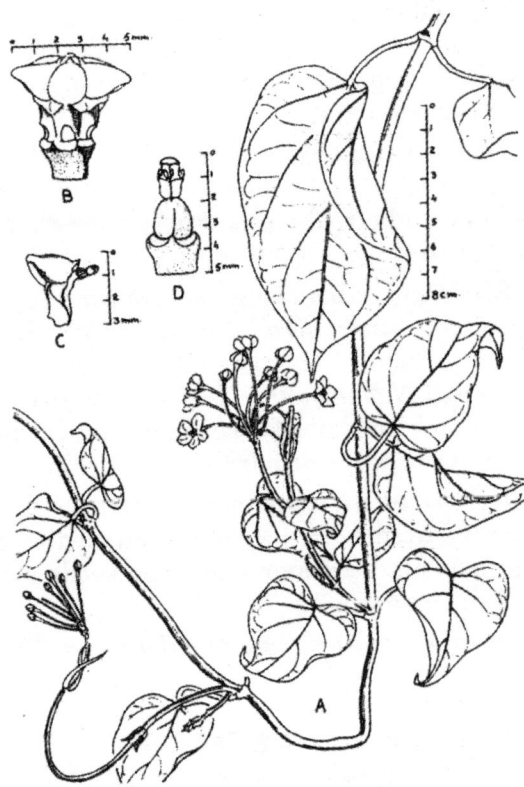

Fig. 5.32: *Wattakaka volubilis* Benth. ex Hook. f. A. flowering twig; B. calyx and corolla removed to show corona; C. a coronal scale and a stamen seen laterally; D. gynoecium.

Distribution: In tropical regions of India. Plains of Tamilnadu, South Maharashtra, Bengal and Assam extending to Myanmar.

Uses: Roots and the tender stalks are emetic and expectorant, while the leaves are used as an application to boils and abscesses.

A glucoside *dregein* has been isolated from the plant.

135. *Woodfordia fruticosa* Kurz.

Family: Lythraceae.

Local names: Dhátaki (Sansk); Dhaur (Himachal). Trade name is Dhatri and different preparations like *Dhatri Loh* and *Dhatri Ásaw* are available in the market.

Description: A pubescent shrub with long spreading branches. Leaves opposite, sometimes in whorls of 3, sessile, surface green, lower white, black dotted. Flowers clustered, shortly stalked. Calyx tubular, bright red, teeth 6 short alternating with 6 minute accessory teeth. Petals 6, red, hardly longer than the calyx teeth. Stamens 12, filaments long, red, far protruding, inserted below the middle of calyx tube. Ovary oblong, 2-celled, style far protruding. Capsule enclosed within the calyx tube, opening by 2-valves (Fig. 5.33).

Distribution: Throughout India ascending to 1700 m in the Himalayan region; also reported in valleys of the Himalayan region: Kumaon and Garhwal.

Uses: The dried flowers are astringent, used in dysentery, dearrangement of liver, disorders of mucous membrane and in haemorrhoids.

Other uses: The plant is noted for the dye extracted from flowers, twigs and leaves. The dried flowers contain 20% of tannic acid and are used in the same way as *Tamarix gallica.*

136. *Wrightia tinctoria* R. Br.

Family: Apocynaceae.

Local names: Dudhi-Indrjao, Kérna (Rajasthan); Kalá-Kuda, Bhakar-aak Mitha Indrjau (Hindi).

Description: A small deciduous tree; bark scaly, smooth. Leaves variable, nerves 6–12 pairs, petiole very short. Flowers

Stem and rootbark of *W. tomentosa*, are said to be useful for snakebite and scorpion sting (Fig. 5.34).

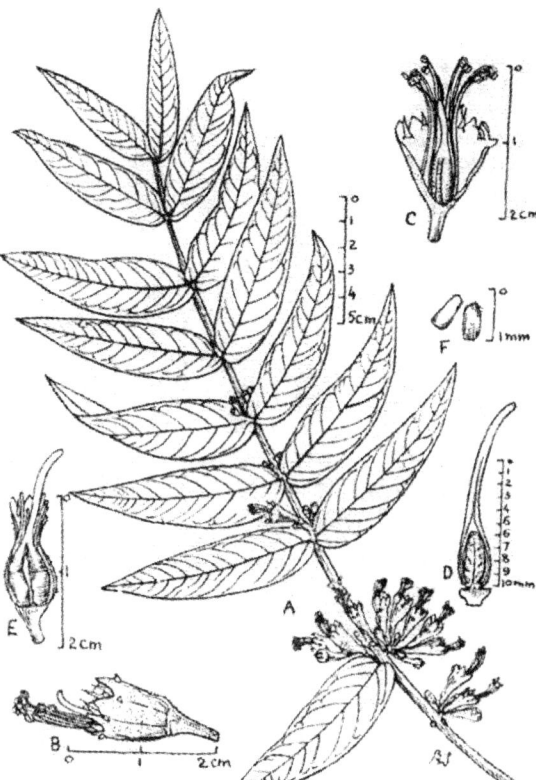

Fig. 5.33: *Woodfordia fruticosa* Kurz A. flowering twig; B. flower; C. flower opened to show the petals, androecium and gynoecium; D. longitudinal section of gynoecium; E. calyx cut open to show the capsule; F. seeds.

white, fragrant, in lax terminal cymes. Fruit has two distinct pendulous follicles.

Distribution: Throughout India, on open deciduous forests on trap formation, extend southwards in considerable abundance to Kerala state. Also on dry sandy soils and in hilly region on rocks in Rajasthan and Banda district of UP.

Uses: The leaves turn black when dry, yield an indigo-like blue dye which is known as 'pala-indigo'.

Unripe fruit is used for coagulating milk.

The seeds are aphrodisiac and anthelmintic, used to cure amoebic dysentery.

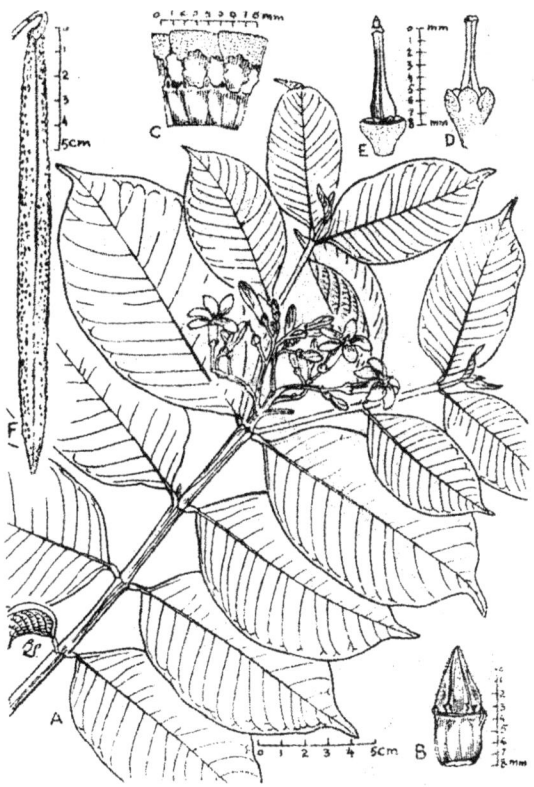

Fig. 5.34: *Wrightia tomentosa* Roem and Schult. A. flowering twig; B. corolla lobes cut away to show androecium; C. part of corolla tube opened and the lobes partly cut away to show the coronal scales; D. fruit.

137. *Zanthoxylum alatum* Roxb.

Family: Rutaceae.

Local names: Témru, Timbur, Timru (Garhwal), Téjbal (Hindi).

Description: A shrub or small tree. Bark corky, strong prickles on the branches, petioles and the midrib of leaflets. Branchlets often rough with raised dry specks. Leaves alternate, imparipinnate, common petiole

and rachis narrowly winged, Leaflets 2–6 pairs, opposite, elliptic-lanceolate, more or less serrate, pellucid-punctate. Flowers yellow, small, usually 1-sexual in dense lateral panicles. Calyx 6–8 fid. Petals none. Stamens 6–8, much exceeding the calyx. Anthers large. Fruit usually of solitary carpel which dehisces ventrally, of the size of peppercorn, tubercled and strongly aromatic.

Distribution: In hot valleys and in forest undergrowth, up to 2000 m in Himalayan region, from Himachal to the Kumaon region and the Khasia hills.

Uses: The wood is used for walking sticks on a large scale.

The fruit resembles that of coriander and used for curing toothache.

Oil from the seeds is similar to eucalyptus oil in odour, which is antiseptic, disinfectant and deodorant. Young stem used as tooth-picks.

138. *Zizyphus mauritiana* Lamk. (Z. jujuba Lam.) cultivated var. Pewandi.

Family: Rhamnaceae.

Local names: Chinese Date, Jujube (Eng.),Bér, Bor, Nadará, Karkándu Kuvala (Hindi).

Description: Moderate-sized tree, almost evergreen, with dark-grey or nearly black bark with long deep vertical cracks, reddish and fibrous inside, young parts rusty tomentose. Leaves variable, elliptic or ovate, rounded at both ends, dark green and often shining above, densely wooly and tomentose beneath, base more or less oblique, strongly 3-nerved, nerves closely penniveined, prickles solitary or in two's. Flowers greenish yellow, in short axillary cymes or fascicles. Calyx glabrous within. Petals concave, reflexed. Fruit globose to ellipsoidal red or orange when ripe.

Distribution: Common and gregarious along foot of the hills and in the arid and semi-arid regions of W. Rajasthan. Also improved varieties are cultivated in Uttar Pradesh and Rajasthan for the fruits.

Uses: Roots as decoction is given in fever and as powder applied to old wounds.

Bark is a remedy for diarrhoea.

Fruits are mucilaginous, pectoral, blood purifier and improves digestion.

Other species: *Zizyphus nummularia* (Burm.f) Wight and Arn. (Vern. Bordi, Jhar-bér-Hindi).

Fruits are astringent, cooling and used in bilious affections. Leaves are used in skin diseases, like scabies also used in cough and cold.

Zizyphus xylopohra Willd. (Vern Gat-wood) Used in Rajasthan as fuel.

C. Herbaceous Plants used as Drugs from Wastelands, Weeds of Croplands and Forest Understorey

5

1. *Abrus precatorius* L.

Family: Papilionaceae.

Local names: Gunchi, Ghumchi, (Hindi), Chirmuti, Ratti, Chanoti, Ratti Kunch Crab's eye, Indian liquorice (English).

The drug Indian liquorice should not be confused with the trade name liquorice used for *Glycyrrhiza glabra* (Muléthi Hindi) which is entirely a different plant species.

Description: A branched twiner with woody stem. Leaves pinnate, about 10 cm long; leaflets 16–40, opposite, oblong, each about 2 cm long. Flowers rose-coloured, in axillary raceme. Fruit a pod with wrinkles, 2–3 cm long. Seeds 3–6 round, hard, white with a black or white spot (Fig. 5.35).

Distribution: Throughout the plains of India and ascending to 1000 m in the Himalayas, from Kashmir eastwards to Assam, and other hills. Sometimes planted in the gardens.

Uses: The medicinal properties of the plant are similar to that of liquorice (demulcent and emollient).

The *root* contains Glycyrrhizin (found in Liquorices), is emetic, alexiteric, emollient and demulcent. A watery extract is used for relieving obstinate coughs. The root is also taken internally for sore throat and rheumatism. It is also useful as an adjuvant for mixtures.

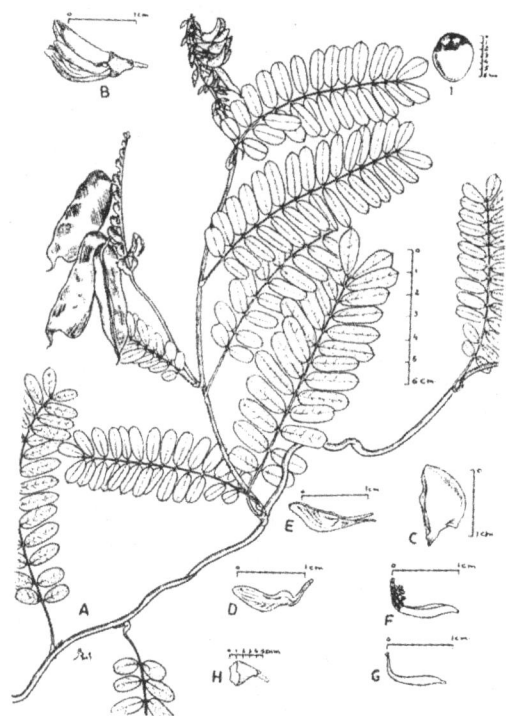

Fig. 5.35: *Abrus precatorius* L. A. flowering and fruiting twig; B. flower; C. standard; D. wing E. keel; F. androecium; G. gynoecium; H. calyx; I. seed.

The *leaves* are sweetish and used as demulcent. Fresh juice of the leaves is used as a cure for hoarseness of voice, dry cough and ardour urine; it is also considered as a good blood purifier. Externally the leaves are used for relieving pain. The juice of green leaves is useful against patches of leukoder-

ma. In Konkan area singers chew the leaves of white seeded variety as a remedy for hoarseness; also chewed with sugar to cure apthae of mouth. In spermatorrhoea with bloody discharges, equal parts of juice of white seeded Abrus leaves and *henna* leaves are rubbed with roots of cumin and sugar.

The *seeds* are purgative and emetic but in large doses poisonous. These are used internally to prevent conception. They lose their medicinal properties when boiled. Powdered seeds boiled with milk are used as powerful tonic and have an aphrodisiacal action on the nervous system. Externally, paste of the seeds made with water is used as rubefacient in sciatica, stiff shoulder, paralysis and other nervous diseases. It is also useful for skin diseases, ulcers, contusions and inflammations. In case of alopecia the paste is rubbed on the bare scalp to encourage growth of hairs. Poultice of seed with water and salt is used over boils to promote suppuration. Seeds without seed coat, finely ground are used for disease of cornea and glandular eyelids. Powdered seeds are used as snuff for relieving headache.

Other uses: Jewellers use the seeds, each unit, weighing 2–3 grams on an average, as unit of weight. Finely powdered seeds are used to increase adhesion when soldering delicate ornaments. The seeds are also used for making rosaries and beads, etc.

Fibre from the stem is woven into baskets.

The seeds contain the poison *abrin* which is made use of for poisoning cattle. The powdered seeds are made into a paste, with water, and the paste is molded into sharp needles which are inserted under the skin of animals.

Extent of utilization: The seeds and the roots are extensively collected and sold in the market as *ratti*.

The root appears to contain sugar and glycyrrhizin. The seeds contain principal poisonous constituent abrin, a *toxalbum* similar to ricin from castor seeds.

2. *Abutilon asiaticum* G. Don.

Family: Malvaceae.

Local names: Kánglu (Hindi).

Description: A small shrub, cultivated for the stem fibre which is used for ropes, etc.

Uses: The leaves are used in gonorrhoea, ulcers, in bladder stones and as an eyewash. These are also diuretic. Bark and root along with leaves are demulcent and diuretic.

Other species

a. *Abutilon avicennae* Gaertn. (Vern. Name: Jaya, Nahni, Khapat) Chinese Jute, Indian Mallow. (Eng.)

An annual herb with more or less cordate-leaves with a long point. Flowers solitary, axillary. Petals yellow. Capsules with 15–20 segments. Seeds many, hairy.

A native of Uttar Pradesh, distributed up to Sindh in Pakistan.

Known for the fibre which has the reputation of being superior than the jute and finer than Manila hemp.

b. *Abutilon indium* G. Don (Karandi, Madmi, Petari, Kanghi) Distributed throughout India (Fig. 5.36).

All the parts used medicinally. The herbs stop bleeding.

Leaves when ground with water, strained and drunk stop bleeding from the piles and also cure hematuria. Twenty one leaves of the plant, seven grains of pepper are ground together and made into seven pills. One pill is swallowed every morning with water in cases of heamorrhoids, whether

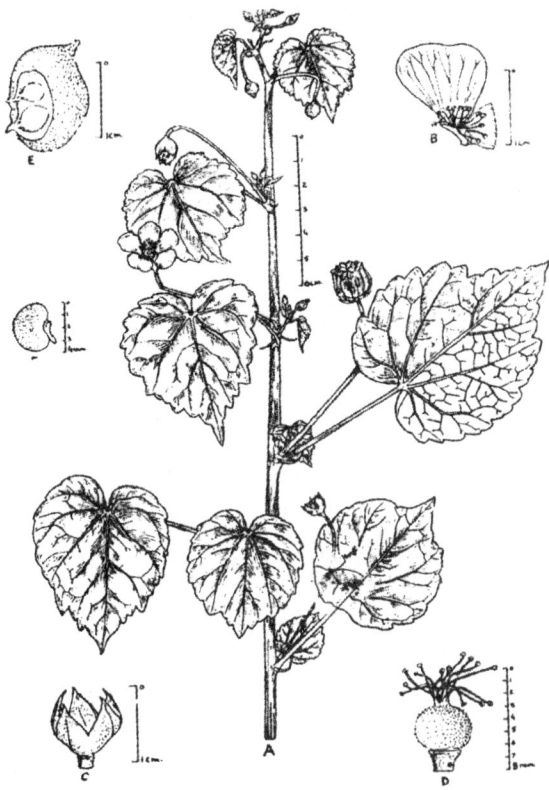

Fig. 5.36: *Abutilon indicum* Sweet A. upper portion of a flowering and fruiting stem; B. androecium longitudinally split open, a petal and base of the other; C. calyx; D. gynoecium; E. fruiting carpel; F. seed.

bleeding or not. In cases of gonorrhoea seven leaves of the herb, grounded with 3 grams of white cumin, with sugar candy and drunk. Mucilage is known as demulcent tonic. Gargle of the leaves cures sore throat.

The seeds also work as aphrodisiac when 3 grams of seeds are taken with milk, after making it in powder.

Decoction of the bark cures diarrhoea and also used as mouthwash to relieve toothache. Bark is mucileginous, considered to be diuretic.

Roots used as pulmonary sedative and an infusion of the roots make a cooling

drink in fevers; also regarded as useful in strangury, haematuria and leprosy. Infusion is a cooling agent called as Balbij. The seeds are nutritive, tonic and demulcent, used as a laxative in piles, as expectorant to relive cough and as an aphrodisiac, and demulcent having reputation of being useful in gonorrhoea, gleet and chronic cystic. Smoke from powdered seeds kills the worms, if the rectum of a child infected with the worms is exposed to the smoke.

c. *Abutilon bidentatuma* A. Rich. (Rotável-Rajasthan)

The flowers are tasty and eaten.

d. *Abutilon glaucum* (Cav.) Sweet (-A. muticum G. Don) (V. Name: Pintári Hindi) is also used in many diseases. Leaves are mucilaginous, also pectoral.

3. *Acalypha indica* L.

Family: Euhorbiaceae.

Local names: Khokáli, Kupi, Vanchi-Kanto (Hindi).

Description: An annual herb, about 30 cm high, with numerous, long, angular, softly hairy branches. Leaves up to 7 cm long, ovate, thin, glabrous, margins toothed. Flowers in long spikes; male and female flowers on the same spike. Male flowers clustered near the summit. Female scattered, bract truncate, 3–5. Capsules small, concealed by large bract, usually one seeded; seeds minute, pale-brown.

Distribution: A common weed throughout the hotter parts of India.

Uses: The juice of the fresh leaves is emetic and purgative. The plant is a popular remedy for bronchitis, pneumonia and rheumatism. Also used as anthelmintic.

As an emetic herb, its action is considered as safe, certain and speedy in cases of cough.

Cotton soaked in the juice of the herb is used for plugging the nostrils to get relief from congestive headache.

Root pounded in hot water is used as a cathartic, in small doses as an expectorant and nauseant.

4. *Achillea millifolium* L.

Family: Compositae (Asteraceae).

Local names: Gandaná, (Hindi) Biran jasif (Trade name).

Description: An erect pubescent herb, about 30 cm high. Stem leafy, grooved, lobes linear; lower leaves stalked, upper sessile. Flower heads radiate, crowded in branched corymbs, bract erect, receptacle flat, covered with branched scales. Flowers white or pale-pink; pappus absent. Corolla of central flowers 5-lobed. Fruit an oblong, flattened achene.

Distribution: In Western Himalaya from Kashmir to Kumaon, 2000–3000 m.

Uses: The plant is diaphoretic, tonic, astringent and stimulant, used in colds, obstructed perspiration, heartburn, hysteria, colic, epilepsy and flatulance. It is said to open the skin pores freely and purify the blood. Leaves are used as tonic. A decoction, when injected into nose, checks nasal bleeding; also injected in bledding piles and vaginal haemorrhages.

Decoction used for sore throat as gargle; hairwash, and lotion for sore nipples. Leaf decoction used as a sure remedy against cold, haemorrhages and nervous affections.

The flowers are stimulant, aromatic, vulnerable; yield an essential oil which is useful as an appetizer and given in disorders of urinary system. The oil has 'Azulene' as the main constituent.

5. *Achyranthes aspera* L.

Family: Amaranthaceae.

Local names: Chirchira, Chirchtra, Latjira, Ohut-Kánda (Hindi), Untaghado, Narkata, Apamarg (Hindi); Prickly. Chaff-flower. Rough chaff tree (Eng.); the plant is an herb and not tree.

Description: A tall, erect, much branched, coloured plant about a metre high. Leaves velvety tomentose, opposite entire, variable in shape. Flowers bisexual, in slender, branched spikes with membranous bracts, persistent but spiniscent. Perianth of 4–5 rigid segments (Fig. 5.37).

Distribution: Common weed, throughout India, on roadsides and field boundaries

Uses: The plant is valued on account of its ash which contains a large quantity of potassium and used in asthma and cough. It is pungent, alterative, antiphlegmatic, antiperiodic, diuretic and laxative, useful in oedema, dropsy and piles, boils and erruption of skin, etc. Dried plant is given to children in colic. Water in which the crushed plant is boiled is given in pneumonia; the decoction being a good diuretic is efficaceous also in renal dropsies.

Infusion of the root is a mild astringent, used in night blindness, cutaneous diseases. Root used as brush for tooth. It is useful in cases of haemorrhoids, leaves and seeds are emetic and used to cure hydrophobia, carminative resolve swelling, digestive and expels phlegm (when salt from root is taken). Young leaves are eaten as spinach.

In case of asthma, the ash of the plant is used and applied externally for ulcers and warts. Mixed with sesamum oil ash is used in the treatment of ear diseases, being poured with sesamum oil into the mealus.

Paste of the root in water is used in ophthalmia and opacities of the cornea. Paste

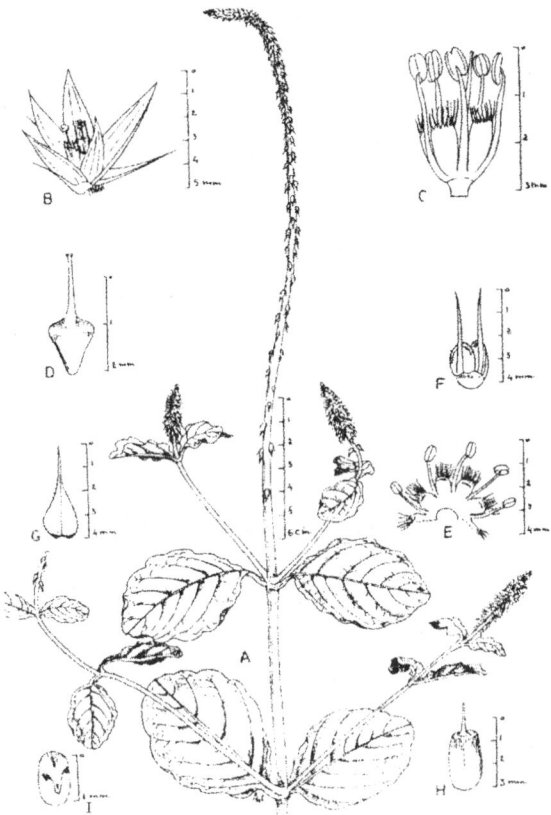

Fig. 5.37: *Achyranthes aspera* L. A. flowering and fruiting branch; B. flower; C. flower with bracts and perianth removed; D. gynoecium; E. androecium opened to show stamens and staminodes; F. bracteoles; G. bract; H. utricle; I. seed.

of fresh leaves, mixed with water, is used for allaying pain from bite of wasps.

Fresh leaf juice with little of opium is beneficial when applied to syphilitic sores.

Other species: *Achyranthes binata* is used as diuretic and as an astringent. Plant seeds are rubbed with rice water and given to cure bleeding piles.

6. *Aconitum* species

Family: Ranunculaceae.

Local names: Atis, Patis (Hindi).

The roots of various species are used in trade and demand considerable attention. Recently efforts have been made to cultivate various species in the farmers field in different parts of the Himalaya, where the different species are found in nature. It has a good price in the market. The active principle is a nontoxic alkaloid *atisine, hitisine* and *hetratisine,* etc. More than 225 kg of dried roots have been estimated to have been marketed from Lahul only, in a single year. Small quantities of *A. violaceum*, fair quantity of *A. chasmanthum* are reported to have been collected from Himachal Pradesh (Chamba district). There are many species in the Himalaya.

1. *A. balfouri* Stapf. occurs in Garhwal Himalaya and in Sikkim occurring (3300–4200 m) in subalpine and alpine Zone in (3500–4300 m). The tubers are poisonous. Exploitation of this plant is banned in Sikkim.

2. *A. chasmanthum* Stapf.ex Holmes (Vern. Banbalnnag-Kashmir). It is found under moist shady slopes and among boulders in Himachal and Kashmir. The root is poisonous and used as substitute for *A. nepallus*.

3. *A. deinorrhizum* Stapf. (Vern Dushia-Bish, Maurabikh, Mohra-Himachal Safed-Bikh Hindi) in alpine belt of Garhwal. A powerful sedative, anodyne, diuretic and virulent poison in large doses. The force of blood circulation is reduced and frequency of respiration diminishes, while in falal dosés there is loss of sight, hearing and feeling followed by convulsions, syncope and ultimately death.

4. *A. elwessi* Stapf. In alpine regions of north Sikkim, Naga hills in east-Himalaya. The roots are poisonous.

5. *A. falconeri* Stapf. (V. Names: Bis, Bikh, Meetha tellia Hindi). In alpine

and subalpine regions of Garhwal hills. Large quantities are collected and sold outside the state. The roots are poisonous and used as substitute for *A. chasmanthum* Stapf.

The root is soaked in cow's urine for a week, then washed, dried and powdered. The powder (0.4 gms) is administered internally in acute non-malarious fever and diarrhoea, when mixed with *Piper longum*. An external treatment named Gul-Dahakarma) is employed by local people to cure rheumatism pain in joints.

6. *Aconitum heterophyllum* Wall. (Vern. Atis, Patis, Mithi-patis). A tall herb with purple marked greenish blue flowers in terminal racemes and spindle-shaped roots in pairs. It is met with among the moist boulders in the subalpine and alpine areas of Kumaon and Garhwal, 3000–3500 m alt. It is the oldest nonpoisonous aconite drug. The roots are antiperiodic, aphrodisiac, astringent, tonic and used to cure diarrhoea, dyspepsia and cough.

The root is said to be aphrodisiac and tonic, checks diarrhoea and removes bile (corrupt). The root powder is administered internally for relief of abdominal pain and diarrhoea, fever of infants.

The tubers are collected during the months of October and November. In Sikkim exploitation of this plant is banned.

The discovery that little alkaloid exists, the drug is not extensively used as it has no antiperiodic value, renders it necessarily to remove it from being considered as a drug, except as a mild bitter tonic.

Atis root of a good quality, should break with a short starchy fracture and present a uniform white surface. The fresh, fully grown root is about 2–4 cm long and about 1 cm thick at its upper extremity. In structure it is of white farinaceous substance within. The TS section seems to consist of 4 or 5 isolated cambium strands, the vessels of which show prominently radiating wedge-shaped formations.

7. *Aconitum laciniatum* Stapf. The species is found in Sikkim region and is poisonous, one of the sources of Calcutta Bish or Bikh.

8. *Aconitum laeve* Royle. A poisonous species in Himalaya from Chitral to Kumaon, 1700–4000 m

9. *Aconitum lethale* Griff. A poisonous species, believed to be the source of poison of the Mishmis in eastern part of Himalaya.

10. *Aconitum luridum* Hook.f. and Th. (Vern. Bish, Mahoor, Butchnap-Hindi). Himalayan region in eastern Nepal to Chumbi.

11. *Aconitum moschatum* Stapf. A poisonous species from eastern Kashmir.

12. *Aconitum nepallus* L. A species widely distributed in Asia, America, Europe.

13. *Aconitum palmatum* D. Don. A nonpoisonous species. The rhizome contains an alkaloid 'palmatisine'. The root is nonpoisonous having tonic, antiperiodic in diarrhoea and used in rheumatism.

14. *Aconitum soongaricum* Staf. The roots contain an alkaloid 'sangorine'; roots are poisonous. Plant of very high altitude.

15. *Aconitum spicatum* Stapf. A plant of alpine areas in eastern part of Himalaya, 3300–4000 m. The roots contain a toxic alkaloid *bikh-aconitine* and is the main species in northeastern markets.

16. *Aconitum violaceum* Jacq. (Vern. Tota, Dudhia). It is common in Ladakh, Kashmir, Spiti, Bashahar, Kunnawar

region; has no commercial value but sold in some areas and can be distinguished from *A. heterophyllum*.

The aconites thus invested with terror, has however been so subdued and reduced to such an extent which is of manageable state, so as to have become a powerful remedy in some of the most troublesome disorders incident to human beings. It is in generally used in intermitent fevers, chronic rheumatism, gout, paralysis, neuralgia, many painful affections of the heart, antiphlogistic in nervous inflammatory diseases as pleurisy, pericarditis, pneumonia, etc. It is also used externally in the form of ointment in neuralgia and rheumatic affec-tions.

Cultivation of Aconites has been discussed by Nautiyal and Nautiyal (2004) Cultivation of High altitude Medicinal Plants, HAPPRC (Bul.No 8, 2004).

7. *Actaea spicata* Linn.

Family: Ranunculaceae.

Local names: Baneberry (Eng.); Mamira (Hindi).

Description: An erect perennial herb, more or less hairy. Stem 20–30 cm high, usually branched. Leaves alternate, divided twice or often into 3 parts (ternately compound); leaflets pointed, often lobed, deeply and sharply toothed. Flowers regular, white, crowded in short, terminal racemes which become longer in fruit. Sepals 4, petal like concave falling early. Petals 4, shorter than sepals. Stamens many, longer than the sepals; anthers small. Fruit a black, ellipsoid, smooth berry, containing about 10 seeds.

Distribution: Temperate regions of Himalaya from Hazara to Bhutan, 2000–3000 m, in shady ravines, woods, pastures and hilly limestone tracts.

Kashmir: Tanmarg forest, in forest north of Hayan pass, Gulmarg, fir forests and amongst low growing shrubs on hill sides asbove 2300 m common. Kishtwar, Hazara, Sonmarg, Kagan, Muzaffarabad, Nagkunda, Shimla, Kulu, Pulga, Ravivalley, Garhwal, Singjavi, Mundali, Harkidoon, Lahul, Chur, etc.

Uses: The *roots* are used in the form of a tincture and act as a powerful nerve-sedative, in neuralgia and rheumatic affections. It is also emeto-purgative, capable of producing dangerous effect, when given in excess. It is useful in various disorders and rheumatic fevers. The decoction of roots is locally used in abdominal pains.

Powdered leaves, stem and flowers are used as insecticide and applied externally for skin diseases.

The *berries* are used internally for asthma and scrofula and externally for skin complaints.

The toxic principal referred is 'oil of ban-berries'.

Extent of utilization: The toxic principle is referred as 'oil of bane berry' seeds contain tannin.

Small quantities of this plant are collected and sold in the plain.

8. *Adiantum aethopicum* L.
(A. emarginatum Bedd.)

Family: Polypodiaceae (Fern).

Local name: Hansraj (Hindi).

Description: A small fern with several fronds.

Distribution: In South Indian hills, at higher elevations in Nilgiris Palney hills and Northern Kanara.

Uses: Infusion of the leaves used as emollient in coughs and diseases of the chest.

Other species

a. *Adiantum venustum* G. Don. (Vern. Hansraj Hindi).

A small plant often seen on shady places. The fronds are tonic, resolvent, expectorant, diuretic and emetic. These are bitter and employed in the form of syrup which corresponds with 'Sirp de Caoillaire' for chest affections.

b. *A. capill-veneris* L. (Vern. Hansraj), *A. flabellulatum* L. and *A. tenerum* Sw. are used for the same purpose.

c. *A. caudatum* L. (Vern. Mayur-rakshika Hindi). Used in skin diseases, diabetes, cough and fever.

d. *A. flabellulatum* L. Used for cough, while the rhizome is anthelmintic.

In eastern parts like Assam, Nepal and Bangla Desh (Sylhet).

9. *Acorus calamus* Linn.

Family: Araceae.

Local names: Bhadrá (S); Bach, Ghorabach, Saféd-bach (H); Vaj (Mumbai) Gandhilovaj (Guj); Vekhand (Mahr). It is *Calamus aromaticus* of earlier writers and *Acorus* of the Greeks.

Description: An aromatic, erect herb; rootstock thick, creeping. Stem flat, 15–30 cm. Leaves radical, tufted, resembling those of *Iris*, margins crimped. Spathe leaflike, long, narrow, not enclosing the spadix. Spadix sessile, 5–10 cm long, cylindric, dense flowered. Flowers bisexual, yellow-green; perianth of 6 segments, persistent, free. Stamens 6 at the base of the segments. Ovary free, oblong, 3-celled, top conical; ovules many. Berries yellow-green, angular from mutual compression, 1–3 seeded.

Distribution: Throughout India, in marshes or on riverbeds, ascending to 2500 m on the Himalaya. It is often associated with some ferns. It is plentiful in the marshy tracts of Kashmir, Sirmur, Manipur and Naga hills. It has been cultivated in Mysore and other regions. In semi-aquatic condition it occurs between 600–2000 m and is found in fresh water swamps of Dehradun (Mathronwala).

Kashmir, Kasauli, Shimla, Kulni, Jaunsar, Naitwar, Tehri Garhwal, Kumaon: Kapkot; Khasia.

Uses: The dried rhizomes of the plant constitute the drug which is used medicinally.

The rhizome, when sucked by children, clears throat and improves the tone of voice. It is useful in children's diarrhoea. Powdered rhizome in water is most efficaceous stomachic. It may be chewed by dyspeptic person and relieve toothache. It is considered as an emetic and a household remedy for flatulant colic. The rhizome, due to the presence of a volatile oil, acts as carminative and is useful in asthma, in curing dyspepsia, remittent fever, dysentery and also as a nerve tonic and insectifuge. It is prescribed for flatulance even in infants.

The leaves and rhizomes are also used for flavouring drinks and for perfumary.

Powdered roots are used as vermifuge. Roots of *Bansa* with green giloe (*Tinospora cordifolia)* and mulheti, (*Glycyrrhiza)* when boiled, strained and mixed with honey cure fever and cough.

Oil from the rhizome is a good nerve stimulant and the essential oil free alcoholic extract shows marked sedative and analgesic properties which justify its use in mental diseases. The antibacterial activity of rhizome has recently been shown experimentally. Oil is also used in liquor and as snuff.

Extent of utilization: In the indigenous system, powdered rhizome is used as an ingredient in the preparation of *Vachadigh-*

rita and *Trishnadivarti*. From Kanatal forest (Tehri Garhwal) alone in 1963–64 about 4 quintals of the rhizome are reported to have been collected (Uniyal and Issar, 1967) and a fair quantity is exported to the plains. Ram Chandra Rao (1975) has shown that a gross income of Rs 12000/can be obtaind with an input of Rs 2000/ha. If distilled for oil, a gross income of Rs 30,000 can be obtained with an input of Rs 12000/ha.

Collection: In May and June this plant is taken out along with the rhizome and dried in partial shade.

10. *Adhatoda vasica* Nees

Family: Acanthaceae.

Local names: Vasáka (S); Adusà, BánÑa, Vasiká (Delhi) Rusa, Adulsaá (Mar); Alduso, Árdusi (Guj).

The trade name *Vasaka* is based on Sanskrit name of the plant; even the scientific name is based on the Sanskrit name.

Description: An evergreen densely branched shrub with smooth black, ash-coloured bark. Leaves opposite, elliptic, 12–20 cm long, pointed. Flowers white with red spots, 2–7 cm long, on axillary, stalked, bractate spikes; bracts leafy, 1-flowered. Calyx deeply divided into 5-lobes, pubescent. Corolla 2-lipped, pubescent outside; upper lip notched, curved, lower lip 3-lobed. Fruit a 4-seeded clavate capsule, pubescent.

Distribution: Common in tropical regions of India from Punjab, Haryana to peninsular India throughout the plains and in hills ascending to 1300 m; it is common near habitations. The plant is exceedingly common in Dehradun and Saharanpur and also in Jaunsar in valleys, on vacant lands.

Uses: The drug *Vasaka* comprises the fresh or dried leaves of the plant. It has the property of healing diseases of lung like cough, asthma, phthisis and tuberculosis. It expels phlegm and is antiseptic and germicidal in whooping cough.

The root is expectorant, antispasmodic, antiseptic, antiperiodic and anthelmintic, used in all forms of cough, catarrhal fever, malaria, bronchitis, cold, asthma, phthisis, diphtheria, febrile disturbances and gonorrhoea.

The leaves containing an active principle *vasicine* are powerful expectorant and antispasmodic, used in diseases of respiratory tract and in tuberculosis, cough, bronchitis and asthma. The juice is also employed in diarrhoea, dysentery, especially in haemoptysis and in bleeding. Powdered leaves are used as febrifuge in malaria. Externally poultice or fomentation of leaves is used for fresh wounds, rheumatic joints, inflammatory swelling; in neuralgia, headache and bleeding from nose. Decoction of the leaves makes good application for scabies and other skin diseases. The chief use of the plant is in the form of juice, syrup or decoction since it softens the thick sputum and facilitates its coming out, thus bringing relief in bronchitis. Larger doses can, however, cause irritation and vomiting. 7 leaves of the plant when boiled in 100 cc. water, strained and mixed with honey cure asthma, cough and phthisis. Syrup is made with sugar. Leaf juice with honey is given to children for cold.

The flowers and fruits are bitter, aromatic and antispasmodic. In the form of infusion they are useful as an anthelmintic and in heat of the blood. Fresh flowers are used in hectic fever, gonorrhoea and ophthalmia. Confection of flowers with sugar prevents spitting of blood with cough in tuberculosis of lungs.

Other uses: The twigs and leaves of the plant are used as green manure in rice fields

and also as weedicide, insecticide and fungicide on account of the alkaloid *vasicine*. As weedicide it is used against aquatic weeds in rice fields, as insecticide in the same way as tobacco leaves and as fungicide they prevent the growth of mould on fruits covered with *vasica* leaves. Often the leaves are used to enhance ripening of immature fruits and development of natural colour by market gardners. The leaves emit an unpleasant smell and are spared from browsing, hence the name *adhatodai* in Tamil, i.e. not touched by goats. The plant may thus be suitable for planting in soil reclamations programme. Wood of the plant gives charcoal for gunpowder and the ash is used by *Dhobis* in place of *Sajji*. Infusion used as remedy again blight on Tea and other crops. The fruits are hung round the neck of children to keep them away from catching cold.

Extent of utilization: The plant is widely used in Ayurvedic and Yunani systems of medicine. The active principle in the leaves *Vasicine* is used in cough syrups.

Studies of Patel *et al* (1982) showed that an improvised dentrifice of a different formula including *Adhatoda vasica* as an ingredient will provide additional benefits by enhancing the total effectiveness, since the extract significantly potentiated the antimicrobial effect of dentrifices. The abrasiveness and pH of these dentrifices reached to the control level according to Federal USA Specifications, when combined with plant extract.

The beneficial effects of this plant on gingival inflammation without producing side effects are toxicity which could prohib-

it its practical applications (Doshi and Patel 1981). Salt of Bansa (2–4 gm) mixed with honey or placed inside betel leaf and eaten relieves cough and asthma.

11. *Agrimonia pilosa* Ledeb. var. nepalensis (D.Don) Nakai (Agrimonia eupatorium sensu Hook. f).

Family: Rosaceae.

Local names: Ratpatia (Kum.). The same name is used for *Senceis nudicaulis*.

Description: A perennial, hairy herb. Stems erect, 30–60 cm. Leaves pinnately compound; lower leaves 10–17 cm; leaflets coarsely toothed, very unequal, larger ones 5–9, ovate, intermixed with a number of much smaller ones. Upper leaves gradually smaller and with fewer leaflets. Stipules adnate to the base of the leafstalk. Flowers yellow, in terminal, spike-like racemes, each flower in the axil of a small, 3-cleft bract and with 2-smaller, 3-toothed brac-teoles at the top of its stalk. Calyx tube top shaped, grooved, bearing outside its mouth a ring of small, hooked bristles; limb 5-lobed. Petals 5, oblong. Stamens 15. Carpels 2, free, enclosed within the calyx tube; stipules thread like, protruding; stigmas terminal, dilated; ovule soiltary. Achenes 1 or 2, enclosed in the hardened, bristly calyx, crowned with a ring of hooked bristles.

Distribution: In the temperate region of Himalaya from Kashmir to Sikkim 1000–3300 m. The plant is the western vicariant of *A. eupatorium*. The latter have small flowers and fruits.

Uses: The root is astringent, tonic, diuretic and useful as decoction in coughs, simple

Patel VK, Doshi JJ and Venkatakrishna Bhatt H. 1982. Effect of *Adhatoda vasica* leaf extracts on efficiency of various dentrifices. Pro. Nat. Acad. Sc. India 52 B1: 21-24.

Doshi JJ and Patel VK 1981. Dental Dealogue.

diarrhoea and relaxed bowels. It is reported to give tone to the system and promotes assimilation of food.

The Himalayan species is not utilized and had been confused with *A. eupatorium* which is reported to contain essential oil and a dye. The plant needs further study from the utilization point of view.

12. *Aerva lanata* (L) Jussieu

Family: Amaranthaceae.

Local names: Gorakhbuti, Chaya, Astmabyda (Sans); Bui, Kapur Jhari (Hindi).

Description: An erect or prostrate herb with many branches from a stout woody rootstock, pubescent or woolly. Leaves alternate, elliptic-obovate or suborbicular, pubescent above and woolly beneath. Flowers minute, greenish white, bisexual, arranged in small dense axillary heads or spikes, sometimes crowded into globular clusters. Perianth oblong, obtuse and apiculate, silky hairs on back. Utricle ovoid; seeds black and shining.

Distribution: An abundant weed over larger areas in Gangetic plain throughout hotter parts of India, up to 1000 m on the hills. It is almost similar to *Nothosaerva brachiata*.

Uses: The plant is anthelmintic and useful in kidney stones and in gonorrhoea.

13. *Ageratum conyzoides* L.

Family: Asteraceae (Compositae).

Local names: Ajganda, Ghandharisedardi, Sahdevi (Hindi); Goatweed Floss flower, White weed (Eng.).

Description: An annual herb about 1 m high, with erect stem, branched, columnar. Leaves stalked, 5–7 cm long, broadly ovate, crenate, hairy on both the sides, stalks about 2 cm long. Flowers heads small, in dense terminal corymbs; bracts striate, sharply pointed. Flowers many, pale blue or white, foul smelling; pappus scales 4–5. Fruit an achene, minute, black (Fig. 5.38 and Colour Plate 5.10).

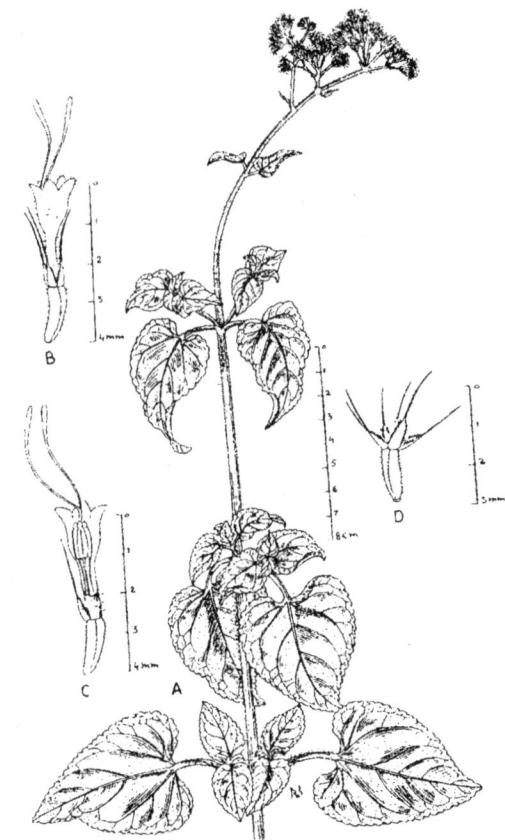

Fig. 5.38: *Ageratum conyzoides* L. A. upper portion of a flowering stem; B. flower; C. corolla partly cut away to show androecium; D. achene.

Distribution: A common weed of the tropics, throughout India, ascending to 1700 m in Himalaya.

Uses: The leaves are applied to cuts and sores. It is given with water, internally, as an emetic. A hot poultice of the leaves serves as remedy in boils, wounds and sores, leprosy, skin diseases, in cuts and sores, etc. Leaves used intravaginally for urine troubles, also considered a styptic and antitetanus when applied to wounds.

As a household remedy, infusion of the plant serves as a tonic in diarrhoea and flatulent colic. Its juice is used and applied locally in prolapse of rectum.

The flowertips are mixed with paddy to repel insect pests.

Root juice is considered as preventive for the development of stone in bladder.

Recently leaves have been used for the manufacture of incense sticks called *dhup*.

14. *Aloe barbadensis* Mill. (-A.vera Tourn. ex L.)

Family: Liliaceae.

Local names: Indian aloes (Eng.) Gritkumári, Ghee Kuvar (Hindi).

Description: A rossette of succulent lance-like leaves with spines on the margins. Flowers orange-red on long stalks. Leaves having a mucilaginous pith which is slimy. *Distribution*: A native of West Indies but grows wild in the south. Planted also in Indian gardens and many of its forms have naturalised in India. Also found in semiarid regions in a semiwild state and has naturalised in India, expanding from dry westward valleys to the Himalayan region up to the southern tip, cape Comorin.

Uses: The plant is stomachic, purgative, emmenagogue, anthelmintic, useful in a number of diseases, particularly in piles. Of late it has been widely used in cosmetics such as face lotions, creams, etc. It has a cooling property, bitter in taste and is anabolic in its action, a fighter of bile (pitta). This property enables it to guard against fever, skin diseases, burns, ulcers, excessive thirst and convulsions.

The anabolic activity energises the body. It contains an active principle *aloin*, responsible for its purgative action, relieves constipation and improves digestion. It acts on the kidneys through its diuretic action. It is used safely in scant urination and burning micturition; reduces fatigue and used as general tonic for general weakness.

It improves vision and reduces redness of eyes, providing a soothing effect to the eyes. Rheumatism and arthritis need a special mention; analgesic and antiinflammatory for joints. It is nontoxic and increases immunity of body.

Aloe leaves are scrapped, sprinkled with turmeric powder, revives swellings of the foot, inflamed eyes. The pith mixed with 'ajowan' is kept in earten pot and used in stomach pain and flatulance.

Its use in cosmetics and other formulations has made the juice available commercially in the market and is much in demand. It resolves swelling of the glands between thighs and armpits.

Other species

a. *Aloe abyssinica* Lam. The leaves are emollient and cathartic. A native of Abyssinia and Central Africa; also reported from Kathiaward coast.
b. *Aloe perryi* Baker. The plant is stomachic, tonic, purgative, useful in dyspepsia, jaundice and amenorrhoea. Reported from island of Socotra.
c. *Aloe socotrina* Lam. Uses are the same as of *A. abyssinica*.

15. *Amaranthus paniculatus* L. (-A. caudatus L.)

Family: Amaranthaceae.

Local names: Chua-mársa, Rajgiri, Kédari-chua, Chuka, Natya; Chaulai, Ramdana. (Hindi), Bustanafroz (Kashmir),

Marsa Himayan region from Garhwal eastwards; Amaranth grain, Velvet flower (Eng.).

***Description*:** A tall robust herb. Leaves long petioled, ovate-lanecolate, tip obtuse, spike in dense soft thyrsis, hardly squarrose. Bracts acicular longer than the ovate mucronate sepals.

***Distribution*:** Cultivated throughout India, asceding to 3000 m in Himalaya; also grown as a vegetable in north India.

***Uses*:** The plant is used as a blood purifier, diuertic in Kidney stone, in strangury, also given in scrofula and urinary disorders and applied to scrofulous sores. It is beneficial in piles.

The seeds are fried and made into *Laddus*. with jaggary, eaten in fast.

***Other species*:** *Amaranthus spinosus* L. (Vern. Jangli-chulái, Katali-chulai, Kanté-math)

A weed (Fig. 5.39), very common on the nitrogen rich soils, on wastelands and in the field. Roots are used in gonorrhoea, eczema and in colic and as lactagogue. Boiled leaves and roots given to children as laxative, applied as emollient poultice to abscesses, boils and burns. Plant is astringent, useful in diarrhoea and dysentery.

A. tricolor L. var. *tristis* (Prain) Nayar (Vern. Chumliság) cultivates as pot-plant.

A. tricolor L. (*A. gangeticus* L.) Vern. Lálság (Hindi). Cultivated for fatty oil and as drug to cure diarrhoea, dysentery, externally used as poultice, in ulcerated throat and mouth.

16. *Anagallis arvensis* Linn.

***Family*:** Primulaceae.

***Local names*:** Jonkmari (H); Loluru, Phopura (Jaunsar)

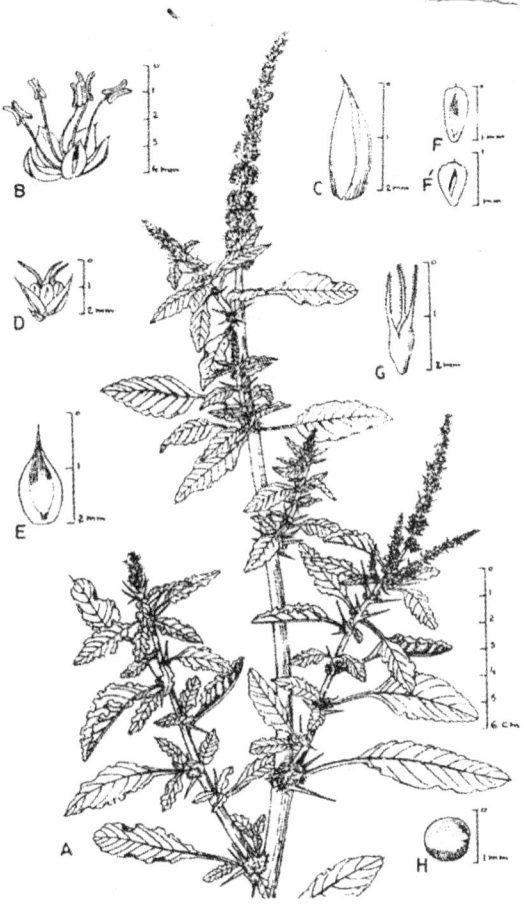

Fig. 5.39: *Amaranthus spinosus* L. A. flowering and fruiting branch; B. male flower; C. one of the perianth lobes from B; D. female flower; E. bract; F and F'. two of the perianth lobes from D; G. gynoecium; H. seed.

***Description*:** A glabrous, gland dotted herb. Stems slender, erect or decumbent, branching from the base; branches 4-angled. Leaves opposite, sessile, broadly ovate, entire, acute. Flowers bright blue, solitary axillary, closing in dull weather; stalks slender, longer than the leaves. Calyx 5-parted; segments narrowly lanceolate, acute. Corolla rotate, about 2 cm in diam, 5-lobbed nearly to the base; lobes glandular-fringed, entire, twisted in bud. Stamens 5,

inserted at the base of the corolla, filaments hairy. Ovary globose; style slender, stigma terminal, simple; ovules many. Capsule small, globose, opening by a circular fissure around the middle. Seeds numerous, minute (Fig. 5.40).

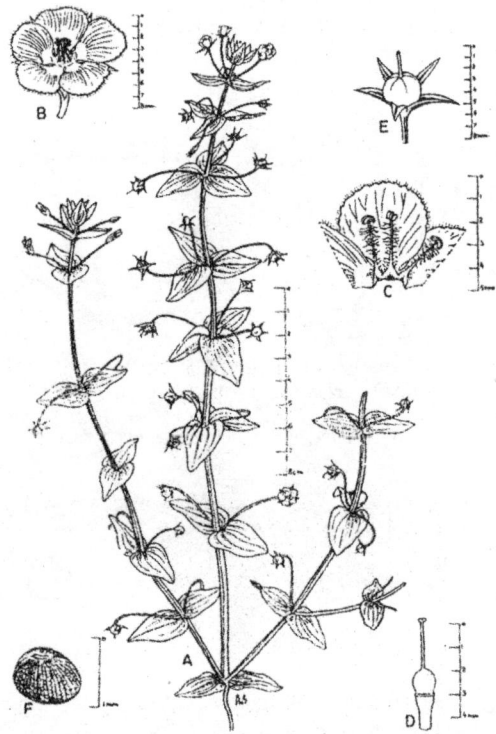

Fig. 5.40: *Anagallis arvensis* L. A. plant; B. flower; C. part of corolla and three of the stamens; D. gynoecium; E. capsule; F. seed.

Distribution: More or less throughout India as a weed, ascending to 2500 m in the Himalaya, on cultivated fields or in the garden during February and March.

Uses: The plant was formerly considered to be detrigent, vulnerary and cephalic. Some have reported its success also in hydrophobia. It is used for curing gout, dropsy, gravel and convulsions. It can expel leaches from the nostril of cattle.

Extent of utilization: The plant is not utilized as a trade commodity.

17. *Anaphalis adnata* DC

Family: Compositae (Asteraceae).

Local names: Chukmak Rui (Kum).

Description: An erect, stout, robust herb. Leaves thick, broadly lanceolate or oblong, often spathulate, acute, basal lobes short, if present. Heads in dense, rounded, terminal and axillary corymbs. Involucral bracts broad, obtuse, erect in flower, spreading in fruit; outer ones usually pale-yellow.

Distribution: In temperate regions from Shimla to Bhutan, 2000–2500 m.

Uses: In Kumaon the flower heads and leaves are dried in sunshine and used as wicks in crude lighter, locally called as *Chakmak*. Hairy pubescent on flower heads used to stop bleeding in Kumaon.

18. *Andrachne cordifolia* Muell. Arg. (in DC. prodr. 17:23 4. 1866) Muell. Arg. (in DC. prodr. 17:23 4. 1866)

Family: Phyllanthaceae.

Local names: Gurguli, Bhatula (Garhwal).

Description: Small shrub, branches round, slender. Leaves alternate, thin, long-stalked, base rounded, rarely cordate; lower leaf surface pale and hairy. Flowers green on slender axillary stalks. Male in clusters of 3–6. Female solitary, both on the same plant. Capsule globose, seeds 6, triangular.

Uses: The twigs and leaves are said to kill cattle, when browsed on empty stomach; leaves are reported to contain HCN.

19. *Andrographis paniculata* (Burm.f.) Wall. ex Nees

Family: Acanthaceae.

Local names: Kiryát (Hindi), Kálmégh (Kumaon); King of Bitters (English).

Description: Erect annual, 30–90 cm high, branches 4-angled. Leaves lanceolate, paler beneath, lateral nerves 4–6 pours. Flowers small, forming large paniculate inflorescence calyx linear. Lanceolate corrolla pink. Capsule tapering on each curl.

Distribution: West peninsula often cultivated in gardens.

Uses: The roots and leaves are febrifuge, stomachic, tonic, alterative and anthelmintic. Expressed juice of leaves with certain spices is dried in sun and made into globules called *Alui* and given to infants to relieve gripping. The plant is also cultivated as an ornamental.

20. *Aneilema scapiflorum* Wight.
(-Murdannia scapiflora Royle)

Family: Commelinaceae.

Local names: Siyá-musli (Hindi, Urdu); Kuréli (Gujarati, Maharashtra).

Description: Herb with elongated roots, finely acuminate leaves, narrowly ensiform. Scape erect with narrow panicle bract large, sheathing, amplexicaul. Capsule ellipsoid, trogonous. Seeds 4–6, superposed in each cell.

Distribution: In subtropical and temperate regions and the upper Gangetic plains, eastwards to Bhutan, up to 2300 m.

Uses: Roots are astringent, useful in headache, giddiness, fever, jaundice and as tonic.

Rootbark dried in shade is useful in asthma; dried powder of the root mixed with sugar is used as an aphrodisiac. Also used in colic, piles, infantile convulsions, asthma and incontinence of urine. With juice of *Tulsi*, it is given in spermatorrhoea.

21. *Angelica glauca* Edgew.

Family: Umbelliferae.

Local names: Chóra, Chorak (Himachal).

Description: A herb, the roots of which are used as spice in Kashmir and Himachal Pradesh.

Distribution: From Kashmir to Garhwal, between 2300–3300 m.

Uses: Plant roots are used as condiment. Contains an essential oil. Seeds produce vomiting and purging.

The plant is stimulant and used in dyspepsia and constipation. The active principle is an essential oil from the roots (1.3%) in dry roots.

Other species: *Angelica archangelica* L. The rootstock and fruits are stimulant, expectorant and diuretic. Dry rootstock contains 0.35% and fruits 1.2–1.3% essential oil and several furocoumarins like *angelicin, bergapeten, xanthotoxin* in addition to *umbelliprenin* and some *phenols*.

22. *Allium sativum* L.

Family: Liliaceae.

Local names: Garlic (Eng.); Lahsun, Lasan, Lasuna, Thum (Hindi).

Description: The genus *Allium* contains over 250 species and is distributed in northern Temperate regions. 27 species are recorded growing wild in India, mainly confined to the Himalayan region. The cultivated types are about 5 (*A. ascalonicum* L., *A. cepa*, *A. tuberosum*, *A. ampelorasum* and *A. sativum* L.). *A. sativum* has flat leaves. Bulb short, compound. Umbels lax, bearing both flowers and bulbils.

Distribution: *A. sativum* is a native of Central Asia, but cultivated widely and exported to other countries.

Uses: The rhizome is used as a spice in vegetable and pulses preparation and has sharp disagreeable smell and taste.

Oil from the tubers is useful in tuberculosis of lungs, cough, asthma and whooping cough. In case of rheumatism, cough and asthma, 2–3 cloves of garlic are given with honey. Externally, it is applied in migraine after grinding and application to the aching side and so are the abscesses and pimples cured.

One clove of garlic is taken with milk to reduce blood cholesterol. Garlic cloves warmed in mustard oil used locally rubbed on the affected parts. Garlic is strongly antiseptic, taken internally it destroys worms, and externally, will rid the in skin of parasites. It has antibacterial properties and used widely both for intestinal disorders and for infectious diaseaes. Antibacterial activity is due to 'allicin' which can inhibit growth of number of microorganisms.

Other species

a. *Allium blandum* Wall. (Vern. Dádu)
The entire plant is used as vegetable; also employed as digestive and carminative. Not utilised commercially except some use in folklore.

b. *Allium stracheyi* Baker (Vern. name: Kéer-Lahul)
The plant is abundant in the *Dangs* (in Gujarat) on moist slopes and on open rocky surfaces in Lahul, Ganga valley, Ralam valley and near Badrinath.
The leaves and bulbs are used extensively in the preparation of *Masalas*. not utilized commercially.

c. *Allium victoriales* L. (Vern.Pángri-Hindi)
Distributed from Kashmir to Jaunsar (Garhwal) Kedarkanta, etc. Var. *angustifolium* occurs at Milam glacier.
The leaves are used as vegetable and bulbs for flavouring; also used as carminative.

d. *Allium ampeloprasus* L. (-*A. porrum* L)
Vern. name: Kirath (Arabic). Paru (Mumbai).
The bulbs are stimulant, expectorant and used to hasten suppuration of boils.

e. *Allium ascolinicum* L. (Vern. names EK-Kandá, Lasum-Hindi, Gundhun-Mumbai).
The bulbs are stimulant, aphrodisiac, also used in earache as the juice of the bulb poured into the ears.

f. *Allium cepa* L. (Vern. name: Palandu (Sansk.) Piaz (Hindi); Dungari-Kanda, Kanda (Mahar.) Onion (Eng).
The bulbs are used as vegetable and are stimulant, diuretic, expectorant, aphrodisiac, emmenagogue, useful in flatulance and dysentery. It is considered as protection from *loo* (during summer heat), plague and cholera. It clears hoarse voice when taken with honey after warming and cures cholera. It cures instantly the hiccups. Haemorrhoids causing pain banadaged with onion baked in ashes. Cold and catarrah, causing congestion of head and nose are relieved after smelling sliced onion. Ointment from onion is used to cure wounds, particularly; the wound of breast.

g. *Allium tuberosum* Roxb. (Vern Bungaghunduna Mahar.
The seeds are given in spermatorrhoea.

h. *Allium schoensparsum* L. The properties are similar to *A. cepa*.

23. *Alpinia galanga* Willd. (-Languas galanga (L.) Stuntz.)

Family: Zingiberaceae.

Description: A rhizomatous plant 2–3 ft. in height. Rhizome, stem and leaves resemble that of sugarcane. It is dried and prepared to make *Soanth*.

Local names: Bará-kulanjan (Hindi), Sugandha-Vach (Sansk); Badi-pan-kijar. (Mah).

Distribution: A pereundal robust herb. Native to Malaysia and Indonesia, distributed in eastern Himalaya and southwest India. Also cultivated in north India for the rhizomes and the fruits which are traded in the market.

Uses: Rhizomes used as spice in curries for flavouring food, and medicine for rheumatism, stomach disorders, bronchial catarrh and as tonic and stimulant; also as substitute for *A. officinarum* (called small galangal).

Rhizome is the source of an essential oil.

Other species: *Alpinia officinarum* (Vern. name, Kulinjan Sugandha bach) is native to China. Cultivated to northern India. Rhizomes used as stomachic, stimulant and carminative, considered as nervine tonic and aphrodisiac.

A. sandarae Sand, and *A. speciosa* (Wendl) K. Schum. are both ornamental. *A. sandarae* is grown for its graceful foliage and *A. speciosa* for its sweet-scented flowers.

A. calcarta Rosc. is used as substitute for *A. galanga* and grown in Konkan area.

A. khulanjan M. Sheriff, gives an essential oil, rhizomes are stimulant, expectorant and carminative.

24. Anemone obtusiloba D.Don

Family: Ranunculaceae.

Local names: Rattanjog, Ratanjota (Kumaon); Kákriya (Garh).

Description: Herb with woody rootstock, fibrous, clothed with old leaf sheaths. Leaves many, heart-shaped, hairy on both surfaces. Flowers white, lower portion tinged with blue purple at high altitudes. Sepals 5. Petals 5. Fruit a head of many achenes tipped by a short style, and not imbedded in wool, but coarsely hairy.

Distribution: In the temperate and alpine regions of the Himalaya from Kashmir to Sikkim, 3000–5000 m. All the varieties of the plant are common; the yellow variety is smaller; the blue and white varieties are succeeded by the yellow one at about 4000 m.

Kashmir: Tanmarg, Khelanmarg, top of Hayan pass, Basam Gali, Tosh maidan, from Toshmaisan to DamamSar, Gulmarg and Aporwat, Kishtwar, Chamba, Pangi, Lahul, Chenab valley, Rotang pass, Dharamsala, Manali, Kokgar, Narkamda, Bhirogtibba (Shimla), Garhwal, Hattoo, Kedarkanta, Deoban, Datmir, Murreehills, Kumaon.

Uses: The pounded root which is acrid, is mixed with milk and given internally for contusions.

The leaves are bitter, emmenagogue and good in complaints of spleen and kidney, remove jaundice. It is also good for sores in the mouth.

The seeds, if given internally, produce vomiting and purging. The oil extracted from them is useful in rheumatism.

Extent of utilization and trade: The air dried plant contains small quantities of a substance with properties similar to *anemonin*. The plant is not much utilised commercially.

Other species: *Anemone narcissiflora* Linn. is said to be poisonous.

25. Argemone mexicana Linn.

Family: Papaveraceae.

Local names: Shialkánta (H), Darudi, Pila-Dhatura, Kantá-Dhotra Siakanta (P); Srigala-Kantaká (S) Tasty food of Camel. Kandar Mexican Poppy, Prickly poppy, Satynashi (trade name).

Description: A glabrous, glaucous herb. Stem prickly, branching, with yellow juice. Leaves thistle like, stem clasping, oblong, sinuately pinnatifid, spinous, veins white (Colour Plate 5.11). Flowers yellow, terminal, on short, leafy branches. Sepals 3, prickly, ovate, produced just below the tip, in a short horn-like excrescence. Petal 6. Ovary prickly, 1–celled, stigma sessile, 4–6 lobed, ovules numerous, borne on the walls of the cavity. Capsule prickly, oblong-ovoid, opening by 4–6 values. Seeds numerous, globose of the size of poppy seeds (Fig. 5.41).

Fig. 5.41: *Argemone mexicana* L. A. flowering and fruiting branch; B. leaf from lower part of stem; C. petal; D. sepal; E. stamen; F. gynoecium; G. capsule; H. seed.

Distribution: Throughout India, weed of waste places both in plain and warm valleys; common in fields and roadsides, ascending to 1500 m, also in warm valleys. An introduced weed, widely naturalised in nearly all tropical countries.

Uses: Herb is used as blood purifier, cure from syphilis, itching foul sores and rheumatism of joints.

Fresh juice of the plant is applied to indolent ulcers to promote healing. It is also applied to the eyes in conjunctivitis. Mixed with milk it is given in leprosy. Externally the juice is applied for relieving blisters and rheumatic pains, healing excortications, indolent ulcers, indolent scabies and herpetic erruptions.

The roots are alterative, used in skin diseases, stimulant and a decoction is used in gonorrhoea, gleet, vesicular calculus and in chronic skin diseases. Powdered root is useful in boils and abscesses. The decoction is also effective as an eyewash and mouthwash for tootache and as lotion for inflammatory swelling.

The leaves in infusion are used in cough.

The yellow juice from the plant contains an alkaloid 'protopine' and 'berberine' (22%), is diuretic and alterative with slight corrosive properties. When fresh, it is applied to indolent ulcers to promote healing. It is used in dropsy, jaundice, skin diseases and gonorrhoea. The juice is widely used as an infusion. When mixed with milk given in leprosy.

The seeds are narcotic, stomachic, emetic, expectorant, cathartic, demulcent and nauseant. Smoke of burning seeds is useful in toothache and pulmonary diseases.

Oil from the seeds is purgative, narcotic and demulcent. It is a powerful alterative in syphilis and leprosy. Used for herpetic and other skin diseases, indolent ulcers, eruptions, headache and conjunctival inflammation. The oil is semi-drying and used as lubricant in soap manufacture and painting. It is preventive of white ants and is medicinal for syphilis and many skin dis-

eases. In India it is used as an adulterant for mustard and other edible oils and is reported to cause blindness. In small doses the oil is aperient and gastrointestinal irritant in large dose producing vomiting and purging and occasional death. It is also given in other inflammatory conditions of the intestine. It can be used as illuminant. Oil mixed with sugar when taken kill intestinal worms. 3 drops of it are administered every two hours to cure cholera.

Extent of utilization and trade: The plant is collected and used on a limited scale by Vaids and Hakims for local use. The yellow juice contains an alkaloid 'Protopine' (argemonine) and 'Berberine'. Prolonged use of the oil produces in man toxic effects resembling those occurring in epidemic dropsy. In USA the oil is used by painters and in S. Africa as a preventive from white ants. The oil is adulterant of mustard oil causing dropsy and purging.

Seeds grounded in water (10 gm), strained and given to drink as a cure for rheumatism. 20 gms of seeds grounded with 7 grains of pepper, strained and given to drink in case of bite by mad dog.

26. *Apium graveolens* L. (-A.dulce Mill.).

Family: Umbelliferae.

Local names: Ajmod, Ramdhuni, Karas, Salári (Hindi); Chanu (Mumbai); Celery (Eng.).

Description: A herb, strong smelling, glabrous, biennial, about 1 m tall, forming a less extensive root system and a short, stubby stem (7.5–15 cm long) with a clump of thick-petioled pinnately compound leaves. The petioles or leafs-stalks constitute the celery of commerce. The greenish-white flowers are borne in small compound umbels, produced on a tall, grooved and pointed flowering stem.

Distribution: A native of Europe, now cultivated in Northwest Himalaya and in hills of Uttar Pradesh, Uttaranchal, Himachal and South India. It is cultivated during winters for its seeds and roots.

Uses: The root is alterative and diuretic. The seeds are stimulant, carminative, nervine and tonic, colic sedative antiseptic, emmenagogue, as antiseptic used in bronchitis asthma and for liver and spleen complaints. Dried ripe fruits are used as spice.

Seed oil is used as fixative and an ingredient in perfumes.

Long leaf stalks are consumed as salád and vegetable.

Other species: *Apium graveolens* L. var. *rapaceum* (Mill.) DC. (syn. *A. rapaceum* Mill.); Salári (Hindi).

Roots are eaten as vegetable. Fruits used for extracting oil, which is antispasmodic, nerve stimulant and used in rheumatism. Seed decoction is household and popular remedy for rheumatism.

It is cultivated in the hills of Uttaranchal and also in south India.

27. *Artemisia nilagarica* (Clarke) Pamp. (-A. vulgaris auct.non L.; A. vulgaris var. nilagirica C.B. Clarke)

Family: Compositae (Asteraceae).

Local names: Indian wormwood (Eng); Nagdona (Hindi); Tarkha (Punjab); Kunja (TG).

Description: An erect, hairy, tomentose shrub like herb; stem much branched. Lower leaves 1–3 pinnately lobed or 1–3 pinnatifid, ultimate segments narrow, rather broad, entire or nearly so, upper surface pubescent or hairy, lower tomentose or hairy, white grey or brown; leafrachis

winged, wings entire. Floral leaves 3-lobed, nearly to the base or lanceolate. Involucral bracts few; pappus none (Fig. 5.42).

Fig. 5.42: *Artemisia nilagirica* Pamp. A. upper portion of a flowering stem; B. leaf from lower portion; C. involucral bract; D. an outer floret; E. disc floret; F. corolla of E partly cut away to show androecium, G. style of E.

Distribution: Abundant on waste grounds on roadsides and near cultivation, in west Himalaya, Sikkim, Khasi hills, Darjeeling, Manipur, western ghats from Konkan southwards.

Uses: The leaves are used locally as antiseptic and expectorant; collected for making incense sticks; also a remedy in skin diseases, used for fomentation.

Leaf juice is useful in infantile diseases and given in affections connected with debility, and asthma. A strong decoction is given as vermifuge.

Boiled leaves are used as poultice in headaches. In case of suppression of mensus and hysteria, it is given as an infusion. It assists parturition and prevents abortion and has use in nervine and spasmodic affections.

The plant yields 2% volatile oil, which is a good larvicide like kerosene oil, also a feeble insecticide.

Other species

a. *Artemisia absinthium* L. Worm wood (Eng.); Vilaiti-afsantin (Hindi); Damar (Sansk).
   A herb. Flowers used as vermifuge, and the flowertops, a source of drug *santonin*, which is used as stimulant, vermifuge and tonic in intermittent fevers. It contains a glucoside 'absenthin' and an essential oil in the external hair-like glands only. In Kashmir 1000–3300 m.

b. *Artemisia dracunculus* L. An aromatic perennial in western Himalaya. Aromatic leaves are stimulant, febrifuge, aperient and stomachic. An essential oil from the leaves and flowering tops is used to flavour wine and other beverages.

c. *Artemisia maritima* L. (-A. brevifolia Wall., A. fragrans Wild.) Wormseed, Santonica- (Eng.) Kirmálá (Hindi) A stout, much branched, perennial aromatic, dried immature leaves and flower heads constitute the drug. Occurs in the Himalaya from Kashmir to Kumaon. Just when the flowers start opening, the 'santonin' is maximum. The drug is used for expelling worms (thread worms and roundworms) from the stomach; also used in fevers and as stimulant. Indian plants besides santonin also contain two more constituents, 'beta santonin'. Out of the 2 varieties,

namely *neercha* and *seski*, the latter yields more oil than the former, (found in July to mid Sept.).

d. *Artemisia pallens* Wall. An aromatic herb known as *Devana*. Cultivated in the south (Karnataka) and near Poona; also on a small scale in Andhra, Kerala and T. Nadu. Leaves and flowers are widely used; the oil having a delicate aroma, used in high grade perfumery and for flavouring beverages.

e. *Artemisia scoparia* Waldst and Kit. V. Name: Barna-Hindi. A perennial herb, twigs made into brooms for cleaning. *Scoparone*, present in this species exhibits marked hypotensive and tranquilizing activities. Also used as a cure for pain in the ear. Seeds and flowering buds contain an essential oil (0.75%) and 0.92% *scoparin* (a crystaline lactone).

28. *Asparagus filicinus* Buch.-Ham.

Family: Liliaceae.

Local names: Allipalli (Kash); Sharanoi, Kauntá (Jaun).

Description: An erect, unarmed shrub. Cladodes 0.75 to 0.5 1.25 cm long, flat, falcate, acuminate. Flowers solitary or in pairs, white; pedicels jointed above the middle, very slender. Fruit a globose berry.

Distribution: In the tropical and temperate regions from Kashmir to Bhutan: Kashmir, Lahul, Shimla, Jaunsar; Balcha, Deoban, Chur, Mundali.

Uses: The tuberous roots are used as drug and are pickled. The roots are considered as tonic and astringent. It is used in cholera and acts as powerful diuretic. It is also used as a cure for rheumatism due to dampness. The shoots are eaten as vegetable.

Time of collection: The plant flowers in May and June.

Extent of utilization and trade: To a limited extent the roots are collected, dried in shade and exported to plains where they are sold in bazaar as *Satavari* or *Bojidan*.

Other species

a. *Asparagus racemosus* Willd.

Family: Liliaceae.

Local names: Shatamuli (H), Satavári (Bo); Satraw-al (Dun) Sharanoi (Jaun).

Description: A much-branched, scandent shrub with terete stem and triquetrous branchlets; spines more or less recurved. Cladodes 2–6 together, narrowly subulate, falcate and divaricate, channelled beneath. Racemes 2–5 cms long, pedicels jointed in the middle, slender below the joint. Flowers white. Anthers purplish. Fruit a globose berry.

Distribution: Throughout the tropical and subtropical regions of India, ascending to 1300 m in the Himalaya, Kashmir eastwards.

Time of collection: The plant flowers in the autumn and fruits in the cold season.

Uses: The roots are bitter and used as astringent, demulcent, diuretic, aphrodisiac, antispasmodic, alterative and antidysenteric. It is reported to increase lactation in animals (dogs). Roots boiled in milk administered to relieve bilious dysentery and diarrhoea and promote appetite.

The drug is collected throughout the country and sold in the bazaar.

The tuberous roots of the following species are also used in medicine:

b. *Asparagus adscendens* Roxb.

Local names: Hazarmuli, Saféd musli (H), Saphetamusli (Mumbai), Jhirna (Garh.)

The roots are used for general debility and in dysentery.

c. *Asparagus officinalis* Linn.

Local names: Halyun (H), Hikua (B).

Distribution: The plant is cultivated.

Uses: The roots are diuretic and an infusion used against jaundice and congestive part of the liver. It contains an essential oil *A. gonocladus* Baker. *A. sarmentosus* Linn. are other species used in medicine.

29. *Asclepias curassavica* L.

Family: Asclepiadaceae.

Local names: False Jpecac; Blood Flower (Eng); Kakatundi, Kaura-dodi (Hindi).

Description: A herb or undershrub, erect, about 60–75 cm high with opposite, lanceolate or oblong lanceolate, glabrous or slightly pubescent leaves, 5–7 cm long and many fld.; umbels of orange coloured flowers. Corolla lobes reflexed , bright orange, of 5 erect spoon-shaped processes, adnate to the stpitate staminal column. Follicle solitary, erect, 7 cm long, narrow-ovoid, flat , winged, comose at the hilum (Colour Plate 5.12).

Distribution: At the foot of hills in the Siwaliks and in grasslands and on sides of canal at Dehradun. The plant has gone extinct near Dehradun, but still found in swampy situations in the forest. Fls in May-June, practically the year round.

Uses: The roots are used as emetic, purgative, used in piles and gonorrhoea. Leaf juice is anthelmintic, sudorific and for gonorrhoea.

30. *Asphodelus tenuifolius* Cav.

Family: Liliaceae.

Local names: Jangli-Piázi, Bokát (Hindi); Dungru (Gujarat).

Description: An annual herb with long terete leaves, sheathing at the base, finely puberulous. Scapes several from the roots, much branched above, 30–60 cm. Flowers white, laxly racemose, solitary in each bract, pedicels jointed below the middle. Bracts broadly ovate, boat-shaped, scarious, with a strong brownish keel. Perianth segments oblong, obtuse, with a brownish costa. Stamens acutely 3-gonous, black.

Distribution: Abundant as a weed of cultivation. Flowers during cold season. In fields on the plains; such as Indo-gangetic plain.

Uses: The seeds are diuretic, applied externally to ulcers and inflamed parts of the body.

31. *Atropa acuminata* Royle ex Lindley

Family: Solanaceae.

Local names: Jhárka, Ság-Angur, (Hindi); Indian Belladona (Eng).

Description: An erect, branched, perennial herb.

Distribution: In altitudes between, 3000–4000 m. Recorded from Kashmir to Himachal Pradesh.

Uses: The roots and leaves are narcotic, sedative, diuretic and used as anodyne; however the berries are poisonous. The alkaloid content in indigenous plants is 0.81% in roots, resemble very much to the European species (*A. belladona*). *A. belladona* and their natural hybrids are grown in Kashmir. The species name is due to the practice among Italian ladies of using belladona to increase apparent blackness of eyes by dilating pupil.

Other species: *Atropa belladona* is cultivated by RRL Jammu in Kashmir. Roots called Belladona radix and leaves *B. folium*, used as sedative and has property of dilating pupils of eyes (atropine).

32. *Bacopa monnieri* (Linn.) Pennell. (=B. monniera Wettst. Herpestis monniera (Linn.) H.B. and K.

Family: Scrophulariaceae.

Local names: Mandukparni, Nirabrahmi, Bahmi (S); Saféd Chamni (H), Jalbrahmi, Jalnervri (Guj); Ghola (Marh); Saumyalata (S) Thyme-leaved gratiola (E).

The trade name Brahmi and Bacopa are based on the local Indian and scientific names of the plant respectively.

Description: A succulent, glabrous, creeping herb, rooting at the nodes; branches ascending, several. Roots arise on the nodes of the stem also. Leaves obovate oblong, entire, lower surface dotted, fleshy. Flowers arise in the axil of the leaves and are borne on short pedicels, blue, generally pale blue, at times white. One of the five sepals is larger than the others. Corolla 2-lipped, 1 cm across. Capsule ovoid, 2 grooved; seeds many.

Distribution: The herb is found in moist or wet places, such as borders of water channels, wells, irrigated fields, etc. in all parts of India. It is sometimes present even in saltish water, ascends in the hills to 1300 m.

Uses: The drug consists of the whole plant. Locally the plant is used as laxative. In medicine the plant is valued for the *Baccosides* as a nerve tonic and is prescribed in nervous disorders, mental diseases, constipation, asthma, epilepsy, insanity, and as diuretic. It helps to improve memory among the aged. This compound has been patented.

The leaf juice is given to infants in bronchitis; the relief is due to the vomiting and purging brought about by the drug. Leaf juice mixed with petroleum, is applied in rheumatism. A poultice made of the boiled plant is placed on the chest of children suffering from cough. The plant is considered as blood purifier.

The alcoholic extract of the plant has tranquilising effect as has been tested successfully on the animals.

Extent of utilization and trade: Very small quantity of the plant used to be collected for local consumption, it was promoted by CDRI, Lucknow, as a master drug.

The plant contains an alkaloid *bramhine*, which is a cardiac tonic; its therapeutic action resembled *strychnine*, but is less toxic.

33. *Barleria cristata* Linn. var. ciliata (B. ciliata Roxb.)

Family: Acanthaceae.

Local names: Jhinti (S); Tadrélu (P).

Description: An erect, hairy perennial herb; stem branching. Leaves shortly stalked, ovate-lanceolate, entire, acute. Flowers liliac or pink, crowded in short, headlike, nearly sessile axillary spikes; bracts none; bracteoles linear hairy. Calyx 4-parted, segments hairy, outer pair lanceolate, spinous toothed, acute, inner narrow. Lower half of the corolla tube cylindric, upper half dilated upwards, limb spreading, lobes 5, ovate, nearly equal. Stamens 2, as long as the corolla acute 2, imperfect, shorter, anther cells blunt. Ovules 2 in each cell, style minutely capitate. Capsule oblong; seeds 4 or fewer, silky.

Distribution: Throughout India, ascending to 2000 m in the Himalaya in the valleys.

Uses: The drug constitutes the roots and leaves.

The plant is bitter and useful in inflammation, fever, bronchitis and diseases of the

blood. The infusion of the plant is given in cough. The roots and leaves are used particularly to reduce swelling and an infusion is given in cough.

Time of collection: The plant flowers from July to Oct.

Extent of utilization and trade: There are two varieties of this plant—var *ciliata* found in H.P., Uttaranchal (Kalsi, Nalapani-DDN) and var *microphylla* collected from Shimla.

Other species: *Barleria prionitis* Linn. (Katseeya, Chabri (H), Karaunta (S). A prickly shrub with yellow flowers is common in tropical regions of the country, especially in the Deccan and Karnataka. The leaves and roots of the plant are used medicinally. Juice of the leaf is used in catarrhal affections of children and applied to the soles of feet to harden them. Roots useful in skin diseases and a paste is applied to boils and glandular swellings. Leaves are chewed to relieve toothache (Fig. 5.43).

34. *Benincasa hispida* (Thunb.) Cogn.
(B. cerifera Savi)

Family: Cucurbitaceae.

Local names: The white Gourd Melon. Ash Gourd (Eng.); Pétha (Hindi).

Description: A cultivated plant, grown for its fruit, which is used both as a vegetable and also candied with sugar in preparation of *Petha* by confectioners, particularly famous for Agra preparations.

Uses: Fruits are edible and so are the seeds. The fruits are laxative, diuretic, specific for haemoptysis and other haemorrhages from internal organs. Juice of the fruit is given in insanity, epilepsy and other nervous diseases. The seeds are anthelmintic and yield a mild pale oil.

The fruit excretes a waxy bloom which, it is said, can be made into candle.

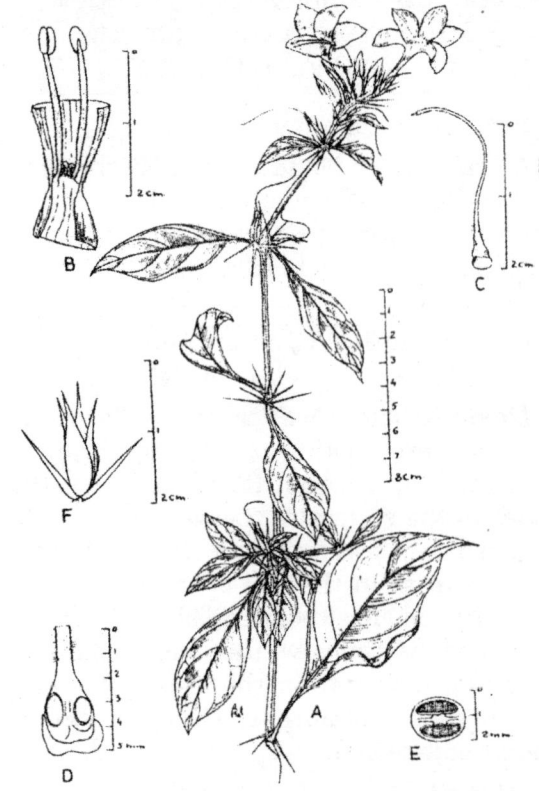

Fig. 5.43: *Barleria prionitis* L. A. flowering twig; B. corolla limb cut away and tube opened to show androecium; C. gynoecium; D. longitudinal section of gynoecium; E. transverse section of gynoecium, F. calyx and bracteoles.

The alterative and styptic properties of the fruit are popularly known as a valuable antimercurial.

35. *Bergia odorata* Edgew.

Family: Elatinaceae.

Local names: Kakria, KarbujaRohwan (Hindi).

Description: A small undershrub, white pubescent, aromatic, bark papery, leaves sessile, thick, elliptic or ovate. Flowers white with purple styles, fascicled, pedicelled 2–8 together varying much in size. Sepals and

petals ovate and obovate respectively. Stamens 10, filaments dilated below. Styles 5, purple, one half the length of the ovary. Capsules 5-celled. Seeds numerous, minute.

Distribution: On rocky places in saline habitats and near cultivated fields; Rajasthan and Gujarat, extending to Sindh.

Uses: Used for cleaning teeth and applied to fractured bones; the leaves are rubbed down in water and used as poultice for sores.

36. *Bergenia ciliata* (Royle) Raizada Bergenia ligulata (Wall.) Engl., Bergenia ciliclata (Ham) Stern. f. ligulata. Saxifraga ligulata Wall.

Family: Saxifragaceae.

Local names: Pakhánbéd (H), Pashánbheda (S), Silparo (Gharwal).

Description: A perennial herb, rootstock stout, woody. Leaves heart-shaped, undivided, entire with margins fringed with hairs, hairless on both surfaces, base of leaf stalk sheathing. Flowering stems leafless, thick, usually reddish. Flowers white or pale rose-pink forming a many flowered corymb or one-sided panicle, flower-stalks hairless. Sepals attached to the ovary, hairless, lobes erect in fruit. Petals 5 rounded. Stamens 10. Carpels 2, sometimes 3, with very long styles. Fruit almost round.

Distribution: In temperate zone growing in rocks in shady places, where the shade is provided by *Abies pindrow, Quercus incana* and *Cedrus deodara* and seen along with *Viola serpens, Valeriana jatamansi* and some ferns. Distributed from Kashmir to Bhutan.

Kashmir: near Shirazia bagh, Aporwat, Gulmarg; Chenab valley, Shimla, Kharamba, Rikshin, and other places in Himachal and Uttaranchal on rocks.

Uses: Local people use the powdered rootstock with *Misri* 'Sugar' as diuretic and useful to break renal calculi.

The roots are tonic, used in fever, diarrhoea, and pulmonary affections, antiscorbutic; bruised and applied to boils and ophthalmia. Rhizome powder is taken with water in dysuria and menorrhagia; useful to break renal calculi.

Extent of utilization and trade: The rootstocks are collected during March and April and cut into small pieces before sent to market for disposal. In 1963–64 about 26 quintals of rhizomes are reported to have been collected only from Kantal forest in Tehri Garhwal (Uniyal and Issar, 1967) and sold at Rs 250/quintal. In the local market both *B. ligulata* and *B. ciliata* are sold as Pashánbed. *B. strachy*, a herb of high altitude, is also used as *B. ciliata* at the place of its occurrence and sold as *Bhotia chai*.

Mixed with the seeds of *Gylycine max*, root powder is given to cattle in diarrhoea. It dissolves gravel in bladder and is used for this purpose.

The rhizome contains tannic and gallic acids, starch, glucose and a red colouring matter, calcium oxalate and mineral salts, etc. no alkaloids have been detected.

Ayurvedic preparartion made out of rootstock are *Pashanbedadikwath, Eladikwath, Trikantakdadikwath, Satawarighrita, Saddhradiya churna*.

The rhizome of *B. ciliata* is said to dissolve gravel or stone in bladder for which it is generally used. It is useful as astringent in diarrhoea and pulmonary affections; it is said to have tonic properties and to act as a antidote to opium. The rhizome is also reported to remove mucous from intestine, useful in hydrophobia, splenic enlargement, menorrhagia, excessive urine, haemorrhage, biliousness, eye sores, ulcers, diseases of lungs and heart. These are considered as absorbent and the powder is applied to boils externally and in dysentery internally. The

roots after rubbing down and mixed with honey are given to children when teething. It acts as antidote to opium.

B. stracheyi, from the alpine zone has rootstocks which are used in urinogenital disorder. The time of collection is March/April, roots are cut into small piecees before marketing. In local market both *B. ciliata* and *B. ligulata* are sold as *Pashanbéd. B. stracheyi*, also used as *B. ciliata* at the place of its occurrence and sold as *Bhutia chai.*

37. *Bidens chinensis* (L) Willd. (-Bidens pilosa L. auct non L. sensu Hook. f. FBI).

Family: Compositae (Asteraceae).

Local names: Phutim (Gujarat).

Description: An erect herb with robust stem, glabrous near base, pubescent upwards. Leaves opposite, 1-pinnate; leaflets 3–5 ovate. Fls. yellow to white. Heads radiate on long stalks, diverging; ligule white. Pappus of 3-barbed bristles. Achenes linear, rough.

Distribution: Throughout India in tropical regions, common on waste places.

Uses: The leaves after pounding are applied over eyelids to cure sore eyes. These are also used as styptic, for stopping the flow of blood and as vulnerary.

Juice of leaves when squeezed into the eyes and ears is reported to cure the eyes and ear complaints. Powdered in water leaf powder is administered as enema for abdominal troubles.

The flowers are reported as a remedy for diarrhoea. Dried flower buds, ground and mixed with alcohol, are used as mouthwash in toothache.

Different varieties reported are var. *bipinnata* (Kumaon and Eastern Himalaya), var. *dubia* (Garhwal, Shimla and Kumaon) var. *radiata* (TamilNadu) and var. *minor* (Assam) [Fig. 5.44].

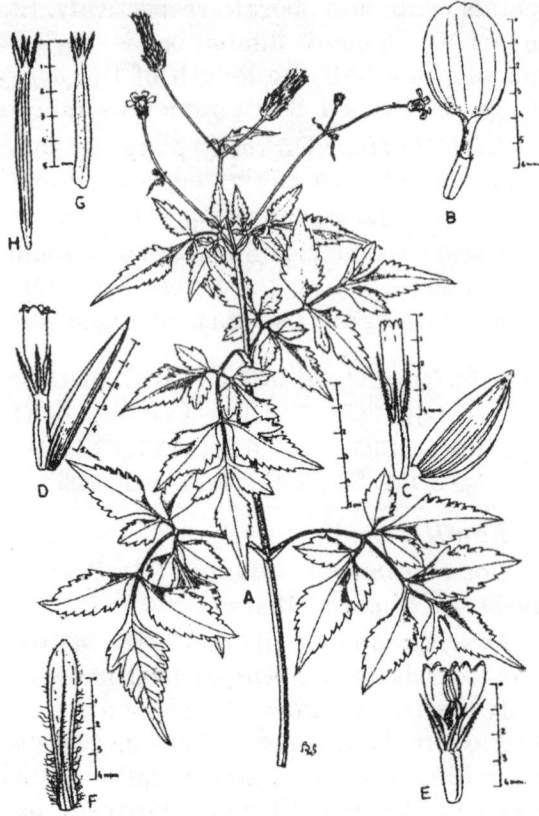

Fig. 5.44: *Bidens bipinnata* L. A. flowering and fruiting branch; B. ray floret; C. one of the outer disc florets with a palea; D. one of the inner disc florets with a palea; E. corolla of C opened to show androecium; F. one of the outermost involucral bracts; G and H. outer and inner achene.

Other species: *B. tripartita* L. reported from central and western Himalaya, from Kashmir to Nepal, 1000–1700 m. In Chinese *Materia Medica* the plant is useful in chronic dysentery and eczema.

38. *Blepharis ciliaris* (L.) Burtt. (-Blepharis edulis Pers)

Family: Acanthaceae.

Local names: Dahshni-Chapár, Uttanjan (Hindi); Shikhi (Sansk.); Karádu (Maharashtra).

Uses: The seeds are diuretic, aphrodisiac, relsolvent and expectorant. *Blepharis sindica* Stocks ex T. Anders.

Local names: Bhoomgri, Jasad, Billi-Khojá (Bikaner).

Description: A dichotomously branched undershrub, greyish pubescent about 30 cm tall. Leaves whorled with about 1 cm long petiole, lamina coriaceous, apex spine tipped. Flowers purplish blue fading to white, apex acuminate with sharp spine at the tip. Flowers in conical hairy spike on woody peduncles. Bracteoles; linear, hairy, shorter than the bracts. Calyx scarious. Corolla tube short, lower lip flat. Capsule 2-seeded. Compressed ovate, acuminate.

Distribution: Northwest India, extending westwards to Pakistan. Frequent on rocky gravels, rarely on sand. Long spikes of the previous years dry on the plant and get blackened.

Germination of this plant is prevented with excess water.

Uses: The plant is boiled in milk and taken as tonic, also fed to cattle for better milk production.

The seed used as cure for earache.

Seeds of *B. ciliaris* contain, crystalline bitter principle and contain, 2.1% *allantoin* and 1.2% bitter glued *blepharin*, catechol, tannins, saponin and glucose.

B. sindica stocks ex T. Anders (-Bhoongri-Rajasthan) is boiled in milk and taken as tonic and fed to cattle for greater milk yield.

Blepharis maderaspatensis is shown in Fig. 5.45.

Blepharis maderaspatensis L. (-*B. boerhaavifolia* Pers.)

A prostrate or suberect herb, much branched, rooting at nodes. Leaves 4 in a whorl, unequal (linear oblong in L. sindica)

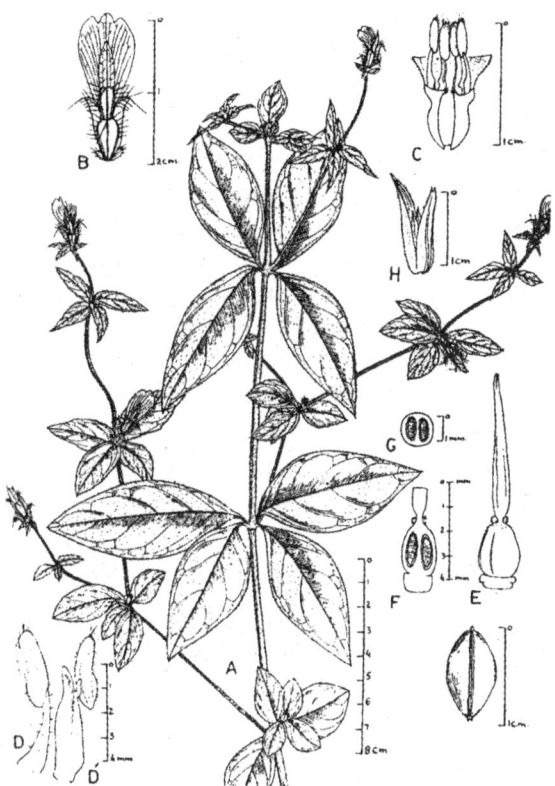

Fig. 5.45: *Blepharis maderaspatensis* Heyne ex Roth A. flowering and fruiting branch; B. flower C. corolla limb partly cut away and tube opened to show androecium; D and D'. an upper and a lower stamen; E. gynoecium; F. longitudinal section of gynoecium; G. transverse section of gynoecium; H. calyx, I. capsule.

with thin elliptic ovate to oblanceolate lamina, hairy on nerves below. Inflorescence an axillary cluster of 2–3 fls. (many flowers in B. sindica have a strobilate spike). Flowers white or purplish pink, bracts in 4 opposite pairs. Capsules ovoid. Seeds 2-flattened. In tropical Africa, Pakistan and India, on gravelly ground.

39. *Blumea lacera* (Burm. f.) DC. ex Wight var. *glandulosa* Hook. f.

Family: Compositae (Asteraceae).

Local names: Kakrondá, Kukurband, Kokarunda, Kakránda (Hindi), Nirundi (Mumbai), Kukkuradry (Sansk), Burabdo (Rajasthan).

Description: Annual herb, hairy, glandular, with a strong odour, much resembling the tobacco plant, but the flowers are yellow; stem very leafy, erect. Leaves obovate. Flowers in short axillary cymes and collected into terminal pisiform panicles. Involucral bracts narrow, acuminate, hairy, corolla yellowish. Achenes 4-gonous, not ribbed; pappus white.

Distribution: Throughout the tropical regions of India. On sandy soils in cultivated fallow.

Uses: The plant is used for curing piles and kills all kinds of worms. It also resolves all kinds of persistent swellings.

Juice of the herb is boiled till it is semi-solid, mixed with equal quantity of *Géru* (red ochre) and made into pills, which are taken twice daily with water to cure piles. The juice of the herb when applied to wounds, the worms get killed.

Other species

a. *Blumea balsamifera* DC. Vern. Names: Kakrondá (Hindi), Kalahad (Gujarat). Warm infusion is sudorific and the decoction expectorant. Plant is a fish poison. Leaves yield crystal fine essential oil. Injection of extract lowers BP, used in treatment of excitement and hypertension.

b. *Blumea chinensis* DC. Leaves are used in Malaysia as stomachic, antiseptic and diaphoretic. Plant is recorded from Eastern Himalaya, Assam, Sikkim and Khaia hills.

c. *Blumea densiflora* DC. Leaves used as sudorific and the juice is insect repellent.

d. *Blumea eriantha* DC. (vern. Nimurdi-Maharashtra) Juice of the plant is carminative, warm infusion is sudorific, cold infusion diuretic and emmenagogue.

e. *Blumea myriocephala* DC. Kukusunga-(Bengal), Kakránda (Hindi).
The plant is bitter, antipyretic, while the leaf juice is anthelmintic, astringent, febrifuge, stimulant and diuretic. An essential oil and camphor is obtained from the herb (0.0.5%), contain *blumea camphor.*

40. *Boerhaavia diffusa* L.

Family: Nyctaginaceae.

Local names: Punarnavá (Sanskrit); Sant (Hindi), Itsit (Jammu), Ghetuli (Bihar) The trade name Punarniva is from the Sanskrit name.

Description: A perennial herb, creeping, with stout rootstock and erect spreading branches. Leaves thick and in equal pairs, rounded at the base, smooth above, margins wavy. Flowers pink, small, in clusters or umbels. Fruit with 5-ridges, glandular.

Distribution: A weed throughout India on fallow lands, ruined buildings and old walls.

Uses: The whole plant, particularly the roots, leaves and seeds constitute the drug, eaten in rainy season as pot herb.

Plant root is used either as a powder or as infusion. It is laxative, expectorant, diuretic, stomachic, diaphoretic and anthelmintic; used in diseases of heart, kidneys, especially in strangury, cough, jaundice, intestinal colic, pleurisy, difficult breathing, gonorrhoea and to relieve intestinal inflammations. The poultice is also used for extraction of gúinea worms.

Root is also considered as a dressing for oedomatous swellings in the form of a paste. A poultice made of the roots by boiling them

in water, is applicable in ulcers, abscesses and fractured bones (personally tested the later application).

The roots are considered as a sovereign remedy for dropsy and the liquid extract is given in chronic Bright's disease. As a dressing for oedomatous swellings the root is used in the form of a paste.

Leaf juice with honey is given to cure liver complaints such as jaundice, also dropped into the eyes in chronic ophthalmia. It has been found effective in regulating the menstural flow.

In mild eases of dropsy, the fresh herb is boiled, salted and eaten with *chapattis*; *Punarnivadi Mandoor* which is used to cure jaundice, stomachache and intestinal inflammation, along with other medicines. The drug is also beneficial in ascites and oedema due to early diseases of liver, kidney, etc.

Infusion of the plant is mild laxative and febrifuge for children.

Extent of utilization and trade: The drug contains an alkaloid 'punarnavine'. The diuretic activity of the drug has been experimentally confirmed long back on the animals. In moderate doses it has been recommended for asthma, while large doses brings vomiting.

Regeneration pattern of this plant has been studied by Pandey et al (2001). Since this herb is encountered in different terestrial habitats, ranging from managed grasslands, wastelands, agro-ecosystems and forest gaps, the regeneration pattern and population structure is diffused in relation to the distribution. However, it has been noted that this plant flourished quite often in the vicinity of species like *Euphorbia hirta*, *Tridax procumbens* and *Vernonia cinerea*.

Regeneration strategy and pattern of regrowth depend on different degrees of disturbances. Repeated clippings and trampling often result into intricate shoot complex borne by simple root system. The effect of intense grazing, fire, and water stress affect the survival pattern (Chandrashekhar and Swamy, 1995). Biology and ecology of a few common species of Boerhaavia has also been studied (Bajpay, 1993). Practices like harvesting of entire plant, trampling and grazing results in stumps of different shapes and size.

Boerhaavia coccinea Mill is depicted in Fig. 5.46.

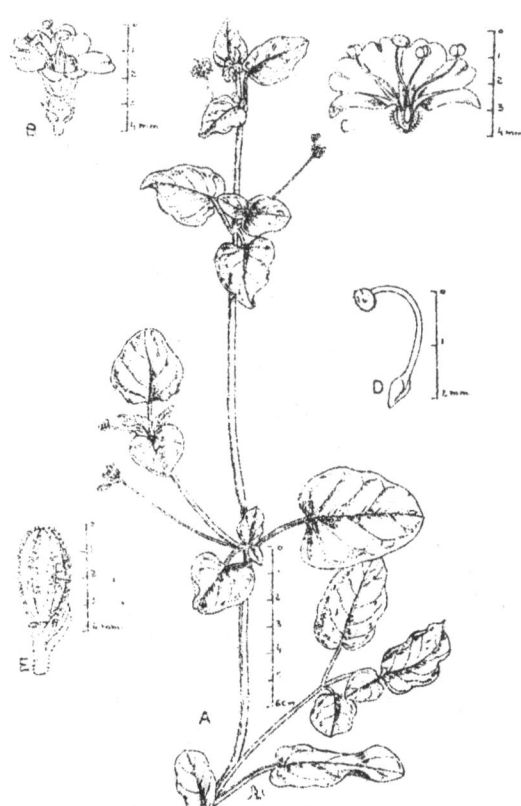

Fig. 5.46: *Boerhaavia coccinea* Mill. A. flowering and fruiting branch; B. flower; C. flower, opened to show androecium and gynoecium; D. gynoecium; E. fruit.

The main sprout arising from stumps often repeats the characteristic architectural model of the parent plant. The distal part of the shoot may die, even in the undisturbed condition during the period of moisture stress and the substitution growth may occur within the same season of growth, if moisture conditions becomes favourable.

The age determination on the basis of number of emerging shoots gets obliterated due to formation of the cavities, nitches, scars and mounds at the base of the shoot system. The age determination on the basis of number of emerging shoots may be quite authentic for younger stumps.

41. *Brassica campestris* L. var. rappa (L.) Hartm.

Family: Cruciferae (Brassicaceae).

Local names: Káli-sarson, Rape seed (Eng).

Description: The crop is grown under irrigation during the 'rabi' season, single or intercultivated with the wheat. Same is true for *B. campestris* var *sarson* Prain. There are other varieties like *L. Brassica campestris* var. *toria* Duth. and Full., *B. hirta* Moench. (*B. alba* (L.) Boiss called white mustard or Safed-rai, *B. juncea* (L.) Czern and Coss, Indian mustard called Rai-Hindi *B. juncea* var. *cuneifolia* Roxb. (Vern. leaf mustard Rai (Hindi), *B. napus* L. (Eng. Rape; (Eng.), *Kali sarsom* (Hindi) *B. nigra* Koch. (Black Mustard-Eng. Kali-Rai -Hindi); *B. oleracea* L. *varacephala* DC. (Borecole, Kale or Karamsag-Hindi); *B. oleracea* var *botrys* L. Cauliflower, Brocoli (Eng), Phulgobhi-Hindi); *B. oleracea* var *capitata* Cabbage-Eng, Bandgobhi-Hindi); *B. oleracea* var. *gemmifera* Zenk. (Brussels sprout, Budbearing cabbage-Eng; Buttangobhi-Hindi); *B. oleracea* var *gangylodes* L. (Syn var. Knolkhol-Eng; Ganthgobhi-Hindi); *B.*

pekinensis (Lour.) Rupr.) Chinese cabbage), *B. rapa* L. (*B. campestris* L. var. *rapa* (Hindi Shalgam; Turnip (Eng).

Uses: The leaf extract of *B. oleracea* var. *botrys* is bitter and used in gout and rheumatic pains.

The seeds are diuretic, laxative and anthelmintic.

The roots and seeds of *Brassica campestris* L. (Vern. Sarson) are used as antiscorbutic and efficient counter irritant in poultice. *Sarsoon oil* is also used in rheumatism, bronchitis and is rubbed against internal muscular injury.

B. rugosa Prain var. *typica* (vern. Rái-Kumaon). Seeds crushed and mixed with turmeric powder in crude salt and smashed gourd are used as anthelmintic. Smoke from the seeds with that of capsicum are used as anthelmintic. Capsicum is inhaled to do away with evil spirits in Kumaon. The practice is done on infants, newly wed brides and even domestic animals who fail to give milk.

42. *Bupleurum falcatum* L. var. marginatum Wall.

Family: Umbelliferae or Apiaceae.

Local names: Kalizewar, Sipil (Punjab).

Description: An erect glabrous herb. Stem corymbose upwards. Leaves linear, sessile, usually curved like a scythe, nerves 5–7. Bracts 2–5, linear-lanceolate, shorter than the umbels. Fruit oblong, ridges distinct. Margin of leaves cartilaginous, middle cauline leaves amplexicaul, narrowed close to the base.

Distribution: In temperate regions from Kashmir to Bhutan, 1500–4000 m.

Uses: Roots with other drugs are prescribed in liver troubles and as diaphoretic. The root causes perspiration and is said to be effective in thorasic and abdominal inflam-

mation and fever. Also used in flatulance and indigestion, in malaria and other fevers.

43. *Bryonopsis laniosa* (L.) Naud.
(-Bryonia lacinosa L.)

Family: Cucurbitaceae.

Local names: Bahupatrá (Sansk), Gargumáru, Ishári, Syapá kariyál (Kum).

Description: It is a perennial, tendril-climbing, dioecious herb with palmately lobed leaves; flowers are in axillary clusters, fruit is a smooth, globular berry.

Distribution: Throughout India.

Uses: The plant is supposed to be efficaceous in snakebite. It is bitter, tonic, used in bilious attacks and in fevers with flatulance.

The leaves are applied to inflammation. The bitter principle is *bryonin*.

44. *Brunella vulgaris* L.

Family: Labiateae (Lamiaceae nom alt.).

Local names: Dháu, Ustékhadus (Hindi); All-Heal, Brunel, Pick Pocket, Pimpernal, West Indian Sumach (Eng). Ustékhadus is the trade name.

Description: A perennial herb with creeping rootstock. Stem bristly. Leaves opposite, ovate, obtuse, hairy, Flowers in whorls of 6; bracts broadly ovate-cordate, with sharp-pointed tips and purple-coloured margins. Corolla deep purple blue or at times white, 2-lipped, upper lip erect, hooded, lower spreading. Fruit of 4, smooth, oblong nutlets.

Distribution: Temperate regions of Himalaya, 1300–2000 m; also in the Nilgiris, Palneys and Kera mountains.

Uses: The herb is stimulant, expectorant, astringent, carminative and antispasmodic. Used in haemorrhages, diarrhoea, relaxed throat when infusion of the herb with honey is given. On sore and relaxed throat,

ulcerated mouth it is used as mouthwash. Also given as an injection for the treatment of internal bleeding piles. For piles the green leaves coated with castor oil are warmed over a fire and applied to the rectum.

45. *Bryophyllum pinnatum* Kurz.
(-Kalanchoe pinnata DC.) B.
calycinum Salisb.

Family: Crassulaceae.

Local names: Ghaimári, Hemságar, Zakhm-é-hayata (Hindi), Air plant, Life plant (Eng).

Description: A perennial, smooth, succulent herb. Stem erect, hollow. Leaves usually simple, opposite, fleshy, ovate or oblong, 8–15 cm long, margins notched, apex rounded. Flowers in large terminal panicles, bisexual, pendulous, cylindric, 5 cm long, Calyx tubular green, tinged with red spotted white. Corolla tubular, tube green tinged with red follicles 4, enclosed within dry calyx and corolla.

Distribution: Throughout India, ascending 1000 m.

Uses: Leaf juice is efficaceous in diarrhoea, dysentery, lithiasis and cholera. Externally leaves are used as styptic, astringent and antiseptic; applied to contused and unhealthy wounds, bruises, boils, fresh cuts, foul ulcers, etc.

46. *Cajanus cajan* (L.) Millsp. (-Cajanus
indicus Spreng.)

Family: Papilionaceae (Leguminosae).

Local names: Arhar, Tuvar (Hindi) Adhki (Sansk.)

Description: A well-known foodgrain used as pulse, usually cultivated in eastern part of UP on field beds. The shrub is up to 3–4 m high with silky sulcate branches. Leaves 3-foliate, gland dotted beneath; leaflets up to 7 cm long, oblong lanceolate, acute,

entire, densely beneath. Flowers in loose corymbose recemes, longer than the leaves, panicled towards the ends of branches. Corolla 3-times as long as the calyx, yellow on veins with red in *C. bicoloor*. Stamens diadelphous. pods brown-tomentose, up to 7 cm long, seeds, topped with the persistent base of the style. Seeds about the size of smal pea, varying in colour from yellow to brown.

Uses: It facilitates the appearance of pustules of smallpox or chickenpox, when water in which the seeds are cooked, is given to the patient.

The seeds also resolve swellings of inner organs and cure inner wounds, like of stomach, liver, intestines, etc. *Arhar* cures alopecia when ground *arhar dal* is applied 2–3 times in a day. Enlarged testicle resume their normal size by application of a warm paste for about 8–10 days.

Green pods and shoots when made into a paste is applied warm to the breast, to relieve their tension due to accumulation of milk.

Swollen gums and wounds in the mouth are cured by using a decoction of leaves as mouthwash.

Juice of leaves and root is dropped into the eyes in case of macula, due to smallpox.

Leaves of the plant and *Neem* tree when ground in water and strained are given to drink in cases of piles, and considered very effective.

The seeds contain two globulins: *cajanin* and *concajanin*.

47. *Cannabis sativa* L.

Family: Cannabinaceae.

Local names: Bháng, Charas, Gánjá, Hashish, Kináb, Sidhee, Vijay (Hindi); Indian Hemp, Canvas (Eng.).

Description: An annual weed of camping grounds and roadsides. The leaves are palmati-partite, lobes finely serrate. Stem slender, grooved, finely tomentose; branches few. The leaves are alternate, palmately divided, toothed, long pointed, dark green and rough above, palédowny beneath; leaflets 7–11, linear lanceolate. Flowers unisexual, male and female on different plants. Male flowers clustered in short axillary drooping panicles. Perianth of 5-parts. Female fls. axillary, sessile, erect, perianth a single, entire leaf enclosing the ovary. Fruit an achene, minute, enclosed in persistent perianth. Fls. July-October (Fig. 5.47).

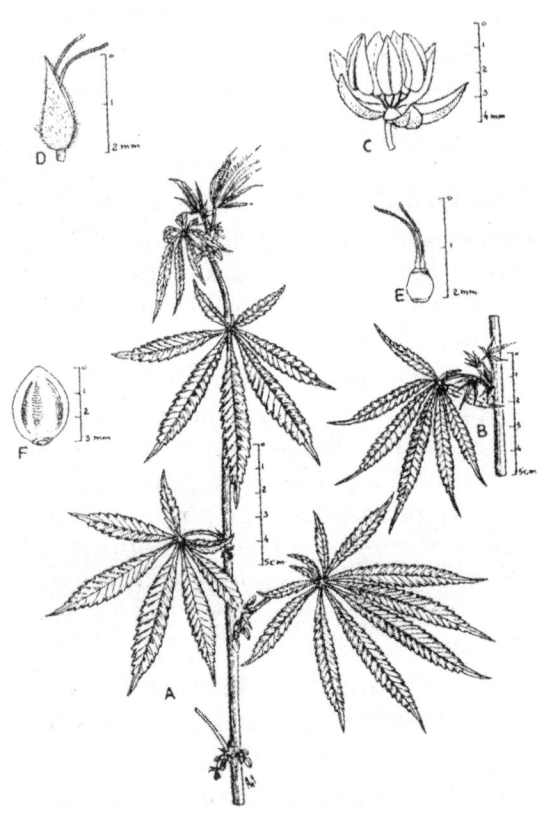

Fig. 5.47: *Cannabis sativa* L. A. flowering branch from male plant; B. flowering branch from female plant; C. male flower; D. female flower; E. gynoecium; F. seed.

Distribution: Throughout India on waste places and by the roadsides, wild in the sub-Himalayan regions, ascending 3000 m, acclimatised or cultivated in the plains. In Dehradun, the plant was abundant on sides of the canal and was being utilized by local people by rubbing the plants between the hands when in flower. Of late, the plant has disappeared from the site. Cultivated for the narcotic drug, *ganja*. Dried flower tops give a resin causing intoxicating effect.

Uses: The plant is used as a tonic, intoxicant, stomachic, antiseptic, analgesic, narcotic, sedative and anodyne. The most important constituents are a resin and a volatile oil; the resin is a potent intoxicant, a narcotic and intoxicant and is utilised in 3 different forms: (1) *Ganja*, (2) *Charas* and (3) *Bhang*.

Ganja is the dried flowering tops of the female plant, covered with a resinous exudation, used as smoke like the tobacco alone or with tobacco. *Charas* is the resinous exudation covering all the aerial parts; the exudation contains a large quantity of a toxic red oil.

Bhang is the beverage prepared from an infusion of the dried leaves and flowering shoots with milk and other ingredients which contains the least quantity of the narcotic drug, charas is practically the drug itself and hence more intoxicating and narcotic than the other two.

Leaves and flowers are placed under the bedsheet to derive away the bugs, used as fish poison and irritant to the skin. In Kumaon hills leaf juice is used to relieve ear trouble when put into the ears. Paste of leaves is used in piles. Juice removes dandruff and vermin from head. Poultice of leaves is used in diseases of the eye with photophobia and in piles.

Seed oil is used also as luminant, while these are given to milch animals to increase flow of milk.

The weed cause intoxication which is exhilirating, inducing laughter and talkativeness. Initially, its use enhances the appetite and sleep. It has a deleterious effect on the mind and the heart leading to insanity.

The leaves are sedative, anodyne, antispasmodic, astringent and diuretic applied after grinding and spreading warm on castor leaf and bandaged. Green leaf juice is instilled into the ear to cure earache; used as a household remedy for dysentery and diarrhoea. It is also used to stop malarial fever when 1 gram of *bhang* is ground and mixed with 2 gram of jaggery and made into 4 pills, one pill is administered every hour. It is also used in case of pains and burning sensation due to piles. Leaves useful in stomachic, relieve flatulance and induce sleep.

48. *Canscora decussata* Schult.

Family: Gentianaceae.

Local names: Shankhpushpi, Shankaoli, Shankhavali, (Hindi); Dankuni (Mumbai).

Description: A spreading herb with leaves of like *doob* (*Cynodon dactylon*) grass. Flowers white, bell-shaped, rather pinkish, flowering during the months of May and June. Since green plant is not available throughout the season, it is uprooted before the rainy season and dried in shade and preserved.

Distribution: Throughout India up to an altitude of 1300 m and growing in moist situations.

Uses: The dried herb purifies blood, strengthen the brain, restores loss of memory, clears the complexion and is used in diseases like spermatorrhoea and diabetes.

Eight gram of the herb is ground in water along with seven grains of black pepper, strained, sweetened and drunk. It acts as prophylactic during smallpox epidemic. Dry herb powdered (3 gram) is taken twice in cases of spermatorrhoea, nocturnal emissions and burning sensation during urination.

'Shankhpushi' is the branded name of a medicine sold in the market, prepared by Baidyanath.

Other species: *Canscora diffusa* (Vahl.) R.Br. is another species which is a nervine tonic and used as substitute for the above species. It is found throughout India.

49. *Cassytha filiformis* L.

Family: Lauraceae.

Local names: Akáshhavalli (Sansk), Amarbéli, Amarvéla (Hindi). The name is also used for *Cuscuta reflex,* a common parasitic herb on hedges and trees in North India.

Description: A parasitic herb spreading on shrubs and trees, rootless and twiner met with throughout India, especially near the coast, similar to the habit of Cuscuta quite glabrous. Stem slender, cord like, long, often intricately matted, branched, dark green. Flowers small in lax-dense divaricated spikes. Perianth twice as long as the bracteoles, tube short, segments 6, in two rows, three outer short, inner longer, concave, valvate. Fruit globose, smooth, white, enclosed in fleshy pertianth tube and crowned by erect segment.

Distribution: Throughout greater parts of India, also in island, South Andaman.

Uses: Used in cases of paralysis, boiled water with the plant stalled to stay in body; kills intestinal worms and useful in cases of gonorrhoea. It resolves swellings, when fomented, after boiling in water. Fomentation with the decoction and bandaging with the residue is used to remove hardness of the muscles of belly. Plant juice mixed with sugar is specific in inflamed eyes. Tonic, alterative, in bilious affections, children dysentery, urethritis and skin diseases: insecticidal, used as hair tonic; with butter and ginger used for cleansing invertrate ulcers.

50. *Capsella bursa-pastoris* (L.) Medic.

Family: Cruciferaea (Brassicaceae nom. alt.).

Local names: Shepherd's Purse (Eng.).

Description: An erect annual herb covered with branched hairs. Roots long tapering. Stem branched, erect. Radical leaves pinnatifid, sometimes lanceolate, terminal lobes triangular, segments entire; upper leaves pinnalifid, lobed at the base, stem clasping, uppermost lanceolate. Flowers white, in racemes. Pods flat, triangular. Seeds many, in two whorls.

Distribution: A weed of cultivation, in temperate regions, above 11500 m. A cosmopoliton weed. Flowers January-December.

Uses: The plant is astringent, used in diarrhoea and as diuretic in dropsy. The fruits are stimulant, externally as rubifacient, used in putrid sore throat, hoarseness, dyspepsia, in diarrohea and piles.

In Europe and America, the plant is of prompt use to arrest bleeding when given, the form of fluid extract which is also given in dropsy as a diuretic, but not of much use in India. It has been asserted to contain *acetylcholine, choline* and possibly *tyramine, fumaric acid* and *inosetol* (in fungoid infected plants only) but reported to be devoid of active constituents (JBNHS 1939, 40. 701). It has, however, been reported to be given in the form of fluid extract or administered intravenously or intramuscularly to control haemorrhage of diverse origins. Extract of dry or green plant exerts great influence as contracting effect on the uterus of guineapigs (Chopra *et al* 1956. *Glossary of Indian Medicinal Plants.* p 50. CSIR, India).

51. *Carum bulbocastanum* Clarke non Koch. (-Bunium persicum Boiss) Fedts

Family: Umbelliferae nom alt. Apiaceae.

Local names: Kála-zirá. Sia-sirak (Hindi), Gunyun (Kash), Umbhu (Ladakh), Black Caraway (Eng.)

Description: A perennial herb. Stem delicate, branched. Leaves pinnate compound, finely dissected, linear umbel with 5–10 fls., petals white pink. Fruit glabrous.

Distribution: In Kashmir it is a weed of cultivated land from Afghanistan westwards in Pakistan and eastwards to Kumaon. It is liable to prove dangerous in fields owing to the fondness of pigs to the roots. It is also existing truly wild on grassy fields, when the shepherds collect it as valuable source of income.

Uses: Since in uses it is identical with *Carum carvi* and collected in large quantities from Himachal and sold as *Kalazira* or black cumin, often adulterated with *Bupleurum falcatum.* Starchy tubers are eaten as vegetable, seeds as spice.

Other species

a. *Carum carvi* L. (vern, Zirá, Shingu-Zirá, Kala-zirá)

A glabrous annual and is being cultivated at many places in Himachal Pradesh, Kashmir, etc. in the subalpine regions. The fruit, its essential oil and aqueous extract is extremly used as mild stomachic and carminative in flatulant colic and to impart flavour to culinary and pharmaceutical products. The seeds are carminative, astringent and stimulant. Fruits are also lactagogue. An essential oil obtained from the fruits is used in medicine and as perfume for soap.

The produce from Lahul forest only is about 35–40 qtls per annum. Though the fruits of *C. bulbocastanum* are also sold as *kalazira*, which is however a distinct plant. The essential oil from the seeds contain *kerotene*, *carvone* and *carvacrol*, etc.

Though the plant is native of Europe and west Asia, it is now being cultivated in Bihar, Orissa, Punjab, Bengal, Andhra, apart from Northwest Himalayan states between 3000–4000 m.

b. *Carum roxburghianum* Benth.and Hook. f. (-Trachyspermum roxburghianum (DC) Sprague

Family: Umbelliferae.

Local names: Ajmod (Hindi).

The plant is cultivated throughout India, the seeds are used in curries and in medicine. As a drug it is carminative and stimulant, useful in curing vomiting and dyspepsia. Also made use of in the preparation of gripe water given to children during summer season.

c. *Carum copticum* Benth. and Hook. f. (-Trachyspermum ammi (L.) Sprague.

Family: Umbelliferae.

Local names: Ajwain (Hindi). The Bishop's weed (Eng.).

It is cultivated throughout India. It is astringent, aphrodisiac, vermifuge and diuretic. A water, locally called *arak ajwain*

is distilled and an oil is obtained from the seeds. The fruits are antiseptic, stomachic, used in diarrhoea, colic, flatulance, indigestion and cholera. The roots are diuretic and carminative. Fruit yields 4–6% oil, an essential oil containing 45–55% *thymol*. Cultivated extensively in the country. The oil is also used and being considered like the fruits. A crystalline substance separated from the oil during distillation is identical to thymol (called *satajwain*) which is mainly antiseptic; besides *thymol* it contains certain hydrocarbons called 'thymene'. *Satajwain* along with peppermint and camphor, are used in formulations for a lotion called *Amritdhara*, a product in use for curing headache, diarrhoea and dysentery throughout the country, perhaps as a home remedy.

52. *Centella asiatica* (Linn.) Urban (=Hydrocotyle asiatica Linn.)

Family: Umbelliferae (nom alt. Apiaceae).

Local names: Mandukparni (S); Brahmi (Garhwal and Kumaon).

Description: A common trailing herb, rooting at the nodes. Leaves many at each node, kidney-shaped, about 5 cm long, palmately nerved. Flowers in clusters of umbels bearing 3 or 4 small flowers without stalks. Fruit very small, longer than broad (Fig. 5.48).

Distribution: Throughout North-west India, ascending to 1000 m, in the tropical and subtropical regions. Collected during Sept-Oct. from humid wild areas like gardens and parks.

Uses: The whole plant is diuretic, tonic and a local stimulant. It has been used as a remedy for skin diseases such as eczema, ulcer, abcesses and scrofula. The plant is also much used for enlargement of glands, chronic rheumatism, chronic nervous

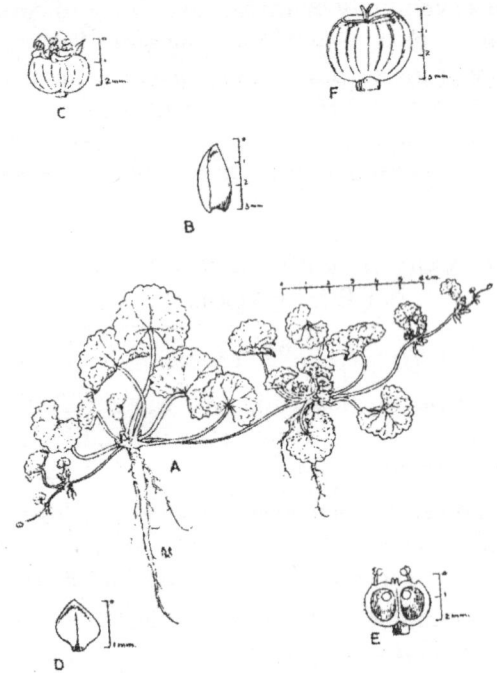

Fig. 5.48: *Centella asiatica* Urban A. plant; B. bract; C. flower; D. petal; E. longitudinal section of flower, petals removed; F. fruit.

disease, amenorrhoea and blood purification. In small doses, it acts as powerful stimulant, especially to cutaneous system. The juice is efficaceous externally in elephantiasis, enlargement of scrotum, diseases of cellular tissues and rheumatic swellings, internally in syphilitic, scrofulous and leprous diseases.

Locally, the leaves are used as brain tonic. In Kumaon the fresh juice of leaves is dropped in eyes, especially in cataract and taken orally in fevers. Externally, leaves are used as powder, juice and ointment. In case of skin erruptions powder of the leaves, in elephantiasis and enlarged scrotum the ointment is used. The leaves are also a household remedy in the treatment of early stages of dysentery in children and bowel complaints. In small doses, the leaves are also administered for improving memory

and mental weakness. They are also useful in various kinds of fever, remittent, intermittent, continued or malarial.

Extent of utilization: Medicinal properties of the plant are known from remote times. Large quantities are collected and sold in the market. Dehradun is a good centre for the trade. The leaves contain an oil and a volatile principle *vallarine*, a word derived from Tamil name of the plant. This plant locally collected by villagers in Dehradun is purchased by traders and sold further to the traders of Haridwar where it is sold as *Brahmi*.

53. *Cardiospermum halicacabum* L.

Family: Sapindaceae.

Local names: Karanasphota (Sansk.); Kanphuli (Hindi); Kapál phodi. Latáphatkari (Mumbai); Baloon vine (Eng).

Description: An annual herb heals, with slender branches. Leaves deltoid, 2-ternate, lanceolate, margin serrate, acute at apex and narrowed at base. Fls. white in umbellate cymes, petals rounded at apex. Capsules trigonous, winged at angles, bladdery. Seed black, globose in hedges.

Distribution: In hedges, tropical and warm areas, throughout India.

Uses: The plant is used in rheumatism, and stiffness of limbs. The roots are diaphoretic, diuretic, laxative, rubifacient and emmenagogue, used in nervous diseases.

Leaves used as rubifacient, as a poultice for rheumatism, leaf juice as a cure for earache.

54. *Centrantherum anthelminticum* (Willd) Kumtze (-Verninoa anthelmintica Willd.)

Family: Compositae (St. Asteraceae).

Local name: Somraj (Hindi); Somraji (Sansk); Kalijira (Mumbai); Kattujirakam (Mal), Sattu shiragam (Tam); Adavijilakara (Tel).

Description: An annual robust herb, erect, leafy: stem 60–90 cm high branched, pubescent. Leaves elliptic lanceolate, acute, coarsely serrate, pubescent on both surfraces, base tapering into the petiole. Flower heads sub-corymbose, many flowered, with linear bract, near the top of the peduncle. Outer involucral bracts linear. Pappus reddish achenes 10 ribbed, pubescent.

Distribution: Throughout India up to 2700 m in the Himalaya, Kahsia hill and black cotton soils of the Deccan region. Often cultivated.

Uses: The seeds are anthelmintic, used in skin diseases, tonic, diuretic and used for destroying pediculi, in scorpion sting.

An active anthelmintic principle has been isolated yield 1% as a bitter resin which is diuretic, antiseptic and stimulant. The achenes have been reported to have essential oil and resins.

55. *Chenopodium album* L.

Family: Chenopodiaceae.

Local names: The white Goose foot (Eng.); Bathuá (Hindi); Vastuk (Sansk.); Chandanbestu, Bethuság (Hindi).

Description: An erect herb up to 3 m in height, green, more or less coated with white mealy pubescence, inodorous. Stem angled often tinged with red purple. Leaves variable in size and shape, entire, petiole slender, often equalling or longer than the leafblade. Flowers in cluster, often forming compact or loosely panicled spikes. Sepals oblong, keeled on the back. Stigmas 2. Seed compressed, orbicular with an acute margin, black and shining.

Distribution: A common weed of the *rabi* season crops, like wheat, and gram; collected regularly as a pot herb and green vegetable. In plains usually but occurs in the Himalayan region up to an altitude of 4000 m.

Uses: The leaves are rich in potash salts. It is used in cleaning copper vessels, preparatory to tanning them.

The plant is laxative and anthelmintic, used for killing intestinal worms. Useful in fevers and disorders of the liver. It has also been found efficaceous in leukoderma, when the leaf juice is applied 4-5 times a day on the white patches and the vegetable eaten along with bread. About two months of continuous use makes spot disapper to its natural colour.

The essential oil is reported to contain *carotene* and vitamin C. The plant growth is stimulated by magnesium, a field indicator.

56. *Cichorium intybus* L. Chicory (Eng), Kásni (Hindi)

Family: Compositae (Asteraceae nom. alt.).

Local names: Chicory (Eng); Kasni, Kaskui (Hindi); Kasni-Trade name.

Description: A hispid herb, more or less erect with bright blue or violet, ligulate flowers, arranged in terminal and axillary heads.

Distribution: It is cultivated to a certain extent in Gangetic plains, in Punjab, Haryana and is occasionally met as escape. It is truly wild in Europe and is believed to be indigenous to western Himalaya, asending 2000 m.

Uses: Used as vegetable. Roots when dried roasted and ground mixed with coffee.

In combination with other drugs it is used for stomach diseases (Mako-*Solanum nigrum*,

Atis (*Aconitum*), Kakra-singi (*Pistacia integerima*), Pipal (*Piper*) and *Nagarmotha* (*Cyprus scariosus*).

It is bitter in taste but very helpful to cure liver problems.

Other species: *Cichorium endivia* L. Vern: Kasni (Hindi).

Native to mediterranean region, cultivated mainly in north India. Leaves are eaten as vegetable. The plant is used as resolvent and used in bilious complaints, tonic, demulcent, in dyspepsia and fever.

The fruit is cooling, in fever, headache, bilious complaints and jaundice.

57. *Cissampleos pariera* Linn.

Family: Menispermaceae.

Local names: Páthá (S); Harjori, Dakhnirbisi (H); Akandi (H); Pardhi (TG); Párhe (Dun); Pari (Kum.), Ambashta (Sans.) Dakh-Nirbis. Arjon (Garh); Ice vine (Eng).

Description: A slender climber with perennial rootstock which throws out long twining branches annually. Leaves peltate. Flowers minute, yellowish and visible during rainy season. Male flowers pedicelled in nearly axillary cymes. Sepals 4. Filaments longer than corolla. Female flowers in elongate axillary cymes. Sepal 1. Petal 2. Fruit subglobose, drupe red, transversely ridged.

Distribution: Commonly distributed throughout subtropical and tropical India, ascending to 2000 m. In hedges at low elevations in the hills and in valleys. Garhwal, Punjab, Himachal Pradesh, Kumaon: Bysani, Almora, Bagehsar valley; Chamba, Garhwal: Tiuni; Shimla, Mussoorie road, Pathankot, Nepal, Assam.

Uses: The root is bitter in taste and a valuable stomachic. It is frequently prescribed in later stages of bowel complaints.

It acts as antiseptic of bladder and is used in chronic inflammation of urinary passages. It is given in dyspepsia, prolapse of uterus, internal inflammation, dropsy and cough. In advanced, acute and chronic cystitis these are used with much advantage. In decoction it is useful in abdominal diseases and in colic pains.

Dried stem and leaves contain alkaloid *berberine*.

The leaves have cooling properties and are used externally for unhealthy sores and sinuses. In Kumaon the juice of the fresh leaves is used in eye troubles. Crushed leaves are used for applying to pupils, boils, and burns and wounds. The leaves are also used in urine troubles.

Time of collection: It flowers during the rainy season.

Extent of utilization: Small quantity of the plant is cultivated and consumed locally. The root contains an alkaloid *Berberine*.

58. *Cimcifuga foetida* L.

Family: Ranunculaceae.

Local name: Jiunti (Hindi).

Description: A perennial, more or less pubescent herb. Stem erect, leafy, branched. Leaves pinnately compound; leaflets ovate or lanceolate, deeply toothed, terminal leaflet 3-lobed. Flowers white, crowded in racemes, solitary in the axil of upper leaves and combined in a terminal, sometimes large spreading panicle. Sepals and petals 7, with no clear distinction between them, ovate, concave, one or two of the inner ones deeply 2-lobed. Stamens numerous, ultimately longer than the sepals. Ovaries 2–5, rarely more, many ovuled, style short, stigma pointed. Follicles flat, tipped with persistent styles. Seeds 6–8.

Distribution: Temperate Himalaya, from Kashmir to Bhutan, 2300–4000 m.

Uses: The root contains a resinous substance called *cimvigugni* and is different from that of *C. racemosa*. It is employed in rheumatic affections, dropsy, early stages of phthisis and chronic bronchial diseases.

In Siberia the plant is used to drive away the bugs and fleas.

59. *Cissus quadrangularis* L.
(Vitis quandrangularis Wall.)

Family: Vitaceae.

Local names: Ashtisanhára (Sansk); Harjora (Hindi); Kandivél (Mumbai).

Description: A common succulent climber, stem 4-angled, succulent, leafless when old, quadrangular, young branch with wings. Leaves 3–7 lobed cordate. As in cymes, peduncled. Calyx cup-shaped. Petals 4, hooded at apex. Berry globose, 1-seeded.

Distribution: Throughout the hotter parts of India.

Uses: Juice of the stem reduces healing period of fractured bones.

The leaves and young shoots are alterative and used in powder form in digestive troubles.

Juice of the stem is also used in irregular menstruation and in scurvy.

Other species

a. *Cissus adnata* Roxb. (-Vitis adnata Wall.) Vern. Kolézan. (Maharashtra). The decoction of tubers is diuretic, alterative and blood purifier. Roots powdered and applied to wounds.

b. *Cissus pallida* Planch. (-V. pallida Wight and Arn.) Vern. Kondage-Canara) Bruised roots applied to rheumatic swellings in Garhwal and western ghats.

c. *Cissus repens* Lamk. (Vitis repens Wight and Arn) Vern. Pureni-Nepal. In

East Himalaya, including Assam and western peninsula. Young shoots are edible and acidic in taste. The plant is applied to foetid ulcerations, to boils and small abscesses as maturant.

d. *Cissus setosa* Roxb. (Vitis setosa Wall. Vern. Harmal-Hindi)

The name is also used for *Peganum harmala* in Punjab. Leaves used in indolent tumours and applied externally to expel guinea worms.

60. *Citrullus colocynthis* Schrad.
(-Citrullus vulgaris (L.) Schrad.)

Family: Cucurbitaceae.

Local names: Mahéndra-varuni (Sabsk); Indráyan (Hindi), Makhal (Mumbai); The Colocynth. (Eng).

Description: A creeping herb with fruits resembling watermelon, perennial, scabrid, monoecious. Roots very deep, white in colour. Tendrils simple or 2-fid, slender, hairy. Leaves deeply divided, 5–7 lobed. Flowers yellow. Male fls. in villous peduncle, calyx hairy. Female fls. Ovary ellipsoid, densely hairy. Fruits up to 8 cm in diam, variegated green and white, intensely bitter (Fig. 5.49).

Distribution: On sandy places where large patches are seen often covering the sand.

Fig. 5.49: *Citrullus colocynthis*, fruits in Thar desert near Jaisalmer (Rajasthan) (Photo R.K. Gupta).

After the rains, new plants come up both by seeds and by perennating rootstock. Though the species is characterised by the absence of xeromorphic structures, in the leaves maximum vegetative growth takes place till winter starts. During the dry season, January onwards, the plants reduce transpiring surface by decreasing the surface area, curling of lamina upside down, thickening of epidermis, drying and falling off old leaves.

The plant show luxuriant growth during the rainy season and is not adapted to the xeric conditions, thus disappears with the change of season.

A common plant of the wasteplaces in northwest India, west Asia, Africa, Arab countries, frequented on sandy places; one of the species as a pioneer to establish on the (fresh) sand dunes.

The branches radiate on all sides with a tap root. The male flowers are produced in large profusion but the female prosper after the rains.

The germination mechanism of the seeds is interesting, since the intact seeds from mature seeds failed to germinate under a variety of experimental conditions in the laboratory. No inhibitors were recorded in the laboratory; pretreatments including some which modified or even pierced the testa failed to affect germination. How this inhibition is overcome under natural conditions is yet unknown. Under laboratory condition germination is possible by mechanically forcing open the testa or by its complete removal.

It is nowhere cultivated except on an experimental scale along with *Ipomoea pescarpae* near Surat and a few coastal areas that too, to prevent sand drifts.

Uses: The roots are purgative, used in ascites, jaundice, urinary diseases and in rheumatism.

Both the pulp and the roots are cathartic and used for purging in cases of rheumatism, paralysis, leprosy, elephantiasis and also as anthelmintic. Equal weight of colocynth root and long pepper are powdered, sifted and mixed with equal weight of jaggery. 10 grams of this powder is taken to relieve pain and swelling in rheumatism. The root, rubbed in water, is applied to piles to relieve the pain.

Colocynth pulp (2–3 grams) is boiled in 25° C of cow's milk, mixed with 10 grams of cow's ghee (butter) or 6 grams of almond oil and drunk. It relieves constipation. The pulp of the colocynth is boiled in water, expressed and strained and boiled further till it is almost solid. 20 grams of this solid, 3 grams of 'gum tragacanth are added and made into pills of about 2 grams each. One or two pills are taken to relieve constipation; if purging is required 4–5 pills are taken. It is also useful in cases of rheumatism and syphilis.

Dried pulp of the ripe fruit also constitutes the commercial drug *colocynth*, which is a drastic hydragogue providing large watery evacuations. In large doses, it causes violent gripping, prostration and sometimes bloody discharges. Small fruits are taken out during the rainy season and stuffed with common salt and *ajowain* which is used as a cure for acute stomachache.

Colocynth seeds, along with those of 'Malkangni (*Celastrus paniculatus*) each 200 gram are crushed together and boiled in three litres of water on slow fire. When one litre is left, half litre of 'seasmum oil' is added and boiled further till all the water has evaporated and only oil is left. The oil is used as hair oil and a drop of its introduced in the nostrils at a time, cures catarrah. In case of congestion and phlegm has putrified, it also gets expelled. The *oil of colocynth* is obtained by boiling 200 grams of fruits in half litre of water, when one-fourth is left, it is strained and mixed with half litre of castor oil and boiled again till all the water has evaporated and only oil is left.

***Extent of utilization and trade*:** Fresh fruits are collected and sold in the market; it is also reported to have been cultivated as a commercial crop. Intensively bitter taste of the pulp is due to an amorphous glucoside which is yellow in colour (0.6%) but not in seeds. The yield is expected to be about 60 kg; oil in the seeds is from 15–16%. Bitter substance is *colocynthin, colocynthenin*. Roots contain *alpha elatrin, hentriacontane*, and *saponins*, while the seeds contain fixed oil, a *phytosterolin*, alkaloids, glycosides and tannin. The pulp contains *alphaelterin*, hentriacontane, a phytosterol' and a mixture of fatty acids.

Other species

a. *Citrullus fisulosus* Stocks (-*C.vulgaris* var. fistulosus (Stocks) Duthie and Fuller); Colocynthis citrullus (L.) O. Kuntze var. fistulosus Chakravarty in Rec. bot. surv. India. 17 (1) : 1959.
   ***Local names*:** Tinda, Tendu (Hindi).
   The plant is cultivated throughout the country for its fruits, used as vegetable, recommended by ayurvedic physicians in diets. The seeds are medicinal.

b. *Citrullus vulgaris* Schard. (Vern. name. Tarbuj-Hindi, Watermelon-Eng.).
   A climber cultivated throughout the country but supposed to be indigenous in tropical Africa.
   The fresh fruit is used, since it is cooling during summers and is diuretic. There are many special forms which vary in colour and flavour of the pulp and in the sizes as well. The wild plant may be either bitter and sweet, without any observable structural differences. The bitter form very close to *C. colocynthis* in Sindh is known as *kirbut*. A small

variety of *tarbuj* is now grown and is in the market during the summer season and in south even during the winter season (var. babysugar).

The pulp is edible and given as a medicine for strength and vitality. The seeds are collected for a limpid oil. In times of scarcity these are pulverised and baked into bread. There is considerable demand on account of their cooling, diuretic and strengthening qualities and so constitutes one of the four seeds (called *Chár-Magz*) sold in the market (Kahrbuza, Khira, *Tarbuj* and Kaddu-Sriphal). The oil of terbuz is sometimes substituted for almond oil.

c. *Citrullus prophatrum* L. (Vern. name. Kantala, Tndras-Hindi.) Small procumbent, trailing herb; on sandy/gravelly soils. The plant is emetic and purgative. Root used in indgestion. The fruit pulp contains bitter resins producing nausea and is purgative.

61. *Clematis buchaniana* DC

Family: Ranunculaceae.

Local names: Kauni-báli (Jaunsar).

Description: A climbing shrub, shortly hairy. Stem grooved. Leaves palmate, the bases of the opposite stalks more or less united; leaflets broad ovate, cordate, often lobed; coarsely and irregularly crenately lobed. Flowers pale-yellow, in long leafy panicles. Sepals tomentose outside, pubescent within, ribbed narrowly oblong, erect, tip recurved and pointed. Filaments hairy throughout. Achenes hairy.

Distribution: In temperate regions from 1500–3300 m.

Uses: The root decoction is used in menorrhagia.

Extent of utilization: Not commercially utilised.

1. *Clematis barbellata* Edgew.
   Local names: Lungéru (TG); Kauni (Jaunsar).
   A climbing shrub, stem terete. Leaves of 3-leaflets, stalked and clustered at the nodes; leaflets ovate lanceolate, often lobed irregularly, sharply toothed. Buds ovoid, acute. Flowers dull-purple, soliary on axillary stalks, at first shorter than the leaves, lengthening in fruit. Styles pubescent, long pointed, spreading. Filaments usually fringed with long hairs. Anthers hairy on the back. Achenes glabrous. In temperate areas on shady aspects in outer Himalaya. Uses same as the above species (in menorrhagia).

2. *Clematis montana* Buch-Ham. Vern. name: Ghantiali, Kariguli (Hindi) Kaunia bali (Jaunsar).
   Climbing shrub; stem terete. Leaves of 3-leaflets, clustered at the nodes. Leaflets narrowly ovate, acute, margins more or less toothed towards the tip. Fls. white, solitary on axillary stalks. On open hill sides, 2000–3000 m; conspicuous during May because of beautiful spray of large white flowers. Leaves act deleteriously on the skin.

3. *Clematis nepaulensis* DC. (Vern. name Ghantiali, Kanguli)
   The leaves act deliteriously on the skin.

4. *Clematis orientalis* L. (-C. graveolens Lindl.)
   The plant is poisonous and contains HCN.

5. *Clematis triloba* Heyne ex Roth. Vern. name: Murhári (Hindi); Laghuparnica (Sansk).
   The plant is applied to boils and itch, used in leprosy, fevers and blood diseases.

62. *Clerodendrum indicum* (L.) Kuntze (-C. siphonanthus (R.Br.) C.B.Clarke).

Family: Verbenaceae.

Local names: Chingári (DDN); Bháringi (Hindi), Turk's turban, Tube flower (Eng). Bamunhati (Mumbai) Arni (Rinj).

Description: A shrub, 1.5–3 m high with herbaceous, fluted hollow stems. Leaves in whorls of 3–5, narrow lanceolate, subentire glabrous, rather hard. Flowers white fading to yellow in rigid terminal panicles. Calyx dark red and enlarged in fruit, segments oblong. Calyx tube drooping, lobes ovate-oblong. Drupe ovoid, dark blue, supported by spreading red calyx.

Distribution: Common in grasslands; often cultivated in gardens for its flowers.

Uses: The roots are useful in asthma, cough and scrofulous affections.

Juice of the leaves is applied to herpetic erruptions, and are vermifuge, bitter and tonic. Anthelmintic property is due to bitter principle in the leaves.

Sections of the hollow stem are tied round the neck as a charm against various ailments in Bengal. The plant is also ornamental.

Other species

a. *Clerodendrum inerme* (L.) Gaertn. (syn Volkameria inermis L.)
Lanjai (Hindi), Kundali (Sansk). A straggling, trailing shrub; also grown as a hedge plant.
Leaves used as poultice, in juice is alterative and febrifuge. Medicinal properties resemble like those of Swertia chirata. Roots when boiled in oil work as liniment. The leaves contain an amorphous bitter principle, resin and gum.

b. *Clerodendrum infortunatum* Gaertn. Vern. Káru (Hindi) Bhánt (Hindi), Bhantáka (Sansk.) A deciduous shrub, up to 4 m high, bark with grey raised corky meticels. Leaves opposite, broad-ovate, slightly cordate. Panicles terminal, branches and calyx redening in fruit. Corolla white, tinged with red. Drupe fleshy, tetragonal, bluish black when ripe, enclosed in leathery calyx. Common in 'sal' forests. Roots and leaves used externally for tumours and skin diseases. Leaves are used as tonic, like 'Chirayata'. Fresh leaf juice works as vermifuge and as febrifuge in malarial fever, especially for children.

c. *Clerodendrum phlomoides* L.f. Arni (Hindi) Vatághani (Sansk).
The roots are given as tonic after measles attack. Leaf juice is given in neglected syphilitic complaints.
The plant is given to cattle in diarrhoea.

d. *Clerodendrum viscosum* Vent. Paste of the root of this pant and 'rye' is given twice in cases of blood dysentery (Sambhalpur dist. in Orissa). Root paste is heated and applied to the chest to relieve pain. 3–4 leaves of the plant are added to fomenting liquor for enhancing fermentation and reduce total fomentation period to about half, besides increasing the intoxicating effect of the liquor. (in Ganjam district of Andhra).

63. *Cleome brachycarpa* Vahl. ex DC

Family: Capparidaceae.

Local names: Panwár (Hindi); Raktágarba (Hindi), Nodi, Navli, Khirnamar (Rajasthan).

Description: Perennial, small, glandular herb, branched from the base. Leaves 3–5 foliate, upper simple; leaflets obovate or oblong. Flowers yellow, in axillary, leafy, long-pedicelled racemes. Capsule scaberulous, inflated. Seeds minute, smooth. Fls. July-October.

Distribution: In northwest India from Agra in UP, westwards to Pakistan (Peshawar and Sindh). On sandy soils and rocky regions and gravelly habitats in association with *Heliotropium rariflorum,* as weed on wastelands.

Uses: The seeds are used in scabies and leukoderma. Leaves are crushed in oil and rubbed in skin diseases. In villages, poor people cook the leaves for vegetable and reports that the vegetable works as prophylactic during an epidemic like plague.

The plant is also used to cure worms in camels nose also in seabees, rheumatism and inflammation. Powdered seeds when taken cures phlegmatic cough when taken with equal quantity of sugar (5 gram each) for about five weeks. The seeds are used externally as powder and steeped in 100 gram of curd and applied for 2–3 days, after the seeds have swollen, to ringworms.

Young leaves and twigs (of *Cassia tora*) cooked as vegetable and eaten with rice in Bastar and Mandla regions of MP. *Baigas* tribals take its tea to keep the body warm in winters. *Jhabua* Bheels crush the seeds with water and apply on the skin to cure eczema. *Boxa* tribals eat young leaves to cure rheumatic pains.

Other species: *Cleome viscosa* L. (Polanisia isocandra (L.) W. and A.) Vern. Names: Pivli-Tilwan, Kalo, Handi-bagro, Bagro. Jangli-Hur, Kanphutia, Tilwan (Hindi).

Erect hairy herb with 3–5 foliate leaves; leaflets ovate or obovate. Fls. yellow, long pedicelled, in axillary racemes. Stamens 10–20. Fruit capsular, 5–6 cm long, striated, glandular pubescent, terminated by a style.

A powerful irritant, cause vesication and redness when applied to skin. The juice mixed with oil is put into the ears to prevent discharge.

Pulp of the plant is a guarded medicine for plague and fever. Leaves are used for dysentery, paratyphoid, bronchitis and gonorrhoea. The seeds are used as a anthelmintic in piles and for removing worms, germs and roundworms.

Cleome gynandra L. is shown in Fig. 5.50.

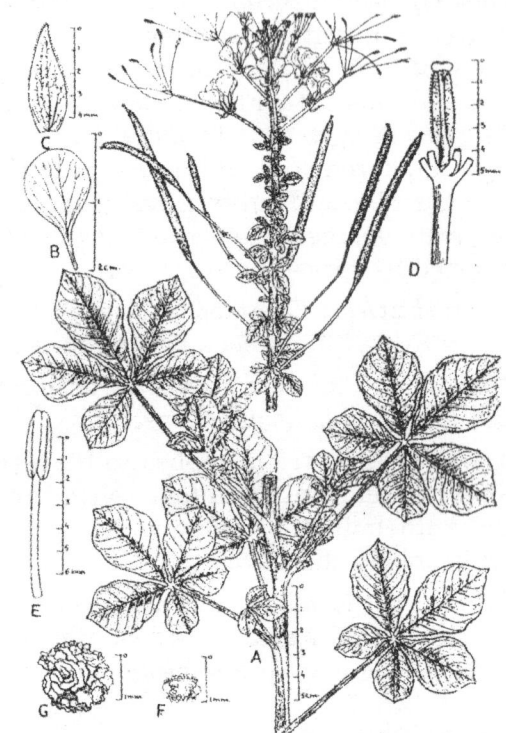

Fig. 5.50: *Cleome gynandra* L. A. flowering and fruiting branch; B. petal; C. sepal; D. gynoecium and upper part of gynophore (stamens cut away); E. stamen; F. ovary in transverse section; G. seed.

64. *Coccinia indica* Wight and Arn. (-C.cordifolia Cogn.)

Family: Cucurbitaceae.

Local names: Kandoori, Kunduri, Kundru, Ghol, Golan, Golends, shivlingi (Rajasthan); Bimbi, Kaidonda (Hindi); Bimba (Sansk); Telakucha (Mumbai); Kovai-fruit (Eng).

Description: A climber of rainy season with beautiful white fls. with green fruit turning to red. Stem grooved, slender; tendrils simple. Leaves 5-angular, occasionally 5-lobed, paler beneath, studied and sometimes rough with papillae. Palmately 5-nerved from a cordate base. Flowers white. Male peduncle 2–5 cm, jointed below the flower. Female shorter, about 1 cm. Fruit bright scarlet, ovoid or oblong.

Distribution: Throughout India, extending to Africa, Malaysia and SriLanka.

Uses: The plant is used internally in gonorrhoea. Leaves applied externally in skin erruptions. Juice from the roots and leaves is used to cure diabetes.

The unripe fruits are cooked as vegetable. Seven leaves of this plant are ground in water along with 7 grains of pepper, strained and drunk to reduce sugar passed with urine.

Coccinia grandis (L.) Voigt (*C.cordifolia Cogn.*). A perennial climber, stem grooved tendrils simple. Leaves 5. angular paler beneath, 5 nerved from cordate case, Flowers white. Fruit bright scarlet. Most common on hedges in Rajasthan (Vern. Shivlinge, Golaw, Golenda) (Fig. 5.51).

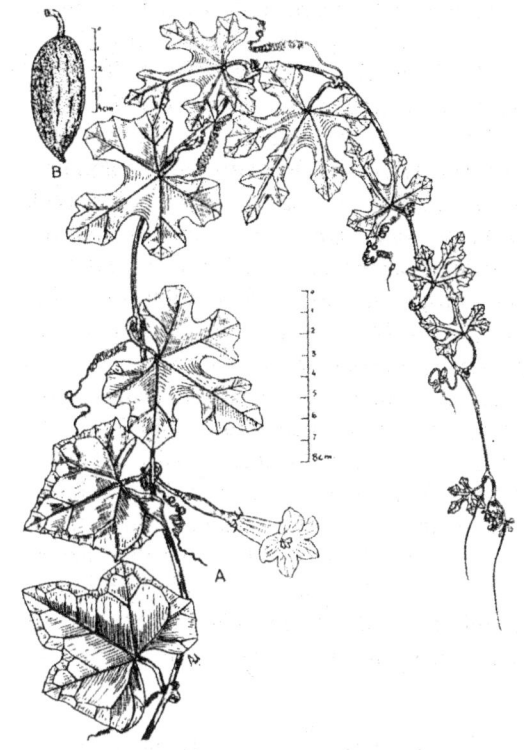

Fig. 5.51: *Coccinia grandis* Voigt A. upper portion of a female flowering stem; B. fruit.

Distribution: Foot of the Himalaya and Shiwalik hills, in tropical regions of the country; found largely on *Capparis* or *Gymnosporia* bushes in Rajasthan.

Uses: The roots are used in the treatment of intermittent fevers and as tonic. It is reported also useful in chronic rheumatism and veneral diseases. Juice of the leaves when mixed with water, forms a jelly which is taken as a coling medicine for gonorrhoea and used externally for eczema, prurigo and impetigo.

65. *Cocculus villosus* (-C. hirsutus (L) Diels.

Family: Menispermaceae.

Local names: Bagar-bél (Rajasthan); Amti-ki Bel (Hindi); Vasnavel, Garudi (Sansk).

Description: A climbing undershrub, with tomentose branchlets. Leaves very variable in size, ovate or ovate-oblong, mucronate, clothed with grey tomentum, petiole up to 1cm. Male fls. in axillary panicles. Female 1–3 together, on short axillary pedicels. Drupe 0.5cm in diam., dark purple.

66. *Colchicum luteum* Baker

Family: Liliaceae.

Local names: Jangali-Kachálu, Hirantutiá (Hindi), Virkum (Kash), Hiranyatutha (Sansk); Golden collyrium (Eng) Trade Name: Suranjan.

Discription: A perennial herb usually the earliest spring flower. Leaves large, oblanceolate Fls. linear, oblone, perianth golden yellow.

Distribution: Temperate Himalaya from Kashmir to Himachal. Used as a substitute for *C. acutumnale.* which does not occur in India. Indian market has two forms, the bitter and sweet, the latter is imported from Persia; though bitter one possesses the same properties as true colchicum. It is also traded as a medicine under Trade name *Colchicum* used for gout and arthritis.

Uses: Corms used as carminative, laxative, aphrodisiac, alterative. Also used in diseases of liver and spleen apart from gout and rheumatic pain, also for external appli-cation in gout to lessen inflammation.

67. *Commelina obliqua* Buch.-Ham.

Family: Comelinaceae.

Local names: Kanjura (Hindi); Jata-kanchura (Mumbai), Anjura (Kumaon).

Description: A succulent herb. Stem branched, procumbent. Leaves sessile, lanceolate, acute or long pointed, sheathing fringed. Spathes usually in a terminals head, partially enclosed by a pair of opposite leaves, usually filled with a clear glutinous substance. Petals pale-blue, one of them nearly white. Ovary 3-celled with a single ovule in each cell. Capsule 3-seeded; seeds minutely dotted.

Distribution: Throughout India, in lower moist parts; also in lower Himalaya.

Uses: The root is useful in vertigo and bilious affections. It is refrigerant and laxative, useful in strangury and costiveness.

Other species

    a. *Commelina benghalensis* L. Kanchárá, kanakooá (Hindi) Kancháta (Sansk.)

The plant is bitter, emollient, demulcent, laxative and useful in leprosy. The leaves are eaten as *Pakori*, fried with gramflour. Throughout India.

    b. *Commelina nudiflora* L. (Commelinaceae) Kanshurá; Koshpushpi (H). Bruised plant is applied to burns, itches and boils. Leaves are used for poulticing sores.

    c. *Commelina salicifolia* Roxb. Jalpipári (Hindi), Panikanchira (Mumbai). The plant is used in dysentery. *Commelina suffruticosa* Blume (Vern Dareorsá-Santhal) The roots are applied to the sores. Tropical regions of India in east from Nepal to Sikkim, Central India till Western Rajasthan.

68. *Convolvulus arvensis* L.

Family: Convolvulaceae.

Local names: Hiranpádi, Hiranphuri (Trade name); Gondal (Bihar) Bhadrabála (Sansk).

A herb with slender, creeping root-stock. Stems many, trailing or twining, angular. Leaves petioled, ovate or oblong-lanceolate, apiculate at the apex, entire, sometimes lobed. Peduncles solitary, slender, with a pair of linear bract; pedicels up to 2–3 cm long, except primary, 2-bracteolate beyond the middle. Corolla widely funnel-shaped, pink or white, with a pale-yellow centre. Capsule globose, dark reddish (Fig. 5.52).

The roots are cathartic and produce marked gastro-intestinal irritation.

Other species: *Convolvulus glomeratus* Choisy ex DC. (Vern. Runchhalivedic-Gujarat) is a purgative.

C.conglomeratus var. *volubilis* (Vern. Ratanjot, Rothaval). The plant soaked in water is given as a cooling drink.

medicinal roots and a yellow dye in America. It may be confused with *Picrorhiza kurroa*, which is similar in appearance.

The plant is used as collyrium, the roots make a valuable tonic in debility, following fever, but not as febrifuge. There is considerable demand for this plant. Locals gather the roots towards the end of rainy season in India and sell it in the market. *Mishmi* poison consists of a mixture or powdered testa roots with a pulp of an acid glutinous fruit called *Dillenia speciosa*.

70. *Corchorus depressus* (L.) Christensen.

Family: Tiliaceae.

Local names: Hadé-Ka- Khet, Báphuli (Hindi), Bahuphali (Guj).

Description: A spreading, much branched, woody perennial with prostrate branches. Leaves small, roundish, shortly petioled. Flowers yellow, small, numerous in leaf-opposed cymes, peduncle short, stout. Capsule cylindrical, 4–5 valves, beaked.

The plant has tonic properties, given as cooling medicine in fevers. Mucilage is used in gonorrhoea. The seeds in decoction with milk given as tonic. Roots and ripe fruits are used in diarrhoea. Leaves in decoction are efficaceous in cases of skin erruptions. Leaf infusion is demulcent, laxative, stomachic, increase appetite and bitter tonic given in fever, dyspepsia and liver disorders.

Other species: Corchorus fasicularis Lam. (Vern name. Khetápat, Hirankuri, Chánchu, Kalábi, Janglipat, Chhunchhadi). Trade Name: Bhaphali. A small climbing herb growing wild as weed in wheat and gram fields. Slender branches exude white milky fluid on breaking. The flowers are white with rosy tinge. Also known as *Karvibel*, since it is bitter in taste. About 10 gram of the herb is ground with 7 grains of pepper, strained and

Fig. 5.52: *Convolvulus arvensis* (Vern. Hirankhuri).

69. *Coptis teeta* Wall.

Family: Ranunculaceae.

Local names: Mamirá(Hindi); Titá (Bihar and Assam); Mamiran (Mumbai) also the trade name. *Mamira* name is given to Thalictrum also.

Description: A small stemless herb with persistent rootstock.

Distribution: In temperate regions of east Himalaya (Mishmi hills). The plant grows over the ground among the moss round the trees. Also on the northern frontiers of Assam.

Uses: The rhizome is a bitter tonic, stomachic, efficaceous in debility and mild forms of intermittent fevers.

Useful as a salve for eyes. In China used as antidiabetic.

Coptis anemonoefolia root is medicinal in Japan, while *C. trifolia* yields both the

given for 2–3 weeks to cure diseases like abscesses, ringworm, scabies and gonorrhoea. The herb is also applied externally. The plant is mucilaginous and restorative.

71. *Coriandrum sativum* L.

Family: Umbelliferae (-Apiaceae nom alt.)

Local names: Dhainá, Dhaná, Dhaneyoka (Hindi), Dhanyaka (Sansk.); Kasneez (Urdu), Kothibir, Kotmir (South).

Description: A smooth erect herb, up to 40 cm high, giving out an unpleasant odour when crushed. Leaves alternate, pinnately divided; leaflets of the lower leaves broadly ovate, lobed toothed; of the upper leaves linear. Flowers in branched umbels, very small, white. Fruit spherical.

Distribution: Cultivated throughout India, particulary on heavy soil. Guna is the famous in MP and the variety is sold at a premium.

Uses: The seeds of this plant, used as spices, have many virtues. It is used in headaches due to heat, vertigo and diarrhoea due to poisoning from croton seeds.

The seeds ground with water are applied as paste to the forehead to relieve pain.

Green leaves of *Dhania,* cucumber and vineger are mixed and placed in a phial, which if inhaled cures fever with delirium. In case of vertigo, 9 gms each of dry *Dhania* and dry embelic myrobalan (*Phyllanthus embelica*) are pounded coarsely, steeped overnight in water; in the morning water is strained, sweetened with sugarcandy and drunk. Seeds of *Dhania*, chewed after meals, help in digestion;

Dhania seed powder mixed with sugar is taken with water to stop excessive nocturnal emissions and moderate sexual desire. According to Unani system the infusion is useful in preserving the sight in smallpox.

The volatile oil obtained from fruit distillation is efficaceous in flatulent colic, in rheumatism and neuralgia, also for correcting the gripping effects of purgative.

72. *Costus speciosus* (Koen.) Smith.

Family: Zingiberaceae.

Local names: Keu (Bihar, Hindi); Kushtha (Sansk.); Kemuka (Mumbai).

Description: An ornamental herb.

Distribution: Widely distributed in Assam Khasi and Jaintia hills, Uttaranchal, Himachal sub-Himalayan tracts. More or less throughout India up to 1300 m. Cultivated as ornamental.

Uses: Tuberous rootstock is the raw material for production of diosgenin. The tribals of Baster wash the tubers, bake them in hot ash, pounded and eaten as *chutney*; also pounded and boiled to make full diet. These are also pickled with lemon and chillies. In Orissa, juice of the rhizome is applied to head for cooling and relief from headache. Roots are rich in starch, bitter astringent, purgative, stimulant, tonic and anthelmintic.

73. *Crotalaria spectabitis* Roth. L. *Crotalaria sericea* Retz.

Family: Papilionaceae.

Local names: Jhunjhunia (Hindi), Pipuli Jhunjhun (Mumbai), Ghantarava (Sansk.)

Description: An erect, silky pubescent shrub, 1–2 m. Stem robust, grooved. Leaves simple, nearly sessile, obovate, narrowed to the base, tipped with a minute bristle, uppermost leaves lanceolate, acute, upper surface glabrous, lower finely pubescent; stipules small. Flowers yellow in terminal racemes. Calyx pubescent. Petals nearly twice as long as the calyx. Pod oblong, nearly glabrous; seeds many.

Distribution: Throughout India, in tropical regions ascending to 1300 m in the valleys.

Kashmir, Punjab plains, Shimla, Dehradun, Kumaon; Bageshar, Sikkim, Sylhet, Central India, Bombay, Bengak, Madras.

Uses: The plant is used to control scabies, poisonous to livestock, seeds, leaves and stem contain monocrotaline; lowers BP in dogs and lethal to livestock.

74. *Cuminum cyminum* L.

Family: Umbelliferae (Apiaceae nom alt.)

Local names: Jeera, saféd-Zira (Hindi), Jiraka (Sansk).

Description: A glabrous annual herb, 30–90 cm high, roots long, slender, Stem arising almost from the base, branches forked. Basal leaves soon falling off, short stalked, feathery, divided into very slender lobes, lobes running parallel to the main axis, about 5 cm long. Higher leaves almost similar having short, sheathing winged stalks. Flowers in large, long-stalked branched umbels, bracts of main umbel about 4 cm long, secondary umbels 2–6, less than 3 cm long, few fld. Capsule cylindric.

Distribution: Native of Arabia, Egypt, Syria but cultivated mainly in Rajasthan (Ganganagar district) UP and Punjab.

Uses: Fruit is stomachic, diuretic, carminative, stimulant, astringent and emmenagogue, useful in dyspepsia, diarrhoea and horseness of voice. The fruit is also given to pregnant women for checking bilious nausea, for promoting secretion of milk after childbirth. Cumin seeds form an important ingredient of spices, used as carminative and sharpens the appetite.

The seeds contain an essential oil.

Seeds mixed with a little clarified butter are smoked like tobacco in a pipe to relieve hiccup.

A poultice is applied to painful swellings on the breast and testicles.

An ointment of powdered fruit with honey, salt and calrified butter, is applied for relief of pain from scorpion stings.

75. *Curcuma longa* L. (-Curcuma domestica Valeton).

Family: Zingiberaceae.

Local names: Turmeric (Eng), Haldi (Hindi), Haridra, (Sansk), Halafa, Manjal (Gujarat).

Description: About 25 species of curcuma are described and some 10 are of economic importance. The underground stem or rhizome is used medicinally.

Distribution: It is cultivated in almost in all the states in India, particularly in Tamil Nadu, Maharashtra, Bengal, Uttaranchal, etc.

Uses: The paste of the rhizome made in water is applied to swellings, insect bite and on wounds. The green rhizome is given to infants as a cure on simple whooping cough. *Haldi* powder with milk is given after injury to help up the wounds and as antiseptic.

An ingredient of spices, it is used in curries, vegetable and pulses and imparts colour, prevents flatulance. It is useful in cough when taken with lukeworm water. It relieves pain and swelling due to hurt and used externally mixed with lime. In cases of gonorrhoea, equal weight of turmeric and *Anwla* are taken after grinding, sifting and mixing with equal weight of sugar, with milk. It also kills intestinal worms, stops bleeding. Green rhizomes made in *laddu* (balls) are taken during winters.

76. *Cuscuta reflexa* Roxb.

Family: Convolvulaceae.

Local names: Akashvél (Hindi), Amarvéla (Sansk), Algusi (Bihar), Nirmula (Mumbai). The Dodder (Eng.).

Description: A well-known leafless greenish white or yellow flowering parasite. Stem twining, succulent, densely interlaced over small trees and shrubs; attached to the plant on which it grows by means of minute adhesive discs. Flowers fragrant, waxy, white in clusters or racemes. Calyx shorter than corolla. Corolla tubular, lobes shorter, triangular, reflexd. Ovary 2-celled. Stigma 2, diverging, acute. Capsule globose, opening by horizonal line near the base.

Distribution: Common throughout the plains of India, ascending in hills up to 2300 m.

Uses: Stem in decoction is useful in constipation, flatulance, liver complaints and bilious affections. Seeds of this plant with that of *Smilax aspera* (sarsaprilla) are used to purify blood.

Powder of the seeds in Kumaon region is given as stomachic. Juice of the stem kills lice of headhairs. The plant is used for fomentation in pains. It is purgative, used for blood purification, bilious disorders and protracted fevers.

The seeds are carminative, anthelmintic, alterative and yield a semidrying oil.

Boiled residue of the plant is used in chest pain.

The plant contains *cuscutalin* and *cuscutin* and the seeds contain pigments *amberlin* and *cuscutin*, a wax and semidrying oil.

Other species

a. *Cuscuta hyalina* Heyne ex Roth. Vern. Amarbél

Found in northwest India on numerous hosts. Extract of plant boiled in water used against chest pain.

b. *Cuscuta chinensis* Lamk. Vern. Amárbel.

It forms closely twining tangled masses on a number of hosts in arid areas. The seeds are tonic, diaphoretic and demulcent. In western peninsula, on trees.

77. *Cyamopsis tetragonoloba* (L.) Taub.

Family: Papilionaceae (Leguminosae).

Local names: Gowár (Hindi); Hauri (Mumbai); Bakuchi (Sansk.).

Description: An erect annual herb. Leaves trifoliate. Flowers violet to purplish, in axillary closed racemes. Pods elongated, compressed and fleshy.

Distribution: Indigenous to India, but never recorded wild. Cultivated throughout the drier regions of India such as in Rajasthan.

Uses: Green tender pods are used as vegetable. Seed coat is tough. Highly nitrogenous protein content about 30 percent can withstand storage. Seeds produce mucilaginous guargum.

The fruits are laxative, used in biliousness. It is a good vegetable.

78. *Cynodon dactylon* (L.) Pers.

Family: Gramineae (Poaceae nom alt.).

Local names: Dub, Durbá, Hariali (Hindi), Dhruva, Haritali (Sansk) Bermuda (Eng).

Description: A perennial glabrous grass with slender, creeping stem, rooting at the nodes, developing matted tufts with short erect branches. Leaves long, very narrow, linear or lanceolate, flat, narrow pointed, smooth, glaucous, arranged in two rows.

Inflorescence of 2–6 spikes, rediating from a slender, 2–5 cm long stalk, green or purplish; stalk of spike slender, compressed or angled.

Distribution: Throughout India, ascending to 1700 m in the hills of the Himalayan region, the grass is considered sacred and used during *Poojas*. Native of the tropics of old world.

Uses: The entire plant is used in medicine which is diuretic, astringent and styptic; fresh juice is used in dropsy as a diuretic, haematuria and vomiting. As an astringent, it is given in chronic diarrhoea and dysentery. It has been found useful in hysteria and insanity.

A cold infusion with milk is prescribed for the bleeding piles, irritation of the urinary path or anus and for checking vomiting.

Externally, the juice is used for checking bleeding from nose, catarrhal, ophthalmia, fresh cuts and wounds; paste of the fresh grass used as styptic, applied locally in gout and rheumatism.

Decoction of roots is diuretic and useful in the treatment of vesical calculus, bleeding piles and dropsy, also in genitourinary diseases.

Crushed roots mixed with curd used in chronic gleet.

It can withstand moderate grazing giving rise to coarser shoots with trampling and overgrazing. It is liked by horses and highly esteemed as lawn grass.

If allowed to wilt, it develops hydrocyanic acid and may be injurious to animals feeding on it.

Extent of utilization: The plant is utilised by Vaids and Hakims. The most common grass, makes a good hay and if carefully staked will keep for years. Can be propagated by cuttings, simply stuck in the ground. It is said to be a persistent weed of black cotton soils and require deep hand digging.

79. *Cyperus rotundus* L.

Family: Cyperaceae.

Local names: Bhadra-musté, Mostaká, Mothá, Nagarmothá (Hindi) Nutgrass (Eng.).

Description: Glabrous sedge with slender stolons, hardening into wiry roots, thickened into woody ovoid tubers. Stem subsolitary at top, triquetrous. Leaves long, often overtopping stem. Umbel frequently compound. Spikes loosely spicate of 3–8 spikelets, but umbels sometimes large, reduced to heads.

Distribution: All warm countries, ascending in hills up to 2000 m. A frequent pest in the garden and cultivated places.

Uses: The rhizomes are fed to animals during scarcity period after boiling and also form ingredient for *agarbatti* (incensce stick).

The roots are pungent, aromatic, acrid, diaphoretic, diuretic, astringent, anthelmintic, vulnerary, stomachic, emmenagogue and lithotritic. 20 gram of roots is administered in diarrhoea, dysentery, ascites, dyspepsia, vomiting, cholera and fevers. Roots are also used in urinary disorders, leprosy, thirst and blood diseases.

Roots in large doses are given for roundworms and are ingredient of many compound medicines

Externally warm plaster of fresh roots is applied to breasts to promote flow of milk. Paste of roots heal wounds and sores.

Roots are also used as sweat and urine producers. Roasted and pounded tubers are used to produce an aroma to certain fabrics.

Other species

a. *Cyperus scariosus* R. Br. Vern. Nágarmotha (Hindi), Nagarmusta (Sansk.), Lawala (Maharashtra).

The tubers are aromatic, stomachic, dessicant, diaphoretic, astringent, useful in diarrhoea in the form of decoction in a compound of *Atis*, *Pipal*, *Nagamotha* and *Kakrasinghi* under acute and critical condition and has practically been tried on the author.

b. *Cyperus esculentum* L. Vern. Kaséru (Hindi), Kaseruka (Sansk.)

Tubers are edible and eaten, contain 20–8% oil with pleasant taste and odour. Flour of the tuber is a high calorie food. In uppergangetic plain, Haryana, up to Nilgiris I and Annamalai hills.

c. *Cyperus stoloniferous* Rtetz. Vern. Jatámási (Maharashtra).

The tubers are stomachic, stimulant for heart. Reported on sea sands on shores of India and tidal muds.

The genus of sedges (Cyperus) contains some of the Indian species and most of these are valuable fodder plants, especially when young, others are dangerous weeds of cultivated lands and a few yield culms and leaves that are employed in thatching, in grass matting; while still others afford tuberous rhizomes, that are either eaten, especially in times of scarcity or are collected and sold as perfumes or medicines. *Cyperus rotundus* (Vern. Motha) and *C. scariosus* (Nagarmotha) are very common and found near the water sources. The tubers are used in disorders of the stomach and as intestinal sedative. *C. tuberosus* is aromatic and used medicinally. These have been regarded as diaphoretic, stimulant and astringent. The scythiens have employed them in their embalming preparations. These are also used in perfumery, particularly in giving a required aroma to certain fabrics and in preparation of incense sticks (*agarbatti*). These are greedily eaten by pigs in times of scarcity and also by the human beings. The other fibrous or matmaking forms are *C. corymbosis*, *C. malecensis* and *C. tagetum*.

d. *Cyperus iria* L. (Vern. Burachucha-Bihar) found throughout India in the rice fields. The plant is tonic, stimulant, stomachic and astringent.

e. *Cyperus longus* L. The tuber is bitter, aromatic and emmenagogue, it is reported from Mt. Abu.

f. *C. platyphyllus* Roem. and Schult. The tuber is tonic and stimulant, recorded mainly from the Deccan peninsula.

80. *Delphinium denudatum* Wall.

Family: Ranunculaceae.

Local names: Nirbisi, Jadwar (Hindi).

Description: An erect herb. Stem up to 1 m, branched. Radical leaves orbicular, long stalked, divided nearly to the base, narrow, pinnately lobed, often toothed. Stem leaves few, shortly stalked, upper sessile more or less deeply 3-lobed, lobes narrow, mostly entire. Flowers few, scattered, spur cylindric, nearly straight. Sepals spreading, varying from deep blue to faded grey. Petals blue, the lateral ones 2-lobed, hairy. Stamens many, filament flattened at the base, tapering upwards. Ovaries 3, sessile, many ovuled, style short, recurved. Follicles many seeded, tipped with persistent style.

Distribution: In wetsren Himalaya, 1500–4000 m

(Kashmir: Pahalgam, on way to Amarnath, Kishtwar, Baramula Kagan valley,

Chitral, Upper Chenab valley; Shimla: Shur, Jako Lahul; Kardang, Kunawar, Kulu, Chamba, Khullil, Pangi; Chakrata, Joshimath, Ranikhet, Bhimtal.

Uses: The plant has the property resembling those of *Atis* (aconites) and is often classed among non-poisonous aconites.

The root is bitter, cooling, vulnerary and alexiteric; it cures cough, diseases of blood, stimulant in conditions of debility. It is used as an appetiser, brain tonic and in insanity. It is good in toothache and in painful piles. The roots are used commonly as an adulterant for aconite.

Extent of utilization: The roots are collected and sold in the market on a limited scale. There is a considerable scope for collection of the roots in the Himalaya.

81. *Dactyloctenium aegyptium* (L.) P. Beauv. (-*Eleusine aegyptiaca* Desf).

Family: Gramineae or Poaceae nom alt.

Local names: Makrá, Kuril, Manchi (Rajasthan); Mhar (Mumbai), Madana (Punjab).

Description: Annual herb about 45 cm high stoloni ferous. Spikes 2–5 cm long; tip of rachis shortly produced, up to 20 cm, long.

Distribution: Throughout the tropical regions, frequently on sandy soils, sometimes also on salt impregnated soils.

Uses: The grains are used after parching, by women suffering from bellyache after childbirth. Seed decoction is used as an alleviator of pain in kidney region (in Africa)

82. *Datisca cannabina* L.

Family: Datiscaceae.

Local names: Akalbir (Hindi), Kalbir, Bhanjhála (Punjab).

Description: A tall, erect herb, resembling hemp.

Distribution: In subtropical and temperate regions from Kashmir to Nepal, 1000–2000 m.

Uses: The roots are used extensively as a yellow dye for silk. The herb is bitter, purgative, febrifuge, diuretic and sedative in rheumatism.

83. *Descurainia sophia* (L) Webb. (-*Sisymbrium sophia* L.)

Family: Cruciferae or Brassicaceae nom. alt.

Local names: Khubkákan (H), Maktrusa (Pb). Khaksi (Persian).

Description: A finely pubescent herb with numerous leaves, sessile, twice or thrice pinnatisect, segments short, thread like. Flowers pale yellow. Pods glabrous, slightly flattened, curved, erect or spreading.

Distribution: In temperate Himalaya, Kashmir to Kumaon, also in the eastern parts of the Himalaya. A common weed.

Uses: The plant has a pungent odour when rubbed and an acrid bitter taste due to a volatile alkaloid. The flowers and leaves are astringent, antiscorbutic. The seeds are expectorant, restorative, tonic and useful in fever, bronchitis, also given to expel worms and in calculous complaints. The water boiled with seeds given to patients suffering from smallpox.

84. *Dicoma tomentosa* Cass.

Family: Compositae or Asteraceae nom alt.

Local names: Bajrandanti (Hindi) also is the name for *Potentilla fulgens* in Kumaon. Choloharnacharo (Guj), Navananjichapala. (Maharashtra).

Description: An annual herb or undershrub, much branched, clothed with white cottony wool. Branches terete. Leaves

narrow, nerves obscure. Heads subsessile, numerous, glabrous. Involucral bracts terminating in long yellow spines. Achenes densely silky. Pappus shining.

Distribution: In northwest India, south to western peninsula. Common spiny annual on the gravelly plains.

Uses: Bitter, febrifuge, used in febrile attacks after childbirth. Also used as local application to putrescent wounds in Africa.

85. *Dolichos biflorus* L.

Family: Papilionaceae.

Local names: Kulthi, Kalath, Hulaga (Hindi), Kuláytha (Sansk.).

Description: A twining or suberect annual with leaflets 2–5 cm long. Flowers yellow, 1–3 in the axil. Pods 5–6 seeded.

Distribution: Extensively cultivated in south on poor soils. Usually sown in June-July and harvested in Sept.-Oct. In districts with good rainfall it is sown in Nov., after paddy is harvested and reaped in March.

Uses: The grains when boiled and mixed with gram are a good fodder for horses. A favourite crop on riverbeds and on sloping land.

Leaf decoction is useful in leukoderma, menstrual disorders. Seeds are used as cure for kidney stones and are a mild purgative; also used in piles and in reducing body weight. The seeds alleviate *Kapha* and *Vata*, useful in asthma, bronchitis, cough, colic, dropsy and abdominal tumors and headache. It is useful in nasal catarrah, renal calculi and menstrual disorders like amenorrhoea. The seeds are the rich source of protein.

86. *Drosera lunata* Buch.-Ham.

Family: Droseraceae.

Local names: Mukhájali (Hindi).

Description: Small herb, often branching. Leaves alternate, halfmoon shaped, peltate, upper surface and margins beset with viscid glandular hairs; radical leaves smaller, resolute, soon disappearing. Flowers small, white in terminal branching racemes. Calyx 5-parted. Stamens 5. Ovary ovoid, 1-celled. Capsule encised within the persistent calyx and corolla. Seeds minute, attached to the valves.

Distribution: An insectivorous plant on pasturelands in temperate regions, 1300–2500 m. The plant has been recorded near the temple on Mussoorie-Chamba road, but now almost absent from Surkhanda region.

Uses: Leaves are bruised and mixed with salt and used as blister in Kumaon. In Ayurveda, the plant is used to reduce gold in powder form. Utilized on a limited scale.

When placed in milk, the plant curdle it, a property attributed to a ferment.

87. *Echinops echinatus* Roxb.

Family: Compositae, nom alt. Asteraceae.

Local names: Utakantá, Gokru (also named for *Tribulus terrestris*); Kantálu, Utáti (Sansk); Utánti (Maharashtra); Untkantalo (Rajasthan).

Description: A thistle like annual herb, rigid with white cottony pubescence, branched from the base. Leaves pinnatifid, lobed spiniscent, white beneath, sessile, oblong. Heads white, globular; flowers intermixed with long spines. Involucre surrounded by strong white bristles. Outer involucral bract 6–8, oblanceolate, pungent, inner with ciliate, connate tips. Achene silky villous (Fig. 5.53).

Distribution: More or less throughout India; in the Himalayan region found as undergrowth in *chirpine forests*, upto 1700 m.

Fig. 5.53: *Echinops echinatus* on the gravely plains of Jaisalmer district in arid Rajasthan (Photo R.K. Gupta).

Uses: The plant is alterative, diuretic, nervine tonic, used in hoarse cough, hysteria, dyspepsia, scrofula and ophthalmia. Powdered roots are applied to cattle for maggots and mixed with *babool* gum to destroy lice.

In Garhwal, the flowers are boiled in water and the decoction is mixed with butter and sugar. The liquid is injected into the nostrils. A yellow liquid comes out from the nostrils, that cures jaundice (based on information collected from Pauri district by the author).

The thorny plant with leaves resemble the *Argemone mexicana* plant in appearance. The juice of the flowers is introduced in the eyes to relieve night blindness, haze, nebula and macula.

Rootbark is powdered, sifted and mixed with honey, is taken for cough and asthma cure. It is also used in sexual debility, when about 10 gm of the roots are crushed and boiled in a mixture of 500 cc of milk and an equal quantity of water along with 5 date-palm fruits. The decoction is taken after the water has evaporated, the bag of cloth containing the roots is removed, the dates may also be eaten.

The leaves and fruits of this plant (10 gm) along with pepper 7 (grains) are boiled in water, strained and taken for curing phlegmatic fever.

88. *Eclipta alba* Hassk. (Eclipta prostrata Roxb.)

Family: Compositae (Asteraceae nom alt).

Local names: Bhángra (Hindi); Bhringráj (Sansk), Kedaraja, Kesuti, Kesaria (Bihar); Jalbhangra (Rajasthan).

Description: An erect or prostrate herb, strigose. Leaves variable in form and look, sessile, narrowed at both ends. Flowers in solitary heads or two together in short, unequal axillary peduncles; involucral bracts ovate, obtuse, equalling or exceeding flowers.

Distribution: A cosmopolitan weed, in wet places near the ditches throughout India. It appears that there are two forms of this plant—one prostrate and the other erect. Also found on dry riverbeds and ponds.

Uses: The plant is considered in Ayurveda as very valuable hence called *Bhringráj*. It is used in the treatment of cough, asthma and anaemia, deobstruent in hepatic and spleen enlargements.

Plant juice in combination with aromatics is used in jaundice. Leaf juice along with honey used as remedy for catarrah in infants.

Roots are emetic, purgative, applied externally as antiseptic to ulcers and wounds in cattle.

Oil of the plant called *Bhringraj tel* is a brand name and sold in market.

The plant is used by local vaids and hakims.

89. *Elsholtzia incisa* Benth

Family: Labiatae.

Local names: Bindá (Kumaon).

Description: A pubescent slender herb up to 1 m in height. Leaves fragrant, long-stalked, ovate, coarsely toothed, base tapering, entire, lower surface gland-dotted. Spikes cylindric, flowers white; floral leaves bract like small, linear. Calyx glandular.

Distribution: In open moist edges in sub-temperate and temperate forests between 1900–2200 m from Jammu to parts of west Nepal.

Uses: Dried leaves and flowering tops on hydrodistillation yield 0.88% pale yellow oil with a strong smell of thymol. Oil on chilling gives large crystals, a first report of occurrence of thymol in substantial quantities in the genus *Elscholtzia*. (Sarina and Prabhakar, 1966). *Curr*. Sci. **35(23)** 604.

90. *Emilia sonchifolia* DC.

Family: Compositae or Asteraceae nom alt.

Local names: Hirankhuri (Hindi); Sadhi-modi, Sudhimudi (Bihar), Sudhamandi (Mumbai).

Description: An erect herb with radical leaves spreading, usually stalked, more or less pinnatifid; lobes entire or coarsely toothed. Stem leaves few, similar to the radical base often lobed, stem clasping. Flower heads discoid, long-stalked, corymbose; involucral bracts in one series, edges often more or less united, reflexed in fruit. Flowers purple with white pappus. Corolla 5-toothed. Tips of style arms minutely conical. Achenes 5-angled angles bristly.

Distribution: Throughout India ascending 2000 m in the Himalaya, in waste places and cultivated fields.

Uses: Plant decoction is a febrifuge and when mixed with sugar is given to infantile in bowel complaints.

Leaf juice when poured into the eyes for night blindness and as fomentation in inflamed eyes, and acts as cooling as rose water, in decoction the leaves are antipyretic.

91. *Ephedra gerardiana* Wall.

Family: Gnetaceae.

Local names: Tut-gandhá (Jaunsar); Asmania, Budagur (Himachal) Rachi, Khanda-Phág (Bushar-Himachal), Tse (Ladakh).

Description: A small shrub, branches opposite or whorled, green, striate, nodes 30 cm long; bark on old stem grey. Leaves scale like, scales connate into a 2-lobed sheath, yellow or brown. Spikes often in whorled clusters. Fruiting spikes red, often succulent.

Distribution: Dry regions of temperate and alpine Himalaya from Kashmir to Sikkim, 2300–3300 m, on dry southern exposures.

Uses: Liquid extract of the plant is used for controlling asthma. Tincture of Ephedra is cardiac and circulatory stimulant, while a decoction of stem is a remedy for rheumatism and syphilis in Russia.

The berries juice is used in affections of respiratory passages (ephedrine).

E. equisetira is the official species recognised, *E. gerardiana* is used as a substitute and marketed for ephedrine hydrochloride which is a cardiac stimulant in toxic conditions of the heart, produced by infections such as from diphtheria and pneumonia.

The plant is sold in the market as *Somkalpalta* and collected during Sept-Oct. The dried stem is also employed as snuff. *E. geradiana* and *E. major*, both can be distinguished by subscandent growth of the former and upright, rarely ascending *E. major*. The young stem collected from Lahul and Jespa

yielded 1% of alkaloid, calculated as ephedrine. However, alkaloid content up to 1.93% has been reported in Wealth of India in similar samples previously.

92. *Eleusine coracana* Gaertn.

Family: Gramineae (Poaceae nom alt.).

Local names: Mandwá, Rági, Nágli, Mandal, Soma (Hindi), Finger millet (Eng.).

Description: A herb cultivated as food crop in hilly districts of Uttaranchal, in Andhra, T. Nadu, Karnataka, Orissa, Bihar and Maharashtra, as kharif crop under dryland condition.

Uses: The grains are tonic millet. In villages bread is made from the flour of the grain and is a staple food of many rural communities.

It also serves as a medicine in need, since it is useful in dropsy, when the flour is mixed with decoction of Kachnár (*Bauhinia variegata*) leaves and bread made from it. This bread is smeared with butter and is eaten by the patient.

Mandua seeds are ground in water and applied to parts affected by spider when it gets crushed under the skin, which normally gives rise to small vesicles and a burning sensation.

The gains are cooling, useful in biliousness and astringent.

Other species: *Eleusine indica* is an annual herb and *E. coracana* is supposed to be the cultivated form of this species. Stems are tufted, 30–60 cm erect, soft. Leaves flat, sheaths hairy at the mouth. Ligule obsolete. Spikelets green 3–5 fld, alternate, not opposite as in *Panicum*. Crowded in two series along one side of 4–7 simple straight, flattened branches and digitately spreading from near the top of the stem. Glumes boat-shaped. Empty glumes 2, shorter than the lowest fls. glumes, unequal, the lower the smaller, keels roughly hairy. Flg. glumes glabrous, acute. Stamens-3, styles 2, short, distinct, feathery. Grain oblong, free within the persistent glumes, outer coat loose, seed wrinkled.

Throughout India, ascending to 1300 m, in tropical regions of the old world. Name of the genus *Eleusine* is derived from Elusis, an ancient city of Greece famous for the worship of Ceres, the goddess of agriculture.

93. *Eulophia campestris* Wall.

Family: Orchidaceae.

Local names: Sálapmisri (Hindi), Amritá (Sansk.), Trade name Pranda.

Description: A terrestrial, glabrous herb with fleshy tubers rhizomes. Leaves appearing with or after flowers. Usually plicate. Fls. racemose on a tall erect, lateral seafe. Sepals free subequal, lip adnate to base of column.

Distribution: In western peninsula, other species are *E. epidermalla*, *E. herbaria E. nuda* and *E. pratensis (Satavari)*.

Uses: The tuber, *salap* is the product of several species like *E. campestris*, *E. nuda* and *E. virens*. The 'salap' of European commerce is *Orchis mascula*. These are dug up after the plant has flowered; the plump firm ones are washed, set aside, and subsequently strung on thread and dried in sun. commercial article is met in three forms-palmate, large ovoid and small ovoid. Esteemed as tonic, aphrodisac, used in stomatitis, purulent cough and heart troubles.

94. *Eruca sativa* Garsault

Family: Brassicaceae

Local names: L. Taramira (Hindi).

Description: An erect, branched herb. Stem branched. Leaves sessile, pinnatifid, segments coarsely toothed, terminal one broad; upper leaves smaller, sometimes nearly entire. Flowers pale yellow or white in racemes. Sepals often tipped with hairs. Petals veined. Pods closely adpressed to the axis, ovoid oblong, turgid, terete, with a larger ensiform seedless beak, pedicels shorter than the calyx. Valves concave, 5-nerved, twice as long as the broad flattened beak, stigma simple.

Distribution: The plant is cultivated as a field crop for oil and often found as escape. It occurs throughout northern India up to 1000 m and ascends in the Himalaya up to 2500 m.

Uses: Young leaves are stimulant, stomachic, diuretic and antiscorbutic.

The seeds are vesicant and the whole plant is considered aphrodisiac.

Extent of utilization and trade: The plant is cultivated on a large scale for oil seeds and its use in medicine is occasional and limited.

95. *Euphorbia hirta* L. (-E. pilulifera auct non L.)

Family: Euphorbiaceae.

Local names: Dáddar-jari, Dudhi (Kash.), Laldidhi (Hindi), Dudnéli (Guj), Moti-dudhi (Rajasthan). Pusitoa (Sansk.) Trade name based on scientific name.

Discription: An annual herb, ascending or erect up to 50 cm in height. Stem round, covered with yellowish hairs. Leaves in opposite pairs, small up to 4 cm long, dark green on upperside, pale on lower side, margins faintly toothed. Flowers whitish, minute, in small, axillary, stalked clusters. Fruit small, 1–2 mm in diam., hairy. Seeds 3-angled, wrinkled, light reddish brown. Flowers throughout the year (Fig. 5.54).

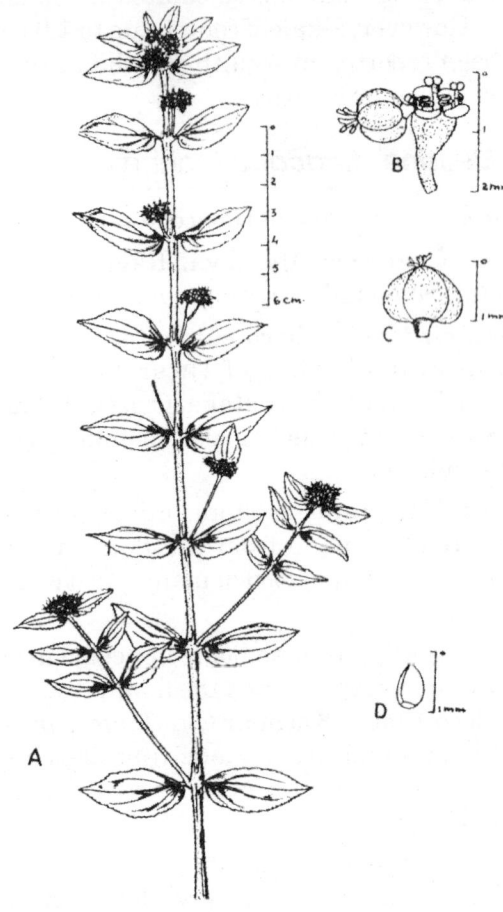

Fig. 5.54: *Euphorbia hirta* L. A. flowering and fruiting branch; B. cyathium; C. capsule; D. seed.

Distribution: Throughout the country in warmer regions and on a variety of soils and moisture conditions.

Uses: Used in bronchial affections, asthma, colic; locally applied for curing ringworms. It is also used in affections of bowel of children and lung complaints.

A fluid extract has been used in asthma and dysentery.

The plant promotes formation and flow of milk in women and is also useful in gonorrhoea and other genitourinary complaints.

Roots of the plant stop vomiting, large doses of the drug, however, cause irritation in stomach and result in vomiting and nausea.

Milky juice of the plant is applied to warts and is known as a vermifuge. The antibacterial and antitubercular activity has been proved experimentally. It contains an alkaloid, effective in respiratory system, dilating on bronchus.

Other species

a. *Euphorbia thymifolia* L. (Choti-dáddar, Nanhi doodhi (Hindi); Raktavinda-chada (Sansk.)

A spreading herb with red branches and reddish green leaves giving a latex. It is used in cases of spermatorrhoea, tenous semen and premature ejaculation. About 7 gms of the herb is ground in water along with 7 grains of pepper, strained, mixed with some sugar and drunk in the morning and evening. In cases of diarrhoea and dysentery, 20 grams of fresh herb or 10 grams of dry herb is ground with water, strained and given to drink in the morning and evening.

b. *Euphorbia hypericifolia* L. Dudi, Haksardána Hazárdana (Hindi), Dugdhika (Sansk.).

Infusion of the dried leaves is astringent, deeply narcotic, useful in diarrhoea and dysentery, menorrhagia and leukorrhoea. Sometimes given to children with milk in cases of colic. It contains a phenolic substance, an essential oil, a glucoside and an alkaloid.

Common in hotter parts of India ascending to 1500 m in the Himalayan region.

c. *Euphorbia helioscopia* L. (Vern. Mahábi, Hirruseah-Hindi).

Roots are anthelmintic. Milky juice applied to erruptions. Roasted seeds given in cholera. Seed oil is purgative.

d. *Euphorbia lathyrus* L. Vern. name: Burg-sadáb (Bihar), Sudab (Punjab).

Leaves are carminative, while the seeds are used in dropsy. A native of Central Europe.

e. *Euphorbia microphylla* Heyne. Vern. name Chotá-Keuree; Dushiaphul (Bihar and Santhal).

The plant is used as a galactagogue.

f. *Euphorbia rothiana* Spreng.The plant is used in the Deccan peninsula and parts of MP and UP from Banda southwards. Used as irritant and the juice is acrid.

g. *Euphorbia rosea* Retz. The plant is used as a drastic purgative in Indo-China.

h. *Euphorbia sanguinea* Hochst and Steud. It is used to soften sore nipples of suckling mothers.

i. *Euphorbia thomsoniana* Boiss. Vern. name Hirtis, Hirvi (Kash.)

Rootstocks are crushed as purgative and detergent for washing hairs.

j. *Euphorbia thymifolia* L. (Vern. name Laghududhika (Sansk.); Chotidudhi (Hindi). Dried leaves and seeds are aromatic, astringent, stimulant and laxative, given to children in bowel complaints.

96. *Evolvulus alsinoides* Linn.

Family: Convolvulaceae.

Local names: Shankhpushpi (Hindi-Trade name), Vishnugandhi (Sansk), Shankavalli (Mumbai), Vishnu karandi (Tam).

Description: A softly hairy, annual/perennial herb with numerous slender branches. Leaves sessile, lanceolate or ovate, acute. Flowers blue or white, axillary, sessile/stalked, solitary or 2–3 together, forming terminal racemes. Corolla funnel-shaped. Capsule lobes folding at the angles, globose. Seeds 4 (Fig. 5.55).

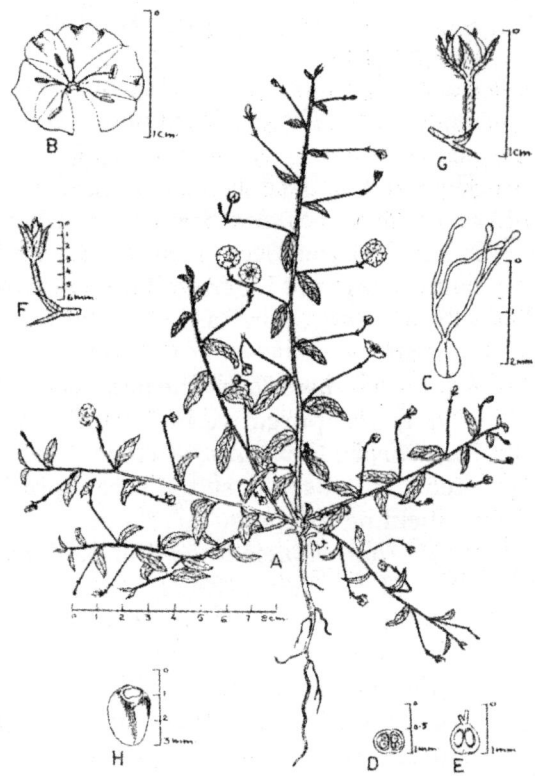

Fig. 5.55: *Evolvulus alsinoides* L. A. plant; B. corolla split open to show androecium; C. gynoecium; D. transverse section of gynoecium; E. longitudinal section of gynoecium; F. calyx; G. capsule; H. seed.

***Distribution*:** Throughout India in open area, gardens, lawns and rocky areas along roadsides ascending to 2000 m. Common in grass fields and on cultivated areas after rains.

The whole plant is used medicinally.

***Uses*:** The plant is bitter, pungent, alexiteric, alterative, tonic and useful in bron-

chitis, biliousness, epilepsy, loss of appetite, dysentery as fibrifuge, leukoderma, teething of infants and believed to brighten intellect. It is used as a laxative and strengthens memory, complexion and appetite. It is used also as a febrifuge with cumin and milk and with oil to promote growth of hair. (Kirtikar and Basu, 1994). It is also reported to be a sovereign remedy for dysentery and useful in internal haemorrhages. In fevers attended with diarrhoea or indigestion. A decoction of the drug with *Ocimum sanctum* is useful (Nadkarni and Nadkarni, 2000).

The leaves are made into cigarettes and smoked in bronchitis and asthma. Also as blood purifier, eye diseases and as memory booster (Girach et. al. 2001).

Important alkolids are: evolvine, betaine and β sitosterol (Chemexil, 1992).

***Extent of utilization and trade*:** The plant is utilised on a commercial scale for the preparation of a ayurvedic medicine *Shankhpushpi* by 'Baidyanath'.

The oil of this plant is extensively used as an alternative to oil for promoting the growth of hairs (Kirtikar and Basu, 1994 ed.). Also in fevers attended with diarrhoea or indegestion.

Application of FYM favours luxurious growth of this plant though it is weed during winter under rabi crops. FYM plus NPK application + hexameal in the field gives best growth.

97. *Fagonia cretica* L. (F. arabica L., F. mysorensis Roth.)

***Family*:** Zygophyllaceae.

***Local names*:** Dhása, Dhamásá, Damahan (Hindi).

***Description*:** A spiny undershrub with glandular hairs. Leaves opposite, small, with a spine at the base of each leaf. Flowers solitary axillary, small, pink coloured. Capsule globular, embedded in a mass of thin straight spines, deeply 5-lobed. Seeds one in each lobe, compressed.

Distribution: Saurashtra, Rajasthan in arid parts, in upper Gangetic plains and Punjab. On sandy soils, loose but mixed with gravel and sand dunes and sandy tracts.

Uses: Leaves, twigs and juice used in medicine. Plant is considered cooling and refrigerant. It is bitter, astringent, tonic, prophylactic against smallpox, dropsy, delirium and other disorders caused by poisoning. It is also used in cough, dysentery, skin diseases. Boiled residue of the plant in water is used for abortion.

The infusion is also useful in fevers for quenching thirst, externally used as a bath for relief of itch and irritability of skin and as gargle in sore throat.

Paste of the plant is applied to tumours as a suppurative; locally over swellings of neck and for scrofula.

Juice of the plant is reputed to prevent suppuration when applied to open wounds. Juice boiled with sugar candy, until thick, is sucked for relief in stomatitis.

98. *Ferula asfoetida* L. (Vern. Hing, Hingrá, Hingru, Hingará)

Family: Umbelliferae.

Description: A perennial, unpleasant smelling herb, leaves downy when young, divided three times, secondary and tertiary divison without stalk, lower leaves 30–60 cm long, ovate stalk sheathing the stem. Flowers arising from the sheaths of lower leaves in simple umbels. Flowers yellow. Fruit very small.

Distribution: Arid hills of west Baluchistan, Afghanistan, Multan and Khorásan.

Ferula is the genus of herbs growing from perennial rootstocks and affords various forms of *asfoetida*. *F. alliacea* is wild in Eastern Persia and prefers a stony soil and elevation of 2300 m and reported to be the chief source of *asfoetida,* while that of European commerce is *F. foetida* and *F. galbaniflua,* is the chief source of the drug called *Gallanum,* a native of Iran-Siraz and Kirman.

F. nartex Boiss. It is distributed in west-Tibet, on the slopes dividing it from Kashmir and once credited as a source of *Tibetan asfoetida*. The resinous mass contains an essential oil and is carminative and anti-spasmodic. *Hing* is used as a substance in Indian culinary.

Uses: The resinous gum from the plant, called *hing,* is used in diseases of the stomach and intestines. It increases appetite, helps digestion of food and expels wind. Asfoetida dissolved in water is placed on the navel with a wad of cotton wool. It is also applied in ringworm, useful in counter-acting opium poison.

In case of toothache, relief is obtained by filling the cavity of the tooth caused by worms, with a little of asfoetida.

Locally, pills of asfoetida are prepared with roasted borax, rind of myrobalan, salt, dried ginger, all powdered and sifted, moistened with lime juice. These allay stomachache, increase the appetite and help food digestion.

An emulsion of the gum is valuable as an enema per rectum in peritonitis, tympanites, hysteria, flatulant colic, etc. Also useful in removing thread worms from rectum and lower bowels.

The gum is also an ingredient of some sedative mixtures and remedies for whooping cough, chronic bronchitis, etc.

Other species

a. *F. galbaniflua* Boiss and Buhse (Vern. Ganda Biroza Hindi; Galbanum-Eng). The oleo-gum resin obtained from stem is used in treatment of bronchitis and asthma.

b. *F. jaeschkeana* Boiss. (-Narthex asfoetida) Hing-Hindi and Kashmir. Roots and seeds produce an essential oil.

99. *Fagopyrum esculentum* Moench.

Family: Polygonaceae.

Local names: Kotu (Hindi), Ogal (Kumaon, Punjab), Phaprá (DDN and HP).

Description: An erect, glabrous, annual herb. Leaves triangular, cordate, acute. Flowers in axillary and terminal peduncled, subcapitate, many flowered cymes. Perianth pale or white. Nut ovate, acute and angled.

Distribution: Cultivated in the hills and often found as an escape. A native of Eurasia, up to 4000 m, also in Nilgiris.

Uses: The grains are used in the same way as of *F. cymosum* Meissn. The seed flour is recommended as a diet in colic and used as an emollient and resolvent. The flour is used during fast.

The plant yields a glucoside, 'rutin'.

Other species

a. *Fagopyrum cymosum* Meissn. (Vern. Banogal-Kumaon).

An erect pubescent herb with long stalked leaves, broadly triangular, acutely pointed, cordate; uppermost narrower and stem clasping; stipules tubular. Fls. white, in racemes, forming stalked panicles. Flower stalk jointed near the middle perianth. 5 stamens with honey secreting glands. Nut ovate, acutely 3-cornered, more than 2 times as long as the perianth, enclosing its base.

In temperate region from Kashmir to Sikkim, 1500–3000 m. Usually in woods. Sometimes the flowers in Fagopyrum are dimorphic, anther short styled with short stamens.

The grains are recommended as diet in colic, cholera and diarrhoea, in fluxes of all kinds and abdominal obstructions. It gives 4–8.5% 'rutin'.

b. *Fagopyrum tartaricum* Gaertn. (Vern. Dárau-Lahul; Kaspat, Kalatrumba, Ugal).

This plant is cultivated at the high altitudes, between 3000–4500 m, from Kashmir to Kumaon.

It is a taller, coarser plant having long grains of black colour, with angles rounded off and keeled towards the top, instead of being sharp.

The plant is cultivated as a sustitute to cereal crops. This variety is richer in 'rutin' than *F. esculentum* (up to 6%). It should be possible to raise exclusive buckwheat crop for 'rutin' production which is very much desired for the high altitude Himalayan region.

100. *Fumaria indica* Pugsley

Family: Fumariaceae.

Local names: Pitpápdá, Paprá, Parpatá (Trade name), Bansulpha, Khét-paprá, Kairu (Kum).

Description: An annual herb, often scandent. Leaves multisect. Flowers small, rosy or purplish, sometimes white in racemes. Sepals 2, small. Petals 4, outer two dissimilar, upper spurred at base, two inner clawed. Stamens 6, diadelphous, with a basal spur. Ovary 1-celled, style filiform. Capsule globose (Fig. 5.56).

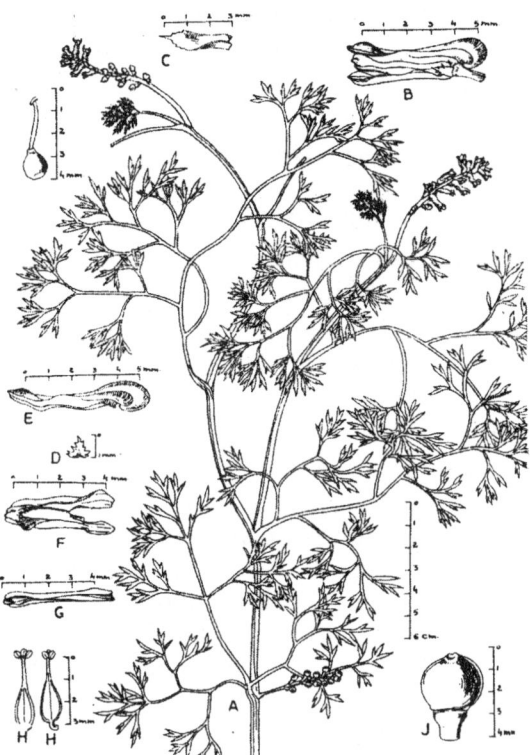

Fig. 5.56: *Fumaria indica* Pugsley A. upper part of flowering and fruiting plant; B. flower; C. bract; D. sepal; E. upper petal; F. lateral petals; G. lower petal; H and H'. upper and lower set respectively of 3 united stamens each; I. gynoecium; J. nut.

Distribution: It grows abundantly in or near cultivated fields as weed, throughout north India in rabi season and on the hills in south India. The leaves resemble very much with coriander leaves.

Uses: The plant is used in fever and influenza.

Infusion from the stem and leaves in Kumaon is used as tonic, diuretic and diaphoretic.

It purifies blood and used in cases of fever, scabies, ringworm, ulcers, boils, gonorrhoea and syphilis. Collected in the month of July.

Extent of utilization and trade: It is good substitute for *F. officinalis*, which contains fumaric acid and alkaloid *Fumarine*. The plant is exploited commercially on a limited scale and is sold in the market under the trade name of *Papra*.

7 gm of the herb, equal quantity of *Emblica officinalis* (*Anwla*) rind and dry coriander are steeped in water overnight, strained in the morning and mixed with *JejubÑ syrúp* to cure irritation of skin.

Malarial fever is cured by drinking a decoction of the plant and green Gilo, each 7 gms (*Tinospora cordifolia*). For intestinal worms this plant (*Fumaria*), *baibarang* (*Myrsine africana*), rind of *Chebulic myrobalan.* (*Terminalia chebula*), each 5 gm are boiled together and a decoction is drunk to kill worms.

A good substitute, *F. officinalis* contains Fumeric acid and alkaloid *fumarine*.

101. *Fritillaria roylei* Hook.

Family: Liliaceae.

Local name: Panja, Garud Panja.

Description: A glabrous, bulbous herb. Stem erect, unbranched, leafy, except on the lower portion. Leaves 3–6 in a whorl or the upper ones sometimes opposite, linear-lanceolate, tips of upper leaves often linear and hooked. Flowers nodding, terminal, solitary or 2–4 in a short raceme. Perianth bell-shaped, segments 6, distinct, yellow green, chequered with dull purple, each bearing a large, viscid gland at the base., tips rounded, not recurved. Stamens 6, at the base of the perianth; anthers linear oblong, attached at the base. Ovary oblong, 3-celled. Style thick, straight, divided at the top in 3 short-pointed lobes or many in each cell. Capsule obovoid, obtusely 6-angled, 3-valved; seeds many, flattened, minutely winged.

Distribution: In temperate and subalipne regions of Himalaya on turfy places. *Nemocharis oxypetala* Balf. (*F. oxypetala Royle*) Vern. *Garud Panja* occurs in Birch forests of Nepal west and Garhwal) Kashmir: Gurais, Kolahai, Astonmarg; Garhwal: Kedrakanta, Harkidun, Chahch-pur peak, Shimla, Lahul.

Uses: The bulbs are eaten as a general tonic. The tubers of *F. oxypetala* are aphrodisiac and tonic.

Extent of utilization and trade: Very small quantity of the drug is collected and sold in the market. (July-August).

102. *Gnaphalium luteo-album* Linn. var. multiceps. (=G. affinis)

Family: Compositae or Asleraceae nom alt.

Local names: Bál-Raksha (Punjab).

Description: An erect, more or less softly woolly herb. Stems often tufted , simple or branched. Leaves crowded or distant, sessile, basal lobes sometimes decurrent, spathulate, both surfaces woolly; uppermost leaves lanceolate, acute. Heads discoid, numerous, in irregularly globose clusters, at the end of corymbose branches. Involucral bract many, scarious, shining, inner as long as the flowers, outer shorter; receptacle flat, flowers bright yellow; pappus hairy. Achenes oblong, slender.

Distribution: Throughout India, ascending to 3300 m. The two varieties recorded are:

var. *multiceps* (*G.affinis*). Shimla, Almora, Kashmir, Deota, and Konain (in Uttaranchal).

var. *pallidum*, differs only in the heads being pale brown, instead of bright yellow.

Uses: The leaves are used as astringent and vulnerary.

Extent of utilization: The drug is used on a very limited scale and not exploited commercially.

103. *Gentiana Kurroo* Royle

Family: Gentianaceae.

Local names: Kutki, Káru-Kutki (Hindi), Páshánved (Mumbai), Nilakant (Punjab). The name 'Pahsanbed' is also used for *Saxifraga ligulata* (Bergenia ligulata Wall.) Engler, also called Patharfod. Himalayan gentian (Eng.).

Description: A herb with thick rootstock. Stem tufted, decumbent, 10–15 cm. Leaves narrowly-oblong; radical resolute; stem leaves narrower. Flowers blue, spotted with white 2 to 5 cm long, 2 cm in diam., solitary or racemose. Calyx, about half as long as the corolla; lobes 5, linear. Corolla 5-lobed. Capsule oblong.

Distribution: In Western Himalaya, from Kashmir eastwards to Kumaon hills. Flowers in September; 1700–500 m.

Uses: Roots are tonic and so the rootstock, stomachic, febrifuge, for urinary affections. The roots are bitter, acrid and in large doses a moderate cathartic. It is also used in fever and dyspepsia and as an ingredient of various purgative combinations.

It contains an active bitter principle, beneficial in certain kinds of fevers and has mild cathartic action.

Other species

a. *Gentiana decumbens* L. Tincture of the plant is used as stomachic in Kashmir.
b. *Gentiana teeta* Fries. Vern. Teeta (Punjab, Himachal).
   The plant decoction is given in fevers. Found in Kashmir and West Himalaya.

104. *Gisekia pharmaceoides* L.

Family: Molluginaceae.

Local names: Balukaság (Hindi), Morang, Sareli, Oreli-morang (Rajasthan), Valuka (Sansk); Valuchi-bhaji (Mumbai).

Description: An annual, diffused or suberect herb, glabrous. Leaves oblong, spathulate, 1.8–5 cm long. Flowers on dense cymes. Fruit of 5 distinct, indehiscent carpels that are covered with papillae.

Distribution: A pantropic weed in tropical regions. Plants of sandy soils most common in Rajasthan, after first showers, completing its lifecycle in 5–8 weeks. Seeds need resting period, 4–6 months, for germination.

Uses: A strong anthelmintic, recommended as specific medicine for tapeworms. The plant is aromatic and aperient. Also used as pot herb in times of famine.

105. *Geranium wallichianum* D.Don ex Sweet.

Family: Geraniaceae.

Local names: Mamiran (Hindi).

Description: A perennial herb with thick hairy rootstock. Stem robust, much branched. Leaves orbicular, palmately 3–5 lobed and irregularly toothed; stipules large, broadly oblong. Flowers blue purple. Sepals long pointed. Petals slightly notched.

Distribution: In temperate regions from Kashmir to Nepal, 2000–3500 m. Kashmir: Gulmarg; Shimla, Mashobra, Dharamsala, Chamba, Murree, Tragbol, Sonmarg, Garhwal, Deoban, Kumaon: Nainital.

Uses: The herb is astringent and is employed as a cure for toothache locally.

The rhizomes are used as substitute for *Coptis teeta* (Maimiran).

Extent of utilization and trade: Commercially the roots are used as an adulterant for coptis.

Other species

a. *Geranium nepalense* Sweet

Local names: Bhand, Bhanda (Hindi and Himachal).

Description: A perennial, pubescent herb with prostrate, diffuse stem, leaves 5-gonal, orbicular, 3–5 palmately lobed, segments equal; stipules narrowly lanceolate. Flowers pale purple. Sepals acute, shortly pointed. Petals slightly notched.

Distribution: In temperate regions, 1500–3000 m, under conifer forests. Also on rocks and walls.

Kashmir: Pahalgam, Murree, Sonmarg, Kagan, Drosh; Pangi, Chur, Upper Chenab valley, Shimla, Waziristan, Darjeeling, Manipur, Lushai, Khasia, Nepal.

Uses: The plant is astringent and used in certain renal diseases; rhizomes used as substitute for *Coptis teeta*.

Extent of utilization: It is not utilised on a commercial scale.

b. *Geranium lucidum* Linn.

An annual, softly hairy, usually glandular herb, often turning to red. Leaves orbicular, 5–7 lobed, lobes 5-fid; stipule acute. Flowers pink, on long peduncle.

It is distributed in temperate regions from Kishtwar to Kumaon, 2000–3000 m, on rocks and old walls.

Garhwal, Nainital, Shimla, Murree, Dalhousiae, Chur, Kunawar, Tehri Garhwal.

The plant is diuretic and astringent.

106. *Geum urbanum* Linn.

Family: Rosaceae.

Local names: Jangli Gunglu, Gogjumul (Kash.).

Description: A softly hairy herb, root-stock perennial. Leaves pinnate; leaflets acutely and irregularly toothed. Radical leaves 10–15 cm; leaflets 6–12, nearly sessile, uppermost pair large, other all small but alternate pairs larger; terminal leaflets stalked, nearly orbicular, often lobed. Upper stem leaves usually of 3-leaflets, the lateral leaflets sometimes merged in the terminal one and forming a 3-lobed leaf. Flowers pale yellow. Calyx lobes reflexed. Style sharply incurved and jointed near the middle, lower portion glabrous, persistent, becoming elongated and hooked in fruit, terminal portion hairy, ultimately breaks off.

Distribution: In temperate regions from Murree to Kumaon, 2000–4000 m. Chitral, Hazara, Murre hills, Shimla, Upper Chenab valley, Chamba, Kulu, Chur, Kumaon, Garhwal, Namik, Mundali, Manali, Bhutan.

Uses: The roots are astringent and antiseptic and are given in the form of infusion for ague and as an excellent sudrofic in chills or fresh catarrah. Constant use of the plant is reported to have highly restorative power in weakness, debility, etc. It is also useful in diarrhoea, sore throat and leukorrhoea.

Extent of utilization and trade: The plant is not utilised on a commercial basis and only exploited on a limited scale.

The root is said to contain *eugenol*, a glucoside *gein* and enzyme *gease*.

107. *Glinus lotoides* L.(-Mollugo hirta Thunb); Mollugo lotoides (L.) O. Kuntze.

Family: Molluginaceae.

Local names: Bakdá, Pada (Mumbai), Gandibuti (Hindi); Bhissata (Sansk.), Dholakani (Rajasthan).

Description: A spreading herb with rounded, villous, small leaves, flowers axillary, pink. Sepals stellately tomentose. Staminoded linear or 0. Capsule a little shorter than the sepals, oblong.

Distribution: Throughout Northwest India, extending to the Deccan peninsula. On cultivated wastelands, in drying depressions locally called *khadins* in arid Rajasthan.

Uses: The dried plant is diuretic, purgative and is a cure for bilious attacks and for wounds and pain in the limbs.

Plant juice in given internally to weak child and for indigestion.

Pounded and boiled in water, it is used as a remedy against urinary troubles, also in indigestion and as purgative in arid Rajasthan.

108. *Gloriosa superba* L.

Family: Liliaceae.

Local names: Shakrápushpi (Sansk.), Kalihári (Hindi), Kariánag (Mumbai), Kariári (Punj), Agnishikhá (Telgu), Ákkinichilam (Tam).

Description: A herb climbing by means of its leaves; root stock fleshy, creeping. Stem leafy, 150–300 cm or more. Leaves alternate, or opposite or in whorls of 3–4 on different parts of the stem, sessile, oblong-lanceolate, tips linear, spirally twisting. Flowers solitary in the leaf-axils, nodding. Perianth persistent, segments 6, distinct, spreading at first, reflexed afterwards, narrowly lanceolate, margins curled and wavy, yellow when young, changing to bright red. Stamens 6, hypogynous, slightly shorter than the perianth, filaments at first green, then yellow and finally red. Anthers versatile, connective green, pollen orange. Ovary oblong, 3-celled, style long, linear, green,

turning to red. Capsule oblong, 45–60 cm, obtuse; seeds globose, numerous (Fig. 5.57).

Distribution: In the Himalayan region, ascending to 1700 m, locally found in Dun forests in Sal region. Also in tropical Asia, Africa (French Guinea).

Uses: The roots are purgative, chola-gogue, anthelmintic, used in leprosy, para-sitical affections of the skin, piles, colic. Starch from the root is given in gonorrhoea. The seeds are being collected for the alka-loids *superbine, gloriosine, colchicine* and other alkaloids.

The plant is also being grown in south India for the seeds on a commercial scale. However, it is reported that the seeds from commercially grown plants in South India

possess lesser amount of alkaloid as com-pared to the naturally found plants in the forests. However, attempts have been made to collect the seedlings from the forests and grown in farms; thus posing a threat to the very survival of plants in nature.

109. *Gossypium arboreum* L.

Family: Malvaceae.

Local names: Nurma, Rui (Hindi), Kapás (Punjab and Haryana), Surropushpa (Sansk.) Cotton plant (Eng).

Descriprion: A shrub, cultivated all over India as a fibre plant, which is obtained from the surface of the seed. The stalks are used for making paper. Stem hairy, erect. Leaves divided to the middle, heart-shaped, with a gland on the under surface, 3–5 or more lobed, broadly ovate, apex pointed, hairy. Flowers solitary, involucre of bracts ovate obtuse. Corolla yellow with a purple base. Capsules ovoid, without hairs, apex sharply pointed. Seeds up to 7, ovoid seedcoat covered with a greyish, firmly adherent down and completly surrounded with white cotton.

Distribution: It is grown throughout the country, considered to be a cultivated form of *G. stocksii.*

Uses: Flowers of the cotton plant when steeped overnight in water and strained in the morning and mixed with a little sugar, when drunk works as tonic for the heart.

Syrup made from the flowers used for the same purpose as above.

Decoction of the rind after the cotton is extracted and the bark of the root is given for inducing abortion and expulsion of pla-centa, and for inducing menstrual flow of facilitating child-birth.

Pith of the cotton seed strengthens human body and acts as an aphrodisiac, also useful

Fig. 5.57: *Gloriosa superba* L. A. flowering branch; B. fruit; C. transverse section of gynoecium.

in initial stages of diarrhoea. The pith (10 grams) is powdered, steeped overnight, strained and drunk in the morning. 30 grams of pith ground in water, strained is given as an antidote to opium and *dhatura* poisoning.

20 gram of *Binaula* seeds (*G. indicum*) are ground with water or with milk, cooked and sweetened with sugar and drunk to fatten the body and improve lactation in feeding mothers.

Other species

a. *Gossypium barbadense* L. (Vern Names: Kapás Hindi)
   The seeds are used as demulcent, laxative, expectorant, galactagogue, employed to procure abortion, but considered nervine tonic. Seed oil used to clear face streaks, spots and freckles. Cultivated mostly in India almost in all states.
b. *Gossypium herbaceum* L. also known as Kapás or Kapasi. The seeds are demulcent, laxative, expectorant, aphrodisiac, employed to procure abortion, considered nervine tonic. Cultivated at Baluchistan in Pakistan.
c. *Gossypium hirsutum* L. The seeds are considered emollient and roots emmenagogue. Leaves and flowers contain a glucoside 'querimeritrin.' Cultivated.

110. *Gymnema sylvestre* R.Br.

Family: Asclepiadaceae.

Local names: Gudumár-buti, Merásingi Medasingi, Grihadruma, adalsinghi, Kakrasinghi, (Hindi); Meshashingi (Bihar).

Description: The plant is a creeper, spreading like *giloe* (*Tinospora cordifolia*) on acacia and other trees. The leaves resemble very much like the bael (*Aegle marmelos*) leaves and exude a white sticky fluid.

More or less a pubescent climber, young stem and branches terete, pubescent. Leaves subcoriaceous, elliptic or ovate, acute or shortly acuminate, cuneate, rounded or cordate at base, often glabrous above, pubescent beneath, especially on veins. Flowers yellow in umbellate cymes, peduncle shorter than the petiole, pedicels slender. Calyx deeply divided. Corolla as long as the campanulate tube. Anthers white. Follicles lanceolate, tapering into a beak. Seeds pale brown.

Distribution: In peninsular India, extending to tropical Africa.

Uses: The leaves of this plant when chewed have the property of temporarily removing the sense of taste. It effectively masks the taste of sweet and is considered useful in cases of diabetes. 9 gm of green leaves are ground with water, strained and given to drink, otherwise 3 gm of powder of sifted dry leaves is taken with water, both in the morning and evening. It also decreases frequency of urination. The 'bubo', due to plague gets resolved by applying a paste of the root of this herb when ground in water.

111. *Gynandropsis gynandra* (L) Briq. (Cleome pentaphylla DC., C. gynandra Briq. G. pentaphylla L).

Family: Capparidaceae.

Local names: Bagrá, Ajgandha, Hurhuirá, Tilavána, Hulhul, Hurhur, Arkahuli (Hindi), Surajavarta (Sansk.).

Description: A tall annual herb, stem striate, hairy. Leaves 5-foliate with a long petiole; leaflets broad-ovate. Flowers white, pale purple, in dense, bracteate cymes. Bracts 3-foliate. Capsule 5–8 cm long, tapering at both ends, nearly glabrous, striated. The plant smells when bruised in hands.

Distribution: Abundant throughout the warmer parts of India and all tropical countries. A herb of rainy season.

Uses: Root decoction is used in fever.

The green leaves are applied to the skin and tied down to form a good blister, used in rheumatism, rubifacient and vesicant. Juice of the leaves is a remedy for otalgia. The leaves are also used in māscular pains and headache (migraine).

Seeds are applied as poultice for sores having maggots and are anthelmintic and rubifacient. Seeds infused in boiling water used as a cure for cough, bruised they are applied as poultice for sores having maggots also given to horses as anthelmintic and against stomachic.

Leaf juice is dropped into the ear for curing pustule, which gets cured or it suppurates or bursts.

For malarial fever, 10 gm of leaves are ground with 20 grains of pepper and made into pills, that are orally administered before the expected time (4 ps) of malarial fever. The leaves also cooked as vegetable and eaten along with rice and cure piles.

4 gm of leaves and *Ajwain* seeds when ground together in water, strained and drunk in cases of dropsy.

10 gm of leaves, 3 cloves of garlic and 3 grains of pepper, ground together in water, strained and given for a few days to person bitten by mad dog to get relief.

112. *Hedera nepalensis* K. Koch.

Family: Araliaceae.

Local names: Mithiári (Jaun), Ivy (Eng).

Description: A large, evergreen climber, adhesing to the trees and rocks by its adventitious roots. Leaves simple, variously lobed, leathery, dark green and shining above, margin entire, base cordate, rounded or cuneate, petiole slender. Flower yellowish green, in pedunculate globose umbels, which again are arranged in subcorymbose panicle; peduncle clothed with minute stellate scales. Fruit yellow, globose, turning black when ripe, shining; seeds 3–4 ovoid.

Distribution: Throughout the Himalaya up to 3300 m. Fairly common on oak, especially common on shady situations. Also found in swamps in the Dun.

Kashmir, Hazara, Murree, Chitral, Chamba, Shimla, Bashahar, Dehradun (Narkanda), Mathronwala, Mussoorie, Nainital, Sarju valley, Pabar valley, Chakrata, Khasia, Lushai, Manipur, Nepal, Sikkim, Bhutam.

Uses: The dry leaves and berries are stimulating, diaphoretic and cathartic. The leaves may be employed for fomentation in glandular enlargements, indolent ulcers, abscesses, etc. Decoction of leaves is externally applied to destroy vermin from the head of children.

The berries are of use in febrile disorders, and infusion is given in rheumatism.

Ivy gum is considered antispasmodic in England.

Extent of utilization and trade: A glucoside, *helixin* has been extracted from the berries and it is due to the presence of this substance that their bitter quinine like taste is due.

The Himalayan Hedera, differs from that of European in the colour of the fruits, having more branches and hairs or scales on the inflorescence.

113. *Hemidesmus indicus* (L.) Schult.

Family: Periplocaceae.

Local names: Hindi Salsa, Anantmul (H); Durivél (Guj), Nagajihva (S); Kalidudhi (MP).

The trade name is Indian Sarsaparilla, since it is a substitute for Smilax (Sarsaprilla).

5

Description: A perennial twiner or creeper, rootstock woody, fragrant. Stem slender, hairless. Leaves vary greatly in size and shape, they are 5–10 cm long but the breadth varies from 0.5 to about 4 cm, dark green with whitish blotches, pale or whitish, hairy on lower surface. Flowers very small, black with a tuft of white hairs at the top. All parts are with milky latex.

Distribution: The plant is distributed throughout India; in north India from Banda to erstwhile Oudh and Sikkim, southwards to erstwhile Travancore.

Uses: The dried roots of the plant constitute the drug. It has long been employed in native medicine. It is useful in fever, loss of appetite, skin diseaes, syphilis, leukorrhoea and other urinary complaints. It is prescribed in the form of a syrup and is demulcent, alterative. The diuretic properties have been shown experimentally. Sometimes the whole plant is pounded or an infusion prepared of the dried leaves. The drug is also largely used as a blood purifier and in rheumatism (osteoarthritis).

Extent of utilization and trade: In Indian commerce the plant is found in the form of little bundles, which consist of entire roots of one or more plants, tied up with a portion of stem.

It is utilized on a commercial scale and is sold in the market.

114. *Heracleum candicans* wall.

Family: Umbelliferae.

Local names: Kaindal (Kash).

Description: A robust herb with a thick, hollow stem, 1–2 m high, densely pubescent in younger plants. Leaves 15–45 long, pinnate or pinnatifid; leaflets oblong 15–30 cm, pinnately lobed and irregularly toothed, upper surface dark green, lower pale pubes-

cent. Umbels large, long stalked, 15–25 cm across. Flowers white, calyx teeth small, linear and acute. Fruit about 1.5 cm long and 4 mm broad, flattened, pubescent and obovate. Root 20–40 cm long, 8–10 cm broad at upper portions, tapering downwards sometimes into a number of branches. When fresh, it emits a strong pungent smell. Colour of the root is purplish which darkens further on drying. The fruit does not shed easily from the plant, it can be easily picked (Fig. 5.58).

Flowers in July-August, fruits end of Sept.

Distribution: The plant prefers moist, exposed, rocky or gravelly localities. In Kashmir it is found near cultivated fields. It is common between 2500–3500 m in Jammu-Kashmir, Chamba, Kullu, Bashahar, Garhwal and Kumaon hills.

Fig. 5.58: *Heraclium candicans* Wall A. aerial portions; B. roots; C. fruit.

Chitral, Aru, Gulmarg, Sonmarg, Pahal-gam, Lahul, Dalhousiae, Shimla, Mussoorie, Ralam valley, Nainital.

Uses: The Indian species do not seem to have been investigated for their essential oils which can be used in perfumery.

The roots are used in incense in Kashmir and also as a cure against some diseases in sheep and goats.

It is reported that application of 20 tones of FYM (Farm Yard Manure) per ha. gives optimum root yield.

Time of collection: The plant flowers during July and August and fruits towards the end of September, when it can be collected. The fruits do not shed early from the plant and can be safely hand picked. Roots can be collected towards the end of September and October; they are very aromatic and laden with moisture when fresh. These should be cut and dried in open when about 60% moisture is removed. It may also lose some essential oil during drying.

Extent of utilization and trade: The roots are not collected but a sizable collection could be made, if there is a demand for oil. The fruits collected from Gulmarg are reported to yield a *furano coumarin*, 'bergapten' and two other crystalline compounds of *lactonic* in nature. A fixed oil of about 9% yield has also been obtained. An essential oil with sweet fragrance (0.1% yield) has also been distilled.

Other species: *Heraceum canescens* Lindl. (Chatrya-TG) is a perennial herb of shady ravines from Kashtwar to Kumaon, 2000–2500 m of which the root decoction is used as vermifuge.

115. *Heliotropium strigosum* Willd.

Family: Boraginaceae.
Local names: Safed-Bhángrá (Hindi).

Description: A small perennial herb, clothed with white appressed hairs. Stem tufted, spreading, much branched. Leaves alternate, nearly sessile, linear-lanceolate, entire, acute. Flowers white or pale blue, in terminal, bracteate spike, lower flowers stalked shortly. Fruit 4, more or less united, minute, 1-seeded nutlets.

Distribution: Throughout India ascending to 1500 m.

Uses: The plant is laxative and diuretic. The juice is used as an application to sore eyes, gum, boils and sores to promote suppuration.

It is also considered as a cure for stinging nettles and insects.

Extent of utilization: The drug is not used on a commercial scale but employed locally.

Other species: *H. eichwadi* Steud. ex DC. The leaves are used for cleaning and healing ulcers, rolled and put into ears to cure earache.

116. *Haloxylon recurvum* (Moq.) Bunge ex Boiss. (Hamada recurva Bunge ex Boiss.) Iljih.

Family: Chenopodiaceae.
Local names: Khár, Khar-Lani (Rajasthan).

Description: A straggling bush, blackish when dry, branched paniculately, branches divaricate. Leaves distinct, trigonous, spreading or recurved, floral leaves about equalling the axillary flowers. Flowers axillary, forming spikes up to 15 cm. Stigma obtuse or obscurely 2-lobed. Fruiting sepals with large erect ovate-obtuse lips and orbicular wings.

Distribution: In flood plains of Rajasthan and Northwest India extending to pen-

insula, westwards to Punjab. Common on the old flood plains of river *Ghaggar*.

Uses: The plant yields the finest type of sajji, inferior quality is believed to be made from various species of *Salsolaa*. The plant ashes are given in water against internal ulcers. It is sometimes also used by washerman to wash the clothes.

117. *Hibiscus cancellatus* Roxb.

Family: Malvaceae.

Local names: Kamblyá (TG).

Description: An annual herb, bristly hairy. Stem up to 1 m, leaves cordate, coarsely toothed or crenate; lower ones long-stalked, orbicular or ovate, more or less 3 or 5 lobed, upper arrow head shaped, lobes long, narrow tapering. Flowers pale yellow with dark purple centre. Calyx an ovate folded spathe, tip obscurely 5-toothed. Bracteoles numerous, linear bristly, curved, the tips meeting over the top of the bud.

Distribution: In the subtropical regions of India, ascending to 7200 m in ravines.

Uses: The root decoction is used in genital disorders.

Time of collection and phenology: The plant flowers during July and August .

Extent of utilization and trade: The plant is utilised on a very limited scale and not exploited commercially.

118. *Hygrophila auriculata* Schum) Heine E. (=H. spinosa and, Astercantha longifolia Nees.

Family: Acanthaceae.

Local names: Tálmakhàná, Kuliakántá (H); Ekhro (Guj); Kokiláksha (S); Talimakháná (Mahr.).

The trade name Talmakhana is based on the local name.

Description: An erect herb, 60–150 cm high; stem unbranched, erect, straight, four angled, hairy, more so below the swollen nodes. Leaves whorled, 6 at a node, each having a sharp yellow spine in the axil. Flowers 8, in a whorl, purple blue, about 3 cm long, 2-lipped; lower lip characteristically folded.

Distribution: Distributed in the marshes and moist places throughout India, often filling the shallow dried ditches and roadside water channels.

Uses: The drug consists of the entire plant including the roots, it is useful in dropsy, jaundice, rheumatism and diseases of urinogenital system. The diuretic activity of the drug is believed to be due to a combination of both inorganic and organic contents of the plant.

Seeds are considered useful in veneral diseases.

Leaves are useful in cough and urethral discharges.

Extent of utilization and trade: The seeds are collected on a commercial scale and sold in bazaars for medicinal purposes.

119. *Hydrocotyle javanica* Thunb.

Family: Umbelliferae (nom alt Apiaceae).

Local names: Looks like *H. asiatica* (*Centella asiatica* called Mandikparni).

Description: A prostrate herb, rooting at the joints. Leaves rough, bristly, orbicular, deeply cordate, 7-lobed, crenate, petiole laxly pubescent. Umbels globose, flowers 30–40 in an umbel, yellow green. Fruit much compressed, not pentagonal.

Distribution: In subtropical and temperate regions from Kashmir to Bhutan, 600–2000 m.

Uses: It is used as a substitute for *Centella asiatica* (*Hydrocotyle asiatica* L.).

The leaves are useful in syphilitic skin diseases, both externally and internally. Also useful in skin diseases, of nerves and blood. Leaves also used to cure indigestion, nervousness and dysentery.

120. *Hyoscyamus niger* L.

Family: Solanaceae.

Local names: Parasikaya (Sansk); Khorasáni aiowain (Hindi); Khorasaniowa (Mumbai), Henbane plant (Eng.).

Description : An erect, more or less hairy and viscid herb with a disagreeable heavy odour. Stem robust 30-90 cm. Radical leaves spreading, stalked, ovate-oblong, coarsely sinuate-toothed. Stem leaves smaller, ovate, irregularly pinnatifid, passing into bracts. Flowers pale-yellow green, veined with purple, darker in the centre, lower ones in the fork of branches, upper solitary in the axils of leaf-like bracts, forming long 1-sided spikes, rollback at the top before flowering, ultimately becoming elongated and straight. Calyx tube ovoid, 5-toothed, limb funnel-shaped. Corolla funnel shaped, lobes 5 slightly unequal. Ovary 2-celled, style longer than the stamens. Capsule enclosed in the globose tube of enlarged calyx, lower part membranous, top-hard, rigid, opening transversely along the constriction between the two portions.

Distribution: A herb, native of Europe and Asia, found in the temperate regions of the Himalaya, extending from Kashmir to the Garhwal Himalaya.

Uses: The plant contains *atropine, hyoscyamine* and *scopolamine* in varying proportions. These solaninaceous narcotics produce an intoxicating effect, when smoked or eaten but continued use leads to disorders of brain. The alkaloid content is 0.4–0.5%.

H. muticus has been wrongly reported to grow in India, however stray plants are met with. RRL Jammu results show 0.35% alkaloids. It is a potent, promisable plant as a source for alkaloid and efforts are needed to be made to develop this plant.

121. *Impatiens balsamina* L.

Family: Balsaminaceae.

Local names: Gulmahéndi (Hindi).

Description: An annual, erect herb. Leaves alternate, lanceolate, deeply serrate. Flowers rose coloured. Sepals minute, standard small, horned, wings broad, lateral lobe rounded, terminal, sessile. Lip small, boat-shaped. Spur short or long, incurved. Fruit a tomentaceous capsule.

Distribution: Widely distributed throughout India, ascending to 1500 m. Common on the borders of rice fields, from Kashmir to Kumaon.

Shimla, Bhimtal, Almora, Kapkote, Jageshar, Murree, Parasnath.

Uses: The flower is tonic and useful when applied to burns and scalds. It is used for pains in the joints. When taken internally it acts as diuretic, emetic and cathartic.

Extent of utilization and trade: The plant is locally used for medicinal properties but not exploited on a commercial basis.

122. *Indigofera pulchella* Roxb.

Family: Papilionaceae.

Local names: Sakéna (Hindi, Kumaoni), Baroli (Maharashtra).

Description: A thinly hairy or glabrous shrub. Leaves 5–15 cm long; leaflets 11–19, opposite, ovate or oblong ovate. Racemes 5–15 cm. Bracts lanceolate, gradually narrowed into a long point. Flowers bright pink fading to violet. Pod glabrous, 2–4 cm, 8–12 seeded, nutlets broad.

Distribution: Plains of north India, ascending to 3500 m. Common in the 'sal' forests and in Siwaliks.

Uses: Decoction of the root is useful for cough and powder is applied externally for pains in the chest.

The plant flowers in the month of November to March and fruits during the rainy season. Flowering varies. The plant is not used on a commercial scale.

Other species

a. *Indigofera articulata* Gouan (I. argentea L.) Vern. Kálá-Klitak (Sansk.) Surmainil (Hindi).
The roots and leaves are bitter tonic and the seeds are anthelmintic.

b. *Indigofera linifolia* Retz. Vern name: Bhángrá(Mumbai), Torki (Hindi).
Plant given in febrile erruptions and used in amenorrhoea.

c. *Indigofera oblongifolia* Forsk. Vern names: Jhilla, Raktapàlaà (Sansk.) Therapeutic properties similar to *I. tinctoria*. Juice of the leaves is prophylactic against hydrophobia, while plant extract is given in epilepsy, bronchitis and in nervous disorders. As an ointment it is used in sores, old ulcers and haemorrhoids.
The plant is grown as a cover and green manure crop.

d. *I. aspalathoides* Vahl. ex DC. Vern. names: Sivanima (Sansk), Sumainil (Hindi).
Leaves, flowers and tender shoots are cooling, demulcent and employed in leprosy and cancerous affections, when taken as a decoction. The plant is rubbed up with butter to reduce oedomatous tumours.

e. *I. tinctoria* L. Vern Guli. Nil, Nilimàndu. Nilika (Sansk).

Leaves useful in hepatitis and juice is given in whooping cough, asthma and diseases of lung, kidneys, enlargement of spleen and liver and palpitation of the heart. Poultice is applied for skin diseases, piles, ulcers, etc. Decoction of roots is given in calculus; an infusion as antidote in cases of poisoning by arsenic (Fig. 5.59).

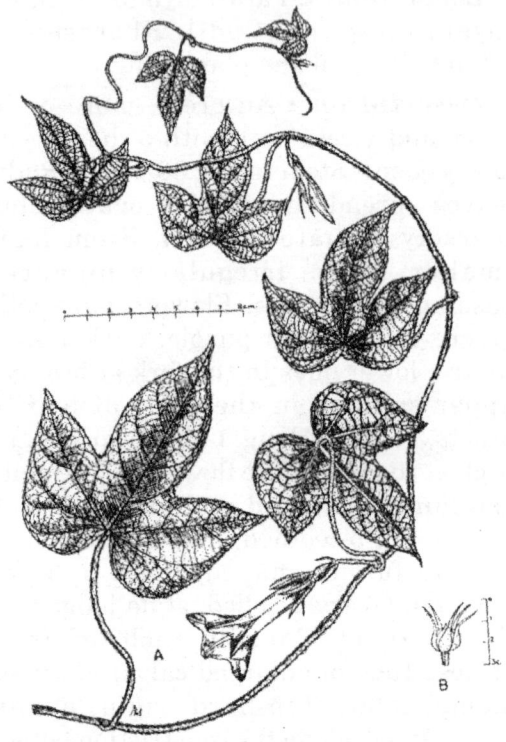

Fig. 5.59: *Indigofera tinctoria* L. A. a branch in flower and fruit; B. flower; C. standard; D. keel; E. wing; F. androecium; G. gynoecium; H. calyx.

f. *I. pulchella* Roxb. Vern. name: Sakéna (Kumaon).
Root decoction is given in cough, while root powder applied externally for pain in chest.

123. *Inula royleana* DC.

Family: Asteraceae.

Local names: Haleem (H); Roylés Inula (E).

Description: A stout herb, more or less hairy and glandular. Stem 30–60 cm high, grooved. Leaves rather membranous, blunt, almost hairless or very hairy above, sometimes thickly woolly beneath, finely toothed, basal leaves egg-shaped or oblong with a winged petiole; stem leaves variable, lyrate with 2 lobes at the base. Heads 8–12 cm in diam. solitary on a stout, erect, hairy stalk, hemispheric. Bracts surrounding the flower-head slender, long pointed. Achene hairless, slender; pappus pale red.

Distribution: Temperate west Himalaya, 2300–4000 m amongst boulders and short grass above 3000 m, common also on open hillsides and *margs, pastures*. Also above Gulmarg, Sonmarg.

Uses: The roots contain 3% alkaloid which produces fall of blood pressure.

Extent of utilization: The plant is not utilised and has vast possibilities of detailed work on the species.

124. *Ipomoea nil* (L.) Roth (=l, hederacea auct non Jacq.)

Family: Convolvulaceae.

Local names: Káládana, Nilkami (H); Káládana (Guj., Beng); Nilpushpi (Mahr); Krishnabij (S); Kana-Kumpan (Guj); Bhorada (NTAL) Jiriki (Tel); Sirikki (Tam); Kanikhondo (Or); Bharar (Kum).

Description: A slender climber with retrorsely hairy stem and ovate-cordate, more less 3-lobed leaves. Flowers blue or tinged with pink on 1–5 flowered, peduncled, long narrow sepals with ligulate tips. Corolla 4–5 cm long, funnel-shaped, tube very short. Ovary 3-celled (Fig. 5.60).

Distribution: Throughout India ascending to 2000 m. Often cultivated but also wild.

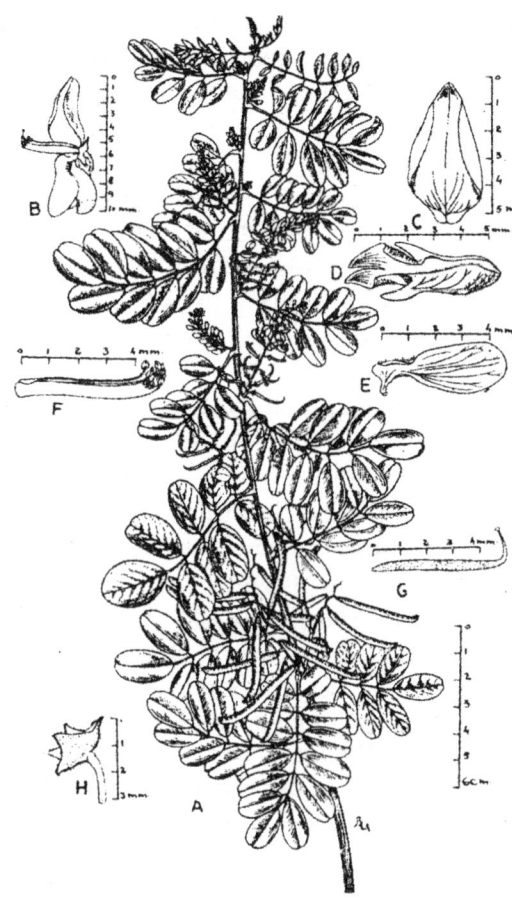

Fig. 5.60: *Ipomoea nil* Roth. A. flowering twig; B. capsule; C. seed.

Kashmir, Bashahar, Punjab, Murre, Shimla, Garhwal: Tons valley, Kumaon: Nainital.

Uses: The seeds known as *Káládána* and *Mirchai*, resemble *Jalap* in their action.

The seeds are laxative and carminative with a bitter taste and cure inflammations, abdominal diseases, fevers, headaches, diseases of head, liver and spleen, pain in joints, leukoderma and remove bad humours from the body. Overdoses of the drug cause irritation. Safe as sure cathartic.

Extent of utilization and trade: Roxburgh was the first to make these seeds

known to European physicians and since then they hold an important position as useful and cheap substitute for Jalap.

It has been recorded that the seeds contain 8.05% of resin resembling *convolvuline*, but in addition are rich in albuminous substances and contain 14.02% of a nauseous fat—a disadvantage for internal administration. Active principle is pale yellow resin, *pharbitisin*, soluble in alcohol.

Other uses: Fresh fruits of the plant are used as vegetable. The plant is often grown for its beautiful flowers.

Other species: Several other species of this genus are of medicinal value. Many of them have laxative properties.

Ipomoea aquatica Forsk. (Vern. Kalmi-Ság; Nari) is an aquatic herb in abundance on the surface of tanks on Gangetic plain and Bengal. Young shoots, leaves and roots are universally taken as a vegetable and the plant for that purpose is cultivated. The juice is believed to have emetic properties and reported to be useful in opium poisoning. The tribals of Bastar crush the flowers and put juice into inflamed eyes.

Ipomoea digitata L.(–*I. paniculata* R.Br.) Rasin similar to 'jalap' resin. (bilaik and, Bhui-kohálá) is cultivated on account of its pink to purple flowers and its tuberous roots. Used in native medicine as aphrodisiac demulcent and purgative.

Ipomoea pestigridis L. (Ghabati, Panchpatri, Baddi-Pasvi). Roots are purgative and applied on carbuncle and other sores as paste.

Ipomoea pescarpae (L.) Sweet, (Dopatilata-H); *I. purpurea* (L.) Roth. (Morning glory); *I. quamoclit* L. (Indian pink; Kamlata-H; Torulatá-Beng) and *I. uniflora* R.and S and seeds of *I. cairica* (L.) Sweet (Railway creeper) are considered to be laxative and purgative.

Ipomoea purga Heyne (*Exogonium purga*) (Jalap-English) is a climber native of Mexican Andes and is cultivated in the Nilgiris, Mussoorie and other parts of north-west Himalaya. It flourishes best on shady woods and with deep rich humus soil. The annual requirements of this plant is estimated to be more than 6 quintals, however, Indian species are satisfactory substitutes.

Formerly, the plant was grown among the chinchona plantations but this system has been abandoned, since while the jalap flourished the cinchona was injured. Plants may be obtained from cuttings set under shade in a moist, sandy soil, but for cultivation on a large scale the smaller tuberous roots may preferably be used. These are placed 30 cm apart and at a depth of 16 cm within trenches filled with FYM. As the plants grow, stakes required to be fixed for them to climb on. A return may be expected in the 3rd year and every third year thereafter. Udhayamagalam (Ooty) on account of jalap yields 250 kg of green tuber or 2000 kg of powder. Drying process is a difficult one and there is frequently loss through mouldiness and fermentation. It may be prevented by cutting the tubercles into slices.

Medicianally jalap is well known as a hydragogue purgative, its action being due to certain resinous principles.

125. *Iris decora* Wall.
(=I. nepalensis D.Don)

Family: Iridaceae.

Local names: Shoti, Sasan, Chiluchi (Hindi).

Description: A perennial herb; rootstock stout, prostrate. Stem slender. Spathe long. Flowers pale liliac, short stalked. Blade of sepal broad, oblong, crest yellow. Petals

oblong. Style arm deeply two lobed, margins toothed. Capsule oblong.

The plant bears flowers during July.

Distribution: In temperate region from Kashmir to Sikkim, 1500–2500 m.

Kashmir, Pindar valley, Deoban, Bodyar, Kumaon, Nainital, Sikkim, Bhutan.

The root is considered to be deobstruent, aperient, diuretic and especially useful in removing bilious obstructions. It is useful externally as an applications to small sores and pimples.

Extent of utilization and trade: The plant is used on a limited scale, only by the local population.

Other species: *Iris kumaonsis* Wall. ex Don is another species of the genus which is used medicinally. The root and leaves are useful in fever. The plant is distributed on grassy slopes, usually above the Fir level in the temperate regions from Kashmir to Kumaon and bear flowers in June.

126. *Kickxia ramosissima* (Wall.) Janchen

Family: Sorophulariaceae.

Local names: Jáncher, Kanodi, Bhintgálodi (Hindi, Gujarati).

Description: A perennial, slightly pubescent herb. Leaves long stalked, lower deeply cordate, angularly lobed; upper narrowly lanceolate, sagitate, basal lobes long, acute, diverging. Flowers yellow, on long slender stalks. Corolla 2-lipped, tube produced at the base in a hollow spur; upper lip erect, 2-lobed, lower spreading, 3-lobed. Capsule ovoid, opening by pores at the top. Flowers from May-June.

Distribution: Throughout India, ascending to 2500 m particularly on calcium rich soils.

Kurram, Kohat, Hazara, Kagan valley, Jhelum valley, Udhampur, Ramban, Jammu road, Chamba, Almora, Nainital and Tehri.

The plant is highly esteemed as a remedy for diabetes.

Extent of utilization and trade: Not exploited on a commercial scale.

127. *Lactuca scariola* Linn. (L. serriola L.)

Family: Compositae.

Local names: Káhu (Hindi), Salád (Mumbai).

Description: A glabrous herb. Stem erect, leafy, branched, usually prickly towards the base. Leaves sessile, pinnatifid, segments toothed, pointing downwards, lower surface usually prickly on the midrib and nerves; stem leaves lobed at the base. Heads erect, flowers yellow. Achenes brown, beak very slender, about as long as the body.

Distribution: In temperate regions, 2000–4000 m.

Uses: The fresh plant is a mild sedative, anodyne, purgative, diuretic and antispasmodic, hypnotic with action like opium but no bad effects. It is useful in treatment of coughs, in phithis, bronchitis, asthma and pertussis.

Decoction of the seed is cooling and is used as demulcent and refrigerant. Boiled and made into confection, the seeds are given in bronchitis.

Leaves are hypnotic and sedative.

Extent of utilization and trade: The drug is used on a limited scale.

128. *Lathyrus aphaca* Linn.

Family: Papilionaceae.

Local names: Jangli Matar (Hindi).

Description: A glabrous annual. Stem slender, wingless, much branched, trailing, rachis ending in a tendril; leaflets abortive; stipules leaflike in pairs, adpressed to the stem, triangular, entire, cordate. Flowers yellow, solitary or rarely 2 at the end of a long axillary stalk. Pods, linear, oblong, 4–6 seeded (Fig. 5.61).

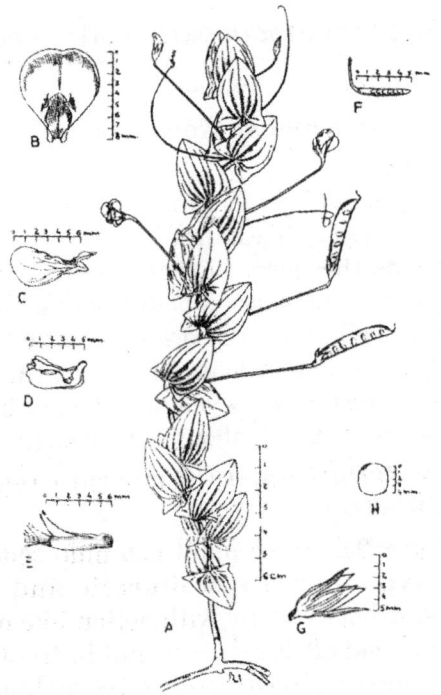

Fig. 5.61: *Lathyrus aphaca* L. A. flowering and fruiting stem; B. standard; C. wing; D. keel; E. androecium; F. gynoecium; G. calyx; H. seed.

Throughout India, ascending to 2300 m.

Kashmir, Punjab, Shimla, Dehradun, Tehri.

Uses: The flowers are resolvent. Ripe seeds are narcotic.

Extent of utilization and trade: It is not utilised for medicinal purposes normally .

Phenology and time of collection: Flowers during March-April.

129. *Launaea asplenifolia* Hook.f.

Family: Compositae (Asteracea nom alt).

Local names: Titlia (Hindi), Tikechana (Mumbai).

Description: A biennial or perennial herb with sessile or shortly petioled leaves, which are sinuately lobed or pinnatifid, lobes toothed minutely, teeth rarely white. Cauline leaves few. Flowering stem 15–40 cm long, many from the root, almost naked, branches dichotomously divaricating. Flower heads terminal, peduncles with usually 1–2 subulate bracts. Involucral bracts glabrous. Achenes columnar, with rough ribs.

Uses: Thr roots in combination with other drugs is used as galactagogue.

Other species: *Launaea chondrilloides* Hook. f. (Vern. name Dudh-Phád). The plant is galactagogue and the boiled residue of the plant in water is taken against constipation.

Launaea nudicaulis Hook. f. Vern. name. Bathal, Dudhlak. Leaves applied to the head of children suffering from fever. Throughout the plains.

Launaea pinnatifida Cass. Vern. name. Bhonpatri (Guj) Bankau (Hindi); Pathri (Mumbai).

Plant is sometimes used as an alternative to *Taraxacum* and given as lactagogue. The juice is sporofic and applied to rheumatic affections. The plant is found on the sandy coastal areas.

130. *Lepidium sativum* L. (Family Cruciferaea or Brassicaceae or Brassicaceae nom alt).

An annual glabrous herb with erect stem 150–40 cm, branched. Radical leaves twice pinnatisect; stem leaves sessile, pinnatifid or lanceolate. Flowers white, in long racemes. Sepals erect. Pods ovate, notched at the tip, margins winged. Seeds one in each cell.

The roots are used in secondary syphilis and also to cure asthma, cough with expec-

toration, and bleeding piles. The seeds administered after being boiled cause abortion, applied to pains and hurts as poultice. Leaves are diuretic, increase milk, useful in scorbutic diseases and blood purifier.

Species like *L. crassifolium* W.K., *L. iberis* Linn., *L. latifoluim L. ruderale* are also used medicinally locally for various diseases.

131. *Leea edgeworthii* Santapa (=L.aspera Edgew.)

Family: Vitaceae.

Local names: Damau (TG).

Description: A large perennial herb. Leaves simply imparipinnate, or the lower often bipinnate; leaflets 2–3 pairs, ovate-oblong, abruptly long-acuminate, obtusely crenate, scabrous, lateral nerves 12–16 pairs, base rounded or subcordate, common petiole not winged, of variable length. Corymbs sessile (so as to appear clustered) or shortly peduncled. Flowers small, green. Anthers not united. Fruit succulent, black when ripe.

Distribution: In subtropical and temperate zone, common in ravines throughout India, ascending to 1500 m.

Shimla, Dalhousiae, Lachhiwala, Kotah range (Garh); Dehradun, Kalimpomg, Chota Nagpur; Parasnath, Sambalpur, north-Canara, Nilgiris, Godavari and Bihar.

Uses: The plant is carminative, febrifuge and stomachic.

Extent of utilization and trade: The plant is not exploited on a commercial scale.

Phenology and time of collection: Flowers during Aug-Sept.

132. *Leucas cephalotes* Spreng.

Family: Labiatae (nom alt. Lamiaceae).

Local names: Guldara (Jammu), Gumma (Hindi); Dronpushpi (Trade name).

Description: An erect hairy, annual plant with shortly stalked leaves, ovate lanecolate, toothed. Flowers white, crowded in globose, terminal whorl, surrounded by numerous lanceolate bracts near the top.

Distribution: A weed of the cropped areas, during rainy season. Both in cultivated land and wasteland. Inflorescence like balls of walnut.

Uses: The plant is reported to have insecticidal properties. The flowers are useful in cough. Both leaf and flowers when boiled in water are given to drink as anthelmintic and in fevers it removes inflation and expels intestinal worms. Leaves are cooked and applied warm as poultice. To reduce swelling, juice of leaves when dropped into the eyes removes yellow-ness due to jaundice.

Extent of utilization and trade: The plant is used extensively for medicinal purposes locally, but it is not much exploited commercially.

Phenology and time of collection: Flowers during Jan.-March.

133. *Lilium polyphyllum* D.Don

Family: Liliaceae.

Local names: Mithávach (Tehri Garhwal).

Description: A bulbous herb, bulbs narrow, of few, long, fleshy scales. Stem slender. Leaves sessile, alternate or nearly opposite or whorled, narrowly lanceolate or linear. Raceme raised on nacked top of stem, 4–10 flowered; bracts leaf like. Perianth green white with purple dots inside; segments obtuse, recurved when fully expanded. Stigma obscurely 3-lobed. Style very declinate.

Distribution: In temperate regions from Kashmir to Kumaon, 2000–4000 m.

Kashmir, Chitral, Chamba, Kullu, Lahul, Shimla, Chakrata, Garhwal, Matkangra, Mussoorie, Below Mundali, Nainital.

Uses: The powdered rhizomes are collected in small quantity for sale and sold as *Mitha-vach*.

Phenology and time of collection: The plant bears flowers during August-September.

134. *Lindenbergia indica* (L.) O. Kuntze (=L. urticaefolia Lehm).

Family: Scrophul ariaceae.

Local names: Gazdar (Mumbai), Bhintachati (Gujarati).

Description: A glandular hairy herb. Stem often tufted leaves opposite, broadly ovate, stalked. Flowers axillary or in small clusters, sometimes forming long leafy racemes. Corolla tube tipped with red. Capsule ovoid.

Distribution: Throughout India on the banks, walls and lower mountain valleys, from Himachal pradesh to Kumaon in Uttaranchal.

Chamba: On Dalhousiae road, Kangra, Shimla, Dehradun, Landour, Rajpur to Mussoorie, Tehri, Deolsari, Almora, Bhagirathi valley, Lohaghat, Jharipani, Pithorgarh.

Uses: The juice of the plant is given in chronic bronchitis and mixed with that of *Coriander* it is applied to skin eruptions.

Extent of utilization and trade: The drug is not used on a commercial scale.

135. *Linum usitatissimum* L.

Family: Linaceae.

Local names: Flax (Eng.), Alsi (Hindi), Atasi (Sansk.), Alasi (Mumbai).

Description: An annual herb with azure blue flowers and capsule having the size of a gram, containing small, flat, pointed dark seeds which also yield an oil locally called *Alsi ka tél*.

The stem is slender, smooth, erect, branched or unbranched up to 60 cm high. Leaves alternate, linear-lanceolate, smooth, pointing upwards, stalks absent. Flowers in loose clusters at the head of the stem, blue-coloured. Sepals 5, ovate, narrow pointed, persistent. Petals 5, larger than the sepals. Capsules roundish, pointed at the apex, divided into 10 cells, each cell containing one seed. Seeds oval, shining, flattened or plump, usually brown-coloured, sometimes white.

Distribution: The herb is cultivated throughout India up to 2000 m in the Himalaya.

Uses: The seeds are reported to strengthen body, cure lumbago, resolve swellings and allay pains. Used as poultice on abscesses, swellings and bursts. An ointment from the oil and lime water when applied to burn relieves pain and dry up the wounds.

Leaves are used to cure gonorrhoea. Flowers as cardiac tonic and dried ripe seeds used as demulcent in the form of poultice. Useful in gouty and rheumatic swellings. Internally used for curing gonorrhoea and diseases of genitourinary system.

The flowers are reputed as nervine and cardiac tonic. The seeds contain almost 30% of a fixed oil which is used both externally and internally. Mucilage from the seeds is taken with honey for cough and cold, externally as an emollient for eye diseases like conjunctivitis. Mucilage preparation takes about five hours by digesting seeds.

136. *Lobelia nicotianaefolia* Heyne ex Roth.

Family: Lobeliaceae.

Local names: Narasal (H), Náli (Guj), Bibhisaná, Devanálá (S), Bantambáku (Beng), Devánala (Marh), Kaduhoge-sopelle (Kan), Kattupukayila (Mal), Kattupukayila (Tam), Adavipogaki (Tel.); Wild Tobacco (E).

The trade name Lobelia is based on the generic name.

Description: A large herb. Stem stout, hollow, 3–5 m high, occasionally branched in the upper region. Leaves very large, up to 45 cm long, smaller upwards, margins not entire, the main nerve whitish. Flowers large, white, in large terminal bunches. Fruit 8 mm, roundish, seeds many, small, yellowish, brown.

Distribution: In the hills of peninsular region and adjoining plains.

Uses: Aerial parts of the plant constitute the drug.

It is used in bronchitis and asthma; causes sweating, nausea and consequently vomiting. The drug is sometimes dangerous and is toxic, if remains in the body for a long time.

The alkaloid *lobeline*, stimulates respiration and is used for reviving respiration in cases where excessive narcotics, anaesthesias and gases have affected the respiration.

The juice exuded from the aerial parts cause blisters on the skin.

Dried plants cause irritation to mucous membranes (in nose and throat).

Extent of utilization and trade: The plant is collected and utilised on a commercial scale. The annual requirements are estimated more than nearly 50 quintals per year.

Phenology and time of collection: Aerial parts are collected from October to November and dried in shade. The coolies employed in collecting the plant avoid approaching areas inhabited with this plant on account of its irritation to mucous membrane.

Other species: *Lobelia inflata* is the true source of 'Lobeline' which is imported. The plant may be cultivated in the Deccan peninsula and Assam. There are other species of *Lobelia* growing wild but are not of much use.

137. *Macrotomia benthami* DC.

Family: Boraginaceae.

Local names: Láljari (TG), Gaozaban (Urdu) (Trade name).

Description: An erect, hispid, perennial herb with alternate leaves, oblong, acute, upper amplexicaul lower petaloid. Flowers blue.

Distribution: In temperate regions of west Himalaya from Kashmir eastwards.

Kashmir: Deosai, Gurais, Pirpanjal, Gulmerg, Ladakh. Garhwal: Harkidon; Rotang in Himachal Pradesh.

Uses: The roots are used locally on cuts and wounds. Useful in diseases of tongue and throat. An important ingredient of *Joshanda* used for cough and cold. Gul é Gaozban (flowers) are also used for Joshanda.

Extent of utilization and trade: Very small quantity of the roots are collected for sale in the market.

138. *Malva neglecta* Wall. M. rotundifolia Linn.).

Family: Malvaceae.

Local names: Khubási, Sonchalá (H).

Description: A much branched, stellately pubescent herb. Stem decumbert, spreading.

Leaves suborbicular, lobed, crenate. Flower small, long stalked; bracteole lanceolate. Petals pale liliac, wedge-shaped, darker streaked, twice the length of sepals.

Distribution: A weed of cultivation, in wet places on the plains of north India, ascending to 3300 m.

Uses: The leaves are mucilaginous and emollient, employed externally in scurvy, reckoned useful in piles. In Europe these are used in the form of infusion for catarrh, dysentery and nephritis. As decoction it is used for fomentation and as poultice is used in sore throat and ophthalmia or for maturing abscesses.

The seeds possess demulcent properties and prescribed in bronchitis, cough, inflammation of bladder and haemorrhoids. The seeds are externally applied in skin diseases.

Extent of utilization and trade: The seeds are extensively used in Indian medicine and the plant is exploited for the seeds which are available in the market.

Malva parviflora L. Vern. Golio. Stellately hairy herb with suborbicular leaves, 5–7 lobed. Flowers purplish or white. Fruits about 6 mm in diam. Carpels wrinkled. Weed in gardens (Fig. 5.62).

139. *Martynia annua* L. (M. diandra Glox.)

Family: Pedaliaceae.

Local names: Kála-Bhédu, Bichu, Hathjori (Hindi), Vinchu (Mumbai), Hathjori (Trade name), Devils Claw. (Eng.).

Description: A naturalised annual, native of America, Mexico, springing on rubbish heaps and in waste places. The plant is more than a metre high with *banyan* like leaves, but dentate. Ripen fruits when burst give seeds having two thorns in their posterior parts, resembling the sting of a scorpion.

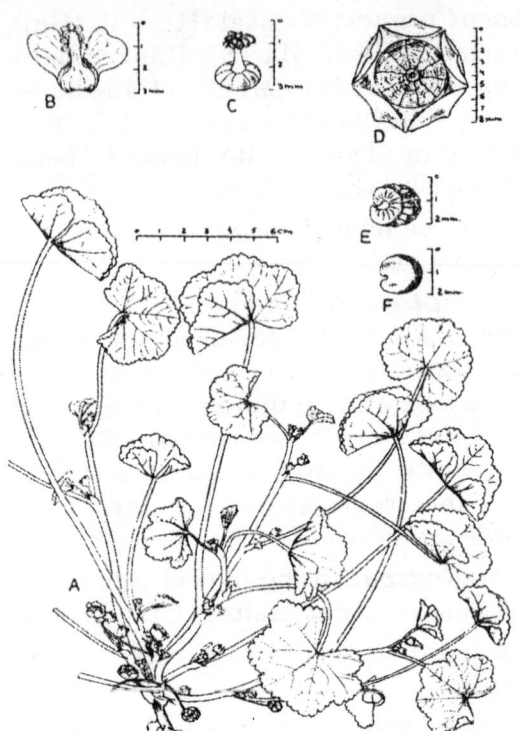

Fig. 5.62: *Malva parviflora* L. A. plant with some of the leaves cut away; B. two of the petals and longitudinally split open androecium; C. gynoecium; D. fruit and persistent calyx; E. fruiting carpel; F. seed.

Distribution: Naturalised in India.

Uses: Leaf juice, fruits are medicinal. The seeds give an oil which is used in cases of paralysis (facial), rheumatism and lumbago. The fruit is alexiteric, useful in inflammation. Also used to cure leukoderma. 1 kg seeds of this plant and 750 gram of Babchi (*Psoralea corylifolia*) seeds are powdered together, sifted and 6–9 gram of the powder is taken daily with water which acts as purgative. Depending on the requirement, the doses can be increased. During the course of the treatment (about 40 days) the patches and the face will darken but should not cause any apprehension. A strict diet regimen needs to be observed, where only wheat

bread with green gram should be taken. In between the mixture of seeds of 'kahu' (*Lactuca scariola*), kasni (*Cichorum intybus*), *Cucumbers*, 'khurfa' (*Portulaca oleracea*), *Coriander* (90 grams each) ground in water, strained and mixed with 20 gram sugarcandy and drunk as *tabrud* (cooling medicine).

A paste of the nut is used as a local sedative and is said to have a curative effect when applied to bite of venomous insects.

Leaves are given in epilepsy, applied to tuberculous glands of the neck. Juice used as gargle for sore throat.

140. *Megacarpaea polyandra* Benth.

Family: Brassicaceae (nom alt. Cruciferae).

Local names: Barámulla (TG).

Description: Perennial robust herb, 1–2 m tall. Stem 6–12 cm, thick at base. Basal leaves pinnatisect, oblong-lanceolate, upper becoming small gradually. Raceme many fld. Flowers 1 cm across, white or creamy yellow. Stamens 8–16. Fruit bilobed at the apex, deeply notched. Seeds suborbicular, brown in colour.

Distribution: In the alpine zone in grassy areas.

Kashmir: Burzil, Sonmarg,; Garhwal Chansil range, under Srikanta, Gori valley, Bayans, Kuari pass.

Uses: The roots are febrifuge and carminative.

Extent of utilization and trade: Very small quantities are collected for trade.

141. *Melilotus indica* All.

Family: Papilionaceae.

Local names: Sénju (H).

Description: Herb with yellow flowers. A pubescent biennial with erect stem.

Leaves of 3-leaflets; leaflet ovate, upper part toothed, veins parallel, running into small sharp teeth. Flowers in axillary racemes. Pod indehiscent. Seeds 1or 2.

Distribution: A weed of the cultivated fields during winter (Dec.-March).

Uses: The plant is useful in bowel complaints and infantile diarrhoea.

Extent of utilization and trade: The plant is not used on a commercial scale but used locally.

142. *Mentha arvensis* L.

Family: Labiatae or Lamiaceae nom. Alt.

Local names: Field Mint (Eng.); Pudiná (Hindi), Pudinah (Mumbai).

Description: A perennial herb which is grown in the western Himalaya from Kashmir to Garhwal. A well-known herb, locally used in the household for making *chutney* during the summer season.

Distribution: Extending from 1500–3300 m elevation in the western Himalaya. The plant has been widely cultivated for its oil, as a plant for cottage industry throughout the alluvial plains of northern India.

Uses: Locally it acts as digestive and carminative and has been used since the earlier times. It has antidotal properties and also used to counteract poisons. In cases of indigestion, dyspepsia, 6 gram of *Pudina* and 3 grams of cardamon are boiled in water (500 cc), strained and given to patients suffering from vomiting and nausea; relieves stomachache and thirst; used as home medicine.

It is recorded that elephantiasis and varicose veins are improved by prolonged use of this plant ground in 100 gram of whey. Application of this plant ground in wine, is taken internally on the face-spots and skin

blemishes are removed. Also a black ring around the eyes vanishes by its continuous application. Instilling juice of green *podina* leaves into the nose or ears; wounds in any other place, the worms infesting it are cleared. It is also used in cases of urticaria, when 6 grams of dry plant boiled with 10 grams of red sugar in water and given to drink.

The fried plant is reported to be antiseptic, carminative, stomachic, refrigerant, stimulant and diurectic (Chopra et. al, 1956)

Other species

a. *M. arvensis* var. *piperascens* Holms (Vern. Japanse Mint) is cultivated for the essential oil, which is the chief source of menthol. It is used in treatment of cold and is an ingredient in the celebrated medicine *Amritdhara*.

b. *Mentha longifolia* (L.) Huds. (-*M. sylvestris* L.).

An erect or diffused herb, strongly scented with small liliac flowers (Fig. 5.63). Found near the water source in Himalaya 1300-4000 m.

The plant is astringent to the bowels, anthelmintic and useful in diseases to heart, bronchitis, loss of appetite, diarrhoea and dysentery. Leaves also used as carminative and stimulant, even when dry. Leaves soaked in water give an infusion which is drunk as cooling medicine. Branded medicine *Pudine Hara* marketed by Dabur is available.

c. *Mentha piperita* L. Vern. Vilayati Pudina-Hindi

A native of Europe but cultivated. The leaves and flowering tops yield an essential oil, used for flavouring foods. The herb is also, stimulant, stomachic, carminative, used for allaying nausea, sickness and vomiting.

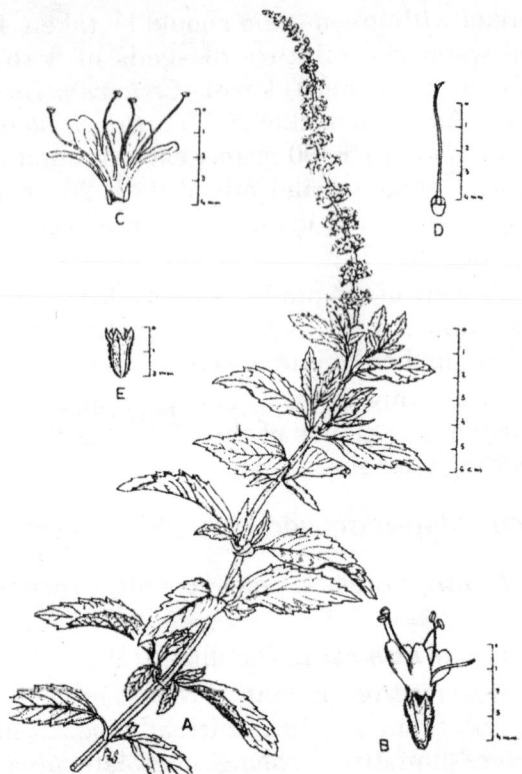

Fig. 5.63: *Mentha longifolia* Huds. A. upper portion of flowering stem; B. flower; C. corolla opened to show androecium; D. gynoecium; E. calyx.

d. *Mentha spicata* L. (-M. viridis L.) Vern. Pahári-Pudina. The seeds are mucilaginous while the leaves are given in fever and bronchitis, as decoction. Used as lotion in aphthae.

The leaves are source of spearmint oil which is used for flavouring food products.

e. *Mehtha rotundifolia* (L.) Huds. (Vern. Apple mint, Round leaved mint). A perennial herb, cultivated in gardens. Used for flavouring food and in confectionary.

f. *Mentha pulegium* L. (European Pennyroyal-Eng.). Native to Europe and west Asia, now grown in Jammu-Kashmir state, introduced by RRL, Jammu.

Leaves and flowering tops yield an essential oil, used in perfumery and in cosmetics.

143. *Meriandra strobilifera* Benth.

Family: Labiatae or Lamiaceae nom alt.

Local names: Kafur-ka Pat (Hindi).

Description: A small aromatic shrub, 60–150 cm high, with opposite leaves, oblong or lanceolate, gradually narrowed from a broad hastate base, upper surface markedly rugose, lower white tomentose. Flowers in dense fld. whorls, crowded in erect tomentose, axillary and terminal cylindrical spikes, often forming panicles; bracts densely woolly outside, but hard in fruit. Calyx 2-lipped. Corolla white, tube as long as the calyx. Fertile stamens only 2, anther cells pendulous from a long connective. Nutlets brown and smooth.

Uses: Leaf decoction is used as lotion for the eyes, ulcers and heals raw abrassions of the skin. It dries up breast milk.

Distribution: In open situations, on stony ground, in western Himalaya, 1500–2000 m.

Other species: *Meriandra bengalebsis* Benth (Vernacular Name Kafur ka Pat (Hindi).

A large straggling shrub with white flowers. It is cultivated in gardens in Upper Gangetic plains and is often known under the name of Bengal Sage. The leaves have a strong camphor like scent and are medicinally used as infusion for application to aphthae and sore throats or arrest the secretion of milk. A native of Abyssinia. Also used for preserving cloth from inect attacks.

144. *Mimosa pudica* L.

Family: Mimosaceae (Leguminoseae).

Local names: Lájwanti, Lajálu, Lajjabatti, Chuimui (Hindi), Lajja (Sansk.).

Description: A small herb with sensitive leaves that collapses when touched, but recovers to its normal position after some time. The rachis 3–4 cm long while the pinnae are 4–5 cm long approximate at the end of the rachis and spreading to look like digitate. Very sensitive. Flower heads long-peduncled. Pods 0.75–8 mm long with densely, prickly sutures.

Distribution: Naturalised, more or less throughout the country; reported to be a native of tropical Africa.

Uses: The herb is used to stop bleeding from intestines and piles; 3 gram of the powder is taken with milk to stop bleeding. About 10 gram of green leaves are ground along with 7 gram of pepper, strained and sweetened with sugar and drunk for stopping bleeding.

Fresh plant is crushed and the expressed juice is drunk as a dose of 30 gram every three hours, in cases of plague and delirium due to high fever.

Stammering followed by an attack of plague or smallpox is cured by drinking for a few days grounded plants in water and strained.

To cure spermatorrhoea and leukorrhoea, seeds of this plant are powdered, sifted and 3 gram of it is taken along with milk, each time in the morning and evening.

145. *Mirabilis jalapa* Linn.

Family: Nyctaginaceae.

Local names: Gul-é-abbás Abási, Krishnakél, (H), 4° clock plant (Eng), Marvel of Peru (E).

Description: A much branched perennial herb; roots tuberous. Leaves triangular

ovate, acuminate. Flowers usually purple but sometimes yellow, crimson or variegated (Colour Plate 5.13).

Distribution: The plant is an ornamental, often cultivated in the house garden and often grows up from self-sown seed. It is indigenous to Central America.

Uses: Root powder if taken internally, acts as aphrodisiac in case of jaundice and dropsy. Leaves cooked as vegetable are eaten.

The leaves have a sharp taste and used to lessen inflammation. They are applied to boils as maturant, used to resolve all kinds of swellings. Even carbuncle is cured by the use of roots of this plant. Equal quantities of roots of karil (*Capparis aphylla*) and this plant with jaggery are taken grounded and cooked. The paste is applied to carbuncle, fresh juice is soothing and used against *urticarea* to allay heat and itching. Some burning sensation is felt on the spot. To remove it 20 gm of Post-Doda (fruits of *Papaver*) empty capsule and 10 gm isabgol (*Plantago ovata*) seeds are cooked and paste applied. 3 gms of camphor powder is sprinkled and used as poultice which relieves burning sensation. 5 gram of powdered flower when given orally benefit haemorrhoids.

Extent of utilistation and trade: The plant is locally used but does not form a trade commodity.

146. *Mollugo nudicaulis* Lamk.

Family: Molluginaceae.

Local names: Rangtia-khár (Rajasthan).

Description: An annual herb, glabrous. Stem erect, many, arising from of tuft of radical leaves, sometimes leafless. Radical leaves spathulate, 2–5 cm long. Flowers in trichotomous cymes; scapes wiry. Sepals oblong. Stamens 3–5. Stigma 3, very small capsule as long as the sepal, somewhat ellipsoid, many seeded. Seeds black.

Distribution: In hot dry parts of India, extending to tropical Africa in south-Sahara. A pantropical weed, on roadsides and wastelands, often in dry rocky crevices.

Uses: The plant is bitter and considered pectoral, used in whooping cough. Leaves are applied to boils to draw out the pus.

Other species

a. *Mollugo cerviana* (L.) Ser. (-*M. spathulifolia* (Fenzl)
(Vern. name: Chirio-ro khét-Rajasthan, Padá (Mumbai); Ghimaság (Bihar).
An annual herb, glabrous. Stem erect, filiform. Leaves whorled, 4–8 in a whorl, radical, tufted. Flowers in umbellate cymes, peduncles trichotomous. Sepals elliptic. Stamens 5. Styles very small. Capsule globose.
The plant is used for promoting flow of lochial discharges and as cure for gonorrhoea. In Rajasthan, it is used for fevers and for promoting lochial discharges as well and blood purification.

b. *Mollugo oppsitifolia* L. (M. spergula L.)
Vern. name: Jima; Phanija-Sansk; Jharási (Maharashtra).
The plant is used for suppression of the lochia, stomachic, aperient and antiseptic. The juice is applied to skin diseases and itch. In combination with castor oil, it is used as a cure for earache.

c. *Mollugo pentaphylla* L. (-Mollugo stricta L.)
Erect, slender, much branched annual. The plant is eaten as a pot herb throughout India. It contains carotene, traces of vit. C. It is stomachic, aperient, antiseptic and emmenagogue; also used in poultice for sore legs. The leaves are bitter and antiperiodic.

147. *Momordica charantia* L.

Family: Cicurbitaceae.

Local names: Karéla, Mokha (Hindi); Bitter Gourd (Eng.); Sushavi (Sansk); Chochhidan (Gujarat).

Description: A slender herb, monoecious. Leaves orbicular, cut into 5-7 subpinnate lobes. Male peduncle slender, bract about the middle of the male peduncle. Female peduncle slender, ebracteate near the base. Fruit ovoid, narrowed at both the ends, many ribbed, covered with triangular tubercles.

Distribution: Cultivated in low hills and plains, throughout the country for the fruits; sometimes found as an escape on the banks of ditches. It extends to Malaysia, SriLanka, China and tropical Africa.

Uses: The fruit has reputation as a cure for the diabetes. Fruit juice mixed with honey is used locally in Kumaon to cure eczema. The seeds are employed in many preparations for diabetes control. Fruit and leaves are anthelmintic, useful in piles, leprosy, jaundice and as a vermifuge.

Roots are useful in haemorrhoids.

Leaf juice is emetic, purgative and given in bilious affections, rubbed in burning of soles of the feet.

Other species

a. *Momordica cochinchinensis* (Vern name: Kakrol, Karkataka (Hindi and Sansk.).
   Seeds used for cough and chest complaints.
b. *Momordica dioica Roxb*. ex Willd. (Kankéro, Jangli Karéla).
   A slender, dioecious perennial with tuberous root. Leaves deeply lobed, petiole without glands. Flowers dioecious, bract near the top of the male pedun-

cles. Flowers large, orange. Fruit ellipsoid, shortly beaked, densely covered with soft spines (Fig. 5.64).

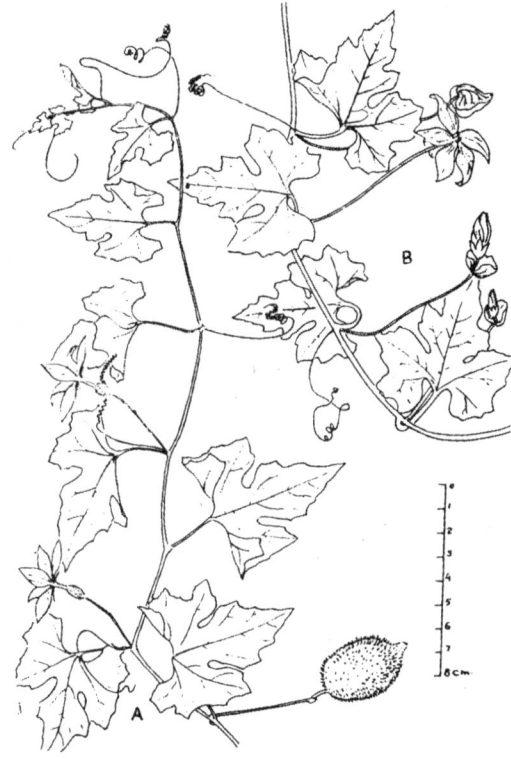

Fig. 5.64: *Momordica dioica* Roxb. A. female flowering and fruiting stem; B. male flowering stem.

Roots used to stop bleeding from the piles, used in urinary complaints, also as sedative in high fevers associated with delirium, also an antiseptic as a paste smeared on body.

Powdered dry fruits produce powerful errhine effect when introduced into nostrils and provoke copious discharge from the mucous membrane.

148. *Morina wallichiana* Royle (M. persica sensu Hook. f.)

Family: Dipsacaceae.

Local names: Békh-akhwár (Hindi).

Description: A prickly perennial. Stem pubescent upwards. Leaves sessile, doubly spinous-toothed, opposite, linear. Flowers white, faintly tinged with pink in elongated spikes. Bracts free, and involucre hairy. Calyx lobes ovate entire. Fertile stamens 2, filaments as long as the corolla lobes.

Distribution: In temperate regions from Kashmir to Kumaon, above 2000 m.

Uses: It is aromatic and used as an incense, and in preparation of *dhup* and *agarbatties*.

Other species: *Morina longifolia* Wall. ex DC; (Vern. name: Biskandrá). A softly pubescent herb with deep pink flowers, crowded in the axil of upper leaves.

The roots are aphrodisiac and tonic. Roots used in the form of poultice.

Small quantities are collected and locally consumed.

149. *Morina longifolia* Wall. ex DC.

Family: Dipsacaceae.

Local names: Biskandrá (TG).

Description: A glabrous, softly pubescent herb. Stem 5–10 cm, leaves sometimes whorled, narrowly oblong, sinuately pinnatifid, prickly; upper ones shorter, sessile, united at the base. Flowers deep pink, sessile, crowded in the axil. Calyx lobes notched. Corolla tube long, slender, limb obscurely two lipped, lobes unequal, ultimately spreading. Filaments much shorter than the corolla lobes. Perfect stamens two. Stigma broad, disc like. Achene small, free within the involucral.

Distribution: Temperate regions of Himalaya from Kashmir to Bhutan. Kashmir, Shimla, Chur, Kunawar, Garhwal, Haunsar: Deoban, Sikkim, Bhutan.

Uses: The roots are aphrodisiac and tonic. Externally they are used in the form of a poultice on boils.

Extent of utilization and trade: Small quantity of the roots are collected and consumed locally.

150. *Murdannia scapiflora* Royle. Aneliema scapiflorum Wight.

Family: Commelinaceae.

Local names: Siyah-Musli, Káli-musli (Hindi), Kureli (Bihar), Sismulia (Gujarat).

Description: An erect, leafy herb; roots of elongated tubers. Leaves all radical, erect, finely acuminate, narrowly ensiform. Scape erect with a narrow panicle, panicle elongate; bracts large, sheathing, lower long, upper small, amplexicaul. Flowers small. Capsule ellipsoid, trigonous. Seeds 3–6, superposed in each cell.

Distribution: In subtropical and temperate regions of Himalaya and the upper Gangetic plain eastwards to Bhutan.

Kalsi, Dholkot (DDN) in Uttaranchal.

Uses: The root is astringent and tonic, useful in headache, giddiness, fever, jaundice and deafness. Root bark dried in shade is useful in asthma. Dried powder mixed with sugar is used as an aphrodisiac.

Extent of utilization and trade: The plant is utilised in trade on a limited scale.

151. *Nardostachys jatamansi* DC.

Family: Valerianaceae.

Local names: Jatámansi (Hindi), Bálchad, Bála-charea (Mumbai); Bhutijatt-Kashmir. Indian Spikenard, Indian Valerian, Musk root, Nard (Eng.).

Description: A dwarf, hairy, perennial herb with long tap root. Stem partly underground, partly above ground. Rootstock woody, stout and clothed with fibres from the stalk of the dead leaves. Stem up to 60 cm in length, upper part covered wth hairs, lower without hairs. Leaves arising from the root stock, 15–20 cm long, 3 cm wide, without stalks, oblong or ovate. Flowers heads. 1–3 or 5, bracts small, oblong, usually hairy, corolla tube hairy within. Fruit small with white hairs, pointing upwards, crowned by calyxteeth.

Distribution: In alpine regions of the Himalaya, extending, eastwards from Kumaon to Sikkim, from 3300–5600 m.; also above 5000 m in Bhutan. On morains and other dry-stony aspects.

Uses: The rhizome is cardiac and respiratory stimulant, nervine tonic, aromatic, carminative, hypertensive, stomatic, laxative, antispasmodic and diuretic, emmenagogue and deobstruent. It is employed for treatment of epilepsy, hysteria and convulsive affection in which it acts as a palliative but not as a cure. It is used in various disorders of digestive and respiratory organs. The root paste is employed as a topical application in haemorrhoids. In mild cases of mental derangement, nervous and convulsive disorders and certain disturbances due to menopause, it is one of the most efficaceous palliative in milder doses. It is used in delirium and for the tranquility of mind. It is much used for cholera, flatulance, palpitation of heart, jaundice, disorders of digestive system, bronchitis, etc.

Also used as vermifuge, in dysmenorrhoea and polysuria. In combiantion with *cinchona* bark it is effective in acute rheumatism.

Locally it is used as hair tonic by ladies and is an ingredient of many hairwashes.

Oil from the distillation of the rhizome is aromatic and used in making medicinal oils. As early as in 1790 the perfume was identified by Sir W. Jones (As. Res. 1790.2: 405–417). Also used as substitute for valerian acrystline acid, *jatamansic acid* has been isolated.

Jatamnsi is also used as infusion prepared by soaking the crushed rhizome and given as vermifuge, given to children along with purgatives like jalap, for treatment of thread worms, an enema of rhizome is given.

A paste made of the rhizome with water is applied to the eyes in stupor and coma.

152. *Nardostachya hardwickii* Wall. (-V. hardwickii wall.)

Family: Valerianaceae.

Local names: Tagar-Hindi.

The plant has properties similar to Nardostachys jatamani, the roots are stimulant and antispasmodic and used in hysteria, epilepsy. As a stimulant in advanced stages of fever and inflammations. A bath is useful in acute rheumatism, the roots give an essential oil.

153. *Nicotiana tabacum* L.

Family: Solanaceae.

Local names: Tobacoo (Eng.), Tambaku, Tamku, Tamak, Tambáku, Tamarkuta (Hindi), Samakhu (Sansk.).

Description: A native of tropical America, now cultivated at many places like Andhra, Maharashtra, Uttar Pradesh and west Bengal. There is a wild variety, *N. rustica* which differs from the *N. tabacum* in its similar stature, its suborbicular leathery leaves and in the greenish yellow flowers, the segments of which are much shorter.

Uses: The leaves are used for smoking and also contain alkaloids which are used as insectides. The leaves are smoked as cigarette or in *hookah*, cigars, *bidi* and pipes.

The oil obtained from the seeds is used as an illuminant and in the manufacture of

paints and varnishes. The oil cake makes a good manure.

The tobacco waste is used in preparation of snuffs, *khamira* tobacco for smoking and manufacture of nicotine sulphate, an insecticide.

In swollen gums, toothache *surti tobacco* (10 gram), pepper (10 gram) rocksalt (2 gram) are finely powdered, sifted and rubbed on the teeth and the gums.

Tobacco is also useful in cataract, when snuff tobacco (10 gram), is triturated in a mortar for 12 hours in castor oil and applied to the eyes daily with a pencil.

Oil obtained from tobacco is useful for wounds and sinus, it also kills headlice.

Nicotiana rusfica L. (vilayat Tambaku) is cultivated in some parts of the country. Use similar to above species. *N. plumbaginifolia* is a wild species found in wasteland in W. Himalaya. A native of Mexico and West Indies.

Nicotiana plumbaginifolia Viv. is shown in Fig. 5.65.

154. *Nepeta elliptica* Royle ex Benth.

Family: Labiatae.

Local names: Tukham-Malangá (Unani-Urdu) also trade name.

Description: An erect herb. Stem densely hairy. Leaves sessile, ovate-oblong, distinctly hairy, teeth small, closely set, angular. Flowers in sessile whorls, crowded in terminal spike, usually interrupted near the base. Flowers pale blue or white. Calyx hairy, teeth linear-lanceolate as long as the tube, than acute. Corolla hardly longer than the calyx.

Distribution: In temperate regions from Kashmir to Kumaon, 1500–2500 m on grassy slops.

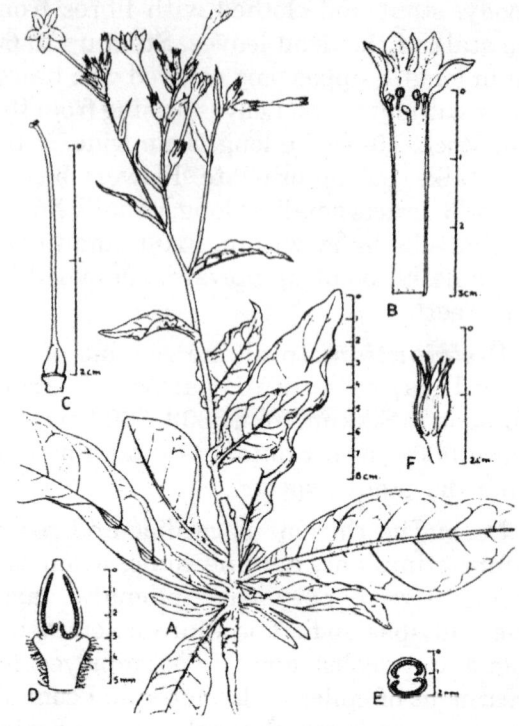

Fig. 5.65: *Nicotiana plumbaginifolia* Viv. A. plant; B. corolla opened to show androecium; C. gynoecium; D. longitudinal section of gynoecium; E. transverse section of gynoecium; F. calyx.

Kashmir, Pirpanjal, Shimla, Garhwal, Jaunsar: Deoban.

Uses: The seeds infused in cold water are used in dysentery.

Extent of utilization and trade: The seeds are not collected on a commercial scale.

155. *Nicandra physaloides* (Linn.) Gaertn.

Family: Solanaceae.

Description: An erect annual herb. Leaves ovate-lanceolate, irregularly sinuatley lobed. Flowers blue, usually axillary. Calyx lobed nearly to the base. Corolla bell-shaped, limbs

spreading, 5-lobed. Filaments hairy, bases dilated, covering the ovary. Ovary 5-celled. Style linear. Stigma 5-lobed, lobes cohering. Berry globose, loosely enclosed by the enlarged, membranous 5 angled calyx.

Native of subtropical America. Commonly found in the temperate regions, 1000–2000 m on roadsides. From Kashmir to Sikkim in subtemperate regions of Shimla, Almora, Mussoorie Nainital, Chakrata, Raniket, Nagaland. Introduced in mountain region of South India.

Uses: The plant is said to be diuretic.

Decoction of the leaf is reported to destroy *Pediculi capitis*.

Fumes of burning seeds when inhaled relieve toothache.

The alcoholic extract of the plant gives 'nicandrin', a bitter principle.

156. *Nigella sativa* Linn.

Family: Ranunculaceae.

Local names: Kalázira (H); Kalonje, Tuk -mogande. Sithulayira, Shuniz (Urdu), Black cumin Small fennel (Eng.).

Description: Slender or stout annual, 15–50 cm high, loosely pubescent, sometimes hairs glandular. Stem erect, finely striate. Leaflobes acute. Flowers single, without involucre. Sepals ovate, whitish, more or less obtuse with short distinct stipe. Follicles coherent throughout the length. Seeds rugose, triquetrous.

Distribution: The plant is native of S. Europe, N. Africa and SW Asia, but extensively cultivated in India.

Uses: The seeds contain 2 kinds of oils; one dark-coloured fragrant and volatile, the other clear, nearly colourless and of about the consistency of castor oil. Medicinally, both the oils are regarded as aromatic, carminative, stomachic and digestive. They are much used in curries, in vinegar and other dishes and are frequently sprinkled over the surface of bread along with 'sesamum oil'.

The seeds resemble onion seeds and used in pickles. These are digestive and carminative, kill intestinal worms and relieve flatulance. These are diuretic and emmenagogne.

Also used in cases of stone in bladder and kidneys, as well as in chronic fevers due to phlegm, rheumatism, gout and sciatica.

50 gm of seeds, steeped in vinegar over night, dried in shade, and ground finely, sifted, powder mixed with honey and preserved. 6–10 gm taken as a dose. In rheumatism, lumbago and pains arising from phlegmatic humours of flatulance; if *kalonji* is taken with other ingredients like *ajwain, sounf* seeds in equal quantity, a dose is of 3–4 gm with lukewarm water.

It also cures leukoderma ringworm, alopecia and acne of the face, when the seeds along with that of fennel (*sounf*) are ground in vinegar and applied. Acute cold is cured by smelling roasted seeds of *kalonji*.

157. *Ocimum sanctum* L.

Family: Labiatae or Apiaceae nom alt.

Local names: Tulasi (Sansk), Tulsi (Hindi); Tulasá (Mumbai), The common Mint.

Description: A much branched herb, young parts pubescent. Leaves elliptic-lanceolate, entire or with shallow teeth, gland dotted; petiole slender and hairy. Flowers in 8–20 cm long spikes, whorls rather close, 6 fld; bracts stalked, elliptic-lanceolate, ciliate, pedicels shorter than the calyx. Calyx pubescent, upper lip rounded. Corolla white, upper lip broadly oblong, 4-toothed at the apex, shorter and broader than upper lip. Nutlets black, ellipsoid (Colour Plate 5.14).

Distribution: Throughout India, cultivated in plains and lower hills. Also in SriLanka, Java and in tropical Africa where it is believed to be indigenous.

Uses: A well-known herb, held sacred and worshipped by Hindus. The leaves give aromatic smell which drives mosquitos. The leaves ground with 'saffron' are given to patients of smallpox or chikenpox, to hasten pustule appearance. It is also given as a drink along with bansa (*Adhatosa vasica*) leaves to get relieved from cough and asthma. Decoction of *tulsi* leaves induces perspiration and relieves fever.

Ringworm is cured by the application of *tulsi* leaves mixed with equal weight of limejuice. *Tulsi* leaves and young leaves of castor plant when applied to mumps behind the ears, these get resolved.

Among the two varieties, the black one is considered to be superior than the white one, which is slightly bitter, pungent, fragrant and appetizing. It cures gas trouble, cough, intestinal worms, skin diseases, kidney disorders, regulates the flow of urine, subdues inflation and restores the balance in the human system by cleansing the body of toxins and strengthening the tone of every organ.

A few fresh leaves can simply be taken every day after washing them clean. Leaf juice can be taken empty stomach in the morning to increase body vitality. The leaves could also be mixed with the leaves of either *bael* (*Aegle marmelos*), coriander, mint, etc. and also with the juices like of carrot, beetroot, spinach, tomatoes and cabbage.

A spoonful of *tulsi* and ginger juice mixed with diluted honey is an excellent medicine for cold, cough, bronchitis, cataract, fever and asthma. The juice and pulp of the leaves can also be applied for skin diseases. Powder of 50 gram leaves of *tulsi* and half the quantity of black pepper, is prepared into pea-sized pellets and dried in sun, if taken can cure all types of the fevers.

Fresh leaves crushed with ginger can cure stomach. For nausea and vomiting some powdered *tulsi* leaves are mixed with milk and taken when slightly hot to cure nausea and vomiting.

As an excellent deworming medicine for children, *Tulsi* leaves juice is recommended along with a pinch of salt. For skin infections like eczema, *tulsi* leaves juice is mixed with lemon juice and applied to the infected parts. *Tulsi* is also good for teeth complaints, like toothache when brushed with *tulsi* powder, it also prevents bad breath and cures pyorrhoea.

Decoction of the seeds and roots is given for curing malarial fever and bring about sweating. *Tulsi* tea is a popular home remedy for cold, cough and fever during the winter season. *Tulsi* leaves mixed with *sandalwood* paste are applied on the forehead to cure headache and to retain coolness.

Tulsi plant has the property and capacity to keep insects and mosquitoes at bay, also called the 'Mosquito plant'. The juice used as an antidote for poisonous bites and stings. Poisonous reptiles, is said, dare not enter the house where the fragrance of *Tulsi* prevails.

Tulsi plant has also been found to be effective in combating pollution, since it emits a peculiar kind of smell, which is reported to purify the atmosphere around us. Thus, the miraculous plant plays a vital role in the scheme of good health and environment for the humanity.

Other species

a. *Ocimum canum Sims*. (-*O. americanum L.*) Vern. name: Kali tulsi (Hindi) Ajaka (Sansk), Bapji (Rajasthan).

The leaves contain an essential oil and made into a paste which is useful in parasitical skin diseases, applied to fingers and toe nails during fever when the extremities are cold. The plant hung in the corner of the room is said to attract mosquitoes and keep rest of the room free of them; seeds are tonic and useful in fever.

b. *Ocimum basilicum* L. Mujarikhi (Sansk), Babuitulsa (Hindi), Sabza (Mumbai).

Flowers are carminative, diuretic, stimulant and demulcent, while the seeds are mucilaginous, given as infusion in dysentery and chronic diarrhoea. It contains an essential oil. Indigenous in lower hills of Punjab but cultivated in other parts of India.

c. *Ocimum gratissimum* L. Vern. name: Vridhatulsi (Sansk); Ramtulsi (Hindi).

Decoction of leaves is used in seminal weakness, while seeds are used in neuralgia. Often cultivated throughout India.

An ubiquitous plant, with which Hindu mythology is replete with legends. It is known as *Vishnupriya,* love of Lord Vishnu, is used as 'Prasad' and partake it, particularly at Badrinath temple in Uttaranchal. The name itself denotes as embodiment of wealth and wisdom, worshipped as *Lakshmi* and *Saraswati.* A mythological legend says that Binda-wife of Sankhachuda, performed penace and pleased, Lord ordinated her to take the form of *Tulsi* plant.

This plant has been popularised as a plant of great devotion, and worship and has found reference in Vedic period. In *Ayurvedic* and *Sidha* system of medicines, the leaves and its juice, the roots and seeds are used to cure various ailments as described above.

158. *Onosma bracteatum* Wall.

Family: Boraginaceae.

Local names: Gázaban (Urdu); Shankphuli, Cholo-Chodhars, Oshthaphala (Hindi). Trade name Gozaban (in Unani system and Gule Gazaban-flowers). The name is also given to *Macrotomia bentham.*

Description and distribution: A herb with hispid parts; hairs setose, usually arising from tubercled base. Flowers bracteate in cymes. Calyx tubular. Corolla tubular. Filaments adnate to corolla tube. Nutlets smooth in Kashmir and Kumaon up to 3700 m.

Uses: The leaves and flowers of the plant are medicinal and used for curing melancholia, insanity and palpitation of the heart. They act as stimulant and cardiac tonic. It is largely used as an ingredient of the common *joshanda,* a decoction which is used for curing cold, catarrh, cough, asthma and chest congestion.

In case of congestion of the nose and chest, due to catarrh, 5 gram of this plant and an equal quantity of liquorice (*Glycirhiza glabra*) and sugar (20 gram) are boiled and drunk. Children suffering from thirst and aphthae get relief when the leaves are burnt, ground and dusted.

Onosma echioides L. (Vern name: Ratanjot, Dhamásá, Dhamani, Maha-rangá Lal-jari Hindi).

A biennial, bristly herb. Leaves alternate, oblong, up to 8 cm long. Raceme elongate, often forked. Corolla cylindric, slightly dilated upwards, glabrous. Raceme in fruit 3–5 cm long; bracts leaflike. Nuts stingy, white, often speckled, shining, smooth. Leaves and branches reddish.

Occurs throughout western Himalaya from Kashmir to Kumaon. Growing under *Capparis* bushes and along the rivers.

The plant is diuretic used in cases of burning sensation, during urination or passing bloody urine. It also purifies blood and in

cases of eczema or burning sensation all over the body, the herb is ground along with 5 grains of pepper, strained, sweetened and given to drink.

The herb is also useful in expelling stones from the kidneys or the bladder.

The root is applied to cuts and wounds.

The leaves are alterative and given as purgative to children, also leaf powder is given for the same symptoms.

159. *Operculina turpethum* (L) Silva Manso

Family: Convolvulaceae.

Local names: Nisoth, Pithori (H), Nashatar (Guj), Nishottara (Mahr), Nisot (Punj) Kalaparni, Triputi (S), Shivadai (Tamil); Tellategada (Tel), Bilialutigadde (Kan), Dudiya-Kalmi (Bihar), Indian Jalap, Turpeth root (E).

The trade name *Turpeth* is based on the scientific name.

Description: A large twining herb with milky juice; roots long, branched, fleshy; stems winged. Leaves 4–10 by 1.5 to 7 cm ovate, cordate. Flowers white, 4–5 cm long, funnel-shaped in few-flowered bunches. Sepals about 2 cm long, but when plant is in fruit, they become much larger and brittle and enclose the fruit.

Distribution: Found throughout India up to an altitude of 1000 m; it is sometimes grown in garden as an ornamental plant.

Uses: The roots (also rootbark) constitute the drug; roots from the white variety should be taken with bark intact.

It is used as a purgative by native practitioners. Out of the two varieties the white is generally preferred since the black is considered to be drastic. The resin is similar to *Jalap resin*. Glucoside *Turpethin* are recorded.

Extent of utilization and trade: The drug contains the *Turpethin*. The turpeth has almost the same properties as the true jalap, obtained from *Exogonium purga* and is a suitable substitute for it. The samples of turpeth, available in the market, have stem and twigs mixed with the roots.

160. *Opuntia dilleni* Haw.

Family: Cactaceae.

Local names: Nágphani, Nigádung, Chappál (Hindi); Vidára (Sansk.).

Description: An evergreen shrub without a main stem; branches jointed, joints flat, succulent, ovate bearing densely woolly hairs with tufts of numerous, unequal, long-sharp spines and bristles. Leaves are generally absent. Flowers arising from the tuft of spines and bristles or margins of the joints, yellow or reddish, bisexual. Calyx lobes numerous on the hollow receptacle. Petals numerous, joined at the base. Stamens numerous. Fruit pear-shaped, fleshy, purple, with spine bearing tubercles near the apex (Colour Plate 5.15).

Distribution: Native of south America but naturalised throughout India, ascending to 600 m in Northwest Himalayan region.

Uses: Milky juice of the cactus is purgative. The pulp from the meshed leaves is applied to eyes in ophthalmia; heated and applied to boils to hasten supperation; baked and given in whooping cough and in the form of syrup used to control spasmodic cough and expectoration.

161. *Orchis latifolia* L.

Family: Orchidaceae.

Local names: Sálp, Sálampanja. Hathjori (Hindi); Munjatak (Sansk).

Description: A perennial herb, glabrous; root tuberous, slightly flattened and divided

into 2 or 3 finger like lobes. Flowering stem 30–90 cm, robust, erect, hollow, leafy throughout or the lower portion bearing a few sheathing scales. Leaves erect, oblong-lanceolate. Flowers crowded, dull purple, lip darker spotted; bracts green, lower much longer than the flowers. Sepals and petals nearly equal, lateral sepals spreading, dorsal forming with petals, a hood over the column. Column very short. Anther adnate to its face, pollinia 2, caudicle attached to 2 small, globose, viscid glands enclosed in minute pouch overhanging the 2-globed stigma.

Distribution: On wet ground from Kashmir to Nepal and west Tibet, 2700–3000 m.

Uses: The tuber *Salap misri* is the product of several species such as *Eulophia nuda* and *E. urens* in addition to *Orchis macula.* The tubers are dug up after the plant has fld. and the plump, firm once are washed and set aside, subsequently strung on thread. These are selected and dried in sun. The commercial article is met in the market in three forms such as 'palmate' called *Salab panja*, large ovoid and small ovoid (*Salb misri*). The tuber is expectorant and astringent.

Other species: *Orchis mascula* L. Vern. Salapmisri (Hindi). Salum (Mumbai). In southern Europe, Asia minor. Uses similar to *O. laxiflora* Lam. Himalaya Drug Company product 'Sperman' for spermatorrhoea is in the market (Medical Digest. 22 sept. 1954).

Orchis laxiflora Lam. Salapmisri-Hindi. A native of Central and Southern Europe till Afghanistan.

162. *Origanum vulgara* Linn.

Family: Labiatae.

Local names: Bantulsi (Kum); Sháhtra (Urdu); Wild Marjoram (E).

Description: An erect herb, clothed with short hairs. Leaves ovate, entire. Flowers pink, crowded in 4-seeded spikes in clusters or heads at the end of branches, sometimes forming terminal panicles. Floral leaves bract like, lanceolate, longer than the calyx, overlapping, often tinged with purple. Calyx bell-shaped, enlarged in fruit, 5-toothed, mouth hairy within. Corolla tube longer than the calyx, limb 2-lipped, upperlip erect, nearly flat, notched, lower spreading, 3-lobed. Stamens 4, in unequal pairs, slightly protruding.

Distribution: Im temperate regions from Kashmir to Nepal and Sikkim, 2000–4000 m Kashmir: Gulmarg, Kunawar, Kagan, Lahul, Murree, Pangi, Shimla, Chamba, Chakrata, Landour, Badrinathm, Kumaon: Almora, Nainital, Milam; Nepal, Sikkim.

Uses: The drug is constituted by the whole plant.

A warm infusion of the plant causes perspiration. Oil from the plant is stimulant and rubefacient and often used as a liniment. It is given as astimulant and tonic in colic, diarrhoea and hysteria. It is also applied in chronic rheumatism, toothache and earache.

In Kumaon the leaves are boiled with tea and used in influenza and fevers.

Extent of utilization and trade: The drug is exploited on a commercial scale and sold in the market.

163. *Oxalis corniculata* Linn.

Family: Oxalidaceae.

Local names: Amrul, Chálmori, Chalmorá Changéri (H): Titpáti (Kum).

Description: A small, procumbent, prennial herb. Stem rooting, pubescent, with appressed hairs. Leaves palmately trifoliate, petiole long, aciduous in taste; stipules small; leaflets obcordate. Flowers axillary. Petals yellow, long stalked, rounded at apex, twice as long as the calyx; bracts setaceous.

5

Capsule tomentose, cylindric, tipped with the persistent style. Seeds several in each cell.

Distribution: Throughout warmer parts, ascending to 2000 m in shady and damp places, Kashmir to Sikkim. Common on the roadsides and fields.

Uses: The herb is bitter and good appetiser, cures dysentery and diarrhoea, dyspepsia and skin diseases. It is used externally to remove warts and opacities of cornea and taken as *salad*.

The leaves are considered cooling, refrigerant, appetizing and stomachic. The fresh leaves are made into a curry and said to improve the appetite and digestion. With water, they are formed into a poultice and when applied over inflamed part cold is produced thereby reducing the pain. Fresh leaf juice is given to relieve toxication from *Dhatura* and is useful in dysentery and prolapsus of rectum.

In Kumoan the juice of fresh leaves is used for application in cuts and swellings and insect bites.

Also used in anorexia, scurvy, skin diseases, tympanitis and thirst. Leaves boiled in butter milk, are used to cure indigestion and diarrhoea and used externally to remove warts.

Extent of utilization and trade: The plant is used extensively by the local people but little exploited commercially.

Other species: *Oxalis acetosella* L. A perennial stemless herb with creeping, reddish, knotty scaly rhizomes and trifolate leaves with long petiole having pink flowers, veined with purple. It is found in the temperate regions from Kashmir to Sikkim, 2000–4000 m in shady and damp places. The plant is refrigerant and taken as 'salad'. It is antiscorbutic and when infused in milk forms whey. The juice of the plant is used as a refreshing drink in fever. Fresh leaves when beaten up with fine sugar, makes a refreshing and wholesome conserve. With advantage the leaves are employed to scrofulous sores.

164. *Paeonia emodi* Wall. ex Royle

Family: Ranunculaceae.

Local names: Udsaláp(H); Chandárya (TG).

Description: A glabrous perennial herb. Stem erect, leafy. Leaves alternate; leaflets 3, usually 3-parted, segments lanceolate, entire, acute or acuminate. Flowers long-stalked, usually solitary in the axil of upper leaves, buds globose. Sepals rounded, outer with a leafy point. Petals broadly ovate, red or white on a fleshy disc. Follicles ovoid. Seeds few. Flowers, May-June.

Distribution: Throughout temperate regions of west Himalaya, 1500–3000 m.

Uses: The tuberous roots form the drug, are used in medicine and are of two kinds, highly esteemed for uterine diseases, epilepsy, bilious obstructions, convulsions, dropsy and hysteria. They are given with milk as blood purifier and in haemoptysis. The roots are also applied to cuts. The root powder is applied to foul ulcers, to kill maggots. Leaves and young shoots are cooked and eaten as vegetable to cure dysentery.

Infusion of the dried flowers is highly valued as a remedy for diarrhoea. The seeds are emetic and cathartic.

Extent of utilization and trade: Small quantity of the roots are collected and consumed locally.

165. *Papaver somiiferum* L.

Family: Papaveraceae.

Local names: Posta, (the product, Afim) (Hindi); Postdhéri, Khaskahsh, Postdana, Poppy (Eng).

Description and distribution: A herb. Native of Asia, now grown in some districts of eastern UP, Punjab, Rajasthan, Madhya Pradesh under license from the respective Govt., since the latex is the source for opium.

Uses: The fruit and seeds of the plant are used in medicine for stopping diarrhoea. The seeds and the *post doda* are available in open market.

The seeds along with almond, and seeds of Pumpkin are taken in cases of insanity and insomnia. The seeds are ground in water, strained and given to drink. The seeds when boiled in water are given to mothers after the childbirth to relieve gripping pain of the womb, and used for fomentation of pelvis. In case of the pain of testicles, 20 gram of seeds and 30 gram of flowers of *Tesu (Butea monosperma)* or *Dhak* are boiled and the decoction is used for fomentation of the parts.

In case of cough, coryza and cold, three grams of seeds and a little table salt are boiled and given to drink. Three gram, *pepper* and three gram of *papaver seeds* boiled in water are given to drink to relieve intermittent fever. Seeds when ground in water, the paste is applied to the forehead to relieve pain.

The seeds are one of the ingredients of the *thandai* (a cold drink), used during summer season to allay effects of *loo* or hot winds.

The *postdana* or seeds are also used during fast and made into sweets.

Latex, a source of opium, is used to induce sleep, relieve pain and relax spasms. The opium contains many alkaloids, chief being *morphine*, *papverine* and *narcotine*.

Morphine is the powerful analgesic, narcotic and stimulant. *Papverine* has little analgesic action and *narcotine* has very wild narcotic effect.

166. *Phaseolus trilobus* Ait (P. trilobatus (L.) Schreb.

Family: Papilionaceae.

Local names: Wildgram (Eng). Pugáni, Mugvan (Hindi), Janglimoth (Rajasthan).

Description: Herb with trailing stems or twining, up to 1 m, hairy leaflets, nearly glabrous, oblong, more or less deeply 3-lobed. Flowers pale yellow, hardly 75 mm long. Pods glabrous, cylindric, curved; seeds 6–12.

Distribution: West Himalaya region up to 2300 m, in west tropical Asia and Africa.

Uses: Leaves are sedative, tonic and used in cataplasm, for eyes. Its decoction is administered in intermittent fever.

167. *Pedicularis pectinata* Wall. ex Benth.

Family: Scrophulariaceae.

Local names: Mishrán (Hindi).

Description: A glabrous herb. Radical leaves persistent, long stalked, lanceolate, pinnatifid. Stem leaves whorled, stalked, lanceolate, pinnatifid. Flowers pink, spicate. Corolla-tube as long as the calyx, beak small, recurved. Stamens attached at the bottom of the tube; filaments hairy. Capsule ovoid, acute.

Distribution: In temperate regions from Kashmir to Kumaon, 2000–4000 m, in meadows and on moist rocks.

Uses: The plant is diuretic. The pounded leaves are used for haemoptysis.

Extent of utilization and trade: The drug is utilised on a limited scale by the local people and not on a commercial scale.

168. *Peganum harmala* Linn.

Family: Rutaceae.

Local names: Gandhyá, Harmal (H), Isband (Beng), Harmar (Guj), Harmal (Punj), Simagoranti (Tel), Simaiyaravandi (Tam), Wild Rue (E). The trade name *Harmal* is based on the local names.

Description: A shrub, about 30–90 cm. Leaves 5–8 cm long, divided into numerous narrow segments. Lowers 2–3 cm diameter, white, single, in the axils of leaves. Fruit capsular, globose, 5–8 mm diameter, deeply lobed. Seeds 2.5 to 4 mm long, brownish, of various shapes and with reticulated seed coat (Fig. 5.66).

Distribution: The plant is distributed throughout the northern and northwestern India and in drier regions of the Deccan peninsula. In Rajasthan, it has been recorded as a common plant of the gypsum deposits in Bikaner, Barmer and Jaisalmer districts and is considered as an 'indicator plant' for high calcium in the soil.

Uses: The dried seeds of the plant constitute the drug as antispasmodic, stimulant and aphrodisiac. The seeds contain several alkaloids and are useful in asthma, hysteria, rheumatism, gallstones, colic pains, fever, jaundice and complaints of difficult and painful menstruation. They are used also as narcotic, anthelmintic and emetic. Seeds are burnt to ward off ill effect.

The alkaloids *harmaline*, *hageine* and *harmine* are all psychomimetics, i.e. act as hallucinogens. The drug stimulates motor tracts of cerebrum and central nervous system. Higher doses are poisonous and bring about severe depressant actions on the nervous system. Recent experiments have confirmed the *bactericidal* action of the drug.

Extent of utilization and trade: The drug is widely used on a commercial scale

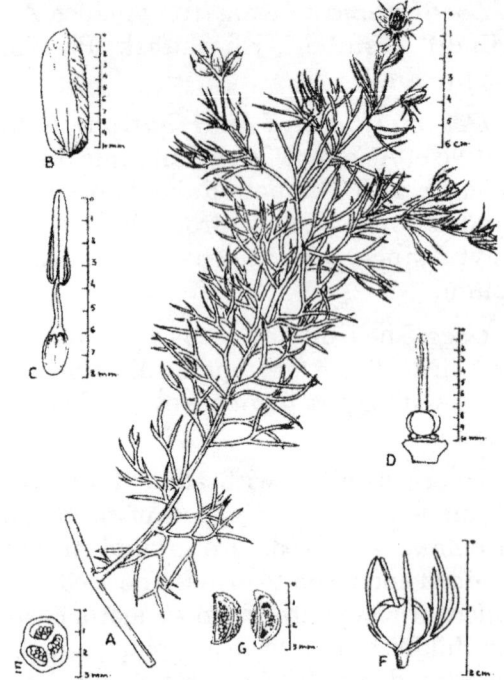

Fig. 5.66: *Peganum harmala* L. A. flowering and fruiting branch; B. petal; C. stamen; D. gynoecium; E. transverse section of ovary; F. capsule and persistent calyx; G. seeds.

and the seeds are sold in the market. The seeds yield a red dye. Powdered roots mixed with mustard oil kill lice in hair. The plant kept in room repels mosquito.

169. *Pergularia daemia* (Forsk.) Chiov. (=P. extensa N. E. Br., Daemia extensa R. Br).

Family: Asclepiadaceae.

Local names: Utran, Utarni, Sadováni (H), Chagulbati (Beng), Utaran (Marh) Chamárdudhi, Nagladudhi (Guj), Karial (Punj), Yugaphalá (S), Uttamani (Tam), Dushtupatige (Tel), Ghol-Lakadi, Kadwadof (Madhya Pradesh).

Description: A twining herb. Stem hairy with milky juice. Leaves 4–6 cm long, some-

times more, ovate or round, hairy on lower surface, deeply cordate at base. Flowers pale or white, small in short clusters. Fruits reflexed in pairs, 5–8 cm long, 1–3 cm broad, covered with spinous outgrowths.

Distribution: Throughout India ascending to 1000 m.

Uses: The entire plant is used in medicine. Juice of the leaves is useful in catarrhal affections and infantile diarrhoea. It forms a constituent of a purgative preparation given in rheumatism and amenorrhoea. It is also useful in curing gynaecological conditions such as excessive bleeding, etc.

Extent of utilization and trade: The active principle of *pergularia* resemble *Pituitrin* in action. The drug is used on a limited scale locally.

170. *Physalis minima* L. var. indica C.B. Clarke

Family: Solanaceae.

Local names: Lakshmirpriya (Sansk); Chirboti (Maharashtra), Tankari (Sansk); Tulatipati (Hindi), Thanmori (Mumbai), Habbikaknaj (Himachal), Kupanti (Telgu).

Description: An annual herb, 30–90 cm with stalked leaves, ovate, sinuately angular, acute. Flowers yellow or blue, single on axillary stalks. Calyx 5-lobed, folding at the angles. Ovary 2-celled. Berry green, globose, about 1.5 m in diam., loosely enclosed in much enlarged inflated, 5-angled calyx.

Distribution: Throughout India, ascending to 2600 m, on waste grounds near the houses. The cape gooseberry *P. peruviana* (locally called *Resbhary*) is cultivated for the ripe fruits which is yellow and sold in the market during winters, which is eaten.

Uses: Fruit is considered as tonic, diuretic and aperient used in gonorrhoea. Leaf juice mixed with water and mustard oil is a local remedy for earache.

Other species: *Physalis minima* var. *indica* C.B. CI. (Vern names: Lakshmi. Priya (Sansk.) Chirboti (Mahr.).

The plant is considered as tonic, diuretic and purgative and is an ingredient of an oil which is given for spleen troubles. It is grown in the garden and is a native of tropical Africa (Fig. 5.67).

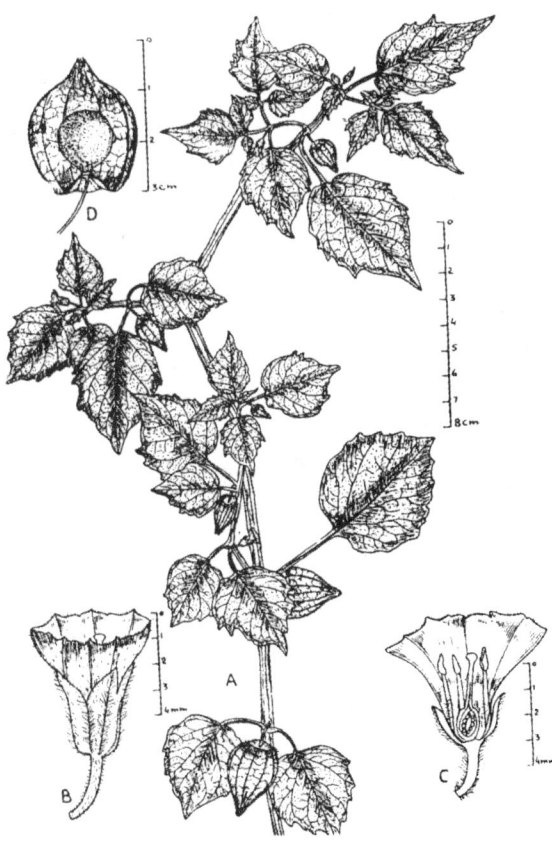

Fig. 5.67: *Physalis minima* L. A. flowering and fruiting branch; B. flower; C. longitudinal section of flower; D. half the calyx cut away to show the berry.

5

171. *Phyllanthus niruri* L.

Family: Euphorbiaceae.

Local names: Bhiamlá (Gujarati), Jar-Amlá (Rajasthan), Bhoniy-mali (Hindi), Bhut-amla (Mumbai).

Description: A much branched, smooth, annual herb. Leaves small, alternate, prominantly arranged in two rows, usually having a fine bloom on the under surface, elliptic obovate, narrow at the base, broad at the apex, size variable, usually less than 1 cm long. Flowers very small, unisexual, both the sexes on the same plant, solitary in the axils. Capsules nearly round having several shallow lobes.

Distribution: A common weed in hotter parts of India from Punjab eastwards to Assam and south to Kerala state.

Uses: The entire plant is of medicinal value, a remedy for dropsy and diseases of genitourinary tract. A bitter neutral substance *phyllathin* has been isolated from this plant. It is diuretic in dropsical affections, and gonorrhoea.

Fresh roots are used as remedy for jaundice, while the leaves are useful as stomachic.

Infusion of young shoots is given in dysentery. Milky juice is given as application to offensive sores.

Powdered roots and leaves after pulverisation are made into a poultice with rice water and used to lessen oedomatous swelling and ulcers.

172. *Physochlaina prealata* (Don) Miers.

Family: Solanaceae.

Local names: Bajarbáng Nándru (Punjab).

Description: A perennial herb with tobacco like leaves.

Distribution: The plant occurs in mass as a weed on the boundaries of cultivated fields, on roadsides. dust heaps near the villages. In Ladakh, and inner dry valleys at about 3000 m.

Uses: The leaves are poisonous and applied to boils. The leaves contain 1% of alkaloids, mainly of *hyoscyamine*, and therefore could form a source of *atropine*. Ladakh and Lahul, being not easily accessible, collection and transport charges make the plant uneconomical. It is being cultivated by the RRL, Jammu. The alkaloid content in the first and second year has been reported to be 0.16–0.24 and 0.74% respectively. It flowers during May and June when leaves can also be collected. The collected leaf material from Lahul, yielded 0.88% alkaloid, calculated as *hyoscyamine*. Smaller collection from the Ladakh area yielded up to 1.02% alkaloid (Chopra 1955). Exploitation of naturally occurring plants can yield up to 100 qtls/acre of leaf material only from Lahul (Sarin. 1967) and if exploited systematically could replace *Saussurea lappa*.

173. *Phytolacca acinosa* Roxb.

Family: Phytolaccaceae.

Local names: Jirrág (Kumaon), Matazor (Hindi).

Description: An erect, glabrous herb. Stem succulent, robust. Leaves alternate, broadly-lanceolate, entire, long pointed, narrowed into a short stalk; stipules none. Flowers pale green, in leaf opposed cylindrical raceme; bracts linear. Perianth of 5-segments. Fruit dark purple, succulent, crowded in an erect raceme; seeds kidney-shaped.

Distribution: It occurs both as wild as well as cultivated plant. An introduced plant from China, often seen as an escape.

Uses: The oil from the roots is used for pains in joints. The plant has poisonous properties but the leaves are cooked and eaten.

Extent of utilization and trade: Locally, the plant is not used in medicine but its power of producing delirium is known. The poisonous properties of the plant are completly destroyed by boiling. A bitter toxie substance *phytolaxa* toxin is present in the plant.

174. *Pimpinella anisum* L.

Family: Umbelliferae or Apiaceae nom alt.

Local names: Shetápushpa (Sansk), Saurif, *Saonf* (Hindi) also name for *Foeniculum vulgare*. Anise (Eng).

Description and distribution: A herb, cultivated in northwest India like UP, Haryana and Punjab also in the south at Orissa. Native of the Mediterranean region.

Uses: The seeds are used to remove flatulance and indigestion. It is reported to increase the flow of urine and milk and is very useful in removing weakness of the eyes; in fever due to disorders of the stomach. The seeds are often taken with sugar after the meals.

Seeds contain essential oil, 90% of which is *anethole* rest being p. methoxypheny lacetone and chavicol.

175. *Piper longum* L.

Family: Piperaceae.

Local names: Longpepper (Eng); Pipar, Piplámul, (roots) Pipli. (fruits) Pimpli, Peeple (Hindi), Pippali (Sansk).

Description and distribution: A creeping or rambling aromatic herb with ovate or orbicular leaves. Dioecious. Male spikes slender, yellow, stamens 2–3. Female spikes up to 3 cm long in fruit. Bracts to male as in female. Stigmas 3–4 cm long, papillose. Fruit a berry. In the warmer states of the country on damp places under shade. Fruit Dec-Jan.

Uses: The fruit is cylindrical, about 2.5 cm long on a climbing herb with granular concretions on its surface, having a bitter taste. It is used in stomachache, increasing appetite and is an aphrodisiac; also useful for the eyes. It is widely used in ayurvedic medicines such as *Churna*. Sifted *Pipli* is taken in phlegmatic cough and asthma with honey. It also forms ingredient of some colyriums useful for night blindness, macula and nebula. Decoction of immature fruit and root is used in chronic bronchitis, cough and cold.

Other species

a. *Piper attenuatum* Macerated roots in water form an excellent diuretic. Distributed in eastern Himalaya and south in the Nilgiris.

b. *Piper aurantiacum* Wall. Vern. name Ranuka (Sansk) Sambha luka (Hindi). In Assam and Nepal. The fruit is bitter and acrid.

c. *Piper betle* L. Támbul (Sansk) Pán (Hindi). Cultivated in hotter and humid regions. Leaf juice makes the common betel leaf, mostly taken with catechu and lime, it is carminative.

Root is used to prevent child bearing. Fruit employed with honey as remedy for cough. Juice of leaves is employed to allay thirst and relieve cerebral congestion.

It contains an essential oil chavicol and enzymes.

d. *Piper chaba* Hunter. Vern. names: Chavika (Sansk); Chab-Hindi, Kankalá

(Mumbai). The fruits are aromatic, used in cough and cold. Cultivated in various parts of the country.

176. *Picrorhiza kurroa* Royle ex Benth.

Family: Scrophulariaceae.

Local names: Kutki, Kadwi, Kédar-Kadwi, Katki, Kuri (Hindi); Katuká(Sansk.), Kalikutki (Mumbai), Karu (Hindi-Punjab).

Description: A perennial herb with elongated, stout, creeping stolon. Leaves almost radical, spathulate, sharply serrate. Flowers white or pale blue purple in a dense terminal spicate raceme. Fruits an ovoid-capsule.

Distribution: In the alpine regions from Kashmir to Sikkim, 3000–5000 m, generally near the springs, on moist rocks and under moist scrubs also under the canopy of *Kharsu* oak (*Quercus semecarpifolia*) and *Rhododendron campanulatum* shrubs.

Uses: The rhizome is a bitter tonic and is a good substitute for *Gentiana kurroa* roots, which is used as antiperiodic and chloragogue (Kapoor and Sarin, 1962). It is also stomachic, purgative in fever and dyspepsia in small doses and cathartic in large doses.

A fairly large quantity of glucoside bitter principle named 'Kutkin', non-bitter product *kurrin, vanilic acid, picrotin*-1 and *Picrotin*-2 are two main active ingredients found in the roots and stolons of *P. kurrooa*.

177. *Podophyllum hexandrum* Royle (-Podophyllum emodi Wall. ex Hook. f and Th. var hexandra (Royle) Chatterji and Mukherji)

Family: Berberidaceae.

Local names: Bakrachimaka, Van-Kákri (Hindi), Bankakri (Himachal Predesh), Paprá (Hindi), Banwangan (Kashmir), Venivél (Gujarat).

Description: A succulent herb with creeping rootstock, 35–60 cm tall with a perennial rhizome. Leaves orbicular, reniform and palmately divided. Flowers white or pink, cup-shaped. Fruit an oblong or elliptic berry, orange or red, with many seeds in the pulp.

Distribution: It grows scattered in *Dangs*, among boulders under the coniferous forests from Kashmir to Nepal, 3000–4200 m; also in Sikkim. It descends to 2000 m in the region of Kashmir in J and K state.

August and September months are best for the collection of roots. In the alpine meadows this plant is sometimes found but the frequency is much less, since it is well distributed in moderately exposed situations, particularly under high altitude conifers and tracts lying between alpine meadows and upper extremity of treeline ranges lying westwards beyond river Sutlej, including Jammu and Kashmir, between 2500–4000 m, and is associated with *Rhododendrons* (high altitude species), *Salix, Juniper* and *Viburnum* species.

Uses: The rhizome and the roots act as hepatic stimulant, chloragogue and purgative. Locally the roots are used as purgative. Ripened fruits are edible.

Podophyllum hexandrum contains not less than 8% resin and 11.52% in one leaf plant and 8.38% in two leaf plants; which is utilised for the preparation of *podophyllum* resin called *podophyllin*. This resin contains several lignins including *podophyllotoxin* from which two drugs, *etoposide* and *teniposide* are synthesized. Both of these analogues are being used in the treatment of lung cancer and other tumours.

Locally the root is considered a chloragogue, purgative, alterative, emetic and a bitter tonic.

Indian plants have much more resin than the American species. The active principle is 10–12% as compared to other samples. *podophyllotoxin* is an antitumour agent.

The demand of this drug, within recent years, has increased manifold. The drug from the Himalaya has not been fully exploited and there is a vast scope for its progressive cultivation and exploitation of naturally occurring plants. The price of the drug is almost about Rs. 60/kg, the drug has a bright future for cultivation as the active principles are reported to have specific anticancerous properties.

178. *Plantago ovata* Forsk.

Family: Plantaginaceae.

Local names: Isabgol, Ispághula, Isbpgul (Hindi); Isadgola (Sansk), Snogdhjeera, Blonde psyllium (Eng).

Description: Herb, almost stemless. Leaves radical, tufted or spreading, simple, prominently ribbed. Scapes nearly erect, naked and furrowed. Flowers numerous, small, green, regular, usually 2-sexual, crowded in terminal cylindric or ovoid spikes. Calyx free, 4-parted, segments nearly equal, usually distinct. Corolla hypogynous, scarious, tube cylindric, about as long as the calyx., limbs 4-lobed. Stamens 4, attached to corolla tube, alternate with lobes; anthers pedulous. Ovary 2-celled, style thread like, divided. Capsule small, opening transversely near the base. Seeds, 2–16, small.

Distribution: Cultivated mostly in Gujarat state and in parts of Rajasthan and Maharashtra.

Uses: The seeds form the mucilage with water. The white husking when removed from the seed is called *Satisabgol* and is sold in the market, both in India and abroad. The mucilage prepared from the seeds when

drunk also relieves heat and thirst due to high fever and in dysentery, when added to curd and allowed to stand for one hour and drunk. It is also useful in cases of dry cough and asthma. Externally, it is applied to whitlow (swelling of finger). Preparation of *isabgol* and *bel* traded as Isabbel in the form of granules by Baidyanath containing *Holarrehna antidysentrica*, *Cyperus scariossus* is patented as a cure for amoebiasis, flatulance, dyspepsia, and bacillary dysentery.

Other species

a. *Plantago amplexicaulis* Cav. Vern. name: Isapaghul, Isabgol. The uses are similar to that of *P. ovata*. Plant has been recorded from Punjab plains and Malwa areas of Haryana (Fig. 5.68).

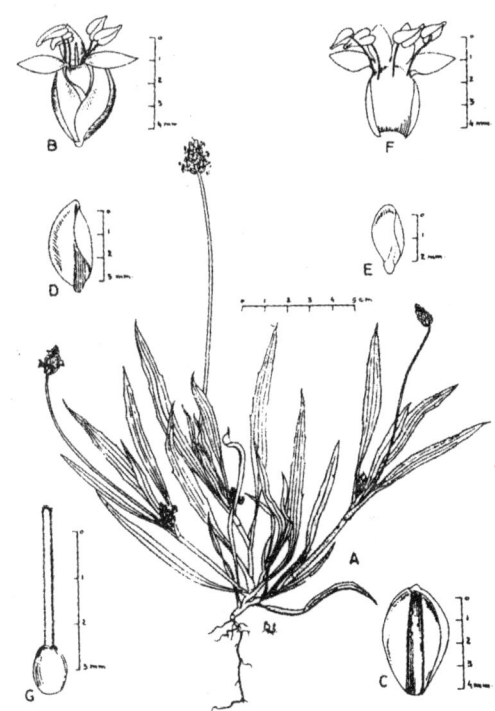

Fig. 5.68: *Plantago amplexicaulis* var. bauphla Pilger. A. plant; B. flower; C. bract; D. outer sepal; E. inner sepal; F. corolla opened to show androecium; G. gynoecium.

b. *Plantago brachyphylla* Roem. and Schult.
The leaves are applied to wounds. Distributed from Kashmir to Kumaon, 3000–4300 m.

c. *Plantago ciliata* Desf. Used as a cure for diarrhoea.

d. *Plantago lanceolata* L. Vern name: Baltangá-Hindi Bartung (Mumbai). Plant with lanceolate leaves which are applied to wounds, sores and inflamed parts. Seeds used with sugar as drastic purgative.

e. *Plantago major* L. Vern name: Lahuria-(Kumaon), Harlang (Mumbai). Seeds are tonic, stimulant and used to cure dysentery; also as substitute to *P. ovata*.

f. *Plantago psyllium* L. Vern name: Fleawort (Eng) In hilly districts throughout India up to 3300 m westwards to Atlantic (Britain). Seeds aperient, used like those of *P. ovata*.

g. *Plantago tibetica* Hook. f. et Thomas. Recorded from Shimla 3100–3300 m on roadsides in Shimla. Seeds used like those of *P. ovata*. Glucoside 'aucubin' is present.

179. *Polygala crotalarioides* Buch.-Ham ex D.Don

Family: Polygalaceae.

Local names: Lil-kanthi (Santhal).

Description: A densely hairy, perennial herb; rootstock woody, often tuberous. Stem short, thick, decumbent, branches long, spreading. Leaves nearly sessile, oblong-ovate. Flowers purple, crowded in axillary racemes. Calyx persistent. Keel of the petals crested. Capsule heart-shaped, fringed.

Distribution: In temperate regions from Chamba to Sikkim, 1300–2500 m. On open grassy hillsides and rock crevices.

Uses: The root of the plant constitutes the drug. The root is chewed to expel the phlegm from the throat, it provokes coughing. The plant is used medicinally in catarrhal affections.

Extent of utilization and trade: The drug is not used on a commercial scale but used locally.

180. *Polygonum plebejum* R, Br. var. brevifolium Hook.f.

Family: Polygonaceae.

Local names: Raniphul (Santhal).

Description: A prostrate herb, branches leafy, finely grooved, smooth, flowering throughout its length. Leaves linear, narrowly obovate; stipules tubular, short, transparent, faintly nerved. Flowers white or pale pink in axillary clusters, half concealed among the stipules. Perianth 4–5 parted. Nut 3-angled, smooth and shining.

Distribution: Throughout India ascending to 2000 m and in the sub-Himalayan tracts.

Uses: The plant is dried, powdered and taken internally in pneumonia. Root is sometimes used in bowel complaints.

Extent of utilization and trade: As an article of commerce it is not much exploited but used locally.

Other species

a. *Polygonum hydropiper* L.

A prostrate herb, rooting at lower joints with lanceolate leaves, glandular dotted and pink or red flowers in slender interrupted racemes. Common throughout the country in wet places, ascending to 2300 m in the hills. The root of this plant is stimulating, bitter and tonic. The dried plant when put in clothes saves them from moth attack. Bruised

leaves are used in place of mustard poultice and are put into mouth to cure toothache.

The leaves contain an essential oil *Oxy-methyl-anthraquinones* having *polygonic* properties, a glycoside which promotes blood coagulation and a *polygonone* containing etherial oil which can lower the blood pressure.

b. *Polygonum amplexicaule* D.Don

Family: Polygonaceae.

Local names: Machran (Lahul); Anjuvar (H); Amli (TG).

Description: A nearly glabrous herb, stem erect, tufted, slender. Leaves few, distant, lower long-stalked, upper stem-clasping, ovate cordate, long pointed, minutely crenate; stipules tubular. Flowers pink or deep red, varying to white, crowded in one or two recent racemes, 5–15 cm long; bracts flat, scarious, glabrous, ovate, acute. Perianth 5-parted. Stamens 8. Styles 3, free, long. Nut 3-angled, smooth.

Distribution: Hooker has reported two varieties and both occur in the Himalaya (in temperate regions) from Murree to Sikkim, 2000–3300 m, on moist shady places in association with *Bergenia ciliata* and some ferns.

Uses: Locally the roots are used in cough (accompanied by blood) and in dysentery. *Unani* hakims prepare 'Sherbat-é-Anjbar', from the roots which is given in acute forms of blood dysentery.

Extent of utilization and trade: Annually about more or less 25 quintals of the root were collected only from Kanatal forest in Tehri Garhwal (Uniyal and Issar 1967) and were sold at Rs 125.00/quintal. Both *P. amplexicaule* and *P. viviparum* are sold as *Anjvar*.

Phenology and time of collection: Flowers from April to June.

181. *Polygonatum verticillatum* Allioni.

Family: Liliaceae.

Local names: Disa (S); Deabrigal, Salam-misri, Devringal (TG).

Description: A glabrous herb; rootstock thick, creeping. Stem erect, sometimes mottled. Leaves in whorls of 4–8, sessile, linear or lanceolate, tips usually acute, sometimes obtuse or slighlty inrolled, lower surface glaucous. Racemes whorled, 2–3 flowered. Perianth white, tinged with green purple. Fruit a berry.

Distribution: Temperate regions from Kashmir to Kumaon, 2100–3300 m, on moist shady places as undergrowth in forests of Fir (*Pieca smethiana*) and Spruce (*Abies pindrow*).

Uses: Locally the rhizomes (root stock) are used as a general tonic and are an ingredient of 'asthavarga'.

Extent of utilization and trade: Very small quantities of the plant are collected and sold in the market.

Phenology and time of collection: The plant flowers in June-July and the best time for collection of roots is August to Sept.

Other species : *Polygonatum sibiricum* Délar (*P. cirrhifolium* Royle).

(Méda-H, Sálam-misri TG). An erect glabrous herb with a creeping tuberous rhizome. Stem terete, climbing by means of the tendril-like tips of the leaves. The rootstock is an ingredient of the *asthavargan* ayurvedic medicine. The plant is found in temperate regions from Shimla eastwards, in moist shady places from 1500–3000 m. The plant is exploited on a limited scale commercially. It flowers from May-June.

182. *Portulaca oleracea* L.

Family: Portulacaceae.

Local names: Kulfá, Khursá, Lunak (Hindi), Loika (Sansk); Purslane (Eng.).

Description: Small prostrate, annual herb, smooth and fleshy; nodes without scales or hairs. Leaves alternate, without stalks, flat, wedge-shaped to linear. Flowers in terminal clusters, yellow, without stalks, Capsules opening transversely. Seeds numerous, kidney-shaped. Fls. all round the year.

Distribution: Throughout India, ascending 1700 m in the Himalaya. Found on old walls, dust heaps and wastelands.

Uses: The herb is a reputed refrigerant and alterative and used as vegetable. It is useful in scurvy and liver diseases; also in diseases of the bladder, kidney and lungs; used in dysuria and haemoptysis. Externally, it is used as dressing over burns, scalds and other skin diseases. Often used as a pot-herb.

The leaves are astringnt, refrigerant, diuretic and emollient. The leaf juice is given in spitting of blood, while half a teaspoonful of the infusion is given in dysuria. Paste of the leaves is applied to temples, as an emollient to allay excessiven heat and pain. It is also applied to the hands and feet when a burning sensation is felt.

The seeds are vermifuge, demulcent, slightly astringent and are used in the same way as the whole herb, used particularly in gripping pains, tenesmus and other painful symptoms in dysentery and mucous diarrhoea; since it has a soothing effect on intestinal mucous membrane. These are effective when given in small doses. Also effective as an anthelmintic and the seeds are used, as vermifuge.

a. *Portulaca quadrifida* L. (Vern. name Bará-Lonia, Chavel-ká-bhaji, Laghulonica, Lonak, Nunisak, Ramghol (Hindi) Upadihy(Sansk). This plant is very much like the other species (*P. oleracea*), however, it differs in the presence of hairs on the nodes. The flowers are solitary, partially sunk in the enlarged end of stalks and surrounded by 4-leaves. Distributed throughout India (Fig. 5.69). The plant is used in the same way as the other species.

b. *Portulaca tuberosa* Roxb. (Vern. Lunuk-Mumbai), jangli-Gajar (Hindi). A plant of the dry districts like in Gujarat and from south Arcot to Kerala (old Travancore). Infusion of the leaves used externally for application in erysipelas.

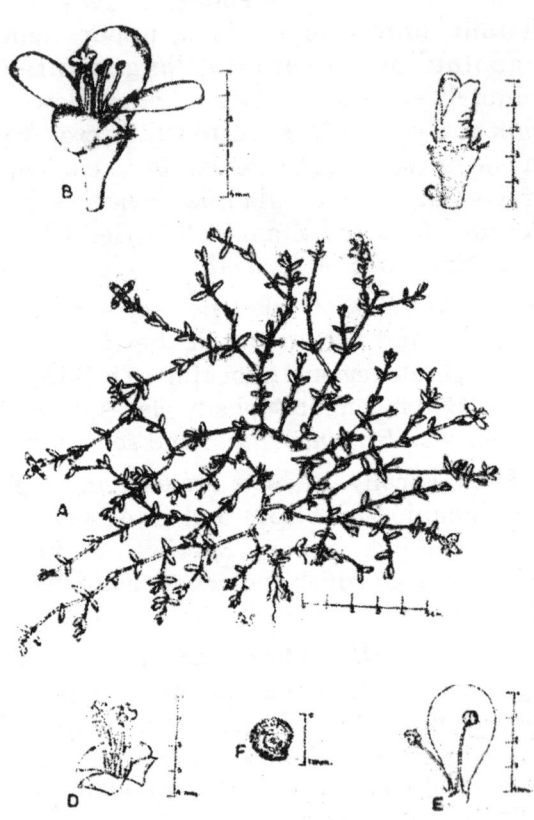

Fig. 5.69: *Portulaca quadrifida* L. A. plant; B. flower; C. calyx; D. petals partly cut away to show androecium and style; E. petal and two stamens; F. seed.

183. *Potentilla fulgens* Wall.

Family: Rosaceae.

Local names: Bajradanti (Hindi, Garhwal and Kumaon).

Description: A perennial herb with robust stem, erect and hairy. Leaves pinnately compound; leaflets numerous, pairs alternately large and small, dimi-nishing in size from the uppermost downwards, ovate, closely and sharply toothed, upper surface green, lower silvery tomentose. Flowers yellow or orange yellow, crowded in terminal corymbs. Petals scarcely longer than the calyx. Achenes glabrous.

Distribution: In temperate regions of the Himalaya, from Himachal-Sikkim 2000–3000 m; now being cultivated on a commercial basis. as a drug in Garhwal Himalaya, Fls. July-October.

Uses: The entire plant is collected, used in scurvy and is widely used in dental diseases, as the name suggests. In Kumaon, rootpowder is used as tooth powder.

Other species

a. *Potentilla nepalensis* Hook. Vern. name. Ratanjot-Hindi. The roots are depurative, and the ashes with oil applied to burns.
   Distributed in west Himalaya from Kashmir to Kumaon, 1700–2700 m.

b. *Potentilla fragarioides* L.
   Leaf infusion is regarded as astringent. Distributed in temperate regions from Kashmir to Bhutan; also in Nigiris in the south.

c. *Potentilla kleiniana* Wight and Arn. (Vern Spangiha-HP, Pinjung-Ladakh, Chinyaphul (Nepal).
   Root and the stem are toxic and locally applied to bites by centipede. The plant is astringent. From Kashmir to Sikkim

in temperate regions, also in east-Himalaya and southern hills like the Nilgiris.

d. *Potentilla reptans* L. Herb used as astringent and febrifuge, while the roots are applied as a lotion. Infusion of plant used for diarrhoea and looseness of bowels.

e. *Potentilla sericea* L. is an astringent while *P. supina* L. roots are used as febrifuge and tonic. Both the species occur in west Himalaya up to 2200 m altitude .

184. *Prangos pabularia* Lindl.

Family: Umbelliferae (nom alt. Apiaceae).

Local names: Kurungás (Kash), Komal (H).

Description: A tall, perennial herb, 1–2 m high. Leaves compound, much dissected into a number of filiform segments. Umbel 20–50 mm across, forming a conspicuous umbrella of small, pale flowers. Fruit 5–7 mm, linear, cup-shaped with undulate, furrowed ridges. Roots up to 40 cm long, when fully mature, tapering downwards into a number of rootlets. The upper portions of the roots representing the subterranean portion are bifurcated into a number of protuberances, each dominating an aerial growth of the preceding year. The whole plant is characterised by a strong spicy odour. The roots have a bitter metallic taste. Fruits August-September

Distribution: On well exposed and dry sandy localities in the Himalaya. It is fairly well distributed in Jammu and Kashmir and Bashahar in Himachal. It is extensively found on the northen slopes of Pirpanjal ranges between an altitudes of 3000–4000 m. The plant grows profusely at Kishtwar, upper Munda (3500 m), Hanjipura in Rajouri and

some parts of Kulgem tehsil and Kamri-pass near Gurez.

Uses: The flower tops and young leaves are employed as an insect repellant, in paddy godowns by local people (Fig. 5.70).

Fig. 5.70: *Prangos pabularia* Lindl A. aerial portions; B. roots; C. fruit.

Fruit of the plant is used as a carminative, stimulant and diuretic in indigenous medicine.

Extent of utilization and trade: Chopra, Handa and Kapoor (1947) reported the presence of 0.65 and 1.02% of essential oil of 1.29 Sp. gr. (at 15° C) in the fruits and roots respectively. The plants of Russian origin yield only 0.26% of oil. The oil was found to contain *myrcene*, *α-pinene*, *camphene*, *borneol* and *dihydrocuminol*. A naturally occurring coumarin *osthol* has been isolated from the roots which is recorded to cause rise in blood pressure and a well-marked stimulation of respiration along with analeptic effect.

Phenology and time of collection: Flowers June, Fruits Aug-Sept.

The fruits should be collected when they are just mature (Aug-Sept), otherwise they are liable to shedding, if allowed to over-ripen on the plant. Umbels are generally cut by hand, dried and thrashed in the sun for two or three days after which these can be safely packed in coarse cloth bags.

Collection of roots is best achieved in October and November when the aerial portions are dry and the root lies in dormant condition. In fields the roots can be easily located by a trained eye by searching for the dried up remains of the stem and also the leaf bases, which form a dense hairy undergrowth around the upper portions. Preferably older roots should be extracted, they can be judged by a large number of protuberances on the upper portions. The roots take about 12–15 days to dry in the sun when about 70–80% moisture is removed. It facilitates drying, if roots are sliced into small pieces.

185. *Psoralea corylifolia* L.

Family: Papilionaceae.

Local names: Bavchi, Babchi, Babéhi (Hindi); Sugandh-Kantak, Bakuchi (Sansk), Lata-Kasturi (Bengal), Bavachi (Guj), Bakuchi (Urdu), Karpokarishi (Tamil), Kusht Nashi-Ayurvedic. Trade name Bawchi and Psoralea are based on vernacular and Scientific names of the plant Kalagija (Telgu.)

Description: An erect herb, branches densely gland dotted leaves round, dotted with black glands on both the surfaces. Fls. small, bluish purple, 10–30 in a bunch in axil of leaves. Fruits black, roundish or

oblong, closely pitted, seed one smooth black, longish round, yields a white Kernel when shelled for internal administration bitter in taste (Fig. 5.71).

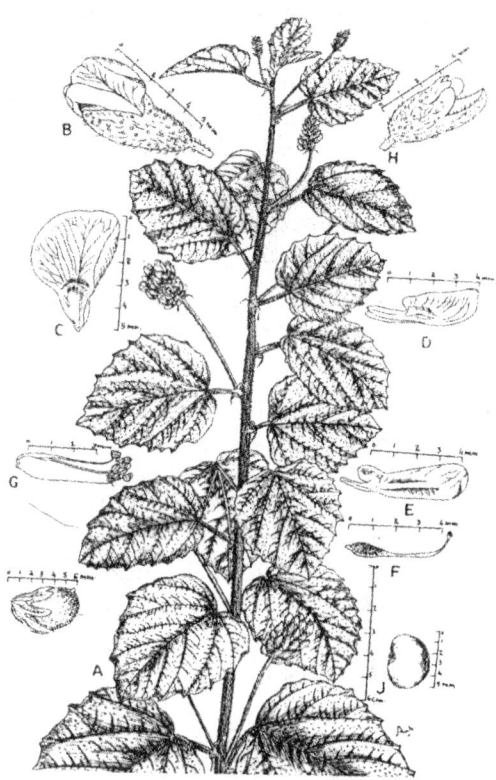

Fig. 5.71: *Psoralea corylifolia* L. A. upper portion of a flowering and fruiting stem; B. flower; C. standard; D. keel; E. wing; F. gynoecium; G. androecium; H. calyx; I. pod; J. seed.

Distribution: Throughout India, a weed of wastelands; sometimes also cultivated.

Uses: The seeds constitute the drug for leukoderma, both for external application and internal consumption. For external application equal weight of the seeds and sulphur (*amalsar gandak'*) are finely ground along with tamarind seeds, which have been steeped for 3–4 days in water, then shelled and crushed. The mixture is applied to white spots which cause itching. It could be repeatedly applied, till the spots become red and liquid starts flowing from them. If the spots do not get redened, *bavchi*, steeped in water overnight, the supernatant is taken in the morning. The seed quantity is increased gradually, till it comes to about 10 grams.

The seeds steeped for a week in ginger juice, changing the liquid daily, are washed and shell removed. Dried in shade, the seeds are taken daily morning (dose of one gram) with water.

The seeds contain an essential oil which is effective on certain bacteria causing skin diseases, like leukoderma and leprosy. Internally the seeds are taken for promoting urination and are anthelmintic.

The oleoresinous extract of the seeds is useful for local application on leukoderma of nonsyphilitic origin.

Plant roots are reported to be useful in caries of teeth and the leaves in diarrhoea.

Extent of utilization: The seeds are used widely in the indigenous system of medicine and are commercially exploited; available in local market.

186. *Ranunculus arvensis* Linn.

Family: Ranunculaceae.

Local names: Butter cup plant (E), Chámbul (Himachal).

Description: An erect annual herb. Stem much branched, hairy above, otherwise-glabrous. Radical leaves long-stalked, wedge-shaped, 3–5 toothed at the tip; cauline leaves short stalked, deeply divided into 3-linear segments. Flowers pale yellow. Sepals adpressed to half the length of petals. Achenes in globose heads, clothed with hooked bristles.

Distribution: In western Himalaya from Kashmir to Kumaon, 1300–2300 m in corn fields, also in moist places.

Uses: In Europe the plant is used in intermittent fevers, gout and asthma but not in India.

Extent of utilization and trade: The plant is not utilised medicinally in India.

Other species

a. *Rauncupus sclcratus* L Vern. name. Shim (Kumaon).

The plant is poisonous and is vasicent, applied to skin to raise blisters. It is poisonous and galactagogue. Contains *anemonin* and 'protoanemonin' considered as the most deadly poison.

b. *Ranunculus aquatilis* Linn. var. capillaceus DC.

In west Himalaya from Kashmir to Kumaon. The plant is used for intermittent fevers, rheumatism and asthma in Europe.

c. *Ranunculus arvensis* Linn. Vern. name. Chambul. (Himachal).

The plant like the above species, is used in Europe for intermittent fevers and in gout. In western Himalaya from Kashmir to Kumon, in Himachal and Rajasthan (Mt. Abu). The leaves contain HCN. Plant contains 'anemonin' protoanemonin. There are other species like *R. falcatus* L. *R. lingua*, L. *R. muricatus* L. and *R. pensylvaticus* L. used medicinally.

187. *Raphanus sativus* L.

Family: Crucifereae, nom alt. Brassicaceae.

Local names: Raddish (Eng.) Muli, Mulo, Muro (Hindi).

Description: An annual herb with pinnate or very deeply-lobed leaves, end leaflet lobe very broad; leaf stalks arising directly from the thick root, spreading in the rosette. Tap root very long, coarse, thick,

white. Flowers in terminal recemes, short at first, much elongated in fruit stage. Flowers white or liliac, with purple veins. Pods not opening when mature, about 4 cm long, columnar, tapering at the apex to a pointed beak, constricted between the seeds.

Distribution: It is widely cultivated throughout India, off season raddish is cultivated in the hilly areas like near Mussoorie which is sold on a premium price.

Uses: Almost every part of the plant is used medicinally, besides being a vegetable.

The leaves are diuretic, antiscorbutic, laxative and lithotriptic. The juice is given in dysuria and some cases of calculus and strangury.

The roots are administered in urinary and syphilitic diseases, piles, stomachic, strangury, dysuria and cases of calculus of the bladder. They are also used as powerful antiscorbutic and administred in small doses and decoction of dried root in 1–2 ounce. The roots are also given in the form of syrup which is effective for treatment of hoarseness, whooping cough, bronchial difficulty of breathing and other chest complaints.

The seeds are peptic, expectorant, diuretic, laxative, lithotriptic, carminative, emmenagogue and stimulant.

Roots also used in urinary complaints, piles and gastrodynic pain.

188. *Rauwolfia serpentina* Benth.

Family: Apocynaceae.

Local names: Chhotáchand, Harkáya, (Hindi), Sarpgandhá (Sansk), Chandrá (Mumbai), Serpentine, Serpentwood (Eng.).

Description: A small herbaceous plant, 15–20 cm or more in height, having a white bark. Leaves in whorls of 3, thin, almost lance-shaped or obovate, oblique, lower end tapering into a short stalk, green when dry,

pale beneath. Inflorescence long stalked, many fld. cyme. Flowers white or pinkish, nearly 3 cm long, with very small stalk, bright red, corolla tube long, dilated, a little above the middle. Drupes obliquely ovoid, purplish black, about 1 cm long.

Distribution: In habit the moist and hot regions of India; abundantly found in Bengal (24 Parganas and Howrah); however, it has almost disappeared from the region of Dehradun, because of excessive exploitation from the natural population.

Uses: The root is bitter in taste and used medicinally. It contains alkaloids of which *ajmaline, serpentine* and *raulfine* are the most important. The roots have febrifuge and emmenagogue properties and have the reputation of acting on the uterus, are used to hasten the expulsion of the foetus. It has marked sedative and hypnotic action and thus a valuable remedy for insomnia, hypochondriasis, insanity, irritable condition of the central nervous system and high blood pressure.

In the northern parts of the country, the roots are so highly valued as a cure for insanity that it started getting locally known as *Pagal-ki-dawa.* Root is also used in intestinal troubles such as diarrhoea and dysentery.

It is also effective in cases of hysteria and epilepsy. Since it induces sleep, it is taken before going to bed.

Bihar variety of roots yields 0.8–1.3% total alkaloids, while Dehradun variety of roots yields 1.0–1.3% of total alkaloids.

Other species: *Rauwolfia canescens* L. The plant grows side by side the above species, inhibiting the same environmental and soil conditions and is found abundantly at many places. The alkaloid 'rauwolscine' in the root bark is 0.1%, while in stem bark it is reported 0.2% and also 0.2% in the leaves.

189. *Reinwardtia trigyna* Planch.
(-R.indica Dum)

Family: Linaceae.

Local names: Bál-basant, Kárkun (Himachal Pradesh); Piunli (Garhwal), Basant (DDN) The same name is used for *Hypericum cernuum.*

Description: A glabrous plant, stem about 30–60 cm, erect or ascending. Leaves entire, ovate-lanceolate, 5–10 cm, narrowed to a slender stalk, minutely mucronate, lower surface pale. Flowers axillary-solitary or in small clusters. Sepals lanceolate, acute. Petals primrose yellow, much longer than the calyx. Stamens usually in two sets, 3 long and 2 short. Styles usually 3, shorter than the stamens, more or less united. Capsule globose separating into as many valves as there are styles (Fig. 5.72).

Fig. 5.72: *Reinwardtia indica* Dumort. A. upper part of flowering stem; B. calyx and corolla removed to show androecium; C. gynoecium; D. calyx; E. transverse section of ovary.

Distribution: In the Himalaya, from Indus eastwards to Sikkim up to 2000 m; also in the hills of Rajasthan (Mt. Abu), western ghat forests and in Karnataka, perhaps the first plant to flower in spring and very prominent.

Uses: Crushed flowers are mixed with mustard oil and applied to wounds and boils.

Also used as medicine for 'Founder 2' in cattle.

190. *Rheum emodi* Wall. ex Missn.

Family: Polygonaceae.

Local names: Revandchini (Hindi); Gandhni, Archa (Garh.), Ladakhi revandchini (Mumbai), Rewandchini (Punj); Archoo (Lahul), Archa (T. Garhwal) Trade Name-Rewatchini.

Description: A perennial stout herb, 1.0–3.0 m in height with very stout roots. Leaves radical, long petioled, very large, 30–40 cm. Flowers small, dark purple or pale-red in tall axillary panicles.

Distribution: Alpine and subalpine regions of the Himalaya in Gharwal reported between 2800–3600 m altitude; also distributed in temperate and subtropical regions. Drought resistant, on partial sunny sites sands and porous soil.

Uses: *Rhubarb* is a mild anthraquinone purgative, having properties similar to that of *senna*. However, rhubarb may exert an astringent action after purgation. *Emodin, rutin, chrysophanol* and *chrysoohenic acid* are the four chief active constituents found in this species. Out of which *chrysophanol* is in higher concentration. It is used as an astringent tonic, its stimulating effect combined with aperient properties, renders it useful in dyspepsia. The powder is also sprinkled over ulcer for healing. Powder of the roots is also used in cleaning teeth. Cost of the drug is about Rs. 80/kg.

Commercial rhubarb is, however, obtained from *R. officinalis* and *R. palmatum* which grow in southwest Tibet and northwest China. The most useful part of the plant is decorticated as dried root or rootstock. Himalayan rhubarb is, however, of little commercial value and is of inferior grade; however considerable quantities are do collected and sold. In addition to its use in Indian medicine, it is employed for the production of certain *anthraquinone* compounds employed as purgatives. It is normally marketed by Kutch Grower's cooperative and not much concerted efforts have been made to collect plant from different locations, except some in Lahul.

191. *Roscoea procera* Wall. (R. purpurea var. procera Wall.)

Family: Scitaminaceae.

Local names: Saféd Musli (TG).

Description: An erect herb. Stem robust, leafy. Flowers liliac, several faintly streaked and tinged with pink. Corolla tube longer than calyx, margin inflexed, forming narrow, flattened, pointed hood; lateral lobes linear-lanceolate, lower obscurely 3-lobed, spreading, notched at the top.

Distribution: In temperate regions from Shimla eastwards, ascending to 3500 m in Kumaon and Sikkim region.

Uses: The roots are used as tonic.

Extent of utilization and trade: Small quantities of the plant roots are collected for sale in the market as *Kakoli*.

Phenology and time of collection: Fls. July-August. The roots are collected from September to October and dried in the sun before being marketed.

192. *Rubia manjith* Roxb. ex Fleming (Rubia cordifolia sensu Hook.f. FBI, R. cordifolia var. munjistha.

Family: Rubiaceae.

Local names: Manjistha (S), Majith (H), Leeckbro (TG). The trade name Majith is based on the Hindi name.

Description: A climbing perennial herb. Stem and branches elongate, ridged, 4-angled, angles minutely prickly. Leaves in whorls of 4, long stalked, cordate-ovate, nerves and margins prickly; stipules none. Flowers dark red in cymes forming large bracteate panicles. Fruit succulent, small, globose. The plant is variable since 2 forms: *cordifolia* proper with 5 leaves rearly 3 costate, veins impressed and *Khasiana* with 3 leaves, rarely 5 costate, veins not impressed Kurz.

Distribution: Hilly districts of India, ascending to 2500 m.

Uses: The root forms the drug. The root is used internally in menorrhagia and has a bitter taste. It is laxative, analgesic, lactagogue and used in eye sore, paralysis, lethargy, liver complaints, enlargement of spleen, pain in joints, rheumatism, leukorrhoea, leukoderma, dysentery, uterine pains and for increasing appetite.

The fruit cures diseases of the spleen.

Decoction of the leaf is used for pleurisy and other inflammatory conditions of the chest.

Extent of utilization and trade: Very small quantities of the plant are collected and sold in the market.

Phenology and time of collection: Fls. July-Oct. The roots are collected during the months of Sept-Oct. and dried in sun before marketing.

Other Species

Rubia tinctorium: The European madder, is cultivated in Kashmir and Afghanistan, distributed westwards from Persia to Spain. De Candolle, considered its original habitat to be west temperate Asia and south east Europe. *R. sikkimensis* is a handsome creeper in east Himalaya and is a source of brilliant red dye. Formerly *R. munjith* was also much employed for extract in dyeing cotton fabrics to various shades of scarlet.

193. *Rumex hastatus* D. Don

Family: Polygonaceae.

Local names: Bhilmorá (Kumaon).

Description: An erect herb with slender stem and broadly triangular leaves, long pointed or 3-lobed. Flowers polygamous, in small whorls, racemed and forming panicles, often dense in fruit. Fruiting sepals orbicular, not fringed, notched at both ends.

Distribution: In the sub-Himalayan tracts from Kashmir to Kumaon, 600–2500 m, on calcareous soils near roadsides, often eroding, leading to small slides.

Uses: The crushed leaves are applied to boils and a sauce is prepared from the leaves, that are acidic.

The leaves are used to clean copper utensils in villages, also used as condiment.

Other species

a. *Rumex acetosa* L. Vern. name: Dock Sorrel (Eng.), Chukaa-Khatta palak (Hindi). The plant is used in scurvy, while the leaves are refrigerant, used as a cooling drink in febrile diseases in Europe.

b. *Rumex acetosella* L. Vern. name: Chutrika (Sansk); Chukhapalam. (Bihar).
Fresh plant juice is refrigerant, diaphoretic and diuretic; used in Europe. Fruits used as poultry feed. Occurs in East Himalaya, Sikkim and Nilgiris.

c. *Rumex crispus* L. Vern. name Chukh (Hindi), Yellow Dock, Curled Dock (Eng). Amla-betasa (Sansk.). Perennial herb, though native of Europe but recorded from Mt. Abu. Leaves used as vegetable, while these roots are used in homoeopathy, mildly laxative and astringent.

d. *Rumex dentatus* L. Hindi-Lalbibi, Amrule, Ambaráh (Hindi), Changéri (Sansk). NW Himalaya, Kumaon and in south India, Maharashtra.
Roots are the source of red dye, also used for cutaneous disorders. The dye is used by local residents in Kumaon for dyeing wool for *Shawls* (Fig. 5.73).

e. *Rumex maritimus* L. Vern. name, Golden Dock (Eng.) Jag-palum, Jangli-pálak (Hindi).

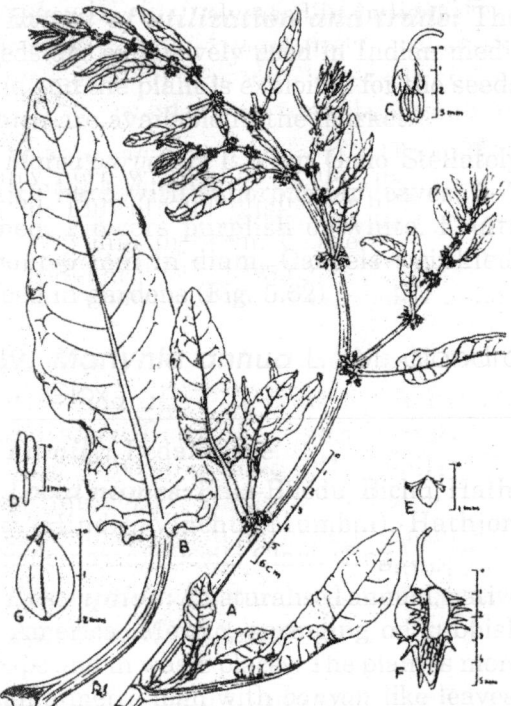

Fig. 5.73: *Rumex dentatus* L. A. flowering and fruiting branch; B. lower leaf; C. flower; D. stamen; E. gynoecium; F. fruiting perianth; G. nut.

The plant is cooling and the leaves are cathartic, applied to burns. Seeds are aphrodisiac. In temperate east Himalaya and the Nigiris in south.

f. *Rumex nepalensis* Spreng.
In temperate regions of Himalaya, 1300–4000 m. The roots are purgative and used as substitute for *rhubarb*. Also used there for venereal diseases.

g. *Rumex patientia* L. (*Rumex orientalis* Bernh). Patience Dock (Eng).
Though native to Europe but found in west Himalaya. The roots are edible, used as purgative.

h. *Rumex scutatus* L. Vern. name: Ambávati (Hindi), French Sorrel (Eng). The herb is refrigerant, astringent and given in diarrhoea, also in dysentery. Also used for salád, soups and sauces.

i. *Rumex vesicarius* L. Vern. name. Chuká, Ambar. Palak, Palagsák (Hindi). An annual, native to southwest Asia and North Africa, also in Punjab and trans-Indus hills in Pakistan. Sometimes cultivated.
Leaves are cooling, diuretic and astringent. Seeds are roasted and given in dysentery.
Leaf juice is cooling, check nausea and given in stomach diseases.

194. *Salsola Kalii* L.

Family: Chenopodiaceae.

Local names: Sajji (Rajashthan); Glasswort, Russian thistle (Eng).

Description: A diffusely, branched, prickly herb or spreading bush. Distributed in the Northwest Himalaya including Kashmir.

Uses: The plant is poisonous and is a vermicide, remedy for worms.

Other species: *Salsola foetida* Del. ex spreng. Vern. Motilane, Loonuk (Hindi). A herb,

used as vermifuge. Found in upper Gangetic and Punjab, Haryana plains.

Valued as a source of Sajji and a valuable fodder for the cammels in arid regions.

195. *Salvia aegyptiaca* L.

Family: Labiatae.

Local names: Tukham-malangá (Hindi).

Description: A low, much-branched, straggling undershrub; branches divaricate, quadrangular. Leaves linear-lanceolate, nerves and veins prominent. Flowers in long racemes, whorls of 2–4 fls. Calyx glandular, hairy in fruit, nutlets oblong, ellipsoid, bluish black.

Distribution: Common in streambeds, along ravines, on open grasslands and on gravelly grounds.

Uses: The plant is a heart stimulant and the seeds are used to cure diarrhoea (being demulcent) and haemorrhoids. They produce plentiful mucilage and are mixed with the seeds of *Plantago ovata* (Isabgol).

Other species

a. *Salvia officinalis* L. (Sage-Eng; Salbia sefákuss-Hindi). Leaves and small tops yield an essential oil, used in perfumes deodorant in insecticidal preparations and as carminative.
b. *Salvia moorcrofliana* Wall. (Vern. Kali-Jari Hindi).
   Leaves, roots and seeds are medicinal. Root given in cold and cough, seeds given for haemorrhoids, dysentery while leaves for guinea worms and itch in the form of poultice. Distributed from Kashmir to Kumaon, 2000–3000 m.
c. *Salvia plebeia* R.Br. The seeds are used in diarrhoea, gonorrhoea, menorrhagia and in sexual disorders. Throughout India.

196. *Saussurea candicans* C.B.Clarke.

Family: Compositae.

Local names: Káli-zari, Batulá (Punjab, Himachal).

Description: An erect herb, cottony, often branching near the top. Radical leaves oblong or obovate, narrowed into a short stalk, pinnately lobed near the base, upper surface rough, lower white tomentose. Stem leaves usually few, lanceolate, usually smaller than the radical leaves. Flower heads long-stalked, solitary or corymbose. Involucral bracts rigid, sharp. Flowers pale red.

Distribution: Kashmir to Butan, 700–2300 m. Flowers April-June.

Uses: The seeds are carminative and considered as a cure for horsebite.

Other species

a. *Saussurea affinis* Spreng. Vern. Gangamula (Assam).
   Root juice along with other drugs given to women as a medicine.
b. *Saussurea costus* (Falc.) Lipsch. (Saussurea lappa C.B. Clarke). Vern. name: Costus (Eng). Kuth (Hindi), Kashmir, Kustam (Tel.), Goshtam (Tam.), Kutki (Garhwal).
   An erect robust herb, 1–2 m high with stout stem. Leaves radical with long lobately winged stalk. Roots stout, carrot like, 40–60 cm long, possessing a charcteristic penetrating odour.
   It is found between 2500–3000 m and has been successfully cultivated in seminatural conditions since long; at least more than 70 years in Tehri state by the erstwhile Maharaja of Tehri Garhwal.
   Roots containing alkaloid 'saussurine' and an oil used medicinally for aromatic properties being tonic,

astringent antiseptic, aphrodisiac, depurative, prophylactic, sedative, stomachic, carminative, stimulant, in asthma, cough and cholera and as alterative in chronic skin diseases and rheumatism. It contains resinoids, essential oil, alkaloid, insulin, a fixed oil and other minor constituents like tannin and sugars. The oil 1% is pale yellow to brownish, viscous. The essential oil is antiseptic, disinfectant (used for pro-tecting woollens). It is carminative, emmenagogue against cough, asthma, swelling and blockage and irregular mensus, pulmonary disorders, difficulty in wallowing and wasting of muscle tissues. Leaves contain an alkaloid *saussurine* and *taraxasyerol*.

c. *Sausssurea obvallata* Wall. Vern. Braham Kamal (Hindi). In western Himalaya at high altitudes, 3000–5000 m (Colour Plate 5.16). It is worshipped to Lord Shiva and given as *prasad* at Kedarnath temple in Garhwal. The roots are applied to bruises.

As a stomachic and tonic it is given in advanced stages of typhus fever, (given as an infusion).

Externally powder of the root is used in treatment of ulcers and skin diseases.

Mixed with mustard oil, root powder is applied to scalp and hairwash and said to have the property of turning grey hairs to black.

The roots contain an alkaloid *Sanssurine* and an oil having disinfectant properties, especially against *streptococcus* and *staphylococcus*.

197. *Scilla indica* **Baker** (Urginea indica (Roxb.) Kunth.

Family: Liliaceae.

Local names: Bhuikandá, Jangli-piaz, Suphaidikhus, Banpiazi (Hindi); White Squill, Indian drug squill, Sea onion (Eng.).

Description: A small bulbous herb; bulb 1.5 cm in dia. or more. Leaves radical. Scapes 30–45 cm, erect; bracts soon disappearing; stalks 3–5 cm, slender. Flowers appearing before the leaves, drooping or spreading, distant in a terminal raceme 15–30 cm, long. Perianth bell-shaped, 6-parted; segments white, with 3 green ribs in the centre, tip rounded. Stamens 6, at the base of segments. Ovary 3-celled, 3 grooved; ovules several in each cell. Capsule 3-valved; seeds black, many, flat.

Distribution: In western Himalaya, Bihar, Konkan and along the Coromandel coast.

Uses: The bulbs are source of raticide. The drug is used as an expectorant, stimulant and cardiotonic in small doses. Squill is used chiefly in chronic bronchitis and asthma; also employed in dropsy, rheumatism and skin troubles. Alcoholic abstracts of the bulb possess anticancer activity against human epidermoid carcinoma.

Young bulbs are planted on ridges, about 2–4 cm below at the surface just before the monsoon in the same way as the onions are cultivated. Bulbs for drug use need to be of minimum size, neither too young nor too old.

The bulbs are oblong and imbricated, while those of *Urginea* are tunicated like onion. The *Urginea* is a plant of the drier hills in lower Himalaya and is an efficient substitute for genuine squill. The other more general substitutes are *Crinum asiaticum, C. latifolium, Dicadi unicolor* and *Pancratium triflorum*. In all the cases the bulbs are expectorant, cardiac tonic and diuretic.

198. *Selinum wallichianum* (DC) Raizada and Saxena (-S.tenuifilium Wall.ex DC).

Family: Umbelliferae or Apiaceae.

Local names: Bhutkéshi (Hindi).

Description: A tall herb, perennial, glabrous; stems erect, finely grooved, hollow. Leaves pinnately divided. Umbels compound, pubescent, long stalked. Bracts 1–8 linear, toothed near the tip, usually fallen off in fruit. Flowers white, polygamous, in umbel. Fruits glabrous, flattened up, ridges all winged. Wings of the lateral side broader than those of closely contagious, dorsal and intermediate ridges.

Distribution: In Himalaya, from Kashmir to Nepal. Fls. August-September; seeds ripen in October and shed when overripe. Roots are best collected in October, since the extraction is easy because of its habitat (in well-irrigated loamy soil).

Uses: The roots are used as an ingredient of incenses.

The smoke of the root is germicidal. The roots yield an essential oil (0.8%) of sweet odour which is hypotensive, sedative and analgesic.

Drying of the roots is done in open when 50% of weight remain, however, it is advisable to expose the roots to sun for a day to ensure against attack by mildew.

199. *Selinum vaginatum* C.B.Clarke

Family: Umbelliferae.

Local names: Also called Bhutkéshi, as the above species

Description: Perennial herb like the above species but the stems 60–120 cm. Leaves 1–2 pinnate; leaflets lanceolate, pinnatifd; segments sharply and irregularly toothed, often lobed fruits oblong.

In west Himalaya, 2000–4000 m, from Kashmir to Kumaon.

The roots which possess sweet and musky odour, are used as incense. These are nervine sedative. The smoke is germicidal.

200. *Seseli indicum* Wight and Arn.

Family: Umbelliferae.

Local names: Vanayamani (Sansk), Vanajowain, Kirminji ajowan (Mumbai).

Description: A small annual herb with villous stem. Leaves minutely pubescent, lower 2-pinnate. Fls. white. Fruit ovate, acute, villous. Seeds dorsally compressed (Fig. 5.74).

Fig. 5.74: *Seseli sibiricum* Benth. A. aerial portions; B. roots; C. fruit.

Distribution: In the sub-Himalayan tracts from Dehradun to Gorakhpur, Bundelkhand, Assam, Bengal and Coromandal coast.

Uses: The seeds are stimulant. carminative, stomachic, and anthelmintic, used as medicine for cattle to expel roundworms.

The fruits contain 1.3% neutral saturated acetone and 0.6% of a compound 'furocoumarin'.

Other species: *Seseli sibiricum* Benth. Bhutkéshi-Kashmir.

The roots are indigenously used in blending beverages and as medicine for the live stock.

Oil from the roots cause fall of blood pressure, vascular constriction and stimulation of respiration. The action appears to be tranquilising in nature.

The fruits contain a neutral unsaturated lactone and a compound of 'furocoumarin'. Large quantity of roots have been collected from JK (lower PirPanjal ranges, Gulmarg and Kokarnag areas).

The plant flowers in August, fruit ripens in mid-October and tend to shed leaves in Nov. when mature. These are dried in sun after cutting the umbel. The roots are also collected by the end of October, when aerial portion starts drying. The collection is a tedious job due to plant habitat which is the rock crevices. The seeds are dried in open sun for about a week with 50% moisture drying up.

201. *Sida cordifolia* Linn.

Family: Malvaceae.

Local names: Kharénti, Kungi (H), Jayanti (S), Bálá (Beng), Baladana (Guj), Chikáná (Mah), Badiananla (Orissa), Katturam (Mal), Arival manaippundu (Tam), Chirubenda (Tel).

The trade name *Bala* and *Sida* are based on the Hindi names and botanical name of the plant.

Description: A small shrub, much branched, star-shaped hairs present all over the body of the plant. Leaves ovate or roundish, thick, margins toothed, petiole shorter than leaves, base cordate (heart-shaped). Flowers small, yellow, one or few together. Fruit divided into 7–10 parts, each strongly reticulated, and with two spiny projections on the tip.

Distribution: Throughout India as a common weed on roadsides, waste places and open scrub forests.

Uses: The entire plant is used in medicine.

It is used as a general tonic and for improving sexual strength. Seeds are aphrodisiac and administered in gonorrhoea. Decoction of the root with ginger is considered useful in certain fevers. Roots are also given in urinary disorders and blood diseases. Powder of the root bark (with milk and sugar or singly) is given in diseases of women such as leukorrhoea, and in nervous diseases. Root juice is used for promoting healing of wounds. The bark of the root with sesamum oil and milk is efficaceous in curing certain types of facial paralysis.

Plant juice with water is effective in spermatorrhoea .

The seeds are used in cystitis, rheumatism and spermatorrhoea and colic pains. These contain 0.085% ephedrine.

Other species

a. *Sida acuta* Burm.f. (Banméthi-H; Bala-S). An undershrub. The roots are useful in nervous and urinary diseases, fevers and stomach complaint. The leaves are also considred to have medicinal properties.

b. *Sida rhombifolia* L. (Mahabala, Svetbarela-Banméth. H, Atibala-Sansk.) Sheriti, Lalhariala Sadev, is an undershrub. The plant is useful in rheumatic pains, pulmonary tuberculosis as a demulcent. The herb grows wild in the fields during summer and the rainy season. The leaves resemble the *tulsi* leaves, flowers small. The herb is used as blood purifier, relieves fever due to bilious and sanguine humours, when ground in water, strained and given to drink. It induces perspiration and lowers fever. It is also given in cases of vomiting blood, which is stopped by drinking it ground in water and strained. In chronic fevers the root, about 10 gram, is boiled with dry ginger (3 gram) in water and drunk. It is also useful in cases of cough and clears hoarseness, when used with kolanjan (Alpinia sp), glyccyrrhiza (mulhati) roots as powder and taken with honey.

c. *Sida spinosa* L. (Gulsakri-H; Nagbálá-S). The roots and the root bark are used as demulcent in irritations of the bladder, in gonorrhoea, fever and as tonic. Leaves of the plant also have similar properties.

d. *Sida grewoides* Guill. (Vern. Dábi) Grounded seeds mixed with jaggary are used to cure lumbago.

Extent of utilization and trade: The plant and the seeds are used in Indian system of medicine and exploited on a limited scale.

202. *Siegesbeckia orientalis* Linn.

Family: Compositae (Asteraceae).

Local names: Lichkura (Hindi).

Description: An erect herb, clothed with crisped hairs; branches opposite, spreading. Leaves opposite, ovate or broadly triangular, coarsely toothed. Heads radiate in leafy panicles. Flowers yellow, sometimes white; pappus one. Involucral bracts in 2 series. Achenes curved, angled, blunt at the base.

Distribution: Throughout India, ascending to 2000 m. A cosmopolitan weed of the warm climate.

Uses: The plant is valuable as a depurative and for healing properties in gangrenous ulcers and sores. Externally, a mixture of equal parts of tincture and glycerine is applied with success in ringworm. Antiseptic properties of the plant have been ascribed only to fresh plant, when applied to unhealthy sores.

Extent of utilization and trade: Not exploited on a commercial scale. It contains *Darutine*, a derivative of salicylic acid. In China (called Kau-Kau) it is known as a remedy for agnea, rheumatism and renal colic. Tincture is a remedy in scrofulous and syphilitic affections. Externally with glycerine it has good effect on ringworm and other parasitic affections.

203. *Sisymbrium irio* L.

Family: Cruciferae or Brassicaceae.

Local names: Khubkalán (Hindi).

Description: A herb, annual, with erect stems. Leaves simple. Flowers white in racemes. Sepals erect, bases equal. Petals clawed. Stigma nearly sessile. Pods linear, seed in one or two in the pod.

Distribution: A weed during the *kharif* crop. The grains are yellow resembling the size of poppy seed. The seeds are available in the market as *Khakseer* or *Khubkalan*.

Uses: The seeds are used in asthma and also yield semidrying oil which is suitable as lubricant. The seeds are used in cases of fevers caused by chikenpox and smallpox. To get relief from the fever about 10 gram of seeds are boiled in about 150 cc of water and given to drink for about a fortnight. In the first week, the quantity of water is grad-

ually increased, so that after boiling 100 cc of water is left, while during the second week the quantity of water is decreased. It is also believed that *Khubkalan* helps for the pox to come out easily without much pain and ultimately dries them.

204. *Solidago varg-aurea* Linn.

Family: Compositae.

Local names: The Golden Rod (Eng).

Description: An erect pubescent herb. Stem rarely branched. Leaves alternate, lanceolate; lower stalked and toothed, upper smaller, sessile, entire. Heads numerous, crowded in leafy terminal panicle. Involucral bracts unequal, narrow, acute, receptacle naked. Flowers yellow, pappus long, rough. Achenes ribbed.

Distribution: Temperate regions from Kashmir eastwards, 1500–3000 m.

The herb is carminative, antiseptic, diuretice, diaphoretic and vulnerary.

Uses: The flowers are tonic, astringent, diaphoretic and carminative.

The dried plant is useful in dropsy.

Extent of utilization and trade: The plant is not exploited on a commercial scale.

Phenology and time of collection: Fls. Sept.

205. *Sonchus oleraceous* Linn.

Family: Compositae (nom alt. Asteraceae).

Local names: Dodak (Punjab), Titália (Bihar), Mhatara (Mumbai).

Description: A glabrous or slightly glandular herb on the upper parts. Stem branched. Leaves thin, lanceolate or pinnatifid, terminal lobes large, lateral lobes pointing downwards, sometimes only one pair, basal

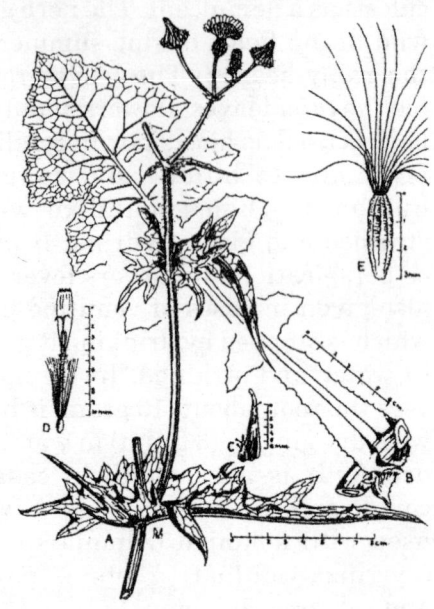

Fig. 5.75: *Sonchus oleraceus* L. A. upper portion of a flowering stem; B. leaf from middle part of the plant; C. involucral bracts; D. flower; E. achene.

lobes acute. Heads yellow. Achenes compresed (Fig. 5.75).

Distribution: Throughout India, ascending to 2500 m. In fields and cultivated places.

Uses: Infusion of the roots and leaves is used as tonic and febrifuge.

The brownish gum formed by evaporation of the latex, when taken internally, behaves as an intensely powerful hydragogue, cathartic and acts powerfully upon liver, duodenum and colon. In its general effects it is said to resemble *elaterium*, producing large and watery discharges so that it has proved a valuable therapeutic agent in ascites and hydrothorax. It, however, requires great care in its administration, since it has the disadvantage of gripping like *Senna* and producing *renesmus* like *aloes*.

Infusion of root and leaves is used as febrifuge and tonic.

Extent of utilization and trade: Not exploited on a commercial scale.

Phenology and time of colletion: Fls. July-Oct.

Other species: *Sonchus brachyotes* DC. (=S. arvensis auct non Linn. sensu Hook.f. FBI)

An erect succulent herb, glandular hairy upwards. Leaves mostly radical, pinnatifid, lobes pointing downwards. Heads yellow, pappus white; ligules long. Achenes flattened and ovate. The plant is bitter, diuretic and good in chronic fevers. The roots are sometimes given in jaundice. Milky juice thickens like fresh opium, the uses are almost similar is that of *Sonchus arvensis*.

206. *Sphaeranthus indicus* L.

Family: Compoistae (nom. alt. Asteraceae).

Local names: Mundi (H), Mundiriká (Sans), Gorakh-mundi (Mumbai).

Description: A nearly erect, softly hairy, shrub like herb; branches long, spreading, winged with the decurrent leaf bases. Leaves alternate, sessile, obovate-oblong, narrowed to the base, toothed. Heads compound, globose, solitary terminal, component heads numerous, very small, crowded, each containing about 12 pink or purple bluish green or bluish red flowers. Involucral bracts nearly as long as the flowers; pappus none. Achenes minute (Fig. 5.76).

Distribution: Throughout India in dry rice fields, ascending to 1500 m. A well-known herb with typical flowers. Flowers and leaves smell like mango fruits.

Uses: The plant is described as pungent, bitter and stomachic, as a remedy for glandular swelling in the neck, urethral dis-

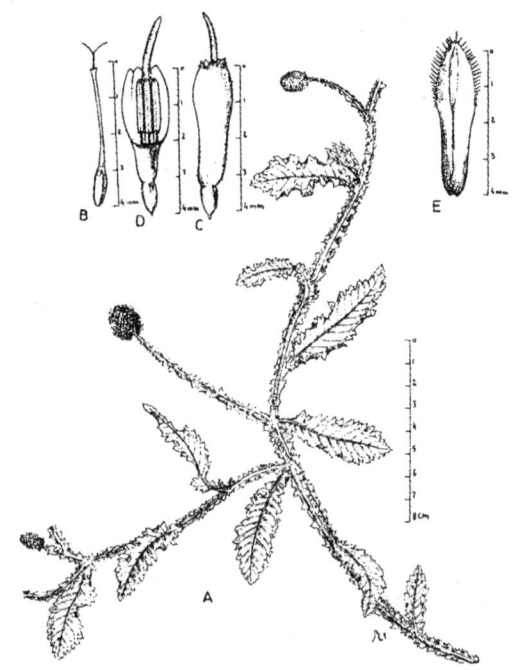

Fig. 5.76: *Sphaeranthus indicus* L. A. a flowering stem; B. female flower; C. hermaphrodite flower; D. corolla of C. partly cut away to show androecium; E. bract.

charges and jaundice. The administration of the drug is recommended in bilious affections and for dispersion of various kinds of tumours. The herb strengthens the heart and brain, purifies blood and increases appetite.

Powder of the root is considered stomachic. Oil from the root is a powerful aphrodisiac.

An essential oil in the herb and an alkaloid has been isolated from the leaves.

Extent of utilization and trade: The plant is extensively used in Indian system of medicine, singly or in combination with other drugs and therefore exploited on a commercial scale. In Bastar young leaves, boiled in water are taken with salt and chillies. It is known that one flower of this herb if swallowed without water in the morning, the eyes will not suffer from inflammation.

The flowers are dried in shade, powdered, sifted and mixed with equal part of sugar, taken 7 grams along with cow's milk in the morning strengthens brain and the eyes, removes general debility and can even check progress of cataract. It purifies blood and cures skin diseases like scabies and ringworm.

Phenology and time of collection: Fls. Dec-April.

207. *Striga asiatica* (Lour) O. Kuntze (-S.lutea Lour.)

Family: Scrophulariaceae.

Local names: Kuránti, Agio (Gujarat), Kuránti (Sansk).

Description: An erect, roughly bristly herb. Leaves sessile, lower opposite, passing into smaller, alternate, floral bracts. Flowers yellow, solitary axillary, sessile, forming termina spike. Corolla 2-lipped, tube cylindric, curved, longer than calyx, upper lip notched, lower large 3-lobed. Capsule oblong, grooved, shorter than the calyx.

Distribution: Throughout India, ascending to 2000 m.

Uses: The plant is bitter, pungent and improves appetite and taste. It is useful in strangury and diseases of the blood.

Extent of utilization and trade: Not exploited commercially.

Phenology and time of collection: Fls. August.

208. *Swertia chirata* Buch-Ham. ex Wall.

Family: Gentianaceae

Local names: Chiráyatá(H), Chiráta (Beng), Kiráta-tikta (S), Nila-vembu (Tam and Tel), Nilabevu (Kan), Chiragita (Marh).

The trade name *Chirayatta* is based on the Indian name.

Description: A robust erect herb. Stem terete, except near the top, 1–5 m high. Leaves broadly lanceolate, in opposite pairs. Pointed at tip. Flowers pale green, tinged with purple, having two glands on each lobe, fringed with long hairs. Fruit 6 mm or more long, ovoid.

Distribution: Temperate regions from Kashmir to Bhutan, ascending to 2500 m. It is mentioned in Sanskrit writings as *Kirata, Tikta*, and often confused with *Andrographis paniculata* and *A. angustifolia*, much inferior in action on human body.

Uses: The entire plant, collected in the flowering stage and dried constitutes the drug.

The drug obtained from dried plant is a bitter tonic, stomachic, febrifuge and anthelmintic. It is antibilious and tends to produce real action, hence much esteemed as a remedy for liver disorders in the form of liquid extract. It causes free discharge of bile and promotes more healthy action. It is laxative and febrifuge, much used in burning sensation of the body, intestinal worms and skin diseases. It is given in fever, diarrhoea and weakness. In the form of infusion or tincture it has been used as a tonic febrifuge and laxative, however, certain experiments did not confirm the febrifuge properties of this plant. The plant contains *chiratin* and *ophelic acid*.

Extent of utilization and trade: Plant is collected when capsules are fully formed, pulled up by roots and bundled. The total annual requirement of this drug is estimated to be more than 400 quintals. Though the whole plant is sold in the market as bundles of dried twigs.

The whole plant is used medicinally, the roots are said to be the most powerful part. It is collected when the capsules are fully

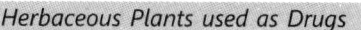

formed. The plants pulled up by the root are tied into flat bundles and are packed in *Bales*. The true *chirata* is sometimes confused with the *Andrographis paniculata* and in some localities *S. angustifolia*, a very inferior to the genuine articles is used as substitute.

Other species: Many other species of the genus from the Himalaya are used similarly as the true chirata. *Swertia angustifolia* Buch-Ham. ex D.Don (Chiraitu-TG) Pahari-chirayata: is one whose properties are similar to *S. chirata* and is taken in Malarial fever as substitute for true Chirayata.

209. *Stephania glabra* (Roxb.) Miers (Stephania rotunda Hook.f. and Thoms. ex parte (non Lour.)

Family: Menispermaceae.

Local names: Gangároo (Kumaon), Parhá (DDN), Gindáru (Garhwal).

Description: Tuberous rooted climbing shrub, twigs ribbed. Leaves 7–16 cm in diameter, broadly ovate or suborbicular, often sinuate, peltate, with 9–11 radiating nerves, pale beneath. Flowers greenish yellow, in axillary compound pedunculate umbels. The outer three sepals linear oblong, the other inner spathulate. Petals shorter than the sepals. Drupe pisiform, red.

Distribution: A common plant in the ravines in western Himalaya from Sutlej eastwards, ascending to 2000 m.

Uses: The root sap is used for massage in headache. The plant is bitter, pungent and improves appetite.

The roots are used in pulmonary tuberculosis, asthma, dysentery and fever.

210. *Striga gesneroides* (Willd.) Vaike. (-S. orobanchoides Benth.)

Family: Scrophulariaceae.

Local names: Missi (Hindi).

Description: A small herb, 15–25 cm high, usually parasitic, rootstock tuberous. Stem slender, with leaves scale like, ovate-oblong, acute, reddish brown as are the stems, passing gradually into floral bracts. Flowers sessile, arranged in erect dense spikes; bracts shorter than the calyx. Calyx 4-ribbed, shorter than the tube, tipped with a bristle. Corolla pink with a white spot at the base of each lobe, tube about 1.5 cm long, hairy in the throat. Capsule oblong, ellipsoid.

Distribution: Parasitic on the roots of a number of plants, like *Lepidagathis, Hamiltonia,* etc. In Rajasthan, Gujarat to South India.

Uses: The plant is useful in diabetes and often chewed to strengthen and colour teeth.

211. *Suaeda fruticosa* (L.) Forsk.

Family: Chenopodiaceae.

Local names: Lunki (Rajasthan), Junak; Morásá (Mumbai).

Description: A much branched, succulent, perennial shrub; branches erect or divaricate. Leaves subsessile, half terete, linear or ellipsoid. Flowers bisexual, axillary solitary or 2–3 together. Fruiting perianth sub-globose. Styles 3 but short.

Distribution: Throughout northwest India, towards Pakistan and Atlantic region in Africa and Arab countries. A common plant of the saline soils. The plant is greyish green in appearance, leaves turn black in drying. The plant is taken by camels and also used as an adulterant 'sajikhar'.

Uses: The plant is edible by camels.

The leaves are used as poultice and when infused in water are used as emetic. The poutice is applied to cure ophthalmia.

Woolly excrescences on the tips of the branches mixed with oil are used for application to camel's sores.

212. *Taraxacum officinale* Weber

Family: Compositae or Asteraceae nom alt.

Local names: Kanphul (Punjab); Bathur (Mumbai), Rasuk (Ladakh).

Description: A perennial herb, juice milky. Leaves all radical, sessile, usually glabrous, variable in shape, narrowly oblong irregularly pinnatifid, lobes linear or triangular, acute, toothed, pointing downwards. Heads ligulate, axillary on a hollow, leafless stalk. Flowers yellow, pappus copious, white, not feathery but soft, ligules 3–5 toothed. Achenes glabrous, flattened, ribbed, narrowed to the base, minutely spiny on the upper half, abruptly contracted into a long slender beak, crowned by pappus.

Distribution: Throughout the Himalya, 300–5000 m in temperate and cold regions.

Uses: The roots are diuretic, aperient and regarded as hepatic stimulant, diaphoretic. It is often prescribed in liver congestion, dyspepsia, dropsy, kidney and skin diseases. It has a specific action on the liver by modifying and increasing secretion; hence used in the chronic diseases of digestive organs, particularly hepatic affections as jaundice, enlargement of liver and dyspepsia, attended with deficient biliary secretion.

Young and tender leaves are used as saláд and in decoction is useful in costiveness and in the chronic inflammation of viscera.

It has been used as a substitute for coffee.

A crystalline principle *taraxacin* and a crystalline substance *taraxaserin*, the *phytosterols tarasterol* and *homotarasterol* are present in this drug.

213. *Tephrosia purpurea* (L.) Pers. (-Caracca purpurea L.)

Family: Papilionaceae.

Local names: Sarpunkhi, Sarphonká (Hindi), Sarpunkhá (Mumbai); Sarmainkh (Maharashtra).

Description: A perennial herb, copiously branched, suberect. Stem glabrescent, terete, finely downy. Leaves short petioled; stipules linear, subulate; leaflets 13–21, green, narrowly oblanceolate, glabrescent above, obscurely silky below. Racemes copious, all leaf opposed. Calyx closely silky. Corolla red, thinly silky. Pods 6–10 seeded, slightly recurved.

Distribution: Throughout India, ascending 1300 m in Himalayan region. Widely distributed on the sandy plains and hills in Rajasthan, sand dunes and other habitats of the desert region.

Uses: The leaves are unpalatable to the livestock in all stages and it covers fallow lands and fields during and after the rainy season. However, in extreme cases of drought we have seen this plant has been browsed by the livestock (stray animals).

The leaves contain *rotenone* and allied substances which had insecticidal and pesticidal properties.

The plant has been described as tonic, laxative, used as anthelmintic for children, used internally as a purifier of blood.

Roots are bitter, given in tympanitis, dyspepsia and chronic diarrhoea. Fresh root-bark is reported to be made into pills after grinding the bark, mixed with black pepper, for relief from obstinate colic.

Glucoside *rutin* has been isolated, while other constituents reported are *tephrosin, deguelin, isotephrosin, rotenone*, etc. Leaves are also reported to contain about 2% of glucoside *osyritin*.

Falciformin, a flavinon has been reported from *T. falciformis* from Rajasthan from the same habitat as of *T. purpurea*. Flavinoids are originally valued as dye stuffs and have become obsolete after the introduction of synthetic dyestuffs but their importance has continued for other reasons, one of these is their *vitamin* property. Precisely *rutin* on large scale is prepared from Buckwheat and Eucalyptus leaves. Similarly, the bioflavidons, herperdin and naringn, obtained as bye products of citrus industry, are employed for the membrane capillatory integrity, control of spontaneous abortion and to aid in surgery and wound healing. Some of the other plants used as Chamaecyparus obtus-Hinokflavone are *Cupressus torulosa, C. sempervirens* (cupreson flavone); *Auracaria cunninghami* and *A. cooki, Garcinia buchanii* and *G. morella.*)

Other species

a. *Tephrosia candida* (Roxb.) DC. Vern. Lashtia (Hindi). Plant used as fish poison.

  The plant is distributed in Himalaya from Garhwal eastwards to Bangla-Desh. It is also grown as an ornamental.

b. *Tephrosia villosa* (L) Pers.
  Herb with downy stem. Leaves sessile with linear stipules. Leaves with 11–19 leaflets grey green, silky beneath. Flowers red in long racemes, lower in distant fascicles. Corolla pale red.
  Throughout Indian plains, extending to tropical Africa. Leaf juice is given in dropsy, also useful in diabetes.

c. *Tephrosia uniflora* ssp. *pterosa* (Blatter and Hallberg) Fillette et Ali. Vern. Bhákar, Biyani (Rajasthan).
  Throughout northwest India extending to the plains of western peninsula.

The leaves are boiled in water and eaten, good in curing syphilis.

d. *Tephrosia falciformis* Ramaswamy Vern, RatiBiyani (Rajasthan). A much branched perennial, branches pubescent or adpressed white silky.
  A source of Flavidon. Branches used for weaving baskets.

214. *Tinospora cordifolia* (Willd.) Miers.

Family: Menispermaceae.

Local names: Giloya, Gilo (H); Guláncha (Beng.), Gulvél (Guj), Gurjá (Kum); Amritavalli (S); Guduchi (Tel), Anebule (Kan); Amrytu (Mal.); Gulvel (Mahr).

The trade name *Gilo* and *Tinospora* is based on the Hindi and English names respectively.

Description: A large climber, having leaves like the betel leaf, stem smooth and shining, lenticelled, with light-coloured pappery bark; branchlets pubescent. Leaves broadly ovate or orbicular, deeply cordate, 7-nerved, shortly acuminate, pubescent above, tomentose beneath, petiole thickened and twisted at the base. Flower greenish yellow, in racemes usually from the old wood; pedicels slender, usually solitary in the female and clustered in male. Fruit a drupe, red, size of a pea.

Distribution: On trees and shrubs in shady places in Siwaliks it is fairly common, throughout tropical regions of India.

Uses: Stem of the plant collected in hot season and dried with bark intact, constitutes the drug. It is a hepatic stimulant. The drug is useful in chronic fever with enlargement of spleen. Green gilo (10 gm) is crushed and steeped over night and drunk after mixing with sugar. It is valuable hepatic stimulant and is given in jaundice and tropidity of the liver. In case of jaundice or burning sensation in the stomach, 20 gm of juice of green plant is mixed with whey and drunk, if similarly taken is useful in diabetes 4 gms of powdered gilo is taken with water or whey (Fig. 5.77).

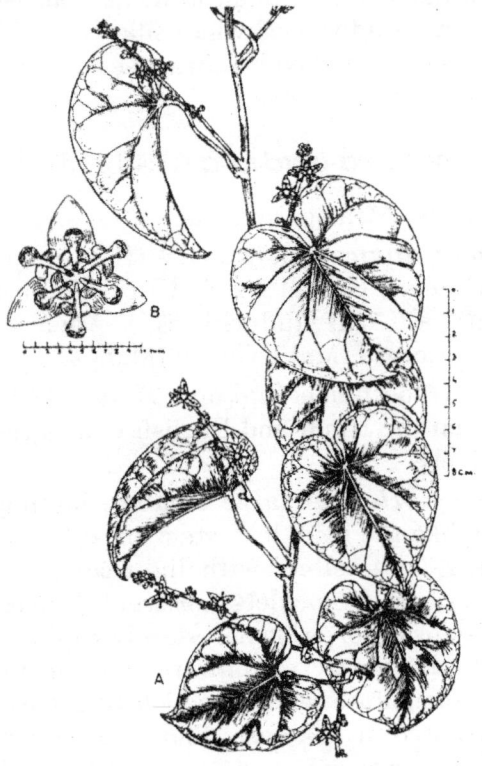

Fig. 5.77: *Tinospora cordifolia* H.K. F. and T. A. flowering twig of a male plant; B. male flower.

It also purifies blood and used in cases of abscesses and boils.

Juice of the leaves is taken to kill intestinal worms. As alterative and tonic it is used in general debility, acidity of urine, skin diseases and dyspepsia. The drug is also believed to be antiperiodic; it is also considered aphrodisiac.

Starch obtained from the roots and stems of plant called "Sat Gilo" is useful in diarrhoea and dysentery; it is also a nutrient.

In Kumaon pith and wood are used in asthma and in fever. A mixture of pith and

wood along with crusted seeds of *glycine max* is given to buffaloes in stomach trouble.

Extent of utilization and trade: The drug is extensively used in medicine and exploited on a commercial basis. It is a tonic, antiperiodic, valuable in general debility after fever, useful in secondary syphilitic affection, chronic rheumatism of mild form. It is also sold as *Sudh Gilo* in powder form. A product called *Amrit Asava* has this plant as the main component.

215. *Trachelospermum fragrans* Hook.f. I (chnocarpus fragrans Wall).

Family: Apocynaceae.

Local names: Dudhi (Kum), Dawarilahara (Nepal), Yokchounrik (Lepcha).

Description: A tall climber, young parts pilose, old stems strongly tubercled. Leaves elliptic-oblong, acuminate, sub-coriaceous, shining above. Flowers white, fragrant, in compound trichotomous corymbose cymes. Corolla salver-shaped with a campanulate mouth. Follicles cylindrical, acuminate, incurved. Seeds compressed having a coma of white hairs.

Distribution: In moist shady ravines, 1000–2000 m, also along the river banks.

Uses: The bark is said to be a powerful tonic. In tincture and in decoction, it is successfully used in malarial fevers. Though it has proved ineffective in treatment of amoebic and bacillary dysentery, it is a good remedy for other types of diarrhoea.

It is sometimes used as a substitute for *Alstonia scholaris* (Chatiun, Saitanka-Jhád, Satwin-H) on account of similar properties.

Extent of utilization and trade: Used on a limited scale and exploited from the Siwalik ranges.

Phenology and time of collection: Fls. April, Fr. Oct-Dec.

216. *Trichosanthes bracteata* (Lam.) Voigt (=T. palmata Roxb.)

Family: Cucurbitaceae.

Local names: Indráyan (Kum); Makal, Panwar, Lal-Indráyan (TG), Chichinda, Makal (Maharashtra).

Description: A large perennial climbing herb. Stem angular. Leaves roughly pubescent, cordate, 3–7 lobed; tendrils branched. Flowers white, male and female usually on different plants. Fruit smooth, globose red, indehiscent, stripped with orange. Seeds flattened.

Distribution: Throughtout India, ascending to 1500 m.

Uses: The fruit is bitter, carminative, purgative, abortifacient and lessens inflammation, cures hemicrania, weakness of limbs, heat of brain, ophthalmia, and leprosy. It is used in epilepsy, rheumatism, and for gargle, etc.

The seeds are emetic and purgative.

Fruit pounded and mixed with coconut oil forms a valuable application to sores under ear and nose. The juice of the fruit or bark, warmed with gingly oil, is used with good effect as a *bath oil* for the relief of long standing or recurrent attacks of headache. The fruit is smoked as a remedy for asthma.

The roots with equal part of colocynth root is rubbed into a paste and applied to carbuncles; combined with equal portions of the three myrobalans and turmeric, it affords an infusion which is flavoured with honey and given in gonorrhoea. In Kumaon the root powder mixed with water is given in fever.

Extent of utilization and trade: The drug is used extensively in the Indian sys-tem of medicine and exploited on a commercial basis. It is sold in the market.

Other species: *Trichosanthes cucumerina* (Jangli-chichonda), a climbing annual has various medicinal qualities and the tender shoots, dried capsules, seeds, leaves and roots are all used in native medicine.

217. *Thalictrum foliolosum* DC.

Family: Ranunculaceae.

Local names: Mamiri, Barmat (Kum); Mamira, Pilijari, Barmat (H.), Chaitra (Kash).

Description: A rigid perennial herb. Leaves extipulate, pinnately compound, petiole sheathing, sheaths auricled. Flowers dingy purple. Anthers beaked. Fruit is a sharply beaked achene, acute at both ends, sharply ribbed.

Distribution: Throughout Himalaya, in temperate regions.

Uses: The roots form the drug of commerce.

The roots combine tonic and aperient properties and have been found useful in convalescence after acute diseases, in mild forms of intermittent fevers and a tonic in dyspepsia. It is also used as a purgative and diuretic. It resembles *turmeric* in appearance and is bitter and pungent, used often as a collyrium in ophthalmia, improves eyesight, in toothache, nail trouble and discoloration of skin.

Ash of the root in Kumaon is used in eye troubles. Paste of the root with *dhatura* seeds is used externally in eczema.

Extent of utilization and trade: The drug is not exploited commercially on a large scale. *Berberine* and *thalictrine* have been isolated from the roots.

Phenology and time of collection: Fls. July.

5

Other species: *Thalictrum javanicum* is used for the same properties as *T. foliolosum*.

218. *Tragia involucrata* L.

Family: Euphorbiaceae.

Local names: Barhanta, Kanchluri, Khajkhoti (Hindi), Vrischikali (Sans.).

Description: A climbing herb, evergreen, clothed with stinging bristles; foliage variable. Leaves simple, thickish, not cordate at base, apex pointed in young leaves, rounded in older ones, oblong-cordate, toothed, covered with stinging hairs. Flowers in long racemes, unisexual, both sexes on the same plant. Male fls. uppermost, female below. Calyx of female flowers with 4–6 pairs of short hairy teeth; petals absent. Capsules round.

Distribution: Throughout India from outer Himalayan ranges in Himachal, extending till Assam and south to Kerala.

Uses: The root is of medicinal value, reputed diaphoretic and alterative. It is given in fevers to promote perspirtation, used in cold extremities and in pains of legs and arms.

In Orissa the roots are used as antidote to snakebite. The roots are also made into a paste with 'mustard oil' and given orally but also applied locally.

Madhya Pradesh tribals use the root and leaf for curing blood dysentery; paste of both root and leaf is orally administered.

Similarly, Santhal and Paharaiya tribals make use of the root for curing dysentery, extract of the root is orally administered.

The drug is effective in itching of the skin, when half teacup of root infusion is given twice a day, for relieving bronchitis; decoction of the root is considered efficaceous.

External application of root paste is used for extracting guinea worms by applying to affected parts. Roots paste with *tulsi* (*Ocimum*) juice also used as dressing for the cure of itching erruptions of skin; this can also be used for external use in leprosy cases.

Fruit when rubbed with little water is applied to head to make the hairs grow and cure baldness.

219. *Trianthema portulacastrum* L. (-T. monogyna L.)

Family: Ficoideae.

Local names: Bishkhaprá, Itsit (Himachal), Lalsabuni (Hindi); Satodo (Gujarat); Punarnavi (Sansk), Sabuni (Bengal). Muchchujoni (Kanara); Sharunnai (Tamil), Galijeru (Telgu).

The trade name 'Biskhapra' is derived from the local name and the plant is sold by this name.

Discription: A succulent spreading herb. Stems much branched, angular. Leaves broader towards the tip. Flowers very small, almost concealed in the base of leaf stalk. Fruits small; like flowers concealed in the base of leafstalks. Seeds black, kidney-shaped, covered with minute overgrowth (Fig. 5.78).

Distribution: Throughout India, common on the grassy lawns after the rains, growing profusely.

Uses: The leaves of the white variety are medicinal and used as diuretic, also used in dropsy and oedema.

The alkaloid from the leaves, called 'Punarnavine' is diuretic and also used in leprosy. It is beneficial in swellings of the body, caused by disorders of liver and kidney; particularly helpful in early stages of the diseases.

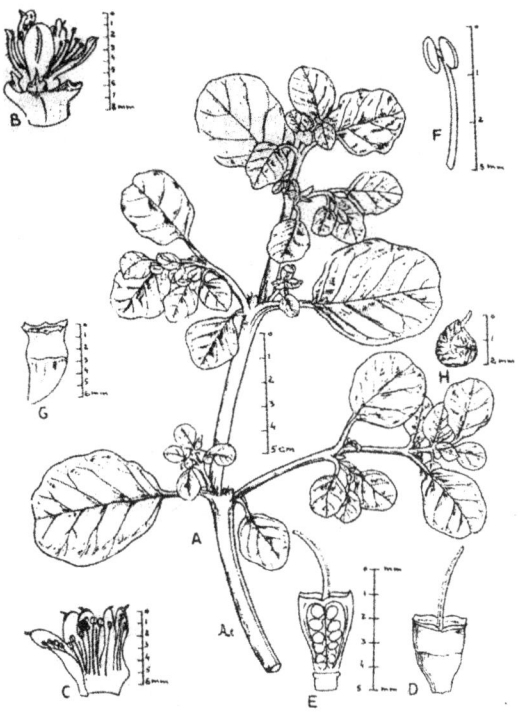

Fig. 5.78: *Trianthema portulacastrum* L. A. upper portion of flowering and fruiting stem; B. flower; C. perianth opened to show androecium; D. gynoecium; E. longitudinal section of gynoecium; F. stamen; G. capsule; H. seed.

The entire plant has been tested for its abortifacient properties; and its dose causes mild contraction of uterus.

Powdered roots are bitter, cathartic and used in amenorrhoea.

Other species

a. *Trianthema pentandra* L. and *T. decandra* L. (Vern. Bishkhprá-Hindi, Itsit (Himachal). The plant having same names as the above species. The roots are believed to be laxative and useful in suppression of mensus and inflamed testicles. In Kumaon, *T. pentandra* is used in abdominal diseases; also as a pot herb. The abortifacient properties of the drug have not been conclu-

sively proved, though it has been reported useful in abdominal diseases and as a cure for pain in bladder.

b. *Trianthema monogyna* L. though a synonym of *T. portulacastrum* (Vern. Punarniva, name same as that of Boerhaavia) having spreading habit and white flowers in rainy season is reported to have medicinal properties. The dried herb powder (20 gram) steeped over night in water, strained, mixed with one gram of *kalmishora* (K_2NO_4) and drunk in cases of dropsy. Dried roots in shade, mixed with turmeric powder and sifted for curing cough and asthma. Leaves dipped in water with pepper, if taken, cause profuse urination.

220. *Tribulus terrestris* L.

Family: Zygophyllaceae.

Local names: Gokhru (Hindi); Laghu-Goshura (Sansk); Kánto (Rajasthan-Ajmer); Negálu (Kanara); Neringil (Maharashtra), Nerunje (Tam); Palleru (Tel.), Trade name Gokhru based on local name.

Description: A prostrate spreading herb, densely covered with minute hairs. Leaves in opposite pairs, 5–8 cm long, compound; leaflets 4–7 pairs, 8–12 mm long. Flowers pale-yellow, solitary, opposite to the leaves or in leaf axils. Fruit spinous, often pricks when picked. Seeds many in each cell of the fruit.

Uses: The fruit constitutes the drug which is widely used. The fruits are cooling, diuretic and used in diseases of urinogenital system and sexual weakness for which the drug is reputed. Fruits used in micturition, calculus affections and in impotence. In the form of infusion used as diuretic in gout and diseases of the kidney; it promotes urination.

5

Extent of utilization and trade: The drug is extensively used in the Indian system of medicine and exploited on a large scale. However, some attempts have been made to grow this plant in the field as natural populations are not able to support the demand without damaging its existence. The fruits are sold in the market and are available in drug shops locally called *Attárs*.

Field trials carried out at Poona by Karnick, 1975, showed profound effect of NPK (10 : 10 : 10) on the growth. Growth regulators as 2, 4D, IAA, induced flower growth and shortened the time of flowering and fruiting of the plants. (Karnick, 1975. Exptl studies on the growth trials of *Tribulus terrestris* L. in Poona, *Proc. nat.* symp. Abstr. 14).

Tribulus alata Delile (vern Badá-Gokhru), gives larger fruits which are used for the same purpose as the above species. Naturally the species also occurs in the desert area of Rajasthan and the fruits can be collected.

221. *Trichodesma indicum* R. Br.

Family: Boraginaceae.

Local names: Salkontá (Rajasthan), Chhota Kulphá (Hindi), Surása (Sansk).

Description: An annual herb, rough with appressed hairs, bulbous based stiff hairs. Stem erect or diffuse. Leaves mostly sessile, ovate-oblong or lanceolate, base narrowed cordate, upper surface closed with stiff hairs, seated on flattened circular tubercles, lower surface harshly hispid, except on the nerves and veins. Flowers pale blue, changing to pink or white. Calyx clothed with long, stiff hairs, segments lanceolate. Corolla limb oblique, funnel-shaped. Nutlets smooth and polished on back, white or bluish when ripe (Fig. 5.79).

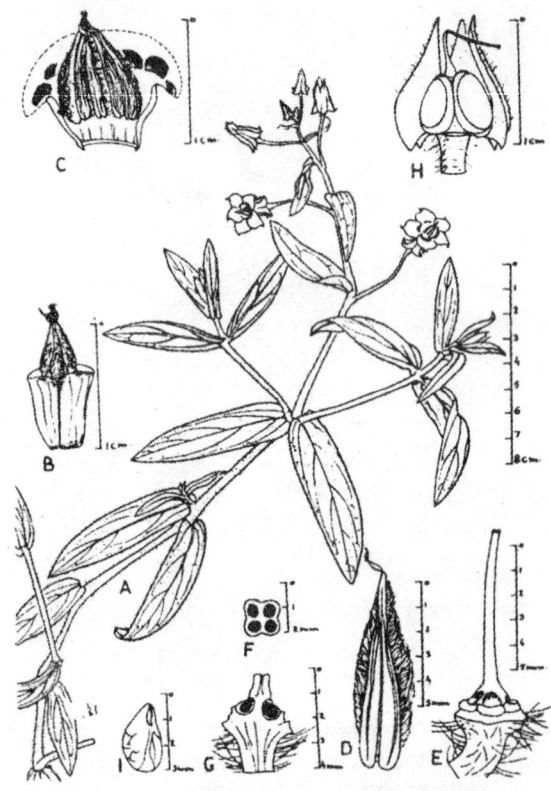

Fig. 5.79: *Trichodesma indicum* R. Br. A. flowering branch, B. corolla cut away to show the staminal cone; C. corolla lobes cut away and tube opened to show androecium; D. anther; E. gynoecium; F. transverse section of gynoecium; G. longitudinal section of gynoecium; H. two of the calyx lobes cut away to show fruit; I. seed.

Distribution: On roadsides and waste grounds, throughout India, ascending 1700 m in the Himalaya.

Uses: The plant is diuretic, used as an emollient and poultice. The roots are pounded and made into a paste which is used to reduce the swelling of joints and given to children as a drink to cure dysentery.

Cold infusion of leaves is a depurative.

222. *Trigonella foenum-graceum* L.

Family: Papilionaceae.

Local names: Méthi (Hindi); Méthra, Muthi (Punj.); Bird's foot, Fenugreek, Greek hayes, Hay-seed (English).

Description: An annual, erect herb, scented, little branched, robust, leaflets 3, lance-shaped, ovate or obovate, toothed. Flowers axillary, 1–2 without stalk, yellow, pretty. Pods sickle-shaped, 5–7 cm long, 10–20 seeded, having an oblique furrow along a part of their length.

Distribution: It is both wild and cultivated in Kashmir to Kumaon, in upper Gangetic plains and valleys; also in the south.

Uses: Though the plant is used as spice in Indian kitchens, and unpleasent in taste, is used medicinally. The author himself is using it as a medicine for the last 30 years.

It is used as tonic, restorative and appetizer, helps in diseases due to cold, like phlegmatic humours, relative constipation and used in cases of cough, asthma and rheumatism.

The dry yellow seeds are used in different forms as powder and the seeds form an ingredient of spices and pickles, taken as such, in *halwa, ladoo* and also drunk after steeping in water over night to cure rheumatism and lumbago. The powder when taken with water, cures diabetes and regulates the blood sugar. Powder when taken with honey is used to improve urination during cold season.

It is reported that the seeds have significant *diosgenin* content, which could serve for demi synthesis of cortisone derivatives.

Other species

a. *Trigonella corniculata* L. Vern. names: Malya (Sansk), Pirang (Urdu).
   The fruit is bitter and astringent, applied to swellings and bruises.

From Kashmir to Kumaon, 2700–4000 m.

b. *Trigonella oculta* Delile
   The seeds are used in dysentery in Gangetic plains.

c. *Trigonella polycerta* L.
   The seeds are used in diarrhoeic condition.
   In western Himalaya up to 2000 m.

223. *Trillium govanianum* Wall.

Family: Liliaceae.

Local names: *Lálri* (Kashmir); Makatoomchi (Lahul).

Description: Glabrous herb, thick, creeping; stem erect, unbranched. Leaves 3, shortly stalked, broadly ovate, arranged in a whorl at the summit of the stem with a solitary, stalked flower in the centre. Perianth purple with 6 segments, narrowly lanceolate, spreading in the flower, reflexed in fruit. Stamens 6, attached to the base. Anthers basifixed. Ovary 3-celled, style purple, divided to the base in 3 long linear arms. Fruit a globose berry; seeds many, ovoid, having pulpy lateral appendage.

Distribution: In the temperate regions of the Himalaya, 2600–3700 m. Associated with *Podophyllum, Aconitum,* etc. in the subalpine zone, but of the two species *T. govanianum* and *T. tschenoskii* Maxim, the former is abundant in some locations of Nepal between altitudes 3500–4000 m having moist soil situations, rich in humus, as undergrowth of high altitude conifers.

Uses: The rhizome is applied in the form of a paste in various diseases of eye, while in Lahul it is used extensively in the form of a paste for healing of cuts and wounds.

The plant possesses similar chemicals and phrmaceutical actions to the American species *T. erectum* and *T. pendulum* Willd.

The Indian species yields a saponin which on complete hydrolysis with hydrocholric acid gives about 2.5% *diosgenin*, an important starting material in the synthesis of steroids (Indian J. Pharmacy, 1965.27 : 106–110,1965).

The alcoholic extract of rhizome is reported to have marked activity on guineapigs uterus in doses of 15–50 mg. On the isolated heart of frog, the extract showed slight stimulation in smaller doses but larger doses depressed it. Isolated rabbit intestine was not much affected.

The drug can best be collected during July-August when in fruiting. Loss of moisture in rhizome when dried in shade is up to 70%, when it gets stony and dark brown in colour.

224. *Tridax procumbens* L.

Family: Compositae or Asteraceae nom alt.

Description: A procumbent, weak herb, stem hairy with ascending tips and very long erect peduncles. Leaves in distant pairs, 2–5 cm long, coarsely toothed or lobed, lobes serrate, both sides hairy, petioles short. Flower heads solitary on long, slender peduncle, about 35 cm long, 2–3 times as long as the stem. Involucre hirsute. Ray florets with a slender hairy tube and spreading, deeply 3-lobed. Achenes silky. Pappus shining. Fls. August-Sept (Fig. 5.80).

Distribution: Native of south America but common throughout the tropical regions of India. Abundant in pastures, banks and waste grounds. Flowers and fruits throughout the year.

Uses: The plant is medicinal and has been personally tested on various occasions in the field both in Rajasthan and Uttaranchal.

Juice of the plant along with leaves is used to stop bleeding from cuts and wounds.

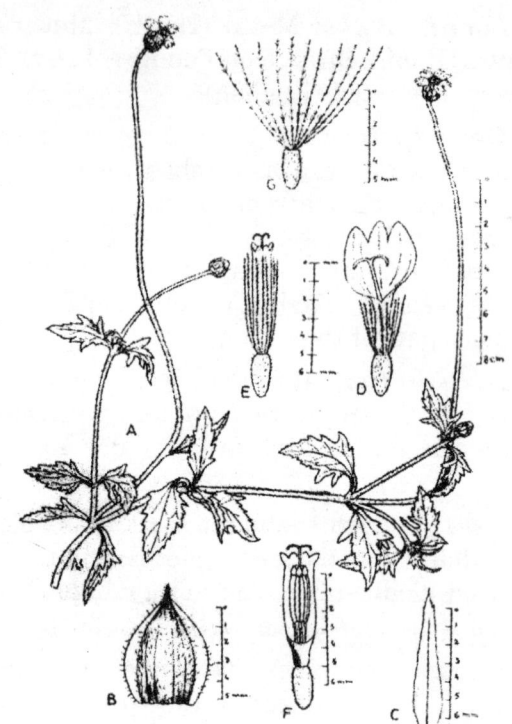

Fig. 5.80: *Tridax procumbens* L. A. upper part of a flowering stem; B. involucral bract; C. receptacular scale; D. ray floret; E. disc floret; F. corolla of E partly cut away and pappus removed to show androecium; G. achene.

225. *Tylophora asthematica* Wight and Arn (-Asclepias asthematica Roxb; Tylophora indica (Burm.f.) Mery

Family: Asclepiadaceae.

Local names: Antamul, Jangli Pikvan (Hindi), Ipecacuanha, Vomiting Swallowwort (Eng), Vallipala (Mal), Pitakari (Marh), Méndi.

Description: A twining perennial, much branched with numerous fleshy roots. Stem elongated and smooth, hairy. Leaves variable, in opposite pairs, 5–10 cm long ovate to elliptic-oblong, upper surface smooth, while lower hairy, rounded or cordate at the base usually pointed at top. Flowers in um-

bellate cymes, stalks bearing 2–3 umbels at the apex. Corolla greenish yellow or greenish purple in short clusters. Follicles 5–10 cm long, in pairs ridged. Smooth, lance-shaped, sharp-pointed. Seeds ovate.

Distribution: In the eastern parts of India such as Assam, Orissa, Konkan, Canara and Tamilnadu plains, ascending to 1000 m.

Uses: The leaves and roots contain the alkaloids *tylophorin* and *tylophinine*. Dry plant yields the alkaloids containing substances with emetic properties.

The leaves are emetic, expectorant, diaphoretic, alterative, blood purifier, aromatic and stimulant. Also prescribed as a decoction to increased lochia in parturient women, and given in rheumatism, including syphilitic rheumatism. Externally the leaves are used to relieve gouty pain.

Roots are used in the same way as the leaves, particularly as diaphoretic and in dysentery. Hot infusion with milk and sugar is given in asthma and bronchitis, as alterative and tonic in chronic cough and diarrhoea for children. It is good for bringing about vomiting.

Extent of utilization and trade: The plant is very much used locally for medicinal purposes. It has been reported that the plant is just on the verge of extinction on account of its over exploitation.

Other species

a. *Tylophora tenuis* Blume
   The plant is used in west coast and Tamil Nadu as a cure for urticaria, smallpox and an antidote for arsenic poisoning.
b. *Tylophora fasciculata* Buch-Ham. ex Wight.
   Juice of the roots and leaves used medicinally. Pounded leaves applied to unhealthy sores for inducing healthy granulation.

The leaves contain an alkaloid and the root powder is given with milk as tonic. Found in upper Gangetic plain and eastwards to Khasia hills through Madhya Pradesh, Konkan, to North Kanara.

226. *Urginea indica* Kunth.

Family: Liliaceae .

Local names: Jangli Piyáz, Jangli-Kánda (Hindi), Ban-Kánda Vaná-Palandam (Sans) Indian Squill (Eng).

Description: A bulbous plant, with thick, white, ovoid bulbs, much like the onion. Leaves strap-shaped, narrowly pointed, 15–40 cm long and 3 cm wide. Flowers drooping or spreading in slender racemes. Perianth bell-shaped, 6-parted, segments white or greenish white, with 3 green ribs, stalk about 3 cm in length. Capsules ellipsoid, 3-celled. Seeds 6–9 in each cell, black, elliptic but flattened.

Distribution: Western Himalaya, Bihar, Jharkhand and to western costal areas; sometimes on sandy shores of the sea.

Uses: The bulbs are medicinal, used particularly those intact in their membrane; those freshly dried are medically active. The properties are similar to that of *Urginia scilla*.

The bulbs are cardiac-stimulant, diuretic, in the form of syrup used as an expectorant in bronchial catarrh and chronic bronchitis.

In large doses it acts as poison and is emetic.

As a household remedy it is used in asthma, dropsy, rheumatism, calculous affections, leprosy and in skin diseases. A confection of the bulb, made with *dryfigs, blackgrapes* and *honey,* is taken in acute bronchitis; juice of the bulb is used with honey and sugar in bronchial catarrh and chronic bronchitis (Fig. 5.81).

The bulbs are used externally to remove warts and corms; for removing warts powdered bulb is locally applied, while for removing corms roasted bulb is crushed and applied.

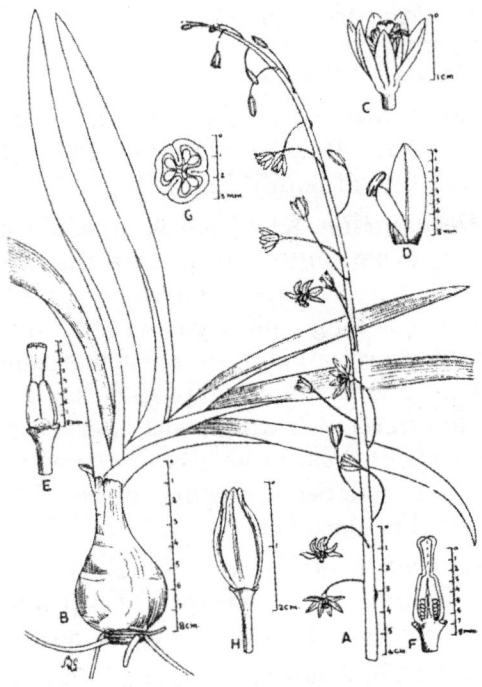

Fig. 5.81: *Urginea indica* Kunth A. plant; B. inflorescence; C. flower; D. a perianth segment and stamen; E. gynoecium; F. longitudinal section of gynoecium; G. transverse section of gynoecium; H. dehiscing capsule.

227. Urtica dioica Linn.

Family: Urticaceae.

Local names: Shisoon (Kum). Bitchugháss (H); Kandár (TG).

Description: Erect herb, clothed with stinging hairs. Stem 1–2 m, often robust, grooved, inner bark containing tough fibres. Leaves opposite, long stalked, ovate or lanceolate, usually cordate, long petioled; stipules free. Flowers small, green, 1-sexual on separate plants, clustered on the branches of loosely spreading, axillary panicles. Achenes flattened, embraced by persistent perianth.

Distribution: In temperate regions, 2300–3300 m.

Uses: The root is diuretic. Juice of the plant is used as an external irritant. In decoction the plant is used in nephritic troubles, in haemorrhages from kidneys, jaundice, etc. Nettle tea is curative for feverish gout as well as intermittent fever, for sciatica, incipient washing, for difficult breathing, heart troubles, rheumatism and lack of muscular energy. Urtication is said to be an invaulable remedy.

Pith of the plant is used as a suppository. The plant is also supposed to do away with evil spirits and it is a practice to move away with a branch of this plant so as to keep away 'evil spirits'. Vegetable from young leaves is efficaceous in sciatica, gout and rheumtic pains. Branches with leaves are used externally on sprain and swellings as the stings irritate the skin.

Extent of utilization and trade: The plant is not exploited commercially, though used extensively by the local people.

Phenology and time of collection: Fls. June-July.

Other species: *Urtica parviflora* Roxb. differs from the above species in having stem obliquely angled, leaves usually long pointed and stipules united. Male and female flowers are on the same plant. Decoction of the plant is useful in fevers.

Urtication is prickling with a bundle of nettle twigs for one or two minutes.

228. Urena lobata L.

Family: Malvaceae.

Local names: Van-bhéndi (Hindi), Aramina, Cadillo (Eng).

Description: A hairy herbaceous plant with cordate, 5–7 lobed leaves, having 1–3 glands on the under surface. Flowers pink, in clusters. Fruit densely pubescent, with hooked bristles, not opening when mature.

Distribution: Throughout the hotter parts of India.

Uses: The leaves are employed to adulterate *patchouli* (Pogostemon spp).

The seeds are mucilaginous and used to make cereal foods.

229. *Valeriana lescehnaultiis (-V. brunoniana Wight. and Arn.) DC.*

Family: Valerianaceae.

It is another species that occurs in the Nilgiris above 2000 m on downs and margin of shota forms and Coorg region. The roots are used as substitute for those of *V. officinalis*. A perennial herb. Leaves simple with one or two leaves slightly pinunate on moist places small leaflets below. The end one, sparingly pubescent. Fls. pink.

Veleriana officinalis L. Vern Bililotan-Rajasthan; Kalvala (Maharashtra).

The roots are stimulant, carminative, antiseptic, useful in hysteria, epilepsy, shell shock and neurosis.

230. *Vanguieria pubescens* Kurz (- *V. spinosa* var. mollis. Mainphal, Moyena-Hindi, Monphal (Bihar), Boi-bindi-sans.).

Family: Rubiaceae.

Description: A large shrub, armed with long thorns. Leaves ovate elliptic 15 cm × 7 cm. Flowers smell, green, subglobose in axillary cymes. Corolla green, subglobose, tube with prominent rubs. Style long, stigma grooved.

Distribution: Central and southern tracts of Bihar near ravines and nalas. Leaves turn yellow in January.

Uses: The tribals (Mundas, Oroans, Hos, Santals, Ladha and Kondhs) of Bihar, Orissa and west Bengal use young roots as vegetable.

The leaves are also taken is powder form during food scarcity.

The Mundas use the leaves to cure dysentery.

231. *Verbascum thapsus* Linn.

Family: Scrophulariaceae.

Local names: Gidhartambáku (H), Ban. Tambáku (Punjab), Jangli-tambáco (Urdu).

Description: An erect herb, closely covered with soft grey yellow stellate hairs, winged with the prolonged leaf bases. Leaves entire, radical short stalked, ovate; upper sessile, oblong-lanceolate. Flowers yellow, nearly sessile, crowded in terminal spike. Corolla woolly outside. Capsule ovoid tomentose (Colour Plate 5.17).

Distribution: Temperate regions from Kashmir to Kumaon, 2000–3000 m.

Uses: Leaves are narcotic to fish but useful in chest complaints, gout and rheumatism. In trans-Indus region, the herb is employed for treatment of asthma. Leaves warmed and rubbed with oil are employed as an application to inflamed parts. A tincture from the leaves has been found useful for migraine of long standing with oppression of the ears. Leaves boiled in milk is a remedy for phithisical cough. A conserve of the flower is useful against ringworm. Infusion of the flower is diuretic. Dried leaves, if smoked in an ordinary tobacco pipe, compltely controls hacking cough.

The seeds are aphrodisiac and narcotic, used in asthma and infantile convulsions.

The leaves contain bitter substances, *saporine* and *croectin*.

Extent of utilization and trade: In Europe it is reputed for curing pulmonary diseases of cattle while in Germany the plant is placed in greenery to drive away mice.

The drug is utilised on an extensive scale locally but not exploited much on a commercial scale.

Phenology and time of collection: Fls. July.

232. *Verbena officinalis* Linn.

Family: Verbenaceae.

Local names: Karaita (Maharashtra), Faristarium (Urdu).

Description: An erect, nearly glabrous, perennial herb. Stem 4-sided, branching. Lower leaves stalked, oblong or ovate, pinnatifid or coarsely toothed; upper sessile, usually 3-parted. Flowers liliac, sessile, in long, slender, bracteate spikes. Fruit ultimately separating into 4 one-seeded nutlets.

Distribution: In temperate and sub-tropical regions throughout India, ascending to 2000 m.

Uses: Fresh leaves are used as febrifuge and as rubefacient in rheumatism and diseases of joints. The drug is employed in early stages of fevers, colds and in treatment of fits, convulsions and nervous disorders.

Roots are a remedy for scrofula.

Extent of utilization and trade: The drug is not used on a commercial scale though locally it is much used.

233. *Vernonia cinerarea* (L.) Less.

Family: Compositae, Asteraceae nom alt.

Local names: Sandári (Rajasthan), Sahádevi (Hindi).

Description: An erect, hoary pubescent annual herb. Stem stiff, cylindric, slightly branched. Leaves petioled, ovate or lanceolate. Flower heads small in lax-terminal corymbs, having about 20 fls. Involucral bracts linear-lanceolate. Achenes white.

Distribution: One of the hardiest plant, a common weed throuhout India. It grows under varying conditions of soil and moisture.

Uses: The plant is diaphoretic and used in decoction to promote perspiration in febrile conditions. Also used as a remedy for spasm of the bladder and strangury. Juice of the plant is given in piles.

Roots are given in dropsy with quinine.

Seeds are anthelmintic and used as a constituent of *masala* for the horses.

Other species

a. *Vernonia roxburghii* Less.
   Properties and uses are as above. Found in upper Kumaon Gangetic plain and in western parts of India.
b. *Vernonia teres* Wall.
   The plant is a medicine for luxations, ulcers and wounds and used in dropsy. Flowers heads are ascaricidal. From Kumaon eastwards to Sikkim, Bihar and Madhya Pradesh.

234. *Veronica anagallis-aquatica* L.

Family: Scrophulariaceae.

Local names: In west Himalaya from Kashmir to eastern Himalaya, till Bhutan and Assam; also in western peninsula.

Description: Perennial herb with hollow, erect succulent stems leaves stem clasping, oblong-ovate, entire or nearly so. Flowers pale purple or white, in axillary racemes, bracts shorter than the flower stalks. Capsule notched.

Distribution: Valleys from Kashmir to Assam and western peninsula, ascending up to 1000 m.

Uses: Plant is used for scrofulous affections of the skin. Bruised plant is applied for healing ulcers, burns and for the mitigation of swollen piles. The properties are almost similar to that of *V. baccabunga*. It contains a glucoside *Aucubin*.

Other species: *Veronica beccabunga* L. Beccabunga (Eng.).

Plant is alterative, diuretic and antiscorbutic, given in scurvy, impurity of blood; remedy for scrofulous affections of skin. Externally bruised plant is applied for healing burns, ulcers, whitlows and swollen piles. Glucoside *rhianthi* (aucubin) has been isolated.

235. *Vetiveria zizanoides* (L.) Nash

Family: Poaceae.

Local names: Khas-Khás (Hindi), Lavancha, Nash, Panni, Ushira (Sansk), Vetiver (Eng).

Description: Densely tufted grass, perennial; underground stem branching with spongy and aromatic roots. Aerial stem tall, smooth, glabrous. Leaf sheaths covering the stem, very much compressed. Leaf bases 30 cm to 1 m long, rigid, somewhat spongy, usually pale green and glabrous. Flowers in conical panicles, 30 cm long, branches whorled, having racemes of spikelets towards the tips; spikelets pale or reddish brown or purple.

Distribution: Wild throughout the Indian subcontinent and Pakistan. On moist heavy soils, lake or steam margin.

Uses: The grass yields an oil called *khus-khus* oil, which is an essential oil and is commercially available. It is used in perfumery and flavouring *sherbats*.

Infusion of the root is febrifuge, refrigerant, diaphoretic and made into a paste in water for use. External paste in fever as coolent, and removes excess heat. An aromatic cooling bath is made by addition of the oil to a tub of water.

Powder of the root along with of *Pavonia odorata* and wood of *Prunus paddum* is added to water tub.

Grass is made into curtains and is in universal demand for thatching.

'Ketone' fraction of the essential oil varies from 20–90% and yields 'beta vetivone' (2–49%).

236. *Viola canescens* Wall. ex Roxb.

Family: Violaceae.

Local names: Banefshá (H), Gul-banefshá (Unani).

Description: A softly pubescent herb. Stem very short, always producing runners. Leaves tufted, ovate, deeply cordate, crenate; stipules fringed. Flowers liliac. Sepals produced at the base. Capsule 3-valved.

Distribution: Throughout hilly districts, 1000–2300 m.

Uses: The plant yields *Banefshá* of trade which is diaphoretic and antipyretic. It is made into a syrup *Sherbat Banefsha* and used to relieve febrile symptoms. The plant cures malaria, bronchitis and asthma.

The roots are emetic and purgative. It is a good febrifuge, tonic, diuretic; alleviates thirst and removes inflammation. In Punjab an oil *Raugan é Banefsha* is prepared. The juice causes nausea and vomiting.

The leaves are a reputed drug for cancer and are considered of great value for relieving pain, due to cancerous growth in throat. Externally poultice of leaves is applied.

The flowers called *Gul Banefsha* possess laxative, emollient, demulcent, astringent and diaphoretic properties. These are used for restraining suppuration. In syrup it is given in bilious affections, epilepsy, nervous disorders, prolapsus of rectum and uterus. The flowers are also a household remedy for treatment of cough, sore throat, kidney trouble and liver disorders. A preserve known as *violet sugar*, made from the flowers, is of great value as an excellent cure for consumption (Fig. 5.82).

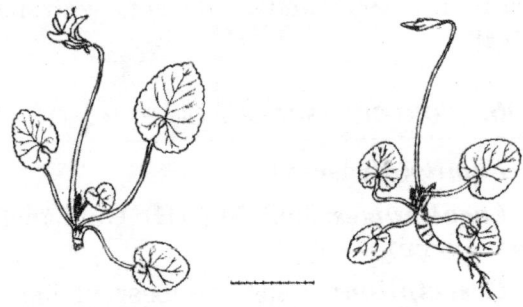

Fig. 5.82: *Viola canescens* Wall.

After boiling the plant in water, it is drunk and relieves fever, cough, sore throat and applied to forehead. A paste in water with coriander seeds is used for curing insomnia due to heat.

The seeds are purgative, are given in urinary complaints and as useful corrective of gravel.

Extent of utilization and trade: The flowers are extensively collected and sold in the market. They are sometimes mixed with the leaves also in order to make them cheap. An important ingredient of *jaushanda* sold in market as a cure for cough and cold.

Phenology and time of collection: Fls. April.

The flowers are collected during Sept-Oct and dried for sending to the market.

Other species

a. *Viola serpens* Wall. (Banafshá-H; Dundibirali-TG, Thungtu-Kum). A stoloniferous, glabrous herb. Stem long. Leaves ovate-cordate; stipules toothed. Corolla pale-lavender, lower petals streaked darker violet purple, rarely all white. It is febrifuge, diaphoretic and antipyretic. The root is emetic and contains glycoside *methyl salicilate* an alkaloid *violaquercitrin* identical with Rutin and a Saponin. Abundant in subtropical and temperate region, 900–2700 m. The plant has the same properties as the above species.

b. *Viola odorata* L. (Violet flower (Eng) Banfshá, Gulbanafsha). The flowers possess slighlty laxative propetries. The best form is syrup to be given as a laxative to children. It is also employed for its mucilaginous, demulcent and expectorant properties. The roots, and seeds are emetic and purgative due to the presence of *violin* and *violia*. Effective remedy in fevers, cold, catarrh pleurisy, pneumonia and cough. Ingredient of *Joshanda* for catarrh and fevers.

The roots contain *violine* and *violia*. The whole plant is collected in February and March. About 200 kg are reported to have been collected only from Kanatal forests in Tehri Garhwal (Uniyal and Issar, 1967) and sold at Rs 12 per/kg. In the local market all the species are mixed up and sold as *Banefsha*. Sometimes the leaves are adulterated with the flowers and sold as *Gulbanefsha*.

c. *V. biflora Violet* is much used in Spain and the roots are given as emetic; the flowers as an emollient, pectoral, diaphoretic and antispasmodic; the leaves used as an emollient and laxative.

237. *Viscum album* L.

Family: Loranthaceae.

5

Local names: Bándá (Hindi), Chulu-ka bándá (Jaunsar), Rini (Kulu), Kahabang (Punjab); Hurchu (Nepal). The Mistletoes (Eng.).

Description: A large parasitic shrub, green all over; branches dichotomous or whorled. Leaves sessile, cuneate-oblong or oblanceolate, with 3–5 longitudinal basal nerves. Flowers dioecious, sessile, in clusters of 3–5, supported by concave bracts. Perianth segments 3–4, triangular, deciduous. Fruit ellipsoid white, smooth almost transparent.

Distribution: Parasite on a number of plants, like rosaceous shrubs. Walnut and Willows, extending to an altitude of 2000 m Kashmir to Nepal.

Uses: The berry is sweet and given in enlargement of the spleen and in cases of tumour. It is laxative, tonic, aphrodisiac, diuretic and cardiotonic.

It contains two active principles; one depresses the heart and contracts isolated intestine and uterus, while the other produces fall of blood pressure.

Other species

a. *Viscum articulatum* Burm. Vern names. Pudu (Hindi); Kamini (Sansk).
   The plant is given in fever attended with aching limbs.
   In Himalayan regions, east of Nepal such as Assam, Khasi hills and in central India, (Madhya Pradesh), western peninsula. In Himalaya it has been reported from the Siwalik region up to 2000 m in outer Himalaya on *Cordia vestita, Cornus capitata, Pyrus pashia,* etc.
b. *Viscum monoicum* Roxb. Vern Bánda, Kuchlé-ka-malang. (Hindi).
   The plant is used as a substitute for *Strychnos nux-vomica.* The dried leaves are powdered and used as a substitute for *strychnine* and *brucine,* and the alkaloid is taken over from the host plant *Strychnos nux-vomica.*

238. *Withania somnifera* Dunal.

Family: Solanaceae.

Local names: Ashwakandica (Sansk); Ashwagandhá(Hindi), Asgandh (Trade name; Winter Cherry (Eng).

Description: An undershrub, erect, 30 cm to 150 cm high branching, all parts clothed with white, stellate hairs. Leaves ovate, up to 10 cm long, a little less in breadth. Flowers bisexual, greenish or lurid yellow, in cluster of about 25 forming umbellate cymes. Berries covered by enlarged inflated calyx, red, globose, smooth; part outside the calyx pubescent. Seeds yellow somewhat scurfy (Fig. 5.83).

Distribution: In drier parts of India, ascending to 1700 m in the Himalayan region.

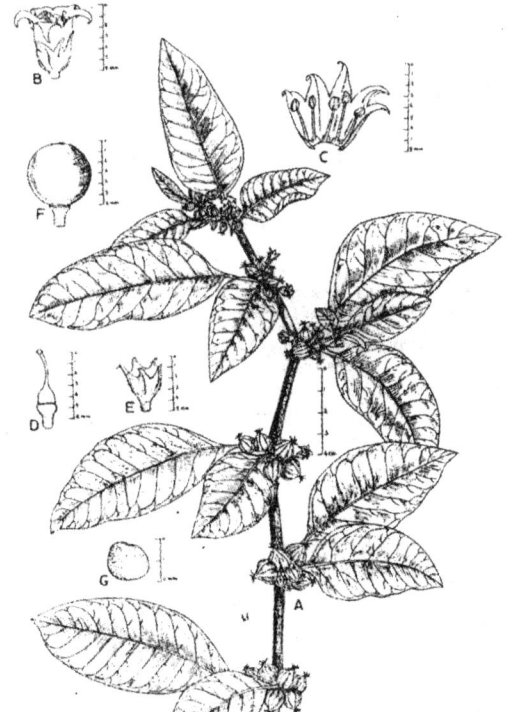

Fig. 5.83: *Withania somnifera* Dunal A. flowering and fruiting branch, B. flower; C. corolla opened to show androecium; D. gynoecium; E. calyx; F. berry; G. seed.

Uses: Roots and leaves of this plant are used locally. The leaves resemble *vasaca* leaves. In case of abscess and swelling of the joints, the leaves are smeared with the oil, warmed and bound on the affected parts to relieve pain and swelling. The leaves are bitter and given as infusion in fever. In east Africa it is regarded as narcotic and anti-epileptic.

The roots are tonic, diuretic, narcotic, abortifacient, used in rheumatism, consumption, debility from old age. The roots cure lumbago, give strength to the body and increase sexual power.

Ashwaarisht, an Ayurvedic medicine, is prepared and traded commercially.

It has also been reported that the roots in the powder form, when used can help in conception, if taken before menstruation along with the cow's milk.

As an insecticide, the leaves are used for killing lice infesting the body. A paste of the leaves is used for the treatment of carbuncle, erysipelas and syphilitic sores. An ointment prepared by boiling leaves in fat is useful in bed sores and wounds, while fresh leaves juice is applied to anthrax pustules.

Bark infusion is given is asthma.

The alkaloid *solanine* possesses the toxic properties, characteristic of the saponins. When introduced in blood stream they produce destruction of RBC, vomiting, diarrhoea, paralysis and coma: sometimes preceded by violent convulsions. The physiological effects of *solanine* and *solsomine* when compared should equal haematomic properties. Both are stimulating to non-striated muscles and in large doses cause spasm of the muscles. *Solanine* stimulates the heart, while *solasonine* depresses it.

Root contains *somniferine*. Root is a powerful tonic and is indigenous medicine given in general debility, dyspepsia, rheumatic affections, etc. (in doses of 30 grams).

The berries and seeds of *Withania* are diuretic and used in chest complaints. Paste of berries is an effective ointment for ringworm. Dried fruits are used in coagulating milk. These are also used in flatulant colic, dyspepsia and intestinal affections.

239. *Xanthium strumarium* L.

Family: Compositae, nom alt. Asteraceae.

Local names: Chhotá-gokhru (Hindi); Arishta (Sansk); Dhupa, Shankeshwar, Shanleshvara, Skankhahuli (Mumbai).

Description: A coarse and rough annual herb. Leaves alternate, heart-shaped or 3-lobed, clothed with rough hairs, irregularly toothed. Male and female fls. on the same plant. Flower heads bisexual, but sterile flowers in upper axils. Male flower heads consist of many dull purple tubular florets. Female flower heads below the male ones, bearing a floret without corolla. Achenes 2, in each head, without a tuft of hairs, surrounded by the 2-beaked hardened covering of bracts and clothed with strong hooked spines (Fig. 5.84).

Distribution: Throughout India, in western Himalaya, ascending to 2000 m.

Uses: The plant is astringent, emollient, sedative, diuretic, sudorific, antiscrofulous, diaphoretic. In decoction (1 part in 10 parts of water) it is given in malaria fever, in gleet, leukorrhoea, menorrhagia, urinary and kidney diseases.

Leaves powder given internally in herpes and scrofula.

Roots are bitter tonic, useful in cancer and also in scrofula. Extract root used locally in ulcers, boils and abscesses.

Fruits are tonic, diuretic, sedative and cooling. These are also given in smallpox. Seeds

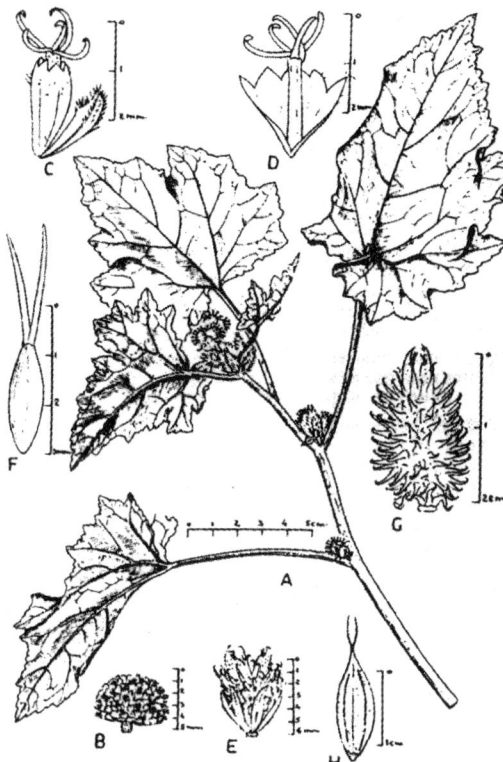

Fig. 5.84: *Xanthium strumarium* L. A. flowering and fruiting branch; B. male head; C. male flower; D. same opened; E. female head; F. female flower; G. fruiting involucre; H. achene

used for resolving inflammatory swellings. They contain a glucoside *xanthostrumarin* and oxalic acid.

240. *Zingiber officinale* Roxb.

Family: Zigiberaceae.

Local names: Adrak (Hindi); Ginger (Eng.).

Description: A perennial herb with a horizontal rootstock, aromatic. Stem elongated, leafy. Leaves linear, without stalks, lower part surrounding the stem, upper apex pointed, smooth, leaves long, about 3 cm in width. Flowers rarely produced; spikes ter-minating the leafy system, oblong cylindric, up to 7 cm long with slender stalks, enveloped by a membrane. Corolla greenish yellow, lips dark purple, often spotted with yellow, 3-lobed. Fruit a capsule.

Distribution: Cultivated throughout the country from Himalaya to Kerala, run wild in many places.

Uses: The underground stem or rhizome, both fresh and dried are used medicinally; fresh rhizome known as *adrak*, while the dried one called *sinth* or *sunth*.

The rhizome is used as a stimulant, carminative favouring agent. It is given in dyspepsia and flatulant colic. The rhizomes are prescribed as an adjunct to many tonic and stimulant remedies.

It is a valuable drug in dyspepsia, vomiting, spasm and painful disorders of stomach, rheumatism, piles, bilious throat, dropsy, tympanitis and other bowel complainst. It is in powdered form prescribed with other drugs as a cure for rheumatism, dysentery, cholera piles, etc. Dried root powder when kept in mouth, stimulates saliva flow, warms and tones stomach.

Ginger tea is pleasent and effective drink during winter season. It promotes mensus in amenorrhoea due to cold.

Enclosed in *tulsi* (*Ocimum* sp.) leaves is used for relief in toothache.

As a paste in water, ginger is locally used as stimulant and rubifacient in headache, shortness of sight due to deficient contractile power of the iris.

Rhizome yields about 1–3% of volatile oil containing, *camphene*, *β phellandrene* and *zingiberene*, *gingerol* and pungent constituents, while other constituents are given as above.

Dried ginger is a well-known domestic medicine that helps in food digestion, expels

wind from belly, increases appetite, intellect and memory. Locally called as *sonth* when ground with black salt (kala-namak), moistened with lime juice and dried (process repeated thrice), if taken 1 gram after meals helps in the digestion stomachache and flatulance. Dried ginger (50 gram) finely ground and mixed with honey (150 gram) is taken two times daily morning and evening, relieves lumbago, loss of memory and pain during cold.

Other Species

a. *Zingiber cassumunar* Roxb. is the wild species called 'Vanardraka' in Sanskrit. The rhizome has same properties as the domesticated species.

b. *Zingiber zerumbet* Roxb. ex Smith, Vern. name Sthulgranth (Sansk), Mahabaribach (Hindi) is also reported but having the same uses as *Z. officinale* Roscoe.

241. *Zygophyllum simplex* L.

Family: Zygophyllaceae

Local names: Vern. name Lani, Lunwo (Rajasthan).

Description: A prostrate, glabrous, much branched, annual herb. Leaves simple, small, sessile, cylindric, fleshy, obtuse; stipules lanceolate and acute. Flowers on peduncles as long as the ovate, oculate sepals. Petals spathulate. Capsule deflexed, rough, of 5 compressed, 2–3 seeded cocci.

Distribution: Distributed commonly in salt districts in desert areas, also in tropical Africa, Central Sahara, West Asia and Sindh province in Pakistan.

Uses: The seeds are collected and eaten under the name *Alati* by the nomads of *Multan*. Also fodder for camels.

The seeds are considered anthelmintic.

References

Abrol BK Chopra IC. 1962. Some vegetable drug resources of Ladakh (little Tibet). Pt. *I.Curr. Sc.* **31(8):** 324–326.

Acosta C. 1578. *Tractodop de las drogies y medicinas de las Indas orient,* Burgos.

Acharya (Yadav ji Trikamji) 1950. *Dravyaguna Vignan.* Niranayasagar a Press, Bombay. pp. 1–264.

Agarwal, VS. 1975. Fruits of Indian plants as a source for medicine. *Proc. National Symp.* (Abstract) p.41.

Agarwal, M. 2002. Medicinal Plants. 9. pp. 100.

Agharkar SP. 1953. *Botany* Pt. 1. *Medicinal Plants.* Revised edition of Bombay Gazeteer, Bombay.

Agharkar SP. 1991. *Medicinal Plants of Bombay Presidency.* 8, 250 pp.

Agnivesh Charkasamhita, *Chikitsa*: 3.267. Niranayasagra Press. Bombay. pp. 422–423.

Ahluwalia KS. 1952. Medicinal Plants of Kangra valley. *Indian For.* **78**: 188–194.

Ahluwalia KS. 1968. Medicinal Plants of Kerala. *Nagarjun* **11**(12): 629–37.

Ahluwalia KS. 1965. Medicinal Plants of Hardwar, UP. Nagárjun 8(11): 665–678.

Ahuja GM,Boyd MR and Cardellineli JH. 1994. *Ethnobotany and the search for new drugs.* pp. 185.

Ahuja BS. 1955 (Reptd 1993). *Medicinal Plants of Saharanpur.* 102 plus 95 pp.

Aitchinson JET. 1869. Lahul, its flora and vegetable products, from communication received from Rev, Heinrich Jaesekhe of the Moravian mission. *J. Linn. Soc.* (bot) **10**: 89–101.

Aitchinson JET. 1881. *Memoirs of the economic products of NWest Provinces of India.* Allahabad.

Ainslie, W. 1813. *Materia Medica of Hindoostan and Artisans and Agricultural Nomenclature.* Madras.

Ainslie, W. 1826. *Materia Medica of Hindoostan and Artisans and Agriculturists Nomenclature.* London. 2 Vols.

Anonymous 1949. *The British Pharmaceutical Codex.* London.

Anonymous 1955. *The India, Pharmacopoea* Manager of Publications. Delhi.

Anonymous 1953. *Pharmacognosy of Ayurvedic Drugs of Travancore-Cochin*, Trivandrum Sr. No 1(1951.No 2 (1953).

Anonymous 1960. List of Medicinal Plants deposited in various herbaria of the Botanical Survey of India. *Bull. bot surv. India* 2 (1 & 2): 180–273.

Anonymous 1975. Recent Advances in the Development, Production and Utilization of Medicinal and Aromatic Plants of India. Abstracts. CIMPO (CSIR) Symp. Lucknow.

Anonymous 1976. *Medicinal Plants of India.* Vol 1. ICMR (CSIR) New Delhi.

Anjaria JV., Prabia M, Bhatt G and Khamar R. 1997. *Nature Healing.* A glossary of selected Indigenous Medicinal Plants of India. Sristi Innovations, Ahmedabad. pp. 1–50.

Anshutz EP. 1974. Therapeutic Byways. The *Homeopathie Bull.* **47**: (3): 53–65.

Arjun S. 1879. *Catalogue of Bombay Drugs including a list of Medicinal Plants of Bombay used in the fresh state.* Bombay.

Arora RB. *et al* 1958. Antiarryhthmic and anticonvulsion activity of Jatamansi. *Indian J. Med. Res.* **46(6)** 782–91.

Arora RB. 1965. *Cardiovascular pharmatherapeutic of six medicinal plants, indigenous to India.* Hamdard National Foundation, New Delhi.

Arora RB. 1965. *Nardostachys jatamansi-a chemical, pharmacological and clinical appraisal.* ICMR, New Delhi.

Atkinson ET. 1882. *Gazetteer of Himalayan districts of northwest provinces of India.* Allahabad.

Baden P and Baden H. 1868–1872. *Handbook of Economic Products of Punjab.* 2 Vols. Roorkee.

Badhwar RL and Chopra RN. 1940. Poisonous Plants of India. *Indian Journ. Aric. Ser.* **10**: 1–44.

Badhwar, RL, Chopra IC and Nayar SL. 1946 Reputed abortifacient plants of India. Indian *Journ. Agric Sc.* **16**: 342–355.

Badhwar RL, Chopra IC and Nayar SL. 1941. Insecticidal and Pesticidal plants of India. *Journ. Bombay nat. hist Soc.* **42**: 854–902.

Baden P. 1975. Kashmir Products. *Indian For.* **18**: 5 1–53.

Bajpay A. 1993. Ecological Studies of *Boerhaavia verticillata* with special reference to phytochemical and Therapeutic Importance. Ph. D. Thesis, BHU, Varanasi.

Bal SN. 1932. *Catalogue of medicinal plants exhibited in Industrial section of the Indian Museum.* Calcutta.

Bal SN. 1940. *An outline of Pharmacopoeal drugs of vegetable origin.* Calcutta.

Balapure KM. 1975. Observations on the availability of medicinal plants on way to Rupkund (District Chamoli) *Proc. Nat. Symp.*

Balasubramanian AV. Vijaylakshmi K and Sridhar Subhasini and Arumugaswamy 2000. Vrkshyaayurveda Expts. linking ancient texts and farmers practices. *LEISA, INDIA* **16(3)**: 31–34 (Sept).

Bans R. 1979. Herbs for health and Beauty. *Illustrated Weekly of India.* February issue.pp. 18–24., 39. Bombay.

Barnard C and Finnemore H. 1945. Drug Plant investigations. *J. Council of Sci. and Industr. Res.* **18**: 277–285.

Basak SL. 2004. *Nutritional and Medicinal value of Plant Foods.* pp. 272.

Basu BD and Kitikar KR. 1981. *Indian Medicinal Plants* 2Vols Alld. (revised 1933 by E. Blatter, JF Caius and KS Bhaskar). Reptd. 1994 (4 vols. Text and 4 vols. Plates).

Baxi N and Advani CH. *Alternative medicines-Acupuncture.* The needle that heals all ailments.

Bedi R. 1961. *Sarapgandha.* Atma Ram and Sons. Delhi.pp 1–39.

Behl PN and Behl A. 1960. *Skin irritant and sensitive plants found in India.*

Bently R and Trimen *Medicinal Plants.* Vols I-IV.

Bennett AW. 1875. *On the medicinal Products of India. Simarubaceae and Burseraceae.* London.

Bhagwan Dash and Lalitesh Kashyap. *Diagnosis and treatment of Diseases in Ayurveda based on Ayurveda, Sukham of Todarmanda* Pt. I.

Bhagwan Dash. *Fundamentals of Ayurvedic medicines.*

Bhagwan Dash and Acharya Menfred M Junius. *A handbook of Ayurveda.*

Bhakuni DS, Dhar ML, Dhar NM, Dhawan N and Mehrotra BN. 1969. Screening of Indian Medicinal plant for biological activity. Indian *Journ. Exptl. Biology,* **7**: 250–260.

Bhandari MM. 1954. On the occurrence of Ephedra in the Indian Desert. *Journ. Bombay nat. hist Soc.* **52(1).**

Bhardari CR. 1950. *Banaushadhi Chandrodaya. An encyclopaedia of Indian Botany and herbs.* Vols 1–10. Gyanmandir Bhanpura, Indore state.

Bhattacharya MN. *Different Diagnosis in medicines.*

Bhattacharjee SK. *Handbook of Medicinal Plants.* Pointer Publishers, Jaipur

Bhattacharjee SK. *Handbook of Aromatic Plants.* Pointer publishers. Jaipur.

Bhattacharjee SK. 2004 (Edt). *Handbook of Medicinal Plants.*

and the Sikkim Himalaya. Bengal Govt. Press, Calcutta. pp 157.

Birdwood GCM. 1862. *Catalogue of the economic products of the Presidency of Bombay, being a catalogue of Govt. Central Museum Division; Raw products.* Calcutta.

Birdwood GT. 1936. *Practical Bazar Medicines.* Thacker Spink. Calcutta.

Bisht BS. *et al* 1960. Pharmacognosy study of the rhizome of *Curculago orchioides* Gaertn. *Ibid.* **19(c)**: 252.

Bisht BS. *et al* 1957. Pharmacognosy of roots and leaf of Coccinea indica Wight and Arn. *Journ. Sci and Industr. Research.* **16**: 46–51.

Biswas K and Chopra RN. 1956. *Some medicinal plants of Darjeeling*

Borthakur SK. 1976. Less known uses of plants among the tribes of Karbi-anglong (Mikir). *Bull. bot. Surv. India,* **18**(4): 16–171.

Borthakur SK 1987. Less known medicinal uses of plants among the tribes of Karbi-anglong (Mikir) Assam. *Bull. bot. Surev. India.* **18(4):** 166–171.

Boyko H. 1938. *Species list of potential plant sources for raw materials in Palestine,* Hadar. Tel Aviv.

Bradu, BL and Atal, CK. 1970. Cultivations of *Ammi majus* L. in Jammu. L. *India Journ. Pharm.* **32(6):** 165–167.

Bremer H and Karel G. 1947. Progress in Phytopharmacy. *Agriculture Review, Ankara.* vol **8:** 20–28.

Brij Lal. 1988. Ethnobotanical studies of the Baiga tribe of Madhya Pradesh. *Bull. Trib. Research and Development.* **16**(3): 23–28.

Brown E, Mathews WSA 1955. Notes on the aromatic grasses of commercial importance. *Pl. Animal Prod.* **2:** 174–187.

Budhiraja KL and Griffithal. 1943. Some common latex bearing woody plants of India. *Indian For.* **69**: 305–310.

Burkill IH. 1935. *Dictionary of the economic products of Malaya peninsula.* 2Vols. London.

Caius JF and Radha KS. 1939. The gum arabica of the bazars and shops of Bombay. *Journ. Bombay nat. hist. Soc.* **41**: 261–271.

Caius JF and Mahskar KS. 1930. *Therapeutic notes on some plants used medicinally in India.* Bombay.

Caius JF. 1936. The medicinal and poisonous plants of India. *J. Bombay nat. hist. Soc.* 37: 917–941.

Caius JF. 1941. The medicinal and poisonous Labiatae of India. *J. Bombay nat. hist. Soc.* 42(2): 380–420.

Cardoso, J. 1905. *Subsidious para a materia medica therapeutica das possesseos ultramarines Portuuesas.* 2 Vols. Lisbon.

Chacko TN. 1993. *Medicinal Plants of East Godavari District.* 2, 56.

Chakravarty HL. 1938. Bengal Medical flora (Compositae). *Sc. and Cult.* **2**: 300–302.

Chakravarty HL. 1942. Identity of Punarnaba. *Journ. Indian bot. Soc.* **21**: 87–89.

Chakravarty HL. 1945. Studies on medicinal plants. *Aristolochia indica. Indian Jour. Pharmacy,* **9**: 96–101.

Chakravarty P and Chakravarty HL 1946. Indian Aconites. *Economic Bot.* **8**: 366–370.

Chakravarty HL. 1975. Herbal Heritage of India. *Journ. bot. Soc. Bengal* **29**: 97–103.

Chandrasekharan S and Swamy RS. 1995. Changes in Herbaceous Vegetation following disturbances due to biotic interference in natural and manmade ecosystems in Western ghats. *Trop. Ecology.* 30: 411–424.

Chaudhury II. 1956. Medicinal Plants of W.Pakistan. Podophyllum emodi L. *Science.* **8(5)** : 230–233.

Chaudhury II. 1961. The medicinal plants of arid regions of west Pakistan. *Pakistan Agriculture.* 1961.

Chaudhuri Rai, HN and Kayal RN. 1971. Pharmacognostic studies of stem bark of 4 species of Cinnamomum. *Bull. Bot. Surv. India,* **13** (1–2): 94–104.

Chaudhuri Rai HN. Notes on the magico-religious beliefs about plants among the Lodhas of Midnapur, Bengal. *Vanyajati,* **23** (2–3): 20–22.

Chaudhuri Rai HN. and Tribedi 1976. On the occurrence of some medicinal plants in 24-Pargana, West Bengal. *Bull.Bot.Surv. India,* **18**; **(1–4)**:161–165.

Chaudhuri Rai HN, Pal DC and Tararafdar CR. 1975. Less known uses of some plants from tribal areas od Orissa. *Bull. Bot. Surv. India* **17(1–4)**: 132–134.

Chaudhuri Rai, HN, Pal DC and Tarafdar CR. 1984. Less known uses of some plants from the tribal areas of Orissa. *Bull. bot. surv. india.* **17**: 132–136.

Chopra RN.1932. *Indigenous Drugs of India.* Calcutta.

Chopra RN. 1933. *Indigenous drugs of India. Their medical and economic Aspects.* Calcutta.

Chopra RN. 1940. Medicinal Plants. *Sc. and Cult.* 5: 620–621.

Chopra RN. 1941. Insecticidal and Pesticidal Plants of India. *Journ. Bombay nat. Hist. Soc.* **42**:

Chopra RN. 1949. *Poisonous Plants of India.* Vol. I. ICAR Scientific Monograph. No **17**, Govt Press. Calcutta.

Chopra RN and Chopra IC. 1955. *Review of work on Indian Medicinal Plants.* ICMR. Publication. New Delhi.

Chopra RN, Chopra IC, Handa KL and Kapoor LD. 1958. *Indigenous Drugs of India.* Calcutta, 2nd edition.

Chopra RN and Nayar SL. 1956. *Glossary of Indian Medicinal Plants.* CSIR, New Delhi, pp. 1–330.

Chopra RN, Chopra IC and Varma BS. 1969. *Supplement to Glossary of Indian Medicinal Plants.* CSIR. New Delhi.l pp. 1–119.

Chopra RN, Chopra IC, Handa KL and Kapoor LD. 1961. *Review of Research on Indian Medicinal and Allied Plants of India.* Vols 1 and 2. ICAR Publication. New Delhi.

Chopra RN, Bhadwar RL and Ghosh S. 1962. *Poisonous Plants of India.* Vols 1 and 2. ICAR, New Delhi.

Chopra RN., Badhwar RL and Ghosh. S. 1965. *Poisonous Plants of India.* Vols 1 and 2. 2nd Revised edition. ICAR. New Delhi.

Chopra IC and Abrol BK. 1964. Some medicinal Plants suitable for cultivation in Indian Arid Zone. *Symp. Problems of Indian Arid Zone* pp. 148–152. Mimeo.

Chopra IC. and Kapoor LD 1968. Steroid sapogenin-bearing plants of India. *Indian For.* **94**(8) 620–630, B*ull. Regional Research lab. Jammu,* 1963: 183–189.

Chopra RN. Kapoor LD, Handa KL and Chopra IC. 1947. Vegetable and drug resources of Jammu and Kashmir. *Journ. Sci. and Inustr. Res.* **6**(12): 480–484.

Chopra RN. 1958. *Indigenous Drugs of India.* Calcutta.

Chopra RN, Handa KL and Kapoor LD. 1946–1949. Notes on the essential oil bearing plants growing in Kashmir. *Indian Journ. Agric. Sci.* **16**: 302-305., 1946, *Ibid,* **17(3):** 100–103, 389–392. 1947.

Chopra IC, Handa KL and Kapoor LD. 1951. Drug Resources of Himachal Pradesh. *Indian Journ. Pharm.* **13(5)**: 118–119.

Chopra IC and Handa KL. 1963. Utilization of some vegetable drugs of northwest Himalayan regions for Cortisone and other steroid sex hormones production. *Bull. Reg. Res. Lab. Jammu* **1**: 192–194.

Clusius C. 1567. *Aromatum er Simplicum alliqd medica mentonum apud Indos nascentium historia*. Antwerp.

Conway. D. 1973. *The Magic of Herbs*. Jonathen Capc. London.

Cruse Robert R. 1949. Chemurgie survey of desert flora in the America SW. *Econ. Bot.* **3**: 111–131.

CSIR 1988. Medicinal *Plants Bibliography of CSIR contributions*, 1950–1987. CSIR. New Delhi.

CSIR 1998. *A dictionary of Indian Raw Materials and Industrial Products*. In: The Wealth of India. CSIR, New Delhi.

CSIR The Wealth of India. A dictionary of Indian Raw Material and Industrial Products. New Delhi.

Dadlani, SA, Singh BP. and Mansoor K. 1971.Chinese Gooseberry. a new fruit plant. *Indian Hort.* **16**(1): 13–15

Dahanuk S, Boyd Me, and Cardellinali JH. 1994. *Ethnobotany and Search for New drugs*. pp.185.

Dalbir Singh. 1964. Some useful Cucurbits for the arid zone in Rajasthan. *Symposium Problems of Indian Arid Zone*, pp 135–139.

Dalziel 1948. *The Useful Plants of West Tropical Africa*. London

Dandiya PC *et al* 1958. Studies on Acorus calamus. I. Phytochemical investigation. *Canadian Pharm Journ.* 91(10): 607–610; II. Investigation of volatile oil. *Journ. Pharm. London* II: 161-163. III. Investigation on mechanism of action in mice. *J. Pharamacology, Ex-ptl.Therapy* 126(4): 334–336.1959. V. Pharmacological action Asarone and beta asarone on central nervous system. *Indian Journ.* Medical Research 50(1) 46–60. 1962. Effects of asarone and beta *asarone on con-ditioned* responses fighting behaviour and convulsions. British J. *Pharmacol. Chemotherapy* **20**(3): 439.

Daniel, Sunderraj and Balsubramanyam 1959. *Guide to the Economic Plants of South India*, Madras.

Dash B *et al* 1965. Preliminary report on chemical, Pharmacological and clinical screening of Saptrangi. Presented at the *Scientific seminar held under the auspicieux of Central Council of Ayurvedic Research held at BHU*, Varanasi on 29th August, 1965.

Dastur JF. 1968. *Medicinal Plants of India and Pakistan*. Bombay.

Deb DB. 1975. Impact of Taxonomy on the study of medicinal Plants. *Proc. National Symp.* pp 41–42.

Den Hertog WH and Wiersum KF. 2000. Timur (Zanthoxylum armatum) production in Nepal. Dynamics in nontimber forest resource management. *Mountain Res, and Development* **20(2)** : 136–145.

Deolia RK. 1994. *Medicinal Floral Ecology in Central India.*

Desai VB *et al.* 1966. Antimicrobial activity of *Abrus precatorius. Indian Journ. Pharmacy* **28**: 164.

Desal VB *et al.* 1964. The effect of *Abrus precatorius* on pregnancy of mice. *Curr. Sc.* **33(9)** : 585.

Deshmukh VK, Godbole SR and Mahabale TS. 1957. *Studies on Pharmacognosy of some Indian Medicinal plants*. Indian Drug Res. Assn. Poona Pub. I.

Devagiri GM, Dhiman RC, Kumar S. and Paul V. 1998. Cultivation and genetic improvement of *Heracleum candicans*: In: *Prospects of Medicinal Plants* (eds) Gautam PL, Rama R., Srivastava U and Singh BB. NBPGR, New Delhi.

Devaraj TL. *Speaking of Ayurvedic remedies for common diseases simple remedies based on herbal medicines.*

Dey KL 1895. *The Indigenous Drugs of India*. Calcutta.

Dey AC, Saxena HO, Uniyal MR. 1969. Botanical Exploration in the Bhagirathi valley with particular reference to the medicinal Plants. *Indian For.* **95(3)**: 190–207.

Dey AC. *Indian Medicinal Plants used in Ayurvedic preparation.*

Dhanuk S and Thatte U. 1996. *Ayurveda.* NBT. New Delhi.

Dhiman AK. 2004. *Medicinal Plants of Uttaranchal state.*

Dhiman AK. 2003. *Sacred Plants and their medicinal uses.* 11. plus 237. pp. 36 colour Plates.

Dobriyal RM and Narayana BH. 2000. *Medicinal Plants Resource Mangement. Current trends and future prospectives in Harvesting herbs* 2000. *Medicinal and Aromatic Plants* (eds AR Nautiyal, MC Nautiyal and AN Purohit). Bishen Singh and Mahendral Pal Singh. pp. 135–138.

Dymock W, Warden CJ and Hooper D. 1890–93. *Pharmacographia Indica.* London. Truber and Co. Vols 3. *A History of Principal Drugs of Vegetable orgin met within British India.*

Edward John Warming. *Bazar medicines and Common Plants of India with a full index of Diseases, indicating their treatment* by these and other agents procurable throughout India.

Engineer Mr and Gajjar MJ. 1919. *Bibliography of the Development of Indigenous drugs.* Monograph No **3**. Chemistry and Technical Lab Gwalior.

Farooque NA and Saxena KG. 1996. Conservation and utilization of medicinal plants in high hills of the central Himalaya. *Env. Cons.* **23**: 75–80.

Farooq S. 2004. *A handbook of Medicinal Plants of Himalayas.*

Firach RD, Brahanan M, Misra MK and Ahmed M. 2001. *Some less known medicinal plants in relation to Unani system of medicine from District Bhadrak, Orissa.* Hamdard Medicines, 44: 50–51.

Fleming J. 1810. A Catalogue of Indian Medicinal Plants and Drugs with their names in Hindostani and Sanskrit languages. *Asiatic Rec.* 153–196.

Flukiger FA and Handbury D. 1874. *A History of the Principal Drugs Pharmacology and Vegetative Origin met within Great Britain and British India.* London.

Gadgil M. 1991. Diversity: Cultural and Biological *Trends in Ecology and Evolution* **2**: 369–373.

Ganguli BN and Kaul RN. 1965. Utilization Potential of a few important medicinal Plants of Arid Western Rajasthan. *Symp. Recent Adv. in the Development and Utilization of Medicinal and Aromatic Plants of India.* RRL Jammu and Kashmir.

Gaur RD. 1977. Wild edible fruits of Garhwal Himalaya **1(1)** 66–70.

Gaur, RD. 1977. Edible Plants of Garhwal hills. Himalaya, **1(1)**: 66–70.

Gazetteer of Bombay State (Revised edn) 1953. *A. Botany,* Part *Medicinal Plants.*

Gedon J and Kinel FA. 1953. Steroid sapogenins aur Indischen Agave arten *Arch. Pharm Berlin.* **286**; 317–319.

Ghosh TP, Gupta BL and Krishna AS, 1929. Ephedra. *Indian For.* **55(4)**: 215–226.

Ghosh GK. 2000. *Herbs of Manipur.* 2 vols. 22 plus 16 and 1164 pp.

Gopal P, 1874. *A Catalogue of Drugs Indigenous to Bombay Presidency.* Bombay

Gopalachari R. *et al* 1952. Chemical Examination of the seed of *Achyranthes aspera L. Journ. Sc. Industr. Res.* **11(B)**: 209.

Gruaham HG. 1999. *Materia Medica. Being an account of the more important Crude Drugs of Vegetable and Animal origin.* 12 + 568 pp.

Guha Bakshi DN and Sensarma P. 1999. *A lexion of Medicinal Plants of India.* Vol **1**. 1999. pp 552. Vol **2**. 2001. pp 408, tals. bib.

Gujaral ML. *et al.* 1957. An experimental investigation of antiarthritic effect of some indigenous and modern remedies. *Journ. Med. Sci.* **44**: 657.

Gujaral ML. 1960. Antiarthritic and antinflammatory activity of gum Guggulu. *Indian Journ. Physiol. Pharmacol.* **4**: 267–273.

Gupta AK, Vats SK and Lal B. 1998. How cheap can a medicinal plant species be. *Curr. Sc.* **74**: 565–566.

Gupta AK. 2003. *Quality standards of Indian Medicinal Plants.* Vol 1 16. 163 pp.

Gupta B. Some problems on the collection of Indian plants for screening of their biological activities. *Proc. National Symp* (Abstr) p.7.

Gupta B and Bal S.N. 1952. Pharmacognosy of Indian Ephedra. *Journ. Sc.Industr. Res.* **2**: 4.

Gupta NG. *The Ayurvedic system of medicine-an exposition in English of Hindu Medicines.* 2 Vols.

Gupta KC. *et al* 1955. Antibacterial substances from *Ocimum sanctum. Antibiotics and Chemotherapy.* **5**:22. 1955.

Gupta R. 1964. Survey Record of Medicinal and aromatic plants of Chamba Forest Divison. HP. *Indian For.* **90**: 459–468.

Gupta, R. 1971. Senna has a growing export market. *Indian Farming* 21(8): 29–32.

Gupta, R. 1977. Medicinal Plants components of Indian forests. *Indian Farming* **26(11)**: 96–98.

Gupta, R. 1972. Vital drugs and essential oil. bearing plants as future cash crops in India. *Indian Farming* **22(6)**: 67–75.

Gupta RK. 1970. Resource Survey of Gummiferous Acacias in western Rajasthan. *Tropical Exology*, **10(2)**: 148–161.

Gupta RK and Saxena SK. 1968. Resource survey of Salvadora oleoides and S. persica for nonedible oil in Rajasthan. *Tropical Ecology*, **9(2)**: 140–152.

Gupta RK 1962. Medicinal Plants of west Himalaya. *Journ. Agric Trop. et bot. appl.* **9** (1 and 2): 1–54.

Gupta RK.1960. Some useful and medicinal plants of Nainital in Kumaon Himalaya. *J. Bombary nat. hist. Soc.* **57**: 309–324.

Gupta RK, Dabholkar MV and Tejomurty PS. 1959. Some medicinal plants in and around Pondicherry. J. *Bombay nat. hist. Soc.* **56(2)**: 235–249.

Gupta RK, and Marlange M. 1961. *Le Jardin Botanique de Pondicherry, Institute Franscais de Pondicherry. Trav. sec. sc. et Techn.* 3(1): 133.

Gupta RK, Gaur YD, Malhotra SP and Dutta BK. 1966. Medicinal Plants of the Indian Arid Zone. *Journ. d'agric. trop. et bot. appl.* **13(6–7)**: 247–288.

Gupta RK, Shukla D and Nambiar KTN, and Ghosh SP. 1982. Evaluation of Turmeric varieties for Doon valley. I*ndian J. Soil Conservatin* **10(2)**: 101–103.

Gupta SS. 1962. Experimental studies of pituitary diabetes. Pt. I. Inhibitory effect of few ayurvedic antidiabetic remedies on interior pituitary-induced hypoglycaemia in albino rats. *Indian Journ Med. Res.* **50**(1): 73–81.

Hamdard, Delhi 1950. *Village Physician.* 2 Vols. Hamdard Publication Delhi.

Hamilton F. (Buchnan) 1904. An account to the genus including the herba toxicaria of the Himalayan mountains or the plants with which natives poison their arrows. Edinburgh. *JSC*: **249**, 1824. *Trans. bot Soc.* 5. *Annals of Royal bot Gardens.* Calcutta **10(2)**: 1904.

Handa KL and Rao PR. 1970 Xantho toxin from *Heracleum candians. Res and Ind.* **15(3)**: 164–165.

Handa KL, Chopra IC and Sobti SN. 1957. Aromatic Plant Resources of Jammu and Kashmir. *Journ. Sc. and Industr. Res.* **16(5)**: 1–28.

Handa KL and Kapoor, LD. 1951. Drug Resources of Himachal Pradesh. *Journ. Pharmacy.* **13(5)**: 118–119.

Hardman, R. 1969. Pharmaceutical products from plant steroids. *Trop. Sci.* **11(3)**: 196–228.

Hass P and Hill TG. 1921-22. *An Introduction to the Chemistry of Plant products.* 2 Vols.

Hazelton LW *et al.* 1948. Studies on the pharmacognosy of *Euphorbia pilulifera. Journ. American Pharm. Assoc.* (science edn) **37**: 491.

Hiralal and Gupta R.K. 1988. Ginger a rainfed crop for hill farming. *Indian Farmer's Digest.* **21**(5): 17–20.

Hocking GM. 1952. *A study of medicinal plants of West Pakistan.* Rome pp. 38.

Hocking GM. 1955. A Dictionary of terms in Pharmacognosy. Illinois. USA.

Hodes ME *et al* 1960 *Vinca lleucablastine*: A preliminary clinical studies. *Cancer Res.* **20**(7): 1041–1049.

ICMR (Indian Council of Medical Research). 1976. Medicinal Plants of India. Vol. 1. ICMR. New Delhi.

ICMR. 1976. Medicinal Plants of India. Vol. I. Indian Council of Medical Research, New Delhi.

International Library Association 1998 *Medicinal Plants Resource Book-India.* pp. 598.

IUCN 2000. *National Register of Medicinal Plants* pp. 161.

Jain SK. 1963. Observations on the ethnobotany of tribals of Madhya Pradesh. *Vanyajati,* **11(4)** 177–183.

Jain SK. 1963. Studies in Indian ethnobotany. Plants used in medicine by tribals of M.P. *Bull. Regional Research Lab, Jammu,* **1:** (1and 2): 126–128.

Jain SK. 1963. Studies in Indian Ethnobotany. Less Known uses of fifty common plants from the tribal areas of MP. *Bull. bot survey of India,* **5(3and 4):** 223–226.(b.) *Bull. Regional Res. Laboratory* 1: 126–128.

Jain SK. 1965a. On the prospects of some new or less known medicinal plants resources. *Indian Med. Journ.* **59(12):** 270–272.

Jain SK. 1965. Medicinal plant lore of the tribals of Bastar. *Economic Bot.* **19(3):** 236–250.

Jain, SK. 1964. Wild plants of tribals of Bastar, MP. *Proc. nat. Inst. Sci. India,* **30B:** 56–80.

Jain SK. 1968. *Medicinal Plants.* National Book Trust. New Delhi. 2nd edn. 1975.

Jain SK. 1967. Ethnobotany, its scope and study. *Indian Mus. Bull.* 2(1)79–80.

Jain SK and Tarafdar CR. 1963. Native plant remedies for snakebite among the Adivasis of central India. *Indian Medical Jour.* 57**(12):** 307–309.

Jain SK and Dey JN. 1966. Observations on the ethnobotany of Purulia District in West Bengal. *Bull. bot. surv. Ind.* **8.**

Jain SK and Tarafdar CR. 1970. Medicinal Plantlore of the Santhals. *Economic Botany,* **24(3):** 241–278.

Jain SK 1992. *Ethnobotany in India.* Proc. seminar. Current Tends in Plant Science. Delhi. Abstr. 70–80.

Jain SK 1981. *Glimpses of Indian Ethnobotany.* Oxford and IBH Publisher. Co. New Delhi.

Jain SK. 1981. Observations on the ethnobotany of the tribals of Central India. In SK Jain (ed). *Glimpses of Ethnobotany.* pp. 191–198.

Jain SK and Saklani A. 1991. Observation on the ethnobotany of the Tons valley region in the Uttarkashi district of the northwest Himalaya, India. *Mountain and Development* **11:** 157–161

Jain SK. 1991. *Dictionary of Indian Folk Medicine and Ethnobotany.* 18, plus 311 pp. Colour plates 16, 433, line drawings.

Jain SK. 1996. *Ethnobiology of Human welfare.* Proceedings of 4th International Congress of Ethnobiology. 1996. 15 plus 519 pp.

Jain SK. 1995. *A Manual of Ethnobotany.* 2nd edn. 12 plus 193 pp.

Janardan KP. 1963. An enumeration of the medicinal plants of Khéda taluka (Maharashtra state), *Bull. bot. surv. India.* **5(3and 4):** 363–374.

Joshi P, Pangti, YS, Joshi, DR and Danl, DD. 1989. *Uttarakhand Ke Vanya, Kand, 'Mool, Phal'*(Hindi). Uttarakhand Sewa Nidhi, Almora.

Joshi SG. 2000. *Medicinal Plants.* 35 plus 491 pp. line drawings.

Joshi P. 1947. *Monograph of Himalayan Drugs.*

Juyal P and Uniyal MR. 1966. Medicinal plants of commercial and traditional importance in Bhillangna valley of Tehri Garhwal. *Nagarjuna* **10(1):** 26–36.

Kala CP. 1998. *Ethnobotanical survey and propagation of rare medicinal plants for small farmers in the bufferzone of valley of flowers.* National Park. Garhwal. Himalaya. Dehradun. Wildlife Institute of India.

Kala CP. 2003. *Medicinal Plants of Indian Trans-Himalaya.* 10 + 141 pp. (ISBN: 81–211–0151–4)

Kamathy RV. Chaudhri Rai HN and Kayal RN. 1971. Pharmagonostic studies on the genus Solanum. *Bull. bot surv. India. pt* 1.**13**(3–4): 224–235.

Kapoor LD, Chopra RN and Chopra IC. 1951. Survey of economic vegetable products of

Jammu and Kashmir, Sindh Forest Division. *Journ. Bombay nat. Hist. Soc.* **50(1)**:101–127.

Kapoor LD. Vegetable raw material yielding Tropane in India. *Indian Drugs* **3**.

Kapoor LD. and Sarin YK. 1962. Indian Podophyllum; its distribution and availability in northwest Himalaya. *Regional Research laboratory,* 1(1): 38–39.

Kapler FT. 1974. Drug action and Reaction. *The Homeopath Bull.* **47(6)**: 101–105.

Karandikar GF *et al.* 1960. Antiinflammatory activity of some ayurvedic remedies and their influence on the hypophysioadreno cortical axis in white rats. *Indian J. Med. Res.* **48(4)**: 482– 487.

Kasera PK, Shukla JK, Prakash J, Sahran P and Chawan DD. 2000. Biology, conservation and mediculture of important medicinal plants in Indian Thar Desert. In: *Journ. Medicinal and Aromatic Plant Sciences. Proc. Nat. seminar on Frontiers of Research and Development in Medicinal Plants* (eds) Sushil Kumar, SA Hasan, Dwivedi, Kukreja HK, Sharma A, Sigh AK, Sharma S and Tiwari R. CIMAP, Lucknow.

Kasera PK, Shukla JK, Prakash J, Saharan P and Chawan DD. 2002. *Propagation and Production Techniques for five important medicinal plants of the Thar Desert Environment.* CAZRI, Jodhpur.

Kaul MK. 2001. *Medicinal Plants of Kashmir and Ladakh.* 173 pp.

Kaushik P. 1988. *Indigenous Medicinal Plants including Microbes and Fungi.* PP. 250.

Kaushik P. and Dhiman AK. 2000. *Medicinal Plants and Raw Drugs of India.* 11 plus 623 pp. Colour photo 80, ISBN: 81–211–0174–3.

Kempana, C. 1974. Prospects of medicinal plants in Indian agriculture. *World crops,* **26(4)**: 166–68.

Khan MA, Chanrackaran I and Ghanim A. 1986. Falciformin-a flavanone from pods of *Tephrosia falciformis. Phytochemistry,* **25**: 3.

Khorana, ML. *et al.* 1949 Anthelmintic activity of 'Kamela' and its constituents. *Indian Journ. of Pharmacy* **11**: 37–43.

Khory RN. 1999. *Materia Medica of India and their Therapeutics.* 5 plus 809 pp.

Kirtikar KR and Basu BD. 1935. *Indian Medicinal Plants.* Vols 4. Allahabad. 4 Vols.

Komar MC. 1920. Report on the investigation of Indigenous Drugs. Madras.

KrishnaMurthy, N, Mathen AG, Nambudri ES, Shivsankar, S. Lewis YS and Natrajan CP. 1976. Oil and aoleroresin of turmeric. *Trop. Sci* **18(1)**: 37:45.

Krishna S and Ghosh TP. 1938. Indian *Tephrosias* as a source of Rotenone. *Curr. Sci.* **6**: 456.

Krishna S and Bhadwar RL. 1949. Aromatic Plants of India. *J. Sci. and Industr. Res.* **8(11)**: 171. Pt. IV. 6(5): 63–76. (Malvaceae to Geraniaceae)

Kurup PA. 1958. Antibacterial activity of *Withania somnifera,* isolation and antibacterial activity. Antibiotics and Chemotherapy, 8,10: 511–515.

Kumar S. 2000. *The Economic Plants of Northeast India.* 123 pp.

Kumar S. 2002. *The Medicinal Plants of Northeast India.* 23. plus 212 pp.

Lamba SS. 1970. Indian piscidal plants.*Econ. Bot.* **24(2)**: 134–136.

Lalramnghinglora H. *Ethnomedicinal Plants of Mizoram.* 19. plus 333 and 66 colour photos.

Lindley J. 1998 . Flora Medica: 15. plus 658 pp. A botanical account of all the more important plants used in medicine.

Madan CL. Gupta US, Prabhakar VS and Kapur BM. 1965. The location of alkaloid in the capsules and seed portion of some Datura species and varieties. *Die Natureiss.* **4(109)** 1–2.

Mahabale TS. 1956. *Studies on Pharmacognosy of some Indian medicinal Plants.* Introduction. Drug Research Association, Poona. No I.

Maheshwari P and Singh Umrao. *et al.,* 1965. *Dictionary of Economic Plants in India.* ICAR. New Delhi. pp197. 2nd Ed. 1983 pp. 288 (Singh Umrao, Wadhwan AM and Johri BM).

Maheshwari JK, Kalakoti BS and Brijlal. 1986. Ethnomedicine of Bhil tribe of Jhabua distirct. Madhya Pradesh. *Ancient Science of Life.* **5(4)**: 255–261.

Maheshwari JK. 2000. *Ethnobotany and Medicinal Plants of Indian subcontinent*. 11 plus 659 pp.

Maiti PC. 1968. Phytochemical screening. *Bull. bot. Surv. India* **10(2)**: 11–122.

Malkhuri RK, Nautiyal S, Rao KS and Saxena KG. 1998. Medicinal plants cultivation and biosphere reserve management: a case study from Nanda Devi Reserve, Himalaya. *Curr. Sci.* **74**: 157–163.

Manga VK and Sen DN. 1945. Influence of seed traits on germination in *Prosopis cineraria* Mac Bride. *J. Arid Environments*. **3**: 371–375.

Masters JW. 1848. A memoir of some of the natural products of Angami, Naga hills and other parts of Assam. *J. As. hort. Soc. India*, **6**: 34–55.

Mehrotra BN. 1975. Survey of medicinal plants around Kedarnath shrine of Garhwal Himalaya. Proc. nat. *Symp. Abstr, Indian For.* **105**: 788–799. 1979.

Mehrotra BN, 1975. Processing of plant samples for chemical and biological investigation. *Ibid*. p27.

Mehrotra BN and Aswal BS. 1987. *Companion to Chopra's. Glossary of Indian Medicinal Plants*. 8 plus 162 pp. (ISBN 81–240–0014–3)

Menon, KPG. 1975. Indian turmeric wins World market. *Indian Spices* **12(2)** : 6–13.

Menon RK. 1960 *Indian essential oils*: A Review. CSIR. New Delhi.

Mertia RS and Kunhamu TK. 2000. Seed germination trial on *Salvadora oleoides*. *Journ. Trop Forestry*, **16**: 50–52. of *Acacia leucophloea* Willd. *Journ. Sci. Industrial Res*. 11**B**. p125.

Mhaskar KS and Cains JF. 1931. Indian plant remedies used in snakebites. *Indian Med. Research Memoir* No **19**.

Mohammed S. 2001.*Biodiversity and ecological adaptation of arid zone Plants*. UGC Final Techn. Progress Report. Govt. College, Churu.

Modak AT *et al* 1966. Hypoglycaemic activity of nonnitrogenous principle from the leaves of *Adhatoda vasica* Nees. *Indian Jou Pharm*. 28: 105.

Moss NS. 1953. *Ayurvedic Flora Medica*. Kottayam.

Mukerjee SK and Murthy VVS. 1952. Presence of *Myricarin* in fresh flowers

Mukherjee SR *et al*. 1962–63. A problem in the act of learning and effect of Marsilin on rehabilitation of epileptics *Physiol. Exp. Med. Sci*. **4**: 337–367.

Murray GA. 1881. *Plants and Drugs of Sindh*. London. pp 1–219.

Nadkarni RN. 1954. *Indian Materia Medica*. Popular Book Depot. Bombay

Nadkarni AK. 1954. *Dr. KM Nadkarni's Materia Medica*. 1927. Revised and enlarged. Bombay

Nadkarni KM. 1927. *Indian Plants and Drugs with their properties and Uses*, Bombay.

Nadkarni KM and Nadkarni AK. 2000. *Indian Materia Medica* (eds) Vols I,II,III Popular Prakashan Pvt. Ltd. Bombay. pp. 1–591.

Nadkarni AK. 2004. *Indian Plants and Drugs*: *with their Medical Properties and Uses*.

Nagarajan M and Mertia RS 2001. Seed germination of *Tephrosia falciformis*. *Ann. Arid Zone*, **40(1)**: 95–96.

Namjoshi AM. 1955. Studies on the Pharmacognosy of *Tinospora cordifolia* (Guduchi). *Bull. Nat. Inst. Sc. of India*. No **4**.

Narayana UB. 2003. *Indigenous Medicinal Specialities*. 32, 488 pp.

Nautiyal MC. 1996. Cultivation of medicinal plants and biosphere reserve management in alpine zone. In: Ramakrishnan PS, Purohit KG, Rao and Malkhuri RK (eds). *Conservation and Management of Biological Resources in Himalayas*. GBPIHE. Oxford and IBH Publishing Co. New Delhi.

Nathawat GS and Deshpandya BD. 1960. Plants of economic importance from Rajasthan. Series I. *Proc. Rajasthan Acad. Sci.***7**

Nayar SL. and Chopra RL. 1951. *Distribution of British Pharmacopoeial Drugs Plants and their substitutes growing in India*. CSIR. Delhi. pp 56.

Nayar SL. 1954. Poisonous seeds of Indian Plants. *Journ. Bombay nat. hist. Soc*. **12(1)**88; **(3)**:515.

5

Nayar SL. 1964. Medicinal plants of commercial importance found in Uttar Pradesh and their distribution. *Journ. Bombay nat. hist. Socn.* **61(3)**: 651–661.

Oaris R and Durand N. 1958. A propos de l'essai des Aloes-dodage Photometrique de l'aloine. *Ann. Pharm. Francsais.* **14**: 755.

O'Shanghnessy1841.*Bengal Dispensatory and Pharmacopoea.*

Pal DC. 1981. *Plants used in treatment of cattle and birds among the tribals of Eastern India.* in SK Jain (ed). Glimpses of Indian Ethnobotany. pp. 245–257.

Pal DC. and Jain SK. 1998. *Tribal Medicine.* pp 336.

Pant,P. 1991, *Kumaon Ki Upyogi Aushadhiya Vanaspati* (Hindi) pp35 Uttarakhand Sewa Nidhi. Almora.

Pandey Sanjay K. and Shukla RP. 2001. The Regeneration Pattern and population Structure of *Boerhavia diffusa* in Northeastern UP. *Trop Ecol.* **42 (l)**: 137–140.

Pandey G. 1996. *Medicinal Plants of Himalaya* vol 1. and 2. (2000). 9. pp 351.

Pandey G. 2000. *Medicinal Flowers Pushpayurveda. Medicinal Flowers of India and Adjacent Regions.* 9 plus 209 pp.

Pareek SK, Srivastava VK and Gupta R. 1989. Effect of source and mode of nitrogen application in Senna (*Cassia angustifolia*) *Trop. Agr. Trinadad.* **62**: 69–72.

Patel RP *et al.* 1958. Antibacterial activity of *Daemia extensa* R. Br. *Indian Journ. Pharm.* **20**: 328.

Patil RP. 1968. Medicinal Plants of Today and Yesterday. *Indian Mus. Bul.* **1. 3(1 and 2)**: 10–13.

Prakasa Rao RS. 1968. Medicinal Plants of Cuddapah District. Andhra Pradesh. *Nagarjun.* **11(12)**: 639–645.

Prakasa Rao Rs. 1964., Medicinal Plants of Hyderabad City and neighbourhood. I. *Nagarjun.* Nov. pp. 181–204, pt 23. December: 261–283.

Prakasa Rao RS 1968, Medicinal Plants of Cuddapah District. Andhra Pradesh. 1. *Nagarjun* 21(7): 358–362., 2 *Ibid.* 383–391, 3**(9)**: 468–473; 4,**11**(10): 532–538; V, Ibid. **5**(11): 590–598; 6(12): 639–345. 7(2): 13–16.

Prasad BN. 1962. *Alectra* parasitica A. Rich.var. *Chitrahutensis,* an indigenous drug in the treatment of leprosy in Bihar. Preliminary observation. *Quarterly publication of British Leprosy Relief Assn. Rev***33**(3).

Prasad G. 2003. *A Manual of Medicinal Trees.* 9 plus 132 pp. 34 colour plates.

Prajapati ND. 2003. *Colour Atlasx of Medicinal Plants.* 19 plus 268 pp. colour plates 134.

Prajapati ND. 2003. *A Handbook of Medicinal Plants. A complete source book.* 53 plus 1008 pp. 134 colour plates.

Prajapati ND. 2003. *Agro's Dictionary of Medicinal Plants.* pp. 398.

Pullaiah T. 2002. *Medicinal Plants in Andhra Pradesh.* 265 pp. 123 gig. 14 colour plates.

Pullaiah T. 2003. *Medicinal Plants in India.* 2 vols. 11 plus 580 pp.

Puri GS and Jain SK. 1960. Survey of some oil yielding plants of West India. *Bull. bot. surv. India* **2**(1–2): 95/098.

Puri HS. Pharmacognosy in Ancient India. *Everyday Science.* **15**: (1) 23–25.

Puri HS. 1975. Phytochemical survey of some plants giving steroid, alkaloid, saponin and Tanin. *Indian Drugs.* pp. 7–10.

Quisunbing E. 1951. *Medicinal Plants of Philippines.* Manila.

Radwanski, S.1977 (abc) Neem Tree. *World Crops* **29(3)**: 62–66**(3)** 111–113, **(4)**: 167–168.

Rai LK and Sharma E. 1994. *Medicinal Plants of Sikkim Himalaya. Status, Usage and Potential.* Bishen Singh and Mahendra Pal Singh. pp152.

Rai LK. 1994. *Medicinal Plants of Sikkim Himalaya.* 8 plus 152 pp. (ISBN: 81–211–0110–7).

Rajpurkar MV *et al.* 1927. Some Pharmacological properties of *Butea frondosa* seeds. *Indian Journ. Med. Sci.* **15(5)**: 353–58.

Rajabhandari KR. 2001. *Ethnobotany of Nepal.* 14. plus 189 pp.

Rau MA. 1961. Recent Survey of Medicinal Resources of Lahul Valley. *Pakistan Journ. Sci. Res.* 4: 202.

Rawat MS. 1998. *Ethnomedico Botany of Arunachal Pradesh* (Nishi and Apatani Tribes) 9 pp. 206. (ISBN 81–211–0153–0).

Rawat RS. 2002. *Nature's Pharmacopoea: Medicinal Plants. Diversity in Doon valley*, Uttaranchal. 6 plus 148 pp.

Ray, PG and Majumdar, SK. 1976. Antimicrobial activity of some Indian Plants. *Econ. Bot.* 30(4): 317–320.

Ray BR and Sharma BK 2004. *Medicinal Properties of Plants. Antifungal, Antibacterial and Antiviral Activities.* 7 plus 600 pp.

Robert B. 2002. *Medicinal Plants. Being Descriptions with Original Figures of the Principal Plants Employed in Medicine and Account of the Characters, Properties and Uses of their Parts and Products of Medicinal value.* 4 vols.

Roberts E. 1984. (Rptd) *Vegetable Materia Medica of India and Ceylon.* 4 + 437.

Rustomjee Nascowajee Khery MD. *The Bombay Materia Medica and therapeutics.* Bombay.

Ophthalmology in Traditional Medicine. CFIKS, Chennai

Sadhale Nalini. 1999 *Surpalas Vrikshayurveda.* Asian Agri-History Foundation.

Said M. 1963. Minor Forest Products of west Pakistan. *Pakistan J. For* 13(3): 276–286.

Samant SS. 2002. *Himalayan Medicinal Plants. Potential and Prospects.* 19 plus 435 pp.

Samant SS. 2002. *Himalayan Medicinal Plants.* Potential and Prospects.

Sanyal. 1924. *Vegetable Drugs of India.* pp. 9. plus 435 pp.

Sarin YK. 1961. Medicinal plants of Bhadarwah and Kishtwar forest Divisions in JK. State. *Symp. Proc. Utilization of medicinal and aromatic plants in India.* CSIR.

Sarin YK. and Kapoor LD. 1962. Survey of medicinal weeds. Bull. Regional. *Res. Laboratory, Jammu.* 1: 25–29.

Sarin YK. 1967. A survey of vegetable raw material resources of Lahul. *Indian For.* 93(5) 489–499.

Sarin YK. 1965. Medicinal, quasi-medicinal and economic plants of Bhadarwah Forest Divison. *Indian For.* 91: 559–572.

Sarin Y.K. and Atal, CK. 1970. Indian Buck wheats as a commercial source of Rutin. *Res. and Ind.* 15(2): 88–90

SarinYK, Kapoor LD, and Chopra IC.1963. *Psycholaina prealata* (Don), A. hyposcyamine yielding plant in Lahul economy. *Indian For.* **89**: 610–611.

Sarin YK and Prabhakar 1966. *Elscholtzia incisa* Benth. A new source of Thymol in northwest Himalaya. *Curr. Sc.* **35: 23.**

Sarin YK and Kapoor LD. 1965. Pharmacognostic study of *Trillium govanianum* Wall. *Indian Journ. Pharm.* **27: 6**–10.

Sarin YK. 1975. Commerce in India-Vegetable raw material-Status and Problems. *Proc. National Symp.* (Abstract) p 43.

Sarup S. *A Pharmacological study of Ephedra.*

Saxena HO, Brahmam M and Dutta PK. 1981. *Ethnobotanical studies in Orissa.* In Jain SK (ed) Glimpses of Indian Ethnobotany. pp. 232–244.

Sehgal, CA. 1954. The blood pressure lowering activity of some fractions of *Rauwolfia serpentina* Benth. *J. America. Pharm. Assn.* **43**: 505.

Schultes RE. 1960. Tapping our heritage of ethnobotanical lore. *Econ. Bot.* **14(4)**: 257–263.

Schultes RE. 1960. Tapping our knowledge of Ethnobotany. *Econ. Bot.* **14(4)**: 257–263.

Secunderabad. atuim. AP. SaiRam, TV. 1997 *Home Remedies* pp. 333. Penguin Books, New Delhi.

Sen SP. 1966. Studies on active constituents of *Cissus quadrangularis. Curr. Sci.* **35**: 317.

Sen, S.P. 1966. Studies on the active constituents of Cissus quadrangularis II. Curr. Sci. 35: 317.

Shabnam SR. 1964. Medicinal Plants of Chamba. *Indian For.* **90(1)**: 50–63.

Shah NC. 1968. The Alpine Medicinal Herbs. *Indian Drugs* **6(2)**: 13–16.

Shah NC. 1975. Prospects of botanical drugs from hill district of UP *Proc. National Symp.* Abstr.pp 39.

Shah and Krishnan 1962. Identity of the Sweta *Punarnava*, Boerhaavia Punarnava sp. nov. of the Ayurveda. *Journ. Sc. and Industr. Res.* **21**(C).249.

Sharma AK. 2001. *A handbook of organic Forming. Agrobios* (India). Jodhpur. pp. 627.

Sharma R. 2003. *Medicinal Plants of India. An Encyclopaedia.* 31 plus 302 pp.

Sharma UK. 2004. *Medicinal Plants of Assam.* ISBN 81–211–0406–8.

Sharma BM and Kachroo, P. 1983. *Flora of Jammu and Plants of Neighbourhood* (illustrations). Bishen Singh Mahendrapal Singh. DehraDun.

Sharma E, Rai LK, Lachungpa S and Awasthi RS. 1995. Status of medicinal plants and their cultivation potential in Sikkim Himalaya. pp 43–51. *Cultivation of Medicinal Plants and Orchids in Sikkim. Himvikas.*

Sharma PC, Srivastava SC and Kapoor LD. 1970. On the identity and chemistry of Mirchakand or Keerkand (*Corallocarpus epigeous* Benth. ex Hook). *Indian For.* **96(9)**: 678–679.

Sharma OP. 1998. *Economic Botany, Medicinal Plants.* 2nd edn. pp. 1-300.

Sheshadri,TR. 1969. A useful development in the chemistry of Flavinoids as natural products. *Proc. nat. Inst. sci.* **35B**: 438–452.

Sholto Douglas, J. 1971. Growing medicinal plants for essential oils and flavours. *Flavour Ind.* **2(12)**: 697–698.

Shrott, J. 1877. List of wild plants and vegetables used as food by the people in famine time. Madras. *Indian Forester* 3: 232.1888.

Shukla M, Baxi D Chatterjee UN. 1965. Ecophysiological studies in arid zone plants. I. Phytotoxic effects of aqueous extracts of Mesquite, *Prosopis juliflora* L. *Curr. Sci.* **34**: 612.

Shukla SD *et al.* 1973. Some Pharmacological action of an alkaloid fraction from *Cardiospermum halicacabum. Indian Journ. Pharm.* **35(1)**: 40.

Shukla JK, Kasera PK and Chawan DD. 2000. Effect of different Index of *Prosopis cineraria* (L.) Druce. Haryana *Agric. Univ. Journ. Res.* **30**: 65–70.

Shyamsundar. *Treatment of Poison in Traditional Medicine.* Centre for Indian Knowledge systems. Chennai.

Singh MP. 1975. Herba; Flora of Nepal. *Proc. National Symp. Recent Advances in Production and Utilization of Medicinal Plants.* pp 44.

Singh D. 1964. Some useful Cucurbits in the arid zone of West Rajasthan.*Symp. Problems of Indian Arid Zone.* pp. 135–139.

Singh J. 1962. Medicinal and aromatic plants of Punjab state and plan for their development. *Indian For.* **88(12)**: 907–911.

Singh, AP. 1981. Indiginous Medicinal System in Uttarakhand Himalaya. *Man and Nature* **5(6)**: 10–12.

Singh V and Pandey RP. 1998. *Ethnobotany of Rajasthan. India.* Scientific Publisher (India), Jodhpur. pp.367.

Singh U, Wadhwani and Johri BM. 1983. *Dictionary of Economic Plants of India.* ICAR. New Delhi. pp.288

Singh DN. 1995. Use of medicinal plants of Sikkim in Ayurvedic medicine: *Cultivation of Medicinal Plants and Orchids in Sikkim Himalaya.* pp. 65–68.

Singh RP. 1981. Indigenous medical system in Uttarakhand Himalaya *Man and Nature* **5(6)**: 10/012.

Singh MP. 2003. *Indigenous Medicinal Plants, Social Forestry and Tribals.* 12 plus 505 pp.

Singh KK and Kumar K. 2000. *Ethnobotanical Wisdom of Gaddi tribes in West Himalaya.* 10 + 139 pp.

Singh KK and Prakash A. 2003. *Tribal Wisdom on Medicinal and Economic Plants of Uttar-Pradesh and Uttaranchal.* 11 plus 183 pages. Colour plates 32.

Singh PB. 1999. *Illustrated Field Guide to Commercially Important Medicinal and Armatic Plants of Himachal Pradesh* (with special reference to Mandi District.) 10 plus 117 pp Society for Herbal Medicine and Himalaya. Proc.

Sigh VK. 1998. *Herbal Drugs of Himalaya.* Medicinal Plants of Garhwal and Kumaon Regions of India. pp 275 (Aspects of Plant Sciences Vol. 15.)

Singh V. 2003. *Ethnobotany and Medicinal nts of India and Nepal.* 2 vols.

Sinha RL 1975. Distribution, availability and industrial potential of Indian medicinal plants in the hill districts of UP. *National symposium on Recent Advances in the development, production and utilization of medicinal and aromatic plants.* Abstr.pp.46.

Sinha RL. 1975. Protection and Planning exploitation of Indian medicinal plants in the hill districts of UP. *Ibid.* pp.46.

Sinha Alka and Kumar Niraj 2002. Role of Biodiversity and Relation of status of Health of the Koltribes of Vindhya region. *Proc. National Acad. Sci.* 71st session, Biological Sciences. Poona.

Sinha GP and Chauhan AS 2001. Ethnobotanical studies on the Lepchas of Sikkim. *TIME* **1**: 25–27.

Srivastava N and Phadya MA. 1995. *Punarnavine* profile in the Regeneration Roots of *Boerhaavia diffusa* L. from leaf segments. *Curr. Sci.* 68: 653–656.

Sood SK. 2001. *Ethnobotany of Cold Desert tribes of Lahul Spiti.* (northwest Himalaya). 10 plus 228 pp.

Srivastava RC. 1988. *Drug Plant Resources of Central India* (An Inventory). pp. 250.

Srivastava GN. 2000. *Indian Traditional Veterinary Medicinal Plants.* pp. 581.

Sushruta Samhita, Uttarantra. **60**: 47. Nirnayasagna Press, Bombay.

Swami B. 2000. *Common Medicinal Plants of India.* A Complete Guide to Home Remedies. pp 329.

Tarafdar CR and Rai Chaudhuri HN, 1981. *Less known uses of medicinal plants among the tribals of Hazaribagh district of Bihar* in SK Jain (eds). Glimpses of Indian Ethnobotany. pp. 208-217.

Tripathi, Pdtt. Ganga Prasad, Dadhich 1908 *Anubhut Chikitsa Sagara* 2 parts. Ajmer.

Tsrong Tsewang Jigme 1995. Tibetan Medicinal Plants: Agenda for cultivation. In *Cultivation of Medicinal plants and Orchids in Sikkim Himalaya.* pp. 75–79.

Uniyal MR. 1968. Medicinal Plants of Bhillangna valley, Tehri Garhwal, UP. *Nagarjuna.* **11(9)**: 478–484., **11(10)** 516–518.

Uniyal MR. 1968. Medicinal Plants of Bhagirathi valley in UP Forests. *Indian For.* **94(5)**: 407–420.

Uniyal MR. and ISSAR RK. 1967. Commercially important medicinal plants of Kanatal forest, Tehri Garhwal. *Indian For.* **93**: 107–114.

Upholf JC Th. 1959. *Dictionary of Economic Plants.* Weinheim.

Varghese E. 1996. *Applied Ethnobotany. A case study among the 'Kharias' of Central India.* 19 pp. 307.

Varier VPS. 1994–96. *Indian Medicinal Plants. A compendiu of 500 species* Vol 1. pp 416. Vol. 2. 1994. pp 416. Vol 3. pp 423. 1994. Vol 4. pp 444. 1996. Vol. 5. pp 592. 1996.

Ved BG. 1960. Place of Momordicacharantia in the treatment of diabetes melitus. *Maharashtra Journ.* **6(12)**: 733–745.

Ved BG. 1962. Place of *Semecarpus anacardium* in the treatment of Cancer. *Journ. University of Bombay,* **30**(3and 5): 85–110.

Vedavathy S. 1997. *Tribal Medicine of Chitoor District.* AP.

Verma DR *et al.* 1959. Oral contraceptives. Pt IV. Hormonal and anti-Hormonal effects of *Rollerin indica. Journ. Physio-pharmacology,* 34: 240–254.

Verma N. 1970. Opportunities for the large-scale cultivations of aromatic and medicinal plants in the province of Uttar Pradesh. *Flavour India* 1(5): 332–334.

Vertika, Singh KP, Sinha Alka and Niraj Kumar. 2004. Admisture dependent doses prepared from medicinal plants by Kol tribe for treatment of Diabetes. *National Acad. Sci. Letter,* **27**: 251–260.

Vertika Neraliya S and Niraj Kumar 2002. Role of biodiversity in prevention and cure of diseases in Kol tribe in and around Allahabad. *Nat. Acad Sci.* 72nd Session B. Biology section. Shillong.

Vaidya Shrikar D. 1999. *Instant and Fast Acting Ayurvedica Treatment. Drug Formulas and Therapies.* Asukari Chikitsa in Ayurveda Jalukar. 10 plus. 206 pp.

Volk OK. 1961. Survey of Afghan medicinal Plants. *Pakistan Journ. Sei. Research.* **4**: 232.

Vyas LN and Gupta RS. *List of medicinal Plants of Alwar.* Sr. I.

Wahid MA and Kazmi MA. 1961. Study of some indigenous pharmacopoeal herbs. *Pakistan J. For.* **11(4)**: 384–389.

Wallis TE. 1946. *Text Book of Pharmacognosy.* J.A. Churchil Ltd. London.

Wallis TE. 1953. *Practical Pharmacognosy.* London.

WHO *Indias medicines for whole SE Asia Region.*

Warman CK. 1999. *Trees of India. Medicinal, Commercial, Religious and Ornamental.* 11. pp. 246.

Waring EJ. 1982. *Remarks on the Use of Some of the Bazaar Medicines and Common Medicinal Plants of India.* 16 plus 292 pp.

Watt G. 1889-1896. *Dictionary of Economic Plants of India.* Calcutta Vold 1–6.

Zaffar R. 2000. *Medicinal Plants of India.* 5 plus 132 pp.

Zhou, S. 1993. Cultivation of *Ammomum villosum* in tropical forests. *Forest Ecology and Management.* **60**: 157–162.

Colour plate 5.1: *Aegle marmelos* (Bael) (Photo R.K. Gupta)

Colour plate 5.2: *Salmalia malabarica or Bombax malabaricum* (Semul) flowering during spring with attractive flowers (Photo R.K. Gupta)

Colour plate 5.3: *Salmalia malabarica* or *Bombax malabaricum* flower in close view (Photo R.K. Gupta)

Colour plate 5.4: *Capparis decidua,* a shrub of arid areas (Photo R.K. Gupta)

Colour plate 5.5: *Cassia fistula,* the common Amaltas in flower (Photo R.K. Gupta)

Colour plate 5.6: *Emblica officinalis* (vern. Amla) (Photo R.K. Gupta)

Colour plate 5.7: *Punica granatum* (vern. Anar) is both a fruit and a medicinal plant (Photo R.K. Gupta)

Colour plate 5.8: *Rhododendron arboreum* (vern. Burans) is an attraction during the flowering season (Photo R.K. Gupta)

Colour plate 5.9: *Taxus baccata* L., a high altitude shrub, is a potential source of taxol, being cultivated for experimental research

Colour plate 5.10: *Ageratum conyzoides,* a weed as forest undergrowth in humid areas
(Photo R.K. Gupta)

Colour plate 5.11: *Argemone mexicana* (Photo R.K. Gupta)

Colour plate 5.12: *Asclepias curass-avica* L. in flowers and fruits (Photo R.K. Gupta)

Colour plate 5.13: *Mirabilis jalapa* (Photo R.K. Gupta)

Colour plate 5.14: *Ocimum sanctum* (vern. Tulsi) (Photo R.K. Gupta)

Colour plate 5.15: *Opuntia dilleni,* the common cactus, with flowers in arid and dry habitats

Colour plate 5.16: *Saussurea obvallata* (vern. Braham Kamal) (Photo R.K. Gupta)

Colour plate 5.17: *Verbascum thapsus* (Photo R.K. Gupta)

Colour plate 6.1: *Taxus baccata,* Himalayan yew (Courtesy: S.Chandola, PCCF)

Colour plate 10.1: *Cassia angustifolia* (senna) in flower (Courtesy: S.Chandola, PCCF)

Colour plate 10.2: *Catharanthus roseus* (periwinkle) with flowers

Colour plate 10.3: *Chrysanthemum cinerarifolium (Pyrethrum cinearifolium)*
(Courtesy: S.Chandola, PCCF)

Colour plate 10.4: *Crocus sativus* (saffron) grown on experimental basis at Dehradun
(Photo R.K. Gupta)

Colour plate 10.5: Standing crop of *Cymbopogon citratus* (the lemon grass)

Colour plate 10.6: Cultivation of *Cymbopogon citratus* after the first cut

Colour plate 10.7: *Datura stramonium* in nature

Colour plate 10.8: *Gloriosa superba* being cultivated extensively as an alternate source of diosgenin

Colour plate 10.9: *Ephedra gerardiana* (vern. Somlata) (Courtesy: S.Chandola, PCCF)

Colour plate 10.10: *Hyoscyamus niger* (Khorasani ajwain) (Courtesy: S.Chandola, PCCF)

Colour plate 10.11: *Pelargonium* (Geranium)

Colour plate 10.12: *Podophyllum hexandrum* (Ban kakri)

Colour plate 10.13: *Rauwolfia serpentina* (Sarpgandha) (Courtesy: S.Chandola, PCCF)

Colour plate 10.14: *Saussurea costus* (Courtesy: S.Chandola, PCCF)

Colour plate 10.15: *Tagetes minuta* (marigold), apart from ornamental is also medicinal (Photo R.K. Gupta)

Colour plate 10.16: *Vetiveria zizanoides* in nature

PART
II

Potential Drug Plants, Quality Control Assessment, Commercial Exploitation and Biodiversity Conservation

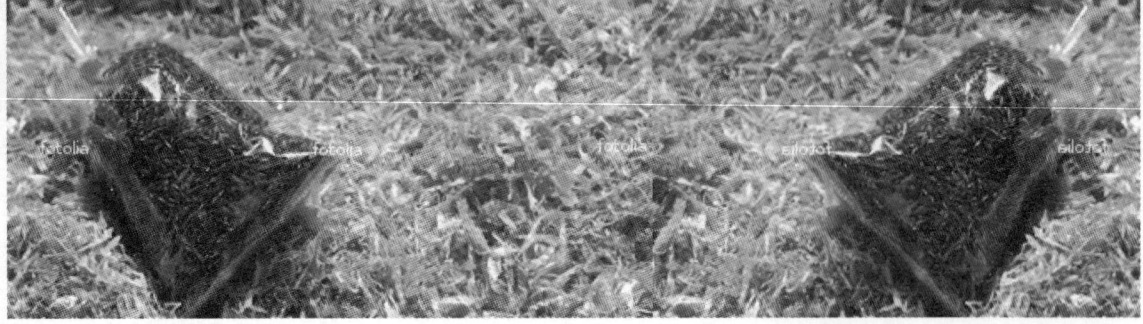

Definition of Drug

1. All medicines for internal or external use of humans or animals and all substances intended to be used for or in the diagnosis, treatment, mitigation or prevention of any disease or disorder in humans or animals, including preparations applied on human body for the purpose of repelling insects like mosquito.

2. Such substances (other than food) intended to affect the structure or any function of the human body or intended to be used for the destruction of vermin or insects which cause disease in humans as may be specified from time to time by the Central Govt., by notification in Central Gazette.

3. All substances intended for use as components of a drug including empty gelatin capsules, etc. or such devices intended for internal or external use in the diagnosis, treatment, mitigation or prevention of the diseases or disorders in humans or animals, as may be specified from time to time and by the Central Govt. in official Gazette, after consultation with the board (Ayurvedic medicine, Homeopathic medicine and Allopathic medicine).

Ayurvedic, Sidha, Unani drugs, include all medicines intended for internal or external use for or in diagnosis, treatment, mitigation or prevention of disease, disorder (in humans or animals) and manufactured exclusively in accordance with the formula prescribed in the authoritative books of Ayurvedic and Unani Medical systems (specified in the first schedule).

Ayurvedic and *Yunani* drugs can be classified into the following categories:

1. *Scientifically studied drugs*: Among such drugs are included drugs like *Sarpagandha* (*Withania somnifera*), *Gul-Banafsha* viola sp. and *Yograj Guggal* including types of Guggal's, like *Tridashang, Singhnad* and *Gokshuradi*, which have been scientifically studied and their therapeutic claims have been justified. There are recently a few additions to the category, just like Bach (*Acorus calamus*), since a number of pharmaceutical companies, the Himalayan Drugs, CDRI, Lucknow, Hamdard Dawakhana, etc. and many are currently working on these drugs and have patented some formulations.

2. *Popular non-toxic drugs*: These are the medicines which are popularly known for their therapeutic value and at the same time are non-toxic, such as 'Chyavanprash', where *Emblica officinalis* (anwla) is the main ingredient and is useful in chronic diseases of the lungs

and also as a general tonic during winters.

3. *Effective but toxic drugs*: These are the medicines having severe toxicity but are of known therapeutic value and used injudiciously like *Bhallakaveleh* where *Semecarpus anacardium* is the important ingredient and is useful in obstinate and chronic diseases.

4. *Hereditary and patent drugs*: Some physicians have specialised in curing certain diseases and the formula, methods of preparation are known to them or trusted members of their family only. Some of them might not be effective as they claim, while in other cases that may not be so effective. Consent for the disclosure of the formula is either denied or is associated with a demand for heavy financial compensation.

5. *Drugs of rare use*: These drugs are not used extensively and only physicians of certain regions use them and claim their efficiency. Such a situation is in the remote areas of the Himalayan region (both eastern and western) and regions in tribal-populated areas.

Rigveda, perhaps is the oldest repository of human knowledge (4500–1600 BC). In *Atharvaveda* which is considered as *Upveda*, definite properties and use of plants as drugs have been given in more details. In early part of the 19th century, a perceptible change from herbal medicine to synthetic chemical drugs has taken place. In spite of the phenomenal growth of modern pharmaceutical industry, the popularity of the herbal medicines has been greatly affected but the traditional folk-medicines still remain greatly as cultural-heritage in different civilisations. Traditional system of medicines still continue to cater to the medicinal needs of the 3rd- world countries which are inhabited with approximately 80 percent of the world's population. Even the WHO (World Health Organisation) has realised the potential of traditional medicines in the management and self-reliance of medicinal care system. The organisation has encouraged inclusion of herbal drugs in "National Health Care Programme", because of their easy availability at a normal price, that is within the reach of a common man and are not only time-tested but safer than the synthetic drugs. In India, not only Ayurveda but Sidha, Unani and Homeopathy are predominantly dependent upon plants and are utilised in the total health care system.

Traditional indigenous system of medicine has been a part of the tradition of each country and employs practices that have been handed over from generation to generation. An important feature is the preference for compound prescriptions over single substance drugs. Earlier in India, medicines used in Indian System of Medicine were generally prepared by practising physicians by themselves which has now been largely replaced by indigenous drug industry. Some of the well-known names are *Zhandu Pharmaceuticals, Dabur Pharma, Baidyanath, Divya-Jyoti Pharma* and many others. It would be surprising to note that now there are more than 7843 licensed pharmacies of Indian System of Medicine (ISM) in India (Ayurvedic, Sidha, Unani) in addition to 857 or more in Homeopathy and a number of unlicensed small scale processing units that manufacture drugs. About 1.5 million practitioners of ISM and Homeopathy use medicinal plants in preventive, promotive and curative applications. In Ayurveda 1000 single drugs and 8000 compound formulations of recognised merit are in vogue. As per one estimate of WHO (the World Health Organisation) global market of medicinal herbs and herbal products is about 62 million US dollars and in 2005 hit 5 trillion US dollars. Presently, India is exporting herbal medicines and material of more than 550 crore rupees, which is expected to increase manyfold within 4–5 years.

Herbal Crude Drugs

Most of the medicinal and aromatic plants or crude drugs (roots, stems, leaves, flowers, seeds, fruits and whole plant) are handled by traditional crude herbal drug dealers in the north and *Pachamarunna Kadi* in the south, who sell drugs under Ayurvedic, Unani and local names. The suppliers are locally traditional, without caring much for the quality (date of expiry, mixed with extraneous substances). However, there is no gradation and standard of quality (ISI/ISO). Sometimes, different species of the same genus are supplied, since they have no knowledge about specific diversity. There are certain known drugs, where confusion prevails with regard to their identity due to the fact that descriptions available in Ayurveda literature are meagre and vague. Different drugs are being sold in different states, under exactly the same name, e.g. Brahmi *Centella asiatica*, *Hydrocotyl asiatica* (UP), *Bacopa monieri* (Bengal), *Hydrocotyle rotundifolia* (Uttaranchal), *Merremia emarginata* (Gujarat), *Hydrocoyle javanica* are used in some parts as Brahmi. Similarly, *Chaerophyllum villosum* is sold in place of *Aconitum heterophyllum* as *Atis*. *Polyalthia longifolia* and *Shorea robusta* in place of *Saraca indica* (Ashoka), Nagkésar (flower buds of *Mesua forea*), *Ochrocarpus longifolia*, *Calophyllum inophyllum*, *Cinnamomum tamala*, *Dillenia pentagyna* are some other examples.

Some Aspects of Cultivation

Ninety percent of the medicinal plants, extensively used in traditional system, are obtained from wild resources such as natural forests, uncultivated land as weeds and wastelands. Many drug plants have become threatened due to excessive exploitation from their natural habitat. Plants like *Rauwolfia serpentina*, *Swertia chirata*, *Aconitum heterophyllum*, *A. dienorrhizium*, *Colchicum luteum*, *Gentiana kurroa*, *Commiphora wightii*, *Podophyllum hexandrum*, *Nardostachys jatamansi (grandiflora)*, *Coptis teeta*, *Achyranthes aspera Cuscuta reflexa*, *Boerhaavia diffusa*, *Evoluvlus alsinoides*, etc. are a few examples.

A few species that are cultivated on a large scale are *Isabgol* (Psyllum), Ahipen (Poppy), Svaranpatry (*Sena*), Chincona, Ashwagandha, *safedmusli*, *Sadabahár*, Solanum species, *Sarpagandha, Ipecae,* etc. Others are cultivated on a large scale as spices or condiments used as medicine like *Kalimirch (Piper nigrum), Haldi (Curcuma longa)*, Adrak (*Zingiber officinalis*), *Dhania (Coriander), Saunf (Feoniculum vulgare) Ajwain (Trachyspermum)*, and *Methi (Trigonella foenumgraceum)*, etc.

The cultivation of plants permits uniform production of quality raw material, whose properties are standard and from which quality crude drugs can be obtained.

Concern about the depletion of the biodiversity as a result of excessive exploitation of drugs plants from natural habitats, and agro-techniques for the cultivation of some of the drug plants have been discussed in this work. However, cultivation of such plants permits production of uniform quality raw material whose properties are standard and from which crude drugs can be obtained unadulterated.

11th Indian Standards Convention Chandigarh 1967. ISI. pp. Doc; S-1/1 to Doc S-1/26.

		Collection from wild	v/s	Cultivation on drugfarms.
1.	Availability of supply	Decrease		Increase
2.	Quality control	Poor		High
3.	Fluctuation in supply	Unstable		Controlled quality
4.	Quality control	Poor		High
5.	Botanical indentity	Not reliable		Not questionable
6.	Genetic improvement	No		Yes
7.	Agronomic manipulation	No		Yes
8.	Post-harvest handling	Poor		Good usually
9.	Adulterants	Likely		Relatively safe
10.	ISI/ISO standard	No		Possible

Merits and demerits of collection of drug plants from wild and under-cultivation have been given as above:

Plants role as raw material for isolates and characterization of new drug molecules for cure and prevention for diseases like cancer and AIDS have created resurgence of interest in them all over the world. It is interesting to know that about 60 percent of the antitumour or anti-infection agents that are commercially available at the advance stage of clinical trials are of natural origin. Presently, there are about 125–130 clinically useful prescription drugs that are derived from about 100 species of higher plants. It is also estimated that about 5000 species of higher plants worldwide, have been studied in some detail as possible source of new drugs, In India, more than 3,300 plants that have been screened for their biological activities.

Medicinal Plants and Phytopharmaceuticals

World's pharmaceutical industry is unique, extremely competitive and highly research based. It is characterised by very special and striking features like, high rate of obso-

lescence of drugs and technologies; requirement of high quality products, good manufacturing practices (GMP), good laboratory practices (GLP), environmental and regulatory requirements, etc. It is a complex, multidisciplinary and time-consuming activity. Normally 12 years of research efforts from discovery to marketing with an average cost 250 million US dollars are required for a drug molecule to be used as therapeutic agent. Though, this long time taking period has been drastically reduced by merging approaches such as advanced genomics, high-through pool screening, combinational chemistry, biology and computer assets, *de novo* drug design. Most of the pharmaceuticals have set a goal of 6–7 years from molecule to marketing.

During the past one decade, bulk production of plant based drugs has become an important segment of Indian pharmaceutical industry. Some of the phytopharmaceuticals which are produced in India at present include morphine, codeine, papverine, quinine, quinidine, chinconine and cinchonidine (Chincona sp. *C. calisaya, ledgeriana, officinalis* and *C. succirubra*); hyoscine and hyoscyamine (*Hyoscyamus*

niger, and *H. muticus*), colchicine (*Gloriosa superba, colchicum luteum* and *Iphigenia stellata*); cephaeline and emetine (*Cephalis ipecacuaha,* cennoside A and B (*Cassia angustifolia* and *C. acutifolia*), rescepine, rescinnamine, ajmalie and ajamalicine (*Rauwolfia serpentina*); vinblastine and Vincristine, ajalmacine, rauacine (*Vinca rosea* or *Catharanthus roseus*), Guggul-lipid (*Commiphora wightii*), taxol (*Taxus baccata*), artemisinin (*Artemisia annua*), etc.

Over the years four principal routes by which plants have led to new therapeutic agents are as below:

1. Chemical constituents isolated from plants used directly as therapeutic agent-digitoxin, strophanthin, morphine and atropine.

2. Plants constituents used as starting material for synthesis of useful drug. Adrenal cortex and other steroid hormones synthesised from plant steroidal sapogenins.

3. Natural products which serve as models for pharmacologically active components in the field of drug synthesis.

Frequently the side effects of a natural product often prevent its use in medicine and resolved only by a preparation of synthetic derivative, e.g. cocaine which led to the development of modern local anesthetics. Modification of cocaine and of podophyllotoxicin to obtain antitumour preparation.

4. Plant constituents which demonstrate a mode of action which is then copied as a muscle relaxant from curare.

Recently, plant constituents have been used as research chemicals, particularly in the area of molecular biology. More than 50 anticancer drugs are not marketed because of their side effects, but are used widely in research. One of the major criteria for selection of the plants for discovery of new lead molecule is the utilisation of available knowledge in coded or uncoded form in traditional or indigenous system of medicine, including folklore or tribal medicine. The past experience in areas of drug discovery have proved that the success rates are very high with such plants as they provide a fertile hunting ground for search of new therapeutic agents.

In this context, the development of new drug; viz. resperine and related alkaloid for hypertension and in certain neuropsychiatric disorders from *Rauwolfia serpentina*, Guggul-lipid for lowering cholesterol from *Commiphora wightii*, Forskohlin (Coleonol) and antihypertensive agents from *Coleus forskohlii*, bacocide for memory enhancer from *Bacopa monnieri* are some of the interesting examples of how observation is recorded in ancient medical texts of Ayurvedas or Unani, a folk medicine, when investigated by a modern scientific method led to successful development of new drugs and provided a lead compound for further modification. However, detailed studies on traditional and folklore plants have been given based on personal experience and from available literature. It is hoped that the information would serve a positive role in identifying new medicines for the emerging complicated diseases which are still incurable through allopathic and other systems of medicine.

Significant medicinal drug plants having leading role in phytopharmaceutical industry are listed in Table 6.1.

Table 6.1: Significant medicinal drug plants having leading role in phytopharmaceutical industry

1.	Morphine, codine, papaverine	*Papaver somniferum*
2.	Quinine, chinconine, cinconidin	Cinchona (*C. calycea, C. oficinalis*)
3.	Hyocine, Hyocyamine	*H. yoscymus niger* and *H. muticus*
4.	Colchicin	*Gloriosa superba, Colchicum luteum*
5.	Taxol	*Taxas baccata (T. wallichiana)*
6.	Artimicin	*Artemisia annua*
7.	Resperine	*Rauwolfia serpentina* BP
8.	Gugulipid,	*Commiphora wightii*
9.	Forkolin	*Coleus forskohlii*
10.	Bacocide (intelligence)	*Bacopa monerri*
11.	Acoron	*Acorus calamus*
12.	Androgracolide, neoandrogracolide	*Andrographis paniculata*
13.	Sapogenin, Glycoside	*Asparagus recmosus*
14.	Azadirictin oil	*Azadirachta indica*
15.	Cymopapain	*Carica papaya*
16.	Asiaticacide	*Centella asiatica*
17.	Cisampelin	*Cissampelos praiera*
18.	Alkaloid (berbery)	*Cosinium fenestrum*
19.	Curcumine	*Curcuma longa*
20.	Colchicine, thiocolchicine	*Gloriosa superba*
21.	Gossiol	*Gossypium*
22.	Kemtothesin	*Mappia foetida*
23.	Eldopa	*Muluna purpurea*
24.	Glycoside	*Piccorhiza Kurroa*
25.	Lignans	*Phyllanthus emblicus*
26.	Celmarin	*Cilvium marinum*
27.	Flavinoids, alkaloids	*Tephrosia purpurea*
28.	Cilarin A and B	*Urginea indica*
29.	Valeopotratus	*Veleriana wallichii*

* The list is not exhaustive but representative.

The Collection and Maintenance of Germplasm of Drugs and Essential Oil Yielding Plants.

There is still, not a serious concern for the collection and maintenance of the germ plasm of medicinal and aromatic plants, though much of the material is widely available in various parts of the country or are under cultivation for a pretty long time. Since it is imperative to increase yield for these crops, shorter growing period, increased active principles, resistance to pests and diseases, collection from different areas is required to be made for the desired traits in parent plant material. Variability in the germplasm is the prime necessity for an improvement programme which is needed on priority to be built up.

There is also an urgent need for agencies responsible for commercial production, testing, certification and distribution of superior, variable seed and planting material. The position is still difficult where live planting material (rooted slips, stolons, tubers, etc.) are required and brought from long distances. CIMPO has been doing this work, for selected species, grown at their farm.

In the universities and colleges the study of medicinal and aromatic plants do not still form the part of the curriculum for graduate and postgraduate courses. Also there is dearth of trained manpower for supporting these plantations and undertaking the research work. However, the national Commission on Agriculture should take up this subject on priority and recommend specific courses on aromatic and medicinal plants in various universities.

Since the aromatic and drug plants have an element of risk, because of the presence of active principle, lack of testing facilities makes farmer apprehensive of the price and long storage sometimes invites loss in its value. Added to it is the attitude of user agency in asking for larger lots. Therefore, a central agency providing speedy testing facility for the produce at nominal price needs to be established with additional

functions for having revolving funds for purchasing from small farmers at a prefixed price.

Resources Survey

The acquisition of new plant material as source of industrial raw material is of real interest to scientists and industrialists. It is easier to develop a natural plant extract than to synthesize a completed molecule in the laboratory. It is well known that the natural material still serves as a better model even for new and possibly improved synthetic drug. Extensive surveys of potentially rich geographical regions within a country and outside are of interest to pharmaceutical firms and Governments, and millions of dollars are spent annually for the screening of plant resources and the results have been rewarding. Scarcity of belladona alkaloids in forties of this century compelled commonwealth countries to search for an alternative source; which ultimately led to the discovery of *Dubosia myropoides* and *D. leichardii,* in rainforests of Australia. *Physochlaina praealata* is another plant from Ladakh which is found to give similar alkaloid. The discovery of cortisones as a remedy in rheumatoid arthritis led to worldwide search for plants containing cortisones or other similar compounds which could be converted into it. The search led to discovery of the cortico-steroids containing *Dioscorea composita* from Mexico and of *D. deltoidea* as a very rich source of diosgenin (3–5 and up to 10% in western Himalaya and *D. prazeri* from eastern Himalaya). Another Indian plant *Solanum khasianum* var. *chatterjeanum* has also been discovered to contain up to 5% of Solassodin, which could be converted into steroid hormones. Hecogenin (for production of steroid hormones) was discovered in leaves of *Agave veracuz, A. sisalana* and *A.*

cantal which could be grown on alkaline soils having pH up to 8.5 or on very poor soils of wastelands.

Thus, the search for new drugs pays rich dividends. The discovery of resperine group of alkaloids from *Rauwolfia serpentina* as a cure for blood pressure and certain forms of insanity in 1952 has been remarkable with the result that over-exploitation of Rauwofia for roots obliges American and Swiss pharmaceuticals to make a worldwide search for Rauwolfia species, which led to pinpoint *R. vomitoria* from Central Africa, and later on from Senegal, Uganda and Congo (with alkaloid content from 0.2 to 3.0%. Similarly, *Carthamus roseum* (rose periwinkle) from Westindies contains several tumour suppressive compounds and atleast one is marketed for use in certain forms of human cancer. Antileukemia properties have been located from alkaloids extracted from *Podophyllum peltatum* and *P. hexandrum*. An African plant, *Phytolaca decandra,* is reported to have yielded a new snail-killing compound. Berberine salts are new addition to modern medicine and Indian firm is manufacturing berberine sulphate and allied salts with Japanese collaboration from several species of Berberis which grow wild in Kumaon and Garhwal. The extracts of *Centella asiatica, Swertia* and *Sopora* are popular in hair tonics. These examples fairly show that enormous potential of trade exists in herbal formulations, even in most advanced countries of the world.

Indian scientists isolated a cardio-gly-coside peruvoside from the leaves of *Thevetia nerifolia,* considered in oral prescription. Hyathine has been isolated from *Cissampelos pareira* which is reported to be a good substitute of tubocuraine (muscle relaxant) imported in the country. Isolation of fata-mansone from *Nardostachys jatamansi* is a remarkable discovery for prolonged hypothe quinine, the active anti-malarial component, has better drug value than its synthetic replica. Use of synthetic drugs is now being discouraged on account of their toxic side effects. It is for this reason that quinine has now become one of our export commodities. The quinine present in *Cinchona febrifuge* is used as the best heart regulariser all over the world. There is possibility of conversion of quinine to quinidine by suitable chemical methods. Hence, the cinchona plantations in India, which had once been closed up have already been reorganised. However, improved methods of cultivation are required in order to increase the production.

Anticancerous: *Vinca rosea*, popularly known as *Nayantara*, has already earned much reputation as a potential anti-cancerous drug. The dimeric indole alkaloids, vincristine and vinblastine of Catharanthus roseus (*Vinca rosea*) has found successful application all over the world as curative for cancer. Recently 16 methoxytabersonine from the same source has been found to be a better anticancer drug, as it is free from any toxic effects.

Semecarpus ancardium L.f. on a preliminary chemical screening has been found to be useful in cancer cases where modern therapeutics have been found unavailing. The following local drugs are said to be useful.

Commiphora wightii, Azadirachta indica, Trichosanthes dioica, Curcuma longa, Pongamia pinnata, Saussurea lappa, Alstonia scholaris, Euphorbia neriifolia, Albizzsia lebbek, Ichonocarpus frutescens, Calotropis gigantea, Psoralea corylifolia, Acorus calamus, Kamjeria rotunda and *Nerium indium.*

An anticancerous bioscreening unit at CDRI, Lucknow, has made it possible not only to carry on the existing plant experiments but also develop synthetic compounds likely to aid anticancer research. Under this programme, if the original extract is found to possess antitumor activity, it is fractioned up to four fractions and in the final process the crystalline material in the selected fraction is isolated for preclinical tests.

Tranquilisers and sedatives: A number of indigenous drugs have shown promise as tranquilisers. The use of *Rauwolfia serpentina* (chota chander) and *R. canescens* (bara chander) is well known for their tranquilising principles *resperine, reseinn-amine* and *deserpidine*. Although, the application of reserpine has considerably reduced because of its toxic manifestations, deserpidine is still recommended as a good tranquiliser. Besides, it is also possible to advise chemical methods for the conversion of inactive rauwolfia alkaloid, rauwolscine into deserpidine yohimbine, serpentine and serpenrinine. Three other rauwolfia alkaloids were known to have considerable tranquilising property. Along with various synonyms *R. serpentina* Benth. ex Kurz has been described in ancient ayurvedic classics

like Charaka and Sushruta. This drug has been scientifically studied by a number of workers. Because of its potensive hypotensive actions it has found place in foreign markets.

Recently, alkaloid glycosine isolated from *Glycosmis pentaphylla*, is shown to possess excellent tranquilising property with least toxic effect. Similarly, active principles isolated from *Acorus calamus* are found to be potent tranquilisers.

Alstonia acholaris has given more than a dozen alkaloids. The plant contains reserpine as one of the major alkaloids and another venserpine which has a better tranquilising property than the reserpine.

Nardostanchys jatamansi has been found to be a potential sedative, hypotensive and antiarrhythmic tranquiliser.

Oral contraceptives: Family planning is being treated on a priority basis in the development of plans of the country. The population crossed to a billion mark in 1999. This may be the situation in many developing countries of the world. Oral contraceptives and other devices are being also imported. In ancient literature a number of single drugs and combined preparations are mentioned for birth control. Some methods of sterilisation and sex-determination have been cited in *Athervaveda* and in *Brhad Aranyakopanishad*. For the sterilisation of women with medicinal herbs, two hymns with an obscure sense are found in Athervaveda which though not clearly indicates this purpose, yet simplifies the sense in a roundabout way. Probably, these hymns suggest the sterilisation of women with the plants described as *Tstavandana*, i.e. a rough creeper and Visa Visataki, i.e. poisonous plant (not identified). The

Sushruta refers the first plant as 'Banaparni' which *Darlia* explains as 'Sarpunkha', identified as *Tephrosia purpurea*.

Mollotus philippense (Kaméla) has been recorded to reduce fertility of animals. Three important forms of ayurvedic medicines, popularly used for birth control are 'Pipaladichurna' (mixture of *Piper longum* roots 1 part, *Embelia tribes* 1 part and Boras 1 part) and 'Kaphaladichruna' (mixture of *Myrica* nagi 1 part *Hedychium spicatum* 1 part, *Mesua ferea* 1 part, *Cuminum cynum* 1 part) and 'Chitrakadichurna' (mixture Plumbago roots 12 g, *Piper nigrum* 2 g, *Curcuma longa* 1 g). Other drugs mentioned as oral contraceptives are *Clerodendrum phlomides*, *Ziziphus mauratiana*, *Michelia longa*, *Plumbago zeylanica*, *Sesbania sesban*, *Anthocephalus cadamba*, *Butea monosperma*, *Euphorbia neriifolia*, *Abies webbiana*, *Cucumis sativa* and *Amaranthus spinosa*. Physicians normally advise to avoid taking of *Carica papaya* fruits and Pineapple during the pregnancy because of the fear of abortion.

Emetic: The alkaloid emetine from *Ipeca vauha* (an antidysenteric drug) is being exported in considerable amount.

Antitubercular: Petrol extract of unripe fruits of *Piper longum* has been found to be highly effective against strain of human tuberculous bacilli (*Mycobacterium tuberculosis*). Other medicines in common use are *Adhatoda vasica*, *Moringa oleifera* and *Allium sativum*. *Ephedra vulgaris* is reputed for manufacture of ephedrine, a drug for asthma. *Adhatoda vasica* for sascine, curative for cough.

Antiasthmatic: Pharmacological and clinical studies on *Saussurea lappa* have proved it to be bronchodilator and an effective cure for bronchial asthma. Other ayurvedic drugs known to have antiasthmatic properties are *Piper longum*, *Costus speciosus*, *Hedychium spicatum*, *Aquilegia agallocha*, *Phyllanthus fraternus*, *Adhatoda vasica* and *Solanum surattense*. Reaction of bronchial asthma to treatment with different ayurvedic drugs is given in Table 6.2.

Cardiotonic: The glucoside *digitalin* from *Digitalis purpurea* is a well established cardiotonic drug. The alcoholic extract of defatted leaves of the red variety of *Nerium indicum* (raktkarvi) contain digitalis-like glucoside having cardiotonic properties. Other potential plants are *Terminalia arjuna*, *Asparagus racemosus* and *Tinospora cordifolia* (protection against gamma radiation of liver).

Purgative: The anthraquinone glycoside, *senna* from *Emodia fraxinifolia* and *Cassia senna* are good purgatives. *Dicumarol* from spoiled sweet clover and *xanthotoxin* separated from *Aegle marmelos* and *Luvangas scandens* are used as curative for thrombosis. Psoralin from *Psoralea corylifolia* is a potent antileukemic drug. Other vegetable drugs like *khirkakli* and *babhachi* could be used against leucoderma.

Antiepileptic and mental diseases: The total extract of *Marsilea minuta* and *M. rajasthanensis* has been shown to be highly efficient in the treatment of acute epilepsy and other mental disorders. The curative principle is shown to be a macrocyclic ketone designated as *marsilin*. Extract of 'Brahmi' was also found to have excellent sedative properties and could be used in several types of mental diseases. Other drugs reported to be useful are *Celastrus paniculatus*, *Allium sativa*, *Acorus calamus*. Plants used as tranquiliser and sedative are *Nardostachys jatamansi*, *Bacopa monnieri*, *Centella asiatica*, *Withania somnifera*, *Celastrus paniculata*, *Evolvulus alsinoides*, *Clitorea ternatea*, *Corchorus sativus*, *Myristica*

Table 6.2: Reaction of bronchial asthma to treatment with different ayurvedic drugs

Age yrs	Incidence of disease in family	Duration of disease (yrs)	Drug used	Dose (mg/kg) body weight	Result	Toxicity
41	Father	5	*Saussurea lappa*	65.5	Relieved	Dry mouth
50	Father	2	do	112.5	No relief	do
50	Father	1	do	69.0	Relieved	do
35	Mother	10	do	80.0	Relieved	do
28	Mother	8	do	73.0	No relief	do
35	Mother	7	*Inula racemosa*	76.5	Relieved	do
35	Father	1	*Inula racemosa*	80.0	Relieved	do
55	Father	1/12	*Costus speciosus*	80.0	Relieved	do
41	Father	1	*Costus speciocus*	76.0	Improvement	do

fragrans, Vitex negundo, Valeriana wallichii, Butea monosperma and *Melia azaderach.*

Antidiabetic: *Salacia chinensis* growing wild in the Konkan areas of western ghats, is being used locally in the treatment of diabetes mellitus. The drug is reported to be a potent antidiabetic drug. Other drugs used are *Aegle marmelos,* (leaves), *Momordica charantia,* (seeds), *Coccinia indica, Trigonella foenumgraceum,* (Jamuna seeds), Gymnema sylvestre (gurmár buti), *Ptero-carpus marsupium, Syzygium cumini* (seeds) and *Azadiracha indica Salacia oblonga* (guduchi).

Table 6.3: Effect of Salacia treatment in initial and final blood sugar levels (Hospital Reports: Anonymous)

Age yrs	Initial blood sugar level in mg/100 cc blood		Observation in weeks	Final blood sugar level in mg/100 cc blood	
	Fasting	After		Fasting	After meals
32M	140	200	8	90	145
48M	150	240	8	125	155
42M	140	180	6	120	165
35M	250	300	8	185	210
35F	150	180	6	122	150
53F	180	220	6	150	198
56F	210	330	6	170	198
40M	150	200	8	130	170
65M	180	200	6	164	282
40M	154	192	6	140	164
66M	150	190	6	140	165
36F	140	175	8	120	158
50M	160	190	6	80	100
56M	190	220	8	148	184
65F	160	200	6	136	172

F = Female, M = Male.

Effect of *Salacia chinensis* treatment in the initial and final blood sugar levels of diabetic patients of different age groups is given in Table 6.3 (after Hospital Reports, Poona).

Antiarthritic and anti-inflammatory: Some of the compound preparations like *Maharyograj Guggal, Kachnar Guggal, Singhnad guggal,* Triodashang guggal and R-compound, *Samirapannaga Ras,* Vatkulan, Vat. tak, Ras Raj Ras. Vrihool Vat. Chintamani Ras and *Maharasnadi Kwath* are effective in arthritis. Guggulu (*Commiphora wightii*) has antiarthritic properties and is used extensively along with gum from *Boswellia serrata* and seeds of *Vitex negundo.*

Diuretics: Non-mercurial diuretics are normally imported, while several plants like *Beorhaavia diffusa, Tribulus terrestris, Vanda roxburghii* and *Bergenia ciliata* are promising as diuretics.

Antileprotic: Various plants like *Hydnocarpus latifolia, Semecarpus anacardium, Alectra parasitica* var. *Chitrakutensis* are good as antileprotic.

Antiamoebic: Plants like *Holarrhena antidysenterica, Aegle marmelos, Syzygium cumini* and *Mangifera indica* are used as antiamoebic.

Anthelmintic: A number of plants are used as anthelmintic in indigenous system of medicines. The most common are *Artemisia nilagarica, Mallotus philippensis, Butea monosperma, Embelia ribes.*

Antibiotic: *Withania somnifera.*

Sources of Vitamins

Vitamin A and B: *Averrhoea carambola, Beta vulgaris, Benincase Ahispida B$_1$, Arachis hypogea* (B$_6$), *Borassu flabelliformis* (B complex), etc.

Vitamin C: Cultivation of Amalaki (*Emblica officinalis*).

Vitamin E: *Cicer arietinum.*

In many cases, it has been found that total extracts of the fresh plant materials are more efficacious than pure components. To cite with a few examples it may be mentioned that total extract of *Prasararni* (Gyandal or Gandh Madal), *Sindhuber* (Nishginda), *Sushni, Brahmi, Salparni,* leaves and roots of *Palta, Vashak, Gulancha* and *Asoka* are more potent than undivided fractions after chemical processing. Sometimes, admixture of crude extract of different plants increases the potency of drug, this is called synergistic action.

Insecticides: Gynocardia odorata (fruit pulp), *Hydnocarpus wightiana* (fruit), *Eupatorium odoratum* (plant), *Rhododendron falconeri* (flower bud), *Gnetum scandens* (leaf), *Eremostachya scandens* (plant), *Tephrosia candida, T. purpurea* (root-bark and seeds) *Derriseliptica* (stem, seeds), *D.ferruginea* (roots), *D. grandifolia* (aerial parts), *D. melaleucensis* (roots), *D. seandens* (roots), *D. polyantha* (roots), *Milletia pachycarpa, P. trifoliata* (root and stem), *Mundulea suberosa* (root).

Pithecolobium bigeminum (leaf-bark, seed), *Pongamia glabra* (roots), *Acacia pennata* (fruit powder), *Albizzia procera* (Bark). Walsura *piscidia* (bark, fruit pulp), *Anamirta cocculus* (seeds, berries), *Myrica nagi* (bark), *Maesa indica* (leaves), *Barringtonia racemosa, B. speciosa* (bark, root), *Anagallis arvensis* (bark, root), *Cyclamen persicum* (roots), *Anagallis arvensis* (bark, root), *Cyclamen persicum* (roots), *Randia dumetorum* (root, fruit), *Zauthoxylum alatum* (bark), *Sapindus trifoliatus, S. muckrosii* (seeds), *Bassia latfolia* (seeds), *Linostoma decandrum* (bark) and *Wikstroemia indica* (bark).

After the brilliant exposition of interaction among the various sapogenins of the Mexi-

Lamba, SS. 1970. *Curr. Sc.* **6**: 456. Krishna and Ghosh.

can plants it was shown that the various derivatives of C21 hydrocarbons, pregnane and allopregnane; C19 and C18 hydrocarbons of androstane and estrone respectively, can be prepared by partial synthesis from various saponins of vegetable origin. These discoveries opened up new vistas for an inexpensive and potentially unlimited supply of the famous drug cortisone and other related therapeutic agents, the dramatic effects of which were demonstrated in the treatment of rheumatoid arthritis after its method of synthesis was worked out from bile acids.

The steroidal sapogenins occur in several species widely separated systematically. Sapogenins found in *Digitalis purpurea L., Periploca gracea L., Strophanthus sarmentosus* DC, *Agave cantala* Roxb. and *A.sisalana* Perr. could not be made use of in industrial production due to difficulties involved in their propagation or because of poor yields of the steroid sapogenin. However, among the natural sapogenins *diosgenin* was found suitable for industrial utilisation in view of its easy conversion to various steroidal hormones. Diosgenin is obtained from various plants of the genus Dioscorea, some of which grow in nature under Indian conditions. Root tubers known as Yams contain the alkaloid. Barua *et al.* 1955, 1956 investigated the diosgenin content in the root tubers of about 19 species in India. Promising types appeared to be *Discorea deltoidea* Wall, and *D. praieri* Prain Burkill, having diosgenin content of 3.55 and 2.1 percent respectively. Chopra and Handa, 1963, made an assessment of the material growing in different northern and northwestern regions of the country and reported methods of cultivation, exact stages of collection of tubers along with method of extract and degradation to the active ingredients. However, it takes three years for the Dioscorea to mature under cultivation

so as to be of commercial value. A better growth of *D. deltoidea* was observed in limited areas where rainfall varies between 100–200 mm and the minimum temperature in winter goes below freezing point and in summer not below 30°C.

Quality Control and Quality Assessment of Drug and Aromatic Plants

Reliable estimates of the total production of drug and aromatic plants are not available for the entire country. In addition to internal consumption, crude drugs and plant materials have been exported in varying amounts to almost all countries of the world. An *ad hoc* expert committee was set up by the export council to form a plan for affective utilisation of facilities available at the Central National Herbarium, Calcutta, and elsewhere in the country for testing crude drugs and medicinal plants reported from India which met in Calcutta as early as in 1967. The need for a comprehensive system of préshipment testing of these products and creation of a suitable set up for the purpose was underlined at a technical session on standardisation of chemical products of plant origin at various conventions of the Indian Standard Institute at New Delhi.

Noting the decision of the export promotion council to bring crude drugs exported from India under compulsory quality control and preshipment inspection, the committee considered ways and means of implementing the same. The authenticated reference and standard sample of various crude drugs and medicinal plants should be made available to any one who desires to have them. In this connection, it was suggested that: 1. Specifications should be laid down for these authenticated samples. 2. An organisation should be recognised for authentication of these samples and for carrying out preshipment work prior to export. Different aspects of testing like identification and determination of active principle should be carried out at suitable stations like CNH (Central National Herbarium, Calcutta), NBPGRI, Lucknow (National Botanical Gardens and Research Institute), Central Medicinal and Aromatic plants Institute, Lucknow, in addition to all the recognised agencies, for carrying out this work. It was, however, agreed that there should be a close liaison among the central institutes which should be entrusted with having most standard samples. Export order should also include the Latin name of the plant along with that in English. 3. The basis of inspection should be simple approved by the foreign buyer. The industrial plantations of crude drugs should be encouraged and CIMPO should supply the authenticated planting material and also necessary guidance for planting and cultivation to interested organisations and individuals.

So far as the plants yielding essential oils of commerce are concerned, it is rather difficult to foreseen their future. The indications, however, are that plant breeders may have to produce new varieties or strains giving higher yields and superior oils. Though some start has been made in some other countries, e.g. roses in Europe, lavender in France, cymbopogon in India, much remains to be done in this field. It is also possible that a better understanding might emerge on the physiology and chemistry of essential oil products in plants, presently very little is known about the fundamental principles involved in the synthesis of essential oils, in plants. One can, however, safely predict a much bright future for synthetic perfumer industry. It is bound to improve the quality and increase durability of its products, as it would be more stable, effective, suitable and economical. The success achieved by present-day

chemists in synthesising nitro-musks has edged them on to explore new horizons, such as the use of terpenes, thymol, carvacol, isoperene, etc.

The utilisation of new reactions of technical processes will unravel the mystery of thousands of odorous substances. Binary and multiple blends will continue to provide basis for creative perfumery and will lead to combinations having most harmonious fragrance, that will be inviting and invigorating, sedative and bold, delicate and subtle but gentle fragrance that will impart honour, life, romance, poetry and aesthetic appreciation for beauty and culture around us.

The aroma therapy, use of plant-based oils to promote well being, suggests six scents to get started. 1. *Bergamot* derived from peel of the fruit has both balancing and uplifting qualities where the scent is diffused by heating a few drops in an oil burner or an electric diffuser. 2. *The Roman Chamomile,* aids digestion, relaxes and calms, lessens the severity of headaches and acts as an antiinflammatory. It is reported to shake away worries and tension by adding 8–10 drops to bath water. 3. *Claryssage scent* is often used to treat premenstrual symptoms and to calm emotional crises quickly. 4–5 drops are diluted in one tablespoon. It has slight narcotic effect. 4. *Lavender* has relaxing effect that can coax you to sleep. The oil is massaged into the palms and sole of feet or a drop is added to pillowcase. 5. *Neroli* made from orange blossoms is rich, heady scent that calms the nervous system when rubbed into shoulders and upper neck. 6. Rose extracted from flower petals is effective for easing emotional traumas. It is put a few drops on a handkerchief or tissue to inhale throughout the day.

6 Import Substitution and Exports Potential

In addition to internal consumption, crude drugs and aromatic plants have been exported in varying amounts to almost all countries of the world. The average annual export in sixties has been around rupees three crores; which has increased manifold. Mumbai and Chennai are still the most important centres. The average unit value of the crude drugs exported was about rupees one per kg. From the export statistics that about fifteen main varieties of crude drugs apart from a large number of medicines, mainly of Ayurvedic and Unani types, were exported.

Through the ages spices of India have not only fascinated social and political history but have also been one of the prime sources of fabulous wealth and prosperity described in the glowing accounts of Pliny, Ptolemy, Marcopolo, Vasco de Gama and other travellers. The Malabar coast of India with its abundance of world's best pepper, cinnamon and ginger, was a vast spice emporium of inland and foreign trade which attracted a host of fortune seekers, intrepid buccaneers, insatiable invaders and ambitious traders who, whosoever helming natures barriers, drained into the sunny realms of India one after the other, to capture her precious spice trade. In their quest to gain complete mastery, they no doubt, brought much misery, but threw the doors of many new and strange countries open, to bring in the fifteenth and sixteenth centuries the great age of discover the world has ever known. Indian spices travelled far and wide to traditional destinations to flavour foods and beverages, to soothe silk and beautify the women of many lands. The aromatic chase of these men eventually led to the conquest and subjugation of Indian traders the absolute masters of the flourishing spice wings of Calicut, Cannanore and Cochin.

Even today, spices continue to be one of the India's significant foreign exchange earners despite the vicissitudes that such trade is poor in world's competition. India is one of the countries that dominate cardamom, both in production and export for world consumer markets. It produces nearly seventy-five percent of the world's cardamom output. India also produces the commercially available true cinnamon as it is sometimes known in the trade. Over thirty-five varieties of herbs and condiments are grown and used in India. Notable among the fascinating array of Indian spices and condiments sought for diversified uses by other countries are pepper, cardamom; chillies, ginger, turmeric, celery, fennel, fennugreek, coriander, cumin, besides curry powders, pickles, chutneys, essential oils of spices, oleoresins and processed spices. India exports most of its spices in the raw form. Among the processed items of export, curry powder and paste, expertly blended in several appetizing combinations, have been enjoying the highest reputation in the sophisticated American and European markets as an Indian speciality. Turmeric, which is recognised as an ideal preservative for pickles and basic ingredient of many species mixers, is yet another item exported as powder. Onion, garlic and ginger are other items which have great export potential in their dehydrated form; sometimes this has led to great price-rise in the domestic markets resulting in political upheavals.

India imported seventy to eighty-five percent of her drug requirements for therapeutic purposes involving an expenditure in foreign exchange to the tune of about one hundred and fifty million rupees even in the sixties. To keep the prices of

medicines within the reach of common man and also to reduce the drain on foreign exchange it is necessary to manage indigenous resources (not exploit) and develop new drugs based on medicinal plants. While there is unmistakable tendency to employ purely synthetic preparations in increasing quantities in clinical practice, today the pharmacopoeia of many countries include larger number of drugs of vegetable origin. Though, with the application of modern medical practice, the possibilities of exploring the Indian medicinal herbs are neglected with the realisation of usefulness of Ayurvedic system on modern scientific lines. Being a necessity, crude drugs have recently received a boost.

Herbal Resources in the Patent Net

Rich biodiversity and germplasm resources in India have been a happy hunting ground for multinationals in the global pharmaceutical market. More than 100 Indian herbs and plants with medicinal properties have already been patented in USA. The list of plants with medicinal properties, which are part of our *indigenous knowledge base* and have fallen under patent net is long and ever increasing—neem, indian mustard seed, *turmeric, bhu amla, jar amla, anar, dushi salai, kareka, rangood-ki-bel, gulmendhi, bagbhernda, vilayti shisham, chakura, arand, kumar ghee, kummari, amaltas, kala-jeera, jangli-ashwagandha, ritha, kalimirch, harad, garden-balsam, shallaki, guruchi, amla* so on and so forth. It is not that India has not fought back to protect its precious biodiversity from falling prey to patents by multinationals, but the success has been few and far between. For instance, India won a crucial battle over turmeric when CSIR succeeded in getting the US authorities to withdraw the patent (no 5401,

504) they had granted in wound- healing properties of turmeric powder to a Mississipi based medical centre. The data are available in our ancient literature, as traditional knowledge which India system of medicine has proved for ages. The European patent office also delivered a favourable interim judgement on the challenge of a European patent on fungicidal effects of Neem oil (Patent no 436 257 B1) owned by WR G. Ace and Co. The reason why multinational pharmaceutical giants have made a beeline for countries like India it is because most of the world's plant species with medicinal value are concentrated in the developing countries. It was estimated that by the year 2000 AD, the value of this germplasm to the pharma industry could be of the order of 47 billion Us dollars. Recognising the potential in herbal drugs Indian companies are also, to some extent, hiking their research and development.

India's geographical position in the centre of the old world, variety of physiographic features, bioclimates, edaphic parameters prevailing in various parts of the country makes her one of the most significant places with rich biodiversity. Some 45,000 plant species have been recorded, given the fact that a number of biologically rich areas exist in India. Certain regions in Northeast India have not yet been fully explored. What is more striking is the uniqueness and the amount of endemism, 33 percent of all flowering plants and 18 percent of all plants (in India). Equally important is the range of domesticated biodiversity: at least 166 species of crops, 320 species of wild relatives of crops are known to have originated here. Within each species, diversity of species is astounding as for example 50,000–60,000 local varieties of rice are reported. O*kra, Citrus* species, banana, mango, jamuna, jute, ginger, turmeric, pepper, cardamom, cinnamon, etc. are a few examples.

The rapid extinction of many species, brought issues related to biodiversity into sharp focus. Nations and Governments realised that individual species were becoming extinct, in addition to large scale habitat's loss, causing mass extinction of species. The conversion on biodiversity (CBD) has agreed upon and initiated work at the earth summit in 1992. The mandate of this convention was to evolve ways and means of protecting and conserving the remaining diversity of life in the world. The equitable use of this collection of living things has yet not been fully resolved.

Economically poor nations are richer in biodiverstiy as compared to economically richer nations. Higher levels of biodiversity are in the rainforests of tropical mountains, tropics and coral reefs of coastlines. India is considered to be one of the 12 mega diversity countries, with 50,000 living organisms, out of 10–30 million organisms of the world.

Pilfering genetic resources of the south to gene-rich countries is taking place. Large scale robbery of genes is being questioned today and governments are clamping down their export. Northern countries naturally want to control and exploit genetic wealth of the tropics and reap benefits that occur due to biotechnical use of these genes. The third world countries are opposing the new regime being thrust down on powerful countries in the forum on Intellectual Property Right (IPR) and Exclusive Marketing Right (EMR).

In India efforts are being made both *in situ* and *ex situ* conservation of biodiversity. About 4.4 percent of the area has been designated as national parks, sanctauries, reserve forest biosphere reserves, botanical gardens, zoological parks, etc. The National Bureau of Plant Genetic Resources (NBPGR) is the nodal agency for the collection and maintenance of Plant Genetic Resources for *ex situ* conservation of plant diversity. The western model of IPR protection is fundamentally opposed to Indian conditions where all knowledge is jointly produced and exchanged in a traditional way to generations after generations. Yet another area of concern for India is the right of its small farmers who ensure India's food security by saving, conserving and freely exchanging seeds. Such an experimentation at the grass root level has helped evolve seeds suited to each ecological niche. This stands threatened today under the TRIPS (Trade Related Intellectual Property Rights) dispensation. Transferring the control of seed production to large seed companies would ensure only monopolistic control.

It is worth noting that in case of *neem*, out of the 26 patents, 20 are in America, two in Japan, three in India and one in Germany. Tumeric, a traditional spice, having many traditional uses was also patented in USA. Similarly, other species like *tulsi* (Ocimum) might be in the pipeline. Ceiba-Geigy has hired local people to collect local useful plants and shiped Indian genetic resources as a major enterprise. Hoechst, India, is setting up a sophisticated screening laboratory at Frankfurt. Surprisingly, the Ministry of Environment and Forests was toying the idea of allowing a US-based company to import tissue culture developed medicinal and aromatic plants in test tubes from India. Considering that, more than two-thirds of the plant species, at least 35,000, are estimated to have medicinal value came from developing countries, the conservation of biological diversity for pharmaceutical use is critical to India, that is home to 7,500

species. An inter-government meeting of the developing countries estimated the value of developing country's germplasm to the pharmaceutical industry at 47,000 million dollars in year 2000. It is primarily for this reason that the US continues to dug its feet when it comes to ratifying the COB (Convention on Biodiversity). The convention recognises that "plants, animals and microorganisms are no longer mankind's heritage but a national property calling for a fair and equitable sharing of the benefits arising out the utilisation of genetic resources". What is required today is immediate policy initiative of how to check the piracy of the biological resources, including plant, animal and human genes. How essential it is to put an immediate curb on this seemingly innocuous practice becomes clear amidst reports that many multinationals are making huge collection of plant species. Among these are the US-based Sabinsa corporation, Phytera Inc., Ecopharm and Phyton Catalytic Inc; the UK-based Xenova Ltd. and the Spanish Pharma Mar. However, the Economic Espionage Act passed by USA takes espionage from military domains to economic domains and redefines 'intellectual property infringement' as a crime. On the contrary, globalisation is being used to force other countries to give up their national interest, national security and national sovereignty. The definition of economic espionage in terms of intellectual property infringement is arbitrary, since IP is being expanded to other areas as public domain of universe systems as well as collective knowledge heritage of non-western societies. A draft biodiversity proposal, however, envisages creation of stage and strict biodiversity cells to regulate biodiversity use. In addition, each village panchayat is required to maintain biodiversity resgisters for documenting existing knowledge on local use and practices. Foreign access to biological resources will be permitted under strict material and information transfer agreements, Prior Information Consent (PIC) of community and charge of a licence fee.

Though the need for such legislation is well established, the government as usual, has hardly moved forward in this, leaving foreign preposterous to freely exploit the resources. Indian policy makers are grossly undervaluing the country's rich biological wealth which may be ruthlessly squandered for short-term monetary gains under the enthusiasm to welcome foreign multinationals. Many cases like that of 'Picroliv' developed by CDRI derived from roots of *Picrorhiza kurroa,* a potent drug for liver diseases caused by hepatitis virus or excessive drinking could be avoided if timely action is taken. Similarly 'bioperine' is the same compound as piperine that RRL is using in new formulations of tuberculosis drug which has been pirated by Sabinsa Corporations of USA.

Commercial Exploitation and Collection of Drug Plants and its Impact on Local Biodiversity Status

India, like Brazil, is blessed with nature's gifts which are being destroyed and the rarest of the rare plants, yielding numerous life-saving drugs, are being savagely stripped off and smuggled out of the country. If evidences available are of any indication, it is a single largest national pilferage of an unprecedented magnitude. As one of the 12 megadiversity centres, India is the home of over 45,000 taxa-flora species. Wealthwise only Brazil comes anywhere near us with 65,000 species, while the Indian species have a large family of several varieties, the Brazilian species have much smaller families. Today as many as 814 of them are deep into the Red Data Book, which contain the list of endangered species. A large number of them yield numerous life saving drugs including anticancer and antiasthma medicines.

Historically, in India plants have been used for treating various diseases. Take the case of *Coptus teeta*, known as 'Mishmir Tista', which has been used by tribals of Arunachal Pradesh to treat fevers. It contains the vital alkaloid *Berberine* and they would uproot the plant, slice a very small portion of its root and then replant it. The plant then lives on to serve them. Now they are learning to rely on poor substitute, opium. The entire Mishmi Tista harvest is exported to Japan and Switzerland, where the middle men in Calcutta (Kolkata) pay pea-nuts to slice the entire root which they sell for exorbitant price.

With a rich base, India is on its way to become the largest exporter of medicinal plants after China. It is difficult, if not impossible to pinpoint when this destructive commercialisation actually began. It first came to the notice in 1982, when the International Trade Centre carried out a study of the medicinal plants in use in European countries. It revealed that out of 80,758 tonnes of medicinal plants, imported into nine countries, India accounted for 10,555 tonnes. It was the single largest exporter of all the nine countries of Europe. For instance, out of a total of 31,452 tonnes of import by Germany, India supplied 6,929 tonnes. Even today, India continues to be the single largest contributor to the global medicinal plants' trade as per the IUCN (International Union for Conservation of Nature). As for volume it is anybody's guess. However, the global plant based medicine trade has an annual turnover of more than 43 billion US dollars, though many of the plants in circulation are banned and restricted to international trade (1995).

Other poor countries like Brazil and Argentina are the next big suppliers, depending on the buying country. Coming back to the ITC (International Trade Centre) study, it clearly showed the trend. However, we preferred to remain ignorant; for the first time the realisation dawned in India was in November 1991. A set of six plants were put on the protected list and their export was banned. However, the list included only one medicinal plant, *Kuth (Saussurea lappa)*, which by then became gravely endangered. It took almost three years to be notified though a mere enactment of law was not enough to check the trade. However, still we are not fully geared to fight this trade.

While the lackadaisical approach still continues and the ignorance reigned supreme, many multinational and national companies kept plundering the wilds, till the Act on Biodiversity Protection got the approval of the country's Parliament. Meanwhile more than 40 more plants had inched towards annihilation. Although they have been included in the protection list, though evidences revealed that their trade and export still continued, largely illegally, but at times officially too in spite of the ban on exports.

The plants harvested are many and take very long time to regenerate. Their population is not uniformly and widely distributed and are endemic to the Himalayan ranges and the western ghats. The fact that the World Conservation Monitoring Centre (WCMC) in UK had issued the grim warning that "if urgent steps are not taken, some biological diversity of immense economic and ecological value will be lost for ever". The six areas it singled out for protection are the Agastyamalai hills in western ghat, and the Eastern and Western Himalaya,

thereby meaning that the entire region is under threat.

The reckless exploitation and destruction of forest cover is another reason that has led to the extinction of many medicinal plants. Extinction of many valuable medicinal plants has also fuelled adulteration in medicines and has given rise to prices of life-saving drugs. Since demand for drug plants increased, as per the WHO, they were accorded the status of prime ingredients in the primary health care programme due to low cost. About 30% ingredient of all allopathic medicines and 15% of Unani and homeopathic and ayurvedic medicines was reported to come from plants.

In a study in Nepal stakeholders perspective on commercial medicinal plant collection in Nepal is interesting. Views of persons involved in commercial medicinal and alpine medicinal and aromatic plant exploitation showed that the local collectors are gaining important financial benefits from this activity, though 71 percent of the non-collecting villagers believed that the medicinal and aromatic plant resources are degraded. Most of the stakeholders, other than the district forest office staff, favour collection over conservation and find that the collection bans are inefficient indicating the potential for addressing village poverty, for example changing the present centrally based regulation mechanism and handing over some MAP-resources for community management.

The overall gravity of the situation and the Indian approach to the issue is revealing. Take the case of Himalayan Yew (*Taxus baccata*) that grows in the Himalayan ranges in subalpine or altimontane regions (Colour Plate 6.1). It is globally regarded as priceless for the Taxol, which is used to cure cancer, the anticancer medicine, way back in 1960 was extracted from the pacific Yew,

and the plant soon received protection status from the authorities. It requires 9 metric tonnes of bark, leaves or shoots to obtain 1 kg of Taxol. To translate it, it needs 30,000 trees of greater than 25 cm dia. A Yew tree takes nearly 50 years to reach this girth. Since efforts to synthesis Taxol proved futile, large quantities were required and the source had to be shifted. All attentions shifted to the Himalayan region Yew. The global as well as national markets have been buying such large consignments of Yew, both legally and illegally. Since 1992–95, 495.137 metric tonnes of Himalayan Yew derivatives were officially exported form Chennai and Cochin. Delhi accounted for 53.75 metric tonnes, while southern export points were fed by the supplies from northeast regions. Delhi received the bulk of the supply from the west Himalayan region. The collectors were paid a meagre sum of Rs 50/- only for a bundle of 40 kg and the truck finally reached the market (Khari-Baoli) of Delhi.

It was more to follow. In April 1993, the Govt. of Arunachal Pradesh issued permits for the harvest and export of more than 5,000 metric tonnes of leaves. In fact it was all set to issue permits for 10,000 metric tonnes but was prevailed by the Central Ministry of Environment and Forests (MOE and F). Although now banned, the trade in Yew still goes on, where the existing stock to a tune of 2000 MT were supplied. The tree has been uprooted out of existence in a fragile environment, since the collection was done in a highly destructive manner where long-term effect of such an exploitation on a slow growing such as this, is to be greatly feared. However, recent attempts to propagate this species through tissue culture had been encouraging and provides ray of hope for its survival.

Another is the case of Agar tree which is regarded of high value non-timber forest

produce. The source of developing divinely fragrant Agar oil, it also has medicinal value and commands high price in the international market. The exotic tree has been frequenting India for the last 2000 years now, in the northeast. It is, today, on the brink of extinction and according to a study only about 100 trees survive in Darugiri Reserve Forest at Meghalaya. Its population in Nagaland, Tripura, Manipur has been completely stripped off, barring a few sporadic saplings at Malkhang, Pukpui, Lungebi and Sihphirranges in Mizoram. The same is the position in Assam, while in Arunachal Pradesh some worthwhile stock was reported to be surviving.

The Agar tree yields oil when infested with the fungus and it takes an expert eye to detect the tree. However, the healthy trees were destroyed in search of the other indiscriminate cutting for the collection of infested fungal wood and felling of younger, uninfested trees by unskilled labourers has rendered Agarwood vulnerable.

Though banned officially, it continued to be exported and during 1990–95, more than 1,245 metric tonnes of agarwood was exported due to the ambiguous export status covered under item 7 of para 158 of prohibited items. The reckless destruction of the tree almost halted the wheels of 200 big or small units manufacturing Agaroil in and around Hojai and Nilbagan in the Naogaon district.

This is true of all cases where the rarest of rare plants have been put on the negative list and yet the trade goes in. Take the example of *Tita*, which is already on the Red Data Book, is banned for export, yet large quantities of rhizomes continue to be exported every year, *Indian Gentian*, now found only in J&K and Himanchal has probably gone off from Uttaranchal hills. In 1992–93, 3.65 tonnes of Kashmal was

exported. In case of Red sanders (*Pterocarpus santalinus*), the situation is even quiver, since it is highly used for *Santalin*, having market in pharmaceutical, food and dye industry. It has a restricted distribution in Andhra, Karnataka and T.Nadu. Because of its wavy grand timber it is used for making musical instruments and expensive furniture; what has been done to save this tree?

Export from India of Medicinal Plants to Countries of the World

Item / Country

1. Belladona (leaves and roots) (Code 292 4003). UAE.
2. *Chirata* (Code 292–4012) Afghanistan, RE, Bahrain island. B.Desh, Italy, Kenya, Mauritius, Singapore.
3. *Galangal* (rhizomes) (Code 292–4016) Are, Brazil, Italy, Kuwait, Netherlands, S. Arabia, Sudan and UAE.
4. *Glycyrrhiza* (dried rhizomes) (Code 292–4017). Canada, Kenya, UK
5. *Ipecac* (dried rhizomes and roots) Code 292–4018. France
6. *Kuth* (root) (Code 292–4021). Afghanistan, Canada, France, Hongkong, Kenya, Nepal, S.Arabia, Singapore, Sundan, Thailand, UAE, UK, USA.
7. *Stryehnos Nux-Vomica* (dried ripe) Germany (GFR), Hongkong
8. *Psyllium* (husk) (Code 292–4024) Afghanistan, B.Desh, Brazil, Bahrain, Canada, Denmark, Ethiopia, France, Germany (GFR, GDR), Hongkong, Italy, Japan, Kenya, Kuwait, Muscat, Nepal, Newzealand, Netherlands, Pakistan, Quata, S.Arabia, Sudan, Spain, Syria, Singapore, Tanzania Rep., USA, UAE, UK, Zambia.

9. *Psyllium* (seeds) (Code 292–4025) Afghanistan, B.Desh, Belgium, Brazil, Canada, France, Germany (GDR,GFR), Kenya, Kuwait, Muscat, Nepal, Pakistan, Singapore, Syria, Sudan, Sweden, Tanzania, UAE, UK, USA.

10. *Sarsaparila* (Code 292–4028) France, Italy, USA

11. *Rauwolfia serpentina* (roots) (Code 292–4031) Mauritius, Pakistan, Singapore, Srilanka, Tanzania. Rep. UAE, USA.

12. *Senna* (leaves and powder) (Code 292–4032)
   Argentina, Australia, Bahrain isles,, Brazil, Belgium, Burma, Canada, Czecoslovaia, France, Germany (GDR and GFR), Guinea, Hongkong, Hungry, Iran, Iraq, Indonesia, Italy, Japan, Jordan, Kenya, Lesotho, Mauritius, Mexico, Muscat, Nepal, Netherlands, Other EA countries, Portugal, Qatar, S. Arabia, Singapore, Spain, Srilanka, Sweden, Tanzania Rep., Thailand, Turkey, UAE, UK, YAR, Yugoslavia and Zambia.

13. *Tukmaria* (Code 292–4034)
   Bahrain isles., B. Desh, Canada, Fiji isles, Germany (GFR), Hongkong, Iran, Iraq, Kenya, Kuwait, Malaysia, Mauritinia, Muscat, Nigeria, other Eest Arabian countries, Pakistan, S. Arabia, Sri Lanka, Sudan, Tanzania, Trinidad, UAE. Foreign trade in medicinal plants, vegetable saps and extracts, alkaloids, hormones and glycosides is depicted in Table 6.4 showing the over exploitation of medicinal drugs and aromatic plants and concern for the biodiversity protection.

Table 6.4: Foreign trade in medicinal plants, vegetable saps and extracts, alkaloids, hormones and glycocides

Product	Unit	Germany	France	UK	Switzerland	US.	Japan
Import of							
Med. plts	tons	26289	15176	3801	3572	NA	152
Veg. saps and extr.	tons	26839	4187	11220	3987	NA	14252
Enzymes	tons	1163	1083	985	456	NA	352
Alkaloids	tons	489500	504902	484100	114821	NA	352
Hormones and	kg	NA	17808	NA	5110	55744	5715
Glycosides	kg	150800	226702	14400	118886	NA	13889
Export of							
Medicinal Plants	tons	5123	2861	417	163	3085	662
Veg. saps and extr.	tons	4599	5995	3393	1886	7099	842
Enzymes	tons	2385	689	875	156	NA	784
Alkaloids	kg	300213	397500	139612	139612	728846	210575
Hormones and	kg	NA	40245	NA	9043	407735	97
Glycocides	kg	236300	30113	122800	5595	24959	190977

* After Statistics of foreign trade in India. GOI. New Delhi.

Forest—An Ecosystem

Ecosystem structure involves a study of the community organisation (producers, consumers and decomposers), their spatial and temporal attributes and the biomass present at a given time. The non-living aspects of the ecosystem, constituting solids, liquids and gases, many of which enter into the body structure of the biotic component, keep circulating into the system. The third most important element of the environment is the energy, mostly in the form of light energy during respiratory activity and decomposition processes. The solar energy is utilised in the ecosystem for building up the biomass by the primary producers (drug and other plants). The primary producers retain a portion of the solar energy in a potential state in the form of body-building material or transform the energy into a dispersed form, as heat energy. One way flow of energy in the ecosystem is governed by the law of thermodynamics which states as follows:

That the energy is never created or destroyed but may be transformed from one type (light) to the other (potential energy of food) and that no process, involving in transformation, will occur unless there is a degradation of energy from a concentrated form (food) into a dispersed form (temperature).

The functional aspects of the ecosystem involves trapping of solar energy (light) and its conversion into potential form (food, fodder including drug and aromatic plants). The food passes from the producer to the consumer of various levels and is ultimately broken into the component elements by the decomposers for a cyclic movement in the ecosystems. Cycling of certain elements, e.g. Nitrogen, carbon, phosphorus and water,

circulating within the ecosystem through different biotic chains are certain examples.

The ecosystem is a dynamic unit with a lifespan, the biotic and non-living components of the system are constantly changing. New taxonomic entrants are always added up by reproduction and immigration, while some are eliminated as a result of emigration, competition and other population hazards created by the diseases, parasites, radiation and over-exploitation of certain plant species (mainly drug and aromatic plants). At times, the abiotic components of the environment undergo certain changes which are reflected by changes in the composition and structure of the ecosystem. Such long-term changes bring about succession, culminating into a climax stage. This is a stable state of the ecosystem in which the input and output of energy balances each other.

In a forest ecosystem, the biological productivity system in which forest as a group of organism, utilizes the energy, water and nutrients of its environment to produce dry matter. The minor systems, within the broad system are vegetation, soil, fauna and eco-climate. Each of these needs special approach for effective understanding of the forest ecosystem as a whole. Any change in the nature of vegetation, soil structure and texture and fauna brings about instability of system, adversely affecting the climate, biodiversity, including economic plants (like drug, aromatic oil bearing and other economic species), in the ecosystem.

Vegetation is the total plant cover (not only the trees) on the units of landscape with noticeable physiognomy, structure and function; it is diverse in horizontal and vertical space. Vegetation can be further looked upon as a series of continuum or simply made up of distinct communities, on abrupt edaphic discontinuities, under the overall regional influence of climate. A special feature of a forest vegetation is, that the trees being the principal components occupy greater vertical heights at different levels than any other terrestrial ecosystems because of their physiognomy.

A broad correlation of soil is possible between soil groups and vegetation types. Nutrient and water are made available to forest vegetation through the agency of root system. Thus, the habitat management is vital for the survival of both animals and plants. However, too much emphasis is laid about the animals like lion, tiger, elephant, rhinos and there is no talk about the plant. Many species are lost or are under the process of losing every day particularly the medicinal plants, in each biogeographical region. Thus, habitat management is vital for the survival of both plants and animals.

The Plant Diversity Loss

The loss of both plants and animals is a process in operation. In India, during the past and at present, the influencing issues are so complex that it is difficult to define the root problem or to find a quick solution. The tropical subcontinent was comparatively rich in biodiversity, particularly in forest wealth, spanning many bioclimatic zones and extraordinary variety of geographical, geological and soil conditions, India is one of the richest assemblages of plants and animals. Humid tropical to xerophytic and temperate to boreal vegetation occurs within this country in a complex-mosaic across coastal, over valley, montane and high (altitude) montane terrains. Forest communities, where woody erect trees predominate (which we call forest), covered extensive stretches of the subcontinent from seashore to tree line, comprising many valuable drug plants.

A fairly large population existed side by side with this biodiversity rich forests and interacted with it. Traditional agricultural communities depended upon the rich resource base and created in addition to crop varieties and strains of domesticated animals in a vast varied spectrum, apart from creating complex interfaces between cultural and natural landscapes. These 'ecotones' were areas, where guided evolution of semidomesticated plants took place. Intrusion of nomadic-pastoral cultures resulted in deforestation, particularly in hill forests, through excessive grazing, and denudation. Through over-exploitation desertification and replacement of extensive climax forest vegetation, both by mosaic of different stages of secondary successions through shifting cultivation also happened here. Fertile soils, adequate monsoon rainfall, optimal temperature regimes and the time tested land management practices permitted high biomass production. Though the population build up gradually yet the subcontinent was both ripped off its biodiversity and the land did not degrade extensively. Land-use ethics were ecologi-cally same and all forms of life venerated. Forests, in particular had sanctity. The relevance of all other components in the biosphere for all our being was well recognised.

The intrusion of alien culture in the subcontinent in the 16th century, the development of a well knit centralised state and change in the whole pattern of lifestyle followed in quick succession. Arrival of European traders and colonialists was the next crucial event. Within a short time, a diverse area was brought under one governing authority. Timber and other products like pepper, cardamom, sandalwood, resins, etc. were in great demand (abroad) earlier. Even mercantile connections with Chinese, the Jews, the Arabs and the Armenians were well established. It was directly a forest produce such as timber, spices and medicinal drugs which attracted and brought in Europeans, the Portugese, the Dutch, the French and the English starting from late 15th century. Ultimately, the desire to monopolise the trade in forest products (aromatic and drug plants) led to bitter struggle between the superpowers that ended with British supremacy by the early 18th century. Naturally, the attention of the colonial powers shifted from mere collection and export of products of this land to effective, intensive exploitation of all available natural resources, in particular, the more easily available agricultural and forest products.

Exploitation for industrial raw material took a little longer time to organise. Tropical plantation crops like tea, coffee, chincona, rubber, pepper, cardamom, cloves were extensively planted. This phase saw introduction of a great variety of new crops. Some of the richest tropical moist vegetation of forests was destroyed for the plantation crops in the Himalaya, Central Indian hills and the plateau, as well as the peninsular Indian hills. The British interest in natural forests was initially for the timber for ship-building industry and tree species such as teak, sal were much sought after. To harness the resources of the subcontinent and to transport them to the harbours for export and to move troops, extensive network of railways was created. This also demanded enormous quality of durable timber. Thus, the demand resulted in 1. direct exploitation and 2. artificial plantations, for which extensive areas of natural forests were cleared.

The devastating effects of forest reservations and the colonial forest management practices destabilized the complex balanced resource, flow systems and the social regulatory mechanisms. This in turn set off cascading repercussions in every part of the country. The massive felling and conversion in ecologically critical forest areas, in particular along the major river valley was a national disaster.

The frequent forest fires wiped out practically the biological diversity from the maximum extent of forests. What remained was drastically degraded and constituted a significant portion of our present waste-lands. The repercussion of this also has a negative feedback effect on the rural agricultural ecology and economy. Policing, fencing off, punishment for trespassing, etc. were resorted, than coopting people, to share the responsibility of protecting the common wealth. There was no measure to share a small part of the forest resources with the hardpressed people, till the JFM (Joint Forest Management) was recently introduced. Because of the very nature of distribution of forests and settlements in the forests of India, no forest could be effectively and totally protected, keeping out the people.

Cattle grazing, in some parts of India is the most effective destroyer of plant bio-diversity. In most cases, with the grazing pressure, annual fires are also routine; the net result of a severely degraded biodiversity and poor ecosystem.

In India, the areas designated as National Parks, the basic issue is to why animals became so rare. We applied the same policing, punitive measures to out-attempt to protect wildlife. The focus is on few wild animals, rather than on a few economically valuable timber species, as was the case in the forest protection measure.

The real biodiversity of the country with which all the rural population, the forest

tribal societies and all their time tested life-styles together achieved some sort of a stable equilibrium, was never even conceptually accepted nor sought to be reestablished. We only created refuge for tigers, birds and Orchids and so on.

The long tradition of sustainable landuse by the tribals, primarily based on complex agricultural practices, was the repository of crop genetic-diversity, possibly occurring anywhere in the world. The spectrum of natural heritage was sustained by richness of the forest around. The degradation of forests and the over powering interest of extraneous cultivators into the tribal world within a short time practically wiped out irreplacable lifestyle.

Habitat destruction, excessive exploitation and a host of other factors have practically wiped out the biological diversity, which included a host of drug and aromatic plant species. Desiccation, desertification and extermination of plants and animal species is taking place at an alarming rate which is an indication that our land is becoming less and less capable of supporting human population. The most dedicated captive breeding programmes are not even going to touch the top of the iceberg, because of the sheer magnitude of the problem.

Unfortunately, the current understanding of biodiversity by most people is a vague notion of the occurrence of some wonder plant such as containing a miracle drug. To locate this goldpot everyone joins in, including the related departments. In the face of such a stream of thinking the real concept of biodiversity has for long been forgotton.

We are busy in reversing the trend of even inadequate conservation measures. National parks and wildlife sanctuaries are getting developed, reduced in area or opened up for consumptive use. The root-cause of eco-degradation remains and festers. Forest by themselves cannot be created or sustained. They are vital integral part of a healthy living landscape. This alone can sustain healthy happy minds. India with its vast area and geographical and topographic diversity hosts 61% of endemic species, out of which peninsular India alone constitutes 32%. The western ghats with 63% of endemics among the tree flora is one of the most important centre of endemism.

However, the rapid growth of population, its needs of wood and land, have made the flora of high altitudes, in particular, very vulnerable. Many of the lowland forests have also been clear felled or are highly degraded. According to the Red Data Book, over 300 species come in threatened category but have been collected during the last century; while some may have lost for ever. Before it is too late, it is necessary to check this genetic erosion based on a scientific approach since many of them must have constituted the life support system for the coming human generation.

However, maps of endemic species distribution gives a clear picture of their concentration and possible area of active speciation. Superimposition of such information on maps of vegetation and biotic environment shall improve their ecological status, according to which the conservation strategy could be worked out.

Suitable routes could be traced taking into account the optimal ecology of selected species and natural areas of distribution. Determination of sensitive zone where equilibrium between the plant cover and surrounding conditions can be traced, especially where anthropoid pressure is high and the climate is fragile. Further, selection of the zones of interest for gene-pool

conservation can be made by plotting some forest types, that are represented in small areas and are likely to vanish without protection.

Conservation Status of Medicinal Plants

All humans are dependent on plants, maybe for food, fuel, fibre, forest and fruits, in order to meet their daytoday requirements for survival. The contribution of medicinal plants, to meet various requirements, for survival in the health sector cannot be underestimated. It is well known that 85% of the traditional medicinal plants are used globally for the primary health care (Kala *et al* 2004); of the world's 2,50,000 known plant species, an estimated 35000–75000 are currently or have in the past been used in folk medicine (Leaman *et al* 1999). WHO (World Health Organisation) estimated that 80% of the world's population, mostly in developing countries, relies on traditional medicine practices, for their health needs. It is true of the poor sections of the population living in rural and far off areas in jungles from developing countries. Natural remedies are not only cheaper than the modern medicines but often are the only drugs available in remote areas (GTZ 2001). In developing nations of the world, particularly in India, traditional medicine has been and still practised since long, and till today plants are the indispensable sources of prevention and curative medicinal preparations, for not only humans, but also for the animals (Bagine *et al* 1997). In the African continent, too, medicinal plants had been in use from times immemorial.

Traditional medicines, once practised in rural areas, have assumed global interest, which is growing day by day. This is partly due to the interest in complimentary medicines in the industrial sector of deve-loped countries and in part resulting from the interest in pharmaceutical industry. Medicinal plants not only provide relief to economically disadvantaged people, small landholders or landless people (with low cash income) but also to the industrially advanced countries like European Union and the United States of America (USA), for the preparation of medicines in their own country. While large sums are being spent to analyse the active principle and other sources from which costly medicines are prepared, e.g. Vincristine from *Carthamnus rosea*, Taxol from *Taxus baccata* and many more.

In India, traditional medicines from plants mainly, sometimes at the doorsteps make them more accessible than the modern health facilities for most of the population, leaving in far off places from the cities and towns. It is relatively inexpensive and accepted by local comm-unities, as compared to the expensive mode of conventional medicines. As of now, the substantial contribution made by the medicinal plant species is widely appre-ciated and understood as can be seen from various steps and the form of organi-sations working on traditional medicinal plants and the interest shown by both the people related to research and the insti-tutions created, relating to medicinal plants research. Of course there had been and is still growing demand for many of the plant species and an increasing interest in the trade. However, the conti-nued habitat loss with more and more area for urbanization and industries, erosion of traditional knowledge, is endangering important medicinal plant species, populations and creating an urgent need for improved conservation measures and sustainable use of these vital plant resources.

Unmonitored trade of medicinal plant resources, destructive harvesting techniques, over-exploitation and habitat change are the primary threat perceptions for the medicinal plant resources, not only in India and its neighborhood, but also for many developing nations of the world (Schoppe-Guth and Freemuth 2001, Lagos-Witte, 2002).

Since most of the plant species used in traditional medicine, till recently used to be collected from the wild and very few have been domesticated with growing efforts from CIMPO and Medicinal Plants Research Institute (MPRI), Lucknow. There is a real danger of species extinction, which needs priority for collection with conservation, research on propagation and restoration of wild genetic pools as well as investigations into possible modifications in active ingredients due to changes in the growing environment. For conservation of plant species, cultivation is often considered an alternative to wild collection from nature's gift. Agrotechniques with plants changing and loosing medicinal properties, needs to be evolved for as many species as possible (Schopp-Guth and Fremuth, 2001).

Several national and international organisations and the gatherings sponsored by them have brought attention to the importance of medicinal and aromatic plants to health and human well-being and to the urgent need for their conservation and management. The ChangMai declaration of 1988 (organised by WHO) resulted in the establishment of a consultation forum on conservation of medicinal plants. other organisations involved are, UNCN and WWF (Leaman *et al* 1999). Others are the Convention on Biological Diversity of 1992 having the objectives to conserve bio-diversity, use it sustainably and ensure fair and equitable sharing of benefits (CBD secretariat, 2001).

In Africa, á decade for African traditional medicine', the Global strategy for plant conservation developed by CBD during the sixth meeting by the conference of parties in April 2002, and the WHO guidelines, standards and policies on herbal medicines including quality control, safety, efficacy and integration, is a milestone on the subject of conservation of biodiversity of medicinal plants.

Threats or Challenges Facing Medicinal Plants

According to the international criteria, over 60,000 plant species have been evaluated for conservation status, of which 34,000 are classified as globally threatened with extinction (Williams *et al* 2003). Plants are endangered by a combination of factors namely; over-collecting, unsustainable subsistence, agriculture, particularly in the tropics, specially in hilly areas. Forestry practices, urbanization, pollution, landuse changes, spread of invasive alien species, pollution and the climate changes are the main causes. Medicinal plant users like local *Vaids* and *Hakims*, in villages plant collectors for drug plants traders could be of help in determining whether or not some species are becoming vulnerable, threatened or rare. Inventory of *in situ* repositories of wild plants are of course helpful in establishing the conservation status of some plants. The involvement of botanical gardens in vegetation and resource surveys to determine conservation status of medicinal plants is important. The Red Data Book, its constant review with herbalists who walk increasingly greater distances for herbs that were previously collected locally (not by vehicles). As the habitats and over-harvesting of commercially important species particularly in the fragile environments of the alpine and subalpine regions, when collected for commercial use reduce medicinal plant resources, in the wild with the result that there is a considerable drop in the availability of the plants used by traditional medicine practitioners.

Bioprospecting: Since there is a revival of interest in the genetic diversity, with the advancement in the field of biotechnology stimulated interest in medicinal plants and the related traditional knowledge, with the result that many organisations both academic and commercial institutions, increased their investment in the search for new biological and chemical compounds. Specific plants and animals as well as micro-organisms are targeted. There is an increasing interest in plants like Taxus, Vinca, etc. Major markets for the extracted plants are at Delhi, Mumbai, Dehradun and other centres in Uttaranchal and other big cities since their is a considerable potential for the existing and potential future bioprospecting in the country.

Habitat destruction: Habitat destruction is the most important anthropogenic cause for diversity loss. It has been estimated that nearly 40% of the earth's potential net primary productivity is either used by mankind or lost as a result of landuse change (Maunder and Clubbe, 2002). Extensive conversion of wild habitat resulted to changes, both in ecological and taxonomic composition of these areas, with adverse effect on medicinal plant's diversity. The rate of anthropogenic activities and the rate of medicinal plants and other forms of biodiversity loss has increased beyond natural levels.

The impact of human growth has resulted in both increased habitat loss and fragmentation. As a result many of the medicinal plant species resources got deplited upon which people depended for their survival. In Africa and Asia, annual forest clearance had been between 2.3 to 3.7 million hac, due to subsistence farming (Maunder *et al* 2002). The trend is likely to continue, as human population and their associated demands on ecosystems grow.

Dilution of customary conservation practices: Sustainable use of medicinal plants in fragile ecosystems was facilitated by common beliefs and practices of the local people. Taboos and social restrictions on gathering medicinal plants limited their harvesting to a great extent. Some communities in the Himalayan region and the arid areas insisted that certain plants should only be harvested after prayer for which the land was designated to the God (like 'Orans' in Rajasthan and land allotted to temples at high altitude like Badrinath, Kedarnath, Gangotri and Yamnotri, etc.). Taboos against the collection of medicinal plants by menstruating women were followed. Taboos that only lateral shoots or roots must be harvested and soils be returned after harvesting the roots or else sickness would replace, are still in vogue. Harvesting limit (not more than 7 in some cases) was also fixed for collection and it was believed that if more plants are harvested some calamity may fall. No single plant should be cut in traditional worship sites. These practices not only protected the intellectual property rights on the technology, but also inherently limited the number of users.

However, due to changes in attitudes and values (acculturation), commercial interests have weakened customary conservation practices that lead to uncontrollable harvesting of medicinal plants and with

subsequent encroachment of the habitats and loss of their biodiversity. The tendency of young generation to copy western lifestyles has watered down the intrinsic indigenous knowledge of medicinal plants and the expanding gap between the young and the old has resulted to latter dying with the invaluable legacy on medicinal plants, while the former living with no sense for their conservation.

***Invasive species*:** Invasion of certain exotics have become increasingly important as agents of species extinction. Most of the alien species are introduced deliberately for their ornamental, agricultural, medicinal, timber value (Tye, 2002), while some of the species like *Parthenium hystrophorus* in most parts of India have invaded some areas since the seeds got imported along with food grains received from the donor countries. *Lantana* is an example of mass invasion. Disturbance, due to human activities, favour invasion, so that vegetation zones that suffered more threatened in that both introduction rated and susceptibility to invasion are higher in inhabited areas.

The invasive species alter community composition or threaten individual species, adversely effecting its existence. They cause drastic changes in community composition or threaten individual species, adversely affecting its existence. Not only they change habitats, but also form monospecific stands, shading out or replacing native species.

***The climatic change*:** The catastrophic effects on plant diversity worldwide could be those that result from climatic changes induced by mismanagement of the earth (Given, 1994). The present industrial revolution and increased agricultural activities have contributed to increasing levels of "greenhouse gases" in the atmosphere with an impact of global biochemical cycles and the climate change. It is predicted that global temperatures shall rise significantly by 2050. Such changes would lead to impacts on individual organisms, communities, natural ecosystems and global biochemical cycles, with potentially great impacts on biodiversity (WCMC, 1992; Given, 1994, Gupta *et al* 2000).

Govt. Policies on Traditional System of Medicine

The number of traditional medical practitioners (TMPs) is gradually decreasing. Some of these are placed as witchcrafts with the result that many of the genuine practitioners have restricted their activities; since many of them do not have the licence to practise. The access and benefit sharing from medicinal plant resources need to be promoted at multi-levels. The lack of laws on traditional medicinal plants and the political will to support natural systems to cure diseases would certainly conserve medicinal plants. Alienation of the forest dwellers and local communities, the tribals, arising from protected areas created a negative view of forest policy with the result that forest laws are broken and the people force themselves into the forest and protected areas to harvest forest products including medicinal plants uncontrollably from the banned region, not only for their own use but also for trade and commerce.

Plant Variety Protection and Farmer's Rights Act, 2001

The Indian Parliament has finally passed the Plant Variety Protection and Farmer's Rights Act, 2001. A long and arduous struggle for the recognition of the rights of farmers in India's *sui generis* legislation has thus ended and a law to grant plant breeder's right on new seed varieties, for the first time has been recognised. Trade Related Intellectual Property Rights (TRIPS) as ratified by the Urugay GATT Round in 1994 necessitated by the commitments that India made necessitated this action. Article 27.3(b) of Trips which deals with the protection of new plant varieties, offers three options. Protection will have to be granted by a patent, an effective *sui generis* system or by a combination of two. The *sui generis* system refers to the grant of plant breeder's rights, of what kind is not defined, except to say that *it should be effective*. Ultimately, India accepted for the *sui generis* option, but not without a struggle by civil society to stop seed patents. Gene campaign has been in the forefront of this campaign which was against patenting of seeds, considered by some to bring great fortune of farmers and the harbinger of rural prosperity. In March 1993, it was a decessive event which contributed a great deal to abandon its pro-patent stand. Once the *sui generis* path was decided, came the question of kind of legislation required to accord plant breeder's right in India for the first time. The gene campaign's demand has been for a farmer's right that would allow the farming community to retain the same control over seed production and use what they have always had. The gene compaigns maintained that plant back rights were no rights in varying degrees, only exemptions. Some exemptions, sometimes referred to as farmers privilege, were allowed by breeders in the early years of UPOV and were limited to plant back rights in varying degrees. In some UPOV members countries, e.g. France, limited exemptions were granted to farmers, in others like Greece, these were more generous. Exemptions for farmers were retained till the 1978 version of UPOV, but have been considerably diluted since. After the last amendment in 1991, exemptions for farmers are no longer a method of course. They have been made optional and are subject to the consent of the breeder.

Gene campaign has insisted, that Indian law has to grant well-defined rights, not just provide beggarly exemptions to its farmers. These rights have to be recognised because of the past and present contributions made by the farming community to the conservation of agro-biodiversity and their role as a dynamic breeder of new varieties which anchor the food security of the world.

The new law recognizes the farmer, not just as a cultivator but also as a conserver of the gene-pool and a breeder who has bred successful varieties. The act makes provisions for such farmers varieties to be registered so that these are protected against being scavenged by formal sector breeders. The rights of rural communities are acknowledged well. The final version of the clause on what constitutes a farmer's right (Section 39) clause IV, now reads like this.....

The farmer.... "shall be deemed to be entitled to save, use, sow, resow, exchange, share or sell his farm produce including seed of a variety protected under this Act in the same manner as he was entitled before the coming into force of this Act.

Provided that the farmer shall not be entitled to sell the branded seed of a variety protected under this Act.

For the purpose of this clause (IV) branded seed means any seed put in a package or any other container and labeled in a manner indicating that such seed is of a variety protected under this Act.

In India, the farming community is the largest seed producer, providing about 85%

of the country's annual requirement of over 60 lakh tons and if the farmer were to be denied the right to sell, it would result in a substantial loss of income for him. The agro-chemical giants, turned life-science corporations, have recently emerged as the largest seed producers in the industrialised nations where seed production is in their hands. Control over the seed sector was established by the simple expediency of buying up all the smaller seed companies. Weak farmer's rights will allow seed corporations to dominate the seed market and strong farmer's right keeps the farming community alive as well as viable competitors and an effective deterrent to take over of the seed market by the corporate sector.

Convention of International Trade in Enangered Species (CITES)

According to the World Health Organisation (WHO), 80 percent of the world's population will be relying on plant-based medicines for their health-care needs by the end of the last century, while still there is growing resistance world over to chemical medicines because of their "side effects". The "holistic approach" to health maintained by the traditional systems of medicines are making them more popular with people.

While the demand for medicinal plants is growing, their availability is decreasing because of the destruction of the forests habitats of such plants. International Development Research Centre (IDRC) of Canada shared the view and recommended urgent initiatives necessary to conserve the genetic resources of medicinal plants and recommended that "steps should be taken to ensure their continued availability".

In India, more than 500 million people depend directly or indirectly on plant-derived drugs for their health care needs. The socioeconomic significance of medicinal plants in India is high as they provide 32 million man days of employment among tribals and local population.

The demand for medicinal plants from multinational pharmaceutical organisations has also put pressure on India's herb resources as is evident from the fact that the export of medicinal plants increased 7–8 fold during the past decade.

Ayurveda, the most widely practised indigenous medical system, makes use of about 2000 herbs for healing; while some more than 300–400 million people depend solely on ayurveda for their health-care needs. While more than 200 million are estimated to be partially dependent on it. There are more than 4200 recognised ayurvedic pharmacies, while the number of ayurvedic physicians, medicine makers and herb collectors are likely to be much higher. However, shortage of medicinal plants, sometimes, has led to their substitution by counterfeit herbs, resulting in the deterioration of the efficacy of medicines supplied.

Recent attempts in a bid to tap US 65 billion dollars international market for plant-based medicines led to the constitution of a Medicinal Plant Board which was recently constituted to formalize and organise medicinal plants, their marketing and trade and coordinate all those involved in development of herbal products. It was also meant to protect biodiversity (of medicinal plants) existing in the country and to organise development of medicinal plants for enhancing the export revenues on herbal material and medicines. The Medicinal Plant Board's responsibility includes preventing the patenting of the traditional medicinal plants or their derivatives by outside agencies as has happened in the case

of *neem, haldi, brinjal* and *gurmar*. Though China has emerged as the largest supplier of herbal products, with its annual sale crossing US 5 billion dollars, the market for plant-based pharma products is growing at a healthy rate of seven percent internationally.

A note by CITES remarks a tree species endemic to India, is sparsely distributed and is varyingly described as endangered or threatened, and rare. Its wild populations are under pressure from over-harvest. There are indications of demand for increasing its number in countries such as in Japan" (status of a protected species is recorded by CITES in its appendix 2 at a meeting in Fort Landerdale, USA). Status of some threatened medicinal plants of Western Himalayas is given in Table 6.5.

The legal status of the red sander's trade is rather ambiguous. Its export in any form is banned; but a little more than 468 metric tonnes of red sandres was exported officially during a period of 6 years, A passage from the ME and F (Ministry of Environment and Forests) note says that although on the prohibited list of exports, large quantities of sandalwood chips or powder exported regularly from India ....however legislation in India". Commenting on the paradox the CITES noted "however legislation in India regulating harvest and trade is unclear" What is going out "legally" or rather openly is "unclear" and only the proverbial tip of the iceberg "Seizures in India attest to the presence of illegal trade" says CITES. Does not it all amount to a great national farse? We have a law that bans. And the same law that permits (the legal export of the banned items).

Table 6.5: Proposed status of some threatened medicinal plants of Western Himalaya
(mentioned in Red Data Book)

Plant name (family)	Local/trade name	Distribution	Status	uses	Parts used	Proposed cite appendix
1. *Aconitum balfouri* Stapf. (Ranunculaceae)	Gobari, Mithá, Haldu-Gobariya	Subalpine and alpine Uttaranchal.	En	Poisnous	Root	I
2. *A. chasmanthum* Stapf. ex Holms	Mohri, Kash, Banbal-Nag	Subalpine and alpine West Himalaya	En	Rheumatic fever	Root	I
3. *A. deionorrhizum* Stapf.	Mohara, Murabi-Kh	HP and J&K, subalpine areas	En	Rheumatism	Root	I
4. *A. falconeri* Stapf. var *latilobum* Stapf.	Kala-Mohra	Endemic to HP	En	Rheumatism	Root	II
5. *A. violaceum wall.* ex Royle	Patis					
6. *A. heterophyllum* Royle	Atis, Ais, Atces	Subalpine to alpine regions of Himalaya	Vu	Rheumatism	Root	I

Table 6.5: Proposed status of some threatened medicinal plants of Western Himalaya (mentioned in Red Data Book) (*Contd...*)

Plant name (family)	Local/trade name	Distribution	Status	Uses	Parts used	Proposed CITE appendix
7. *Atropa acuminata* (Solanaceae)	Royle ex Lindl.	HP and Kashmir		Asthma, whooping cough	Root and aerial parts	III
8. *Aristolochia indica* L.(Aristolochiaceae)	Isarmul	Throughout India		–	Dried stem and plant parts roots	IV
9. *Angelica glauca* Edgew. (Apiaceae)	Choru	Kashmir and Kumaon	Vu	Dyspepsia & constipation		I
10. *Colchicum luteum* Baker (Liliaceae)	Hirantutiya (K), Irkim, Mppond (Punj) Sauranf (Karvi)	Northwest Himalaya	Vu	Cancer, Arthritis.	Dry corm	I
11. *Gentiana Kurroa* (Royle Gentianaceae)	Kamalphul, Nilkanth kuru, Kutki	Nowthwest Himalaya	Vu	Staintalis appetizer	Dried rhizomes	I
12. *Dactylorhiza hatagirea* (D.Don) Soo (Orchidaceae)	Salampanza, Salap Halfajari	NW and East Himalaya		Medicinal	Tuberous roots	I
13. *Dioscorea deltoidea* Wall. ex Kunth. (Dioscoreaceae)	Kriss, Kins, Kithi	Kashmir to Assam Himalaya	Vu CITES II	Medicinal Diosgenin	Tuberous roots	II
14. *E. phedra gerardiana* Wall ex Stapf. (Ephedraceae)	Ephedraceae	Western Himalaya	Vu	Bronchial asthma pneumonia, heart stimulant diphtheria	Dried stem	I
15. *Gloriosa superba* L. (Liliaceae)	Kalihari, Languli	Sub-Himalayan tract and tropical India	En	Anthelmintic	Tubers	II
16. *Iphegenia indica* Kunth, (Liliaceae)		JK, Siwaliks and Meghalaya	DD	Colic, headache	Corms	I
17. *Meconopsis aculeata* Royle		Himalayan Blue Poppy West Himalaya		Narcotic, poisonous	Roots	IV
18. *Nardostachys grandifloa* DC (Valerianaceae)	Jatamansi	West and East Himalaya	Vu	Tonic, stimulant, heart trouble, fits	Rhizome	I

Table 6.5: Proposed status of some threatened medicinal plants of Western Himalaya (mentioned in Red Data Book) *(Contd...)*

19.	*Rheum australe* D.Don (Polygonaceae)	Dolu	Uttaranchal Kumaon Garhwal	Vu	Purgative, astringent used in dyspepsia	Roots	II
20.	*Physochlaina Praealta* Miers Solanaceae)	Bajarbang	NW Himalaya	Lr	Used as Mydriatic, pupil dilatation	Leaves	II
21.	*Rauwolfia serpentina* Benth. ex (Apocynaceae)	Kurz Sarpgandha	Foothills and Trop. Himalaya	Lr	Control high BP	Roots	II
22.	*Rheum emodi* Wall. (Polygonaceae)	Rubarb, Dolu	Subtrop and alpine Himalaya	Vu	Purgative, analgesic	Rhizome	II
23.	*Berberis aristata* DC. (Berberidaceae)	Rasaut, Daru-Haldi Kingora	Himalaya and Nilgiris	Vu	Ophthalmic, Ulcers	Dried roots	I
24.	*Acorus calamus* L. (Araceae)	Vach, Ghorabach Calamues.	Fresh swamps in foothills		Stimulant, expectorant for memory	Rhizomes	II
25.	*Costus speciosus* Sm (Zingiberaceae)	Keu, Kust	Sub-Himalayan tract and Net India	Vu	Diosgenin	Tubers	III
26.	*Jurinea dolomiea* (Asteraceae)	Dhoop	NW Himalaya		Incense		I
27.	*Hyoscyamus niger* L. (Solanaceae)	Khorsani-ajwain Brose wood	NW Himalaya	Lr	Insomnia hysteria	Dried leaves and flg. tops.	II
28.	*Swertia chirayta* Royle ex Benth. (Gentianaceae)	Chiráyta	Kashmir to Bhutan		Tonic, vermifuge & antipyretic	Entire plant	I
29.	*Urginea indica* Kunth. (Liliaceae)	BanPiyaz, Kunda Jangli-Piyaz	Uttaranchal, Bihar, Goa, Coromandal coast		Cardiac tonic, bronchitis, and diuretic	Bulbs	I
30.	*Saussurea costus* (Falc.) Lipsch (Asteraceae)	Kuth	NW Himalaya and Goa	En	Tonic, carminative, Cardiac complaints, asthma.	Roots	II
31.	*Podophyllum hexandrum.* Royle (Podophyllaceae)	Bankakri, Papri Indian Podophyllum	Subalpine alpine region in Himalaya	Vu CITES-II	Tumorous growth, Cancerous tissue and skin diseases	Dried rhizomes	(CITES I)
32.	*Euphorbia* sp. (Euphorbiaceae)			CITES-II			(CITES II)

Table 6.5: Proposed status of some threatened medicinal plants of Western Himalaya (mentioned in Red Data Book) *(Contd...)*

Plant name (family)	Local / trade name	Distribution	Status	Uses	Parts used	Proposed CITE appendix
33. *Taxus wallichiana* Zucc. (Coniferae)	Yew	Temp. and subalpine Himalya	Vu	Cancer	Leaves, bark	(CITES II)
34. *Picrorhiza kurrooa* Royle ex Benth. Scrophulariaceae	Kutki, Tinami, Katki	Kashmir to Sikkim	Vu	Secretion of bile and stimulant for digestive system	Rhizomes	(CITES I)
35. *Arnebia benthamii* (Wall. ex G. Don) Jhonston (Boraginaceae)	Ratanjot, Balchar	Kumaon to Uttaranchal	Lr	Cardiac disorder, hair tonic and tongue diseases	Plant	I
36. *Bergenia stracheyi* (Hook.f.and Th) Engl. (Saxifragaceae)	Pashanbed, Patharphorid	J&K to Uttaranchal	DD	Kidney stone	Root	I
37. *Hedychium spicatum* Ham.(Zingiberaceae)	Sitruti Kapurkachri	Subtropical Himalaya	Vu	Carminative	Rhizome	II

* The classification is based on the proceedings of meeting held by BSI, Dehradun, proposing the CITES. Categories for the W. Himalayan region.

Chapter XVI, Part of the Export-Import Policy for 1992 and 1997 says "all wood and wood products in the form of log, chips, powder, flakes, dust, timber are prohibited in the same chapter. Serial no 34 and 44 of part V states plant portions and derivation obtained from the wild and processed timber of all species excluding sandalwood and Red Sanders may be exported subject to specified terms and conditions". Can we guard the vast natural wealth of this country with this piece of law. This is the legal situation. Now one look at the illegal one. We have four regional Directors to check the export of protected species. But there are 14–15 points from where the agricultural products are being exported. So some amount of smuggling will always go on through misdeclaration. But had the amount been just 'some' the country's plant bank would not have been in a state of bankruptcy.

Indian Plants in CITES

Appendix I of CITES includes highly threatened species, deemed to be threatened with extinction and may be affected by trade (Table 6.6); trade strictly regulated in wild; trade permitted in special circumstance only in artificially propagated plants/flasked seedlings; both import and export permits are required (ca.675 species in the world).

Appendix II: Species though threatened or become so, if trade is not properly controlled; trade allowed both in wild and artificially propagated plants subject to licensing; only export license is required (ca 21000 species in the world) Table 6.7.

6

Table 6.6

Family name	Species name	Common
1. Asteraceae	*Saussurea costus*	Kuth
2. Cycadaceae	*Cycas beddomei*	Beddome's Cycas
3. Nepenthaceae	*Nepenthes khasiana*	Pitcher plant
4. Orchidaceae	*Renanthera imschootiana*	*Red vanda*
	Vanda coerulea	Blue vanda
	Paphiopedilum	Lady's slipper

Table 6.7

1.	Apocynaceae	*Rauwolfia serpentina*	Sarpgandha
2.	Asclepiadaceae	*Ceropegia spp.*	–
3.	Araucariaceae	*Araucaria araucana*	Monkey Puzzle tree
4.	Berberidaceae	*Podophyllum hexandrum*	Indian Podophyllum
5.	Cactaceae	*Cactus sp.*	Cactus
6.	Cyatheaceae	*Cyathea sp.*	Tree ferns
7.	Cycadaceae	*Cycas spp.*	Cycads
8.	Dioscoreaceae	*Dioscorea deltoidea*	Elephant's foot
9.	Euphorbiaceae	*Euphorbia spp.*	Euphorbias
10.	Liliaceae	*Aloe spp.*	Aloes
11.	Orchidaceae	*Entire family*	Orchids
12.	Fabaceae	*Pterooarpus santalinus*	Red sanders
13.	Thymaleaceae	*Aquilaria malaccensis*	Agarwood
14.	Taxaceae	*Taxus wallichiana*	Yew
15.	Valerianaceae	*Nardostachys grandiflora*	Jatamansi
16.	Scrophularia-ceae	*Picrorhiza kurooa*	Kuru

Appendix III for Napal: A support mechanism to domestic legislation; countries can ask other parties to monitor trade in those taxa not listed in appendices I and II (Table 6.8).

Table 6.8

1.	Gnetaceae	*Gnetum montanum*
2.	Magnoliaceae	*Talauma hodgsonii*
3.	Podocarpaceae	*Podcarpus nerifolius*

An orchid, lady's slipper, is spectacular and of awesome beauty. Today, it is being ravaged by collectors and exported for the last 150 years. Worldwide 3 million orchids are estimated to be traded annually. In Red Data Book, the entire orchid family has been included. It grows in the Himalayan region and western ghats. South India lost three beautiful flowers decades ago. Over 3000 plants are plucked. The lady's slipper is a difficult and expensive plant to grow in captivity. Collection from the wild is the cheap alternative, hence, the illegal trade. Species collected from wild cannot be exported legally. In fact Kallimpong, Shillong and Thiruvananthapuram are the main centres for cultivation and the orchids are exported to Japan and USA not through regular routes.

Moreover, the note of ME and F at every step recorded the felling of endangered plants and wood as destructive. Even the Dr. Subramaniam Committee Report set up to evaluate the trade in wild species has remarked "the trade in fauna is largely undocumented though large".

Almost all the scientific studies have hailed India as the custodian of the richest biodiversity on planet earth. But knowingly or unknowingly, openly or surreptitiously, legally or illegally, we are destroying our own wealth. Should not we as ourselves ask the question: What will we export or smuggle once the species is finished? Also are we not killing the very healer that heals us?

Finally, the Indian Parliament passed the Plant Variety Protection Bill and Farmers Right Act in 2001 and with this has ended the long and ardous struggle for the recognition of the rights of farmers in India's *sui*

generis legislation. The legislation was necessitated by the commitment that India signed on Trade-Related Intellectual Property Rights when it ratified the Uruguay GATT round in 1994. Article 27.3(b) which deals with the protection of new plant varieties offers 3 options. Protection will have to be granted by a patent, an effective *sui generis* system or by a combination of the two. The *sui generis* system refers to the grant of Plant Breeder's Rights of what kind is not defined, except to say that it should be "effective". India ultimately opted for the *sui generis* option but not without a deter-mined struggle by civil society to stop seed patents.

In its more than 20 years of existence CITES has achieved many successes such as Training. Programmes for co-operation between CITES and Customs co-operation Council and INTERPOL for species under very exceptional circumstances. Under appendix II trade is permitted with proper permits which provide ideal records for those who monitor and analyse international wildlife trade. This is regulated within countries and for which the co-operation of nations is sought.

References

Atal CK and Kapur BM, Cultivation and A Utilization of Medicinal Plants (RRL, Jammu)

Anonym 1976. Foreign exchange saved by Indigenous Medicinal Plants. *East Pharm.* **19**(216): 50

Anonymous 1996. Annual Statistics of Foreign Trade in India. Ministry of Commerce GOI. New Delhi.

Anonym. 1980. Seminar on export potential of Medicinal plants, phytochemicals and bulk drugs. *IDMA Bull.* **11**(13) : 154–156.

Bagine R, Muthoka P and Mungai G. 1997. Conservation of medicinal plants in Kenya. In *Conservation and utilization of Indigenous Medicinal Plants and Wild Relatives of Food crops* (eds A. M. Kinuya and WM kofi-Tse kpo). pp : 81–84. UNESCO, Nairobi, Kenya.

Balick JM. 1995. Medicinal Resources of Tropical Forest: Biodiversity and its importance to Human Health. Rep. 202. (ISBN: 81–211–0327)

Barua AK, Chakravarty D and Chatterjee D. 1955. Steroid hormones from Indian Dioscorea plants. *Bull. Inst. Sc. India* 1955. 4: 15–20. Barua Ak, (MRS Chakravarty D and Chakravarty R.N. 1956. J. Indian Chem. Soc. 33: 799

Barua AK, Chakravary D and Chatterjee 1955. Steroid hormones from Dioscorea plants. *Bull. Inst. Sc. India.* 1955.4: 15–20

Barua AK Chakravarty D (Mrs) and Chakravarty RN. 1956. *Journ. Indian Chemical Soc.* **33**: 79.

Bedoukian PZ. 1987. Perfumery and Flavour Materials. *Perfum. Flavour* **12(2)**: 1–30.

Chopra IC and Sobti SN and Handa KL. 1962. Cultivation of medicinal Plants in Jammu and Kashmir. *ICAR Res. Series.* No **13**. pp.1–75.

Chopra IC and Handa KL 1961. Review of Research on Indian Medicinal and Allied Plants. *ICAR Review Series.* no 34.pp.1–60.

Chopra IC and Handa KL. 1963. Steroid sapogenin bearing plants of India. *Indian For.* 94(8): 620. ; *Bull Reg. Res. Lab, Jammu* 192: 183–189.

Dery BB, Otsynia and Nglatigwa (eds). 1999. *Indigenous uses of Medicinal Trees and Setting Priorities for their Domestication in Shinyanga region.* pp. 1–2. Tanzania, International Center for Research in Agroforestry, Nairobi, Kenya.

Dhar U, Rawal RS Samant SS, Aori S and Upreti J. 1999. People's participation in Himalayan biodiversity conservation. A practical approach. *Curr. Sci.* **76**: 36–40. Farooquee NA and Saxena KG.1996. Conservation and utilization of medicinal plants in high hills of central Himalayas. *Environmental Conservation* **23**: 75–80.

Dobhal R. 1999. Recognising biodiversity and indigenous knowledge system under new intellectual property regime. *Curr. Sci.* **76**: 1063–04.

Germany Technical Co-operation (GTZ). 2001. Medicinal plants: Biodiversity for Health care, Issue Papers. Biodiv. GTZ. Germany.

Given DR. 1994. *Principles and Practice of Plant Conservation.* Chapman and Hall. London.

Gupta RK, Dabral BG, Meher-Honji VM and Puri, GS. 2000 *Forest Ecology: Environment, Forests and Rainfall.* Vol. 3 pp. 1–458. Oxford and IBH. Pub. N.Delhi.

Gupta SS. 1962. Experimental studies of Pitutary diabetes. Pt. I. Inhibitory effect of a few Ayurvedic antidiabetic remedies on Interior pituitary extract induced hypoglycemia in albido rats. *Indian Journ. Med. Res.* **50(1)**: 73:81.

Gupta AK, Vatas SK and LAL B. 1998. How Cheap can a medicinal plant species be? *Current Scince* **74**: 565–565.

IUCN (International Union for Conservation of Nature) 2002. *Medicinal Plant Conservation;* News letter of the medicinal plant specialist group of the IUCN survival commission. Vol 7.

IUCN 2001. Medicinal Plant Conservation: News letter of the medicinal Plant specialist group of the IUCN survival comission. Vol **8**.

Jain S. 1980. Current market situation of various essential oils. *Indian Perfum.* **24(1)**: 20–21.

Kala CP. 1998. Ethnobotanical survey and Propagation of rare Medicinal Herbs for small farmers in the buffer zone of valley of Flowers., National Park, Garhwal Himalya, Dehradun. Wildlife Institute of India.

Kala CP, Farooquee NA and Dhar U. 2004. Prioritization of medicinal plants on the basis of available knowledge, existing practices and use value status in Uttaranchal, India. *Biodiversity and Conservation.* **13**:453–469.

Kamatenesi-Mugisha M and Bukenya-Ziraba R. 2002. Ethnobotanical survey methods to monitor and assess the sustainable harvesting of medicinal plants in Uganda. In *The Royal Botanic Gardens, Kew. Plant Conservation in the Tropics: Perspectives and Practice* (eds. M. Maunder, C. Clubbe, C. Hankamer and M. Groves. PP. 467–482. The Cromwell Press, Ltd. UK.

Lawrence BM 1985. Review of the World Production of Essential Oils *Perfum. and Flavour.* 10(5): 1–16.

Leaman DJ, Fassil H and Thormann, I. 1999. *Conserving Medicinal Plants Species: Identifying the Contribution of International Plant Genetic Resources Institute* (IPFRI) Report.

Lagos-Witte S. 2002. Conservation of Medicinal Plants in Central America and the Cari bbean: A GEF project begins. In *Medicinal Plant Conservation: Newsletter of the Medicinal Plant Specialist group of the IUCN Survival Commission.* **8**: 21–24.

Madhuri RK, Nautiyals, Rao KS and Saxena KG. 1998. Medicinal Plants cultivation and biosphere reserve management: a case study from Nanda Devi Biosphere, Himalaya. *Current Sc.* **74**: 157–163.

Madhuri RK, Nautiyals, Rao KS and Saxena KG. 1980 Role of medicinal Plants in the traditional health care system : A case study from Nanda Devi Biosphere Reserve, Himalaya. *Curr. Sc.* **75**: 152–157.

Mc Neel, Miller KR, Reid WV, Mittermeir, Werner TB. 1990. Conserving World's Biological Diversity. Gland, Switzerland: ICUN, WRI,CI, WWF, US znc World Bank.

Maunder P and Clubbe C. 2002. Conserving tropical botanical diversity in the real world; In *Plant Conservation in the tropics : Perpective and Practice* (eds. Maunder M, Clubbe, C. Hankamer and Groves M. pp. 29–48. The Royal Botanic Gardens, Kew, The Cromwell Press Ltd. UK.

Maunder M, Stanley Price M, Soorae P and Mashuari S. 2002. The role of tropical Botanic Gardens in supporting species and

habitat recovery: East African opportunities. In *The Royal Botanic gardens Kew. Plant Conservation in the Tropics: perspective and Practice*. Ibid. pp. 115–134. The Cromwell Press. Ltd. UK.

Nair MP, Sastery ARK 1989. Red Data Book of Indian Plants. Calcutta BSI.

Nautiyal MC. 1996. Cultivation of medicinal plants and biosphere reserve management in alpine zone. In : Ramakrishnan PS, Purohit AN, Saxena KG, Maikhuri RK (eds) *Conservation and Management of Biological Resources in the Himalaya*. Almora. GBP Institute of Himalayan Environment and Development. New Delhi. Oxford and IBH Pub. Co. N. Delhi.

Odera JA. 1997. Traditional beliefs, sacred groves and home garden technologies. In *Conservation and Utilization of Indigenous medicinal Plants and wild Relatives of Food crops* (eds. AM Kinyua and WM Kofi-TseKpo) pp. 43–48. UNESCO, Nairobi. Kenya.

Robin SR. 1983. Selected market for the essential oils of Lemon grass, Citronella and Eucalypus. Report. *Trop. Prod. Inst.* No G 171.pp. 1–91.

Secretariat of the Conventation of Biological Diversity (CBD) 2001. *Handbook of the Convention on Biological Diversity*. Earthscan Publications Ltd. London. UK.

Schopp-Guth A and Fremuth W. 2001. Sustainable use of medicinal plants and nature conservation in the Prespa National Park area Albania. In *Medicinal Plant Conservation: News letter of the Medicinal Plant Specialist Group of the* IUCN *Survival Commission*. **7**:5–8.

Srivastava AK and Hussain A. 1977. Research and Development on medicinal and aromatic plants in India. *J. Ind. and Tr.* 27: (1–12) 33–35.

Sumansahai 2001. Plant variety protection and Farmer's rights Act, 2001.

Sundresh I. 1985. India's position in the world trade of essential oils and future export possibilities. *Perfum. Journ.* **7(2)** : 18–30.

Sundaresh I. Export potential of medicinal plants and their derivatives. *Cult. Util. Med. Plants* : 800–823: 787–796.

Sundaresh I. 1978. Export potential of medicinal plants and their derivatives. *East. Pharmst.* **21**: 63–71.

Tye A. 2002. Threatened species management in an Ocean Archipelago: The Galpogos Islands. In: The Royal Botanic Gardens, Kew. *Plant Conservation in the tropics: Perspective and Practice* (eds. M. Maunder M. Clubbe C. Hankamer and Groves M) pp. 323–347. The Cromwell Press Ltd. UK.

Varshney SC. 1990. Export potential of aroma chemicals—a case of value-added products. *Indian Perfum.* **34(3)**; 233–237.

Williams C, Davis K and Chyne, P. 2003. The CBD for Botanists: An *Introduction to the Convention of Biological Diversity* (CBD) for people working with Botanical collections, Royal Botanic Gardens, Kew UK.

World Conservation Monitoring Centre (WCMC) 1992. *Global Biodiversity: Status of Earth,s Living Resources*. Chapman and Hall.

❑❑❑

Environmental Considerations: Farming of Aromatic and Medicinal Plants

Ecological Evaluation of the Habitat of Drug Plants

The theory of climatic analogues as an explanation of the phenomenon of acclimatization and as a guide for the introduction of drug plant species has serious limitations. The presence or absence of the vegetation associated with the drug plants is determined by the nature and intensity of the habitat factors. Such factors like climate, edaphic, physiographic and biotic, control the ecological aspects of the vegetation and produce and essentially uniform environment for the vegetation to subsist. Climate, however, makes the greatest imprint on the vegetation and serve as causative element in the development and distribution of varied plant life. Evaluation of these factors with the help of ecological formula was proposed by Gaussen (1954).

The Indian subcontinent, except some hilly regions like Himalaya in the north and the ghats in the south, is thinly forested. The proportion of the area under forest to the total area of the land is not sufficient of preserve the essential climatic and physical conditions required for the maintenance of life in the country. The socioeconomic conditions of the people are such that the forests are a part and parcel of their daily life and are regularly exploited for various purposes like fuel, wood, fodder and for cultivation. The anthropogenic influence on the vegetation is so intense that the area under forest stands to only not more that 19% (with 40% or more forest cover) of the total available land (Report Forest Survey of India).

The Indo-Gangetic plains, which are largely under cultivation, have an acute need for raising large plantations of forest species with medicinal uses. Farm forestry and fuel plantations are also extremely important for the well-being of the people. Reforestation of the arid and semiarid areas with suitable, quick growing species, is needed in order to check spread of desert-like conditions to other parts of the country; sand-dune stabilisation and the urgency of providing wood and fuel requirements to the local population is an urgent task.

Though in humid areas, as for example in Himalaya and the Ghats, conditions do not appear as bad as in the arid and semiarid regions, there are extensive areas in the river catchments where forest vegetation has been indiscriminately exploited during the past. Sooner or later the problem of reclothing the cold desert areas of Himalaya like Ladakh and Lahul and Spiti, with a view to reassuring the regular water supply and protection from disastrous floods, and silting of reservoirs

constructed for hydroelectricity, have to be considered more seriously than has been done in the past. Further, the development and utilization of basic natural resources, particularly the drug plants, in the present developing economy is a must and requires special attention.

The present vegetation in most part of the country is not a 'climatic climax' and composes a number of communities preserved as 'seral' stages on account of the special habitat features like special rock, soil type and biota. Thus, the species comprising any such community may not be altogether suitable for replanting these areas as edaphic conditions must have considerably changed for some species against others. The reforestation programme with drug plants can be expected to yield results only, if these are based on ecological study of relict vegetation, patches of which are present in some form or the other in the area. Use of exotic species for reforestation may be favoured in some cases on account of their superiority to the indigenous species, either economically, or in growth and development characteristics. For many years Eucalyptus species for oil and wood have been tried with success in many parts of the country, like *Tarai* in north and Nilgiris in south, now is the main source of Eucalyptus oil. *Casuarina equisetifolia* plantations have been raised in south India along the seacoast as well as on inland (islands) Mediterranean species like *Castanea sativa, Fagus sylvetica, Quercus rubour*, Poplars, Hawthorn, Hazelnuts have been tried along with some drug plants with fair success in various parts of the country. Introduction of *Prosopis julifera* in dry areas has proved very successful but its effects on the natural vegetation have not very encouraging. During the recent years, several Australian Acacias and Eucalyptus have been introduced but it is still difficult

to say what their effect is on long-term perspective on the natural vegetation and how they shall be able to compete with the indigenous vegetation. Moreover, in the second or subsequent rotations the species may not show sufficient growth or may suffer from ill effects of climate, or fungus and insect attacks, as exemplified by several plantations made earlier. In the same way, extension of areas of indigenous species have raised similar problems, though less troublesome. There are examples to show that the trial and error method for introduction of exotics may prove much more expensive than the scientific methods. It would, therefore, be useful to evolve some scientific method for the introduction of exotics (aromatic and drug plants) and as well as extension of indigenous species to areas outside their geographical limit.

Early Techniques of Plant Introduction

In USA and Europe introduction of crop plants from different parts of the world has been attempted since long and one of the methods adopted was proposed by Merriam (1890) in his work on life zones and crop-zones of USA. This method utilized temperature data of the two areas, on which plant introduction was to be planned. Livingstone and Sherve (1921) used the ratio of precipitation to evaporation as an index to tolerance of plant growth to new conditions. Later on Koppen presented a classification of the Climates for the world in which his main climatic groups are based on monthly and annual means of precipitation and temperature. Thornthwaite's recent classification (in 1933) is based on precipitation-evaporation ratio, temperature-evaporation ratio and seasonal distribution of rainfall. Using these data Nuttsonson in 1947 developed agro-climatic analogues or homocline techniques for introduction of crop plants belonging to

different parts of the world into North America. Hanson in 1949 critically examined these studies and suggested that for plant introduction purposes, factors of environment other than the climate and latitude should also be utilized. He states that "if the suitability potential of local environment of one region are to be evaluated in reference to those of another region...it is desirable to utilize all available information from the fields of climatology, soils, topography, latitude, altitude, crop geography, natural vegetation, agricultural practices and the topography of all kinds of crop plants, (including aromatic and drug species), fruit trees, vegetable and grasses". Obviously, to evaluate the habitats precisely where the species to be introduced is indigenous, the correct method is to select the areas nearly approaching in its natural habitat. The solution, however, is not simple, because the interaction of several affecting factors ... the sumtotal of which produce the bioclimate, while several factors would play their part.

Ecological Evaluation of the Habitat

The theory of climatic analogues ... the more similar the climatic conditions, the better the adaptation ... as an explanation of the phenomenon of acclimatisation and as a guide to the introduction of exotics has serious limitations. It has been reported that if necessary protection (from fire, grazing, etc.) and irrigation, cultivation, etc. is given in early stages, many species are capable of widening their habitats and extending their boundaries. Another factor is the *exceptional year*....a species may continue to grow well and may be sufficiently acclimatised and yet it may not have sufficient vitality to survive in the exceptionally unfavourable year, or develop the principal active ingredients required for an efficient drug under its new

environment. Hence, accordingly species of restricted natural range with very exacting requirements may prove successful for plant introduction, based on the theory of climatic analogy but not for plants whose plasticity does not allow them to meet the challenge of new conditions. The presence or absence of vegetation is determined by the nature and identity of habitat factors. The habitat factors such as climate, edahic, physiographic and biotic control ultimately the ecological aspects of vegetation that produce an essentially uniform environment for the vegetation to subsist. Climatic factors, however, make the greatest imprint on the vegetation and serve as causative elements in the development and distribution of varied plant life. The climatic influences may either be regional or local. While the regional climatic influences delimit the climatic regions, local modifications due to topographic variations, interrelation of land and water masses, etc. may have local modifications which may introduce heterogeneity causing microclimatic differences.

Take, for example, in the arid regions the atmospheric conditions which define the regional and local climate mainly comprise temperature, moisture and light which are dependent on and are governed by solar radiation, air temperature, atmospheric humidity, precipitation, wind, lightning and atmospheric impurities which act and interact to produce the arid conditions. The role of any individual element is rather difficult to isolate on account of the complexity of the process. However, it is known that though all phases of environment have atleast potential influence on vegetation, all factors are by no means equally important at any one time. As a rule, each single factor assumes increasingly greater importance and is more limiting in effect, whenever or

wherever it begins to tax the ability of an organism, either to tolerate it in greater intensity or to survive under a lower intensity as shown by *Daubenmire* (1947). For instance, each species of plant has a maximum as well as minimum temperature tolerance and some were well between these extremes lies a range within which variations in temperature are relatively of minor importance for survival. This is the optimum range as far as this factor is concerned. Therefore, the factor, the study of climatic factors is necessary only with regard to optimal ranges of meterological elements. However, some plants are very specific in their climatic requirements and the range of variations in which they can survive, develop and regenerate is very narrow. However, Boyko in 1953 asserted that there hardly exists any better indicator of the climatic conditions of an area than its plant cover.

Biogeographic regions and potential areas of the vegetation types of India are given in Fig. 7.1.

The following ecological methods for assessing meteorological conditions have been used in areas where meteorological data are either wanting or are available for a brief period.

1. Geographical shifts in amplitude in relation to ground water table.
2. Geographical shifts in amplitude in relation to IE factor (yearly insolation due to exposure).
3. Topographical shifts in amplitude in relation to IE factor.
4. Overlapping amplitudes.

All these methods are based upon three fundamental laws, viz.

1. Leibig's law of minimum.
2. The geological law of plant distribution and

3. The biological law of climatic extremes.

Factors Determining the Distribution of Drug Plants

As discussed earlier the main factors that determine the distribution of drug plants may broadly be classified as below:

1. General atmospheric factors determining the climatic types.
2. Edaphic factors bringing about local variations and
3. The biotic factors.

The Atmospheric Factors

The atmospheric factors, viz. rainfall, humidity, insolations and wind combined ... make up what is called the climate. Rainfall determines the amount of water available, humidity regulates to a large extent the loss of water from the plants, insolation provides energy for photosynthesis and determines air and soil temperatures and the wind distributes warmth and moisture and at certain seasons it is a potent factor or in desiccation. It is generally agreed that water is the most important variable atmospheric factor wherever the temperature is suitable for the plant growth. In the altitudes climate is determined by latitudes and is modified by altitude and by the monsoon.

Climate as determined by the latitude: The hills and the plains in the north of the country experience well-defined alterations of winter and summer and such alterations are sufficiently marked in the low-lying Indo-Gangetic plains to produce distinct seasonal variations in the aspect of vegetation. In peninsular India the seasons are not so well defined.

Climate as determined by the altitude: That temperature falls with the increasing elevation is well known. Few places in the

7

I. WET EVERGREEN FORESTS OF KERALA - W. KARNATAKA

Culleania - Mesuua - Palaquium

Dipterocarpus - Mesua - Palaquium

Persea - Holigarna - Diospyros

Memecylon - Syzygium - Actinodaphne & Bridelia - Terminalia - Ficus

Montane (Shola) forest

II. WET EVERGREEN FORESTS - TEAK ZONE ECOTONE

Tectona - Lagerstroemia lanceolata - Dillenia - Terminalia paniculata

Tectona - Terminalia - Adina

Anogeissus

III. TEAK ZONE

Anogeissus - Terminalia - Tectona

Tectona - Terminalia

IV. MISCELLANEOUS FORESTS ZONE FORMING TRANSITION BETWEEN TEAK ZONE AND SAL ZONE

Terminalia tomentosa

Anogeissus latifolia

Terminalia - Anogeissus

Cleistanthus

SAL ZONE

Shorea - Buchanania - Cleistanthus

Shorea - Cleistanthus - Croton

Shorea - Terminalia - Adina

Shorea - Buchanania - Terminalia

Shorea - Dillenia - Pterospermum

Shorea - Syzygium operculatum - Toona

Toona - Garuga

VI. NARDWICKIA ZONE

Hardwickia binata - Albizia amara

VII. ALBIZIA AMARA ZONE OF COROMANDEL - CICAR

Albizia amara - Acacia

Anogeissus latifolia - Chloroxylon - Albizia amara

Manilkara - Chloroxylon

VIII. ANOGEISSUS PENDULA SEMI-ARID ZONE OF WESTERN RAJASTHAN

Acacia senegal - Anogeissus pendula

Anogeissus pendula - Anogeissus latifolia

IX. THORN FORESTS OF SEMI-ARID DECCAN

Acacia - Anogeissus latifolia

X. SEMI-ARID DECCAN - INDIAN DESERT

Acacia - Capparis

XI. INDIAN DESERT

Prosopis - Capparis - Zizaphus - Salvadora - Calligonum

XII. NORTH-WEST HIMALAYA

Subtropical evergreen Sclerophyllous forest

Alpine steppe

XIII. NORTH-WEST HIMALAYA - EASTERN HIMALAYA

Subtropical Pinus roxburghii forest

Temperate mixed Oak and coniferous and Temperate coniferous forest

Subalpine forest

Alpine scrub

XIV. EASTERN HIMALAYA - NORTH-EAST INDIA ANDAMAN - NICOBAR

Tropical wet evergreen forest of North-East India and Andaman-Nicobar

Tropical moist deciduous forest of North-East India

Subtropical broadleaved hill forest

Montane wet temperate forest

Mangrove

Salt marsh

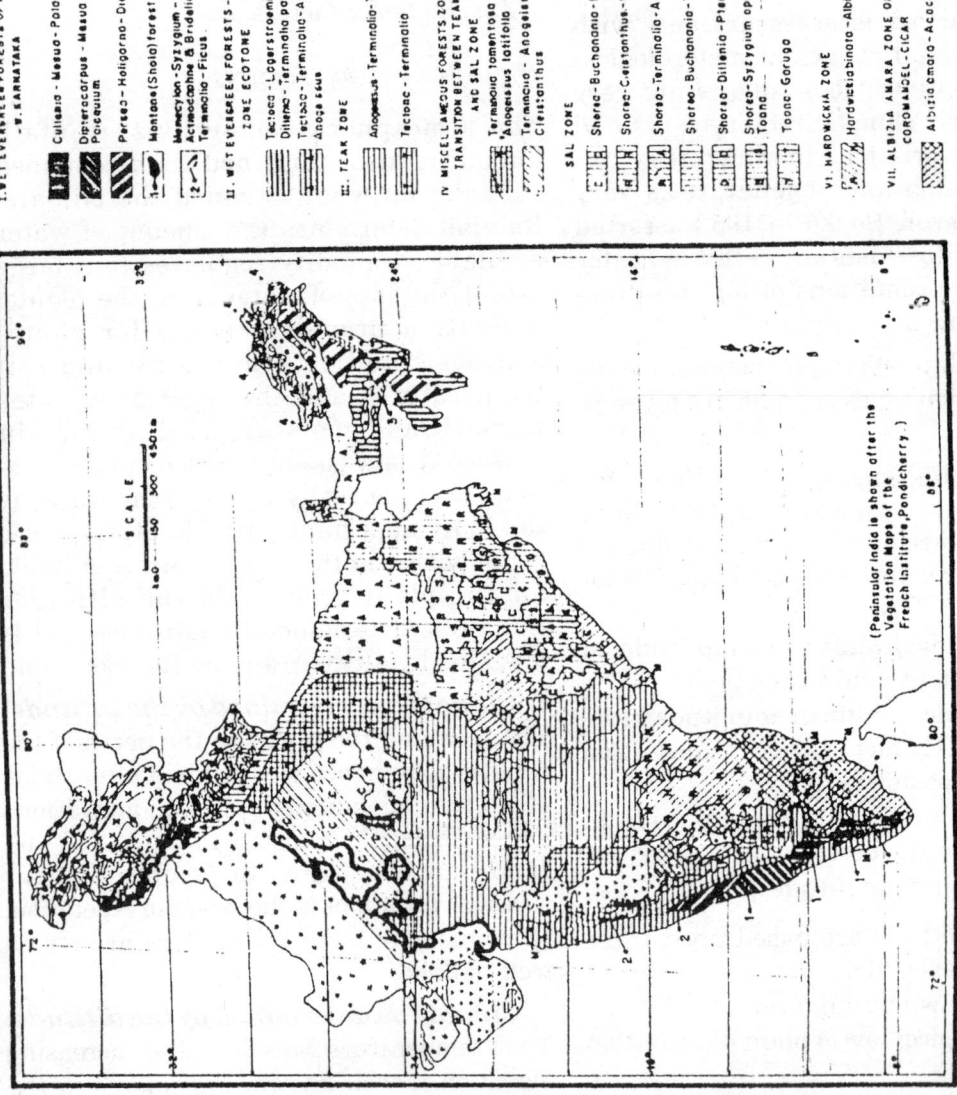

(Peninsular India is shown after the Vegetation Maps of the French Institute, Pondicherry.)

SCALE
150 300 450 km

Fig. 7.1: Biogeographic regions and vegetation types within these regions

world can show much striking changes in the temperature within such short horizontal distances as in the Himalaya. As there are still very few meteorological stations in the Himalayan region we must be satisfied with the available temperature data. Blanford (1884) gives variation in temperature of about 1 °C for each 200 ft rise in west Himalaya. Falling temperatures due to increase in altitude greatly affects the precipitation. Schimper, 1903, says that increasing altitude is associated with increasing rainfall up to a certain level 'while above this level atmospheric precipitation again rapidly diminishes'. He quotes Hill that "the line of the greatest rainfall in the Himalaya lies at 1270 m (14265 ft) above sea level., the rainfall here amounts to 3.7 times much as in the neighbouring plains but at 3000 m (9840 ft) altitude it is only one-fifth of the actual. In, Dehradun the rainfall is 85 inches and it is 119 inches at Rajpur (3200 ft). The moisture-bearing winds from the plains lose much of their water as they drive up the first abrupt slope of the mountain. Passing northwards each successively higher range take the toll, till the last steep ascent of the snowy peaks drains its air currents of most of their moisture. Northern slopes and the east and west valleys receive much less precipitation. In north of the snowy peaks, summer rain is very scanty and the places like in Tibet and Ladakh approaches almost desert conditions.

Relative humidity and the duration of the humid period are also modified by lowering temperature due to altitude. Even with less precipitation, humidity tends to remain higher and to persist longer in the lower temperature of high altitudes. At high altitudes relative humidity is subject to extraordinary and great fluctuations, so that very commonly complete saturation with water vapour and greatest dryness

succeeded one another at short intervals and absolute humidity regularly decreases with falling temperature in passing to higher altitudes. The effect of the vagaries of humidity on the vegetation of the higher altitudes appear not to have been evaluated but is most considerable.

Modification of the Climate due to Monsoons

The monsoon is a great wave of moisture bearing air moving over the peninsula of India during summers. In the plains of north India the result is copious rainfall and high humidity from the middle of June to the end of September. During rest of the year there is little or no rain and except as modified by cooler winter (November to February), humidity steadily falls, March-May is exceedingly dry and hot.

Himalayan regions, lying as they do immediately north of the plains, are subject to the same generally monsoon condition. There is some marked alteration of wet and dry conditions and heavy rainfall during the monsoon season.

The Climatic Factors

Among the factors affecting the distribution of vegetation the most important are temperature, precipitation, atmospheric humidity and the dryness. These have been recorded since a long time and may give some value. For the country, including Pakistan, Maynmar, Sri Lanka homogeneous documentation exists in the Memosri of the Meteorological services. Based on the nature of available information we may classify the types of stations.

The principal stations have only monthly precipitation and the number of rainy days. The duration of these observations very from 50–60 years with an average of 40–50 years.

Most of the stations are in the plains where the region is cultivated and the natural vegetation is much degraded or practically not existent. Documentation for the hill regions is still poor and it may be said that the vast programme of hydroelectric generation and construction of dams is still based on very poor information. Very little studies on the micro-climate have been made which gives exactly an idea of the variations of temperature and humidity in the vegetation cover.

Temperature: From the data available with the meteorological station it may be seen that say, for example, out of 245 stations, only 14 are between 20–25 °C, 18 °C between 15–20 °C, situated in Assam and western ghats, 2 stations at elevation of 2000 m in the ghats and outer Himalaya has an average value between 10–15. Finally 4 stations are situated and have the average between 1–10° C. For 80 percent of the stations the average annual temperature is above 20 °C.

The temperature of the coldest month is important for the vegetation and the three regions are distinguished where lower temperatures play different important roles. In high altitudes it is an important limiting factor for the vegetation. It determines the period of freezing of the snow which arrests the vegetation period and eliminates a large number of less resistent species.

In the average altitude of the extra-tropical zone the risk of freezing during winter is great but the frost is limited to the early hours of the day. In the Siwaliks there are small periods of freezing that eliminate but a number of tropical species are resistent. Such conditions are also found on the summit of South Indian hills like Nilgiris and the Palneys, etc.

In the average altitudes of the intertropical regions low winter temperatures, on the contrary, have a beneficial effect on the vegetation. These do not play the role of a limiting factor but determine the condensation which is abundant and reduce the evaporation. The deficit of saturation diminishes rapidly with the temperature.

The Zonal Concept in the Establishment of Drug Farms

Production of quality drug plants depends on several controllable and uncontrollable factors. The controllable factors, such as high quality seed, fertilizers, pesticides, etc. can be easily managed to favour quality production of drug plants. There are many factors like temperatures, rainfall, relative humidity, wind velocity, etc. which are uncontrollable and thus make the drug production a risky affair. In order to minimise the risk and ensure high productivity of drug plants, there is need to develop technology suited to various ecofloristic zones. To meet this objective, the country has been divided into several ecofloristic regions based on biodiversity and the zonewise establishment of drug farms should be established to ensure quality production of drug plants. Since the forests are typical ecological systems, many of the drug plants from a particular forest-ecosystem are a part of that system and cannot be grown purposefully as an individual species. Though, a zone is a too big an ecofloristic situation, subgrouping with the zone based on specific factors like soil types, rainfall, temperature, flood water loging, forest type and topography is essential. Once these subgroups on ecofloristic situations are identified, one can study the gaps in adoption of early available technology and identify constraints of productivity existing in a specific ecofloristic situation, which will help in formulating the research strategies relevant to the problems

of the specific situation. The technology thus developed, will be less risky, and favourise to stabilise drug plants production.

Clarke (1897) was perhaps the first to define a number of botanical regions in the country which was followed by others like Razi (1954), Legris (1963), Legris and Meher-Homji (1968), Gadgil and Meher-Homji (1982) and Rodgers and Panwar (1988). These phyto biogeographic zones provided regions with a similar phyto/zoo-geographic elements for planning protected area network and conservation of the genetic potential. Previous attempts for soil conservation (Das 1977) where potential evapotranspiration and rilling hazard ratio was used, while Bali (1969), Ahmad (1973), Singh (1971 and Khosla 1949) defined Hydrological and geohydrological regions. Gupta *et al* (1970) defined 20 land resource regions of India. Chaudhry and Sharma (1983) based on remote sensing techniques distinguished 12 ecofloristic regions. Inspite of a number of shortcomings, landsat imageries permitted a relatively rapid and accurate delineation of broad land uses and vegetation types. Though, the landsat imageries required elaborate decoding of features on the ground, the aerial photos give an optical image of ground, while the multispectral landsat imageries display complex representation of features on earth surface. Ecofloristic zones of Doon valley at 1:500,000 were delineated (Gupta and Sharma, 1993).

The criterion to define the ecofloristic zones is the physiognomy of vegetation where constituents of forest types and the land-scape was taken into account in relation to density, continuity of plant cover, height, etc. Thus, the forest formations could be separated into dense primary forest, secondary forest and scattered schrubs. The floristic composition

and the dynamic aspects are intimately related, because of large-scale disturbance due to anthropoid features and erosion. The densities of forest, the continuity and discontinuity can easily be delineated with the help of satellite imageries and aerial photographs. Another important aspect is the phenology and some of the phenological characteristics of vegetation could be studied using aerial photographs and landsat imageries at different seasons. Gupta (1992) provided information on major vegetation types at smaller scales and possible recognition at large scale for the southern part of the country.

Attempts to define the agro-ecological zones during the past two decades delineated crop suitability regions. ICAR (1972) Planning commission (1989) and Sehgal *et al* (1990) provided different regions based primarily on physiographic units, soil units and climatic conditions. Potential farming systems, constraints and improvements in each ecological unit were provided. In most of these classifications, relationshjp between values of PET (potential evapotranspiration) and rainfall temperature (Precipitation $\leq$ 2T) has been used which roughly corresponds to the criteria used by earlier workers. So far the attempts for ecological zoning are primarily for agricultural improvement and has neither formed the basis for integrated landuse planning for biodiversity conservation.

Land resource regions and areas of India are shown in Figs 7.2 and 7.3.

Ecofloristic Zones (EFZ)

All three ecofloristic zones have been described and delineated broadly based on physiographic units, altitude, bioclimatic types and series of vegetation described by

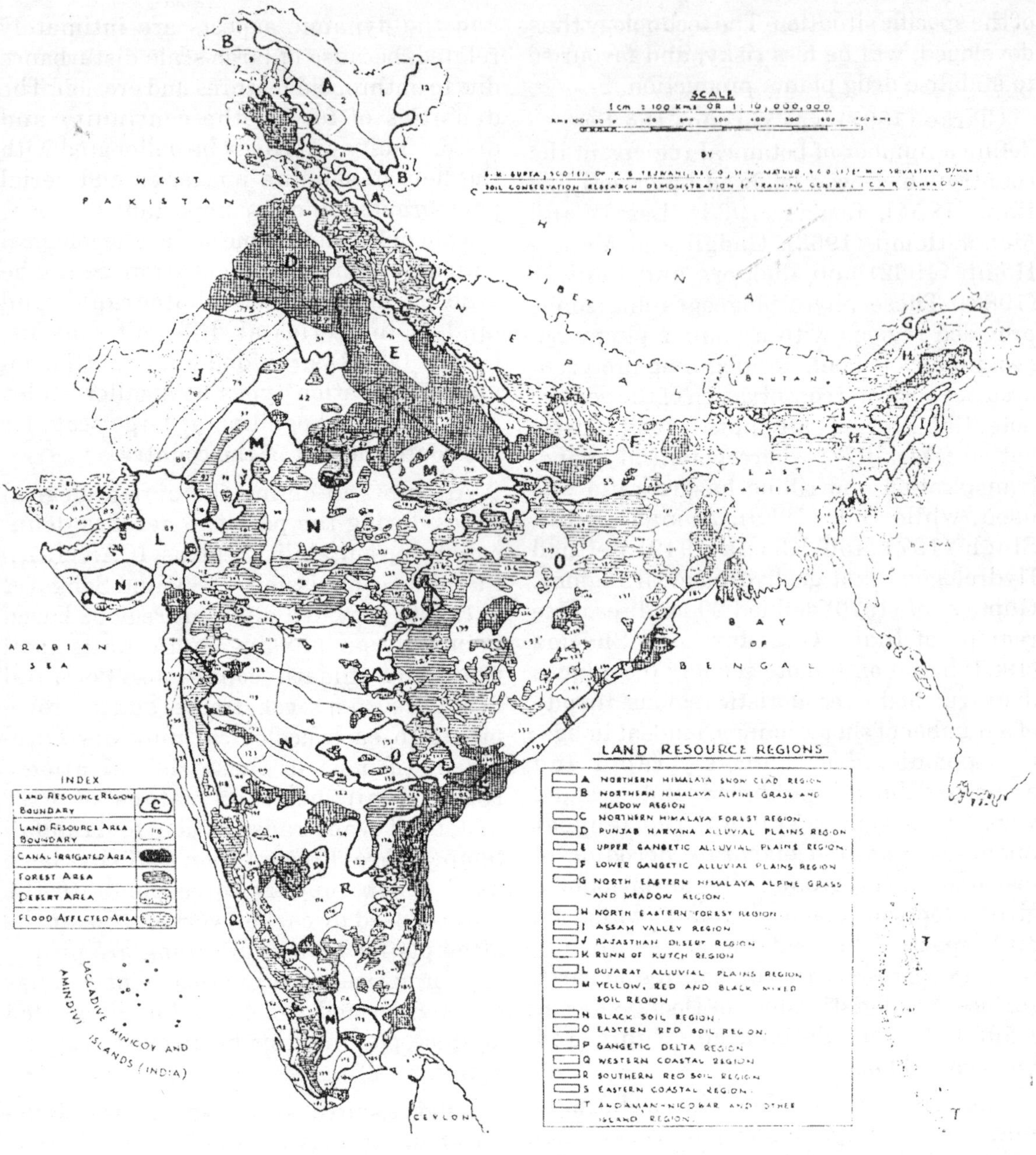
SCALE
1 cm = 100 Kms OR 1:10,000,000
WEST PAKISTAN
TIBET
NEPAL
BHUTAN
EAST PAKISTAN
ARABIAN SEA
BAY OF BENGAL
CEYLON
LACCADIVE MINICOY AND AMINDIVI ISLANDS (INDIA)
INDEX
LAND RESOURCE REGION BOUNDARY
LAND RESOURCE AREA BOUNDARY
CANAL IRRIGATED AREA
FOREST AREA
DESERT AREA
FLOOD AFFECTED AREA
LAND RESOURCE REGIONS
A NORTHERN HIMALAYA SNOW CLAD REGION
B NORTHERN HIMALAYA ALPINE GRASS AND MEADOW REGION
C NORTHERN HIMALAYA FOREST REGION
D PUNJAB HARYANA ALLUVIAL PLAINS REGION
E UPPER GANGETIC ALLUVIAL PLAINS REGION
F LOWER GANGETIC ALLUVIAL PLAINS REGION
G NORTH EASTERN HIMALAYA ALPINE GRASS AND MEADOW REGION
H NORTH EASTERN FOREST REGION
I ASSAM VALLEY REGION
J RAJASTHAN DESERT REGION
K RUNN OF KUTCH REGION
L GUJARAT ALLUVIAL PLAINS REGION
M YELLOW, RED AND BLACK MIXED SOIL REGION
N BLACK SOIL REGION
O EASTERN RED SOIL REGION
P GANGETIC DELTA REGION
Q WESTERN COASTAL REGION
R SOUTHERN RED SOIL REGION
S EASTERN COASTAL REGION
T ANDAMAN-NICOBAR AND OTHER ISLAND REGION

Fig. 7.2: Land resource regions and areas of India

Fig. 7.3: Ecological subdivisions of the Indian subcontinent (After Legris, 1963)

Gaussen *et al* (1961; Puri *et al* 1989), Gadgil and Meher-Homji 1982. These are as follows:

EFZ 1: *Trans-Himalayan Indus catchment*—Steppe region of *Artemisia* and *Ephedra* (3500–4500 m). Rainfall 100 mm, T 10 °C Tm = 0–5 °C. Dry months + 9 + frost.

Soils: shallow stony, skeletal.

Degraded forms: Barley, grazing, off season vegetables and seeds of vegetables. Steppes have scattered *Juniperus* scrubs, patches of millets cultivation with irrigation.

EFZ 2: *North-West Himalayan with Mediterranean tendency*—no tropical or bixeric regime.

A. Steppic with upper region of Sutlej basin (900–2000 m or above)

A1. Alpine dry zone (Kinnaur, Lahul and Spiti.
  Rainfall: 500–100 mm.
  Tm = –5 °C. Dry months 9 plus frost.
  Floristics: Shrubs of Rhod dendron, Juniperus and Lonicera.
  Soils: Shallow, stony.
  Degraded forms: Degraded scrubs, *Salix fruticosa* on riparian sites.

A2. Montane dry (2000–3000 m)
  Rainfall: 1000 mm with winter snowfall. Dry months 4–5.
  Floristics: Pinus gerardiana, Cedrus deodara.
  Soils: Shallow, skeletal.
  Degraded forms: Degraded forests of conifer (*Pinus gerardiana* and *Cedrus deodara*) and scrubs; *Buxus semervirens, Fraxinus* on riparian habitats, also *Pistacia atlantica* is found. Grazing lands for sheep.

A3. Subtropical dry (900–2000 m).
  Rainfall: 1000 mm with winter snow. Dry moths 4–5.

Floristics: Characterised with species like *Olea cuspidata, Fraxinus excelsior* and *Cedrus deodara.*
Soils: Mountain brown hill soil.
Degraded forms: Open woodland in valleys, cultivation of apricots, beans, etc. under irrigation.

B. Region of Kashmir (Karewa plateau and high hills)

B1. Alpine zone (4000–5000 m).
  Rainfall: 100 mm in form of snow. T 10 °C. Dry months up to 9 with frost.
  Floristics: Scrubs of *Juniperus,* Cotoneaster and bent wood of *Betula utilis.*
  Soil: Mountain meadow soils.
  Degraded forms: Scattered shrubs, poor pastures, particularly for sheep and goats.

B2. High mountain (subalpine) 3000–4000 m altitude.
  Rainfall: Up to 1000 mm with snow. T 10 °C. Dry months 56.
  Floristics: Betula utilis, Abies pindrow, Taxus baccata and *Acer caesium.*
  Soils: Brown hill soils.
  Degraded forms: Mixed forest, open woodland and grasslands. Orchards of apples, cherry, etc.

B3. Montane zone (2000–3000 m)
  Rainfall: 1000–1500 mm during winters with snow. T 15 °C.
  Floristics: Pinus wallichiana, Cedrus deodara, Picea smithiana, Aesculus indica, etc.
  Soils: Morainic deposits, glaciated.
  Degraded forms: Degraded conifer forests and pasturelands. Cultivation of maize, buckwheat, amaranth and vegetables. Apricot and apple orchards.

B4. Submontane (Kashmir plateau) 1500–2000 m altitude.
Rainfall: 1000 mm; Dry months 3, T 10 °C
Floristics: Pinus wallichiana, Cedrus deodara, Viburnum cotonifolium, etc.
Soils: Submontane soils.
Degraded forms: Degraded conifer forests and grasslands, on riparian habitats are *Aesculus indica, Celtis caucasica, Alnus* species. Cultivation of paddy and wheat with plantations of Salix (Willow) and Poplars.

EFZ 3: *Western Himalaya including parts of Jammu, Himachal Pradesh and UP hills under tropical monsoonic regime.*

A. Alpine zone (above 4000 m altitude).
Rainfall: 500–1000 mm with snow during winters. T 10 °C. Dry months 7–8, plus frost.
Floristics: Scrubs of *Rhod dendron* and *Junipers.*
Soils: Montane meadow soils.
Degraded forms: Poor pastures with scattered shrubs.

B1. High montane (subalpine) region (3000–4000 m altitude) comprising inner hills from Shimla to Nepal.
Rainfall: 1000 mm with winter snowfall. T 10 °C with high winds.
Floristics: High level oaks (*Quercus semecarpifolia*), *Abies pindrow, A. spectabilis, Betula utilis* and *Rhododendron* scrubs.
Degraded forms: Oak scrubs and ringal bamboo. On riparian sites *Alnus nepalensis, Juglans regia* and maples, etc.

B2. Montane zone (1800–3000 m altitude) with short dry season. Outer slopes of Pir Panjal to Garhwal-Kumaon region.
Rainfall: 1000–2000 mm with snow during winters. T 15 °C Dry months 1–3.
Floristics: Oaks and conifer forests with species like *Quercus leucotrichophora, Q. dilatata, Q. semecarpifolia, Cedrus deodara, Picea smithiana* and *Pinus wallichiana.*
Soils: Brown hill soils.
Degraded forms: Oak scrubs, low thickets. Potato cultivation with amaranth on terraces, fodder trees on terrace riser.

B3. Subtropical zone (1000–2000 m alt.) outer hill ranges and valleys.
Rainfall: 1000–1800 mm T 20 °C. Dry months 7–8.
Floristics: Chirpine with oak and other broad-leaf species on humid situations.
Soils: Skeletal, brownhill soils.
Degarded forms: Open woodland savanna, savanna woodland with shrubs of *Carissa, Euphorbia* and *Rhus* on dry situations, degraded oak scrub with *Lyonia ovalifolia* and *Rhododendron* on humid situations.

B4. Tropical zone at low elevations of up to 1000 m altitude.
Rainfall: 1000 mm. T 20 °C. Dry months 7–8.
Floristics: Shorea robusta with *Anogeissus latifolia, Terminalia tomentosa,* etc.
Soils: Tarai and submon, tane soils.
Degraded forms: Forests degraded to savanna, open woodland overgrazing and fuel-fodder needs of both sedentary and nomadic populations. Cultivation of paddy, maize and wheat. Riparian situations with *Dalbergia sissoo* and *Acacia catechu.* Planations of mango, guava and of eucalyptus, bombax, etc.

EFZ 4: *Northeast India and Eastern Himalaya from Sikkim eastwards.*

A. Alpine zone (above 4000 m altitude), inner ranges of east Himalaya.
Rainfall: 500–1000 mm, T coldest months 0–5 °C. Dry months 5–6.
Floristics: Dwarf scrubs of *Juniperus, Rhododendron, Betula* with *Ephedra* and *Artemisia* on dry slopes.
Soils: Alpine meadow soils.
Degraded forms: Degraded grasslands and scrubs.

B. High montane (subalpine) zone between 3000–4000 m altitude.
Rainfall: 1000–1500 mm. T 5–10 °C. Dry months 3–4.
Floristics: Tsuga dumosa, Cupressus torulosa, Juniperus, recurva, Abies densa, Betula utilis, Picea spinulosa.
Soils: Morainic.
Degraded forms: Degraded conifer forests, bamboo thickets.

C. Montane zone between 2000–3000 m altitude.
Rainfall: 2000 mm. T 10 °C. Dry months 2–5.
Floristics: Quercus species, *Lithocarpus* species, *Larix griffithii, Pinus wallichiana, Picea spinulosa, Machilus edulis, Cinnamomum obtusifolium* and *Acer campbellii*, etc.
Soils: Brown soils.
Degraded forms: Lauraceous forests and pine scrubs. On riparian habitats *Alnus nepalensis* and *Pinus wallchiana.* Zone of shifting cultivation.

D. Tropical moist deciduous zone of northeast between 1000–2000 m altitude.
Rainfall: 2500 mm to 4000 mm. T 15–20 °C. Dry months 3–4.
Floristics: Careya arborea, Dillenia pentagyna, Toona ciliata Aphanamixis polystachys, etc.
Soils: Tarai soils
Degraded forms: Open woodland and savanna, under shifting cultivation.

E. Very moist subtropical zone with moist and semi-evergreen forests between 1000–2000 m altitude between Kohima and Shillong.
Rainfall: 2000–4000 mm. T 15 °C Dry months 4–5.
Floristics: Schima wallichii, Castanopsis indica, Lithocarpus spicata, Phoebe species.
Soils: Brown hill soils.
Degraded forms: Open woodland, savanna, subjected to shifting cultivation; cultivation of citrus, pineapple, paddy and millets.

F. Very moist tropical zone at low elevation below 1000 m altitude.
Rainfall: 2000 m, T 20 °C. Dry months not more than 4.
Floristics: Tetrameles species, *Sterospermum* species, *Messua ferea, Shorea robusta, Dillenia pentagyna*, etc.
Soils: Alluvial soils.
Degraded forms: Open woodlands and savanna. Riparian situations are with *Duabanga sonneritoides, Pterospermum acerifolium; Terminalia* species, *Bombax ceiba, Albizzia procera.* tea, potato, chincona and rubber plantations.

EFZ 5: *Indian desert with long dry season of Western Rajasthan and Kutch peninsula.*
Rainfall: Less than 500 mm. T 10–20 °C. Dry months 9–11.
Floristics: Prosopis cineraria, Capparis decidua, Ziziphus nummularia, Calligonum polygonoides, Salvadora oleoides, etc.
Soils: Mainly alluvium with variations such as saline regosol and wind transported.
Degraded forms: Discontinuous thorny thickets and scattered shrubs. Cultivation of millet, pulses, oil seeds and shrubs of Prosopis, Capparis scattered in cultivation. *Acacia cupressiformis* is typically present in cultivation.

EFZ 6: *Deccan desert, Bijapur to Malegaon, north Gujarat plains and piedmont regions of Aravalli ranges.*

Rainfall: 400–850 mm, T 20 °C. Dry months 7–9.

Floristics: Capparis decidua, Acacia nilotica ssp. *indica.*

Soils: Skeletal, red ferruginous, alluvium black.

Degraded forms: Savanna-woodlands, scrub savanna, discontinuous thorny thicket, scattered shrubs. Cultivation of millet, pulses, cotton, groundnut and tobacco.

EFZ 7: *Semiarid Deccan thorny forests of Maharashtra.*

Rainfall: 600–800 mm. T 20 °C Dry months 7–8.

Floristics: Acacia nilotica, Anogeissus latifolia forests.

Soils: Alluvial black clayey.

Degraded forms: Scrub woodland; cultivation of millets, pulses and cotton.

EFZ 8: *Semiarid forests of Rajasthan. Madhya Pradesh and Bundelkhand region of UP,* up to 1000 m altitude.

A. Forests of Sirohi, Pali, Ajmer, Jaipur, Sikar and isolated hills of Jodhpur and Nagpur districts.

Rainfall: 400–700 mm, T 20 °C. Dry months 8–10 in more than 1000 m altitude.

Floristics: Acacia senegal, Anogeissus pendula, Calotropis procera, etc.

Soils: Brown soils.

Degraded forms: Scrub woodlands and scattered shrubs. Cultivation of millets, pulses and oil seeds.

B. Hills of Rajasthan and Madhya Pradesh up to 500 m altitude comprising hills of Alwar, Sawaimadhopur, Bharatpur, Pali, Ajmer, Tonk, Bundi and Kota districts.

Rainfall: 500–900 mm. T 20 °C. Dry months 8–9.

Floristics: Acacia catechu and *Anogeissus pendula.*

Soils: Brown and black.

Degraded forms: Scrubs and scrub woodlands. Cultivation of millet and plantations of *Acacia catechu.*

C. Hills of Rajasthan up to 700 m (districts of Udaipur, Pali, Ajmer, Sirohi, Bundi and Kota) and Madhya Pradesh (Shivpuri Orcha) and Bundelkhand.

Rainfall: 600–900 mm. T 20 °C. Dry months 9.

Floristics: Anogeissus pendula and *Anogeissus latifolia.*

Soils: Ferruginous brown.

Degraded forms: Scrub woodlands scattered shrubs. Cultivation of millets and oil seeds.

EFZ 9: *Semiarid forests of Coromandel coast.*

A. Coastal plains of Tamilnadu, Andhra Pradesh (Kurnool, Anantapur), Karnataka (Bellary, Chitradurga, Sharwal and Bangalore) 600 m above the sea level.

Rainfall: 500–1000 mm T 20 °C. Dry months 5–9.

Floristics: Albizzia amara, Acacia latronum.

Soils: Ferralitic, sandy loam.

Degraded forms: Shrub Ñavanna, discontinuous thorny thick, etc. scattered shrubs, mined areas for iron and manganese ores. Cultivation of paddy, millets, groundnut, cotton and annona fruits.

B. Interior Karnataka more than 1000 m altitude.

Rainfall: 500–1000 mm. T 20 °C. Dry months 5–6.

Floristics: Anogeissus latifolia, Chloroxylon sweintinia Albizzia amara.

Soils: Red ferruginous.

Degraded forms: Woodland savanna. Cultivation of paddy, millet pulses and cotton.

C. Coastal plains of Coromandel up to 600 m altitude.

Rainfall: 100–1500 mm. T more than 20 °C. Dry months 5–6.

Floristics: Manilkara hexandra, Chloroxylon sweintinia Memecylon umbellatum, Drypetes sepiaria, Diospyros species and *Albizzia amara,* etc.

Soils: Lateritic.

Degraded forms: Lowthorny thickets. Cultivation of paddy, millet, groundnut and castor.

EFZ 10: *Semiarid Hardwickia forests on dry plateau of Karnataka, Andhra, Satpura, Vindhya and dry plateau of Salem up to 1000 m altitude.*

Rainfall: 500–1200 mm. T 15–20 °C. Dry months 6–8.

Floristics: Hardwickia binata and *Albizzia amara.*

Soils: Black clayey, skeletal gravelly.

Degraded forms: Degraded woodlands on skeletal soils. Cultivation of paddy, millets, pulses, castor, tobacco, *Redsandal* and *Annona* fruits.

EFZ 11: *Semiarid Shorea robusta forests on dry plateau and hills of Bihar and Orissa (Chhotanagpur to Khandaman, Keonjhargarh, Simlipal, Dhenkal, Sujauga and Raigarh) up to 900 m alt.*

Rainfall: 1400–2000 mm. T 10–20 °C. Dry months 5–6.

Floristics: Shorea robusta, Syzygium operculatum, Toona ciliata and *Symplocos spicata,* etc.

Soils: Lithosols, ferralitic, laterite, leached sandy loam.

Degraded forms: Scrub woodland, thickets, scattered shrubs and mined areas. Millets, pulses, Turmeric and *Caryota urens.*

EFZ 12: *Sal forests of Puri and Cuttack districts in Orissa.*

Rainfall: Up to 2000 mm. T 20 °C. Dry months: 5.

Floristics: Shorea robusta, Dillenia pentagyna, Pterospermum heyneanum.

Soils: Alluvial, ferralitic, clayey.

Degraded forms: Scrub woodland, thickets and scattered shrubs. Paddy, pulses, jute, casuarina, cashewnut and vegetable cultivation.

EFZ 13: *Semiarid hill forests of Vishakhapatnam, Koraput, Baildala and Papikonda districts up to 900 m altitude.*

Rainfall: 1500–2000 mm. T 15–20 °C. Dry months 5.

Floristics: Toona ciliata and *Garuga pinnata.*

Soils: Sandy ferralitic.

Degraded forms: Scattered shrubs, scrub woodland, thickets; mined areas for alluvium and iron ore; shifting cultivation, paddy, millets, ginger, turmeric, mango, jackfruit, tamarind, *Caryota urens.*

EFZ 14: *Sal forests of Central India, U.P., M.P., Bihar, Orissa*

A. *Sal* forests of central India (Pachmarhi, Chindwara) above an altitude of 1000 m.

Rainfall: 1000–2000 mm. T more than 15 °C. Dry months 5–7.

Floristics: Shorea robusta, Buchanania lanzan, Terminalia tomentosa.

Soils: Red ferruginous sandy loam, lithosol and ferralitic sandy.

Degraded forms: Shrubs woodlands, degraded shrubs. Cultivation of paddy, pulses and plantations of teak.

B. Sal forests on the hills of Balaghat, Bastar, Bilaspur, Mandle, Raipur, Surjuga, Koraput, Palamau, Bharaich and Mirzapur districts up to 1000 m altitude.

Rainfall: 1000–2000 mm. T 15–20 °C.

Dry months 5–7.

Floristics: Shorea robusta, Terminalia tomentosa and *Adina cordifolia.*

Soils: Red loamy, red and yellow soils.

Degraded forms: Scrub woodland, subjected to shifting cultivation.

C. Sal forests of low hills (100–400 m) in Bengal (Birbhum, Dhanbad, Bardhman, Medinapur), Bihar (Balasore, Purulia) and Keonjhargarh in Orissa.

Rainfall: 1000–1500 mm. T 15–20 °C Dry months 6–7.

Floristics: Shorea robustan, Cleistanthus species, *Croton oblongifolium*, etc.

Soils: Red and yellow soils.

Degraded forms: Degraded woodlands, shrubs, thickets and scattered shrubs. Cultivation of paddy, millet, pulses and plantations of Teak.

D. Sal forests on hills up to 1000 m altitude.

Rainfall: 1400–2000 mm. T 15–20 °C. Dry months 6–8.

Floristics: Shorea robusta, Buchanania lanzan and *Cleistanthus collinus*, etc.

Degarded forms: Scrub woodlands, thickets, scattered shrubs. Cultivation of paddy, millet pulses and plantations of teal, sal, mahua and chironji.

EFZ 15: *Miscellaneous forests (Sal and Teak mixed) zone on the hills of Gujarat, Rajasthan, Madhya Pradesh and peninsular hills.*

A. Hills of Mehsana, Junagrah, Bhavnagar (Gujarat), Udaipur, Sirohi, Kota, Mount Abu (Rajasthan), Rew, Tikamgarh, Satna Panna (M.P.) and Jhansi (UP) with pronounced dry period.

Rainfall: 700–800 mm. T 15–20 °C. Dry months 8.

Floristics: Terminalia tomentosa, Anogeissus latifolia.

Soils: Red ferruginous loam, sand, alluvial, black on alluvium, ferralitic.

Degraded forms: Shrub savanna. Cultivation of paddy, millet wheat, pulses and cotton. Plantations of teak.

B. Hills of peninsular India (east of Godavari, Vishakhapatnam, Srikakulum, Mahanadi basin, Wainganga and Chhattisgarh up to 1000 m altitude with short dry period.

Rainfall: 1000–1500 mm. T 15–20 °C. Dry period 5–6 months.

Floristics: Terminalia tomentosa, Anogeissus latifolia and *Cleistanthus collinus*, etc.

Soils: Vertibrown, loamy ferruginous, black alluvial.

Degraded forms: Degraded shrubs. Cultivation of millet, paddy, wheat, pulses. Plantations of teak.

EFZ 16: *Teak forest zone of peninsular India up to Bundelkhand in the north.*

A. Lowhills up to 1000 m altitude from Kanyakumari to Jhansi.

Rainfall: 800–1800 mm. T 15 °C onwards. Dry months 5–8

Floristics: Anogeissus latifolia, Terminalia paniculata and *Tectona grandis*, etc.

Soils: Medium black.

Degraded forms: Shrub savanna, low scattered shrubs. Cultivation of paddy, millets, castor. Plantations of teak, *Pterocarpus marsupium*, Rosewood, Sandalwood, *Boswellia serrata* and *Sterculia urens*, etc.

B. Lowhills up to 1000 m altitude of Andhra Pradesh (Nallamalai, Karimnagar, Khamam, Warangal), Madhya Pradesh (Sagar, Panna, Raipur); Orissa (Koraput, Kalahandi) and Maharashtra (Nagpur, Chanda and Bhandra districts).

Rainfall: 1000–1500 mm. T more than 20 °C. Dry months 5–6.

Floristics: Tectona grandis, Lagerstroemia lanceolata, Dillenia pentagyna and *Terminalia tomentosa*, etc.

Soils: Tropical ferruginous.

Degraded forms: Scrubs and shrub by savanna. Cultivation of paddy, millets, castor. Plantations of teak, Sterculia, Boswellia and Diospyros species.

EFZ 17: *Wet evergreen teak forests ecotone.*

A. Moist deciduous teak forests with short dry season on low hills of Kerala in Westernghats, Maharashtra and eastern fringes of Karnataka state.

Rainfall: More than 2000 m. T more than 15 °C. Dry months 3–7.

Floristics: Tectona grandis, Lagerstroemia lanceolata, Dillenia pentagyna and *Terminalia tomentosa.*

Soils: Tropical ferruginous.

Degraded forms: Woodland tree savanna, shrub savanna, iron ore mined areas. Cultivation of paddy, millets, bamboo and teak plantations.

B. Moist deciduous teak forest with pronounced dry season on low hills of Thane-Nasik range, Dangs, Bhubneshwar and Nagar-Haveli up to 1000 m altitude.

Rainfall: 1800–2500 mm. T 15 °C. Dry months 7–8.

Floristics: Tectona grandis, Terminalia paniculata, Adina cordifolia and *Anogeissus latifolia,* etc.

Soils: Tropical ferruginous, ferralitic.

Degraded forms: Scrub woodland, thickets, scattered savanna. Cultivation of paddy, millets. Plantations of teak, casuarrina, mango and bamboo.

EFZ 18: *Wet evergreen forest of eastern and westernghats.*

A. Very moist evergreen forests with short dry season on western side of the eastern ghats in Kerala and some moist pockets of Tamilnadu states.

Rainfall: 3000 mm. T more than 20 °C. Dry months 4.

Floristics: Cullenia exarillata, Mesua ferrea, Palaquium ellipticum, Poeciloneuron indicum, Gluta travancorica, etc.

Soils: Lateritic loamy.

Degraded forms: Shrub savanna. Cultivation of paddy, tapioca, tea, rubber, cardamom, nutmeg, jackfruit, cinnamon and mixed plantations, area of shifting cultivation.

B. Very moist evergreen to semi-evergreen forests with short dry season on westernghats (between 12–14 degrees north latitude) up to 1500 m altitude.

Rainfall: 2000 mm. T more than 15 °C. Dry months 4.

Floristics: Dipterocarpus indics, Mesua ferrea, Palaquium ellipticum, Kingiodendron pinnatum and *Humboldtia brunonis,* etc.

Soils: Loamy lateritic with humus.

Degraded forms: Shrub savanna and savanna. Cultivation of paddy, sugarcane, shifting cultivation. Plantations of coconut, tea, rubber, cocoa, coffee, mango, nutmeg, Garcinia and Caryota urens.

C. Moist with pronounced dry season on hills of Shimoga, North Canara, Belgaun and Goa up to 1500 m altitude.

Rainfall: 2500–3000 mm. T more than 15 °C. Dry months 6.

Floristics: Persea macarantha, Holigarana arnottiana and *Diospyros microphylla.*

Soils: Lateritic.

Degraded forms: Shrub savanna, mined areas for copper and manganese, areas of shifting cultivation. Cultivation of paddy. Plantation of coconut and *Persea macarantha.*

EFZ 19: *Wet evergreen, montane shola forest region of westcoast westernghats (high hills of Nilgiris, Annamali, Bababudal, Giri and Kundremukh hills above 1500 m altitude.*

Rainfall: 1000–1500 mm. T more than 10 °C. Dry months 0–4.

Floristics: Elaeocarpus serratus, Gordinia obtusa, Meliosma arnottiana, M. wightii, Schefflera stellata, S. wallichiana, etc.

Soils: Lateritic, humic, montane type.

Degraded forms: Scrub woodlands, thickets, scattered shrub fernlands, area of shifting cultivation. Terraced cultivation of paddy, millet, vegetables (potato, cabbage) buckwheat, etc. Plantations of tea, coffee and of *Eucalytus globulus, Acacia mearsnii, A. delabata, Cupressus macrocarpa,* etc.

EFZ 20: *Wet evergreen forests of the westcoast, westernghats with pronounced by season hills from 700–1500 m altitudes.*

Rainfall: More than 3000 mm. T 15–20 °C. Dry months 6–7.

Floristics: Memecylon edule, Syzygium cumini, Actinodaphne hookeri, Terminalia chebula, etc.

Degraded forms: Scrub woodland, thickets, scattered shrubs. Cultivation of paddy, millets. Plantations of *Terminalia chebula* and *Colchicum.*

EFZ 21: *Westevergreen forests of west coast-westernghat, upper slopes of Maharashtra hill, Mount Abu and summit of Ranthambor hills with pronounced dry season.*

Rainfall: 2000–2500 m. T up to 20 °C. Dry months 7–8.

Floristics: Bridelia retusa, Syzygium cumini, Terminalia chebula.

Soils: Strongly ferralitic.

Degraded forms: Woodland and shrub savanna. Cultivation of paddy, millets and cotton. Plantations of Ficus.

EFZ 22: *Tropical evergreen forests of Andaman and Nicobar islands coast to 800 m altitude.*

Rainfall: More than 3000 mm. T 25 °C. Dry months 2.

Floristics: Dipterocarpus pilosus, Artocarpus chaplasa, etc.

Soils: Red loamy and sandy.

Degraded forms: Monoculture conversions, shifting cultivation area. Plantations of teak, oil palm, etc. Cultivation of paddy.

EFZ 23: *Coastal formations of east and west coasts, Godavari, Andaman-Nicobar islands, and Mangrove formations.*

Rainfall: More than 3000 mm. T more than 25 °C. Dry months 2.

Floristics: Avicennia, Rhizophora, Myristica swamps, etc.

Soils: Coastal.

Degraded forms: Scrubs of mangrove, paddy cultivation, fish rearing and pearl culture, etc.

Ecological subdivisions of the Indian subcontinent are shown in Fig. 7.3.

Appendix I

Climatic factors of some stations of the arid and semiarid regions for determining the bioclimates of the region.

Type of bioclimate	Value of factors		Stations
1. Desertic hot (arid) ($x = 200$–350)			

I. With mediterranean tendency.
      (rainfall during short days)
      a) Temperate hot ($m = 10$–$15\ °C$)
	$t_4\,s_6\,x_6$	....	Nokundi, Panjgur.

II. With tropical tendency.
      (rainfall during short days)
      a) Hot ($m = 15$–$20\ °C$)
	$\begin{cases} t_{4/5}\,s_{5/6}\,x_6 \\ t_{4/5}\,s_6\,x_6 \end{cases}$	....
		 ....
			Lasbela; Hydrabad
			 Sakkar.

b) Temperate hot ($m\ 10$–$15\ °C$)	$\begin{cases} t_4\,s_6\,x_6 \\ t_4\,s_{5/6}\,x_6 \end{cases}$	....	Jacobabad. Dera-Ismail Khan, Sibi, Multan, Bhawalpur, Khanpur.
2 (a) Subdesertic Hot Semiarid accentuated ($x = 250$–290)			

I. With mediterranean tendency
      (rainfall during short days)
      a) Hot ($m = 15$–$20\ °C$)
      b) Temperate hot ($m = 10$–$15\ °C$)
	$\begin{cases} t_{4/5}\,s_{5/6}\,x_6 \\ t_4\,s_6\,x_{5/6} \end{cases}$	....
		 ....
			Pasni.

II. With tropical tendency
      (rainfall during long days)
      a) Hot ($m = 15$–$20\ °C$)
	$\begin{cases} t_{4/5}\,s_{5/6}\,x_{5/6} \\ t_{4/5}\,s_5\,x_{5/6} \\ t_{4/5}\,s_{4/5}\,x_{5/6} \\ t_5\,s_2\,x_{5/6} \end{cases}$	....
		 ....
		 ....
			Lasbela.
			 Karach, Badin.
			 Deesa.
			 Bellary.

b) Temperate hot ($m = 10$–$15\ °C$)	$\begin{cases} t_4\,s_{5/6}\,x_{5/6} \\ t_4\,s_5\,x_{5/6} \end{cases}$	....	Kirthar. Layalpur; Khusab, Montegomery.
2 (b) Subdesertic non- attenuated ($x = 200$–250)			

I. With mediterranean
      tendency
      a) Hot ($m = 15$–$20\ °C$)
      b) Temperate hot ($m = 10$–$15\ °C$)
	$\begin{cases} t_{4/5}\,s_{5/6}\,x_3 \\ t_{4/5}\,s_5\,x_5 \end{cases}$	....
		 ....
			Ormara.
			 Peashawar valley.

II. With tropical tendency a) Hot ($5m = 15$–$20\ °C$)	$t_{4/5}\,s_{4/5}\,x_5$ $t_{4/5}\,s_4\,x_5$ $t_{4/5}\,s_5\,x_{5/6}$ $t_{4/5}\,s_5\,x_5$ $t_4\,s_5\,x_5$ $t_4\,s_{4/5}\,x_5$	....	Ajmer, Jaipur, and Mainpuri. Jhansi, Kota, and Mainpuri. Sambherlake, and Jodhpur. Barmer, Bikaner, and Bhuj. Lahore, Hissar. Delhi, Agra.
b) Temperate hot ($m = 10$–$15\ °C$)	$t_4\,s_4\,x_5$ $t_4\,s_5\,x_6$	....	Aligarh. Sri ganga Nagar.

Appendix II

Bioclimatic types of the Himalayan region and their analogous types towards the Alps and Pyrenees for plant introduction (after Gupta, 1964).

1. *Subderertic* Hot (*Hemi-Eremic*).
   Hot and Temperate hot (temperature curve always positive)
   Dry season long, long days dry (m 0–10 °C, P: 100–200 mm), Dry months 9–11
   Chagai mts, northern slopes of Rakosh, north of Kharan, southern chains of Baluchistan (1000–1500 m) and the valey of Gilgit (rainfall less than 1000 mm), Gilgit.
   Dry season short. Long days dry (*m* 0.10 °C. P more than 250 mm). The valley of Gilgit (except the base), High valley of Banno and the high plateaus of Baluchistan.

2. Xerotheric (Mediterranean).
   Sector 1. m 0–10 °C. P: 250–500 mm.
   The extreme northeast and southwest of the valley of Peshawar (less than 500 m) Northeast of the valley of Peshawar.
   Sector 2. m 0–10 °C. P: 250–500 mm.
   Valley of Indus-north of Kohistan, northwest of the valley of Kurram and high plateau of Waziristan (P more than 2000 m). Plateau of Kalat, of Baluchistan and valley of Indus and Kurram.
   Analogous area: South Spain and Caucasus (Baku), Syria. Iraq (less than 500 m). Iran (1000–2000 m); SE border of Caspian sea less than 500 m. Tashkent ES a markand.

3. Thermomediterranean.
   Dry season long *m* 0–10 °C and the valley of Chitral (Drosh), Indus Samarkand (swat), hills north of Peshawar, and mountain slopes on SW of Peshawar valley and Kurram valley, Dras.

Analogous area: Spain, Italy, Greece, West coast of Turkey, Crete, Libya and Israel (Telaviv), Turkey lake of Van, Iran (200–2000 m) and frontiers of Afghanistan and Pakistan.

4. Mesomediterranean.
   Dry seasons long *m* = 0–10 °C.
   Valley of Kashmir roundabout Srinagar, valley of Panjkora of Swat (till 2500 m). Height plateau of Kurram (1500 m).
   Analogous area: South Spain, Portugal westcoast, east coast of Italy, Georgia-region of Tiflis (less than 500 m), southern border of Caspian sea less than 500 m) and Afghanistan, less than 5000 m

5. Submediterranean.
   Transitional *m* 0–10 °C.
   Basin of Kashmir (except base of valley of Srinagar) and valley of Jhelum, more than 1000 m above sea level.
   Analogous area: North of Portugal and Spain, Turkey, Greece and Bulgaria (till 1000 m), Pontic region of Crimea, south west border of Caspian sea (till 1000 m).

6. Temperate *m* 0–10 °C
   Slopes of the valley of Kashmir, receiving 1000 mm of rainfall, outer slopes in Kashmir, slopes shelted by first chains of mountains in Kangra and outer hills of Mandi (P 1000–1500 mm). Mountain zone of Kangra, Tehri Garhwal and Kumaon (P 2000–3000 mm) except small parts having 1500–2000 mm rain). Isotherm of 0 °C for coldest months pass at 1500 m in Pyrenées, while in northwest it is 2500 m, 3000 m in Kangra.
   Analogous area: France-high Pyrenees (up to 1000 m), Italy, part of Lombard plain, prealp till 1200 m, Yugoslavia; Black sea-southern part of the border (up to 500 *m*) eastern coast (region of Batoum in Georgia up to 1200–1500 m).

7. Cold and temperate cold climate (temperature curve negative for a few months of the year).
More than 8 months (dry plus snow)
Metro losy stations are at Skardu, Kargil, Drǝs and Leh. The limit of 4 months of snow pass at 3500 m in northwest and high valleys, 4000–4500 *m* in Himalaya.
Analogous area: Turkistan, Afghanistan (near Kabul, Hajra up to 3000 m.

8. Temperate cold (*Less than 4 months snow*).
In Himalayá isotherm for 0 °C for coldest month pass near 2500 m in northwest and valleys of Kashmir, near 3000 m in Kangra and 3500 m in Garhwal. The upper limit of the climate is found near 3500–4000 m in Kangra and 4500 m in Garhwal. Between this limit is the temperate climate having 4–7 months of snow. Rainfall is 500–3000 mm in the region of Garhwal and Chamba, 1000–2000 mm in Kohistan and 600–700 mm in Kashmir.
Analogous area: Spain, Pyrenées (curve of 0 °C for coldest month is between 1500–2000 m; France-central massif, Alps less than 2000 m), Austria, Hungary, Bulgaria, Rumania, Causasus (less than 2000 m), Georgia, extreme northeast of Turkey.

9. Average cold (3–5 months of snow).
This climate is found above the limit of cold climate which is near 5000–5500 m and the lower limit at nearly 3500–4000 m. Rainfall in the Kumaon region amounts to 3000 mm and in the western part 1500 mm.
Analogous area: Summit of Pyrenées above 2000 m; Alps of Italy, Austria, Romania less than 2500 m, Balkans, Caucasus (2000–3500 m). Elbrouz (below 4500 m and mountains of Kopet-Dagh (2250 m).

10. Very cold (Snow more than 8 months).
Peaks more than 5000–5500 m in Himalayan chain, rainfall extremely variable in different regions (in Kashmir 15000–2000 mm, in Chamba region of HP and Tehri Garhwal 3000 mm; Kangra and north Garhwal 2000–3000 mm. Interior chains in the neighbourhood have a heterogenous rainfall, ranging from 600–1500 mm in the extreme northwest, and 600–3000 mm in SE of the ridge.
Analogous area: In Pyrenees more than 3200 m of altitude Alps: peaks of Pelvoilx, Mount Blanc massif, Cervita massif (3500–4000 m; Swiss Alps (above 3000 m), caucasus mountains 3000–4000 m chains of Hindukush, region of Gardex (more than 4700 m).

11. *Glacial climate:* Thermal curve always negative.
North of the valley of Indus, between Gilgit and Leh, a large zone of Karakoram and the peaks above 6000 m of altitude, e.g. Nagaparbat, Tirichmir, etc.
Analogous area: Absent in Pyrenees, Alpe-mountains Blanc, Grand Oaradis and peaks above 4000 m altitude, Caucasus-Elbrouz 5829 m, Skhara 5,182 m; chains of Kohi Baba and summits of Hindukush (more than 6000 m).

References

Ahmad E. 1973. *Soil Erosion in India*. Asia Publishing House, Bombay.

Bali JS. 1969. Soil Conservation for Large Dam Projects in India. The *Harvester*. 2(1) 3–39.

Bhatnagar HP and Seth, SK. 1960. Indicator species for Sal (*Shorea robusta*) natural regeneration. *Indian For*. **86(9)**: 520–530.

Blanford, H.F. 1884. The theory of winter rains of Northen India. *J. asiatic Soc. Bengal,* 53: 1–16.

Chaudhry AB and Sharma MK. 1988. Information content of Lansat data visual interprelation. Natural Resource Management System. National Seminar. *Mimeo*. pp 1–17.

Clarke CB. 1897. Sub-areas of British India. *Journ. Linn. Soc.* **34**: 1–46.

Das DC. 1977. Soil Conservation Practices and Control in India.

Daubenmire RF. 1947. *Plants and Environment*. John Wiley and Sons. N York.

Gaussen H. 1954. Theorie *et Classification des Climates et micro-climates*. VIIth Congress International Bot. Sec. **7**: 124–130.

Gupta RK. 1960. *Studies on the vegetation of NW Himalaya*. Ph. D. Thesis, Poona University, Poona.

Gupta RK. 1964. Forest types of Garhwal Himalaya in relation to edaphic and geological factors. *Journ. Soc. Indian Foresters*. 4:147–160.

Gupta RK. 1964. The bioclimatic types of W. Himalaya and their analogous. Types towards mountain chains of the Alps and Pyrenees. *Indian For*. 90 (8): 551–557.

Gupta RK. Dabral BG, Meher. Hoinje and Puri, GS, 2000. *Forest Ecology*. Vol 3 pp, 1–454. Oxford and IBH. Publishing. Co. New Delhi.

Gupta RK. and Sharma MK. 1995. Ecofloristic Classification for Sustainable Development of Doon valley. *Journ. Himalayan Geol.* **6(2)**: 81–94.

Gupta RK 1992. Ecofloristic Zones for Conservation and Agro-Silvo-pastoral Management in India. *Farming systems and Integrated Pest Management*. Eds. JP Verma and Anupam Verma. Malhotra Publishing House. New Delhi.

Gupta SK, Tejwani KG, Mathur HN and Srivastava MM. 1970. Land Resource Regions and areas of India. *Journ Indian Soc. Soil Sci.* **18(2)**: 187–198.

FAO 1978. *Report on agro-ecological Zones Project*. Vols. 4. Results of SE Asia World Soil Conservation Report. 48–4. Fao, Rome pp 39.

Gadgil M and Meher-Homji VM. 1982. Indo-US *bionational Workshop on Conservation and Management of Biological Diversity*. Indian Institute ScI, Bangalore. *Conservation of India's Biological Diversity*. pp. 1–24. Nav Bharat Enterprises, Bangalore.

Hill, S.A. 1876. Meteorology in N.W. Himalaya. *Indian Meteoro. Memoirs* pp 378–419.

Khosla AM. 1949. *Appraisal of water resources, analysis and utilization of Resources*. UNESCO Conference of Conservation and Utilization of Resources.

Koppen W. 1900. Ver such einer classification des Climate. Vorzugsweisse nach chren Buziechungen pflanzenwelt. *Geogr. Zeitsch*. **6**: 5936 II and 657–679.

Livingstone BE and Sherve F. 1921. *The distribution of vegetation in United Stated, as related to climatic conditions*. Varngie Inst. Washington. Publication No 284.

Legris P. 1963. La vegetation de lInde, Ecologie et Flore. *Trav. Sec. Sci. et Techn*. Pondichery. Tome 6. French Institute, Pondichery.

Legris P and Meherhomji VM. 1968. *Utilization of data on conservation and utilization of Resources*.

Mohan NP and Puri GS. 1956. The Himalayan Conifers V. The succession of Plant Communities in Chirpine (*Pinus roxburghii*) forests of Punjab and Himachal Pradesh. *Indian For*. **83**-356–364.

Mohan NP, Puri GS and Gupta AC. 1956. The Himalayan Conifers. IV. A case study of some soil profiles under some forest communities in the Bashahar Himalayas *Indian For.* **82(6)**: 295–389.

Mohan NP, Puri GS and Gupta AC. 1957. The Himalayan conifers VII, The succession of forest communities in the oak forests of Bashahar Himalaya. *Indian Forest Rec.* (Ns Silvic. **19(2)** 19–36.

Nutton son My. 1952. *Ecological crop geography and field practice of the Rynkun islands, natural vegetation of Ryukyn and agro-climatic analogues in the Northern Hemisphere.* American Institute of Crop ecology.

Puri GS, Gupta RK, Meher-Homji VM and Puri SS. 1989. *Forest Ecology* Vol 2. Plant form *Diversity, Communities and Succession.* Oxford and IBH and Publ. Co. New Delhi pp. 582.

Razi BA. 1954. Some observations on Plants of south Indian Hill tops and other districts. *Proc. National Inst. Sciences.*

Rodgers WA and Parnar HS. 1980. Planning wildlife protection area and network in India. Field Document. Establishment of wildlife Institute of India. Dehradun. pp. 345.

Sehegal JL. Mandal C and Vedvelus 1990. *Agro-ecological regions of India.* National Bureau of Soil Science and Landuse planning Nagpur.

Schimper AFW. 1903. *Plant Geography, upon a physiological basis.* Vol. 1. pp 839. Translated from German by WR Fischer.

Seth SK and Yadav JSP. 1960. *Soils of the tropical moist-evergreen forests of India.* Tropical Moist Evergreen forest Symposium pp 14. Mimeo.

Singh RL (ed) 1971. *India, a regional geography.* National Geography Society of India.

Thornthwaite CW. 1933. The Climate of the earth. *Geogr. Review*, **23**: 433–440.

Zonal Concept and Ecological Evaluation of the Habitat of Drug Plants

Introduction

Regional Research Laboratories of the CSIR (Council for Scientific and Industrial Research), particularly, the Regional Research Laboratory, Jammu, had been the pioneer in the cultivation of medicinal and aromatic plants, both at Jammu and Srinagar in Kashmir, at their farms and have produced valuable data. The work of late has also been extended to other sites by the Agricultural Universities in the country through the Coordinated Project on Medicinal Plants, by the ICAR in their laboratories and stations under their control. New varieties of Isabgol (HI-5) and Mulethi (MH-1) were released by the HAU, Hissar, for Haryana conditions. Variety MH-1 of Rosha grass was released, which is considered to be the best oil yielder, and in quality and even better than the Chinese and Russian materials. Other plants that received attention are *Vinca rosea* (Sada-bhar), Plantago (Isabgol) Sarpgandha (Rauwolfia), Trigonella (Metha), Lemon-grass, Mentha, Khus (Vetiveria), Dhania (Coriander), Rose (Gulab), etc.

Conservation of herbal plants and their scientific exploitation, particularly in the Himalayan region, has also been initiated since 1980–81 first in the Kashmir region and then in other parts of the Himalaya.

(In Himachal Pradesh), Farmer's Societies were created to explore and exploit medicinal plants from the forest areas, though limited efforts were made to encourage farming of herbal species but it could spread awareness about their use as a low-cost health-care system in remote areas. In parts of UP (now Uttaranchal), establishment of a Herbal Research and Development Institute, at Gopeshwar, (district Chamoli) in the region, was decided on 7.12.1989. A state expert committee with Dr. Satyapal Gupta was constituted as a 4 member group on 1.1.1988 and consequently, in December, a workshop was sponsored by Ayurvedic and Unani Medicine Department of UP, to finalise the report presented by the directorate. A sub-committee constituted for the purpose selected the site at Dunagiri, near Dwarhat in Almora district. Also Mandal village, in Chamoli district, was ready to give 20 acres of land for the Institute. A Govt. order was issued on 13.12.89 to setup the Institute at Mandal. Five regional centres were also recommended for the Institute at Pauri, Pithoragarh, Almora, Chamoli, Uttarakashi, in addition to Dehradun and Nainital being the ideal places. The herbal institute was also entrusted to start diploma and degree courses in Herbonomy. Some of the farms were also recommended to be converted to drug farms. Emphasis

was proposed for the cultivation of a few medicinal plants like *Crocus sativus*, (Jafran or Kesar), *Aconitum* species (Atis), *Pavetta indica* (Kukum Kuku), *Bergia ligulata* (Pashanbed, Silphar), *Tinospora cordifolia* (Giloe), *Semecarpus anacardium* (Bhilmor), *Rumex dentatus* (Changeri), *Adhatoda vasica*, (Vasa or Bansa), *Dactyloctenium aegyptium* (Makra); *Mucuna pruiita* (Kaunch or Kawanch); *Terminalia arjuna* (Arjun), *Achyranthes aspera* (Apamarg, Chirchita), *Vernonia cinerea* (Sahdevi), *Prunus cerasioides* (Pyan, Padamrukh); *Citrullus colocynthis* (Indrayan), *Cyperus sacrious* (Nagarmotha), *Myrsine africana* (Baibharang), *Aconitum ferox* (Vishi), *Aconitum luridum* (Bish), *Aconitum heterophyllam* (Maha Vish), *Delphinium denudatum* (Nirbisi), *Leucas cephalotes* (Gumma), *Ougeinia dalbergioides* (Timni); *Eranthemum roseum* (Desmool) and many more species that are used coonly and are on the verge of extinction.

Similarly, in Central India commercial cultivation of some medicinal plants in vogue like is Aloe vera, *Abelolomoschus moschata*, *Acorus calamus*, *Abroma paniculata*, *Asparagus racemosus*, *Bacopa monnieri*, *Adhatoda zeylanica*, *Argle marmelos*, *Andrographis paniculata*, *Catharanthus roseus*, *Cassia angustifolia*, *Chlorophytum borivilianum*, *Coleus forskohelei*, *Comiphora wightii*, *Cymbopogon flexuosus*, *C. martinii*, *C. winterianus*, *Emblica officianalis*, *Glycirrhiza glabra*, *Gloriosc verba*, *Gymnema sylvestre*, *Jatroph· ·urcas*, *Lepidium sativum*, *Mucuna ·ru ·ens*, *Plantago ovata*, *Phyllanthus amarus*, *Ocimum tenuifolium*, *Rauvolfia serpentina*, *Saraca asoka*, *Siwartia chirayata*, *Solanum nigrum*, *Tinospora cordifolia* and *Withania somnifera* (after Vasih Gaurav).

A very broad classification of the climate of some important drug plants is as follows:

1.	Semiarid condition but dependent on rabi or kharif crops.	Withania, Aloe, Vinca, Isabgol Acorus (Chandrasur).
2.	Not damaged by animals.	Aloe, Thyme, Vinca. (Acorus).
3.	Not much damaged by insect attack.	Withania, Chandrasur, Thyme Aloe and Vinca.
4.	Not much technical. knowledge is required.	Vinca, Withania, Isabgol.
5.	Weak, marginal wasteland.	Vinca, Withania, Rauwolfia and Aloe.
6.	Tolerant to high soil pH (8–5).	Isabgol, Khurasani ajwain (Hyoscyamus niger).
7.	Can grow under humid moist condition.	Chandrasur (Acorus).
8.	Can grow in shady places.	Withania, Rauwolfia and Aloe.

The following pages provide a much detailed account of the factors that affect the quality and distribution of medicinal plants which could form a guideline for exploring ecological conditions for the cultivation of medicinal and aromatic plants in the country.

The Climatic Regime of the North India including the Himalayan ranges

Distinct climatic regimes in northern India is a well-known feature. After the close of rains, at the end of September, the sky is clear, the atmospher transparent and the air permits the free passage of heat from sun to the earth during day and in the calm nights that prevail at this season. The months of October and November are characterised not only by clear skies but by great temperatures range and heavy dews at night. These conditions prevail throughout the greater part of December and towards the end of this month and in the beginning the temperature falls (10 °C), even in the plains. About the end of December and in January and February, however, clouds often interfere with the free

radiation of heat at night and the daily range of temperature is less on the average than that of November. Some rain usually falls at this time of the year, especially in the northern districts. In March and April, temperature rises rapidly, especially at a distance from the mountains and the air becomes extremely dry. Hot winds from the west or northwest blow down the valley of Ganga and rapidly change the appearance of the country from that of highly cultivated plain to that of a parched land, almost only green things left being groves of mango tree. In April, the daily range of temperature over the plains is at the maximum, exceeding 30 °C in most parts of Northwest India. The nights are still cool while during day temperatures as high as 11 °C or even higher are recorded. During May and first half of June, the temperature continues to rise though much less rapidly than in March and April, until by the 15th or 20th of June, if the periodical rains have not commenced. The days in June are thus only a few degrees hotter than those in April; as the rainy season approaches the range of temperature diminishes and the night becomes hot. Rains seldom fall during hot season but do occur during thunderstorms. About the middle of May, the water vapour in the air began to increase rapidly and is brought by the prevailing southwest upper current of the atmosphere which seems to descend gradually until it merges with the surface sea winds of the Bay of Bengal and forms the SW monsoon or prevailing winds of the rainy season. The northern strata of these winds is deflected from their normal course by the mountains and directed towards the sea of highest temperature in Punjab and Haryana, thus appearing as east or southeast and not as SW winds. Along the foot of the hills these easterly winds are felt occasionally by the middle of May when the quantity of vapour in the air first show signs of a rapid increase.

During the later half of June sea winds increase in strength and gradually advance along the foot of Himalaya until by the beginning of July the rains have usually set in all over North India. In ordinary years the rains continue to fall, not every day, but frequent intermissions of break, until about the end of September, when the easterly winds cease except to those to the hills, where they last a month longer and are succeeded by calm or feeble currents from the west.

Himalayan range on account of their nearness from the equator and its elevation, presents many points of advantage, as compared to other mountain ranges like the European Alps and other ranges of the world. Nearly all the snow peaks and most of the passes over the Indian watershed stand above the lower half of the atmosphere and thus completely cut off all communication between Indian and Central Asia, except in upper strata. High levels also provoke low air pressure and temperature and intense radiations, etc.

In Uttaranchal, climate like any other mountain chain varies with the altitude and as the physical features are very much diversified, the lower and outer ranges have a climate similar to the plains, while the higher altitudes are covered with perpetual snow. The valleys along the rivers (called Gagar) are extremely hot and experience climate similar to plains. Here every altitude has its own temperature, from the lower valleys which are hot to eternal frost, but at all elevations the force of sun is excessive. The summer rains, too, gradually diminish in strength, as we move along the chain, from east to west being at their maximum in Sikkim. It has rightly been pointed out that the climate is chiefly decided by the geo-

graphical position, elevation and monsoons. During winters, the wind blows from northwards to eastwards which brings dryness and clear sky and this change between summer rainy season and cold dry season characterises sharply the Indian Monsoonic Climate; however, its influence on the climate depends on the sea level. At Nitipass, between Uttaranchal and Tibet, the precipitation is only 140 mm during July-September, but in winter the entire pass is covered by snow. The winterly precipitation in higher ranges of the Himalaya is important in forming the glaciers out of which the most important rivers originate. Winter snowfall at 4600 m is about 9 m and on higher ranges may amount to 14–30 m, and is due to cyclonic (eastward) storms across the North India. As more and more a place is situated in the Himalaya, the lesser will be the precipitation and the hills near the plains in the south of districts take their first toll and while passing northwards, each successive higher range takes its toll, till the last steep ascent of the snowy peak drains the air currents for most of their moisture. However, the orientation of the valleys and the aspect of the slope plays an important part in the distribution of the rainfall. The northern slopes and east and west valleys receive much less precipitation (pptn) than the southern slopes. As for example, the pptn at Narendranagar Uttaranchal is 2674 mm, while at Tehri it is 794 mm. In north of the snowy peaks, summer rains are scanty and the highland approaches desertic conditions. At fairly exposed stations of nearly same altitude there is a gradual diminution of annual rainfall as one passing from west to east. Thus, at Darjeeling the rainfall is 300 cm, while at Nainital it is 180–280 cm. Maximum rainfall is estimated at an height of 1300 m and varies a good deal with the altitude up to 3000 m and above that there

is more snowfall and the mountain slopes in the inner ranges are also cut off to the summer monsoon due to their height.

The variation of rainfall with season is well marked; November being the fierest month except the higher mountains where winter snow begin from the end of November. For higher mountains the precipitation records are lacking. In the upper regions, the rainfall almost daily is the form of mist. The frequency of snow increases with altitude and above 4000 m most of the pptn comes in the form of snow. There is no sharp altitudinal limit in this respect, sometimes hail is also of frequent occurrence. Mist has proved to be an important source of moisture for the vegetation in many parts of the hilly region but has considerable ecological significance.

Temperature

Every elevation in the hilly region has characteristic mean temperature and the variation is different at different altitudes. On southern slopes the difference of latitude and longitude make little difference in the mean annual temperature. There is a great uniformity of the mean temperature at the same elevation with the difference that in Bengal and Sikkim sunrays when most intense are to a great extent cut off by the clouds, while in north-western part of the Himalaya the winter is almost, if not quite, cloudy as the summer. Thus, Darjeeling has the same temperature in January as Simla, Chakrata and Mussoorie, while in May and June it is 7–8 °C cooler.

The mean annual temperature diminishes respectively with the height and the variation of temperature between the hottest and the coldest months and the daily range of thermometer are also greater, as a rule, in the interior that in the outer hills, owing to a large proportion of cloudy sky and

great humidity of air than in the later regions. Both the diurnal and annual range of temperature decrease on ascent from the plains up to a height of 2000–2700 m and beyond that decreases again. These places, however, lie to the north of Indian water-sheds, where humidity is less than on the southern side of the snowy ranges. Moreover, the annual range in Tibet and Ladakh is greater on the Indian side of the chain, on account of the differences in the latitude. Owing to greater annual range of temperature on plains than on the hills, the diminution of temperature in the first 2000 m is rapid in hottest months and least so in cold months. In the clear still nights of cold weather, especially in November and December, before winter rains and snow sets in, nocturnal loss of heat goes on almost as freely as in the plains, as in mountain peaks. The low temperature of plains in winter is due to the draining of cold air down from mountain slopes through the river gorges. This does not affect the temperature of places at long distances from the mountains but has a considerable effect of the foothills of Siwaliks. From March to June absorption of heat in melting of snow and evaporating of snow, on the outerhills, keep down the temperature in May and first half of June when plains are hottest and decrease of temperature at 2000 m is more than twice as great as in December. A rate of decrease of temperature with elevation at the time of maximum day temperature in the month of January averages 3 °C per 300 m in the western part of the Himalaya and 4 °C per 300 m in eastern Himalaya.

Greater annual range of temperature at more elevated stations, especially such as behind the first snow range, receiving little or no summer precipitation, cause of greater differences in the rate of decrease of temperature with height out in opposite direction. The relation of this to the great height of snowline of the northern side than on southern side of the Himalayan chain is obvious. Though, on the average of the year, the decrease of temperature on ascending is nearly to the height but it has been seen that this rule does not apply to individual months. In cold season the variation is slow up to 1600 m but rapid in higher elevations, whilst in summer there is a sudden decrease of temperature on entering the mountain zone and have any variation at 4000–4300 m. Fall in temperature with an altitude, greatly affects precipitation. The increase in altitude is associated with increase in rainfall up to a certain level above, which the precipitation again diminishes. The elevation at which maximum pptn is recorded at 1270 m above the sea level and the rainfall amounts to 3.7 times higher as in the neighbouring levels. Thus, the average precipitation at Dehradun (678 m) is 2160 mm, Rajpur (970 m), 3040 mm and at Mussoorie (1044 m) it is 2400 mm. The yearly average temperature at Dehradun is 21 °C at Mussoorie, 14.1 °C and estimated temperature between 3350–3650 m would be 6 °C. However, a variation of temperature of about 1 °C for every 100 m rise has been reported; thus the mean annual temperature at 3700–400 m would be about 5 °C (42–45 °F). However, it has been recorded that there is a general decrease of 0.55 °C for every 100 m rise in elevation, which holds good even on high altitudes, if the slope is regular and uniform. This coefficient is an average and is valid for diverse regions, e.g. France 0.57 (Angot), Algeria 0.55 (Seltzer). Switzerland 0.54 (Schrotter) and is in almost every temperate countries. As far as the effect of temperature on different expositions is concerned, various accounts have been published and it is natural that expositions receiving sunrays are warmer than those in the shade. The contrast between the two

aspects can also be quantitatively appreciated by the lower limit of snow that shows a difference of altitude, whose value in Switzerland is between 150–650 m, eastern Alps 200 m, Bernina 200–300 m and Alps Penneines 500 m. The upper limit of the forests show a difference of the same order but less marked (100–200 m).

The upper limit of the natural vegetation in the Himalaya is about 4500 m, where the mean temperature is 9–10 degrees above freezing. In Alps where lower limit of the perpetual snow with natural forest of Pinus caribea, the mean temperature is several degrees below the freezing point. This difference of habit between the Himalaya and alpine Pine is curious.

The effect of different exposures to sun and its importance of valleys and slope winds is interesting. At middle altitudes of Nepal it has been shown that on southern dry and sunny slopes *Pinus roxburghii* forests on north shadowing slopes a mixed forest of deciduous and evergreen species (like Quercus, Castanopsis, Acerpyrus etc.) but in deep and broad valley of east Nepal, there is a remarkable difference in the vegetation between eastern and western slopes. In the early morning, until about 10 o'clock the eastern slopes receive direct insolation for several hours in the afternoon when sunrays are coming from west, the slopes are cloudy, caused by so called slope winds. In this way slopes, exposed to west, get less direct insolation than eastern slopes, where sun is shining before the convection clouds are coming up. A second wind system is directly connected with the cycle of slope winds, especially in deep trans-section valleys. This can give rise through the daily interchange of air between the low-lying plain and the Tibetan alpine plateau (so called compensation currents) to a system of mountain valley winds, which produce a dry climate and xerophyllous vegetation on the floor of the valley and the best known examples of such dry valley is in West Nepal and in Bhutan.

The Radiation

Data on the average intensity of solar radiation and its variation for the Himalayan region is still very scarce. Up to Chakrata, the excess temperature of the solar thermometer does not increase with a fair degree of regularity but it appears to be less at Leh than at Chakrata. There is no regular increase in the heating power of the sun as the season changes from winter to summer. Because of the rarified and clean atmosphere, at high levels less of the incoming radiation is absorbed above the ground so that under a clear sky the intensity of sunshine is greater than in the lowlands. Under the same conditions, the relative amount of ultraviolet light is considerably higher at high altitudes than at sea level. Because of the frequent occurrence of cloud and mist in upper mountains, the intensity of insolation is reduced during part of the day. Although part of the incoming radiation penetrates the clouds as diffuse light, infrared rays, which account for almost half of the insolation seem to be almost completely absorbed. Therefore, the radiation climate of the alpine belt is very variable, displaying a fairly diurnal cycle according to the cloudness. When the sky is clear the intensity of sunshine is remarkable and may prove deleterious to unprotected plants tissue as well as to human skin.

Though, part of the incoming radiation reaching the ground is reflected but much of it is absorbed by the ground and vegetation being mainly converted into heat. Apart from the reflected sunlight the ground and the plant-cover also emit radiation

themselves. In daytime these low have radiation of little importance, being much smaller in amount than the incident short wave radiation, but at night the former is by far the most important and may lead to a pronounced cooling of the soil surface. This long wave radiation from the ground is counteracted by reradiation from the atmosphere, the intensity of which depends upon its temperature, and its contents of water-vapour and carbon-dioxide, especially upon the occurrence of clouds. With increasing altitude above sea level, the intensity of reradiation decreases, and hence the effective long wave radiation from the ground increases. Thus, on very clear night in the alpine belt there is a strong long wave radiation, resulting most often in night frost. This is very important for microclimatic differentiation.

The Clouds and the Winds

During the daytime the mountain slopes are heated by the sunshine and the air near the ground becomes warmer and moves upwards, following the slope. The higher level, the ascending winds are cooled and part of their moisture content is condensed, causing cloud formation. At night, on the other hand, the mountain slopes rapidly loose heat by outward radiation and the air near the ground gets cooler and flows downhill. As in other tropical mountains, this daily cycle of clouds formation is regular, so that from about noon, most of their upper parts are capped in clouds. This cycle of cloud formation creates a pronounced insolation between the eastern and western sides of the same mountain, and hence a different altitude for snowline in east and west. From the ecological viewpoint, this diurnal cycle of cloudiness is very important since the screening of direct sunshine causes rapid decrease in temperature and increase in relative humidity.

Since cloudiness on medium altitudes in tropical and subtropical mountains is a very general phenomenon, it affects the availability of moisture. As an example in Nepal (east) at Thandung 3100 m during monsoon (June-October) the pptn in 1963 was 2737 m and 600 mm resulting from mist and fog. This 600 mm is one of the basic reasons that cloud forests are overloaded by moss, ferns and small epiphytic Orchids. The humid belt of cloud forest is last, not the least very important for the whole economy of lower and drier much populated hill, especially in dry valleys. Hence, most of the perennial rivers rise in heights of moist *Abies spectabilis* forests.

The Snow and Glaciers

Most of the precipitation falling on high altitudes is in the form of snow and unequal distribution of the snow cover in some region furnishes the most important ecological factor that largely governs the distribution of the vegetation. When the snow falls in a calm weather, for some time it may form an even cover but as a rule it is soon swept away from elevations and windward slopes, accumulating in hollows and leeward slopes. Thus, the hill tops and ridges become relatively free from snow-earlier, whereas in hollows the snow drift may be several metres thick. This uneven snow cover leads to a number of edaphic and microclimatic positions within a levelled area. Glaciers at present exist on different mountains and their difference in height is probably due to difference in insolation, precipitation and exposed slopes.

Above altitudes varying from 4000–5000 m (12,000–15,000 ft) snow lies perpetually in the Himalaya, called the snowline. This is the lower limit of perpetual snow in a given latitude and altitude, that constitute

a boundary line to snow, which resists the effect of summer. The altitude of snowline varies also due to geological slopes and aspects being lower on dip-slopes and northern aspects than on escarpments and south aspects. The belt of perpetual snow is about 35 m in breadth and runs along the northern boundary of the chain. Snowline determines the tree limit and any change in snowline brings about a corresponding change in the extension or recession of the tree line. The study of snowline, therefore, is an important ecological factor. The bend that occurs in lower portion of stem, in trees due to weight of snow is well known; snow also damages flowers and fruits.

Avalanche strips are typical in the higher parts of coniferous forest belt and are unknown from high mountain regions of east Nepal. The simple reason, that in west Himalaya the snowfall is more during winters than in the eastern part and the upper forestline on the northern slopes is 200 m in Khumbu region is one caused by low temperature but also increases towards the inner part of the main Himalayan range.

The Microclimate

Superposed upon the general climatic pattern, there are a number of micro-climatic variations in the hilly regions. Whereas, the general pattern of vegetation is governed by the macroclimate, the detailed vegetation pattern is largely determined by microclimatic differences. The climatic experience felt by a living plant is not the same as recorded on an ordinary meterological screen. Because of the rarified air and intense radiation, the local micro-climatical differences present at high altitudes are more extreme than in most low level biotopes. Unfortunately, the micro-climatological observations so far made are very few and fragmentary.

The effects of soil covers of different colours, of a crop cover and of wetting the surface ground, determines the absorbing and radiating power of the surface. The-rmal conductivity of the soil is an important factor in controlling the distribution of temperature in soil as well as the air above it and it varies with the water content. Wind movement in the air layers near the ground are the counter part of the thermal structure. Simultaneous observations of wind velocity at various heights show considerable variations. Moreover, the change from the surface to the climate of open space is not quite gradual. There is a considerable amount of evaporation from the soil during all seasons. It may be expected that during wet season the specific humidity in the air would be plus-minus constant with height above the ground with a tendency to be maximum near the ground during the period of sunshine. Evaporation expresses the combined effect of temperature, humidity and sunshine, etc.

Though a good deal of work has been conducted on microclimatic conditions of agricultural crops, forest vegetation and ground flora composing aromatic and medicinal plants have not received much attention except a few studies by some foresters and forest scientists (Gupta *et al* 2000). However, the results are largely paralleled by those occurring in the Andes of tropical South America and mountains of tropical Africa.

Factors affecting the Quality and Distribution of Medicinal Plants

The quality of medicinal plants refers to the intrinsic value of the drug, that is the amount of medicinal principle or active principles in the form of starches, sugars, acids, bases, gums, fixed oils, volatile oils, resins, tannins,

alkaloids, glycosides, hormones, vitamins, etc. India being a country of varied climate, soil and topography, present almost all the possible combinations of soil and climate. There are regions of extreme rainfall from drought to very humid and extreme heat to freezing temperatures. The quality and potency of a drug much depends upon the conditions of its growth and habitat. In ancient literature, there is sufficient proof to show that a regular study and culture of plant material was under constant review and the branch of the science called "Vrikshayurveda" dealt mainly with the cultivation of plants and prevention of plant diseases. There is also an indication of using "Koonpa-jala" rich with chemical properties during the nursing of plants to get satisfactory medicinal property.

Isolation of the active principles of quite different nature from plants obtained from different sources have been reported by various workers. *Centella asiatica* (Vern. Mandukparni), widely used as a medicinal plant, is recorded to give different glycosides when procured from different sources. A systematic study and quantitative leaf microscopy of this plant obtained from different sources, however, did not indicate any morphological change. The variation is perhaps due to ecological changes. Similarly, some species of Ephedra from Northern India, with an average rainfall of 500–700 mm or less gave a good content of alkaloid, while those growing in the eastern part of the Himalaya under heavy rainfall of 1900 mm or more have little effect or no active principles. *Cinchona succribra,* grown at an altitude of 300–450 m above sea level produces low levels of quinine, though the growth is good, whereas the same plant growing at an altitude of 900–1500 m shows high percentage of quinine in the bark. Digitalis leaves obtained from Ooty (Udhaga-

mangalam) show little glucocidal activity, while those from Mungpo have good results, even better than imported ones, though both the places are at the same level. The total alkaloid content in *Rauwolfia serpentina* roots, obtained from different parts of the country vary considerably, which naturally indicates variations in the biological potency (Table 8.1).

Table 8.1: Chemical composition of roots of Rauwolfia serpentina raised at different localities (Chopra, Sobti and Handa 1962)

Locality	Age (yrs)	Total alkaloid content (%)	Resperine content (%)
Jammu	2	1.73	N A
	3	1.70	...
Jamnagar	2.5	1.85	...
Dehradun	2.0	1.65, 1.78	...
Haldwani	2.0	2.38	0.169
Nilgiris	3.0	2.30	0.2

A good deal of variation in the active principle is in different parts of the plant species at different periods of their growth. Even the same part and at the same time of the year shows remarkable variations in the content of the active constituents, e.g. the young and the old leaves of the plant and unopened and opened flowers differ materially, despite the fact that they are collected from the same plant and during the same season. The seasonal variation of the total alkaloid in *Rauwolfia canescens* has been studied, which indicates its highest alkaloid content during the months of October to December as shown in Table 8.2.

Constituents of **Centella asiatica** from different sources and sites are given in Table 8.3 (Chopra, Sobti and Handa 1962).

Diosgenin content of **Dioscorea deltoidea** from different sites of West Himalaya (in Kashmir and H. Pradesh) is shown in

Table 8.4 (quoted after Chopra, Sobti and Handa, 1962).

The potency of the drug may also vary due to variations in the soil composition from place to place; though the plant material may have been collected at the same altitude and in the same season.

Table 8.2 : Average amount of total alkaloids in *R. canescens* in different months of the year (Chopra, Sobti and Handa, 1962)

Seasons	Percentage of alkaloid			
	Leaves	Stems	Roots	Whole plant
February-March	0.40	0.10	0.06	0.56
April-May	0.50	0.20	0.10	0.80
June-July	0.67	0.25	0.14	0.16
Aug-Sept	0.86	0.28	0.19	1.33
Sept-Nov	1.20	0.29	0.20	1.69
Dec-Jan	1.20	0.29	0.20	1.69

Temperature

Temperature is the most obvious factor of climate and can broadly be related to altitude and the latitude giving a rough differentiation into four zones, viz. tropical (very hot and winterless), subtropical (hot with cool winter), temperate (with a warm summer and pronounced winter) and arctic (with a short summer and long severe winter). The relation of climate to altitude is so greatly influenced by other factors such as distribution of land and sea, altitude, prevailing winds and ocean currents and their direct influences, that no hard and fast lines can be drawn. It is, therefore, necessary to substitute suitable temperature ranges for latitude to make such a classification of any practical value. No single temperature figure such as a mean annual temperature, will meet the requirements. However, the factors having the greatest influence are the two ends of the scale so that at the colder end the period during which the temperature is high enough to permit plant growth is of predominating influence, whilst at the hotter end it may be the existence or extent of a cool period enough for growth to be stopped or slowed down. It is for this reason that extensive studies of the subject made in other parts of the world are of much limited value for India.

As far as the temperature data for India are concerned the climatological atlas and the climatological tables for observation is the main source which has been used. It

Table 8.3: Constituents of *Centella asiatica* from different sources and sites (Chopra, Sobti and Handa, 1962)

Locations / Sources	Active constituents			
	Free terpnis acid	Triperpene	Glycoside aglycone	Sugars or glycoside
Madagascar	...	Asiaticoside	Asiatic acid	Glucose and phamnose
	Centic acid			
SriLanka	Centoic acid Centellic acid	Centelloside	Centellic acid	Glucose and fructose
India	... Brahmic acid Isobrahmic acid Thankunic acid Isothankunic acid Asiatic acid	Indocentelliside Brahmoside Brahmoside Thankunside Insothankuniside	Indocenic acid ... ... Thankuni acid Isothankuni acid	do do do

Table 8.4: Diosgenin content of *Dioscorea deltoidea* from different sites of West Himalaya

Sites	Altitudes in m	Weight of rhizome in gm (average)	Diosgenin percent
Poonch	1800–2000	12	Traces
Sanasar	2000–2400	15	3.8
Satote	1500–2500	18	4.3
Hulmarg	2000–2500	28	3.8
Phalgam	2000–2500	25	4.6
Katra	1500– 1800	45	4.6–4.8
Kulu	1800–2000	42	4.3–5.0
Bhadarwah	2000–2200	52	4.8–8.0

Table 8.5: Division of the country on the basis of temperature

Zone	Mean annual	Mean of January	Winter annual
Tropical	Over 24 °C	Over 18 °C	None, no frost
Subtropical	17–24 °C	10–18 °C	Definite, but not severe, frost rare
Temperate	7–17 °C	–1–10 °C	Pronounced with frost and some snow
Alpine	Under 7 °C	Under –1°C	Severe with much snow

may, however, be noted that mean annual temperature exceeds 24 °C over whole of the country with the exception of the hill areas and the northwest. Variation with latitude is almost completely obscured by other factors, though in the Indo-gangetic plains there is roughly a fall of 0.75 °C for each 1° latitude. In striking contract with the irregularity is mean annual temperature distribution in winter months, January being the representative. The mean maximum isotherms then run very fairly directly east and west over the northern half of the country, the variation of 1° latitude being about 0.75 °C. The 18 °C isotherm for January roughly follows the Tropic of Cancer.

The country can broadly be divided on the basis of temperature as given in Table 8.5.

Such a classification obviously results in certain localities falling in one zone in January temperature but in another on mean annual temperature. Thus, Jodhpur is subtropical on the former criteria and tropical on the latter. The vegetation itself then indicates to which it more properly belongs.

The Altitude

From the known relation between altitude and temperature, the upper altitudinal limits of the zones might be expected to fall considerably higher in peninsular India than in the Himalayan region, but for a variety of reasons this is not very noticeable. Thus, a mean annual temperature of 17 °C is experienced in the Nilgiris at about 1830 m, in the Western Himalaya at about 1700 m and in Assam at 1500 m. The corresponding figures for the January temperature of 10 °C are Nilgiris 2150 m, West Himalayan region 1600 m and Assam 1500 m. The extremely small range of the monthly means in the Nilgiris (12.5 °C to 15.7 °C) is of course an influential factor in this connection. The subtropical zone should take in the great Indo-gangetic plain, running up over the plateau and lower hills; it should be divisible in a general way into warmer lowland and cooler hill subzones. But in India, the former has so much in common with the tropical zone as to render joint consideration a necessary curse. The temperate and alpine zones are only met within India in the higher hills along the northern frontiers.

The effect of altitude varies with geographic conditions, a rise in the hills of altitude 270 m corresponding to a fall of 1 °C in mean temperature upto about 1500 m, above which the fall is more rapid. In general this fall is more pronounced on the leeward side of the hills than on the windward side, thus it is reported to be 1 °C for 100 m rise on the westside and 122 m on the eastside of the Nilgiris. It may also be noted that towards the north, the rate of fall is much more rapid in summer than in winter perhaps ten percent above and below the mean respectively (this difference amounts to 20 percent in Europe).

The diurnal range of temperature does not seem to effect the vegetation, provided tolerable limits are not over-stepped. The range over the country as a whole is about 10 °C in January and July but 20 °C in April and October increasing in a northwest direction from the east coast. The maximum temperatures recorded do not seem to reach a level critical for plant survival, but the minimum is of undoubted importance. Few of the tree species in west tropical forests zone can survive frost and the indications are that many tropical trees are weakened, if not killed back in part or to the ground (by exceptionally high or exceptionally low temperatures), this does not necessarily kill them out or even prevent their forming an important or even dominating part of the vegetation. Although biotic factors may be involved, the principal adverse climatic factor is undoubtedly frost. Where the frost is relatively shallow-ground frost, the regeneration may be protected by the overwood, provided it is dense enough, and conversely, clearings or heavy things may result in frost incidence where formerly there was none. Though analyses have been made by climatologists to see if any cyclic recurrence exists in temperature data, but no statistical basis for such an assumption has yet been found (Paramanik and Jaganathan, 1953).

The Rainfall

There is a wide range of conditions from less than 1500 mm in the Thar desert to nearly 5000 mm in Khasi hills of Assam as far as total annual rainfall is concerned. Though, it is an important factor in determining the natural vegetation, its seasonal distribution exerts a far-reaching influence. Practically, in most part of the country, except for the innermost Himalayan ranges and north-west corner, major part of the total rainfall occurs during summer to autumn months and the rainy season varies in duration with a general increase from Northwest to east so that upper areas of erstwhile Assam state experience the shortest dry season and the northwest, the largest. The southeastern part of the Indian peninsula (on the Karnataka coast) gets relatively little rain during the Southwest monsoon period from June to September, and most of the rain falls during October to November which has resulted in a special type of evergreen vegetation, not met with anywhere in the country. The lengthening of moist season also affects the vegetation, viz. the Nilgiris, westernghats and far north as Kalahandi hills.

The intensity and the distribution of rainy days during the rainy period affect the vegetation growth to a great extent since other conditions being equal, the minimum rainfall necessary for the development of a given type increases with increasing length of the dry period. This effect appears to be less marked, the more xerophytic the type considered; the most pronounced for the type requiring the highest rainfall. There is also a less perceptible correlation with the number of rainy days reflecting the distribution of precipitation within the rainy season.

Relation between the temperature and the Rainfall

If the length of same dry period is experienced, it appears that a greater rainfall is required with a lower temperature than a higher one to permit the development of a given plant type, at least for the pronouncedly xerophytic types. Well-developed evergreen forests occur in westernghats with 2500 mm rainfall and a five-month dry season, but not with the same amount and distribution of rainfall in the cooler North Bengal though it occurs in equally cool Assam with only three dry months. The factors involved, however, are so complex that precise correlation is difficult.

For plant life, available moisture is of critical importance and it has been pointed out that the apparent loss of water by interception is not necessarily total. From a consideration of the energy relations, the evaporation of the intercepted rainfall from wetened surface must reduce transpiration unless the incident radiation is enough to sustain both processes. The determination of evaporation is not as simple as the measurement of rainfall but indirect methods have been worked out which give acceptable approximations. Evaporation is regulated by the rate of air movement. Thornthwaite earlier has worked out precipitation effectiveness (PE), i.e. the ratio of precipitation to evaporation in terms of rainfall and temperature. Bharucha and Shanbagh (1957) suggested that a good approximation for evaporation is obtainable from known values of barometric pressure, relative humidity and wind velocity so that the Precipitation/Evaporation (or P/E) ratio can be determined. They also derived a correlation between P/E and the number of rainy days (D) given by the formula $P/E = 0.065D = 0.0047 D2$. It will be noted later that the main forest types of India fall very fairly within the ranges of P/E.

Thornthwaite (1948) later developed a new approach, which is generally referred as Potential Evapotranspiration (PE) which is the value that would be obtained from a vegetation covered surface if moisture remained in adequate supply. There are conflicting views regarding potential evapotranspiration and much experimental evidence has been brought forward that it does not vary with climate, plant and soil type, although it appears that the effect of crop morphology on evapotranspiration rates are not large and that meteorological factors play a dominant role where soil moisture is not limiting. Potential evapotranspiration has been determined from mean temperature, day length (or latitude), the latter being a measure of radiation or albedo. Knowing PE and precipitation, it is simple to work out geographically and statistically the moisture available for recharging the soil, month by month, the amount that will be withdrawn so that each month will show a surplus (s) or deficiency (d) of water needed for evaporation. Annual values for s, d and PE can be obtained as totals of the monthly figures and a moisture index calculated from an empirical formula (Thornthwaite *et al* 1955, 1957)

In tropical regions of the country it is evident that length of the dry season is very influential in detraining the vegetation and this factor has received the attention of Prof. Gaussen in 1957 (Bagnouls F and Gausseth H. 1957). The effect of temperature is allowed for by varying the limiting rainfall for a month to be counted as dry by mean monthly temperature, e.g. less than 75 mm when mean monthly temperature is about 30 °C and less than 50 mm for mean monthly temperature of 20–30 °C. The values of xerothermic index are obtained by drawing ombrothermic curves, i.e. the monthly values of Temperature (T)

and Precipitation (P) together using a rainfall scale in mm twice that of Temperature in °C. Thus, in a wet climate, P curve is entirely above the T and in a very dry climate the converse will hold good. In most Indian climates the curves intersect in two or more places indicating periods of excess or deficit. However, for drawing the ecological subdivisions of India, (map after Legris, 1963), the following values have been used.

Temperature $t_5 = m: \geq 20\ °C; t_{4/5} = 15 < m \leq 20\ °C; t_4 = 10 < m < 15\ °C.$

$t_{3/4} = 5 \leq m\ 10\ °C; t_3 = 0 < m \leq 5\ °C\ t_{2/3} = -5 \leq m\ 10\ °C$

$T_2 = M > -5\ M \geq +10\ °C\ t_1 = M \leq +10\ °C$

Rainfall (Precipitation) (P) : $S_1 = P \geq 2000$ mm; $S_2 = 1500 \leq P < 2000$ mm.

$S_3 = 1000 \leq P < 1500$ mm $S_4 = 500 \leq P < 1000$ mm $S_5 = P \leq 100$ mm

Xerothermic index : $X_1 = 1$–40 dry days; $X_2 = 40$–100 dry days; $X_3 = 100$–160 days; $X_4 = 160$–230 days; $X_5 = 230$–280 days; X_6 = more than 280 days.

m mean temperature of coldest months M = Mean temperature of hottest month. P = Total annual rainfall, X = Xerothermic index.

Effect of Sea and the Mountains

Proximity to the sea or big lakes both affect the climate as a moderator of temperature and as humidifier of the atmosphere, thus affecting the vegetation to a considerable extent. This is allowed partially in some the classifications of climate. Such an effect may also be masked by other factors such as or geographical rainfall in the elevated landmass of westernghats which cause much heavier rainfall near the foothills than on the plains, in contrast to a very low rainfall over flat delta of Indus is very striking. A coastal situation again influences the wind factor of the climate.

The mountain masses generally generate their own climate such as the high hills of South India and the Himalaya. Mountainous country with its deep gradients generally emphasises the climatic effects of aspect, such as incidental radiation including light and moisture relations, particularly in the soil and exposure to wind-all in addition to effects of altitude on temperature. The general montane climates are more moist than those of adjoining lands. The higher hills of peninsular India are of special interest from the climato-vegetation viewpoint, though the vegetation has generally been affected by human activities, all tending to render moisture conditions less favourable. The considerable difference in conditions (and vegetation) between an isolated hill with maximum exposure to mostly adverse influences, and an extensive landmass of the same altitude is noteworthy.

The Himalayan mountain range is so vast that it plays a great part in regulating the climate over most of the country. Forming, as it does, an effective barrier to moisture-bearing winds from the oceans to the south, it also conditions a dry montane climate in the inner ranges with marked effect on the vegetation. The valley of Kashmir screened by the Pir Panjal range has a climate contrasting with that of the south side of the range though with a very limited flora, the forests are less different than might be expected.

Incident Radiation

The incident light intensity and with it the temperature are influenced by the aspect and the gradient of the micro-relief, just as on the major topography. Full solar radi-

ation reaches the top canopy but diminishes as it penetrates directly or after reflection down to successively lower strata and ultimately to the soil, where part will be absorbed as heat, reacting on roots activity. It has been demonstrated that the total absorption by a full cover of vegetation can vary but little, provided adequate soil moisture is available. When this is not the case, the difference may be reflected by a wide difference in vegetation types, i.e. grassland instead of forest. The characteristic development of epiphytes and their vertical distribution are determined jointly by light and humidity. It appears probable that reduction of light intensity by cloud belts plays a significant role in the process of stunting and low growth rates of vegetation. In the tropical wet-evergreen forests, light intensity at ground level has been found to be less than one percent of that at the top of the canopy, with moving sunflacks during the middle hours of the day of varying intensity up to one hundred percent (Richards, 1952). Satisfactory methods for measuring light intensity are not easy to devise and among the various instruments photoelectric cell alone meets most of the requirements.

Temperature records at different heights above the ground and in the soil provide a measure of the magnitude of differences under different canopy (Krishnaswamy *et al* 1957). A dense plant cover reduces radiation losses of heat at night, but at the same time reduces heat received and stored during daytime installation so that the net effect of the upper canopy will largely depend on other site factors. Puri recorded that soil temperature at all depths under Sal (*Shorea robusta*) forest was higher in the cold weather than under the leafless teak or in the open (Gupta, *et al* 2000).

Winds

The effect of winds on the distribution of vegetation, as such is limited in contrast to its effect on cloudiness, mist and rain. On exposed sites it is having a stunting effect which may be marked on the hill tops and on ridge crest forests. Height is significantly reduced under crowns to be relatively low, branchy and dense. Comparable effects are seen on the leaves which tend to be relatively small and tough. Along the coast, the Mangrove forests do not show such effects but the littoral vegetation which is exposed not only to strong winds but also to flying sand grains and salty spray shows a marked reaction both in the poverty of flora and its morphology. Drying winds of arid regions show effect on the vegetation much more clearly marked and this effect is mainly exerted through moisture relations.

On croplands and drug plantations the effect of wind can be considerably reduced by erecting shelter belts which improved considerably the yields from the cultivated land. However, in sand stabilisation programmes on the coast and inland, in arid regions, the effect of shelter plantations also reduces the flow of sand on the land and on the plantations.

The Topography

In a flat country the factor most effective is the minor differences of levels determining the movement of water, both over and through the soil. Even a difference of a few cm or metres or less may influence the moisture conditions. This feature, though often complicated by other physical differences, is well marked on alluvial plains with terraces of different levels. In the degradation of a plain, erosion often creates a plateau, slope and bottom between the three

sites, enough to cause difference in vegetation types with a respective pattern or catena.

In the hilly country, the factors of topography exerting most influence on vegetation are the aspect and gradient. They are reflected in the variations caused in the altitudinal limits of species and vegetation types, the latter appearing to express the integration of many climatic variables.

The influence of aspect is much simpler within the tropics than in higher latitudes, being largely limited to its relation to wind directions, particularly moisture-bearing winds. Thus, a southwest aspect facing the monsoon winds has a much higher and better distributed rainfall than the opposite northeast aspect, which is in the rain-shadow. The situation is accentuated where a whole range of mountains, such as the westernghats or the Himalaya, lie more or less across the wind, particularly as in the former example it makes the edge of the plateau. In this case the drop in rainfall on the lee side is typically very pronounced and no longer local, as typified by the entire central Indian plateau.

Other major effects of variation in aspect are related to the consequent variation in insolation per unit surface. Southern aspects receive more solar radiation than northern in proportion to their latitudes and gradient. Westerly and southwesterly aspects receive maximum amount in the afternoon when the cooling effects of night conditions have been lost, in contrast to the easterly and southeasterly aspects where insolation is at the maximum in the morning. In higher latitudes, steepness of slope cancels out the effects of lower altitude of the sun on a south aspect but accentuates it on a north aspect. All these differences affect the temperature in the first instance but even more influential on the vegetation is the effect on soil moisture. On steep slopes, the soil is shallow and quickly dries out so that almost universally the vegetation type carried is of drier types, e.g. deciduous instead of wet evergreen in the westernghats and in extreme cases not even a tree type at all, except scrub and/or grass.

The vegetation in the altitudinal range of a forest type on the various aspects is determined by a complex of factors including not only insolation, temperature and gradient with their effects of moisture, but by geology, stratigraphy, soil and adjacent topography, so that there is no rule by which a precise figure can be given for the upper limit of the vegetation type on a north aspect, the limit of the south aspect being known. However, the difference in the Himalaya is of the order of 350 m on an approximately SW/NE axis.

The Soils

As is well known that climate affects the weathering of the rocks and formation of the soil, though the process is very slow but given geological time, the soil should become 'mature revealing this uniformity and stability in equilibrium with the climate—a soil complex comparable with the vegetation climax. As with the climate, many classifications have been devised combining genetic relationships with ease and convenience in field use. As with the vegetation the boundaries are rarely very sharp and for practical application it is difficult to avoid using mixed criteria. It may also be pointed out that most of the studies on soils in India are with an agricultural bias though in recent years some progress has been made in studies of forest soils. A large-scale soil map has been published in 1954 showing

distribution of twenty main soil groups, while the latest classification in 1962 differentiated 27 broad soil classes.

Historically, Wadia et al in (1935) compiled a Soil map of India based on geological formations which was further improved in 1945 in which he classified soils as red soils, black soils (regurs), laterite and lateritic soils of peninsular India, Delta soils, desert soils, bhabar and terai soils and alkali soils of the Indo-gangetic plains. Viswanath and Ukil in 1944 published a soil map by placing the soils of India into different climatic zones on the basis of N.S. quotients. Raychaudhry and Mathur in 1954 divided soils into 16 major soil regions by integrating the effects of climate, vegetation and topography on soil formation. Raychaudry et al in 1963 have divided soils into 27 categories and Govinda Ranjan and Dutta Biswas in 1968 divided the soils of India into 23 major groups on the Map prepared for FAO/UNESCO World Soil Map Project. Govinda Rajan in 1971 further revised the Soil Map of India (scale 1:7 million) and distinguished 25 broad soil classes to which he gave 7th approximation classification equivalents to group of order level depending upon the information available on the soil classes.

The above-mentioned soil maps indicated broad soil classes based mainly on variations in climate and vegetation. As the soil survey work progressed, soil variations related to relief or physiography in different climatic environs were recognised and mapped. The concept of soil series was adopted for mapping and sixty-four soil series were thus identified occupying key interpretive positions in soil classification framework. The series so selected were from the work on survey of soils. Murty and Pandey however, prepared another map in 1983, taking into account the soil-physiography

relationship based on soils as a function of climate, vegetation, relief, parent material and age or time. Mutually, exclusively landforms units consistent with the scale were separated on 1: 2, 50,000 topographical map.

The following legend was developed and used.

I. Siwalik and Himalaya.
    1. Mountain ranges
    2. Hilltops and hill ranges
    3. Dissected hill ranges
    4. Moderately sloping pediments with scattered hills (upper)
    5. Moderately sloping pediments (lower)
    6. valleys.

II. Indo-Gangetic plains
    1. Level active flood plains.
    2. Nearly level recent flood plains.
    3. Old meander plains.
    4. Nearly level old flood plains.
    5. Low-lying old flood plains.
    6. Delata plains.
    7. Swamps and marshes.

III. Desert region
    1. Hilltops and hill slopes.
    2. Pedimonts with scattered hills.
    3. Moderately sloping to level pediments.
    4. Sandy plains and sand-dunes.
    5. Playas.

IV. Peninsular India including coastal region.
    1. Hill ranges.
    2. Highly dissected hilly terrain.
    3. Denuded ridge.
    4. Dissected rolling lands.
    5. Upper pediments.
    6. Lower pediments.
    7. Pediments.
    8. Mesas and buttes.

9. Upper plateau.
10. Lower plateau.
11. Valley bottom and narrow valley.
12. Broad valley.
13. Valleys.
14. Alluvial plains and terraces.
15. Flood plains and delta plains.
16. Coastal plains.

The broad soil orders identified are as follows.

Alfisols: Soils of medium to high base status and subsurface horizons of clay accumulation. Fomative element-alf.

Aridisols: Usually dry soils with pedogenic horizons and very low in organic matter. Formative element-id.

Entisols: Soils without pedogenic horizons that have retained original structure of parent materials. Formative element-ent.

Inceptisols: Soils with weakly developed horizons showing alteration of parent materials. Formative element-ept.

Mollisols: Soils with nearly black organic-rich surface horizon and high base status. Formative element-oil.

Ultisols: Soils with red colour, low base status and subsurface horizons of clay accumulation. Formative element-ult.

Vertisols: Soils with high clays of cracking nature and dark grey colours. Formative element-crt.

By far the greatest areas are occupied by the red soils (totally in the east and south), black (west and centre), alluvial (north) and hill (north) types. The alluvial soils predominate the Indo-Gangetic plains, across the north of the country and are often low in nitrogen and phosphorus, though less so in the forests, where they have been denuded. Potash is usually adequate as also lime which is liable to form a Kankar pan, especially over bands of clay in the older and higher alluvium. These alluvial soils are usually deep, but may show marked stratification of sands, clay and loam and the vegetation they carry is greatly influenced by moisture and soil aeration.

The *coastal sands* are met with at places all round the coast and carry characteristic vegetation. They are mainly quartzite or calcareous sands with low silt and clay fraction. The depth is variable and the water table is characteristically high. The nutrient status is typically low, especially as regards nitrogen and phosphorus.

Saline and *sodic soils* are mainly developed in the drier parts of the country (in the northwest) and over the Deccan plateaux. These soils are generally unfavourable for tree growth and a few species develop under the adverse conditions which are often further accentuated by inadequate drainage conditions. A special class of such soils occurs in the extensive saline flats bordering the Rann of Kutch.

The *arid* and *desert* soils are typical of the western part of the country in the states of Rajasthan and Gujarat. They are often alkaline, usually with adequate mineral status, but with low organic matter and nutrient availability. Typically they are structureless. Climate as well as soil being unfavourable, the vegetation is very poor consisting of dry deciduous forest, at best or open woodland or scrubs.

The *black soils* are mainly developed from basaltic rocks and so predominate over most of the Deccan plateau. They are usually montmorillonitic which crack widely and deeply in the dry season. They are usually neutral to alkaline in reaction, poor in nitrogen and phosphorus but contain adequate potash and lime; in fact concretionary lime is usually present. The dark colour is due to the mineral constituents,

humus content being typically low. Like the lateritic soils they have a marked influence on the vegetation. Teak is the most common tree on such soils and where they are deep, either dry or broadly drained, or highly calcareous inferior non-teak bearing dry deciduous forests and sometimes thorn forest (Acacia species) replace the mixed teak forest.

The *red soils* are mainly associated with the acid granites, gneisses and schists and are feralitic in nature. They are typically coarse, often with ferric-concretions, with a neutral or acid reaction, and poor in organic matter and mineral nutrients. The clay fraction is mainly kaolinitic and they tend to carry the moist-evergreen and mixed-deciduous forest with Shorea robusta (Sal).

The *lateritic soils* are characteristic of the tropics with monsoon rainfall and tend to cap the hills in most part of the peninsula; they develop mainly on the more basic rocks. They are usually acidic, poor in mineral nutrients and have a low silica sesquioxides ratio. In the extreme case, the soil hardens on exposure or forms the familiar vesicular laterite used for building and forming extensive 'pavements' which are very infertile and slow to breakdown. These pronounced physical and chemical properties exercise a marked effect on the vegetation. Thus, Teak is not often to be found but Sal (*Shorea robusta*) is more tolerant and *Xylia* is more characteristic. *Anacardium occidentale* is also markedly tolerant. The influence of predominance of laterite and lateritic soil is not only on the distribution of individual species but affects the vegetation as a whole.

The *foothill soils* have not been investigated in much details. Brown earths are usual, slightly acidic or neutral, with high base status.

The *montane soils* of the Himalaya are derived from a very wide range of parent materials and differ fundamentally from most of those of the peninsula by being more or less immature. They show many of the characteristic of the well-known types of the north temperate zone, though pronounced podosols profiles are rarely found. Brown forest soils predominates, mostly with adequate mineral nutrients and humus for plant growth, which is accordingly regulated more by climate than by soil. The soils are usually acidic in reaction. Addition of raw humus in some of the forests is an hinderance to the regeneration of few dominant species.

The *skeletal soils* are of limited occurrence mainly on high land tops in the drier regions of the peninsula and in the Himalaya. They support markedly xerophytic or degarded communities of vegetation but are not of great importance except locally.

Peaty soils are rarely met within the country, except in local depressions in the wet alluvium of Brahamputra valley and in Kerala where they support a special type of woody vegetation or reeds.

Effect of geology and soil on the vegetation, particularly of the tree and shrub species has been studied by a number of workers and it seems that the physical condition of the soil, above all those affecting moisture supply, root aeration and availability of chemical nutrients are much more important on plant species with limited exceptions than the chemical composition as such. Many types such as the tropical rain forest, chirpine forest, and dry deciduous forest extend across geological boundaries, often involving considerable change of rocks with hardly any discernible alteration in the appearance or composition of the vegetation, though examples can be quoted of the abrupt change but they are exceptional. However, Mohan and Puri (1956, 1957) and Gupta

(1960, 1964) have emphasised the differences in forest types in the Himalaya on dip and scarp slopes which is complicated by the usually different aspect and gradient prevailing on the two sites. Depth and soil moisture are again important.

The lime requirement of the plant species varies and much has been written with respect to forests such as distribution of teak in the mixed deciduous forest types (Griffith and Gupta, 1947). The part played by laterite in influencing the forest type distribution has also attracted the attention of scientists and is well known that teak generally avoids it but sal atleast tolerates it. Of the associated species of *Xylia* and *Cleistanthus* are prevalent on laterite but many others are absent. Physical conditions are liable to be highly unfavourable so that the forest is restricted in its development, even where it remains basically the same as on the adjoining red or black soils. The inter-relationship between the soil of the teak forests (Seth and Yadav, 1959), the tropical evergreen forests (Seth and Yadav 1960) and of the dry zone forests (Seth and Yadav, 1960) have been extensively studied and various profiles have been provided by many scientists.

How far the soil is affected by the vegetation not much information is still available. If the soil improves in depth, physical condition and mineral content in the rooting zone there will be likelihood that a vegetation type with greater demands on the soil and of greater competitive power will displace the original type. That, this can happen is well illustrated by changes occurring on the stabilisation of moving sand or on new riverain gravel bed. One view is that this process should continue until ultimately (provided the soil itself becomes mature), the climate, soil and vegetation are all in equilibrium, the latter in a 'climax condition'. Another view is that such an equilibrium is not attainable, but that cyclic changes will occur. Instances are available suggesting that within a general type there may be minor units disturbed irregularly each succeeding and being succeeded by a different unit. Since the distinction among these vegetation units is ultimately arbitrary and gradual transition is common, it would seem better to envisage a cyclic continuum rather than a mosaic of discrete units.

A method of study of the relationship between plant and soil as regards mineral nutrients that is being rapidly followed is that of foliar analysis, not that restricted to leaves only. The amount of nutrients that have actually been taken up by the plant are assumed to reflect both, what the plant actually requires and what is available to the roots in the soil (regardless of the gross content of different elements partly in non-available form). Selective absorption certainly takes place and where a given nutrient is in small supply, deficiencies may result affecting both species and type. This is particularly important in respect to the occurrence of medicinal plants. The method is a promising one but the greatest care is required in collecting the samples for analysis and for interpretation. Soil moisture is another factor which varies considerably from place to place within general climate and vegetation type. These variations are largely correlated with those in light and temperature. Micro-relief is equally important in determining the movement of moisture through the soil. These effects are more marked where moisture at any time of the year is in limited supply and the ground vegetation (containing mainly herbal medicinal plants) is liable to be more affected than the higher strata which are more widely and deeply rooted. In the wetter climate the same causes may result in local excess of water continuously or periodically.

References

Bagnouls F. et Gaussen H. 1957. Les Climats Bioclimatique et leur classification. *Annals de Geographique.* 355: 193–220.

Bharucha FR and Shanbagh Gy. 1957. Precipitation effectiveness in relation to vegetation of India. Pakistan and Burma. Memoire No 3. University of Bombay.

Boyko H. 1935. The Designation of Plant Indicators. *Applied Ecology and Geography.* Vol. I.pt.2.

Central Soil Salinity Research Institute, Karnal. 1975. *Salt Affected Soils of India.* Map. First Edition.

Chopra TC, Sobti SN and Handa KL. 1962. Cultivation of Medicinal Plants in Jammu and Kashmir. ICAR Research Series No. 13. pp 76.

Daubenmire RF. 1947. Plants and Environment. John Wiley and Sons. N. Tork.

De Martonne E. 1926. Une Novelle founction climatologique l'indicede Aridite. *La Meteorologie.* 68: 449–458.

Diction BT. 1952. Plant Introduction. Proc. International Symposuum on desert research. May 1952. Israel.

Duv Devani 1957. Desert Research for Arid Agroculture. Discovery **18(8).** 330–334.

Eig A. 1935. Plants and Plant Communities as indicators. *Hassadesh. Telaviv.* Vol 15: 1–8.

Gaussen H. 1954. Expression du millieu par des formules écologique. Leur representation cartographique. Les Division écologique d¢ monde. Collq. *International du CNRS.* Paris.

Govindrajan SV. 1971. Soil Map of India. *Review of Soil Research in India.* eds. JS Kanwar and SP Raychaudhri.

Griffith AL and Gupta RS. 1947. Soil in relation to Teak, with special reference to laterisation. *Indian For. Bull. No* 141.

Gupta R.K. Dabral BG, Mehe-homji VM and Puri GS. 2000. *Forest Ecology,* Vol 3 Enviroment, Forest and Rainfall. Oxford and IBH Publishing Co. New Delhi. pp. 458.

Gupta R.K. 1960. Studies on the Vegetation of NW Himalaya; Ph D. thesis. Enviroment, Forests and Rainfall and IBH Publing Co. New Delhi pp. 458.

Gupta R.K. 1964. Forest Types of Garhwal Himalaya in relation to edaphic and geological formations *J. Soc. of Indian For.* 4: 147–160.

Hedberg Olov. 1964. Features of Afro-alpine Plant Ecology. *Uppsala.* pp 114.

Hill SA. 1876. Meteorology of Northwest Himalaya. *Indian Meteor. Memoirs* Vol 1. Pt. 6. pp 378–419.

Krishnaswamy VS, Dabral SN and Nath P. 1957. Forest Influences. Study of Soil, temperature and Humidity in the New Forest, FRI. *Indian Forester.* 83: 416–450.

Legris P. 1963. Itude Ecologique de la zone, mediterranean (map) Carte bioclimatique de la zone meditarranean par Emberger, Gaussen, Kassas, de phillipe. Notice Explicative et 2 feuille de cartea 1;500,000 et 1;10,000, 000. Paris.

Legris P. 1963. La Végètation de l'Inde. Ecologie et Flore. Trav.de la Sec. Sci. et Tech. 6: pp 554.

Meherhomji VM. 1963. Les Bioclimats du subcontinent Indian et leurs analogues dans le monde, Doctorate thesis. pp 253.

Mohan NP and Puri GS. 1956. Succession of forest communities in Chirpine (Pinus roxburghii) forests of Punjab and HP. Indian For. 83: 351–364.

Mohan NP and Puri GS. 1957. The Succession of Forest Vommunities in oak-conifer forests of Bashahar Himalaya. *Indian For. Rec.* (ns) Silvic. 10(2): 19–36.

Murthy RS and Pandey S. 1963. Soil Map of India (sub-order association) Scale 1:6, 300,000. National Bureau of Soil Survey and Landuse Planning ICAR. Nagpur.

Murthy RS and Pandey S. 1978. Soil and Landuses in the Himalayan Region. National Bureau of Soil Survey and Landuse Planning New Delhi. National Seminar of Resourses

Development and Environment in the Himalaya region. pp 9.

Murthy RS, Shankarnarayana HS and Hirekerur LR. 1976. Distribution, genesis and classification of Acid soils of India. *Bull. Indian Woc of Soil Sc.* New Delhi.

Puri GS. 1954. Soil Climate of some Indian forests. *J. Indian bot. Soc.* 33: 394–416.

Prarmanik SK and Jagannathan P. 1953. Climatic changes in Indian. *Rainfall Indian Journ. Meteor. Geophys.* 4: 291–309.

Raychaudhri Sp, Agarwal RP, Duta Biswas NR, Chopra SP and Thomas PK. 1963. *Soils of India.* ICAR New Delhi.

Raychaudhri SP and Mathur LM. 1954. Bull. *Nat. Inst. Sci.* New Delhi.

Richards, PW. 1952. The tropical Rain Forest-An Ecological Study Cambridge University Press pp 152.

Seth SK and Bhatnagar HP. 1959. Characteristic of the soil suitable for Sal natural regeneration. *Indian For.* 85: 631–640.

Seth SK and Yadav JSP. 1960. Soils of tropical moist evergreen forests of India. Tropical Moist evergreen Forest Symp. (Mimeo) pp 14.

Thornthwaite CW 1933. The Climate of the Earth. *The Geographical Review* 23(3): 433–440.

Thornthwaite CW. 1948. An approach towards a rational classification of climate. Geogr. Review 21: 633–655.

Thornthwaite CW and Mather JR. 1955. *The water budget and its use in Irrigation* "Water". The year book of Agriculture. 1955. USA

Thornthwaite CW and Mather JR. 1957. Instructions and tables for computing potential evapotranspiration and water balance. Centreton NJ, USA.

Vaish, US and Krishna Gaurav. 2004. National Acad Sci. 74th Annual session. Dec, 2004. Sec. B. *Biological Science Abstract of Papers* p. 54.

Voelker JA. 1893. Improvment of Indian Agriculture quoted by Vishwanath. UKIL. 1944.

Wadia DN, Krishnan MS and Mukerjee PN. 1935. Introductory note on the geological formation of soils of India. *Rec. geol. surv. India.* 4. p2 63.

Yadav DL and Thakur PC. 1972. Soils of Himachal Pradesh. Soils of India. Fertilizer Assn. of India. New Delhi.

❏❏❏

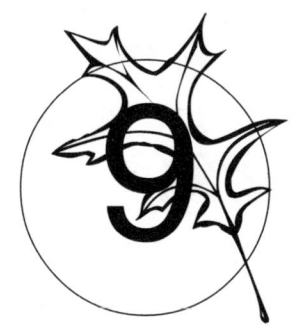

Cultivation and Multiplication of Drug and Aromatic Oil Plants

9

Introduction

Though at present, India is not sufficient in pharmaceutical products, drugs are being imported, as much of the biodiversity in drug plants is being exported at the cost of their survival and existence. In order to meet the public demand for drugs and protect our valuable biodiversity from extinction, it is necessary that utmost attention is given to the production of raw materials for the industry, both for local consumption in the country and for export. The chief source of crude drugs has been primarily through the collection of medicinal plants growing scattered in the wild, in forests and in fields. The process of collection, besides being laborious, is very expensive and consequently manufactured drugs and medicines are highly expensive. The sparse distribution, difficulty of access, cost of transport necessary to reach the natural habitat of the drug plants; indiscriminate collection leading to extermination of the species, ignorance of the correct identification of the genuine plants leading to admixture of the genuine with the spurious and other related (alternate) plants are some of the difficulties faced in the collection of drug plants. This results in sub-standard material and difficulties in meeting the export potential targets. Transport is one of the greatest handicaps in the exploitation of drug plants, since with storage it losses

quality and has, thus, prevented desired exploitation of such fine quality drug plants as *Ephedra, Artemisia, Podophyllum, Psysochlaina, Aconitum, Picrorhiza, Valeriana,* etc. which grow at very high altitudes in far off places in the alpine, subalpine and desertic high altitudes of more than 4000 m.

Excessive extraction from only one area has led to deterioration of the yield and plant quality and may sometimes bring total annihilation and extinction of the particular species. Quite a number of medicinal plants have thus been affected during the past; *Belladona* presents a typical example. In the beginning, tons of belladona were collected annually from the forests of Kashmir, but excessive extraction gradually reduced the yield with the result that this source can no longer be relied upon, to meet the growing demand of the drug industry.

Based on the recommendations of the Pharmaceutical Enquiry Committee in 1955, immediate steps were taken to organise the cultivation of medicinal plants in a scientific manner and sponsor agencies for their proper collection, storage and marketing. This has resulted in improved methods of cultivation of certain medicinal plants and a better percentage of active principles from them. Subsequently, Central Medicinal Plants Organisation (CIMPO)

was created and extensive efforts were made for the cultivation of certain species of drug plants which ultimately led to the establishment of Central Institute for Medicinal and Aromatic Plants (CIMAP) located at Lucknow. Indian Council of Agricultural Research (ICAR), also established a coordinated programme on the medicinal plants with its headquarters at National Bureau of Plant Genetic Resources (NBPGR), Delhi and took positive steps for the cultivation of several drug and aromatic plants in the country.

Though, cultivation of medicinal and aromatic plants attracted the attention of a number of workers, yet much remains to be done regarding their adaptability to various soil and climatic conditions, as discussed in the previous chapters, before sufficient data are available for implementing the cultivation of medicinal plants on a large scale in various phytoclimatic and agro-ecological zones of the country. It is also felt that the medicinal plants should be cultivated on scientific lines on a large scale, in regions suitable for different ecoclimatic and ecofloristic zones. In order to obtain optimum yields and make the country self-sufficient and ready to export quality standardised material in order to gain foreign exchange. Though, a beginning was made by establishing centres for research at Kashmir, HP, Uttaranchal, Assam, Chennai and Bengal, experimental farms were established since 1958 under the supervision of CSIR. Manufacturing pharmaceutical concerns in almost all countries of Europe, USA and the Russia took large initiatives since 1903 when plant introduction service was organised in USA. Some of these organisations in the United States of America also publish statistical information regarding principal markets for such products nearest to the centres of production. It also helped in the procurement of seeds to organisations like the NBPGR (National Bureau of Plant Genetic Resources) New Delhi and other centres concerning with acclimatization of new crops and of their economics and industrial importance.

Cultivation of medicinal plants in the West Himalayan region started in the fifties at three places: Kashmir, Himachal, hilly areas of the then Punjab and UP now including Uttaranchal and H. Pradesh. The inner ranges of Western Himalaya in Kashmir, remains practically dry during the rainy season, thus experiencing a Mediterranean type of climate as compared to the Eastern regions and thus also has a favourable harvesting time. In the Eastern part the heavy monsoon downpour creates problem in harvesting, drying and storage of bulk crops. Nurseries in the Western Himalaya were established at different altitudes starting from 300 m onwards for subtropical, temperate and alpine plants. Propagation of temperate species were conducted at altitudes of over 1500 m above the sea level and in the dry alpine regions. It is of interest to know that more than 50 percent of the plants used in British Pharmacopoeia grow in this region. The chief aims of these nurseries and experimental stations are as follows:

1. Cultivation of drug plants under various altitudinal, climatic and soil conditions and to determine where these can be best propagated.

2. Increase active principle content and the yield per acre (ha) by using suitable manures and fertilizers.

3. Introduce exotic medicinal plants and propagate these under suitable climatic and soil (ecological) conditions.

4. Study diseases caused by various pests, fungi, etc. and to devise measures for their control.

5. Produce good quality seeds through selection and hybridization, and

6. Study the optimum conditions for collection, curing and storage of various vegetable products used in medicine.

The experimental work carried out in the country, since then, indicated great scope for the cultivation of a number of species and varieties of medicinal plants. The number of exotic plants, successfully raised also showed that the active principle content comes up to the required standard and in some cases even better than in their own country. For example, by varying the time of collection of Belladona and Hyoscyamus leaves, i.e. when their photosynthetic activity is the highest, the alkaloidal contents have been found to vary from 0.4 to 0.8 percent respectively. Similarly, marked variations of active principle have been recorded in the case of *Pyrethrum* and *Digitalis*. *Physochlaina praealata*, which grows in the cold desert region of Ladakh, contains over 1 percent of hyoscyamine and can be successfully cultivated in the Kashmir valley. It showed 0.2 percent of alkaloid content in the first year of growth and 0.6 percent in the second year. Experimental work was initially carried out on a number of plants like *Atropa belladona, Hyoscyamus niger, Mentha piperita, M.pulegium, M.arvensis, Chrysanthemeum cinerariaefolium, Digitalis purpura, D.lanata Chenopodium ambrosoides* and *Glycerrhiza glabra*, etc.

The drug farms and the nurseries established by ICAR, CSIR and Agriculture Universities, etc. have also induced the local peasantry, through the efforts of various non-governmental organisations, to take up cultivation of drug plants in their holdings, where food crops are not economical. Thus, the cultivation of drug plants has been established as a small-scale and cottage industry in limited pockets in the regions of Western Himalaya and other places of UP, Madhya Pradesh, etc. The technical advice, seeds and other planting materials of improved nature, need to be supplied for the development of private farms which will go a long way to meet the local and foreign requirements of drug industry. Though cultivation of drug plants is a fascinating subject by employment of techniques such as grafting, selection, hybridization, induction of polyploidy, by colchicine and other hormonal treatments, much can be done to improve the quality of the crop.

In the foregoing pages efforts have been made, based on the data available to the author, to review the work which is just a beginning and a small effort on the subject which needs constant updating. Ecogeographic and climatic diversity in India harbours a rich wealth of medicinal plants. A recent survey of NBRI (National Botanical Research Institute, Lucknow, revealed that about 600 plant species are being extensively used in different medical formulations by almost 160 pharmaceutical concerns in India. Tribals constitute the original source about the medicinal value of most of the plants.

Fruits, berries, roots, rhizomes and leaves of more than 100 wild plants are used as food by the tribals. *Madhuca indica* is the main source of oil and beverages of the tribals. Selection of some plants seems to have been made merely on the basis of "law of signatures", e.g. *Aristolochia indica* is used in snakebite, because the flowers are like the hood of a snake. Central Drug Research Institute, Lucknow (CDRI), undertook the task of scientifically evaluating the pharmacological properties of wild plants (Dhar et al 1968, 1969). The Central Council of Research in Ayurveda, National Institute of Ayurveda, has in hand many projects. Recent Research findings disclosed

that *Pippali* (*Piper longum*), *Amalaki* (*Emblica officinalis*) and *Ashwagandha* (*Withania somnifera*), possess anabolic and immunity-enhancing powers. *Abutilon indicum* (*Atibala*) proved to be the best drug in augmenting the antibody production. *Withania somnifera*, in a clinical study, provided significant relief from anxiety, neurosis, besides quantitative reduction in anxiety level and neuroticism. Significant increase in haemoglobin level and RBC count was also reported. The drug also possessed immune modelling qualities without any side effect. In another study, *Centella asiatica* (*Mandukparni*), administered to human volunteers showed remarkable increase in the serum level.

A study on "Rasayan" effect on normal adults showed that administration of *Punarniva* (*Boerhaavia diffusa*) and *Mandukparni* enhance haemoglobin level and decrease blood urea and acid phosphates. The two compounds Baccocide A and Baccocide B were obtained from Brahmi (*Bacopa moniera*) plant, both of which were found to be equally effective in improving memory enchancer. They are non-toxic. A drug 'Memory plus' was released by CDRI, Lucknow as memory enhancer. Latest research findings on camphor (Vern. *kapoor*), indicates that it has potential use in radiotherapy for cancers; it produces normal cells from the damaged effects of radiation but makes the cancerous cells more vulnerable to ionising rays. The protective effect of camphor is believed to be similar to that of caffeine, which has been demonstrated to protect cells against radiation-induced damage. Various plants have also now been propagated through tissue culture techniques, to evolve superior germplasm of the plant (such as *Chlorophytum*), known for its aphrodisiac properties and is fast becoming extinct. Since, the seed

germination is poor, the technique could be used to produce a large number of 'dropagules' round the year, which could be planted in the field and under greenhouse conditions.

A lactonic glycocide—a good adaptogen which increases the resistance of human-beings as well as the animals against strain and tension was isolated from the plant called *Raktkanér* (*Nerium indicum*).

Antibacterial activity of volatile fractions in the form of essential oils, isolated from the seeds of *Carum carvi*, *C. roxburghianum*, *Cuminum cyminum*, *Foeniculum vulgare*, *Pimpinella anisum* and *Trachyspermum ammi*; has been reported against gram-negative bacteria, viz. *Enterobacter faecatus* and *Escherichia coli*, as well as gram-positive bacteria, *Bacillus cereus*. Effects of plant extract from *Carica papaya*, *Musa paradisiaca*, *Eucalyptus terticornis* and *Parthenium* spp. (Congress grass) have also been recorded.

Medicinal Plants Cultivation and the Sustainable Development

Recent studies (Silori and Badola, 2000) showed that in a village in the buffer zone of "Nanda Devi Biosphere Reserve", 5 villages in the valley at 3000 m are inhabited by *Bhutia* community, subgroup Johari, practising seasonal migration, moving to lower altitudes outside the reserve during winter season (November-April). Such migrations are a common feature throughout the Himalayan region from Kashmir to Kumaon where tribals during the winter months migrate to lower altitudes along with their animals. In general, the number of migrant families have declined for various reasons in the Himalayan region (Prasad, 1989), because of restrictions on cross-border trade. In general, the migrants are older leaving their children and younger in contrast of elderly

in the lower altitudes, where they go to plains in search of jobs. The lack of alternative employment opportunities force the people to migrate to the lower plains.

At high altitude *Bhotiyas* have been known to cultivate Jambu (*Allium wallichii, A. carolianum*) and some other plants have been traded throughout the Himalayan region (Atkinson, 1882). These are propagated through bulbs and their leaves are harvested twice (June-July and Sept-October), though commercial cultivation is a recent phenomenon. Most of the people cultivate on their agricultural fields, on fallow land and the kitchen garden. In some cases, areas surrounding abandoned houses are also used. The villagers are reported (Silori and Badola, 2000) to cultivate a total of 12 species like *Allium wallichii, Saussurea costus, Nardostachya grandiflora, Picrorhiza kurroa, Aconitum heterophyllum, Angelica glauca, Carum carvi, Dactylorrhiza hatagarea, Pleurospermum angelicoides, Podophyllum hexandrum* and *Rheum australe* (Tables 9.1 and 9.2).

Stakeholders' perspective on commercial medicinal plant collection was recently been studied (Larsen and Smith, 2004, Aryal, 1993, Larsen, 2002, Olson and Larsen, 2003) from Nepal. Views of persons involved in commercial alpine medicinal and aromatic plants exploitation have been recorded through open-ended questionnaires. The issues explored are related to striking a balance between poverty alleviation and degradation of medicinal and aromatic plant resources. The main findings showed that plant collection give the financial benefits but majority of non-collecting residents believe that it adversely affects the environment and plant resources. However, stakeholders, other than staff of the Forest Department, are in favour of collection over conservation and find that collection bans are inefficient. The potential for addressing village poverty lies in changing the regulatory mechanism and handing over the resources for community management.

Drug Farms—their Establishment, Maintenance and Economics

The presence of active principle in a drug plant is of interest which should be in constant quantities established in taking repeat samples collected from different age groups and during different places at the same day, should be analysed since the high yields may be due to the locality, growth stage of the plant, season of collection and methods of drying and other factors. The reports published on research work taken on small areas may be misleading, because it may be easy to get higher yields from a small well-looked after experimental plot and the yields thus obtained when multiplied for a large acreage, give generally a rosy picture. Such reports may be misleading for a large drug plant farm raised for commercial exploitation in two ways. Firstly, the industry after a few initial frustrating experiments stops to take notice of such reports and secondly, the confusion and misunderstanding it creates in the mind of planners and policy-makers. If the yields in experimental plots are of erratic nature and have no relation to any of the conceivable factors one has to look for the answer in other fields like detailed taxonomy. Another reason, why a plant may not be of much commercial interest, may be because of the difficult or

(Misra AK, Singh Pooja and Tripathi NN, 2004. *National Acad. Sci.* Abstr. papers, Biological series. pp 65–66. *Nat. Acad Sci*. Abstr. Papers, Biology section. p.50).

Table 9.1: Harvest season, parts used and uses of some Alpine drug plants from the Himalayan region

Name	Harvest season	Part used	Uses
Aconitum hetrophyllum (Atis)	Sept-Oct	Root	Stomach and Intestinal disorders.
Allium carollianum (Pharan)	August	Leaves	Condiment and spices.
A.wallichi (Jambu)	Aug-Oct	Leaves	Condiment and spices.
Angelica glauca (Gandrayan)	October	Root	Whooping cough.
Carum carvi (KalaJira)	August	Seed	Condiment and oil.
Dactylorhiza hatagirea (Hatagari)	October	Root	Urinary disorders.
Nardostachys grandiflora (Mansi)	September	Root	Incense.
Picrorhiza kurroa (Katuki) Kutki	October	Root	Influenza and diorrhoea.
Pleurospermum angeloides (Choru)	October	Root	Fever and headaches
Podophyllum hexandrum (Ban.Kakri)	September	Fruit	Skin diseases
Rheum australe (Dolu)	August	Root	Wounds and cutaneous diseases.
Saussurea costus (Kuth)	August	Root	Gastric pains and disorders.

Table 9.2: Average production and projected earnings

Name	Area (ha) under production	Average Productivity Kg/ha ± SE	Market rate (Rs/kg)	Projected cost (Rs/ha)
1. *Aconitum heterophyllum*	(0.57)	—	250	—
2. *Allium carllianum*	(1.82)	79.3 ± 41.4	55	4,362
3. *A. wallichii* Baker	(2.31)	102.8 ± 36.4	55	5,654
4. *Angelica glauca* Edgew	(1.78)	593.3 ± 122.3	80	47,464
5. *Carum carvi.*	(2.14)	266.2 + 61.3	200	53,240
6. *Dactylocrhiza hatagirea*	(0.37)	144.3 ± 51.7	600	86,584
7. *Nardostachys grandiflora*	(0.15)	—	30	—
8. *Picrorhiza kurroa*	(0.64)	—	250	—
9. *Pleurospermum angeloides.*	(0.12)	.—	30	—
10. *Rheum australe*	(0.62)	—	30	—
11. *Saussurea costus*	(1.31)	190.8 ± 19.8	30	5724
12. *Podophyllum hexandrum*	(0.10)		40	

—Productivity, not recorded as crop was in the field (after Silori CS and Badola 2000)

scarce availability of plant material in required large quantity.

Before taking up large-scale cultivation of drug plants it is necessary to have test plots, in varying ecological situations, to get some indication on the suitability of the site for the pilot scale trials. Some plants may give better yields of the chemical when grown in regions outside their natural habitats. Pilot scale cultivation should be taken in 1–12 acres (about 1 ha) and the total active principle received per acre (ha) should be taken in consideration and not on the total crop. A study of the agronomic requirements in detail, where land is not available for pilot scale trials, should be made in detail. Plot scale cultivation in 10 to 12 acres and total active principle per acre, and not on the total crop is required.

Cost angle, method of propagation, field preparation, transplanting seedlings in the field where direct sowing is not feasible, actual farm maintenance, weeds, diseases, pests and their control, harvest and drying, packaging, storage and transport and cost and economics of large scale cultivation is the prerequisite. However, information on habitat preference for critically endangered species such as *jatapmansi* need he studied in detail (Air et al 2000). Preparation of the field, transplanting of seedlings in the field where direct sowing is not possible/feasible, actual maintenance of the farm, weed, disease, pests and their control, harvesting and drying, packing, storage and transportation and finally cost benefit ratio are the main steps.

Method of propagation: Plants can be propagated either from seed or from stem, root or leaf cutting. Seeds to be used for initial development should normally be from high yielding strains which have already been tested at research station. Too much stress of generic improvement and hybrid cannot be laid at this stage, though the work can be simultaneously undertaken. The life of a drug plant is normally 10–15 years and by the time a better breed is developed the drug may be already replaced by a better new plant or the active principle synthesised in the laboratory. Knowledge on the quality of seed should be used for each experiment and the question of viability and germination should be studied at this stage. Simultaneously, experiment should be carried out on the problem of the best method of storage of seed, if necessary, over a long period. In case where it may be necessary to maintain a nursery for commercial cultivation, the seed requirement should be worked out. Techniques for maintaining nursery and its cost can be studied at this stage. At the

development stage of the work, it is advisable to prepare a detailed proforma, of different operations, man hours involved for each operation and its cost. At the conclusion of experiment on different methods of propagation it should be possible to get best answer for the method of propagation, type and quantity of propagating material that would be required, weeds, pests and diseases and their control at nursery stage, ideal time for transplanting and efficient method of transfer to the field, etc. However, recent advancement of tissue culture techniques have made the task much easier for obtaining true to the type material of common drug plants.

Field planting: Depending upon the habit of the plant, its duration, rate of growth, habit and interculture operation needs, etc. the ideal spacing that may have to be adopted should be determined by trying out various statistical combinations. The spacing to be adopted still have to be correlated to yield per ha. and cost of crop. In climbers various types of support systems shall have to be studied for their suitability and duration. In short-duration crops, possibility of getting two or more crops per ha and per annum should be investigated.

Maintenance of drug farm: Detailed report, on the physical and chemical nature of the soil, shall help in determining ideal manure requirement when the experiment on various types of fertilizers are being undertaken. Different types of ploughing will have certain effect on the crop-growth. Virgin forest-land may give very high initial yield but if soil conservation measures are not taken from the very beginning the land may degenerate quickly. Irrigation and fertilizer requirements should be studied in relation to climatic conditions and nature of the soil. Problem of weeds, diseases, pest

and control need investigations. In case of wild plant taken for cultivation, careful and frequent observations on the occurrence of pests and diseases shall be vital to the success and new methods may frequently be necessary as the new pest control problems arise.

Harvesting and drying: Higher content of an active principle in a plant can be correlated to the stage of maturity of/in fruit, e.g. *Solanum khasiana*, age of crop as in tubers of *Dioscorea* or season of harvest or even weather condition on the day of harvest as in *Mentha*. In some cases, it may be cheaper to leave behind a crop like tuber of *Dioscorea*, instead of trying to collect every bit of it. Various methods of drying like sun shade, partial shade, artificial, etc. should be thoroughly compared. The way a crop is dried, i.e. as a whole, in pieces milled when fresh or after drying will also have some effect on the chemical yield. On *Dioscorea* it has been shown that a harvest after three years growth is feasible, after 4 years is economically slightly better while after four years the cost of crop increases because of the comparatively slow growth in relation to cost of maintenance of the plantation.

Packaging, storage and transport: The problem of proper packing and transport is as important as the cultivation. In most cases, the plants shall be harvested only during a short period of the year. While it may have to be processed during the course of the year. Ideal conditions for packing and storage shall have to be established, so that there is no deterioration of active principle. Type of packing like gunny bags, polythene bags, temperature and light intensity of storage-sheds or godown, nature of packing, loose with aeration or tightly pressed are required for storing the annual herb plus pest problem

in storage shade. In case storing facilities are not available, satisfactory arrangement is required for shipment of raw material. Uninterrupted production shall have to be planned in advance.

Costs and profits: At the conclusion of all the activities a clear picture of various operations is necessary at different stages, the problem involved and the expenses shall emerge. However, it should be cautioned here that the cost, thus arrived, for various stages are not the total cost of the cultivation. These figures will only help in final evaluation to the cost of crop obtained.

A drug farm is a huge establishment with a large area under cultivation. There shall be housing, essential godowns, sheds, farm equipment like tractors, harvesting equipments and irrigation system. The capital involved can be grouped as investment on equipment running cost like wages, seeds, etc. and civil cost-fencing, godowns, etc. The cost should be evaluated from two other variables like lower yield than anticipated because of bad season, breakdown of essential services and increase in the cost, etc.

Farm maintenance: Operation schedule for the farm should be prepared where job-chart for a week, two weeks and a month for distribution of labour, maximum utilisation of farm equipment and procurement before required schedule be made. Soil conservation and soil fertility maintenance measures have to be taken so that constant for better yields are obtained over a period of a number of years. The question of crop rotation or soil improvement by mechanical means need to be studied simultaneously. Efforts for the development of a drug farm should not ease, since brilliant suggestions might come from a labourer working on the farm.

9 Maintenance and Utilization of Germplasm of Medicinal and Allied Plants

Vavilov listed 640 species under his centres of origin excluding the ornamentals. In this list, many of the medicinal plants are considered important as of today, also from the cultivation point of view were not in cultivation even in the early years of the last century. There is a mention of a few species such as Hemp, Poppy, *Mentha piperata, Lavender, Thyme*, some umbelliferous plants and Cocos. Another important publication on the origin of cultivated plants is by Alphonso de Candolle which was made use of also by Vavilov in determining the centres of origin. This publication also does not deal with the origin and cultivation of medicinal and allied plants.

It is during the last few decades that we have come to know a little more about the potential of our medicinal and aromatic plants both from the view point of pharmacognosy and pharmacology. Due to the pioneering works of Col RN Chopra and subsequently by others, we, in India, have now some knowledge of our medicinal plants. In case of medicine the pharmaceutical industry has largely depended on the collection of raw material from the wild growth with the result that several plants are under cultivation not only in India but in other countries as well. In many cases, the active principles originally located in plants have been synthsised.

The Need for Germplasm and Selection of Proper Strains

Improved varieties of agri-horticultural plants resulted in the use of germplasm. Whenever and wherever the extensive germplasm becomes available to plant breeders the results have been distinctly encouraging. The "Green Revolution" is the result of germplasm collected worldwide which reaches through collaboration of scientists worldwide. In case of crops like wheat, rice, maize, millets and more recently in pulses with the result that breeders of agri-horti crops are today better informed of the centres maintaining rich collections of germplasm for purposes of exchange. The situation regarding medicinal plants is different, though in early period, when the science of botany was taking birth, it was the medicinal plants which formed the basic material of the early botanical workers. However, after the development of the science of genetics, agri-horticultural plants have almost completely bypassed the medicinal plants in the field of research relating to improvement through genetic manipulation.

There had been several expeditions by individuals and also jointly though FAO for the collection of germplasm for agri-horticultural plants but very few for the medicinal plants. We may recall the botanico-pharmacognostic exploration undertaken by DR HH Rusby and others for Parke Davis 1885 co., which led to the discovery of *Rhamnus purshiana* in abundant supply in S America, which yields the *Cascara sagrada*. In 1920, Joseph F. Rock found an avenue of *Hydnocarpus* trees in Bangkok and this tree yields oil similar to chalmoongra oil, used in treatment of leprosy; later Rock found the true chalmoongra tree *Tarakatogenos kurzii* in Burma (Myanmar). Whereas extensive catalogues of world germplasm in case of several crops and fruit plants are now available, there are few worthy of mention in the case of medicinal plants, like *Index Seminum* of some botanical gardens though there is a mere listing of species and not of genotypes or chemotypes of the individual species. Even in USA, *Index Seminum of*

Medicinal Plants, issued by the College of Pharmacy, simply listed some 400 species without any reference to genotypes of each species. The germplasm list of All Union Scientific Research Institute of Medicinal and Aromatic Plants at Vilav, (Moscow) is perhaps the most exhaustive and of better quality. There have been quite a few technical conferences on plant exploration and introduction but there is hardly any proposals for augmenting work on exploration for the collection of germplasm in medicinal plants and its maintenance for evaluation and utilisation from time to time. Today, plant breeding aims at increasing the agronomic potential of a base material as reflected in increasing yields, improved variety for diseases and pest resistance, wide adaptability and such other attributes. It is the breeder's natural instinct in striving for the constant improvement of material with which he is working. It is, therefore, necessary to emphasise the need for cataloguing the genetic material of important medicinal plants.

Probably, several world organizations, like the United Nations FAO, have not yet considered to plan and pay necessary attention in this direction as well.

Techniques for the Improvement of Medicinal Plants

One of the first steps in the improvement of plants, be it cereal, fruit or drug, is the exploration of genetic variability. In the narrower sense, it means a close study of the same species in the same locality; in the broader sense, a search in the regions where the plant grows naturally for types that possess better qualities. Plant-breeding methods are the same for drug plants as are for other economic plants, which have not been given the same attention as our agricultural crops. Except for a few genera, the vast majority of medicinal plants of which at least 2000 species have been listed, are collected in the wild. Hence, our first task is the survey of the genus or species. To undertake and succeed in this, genetics must be linked with cytology, for the variabilities of a plant; which may be both environmental and hereditary and the variability of chemical composition may be associated with the chromosome complex and the number, and polyploidy of the plant must be studied. Hence, the necessity to bring together as many geographical races as possible of the plant we wish to improve. In modern genetical language we call this a gene pool, though the ultimate interpretation of the gene action is in the term of biochemistry or chromosome chemistry. Basic chromosome studies give us the clue for parental relationship which is the next step in the genetic improvement of a plant.

The creation of the next genetic variability by hybridization between selected individuals and the concentration of the new genetic variability into desired channels, and finally the efficient selection of the desired genotype and large-scale cultivation constitute important steps.

The scientists devoted to improvement of medicinal and allied plants had been working with narrow genetic base such as small collection of plants like *Rauwolfia, Dioscorea, Datura* and other plants which are available with certain institutions and experimental stations both within and outside the country. However, there is no effective channel in existence even to pool together the limited existing germplasm. Though, there are reports, here and there of certain superior genetic-stock having been located but not much attention is paid to the genetic improvement of medicinal and aromatic plants; though certain attempts by RRL Jammu and of CIMAP are praise-

worthy. The coordinated project of ICAR, located at NBPGR at Delhi, has also made some attempts in this direction which incorporates the medicinal plant unit of the IARI, New Delhi.

Eight exotic strains of *Glycerrhiza* were tried at IARI and one Russian strain Excise EC 21950 gave highest glycyrrhizic acid 2.30% in one year old roots, the lowest of 0.55% was given by *G. foetida*. In *Mentha arvensis* the range of variation in oil content in two cuttings (on dry weight basis) was between 0.67 and 4.13 pc as against 2.76 to 3.73 in Japanese mint and 2.6–4.20 pc in Jammu mint. Sobti (1966) indicated a variation in Jammu mint progeny and analysed 150 seedlings to show genetic variation. There has been a report by Dhar (1965) that total alkaloid content in geographical races of *Rauwolfia serpentina* were genetically controlled. Similarly, variation in fresh root weight per plant (Biswas 1956), fruit set per plant (Chandra 1956) have also been reported in Rauwolfia. Wilson (1966–67) reported diosgenin content in *Dioscorea deltoidea* to range from 3.1–5.0 percent (dry weight), while Barua et al (1955) obtained the maximum of only 3.35 percent. Ilieva (1967) raised the Santonin content in *Artemisia maritima* var. *salina* varying from 3–5% through chemotype selection. Seale (1955) picked up high sapogenin yielding section F16 and F13 from *Dioscorea floribunda* (PI 201783). In case of *Rauwolfia serpentina*, a wide range of variation was observed in almost all the characters in plants of two different age groups. That such a variability was present even in six months old crop indicated that it should be possible to make single plant selections in the first year itself. Progeny of one plant, thus selected, has been found to produce short-stalked peduncles and flower having 26 fruits per spike. In another observation on

150 plants in a year old bulk population. Fresh root weight per plant ranged from 40–150 gm. In case of *Plantago ovata* (Psyllium), ten accessions each from Sidhpur, Mehsana, Banaskanta were studied at IARI Delhi and it was observed that Sidhpur and Mehsana material was significantly superior to that of Banaskanta in tiller number, spike number and yield per plant but no significant difference was observed in spike-length. Chandra (1967) reported from single culture studies that Patan samples (Mehsana district) had a higher yielding potential. On the other hand, Kanitkar and Pande et al (1967) reported that material from Palanpur (Banasakanta) had the better seed quality. These observations indicate the desirability of making intensive surveys of *Psyllium* and other material in their growing district.

In *Datura metel*, the plant height showed positive and significant correlations with primary branches per plant, leaf length, leaf blade width which points towards usefulness of selecting taller plants for more herbage— the main source of alkaloid. In *Atropa belladona*, plant height, tiller number and alkaloid showed higher coefficients of variance (33.2, 48.5 and 42 percent) respectively. In *D. stramonium*, Newyork selection showed significantly higher leaf per plant and fresh weight of leaves per plant. Information on variation amongst yield parameters and their correlation is vital, for any plant-breeding programme. Lack of well-defined genotypes and little effort towards exploitation of natural variability in medicinal plants help in improvement of these plants. Now, efforts are being made in estimating the extent of variation and their correlation amongst some yield attributing traits in *Datura metel, D. stramonium, Atropa belladona* and *Ammi majus*. Exploitation of heterosis is a worthwhile pro-

position in Poppy, since both spontaneous and induced male sterility have been recorded in this crop.

There are reports of new superior materials arising as a result of mutation and polyploidy breeding. The reference relates to Gama radiation, *in Solanum lanciniatum, Mentha piperata* var. *rubersceme* (Ozola, 1969) and in *Digitalia lanata* (Michalsko, 1966).

Polyploids have also been recorded in case of *Mentha arvensis* Janaki Ammal and Sobti et al 1962, *Rauwulfia serpentina* (Janaki Ammal et al 1962, Bhaduri and Biswas, 1965), *Cymbopogon* (Janaki Ammal, 1966), *Datura innoxia* (Singh et al 1967), *Mentha piperata* (Lutkov 1962), *Plantago ovata* IARI and Chandlar Clyde, 1954 in USA. In general, it appears that there is necessities for intensification of efforts in selection work in populations to improve upon their economic and agronomic bases, because in several cases, for one reason or the other, some of these new materials have not been pushed into commercial cultivation.

Work on the improvement of medicinal and aromatic plants has been taken up at a number of places through induced mutation. Late Prof. Janaki Ammal conducted a cytogenetic survey of Indian species of Cymbopogon with special reference to their oil content and commercial exploitation. Survey of 17 out of 24 Indian species of Cymbopogon including 33 races showed polyploidy up to hexaploid level, 2n = 60 exist in the genus. That, in general, the higher polyploids had higher oil content than the diploids, oil content estimation of three species of Cymbopogon (*C.scheoanthus, C.citratus* and *C.rugosa*) showed that they differed in their chemical composition. The allopolyploid nature of hexaploids was revealed in their chemical behaviour and the difference in the nature of the oil constituent. These polyploids were found in regions of higher altitudes where the three meet in their distribution.

The following high-yielding polyploids and interesting diploids are recorded to have been selected for commercial exploitation.

Cymbopogon polyneuros	2n = 20	1.67% mainly limonene
C. ladakensis	2n = 20	1.70 piperitone 40%
C. flexuous (Coimbatore)	2n = 40	2.05%
C. citratus	2n = 40	1.25%
C. stracheyi	2n = 40	2.85%
C. nardus var. *conferiflorus*	2n = 60	2.25% 40–50% citronella
C. pendulus	2n = 60	2.0 70–80% citronella

The presence of extra-chromosome in diploid *C. flexuous* from Keti valley, Nilgiris, and their associates with increase in oil content called for a survey of plants from regions which are under study.

Effect of gama radiation in *C. martini* has produced mutants with much larger leaves. Effect of gama radiation on *C. winterianus* showed that percentage of sprouting, average height of plant and number of suckers per plant decreased with increasing exposure. Percentage of essential oil on dry plant basis also decreased after radiation (1.20 from 1.94%).

Besides X-rays and gamma rays a large number of chemical mutagens, namely ethyl methane sulphonate, ethyleamine and methyl nitrosoguanindine have been utilised. Some of the mutants in *Atropa belladona* have been released for cultivation, while mutants of *Solanum khasianum, Datura metel* and *Mentha arvensis* have also been worked out. Certain studies, however, did not show any increase over control, thus indicating that higher herb yield is manifestation of enhanced tillers, increased

number of leaves formed and bigger leaf-size. There have been reports of new superior materials arising as a result of mutation and polyploidy breeding. The reference relates to gama radiation in *Solanum inciniatum, Mentha piperata* var. *rubrescence, Digiatlis lanata*, etc. Induced mutation in *Datura innoxia* and effects of ionizing radiation on seeds of *Asclepias curassavica*, recorded higher yield as compared to untreated seeds. Polyploids have also been reported in *Mentha arvensis, Rauwolfia serpentina, Cymbopogon* and *Plantago ovata*. However, in general, it appears that there is necessity to intensify selection work in polyploid populations to improve upon their economic and agronomic base, because in several cases, for one reason or other, these new materials have not been very well pushed in commercial cultivation.

Current Needs and Possibility of New Germplasm

There is a general agreement regarding all round need for additional germplasm of important medicinal plants. There is also need to develop contact outside for mutual exchange; but first of all we require to make a good search to locate genetic stock. There are some reports of material which is in the process of development. For example, in Puerto Rico Martin et al have obtained hybrids amongst sapogenin bearing Dioscoreas, viz. *floribunda, spiculifera, compo-sita* and *firesrichsthalli*, the first two having higher sapogenins. The Indian species *D. deltoidea* flowers and tuberises at a much younger stage in the hilly temperate areas where it is naturally growing. Hybridization work can be possibly accelerated to create additional variability for wider adaptability. *Hyocyamus muticus* from the upper Nile region is reported to contain as high as 1.8 percent alkaloid. In Bulgaria, some races of *Atropa belldona* have been differentiated. An atropine race from Samokovand and Scopolanine race from Varua give high yield of alkaloid. New peppermint varieties such as Pirulki 324 from USSR and Marica from Bulgaria, are most resistant to Puccinia menthae.

There is a need for additional germplasm of plants for use in the synthesis of steroid hormones and cortisones, besides *Diocoreas*, already mentioned, are *Solanum*, Hops, etc. Hops are used in medicine for their sedative and soporific properties and also as tonic. Their principal use, however, is in brewing industry. There are also some reports of its early introduction and performance in India. A number of varieties are introduced at Delhi and J and K. These represent some problems of introduction, as they do not grow everywhere and anywhere in the world near than about 36 degrees to the equator.

The roses *Rosa damaskana* largely cultivated in west UP in Aligarh and to a small extent, in Haldighati near Udaipur. The genetic material is distinctly different in several agronomic and botanical aspects. Haldighati type occurs as an off-type in Aligarh plantations. The Aligarh growers consider it as an undesirable off-type because of its short flowering period. The material introduced from Bulgaria by IARI has turned out to be of Haldighati type, which is inferior. The data produced on Aligarh type on an average are 500 flowers per plant over a period of 40 days compared to 100 flowering plants produced in Bulgarian type over a harvesting period of 30 days. The two types, therefore, have different pruning requirements for proper flowering. The number of petals per flower also varies, e.g. 40–50 in local and 25–30 in Bulgarian plants.

Pimpinella anisum forms heterogenous bulk population, raised from various samples. A uniform strain has been developed which meets climatic requirements of Delhi. It was possible to raise a good crop by sowing presoaked seeds by mid-Nov. in shallow furrows under optimum moisture conditions followed by light compacting. A yield of 300–400 kg/acre may be expected. It matures in 175–180 days and oil content as determined is about 2 percent more (vol/weight), it does not suffer from any pest or diseases.

Vital Drugs and Essential Oil-Bearing Plants as Future Cash Crop

India with varied climatic and soil conditions show magnificent variability in plant resources which is distinctly valuable. Though efforts have been made to collect and identify the plant resources of the country during the last two centuries, very little has been done to harness this wealth of plants for industrial development. However, the Council of Scientific and Industrial Research (CSIR), with its regional centres, has made a beginning, yet much remains to be done. The Indian Pharmacopoeia recognized a few plant species, this is too meagre, while considering the fact that the 38 percent of drugs of vegetable origin are listed in National Pharmacopoeia of the eastwhile USSR. Similarly, hardly a few dozen essential oil-bearing plants are made use of by the Indian pefumery industry. The Indian food-flavour industry is also at infant stage and needs specific consideration. Our export trade in essential oils is made up to oils of sandalwood, lemongrass and *Palmarossa* grass, which betrays the naturés vast, treasure of Indian plant resources.

The natural resource of germplasm for all plant industries comes from the wild natural population, while large phytochemical establishments require a regular supply of enormous quantities of raw material, which only commercial cultivation could support. In sixties, the GOI set up a phytochemical project in Kerala for the manufacture of nine basic pharmaceuticals at Narimangalam. This project could not materialise because regular supply of raw material could not be assured. Our total dependence for vegetable drugs on collection from wild became too clear to miss its real significance. Still more painfull is the fact that scientific technology for commercial cultivation of many plants was largely not known. Though experimental cultivation has been attempted during the past in a number of states by the Regional Research Laboratories and their experimental stations and the local forest departments, a comprehensive package of practices has neither been worked out nor tested on large commercial scale so as to form the base of industrial production in future. The CIMPO and its field studies have though come out now with some information for a few species. Still much is required to be done for the commercial and industrial cultivation on this group of plant as a *future cash crop* in the country.

The pharmaceutical discoveries of the last four decades have discovered some new vegetable drugs to give relief to the mankind from many dreadful diseases. There had been expansion and improvement in health services and standards for better medical care, reducing infant mortality and increasing lifespan of mankind. However, the prices of the drugs produced in the organised sector of the industry is a tale of continuous rise; especially with the recent tarifs and World Trade Organization (WTO) has made the drugs beyond the reach of ordinary man. "As a matter of fact, in drugs generally, India ranks amongst the highest priced nations

of the World—a case of inverse relationship between per-capita income and the level of drug price (Kefauver report)". The vegetable drugs, which cater to the needs of the vast majority of rural India have still a role to play in maintaining the cost of health services within the reach of common man. The state and the centre has, thus, the responsibility to develop and perfect the technology from cultivation to manufacture of finished products within the country and make them affordable at nominal price for the rural and weak section of the society.

The sector of essential pills and perfumery covers a wide range of human activity, though only a few specific places are known in the country for the production of essential oils. Industries like soaps, cosmetics, detergents and pharmaceuticals are also sometimes dependent on these basic raw materials. "From dawn until twilight and long into night and from the cardle of the infant to the silence of the grave, we are surrounded by the odorous materials. We take our odours for granted and little realise how much we would be affected, if our lives are deprived of perfumes".

The entire process of collection, processing and the sale of drug plants is in the hands of herb collectors and traders, with only a few items which are in preview of organised sector of the industry. Actual data on the annual production of vegetable drugs and essential oils and their consumption are simply not available and the only information available is the import and export statistics published by the govt. agencies. A few items that need immediate attention could be the plantation of cinchona that need to be increased to meet steep rise in demand with a plant to improve planting and extraction methods with renovation of old inefficient processing factories. *Caffeine* is a basic pharmaceutical, whose production can be raised many times. Tea waste control order of 1959 be amended. Medicinal plants found in wild like *Rauwolfia serpentina*, *Digitalis lanta*, *Psyllium*, *Gloriosa superba*, Indian Celery, *Metha arvensis* var pipera-scens, Geraniums and many others need to be cultivated on commercial scale. Though some steps have already been taken as discussed earlier. However, an attempt has been made to provide the agrotechniques for the cultivation. The information provided is, however, indicative and not exhaustive which needs to be amended for different eco-climatic and soil regions of the country. Supporting research in the improvement of agro-techniques shall improve the crop yield and open new vistas for export. Similarly, techniques for distillation might enable this country to enter into the export market, as our oil samples have been assessed to compare with some of the superior quality oils produced by countries like Japan, China and Tawain.

The Central Indian Medicinal Plants Organization has drawn up a priority list which could be cultivated commercially in the country. The commercial cultivation of medicinal and aromatic plants from the hills has been tried in the subtropical and tropical regions of India with some success. However, plants from the alpine and subalpine regions may not survive in the subtropical and tropical regions for which the floor of the forests needs to be managed, cultivating these plants on a commercial scale. Also the farmers field, where crop cultivation is not very profitable, could also be used for the cultivation of these plants as has been done by Mr. Badoni in Tons Division. Since each individual medicinal or aromatic crop has its own specific requirement of soil, climate, irrigation and fertilizers, this should be given serious consideration while planning their cultivation to obtain expected returns.

Crude drugs and herbal medicines, no doubt, are being steadily replaced by their active constituents. Several thousand patented medicines and gelanicals are being used widely all over the world but invariably contain a number of vegetable drugs each in their formulations. As such total demand of vegetable drugs has been increasing mainfold both in quantity and quality value. Thus, there is no reason for the apprehension of a possible fall in their demand in the foreseeable future.

India's share in the trade of perfumes and pefumary material is not more than 12%. About more than 280 million dollars of natural essential oils and allied perfumery material is purchased annually by developed countries of the world, excluding east European countries. World demand for new perfumery material and essential oils having differences in content and note is rising and these items fetch higher remuneration. Some of the items like Davana (*Artemisia pallens*), Panari (*Pogostemon cablin*), Costus (*Saussurea lappa*), French Basil (*Ocimum basilicum*), Khas (*Vetveria zizanioides*) and floral perfumes like Jasmine, Champaca, Nagkesar, *Kewra* and some other oriental flowers have a wide scope for aggressive marketing and conscious quality control for export.

Cultivation of medicinal plants is an amalgam of traditional and modern technology. One has to make sure the market need and need cooperation of the farmers, who are close to the soil. Hence, a novice has to consult and learn from those who have succeeded in the field and those who have failed. Networking is an important aspect to gain an extra-edge since one should expand the knowledge both from seminars, lectures and literature. However, marketing could be a problem in this field for which *net* could be exploited if one wants to export

the product. As an example, *Aloe vera* is a common medicinal plant, easy to grow but needs to be processed within 24 hours of being cut, else it gets contaminated. The crop is meant to be used for medicine, hence good quality, authentic planting material is the most important to be used, since cheaper planting material may be adulterated and is widely sold in the market which should never be purchased and used.

Table 9.3 lists fungus diseases and insect pests and Table 9.4 describes management of insects and pests in aromatic and drug plants.

Traditional Method of Plant Protection

Farmers use the leaves and latex produced by *Calotropis procera* (vern. *Ákda*) and *C. gigantea*, to control pests. The farmers have wide variety of practical ways of making use of this plant's insecticidal properties. Green leaves of this plant are put in cloth bags and placed at the entrance of an irrigation channel. This method controls aphids and weeds. The leaves are also used to control termites. They are soaked in water for a day and after the liquid is filtered off, it is poured in the termite-infested soil.

The ancient texts include many recipes for crop production and it is necessary to know how these ancient techniques be applied, designed and incorporated. We need to look living folk-lore practices of local farmers to understand and analyse them for improving germination, pest and disease resistance and effect of plant growth regulators. However, experiments on crop plants like paddy have been conducted where seeds have been subjected to different treatments: soaked in water for 24 hours, soaked in a mixture of cow's urine and powdered vacha (*Acorus calamus*) for 24 hours, soaked in milk for 24 hours then rinsed with water, etc.

An experiment with mango-orchard infected with mango leaf showed that removing affected parts and spraying the leaves, affected parts with a mixture of neem oil and (pongam oil from *Pongamia binata*) in a soap solution and fumigation with herbal mixture consisting of *Embelia ribes* and a herb *Daruharida* showed dramatic reversal in the disease.

In another study a custard apple tree that has not yielded fruit for 15 years, showed that application of about a litre of milk together with various pulses, ghee and honey produced a record crop of custard apples, with a particularly delicious taste.

Plant growth regulators like solution of cow's urine diluted with water or a modified *Panchagavya* solution mixture of cow's urine, milk, water and ghee or another regulator consisting of goat flesh extract, black gram powder and sesame seeds showed excellent results when applied to plants.

Use of Saline-Alkali Soils for the Cultivation of some Aromatic Oil-Yielding Plants

Saline-alkali soils can be profitably used for the cultivation of aromatic and oil yielding plants; the main species are as follows.

1. *Cymbopogon martinii* var. *motia* has been successfully grown on the alkaline soils near Lucknow which yielded good quality of palmarosa oil. The composition of the oil based on gas-chromatographic data in order of emergence of peaks has been recorded as Limonene (0.4%), Methyl heptanone (0%), Geranyl acetate (1.1%), Geraniol (88.9%), Farnesol (6.7%) and unknown 0.9%.

2. *Oil of spearmint*, which is highly priced in essential oil industry has been produced from plants raised on alkaline soils near Lucknow. The oil based on gas chromatographic data has been estimated to contain Limonene (13.1%), Phellandrene (3.2%), Cineole (10.0%), Carvone (56.2%) and Dihyrocarveol (13.1%).

3. *Mentha piperata* oil has been evaluated from the plants raised on alkaline soils near Lucknow and has been recorded to contain Limonene (3.3%), 1 Terpinene (3.3%), Methofuran 9.0%, Menthone 9.32% Menthyl acetate (11.7%) and Menthol (42%).

4. *Ammi majus* seed oil with an improved method consisting of proper cleaning and powdering material with solvent oil at 70–100 degrees, separation of a portion of coumarins by crystallisation, treatment of coumarins with alcohol HCl followed by dimethyl-sulphate and finally purification of xanthotoxin.

5. *Matricaria chamomilla* yields blue oil having antiphlogistic properties, used against stomach ailments and as flavouring agent in fine liquors and perfumes. The crop is well suited for saline-alkali oils where 6–8 weeks old seedlings, closely spaced at 30 × 35 cm gave the highest yield. Optimum time of planting is between last week of November and first week of December. The crop also responds well to fertilisation at the rate of nitrogen (50–60 kg/ha and P_2O_5). 62 qtls fresh flowers per ha were recorded to have been obtained with a net income of Rs 1500 to Rs 5600 from the sale of flowers and oil respectively.

Plants of Futuristic Importance for Export and Domestic Use

Papaver somniferum—Morphin, Codin, Papaverin.

Table 9.3: Fungus diseases and insect pests of main drug plants

Name of plant	Insect/fungus	Stage
Withania somnifera Dunal	*Acherontia styx*	Defoliates
Solanum aviculare Forst.	*Agrosti sypsilon*	Cut out young plant
Abelomoschus moschatum Medic	*Alcides affaver*	Grub and adult bore stem
	Althaea rosea Cav.	do
Plumbago zeylanica	*Anisphyra ocularis*	Larvae defoliates
Tephrosia candida(Roxb.) DC	*Aphis craccivora*	Nymph and adult desap
Abroma angustata L. *Buddleja asiatica*	do	do
Dioscorea bulbifera L.	*Aphis gossypii*	do
Plumbago zeylanica L.	do	do
Vinca major L.	do	do
Asclepias curassavica	do	do
Calotropis gigantea	*Aphis nerii*	do
Pergularia daemia	do	do
Nerium indicum	do	do
Anethum graveolens	*Autographa shryson*	Caterpillar defoliation
Salvia moorcroftiana	do	do
Mentha species	do	do
Nepeta species	*Nilgrisigna*	do
Salvia moorcroftiana	do	do
Verbascum thapsus	*Carpocoris* sp	Nymph desap flowers
Mentha piperata	*Chrysolina ewanthematica*	Adult defoalte plant
Verbascum cormandelinum	*Cionus* species	Both larvae and
Verbasdum thapsus	do	adult bore flowers and buds
Chenopodium ambrosioides Rumex acetosella	*Cletus bipunctatus*	Nymph and adult desap flowers
Asclepias curassavica	*Corynodes peregrinus*	Adult defoliates
Calotropis gigantea	do	do
Dioscorea bulbifera	*Crioceris impressa*	Grub and adult bore tender stem and puncture leaves
Carthamus tinctorius	*Dactynotus compositae*	Nymph desap leaf buds
Rauwolfia serpentina	*Deilphila nerii*	Ceterpillar defoliates
Abelomoschus moschata	do	do
Artemisia nilagarica	do	do
Cannabis sativa	do	do
Chrysanthemum indica	do	do
Chenopodium ambrosoides	do	do
Cynoglossum micranthum	do	do
Clerodendrum species	do	do
Helianthus tuberosum	*Diacrisia obliqua*	do
Mentha arvense	do	do
Ocimum species	do	do
Phostemons plectranthoides.	do	do
Plumbago indica	do	do
Reinwardtia trigyna	do	do
Withania somnifera	do	do

Table 9.3: Fungus diseases and insect pests of main drug plants *(Contd.)*

Name of plant	Insect/fungus	stage
Abrus precatorius	Dichrocrocris puntiferalis	Caterpillar bore ripe fruits.
Hyoscyamus niger	Dolycoris species	Nymph desap leaves and buds.
Abelomoschus moschatus	Earias insulta	Caterpillar bore fruit
Abutilon indicum		
Solanum aviculare	Emposca decipiens	Nymph and adult desap leaves.
Datura metel, D. stramoinum	do	do
Solanum aviculare, S.surattense, Withania somnifera	Epilachna decipiens	Nymph and adult desap leaves
Solanum turvum	Eublemna olivacea	Caterpillar defoliates.
Mentha arvensis	Euxia segetum	do
Solanum aviculare	Euzophera perticella	Caterpillar bore root and stem.
Polygonum serratum	Galeurcella placida	Grub and adult
P. hydropiper	do	Puncture leaves
Rumex nepalensis	do	do
Artemisia absenthium	Haltica species	Adult feed on flowers
Hyoscyamus niger	Heliothis armigera	Caterpillar defoliates
Mentha arvensis	do	do
Datura stramonium	do	do
Anethum graveolens	Hyadaphis aoriandri	Nymph desap leaf buds
Barleria dristata	Junonia hierta	Caterpillar defoliates
Mentha arvensis	Laphygma exigna	do
Nepata species	do	do
Pergularia daemia	Lasioderma serricornae	Grub and adult bore dry roots
Solanum ferox	Leucinodes orbonialis	Caterpillar bore stem
Withania somnifera	Lygeus equistris	Nymph and adult desap
Artemisia nilagirica	Macrosiphoniella	Nymph desap leaf buds
A. absinthium	Pseudoartimisia	do
Rosa demascana	Macrosiphum rosaeformis.	Nymph desap leaf and buds.
Hemidesmus indicus	Margaronia bicolor	Larvae defoliates
Ocimum species	Monanthia	Nymph and adult
Mentha arvensis	Globulifera	Suck leaf sap.
Hyoscyamus niger, Kalanchoe pinnata	Myzus persicae	do
Mentha arvensis	Odontotermes obssus	Nymph tunnel roots
Mentha species	Oxya species	Nymph and adult feed on leaves
Abutilon indicum	Oxycarenus loetus	Nymph and adult
Abelomoschus moschata	do	Desap flower and unripe fruit
Anethum graveolens	Papilio machon-asiatica	Caterpillar defoliates
Hieracleum candicans	do	do
Ruta graveolens	do	do
Mentha arvensis	Phytomyza articornis	Larvae mince leaves
Helianthus tuberosus	Protrama penecaeca	Nymph suck root sap
Mentha species	Pyrausta incolralis	do
Desmodium gangeticum	Riptortus linearis	Nymph desap unripe fruit
Abelomoschus moschata	Sylepta derogata	Caterillar roll leaves and feed inside
Mentha arvensis, **Ocimum** basilicum	Syngamia abruptalis	Caterpillar defoliates
Blumea laccera	Vanessa cordrii	Caterpillar defoliates

Table 9.4: Management of insect and pests in aromatic and drug plants

1. *Pod borer:* Apr-Dec. Larva stage (green, brown and yellow organism). Eat leaves and upper part of crop in later stage. Damaging pods, seed and fruits. (Mentha, Muskdana).

Insecticidal	%	Trade name	Method of use
i. Profinophos + Cypermethrin	0.01	Rocket 44	1 ml/lit water
ii. Chloropyriphos 50% EC + Cypermethrin	0.05	Annconda	l ml/lit water
iii. Cypermethrin + Malathion	0.05	Virat	1 ml/lit water

2. *Spotted ballworm:* Throughout the year. Larva green with black spots on the body. Eat succulent stem, flower, pods. The upper part gets brown and hairy (Muskdana).

AI	Humidity	Trade name	Method of use
Lamda cyhalothin 5% EC	0.5	Caral	0.75 ml/lit

3. *Pinkball worm:* May to August. Larva dark rosy. Adult dark grey 8–9 mm (Muskdana-damage seeds, flowers fall early).

i. Prophenophos	0.01	Karina 50 EC	1 ml/lit water
ii. Endosulphan	0.05	Thiodon	2 ml/lit water
iii. Chlorpyriphos 20% EC	0.05	Derswan	1 ml/lit water

4. *Stem borer:* July to March. Larva light rose-coloured, head brown larva bores stem and feed new twigs dry. Citronella-Lemongrass, Palmarosa.

i. Chlorpyriphos 50% EC		Annakonda	1 ml/lit water
ii. Cyper methrin 50% EC			
iii. Chlorpyriphos 20% EC	0.05%	Desr and von	1 ml/lit water
iv. Parathion 20% EC	0.05	Folidoal	1 ml/lit water

5. *Shoot borer:* July March, larva dirty while with 5 long stripes. Enters in lower part of stem by boring and dries plant by eating citronella, lemon grass and palmarosa.

i. Chloropyriphost + Cypermethrin 5% EC	0.05	Annakonda	1 ml/lit water
ii. Chlorpyriphos 20% EC	0.05	Dersvon	1 ml/lit water
iii. Pavathion 20% EC	0.05	Folidol Ektaks 50 EC	1 ml/lit water

6. *Grub white:* June-November. Length 3.25 cm . Head dark brown, strong legs. Adult dirty brown. Comes to notice after, damage is done. In thin roots nods in roots and main root damage is done and main roots. During night eats leaves of plants Citronella, Mentha, Palmarosa.

Forest		Thimet 10° C	800–1000 g/ha
Chlorphyriphos 20% EC	0.05%	or Desvon	1 ml/lit water

7. *Termites:* Throughout the year. Inside soil where cellolosic matter like wood, FYM are in abundance. Termite on ground from hill. Plant root is damaged, stem becomes hollow.

Mint, Citronella, Geranium, Muskdana – It damages woody parts, maximum damage is to root portion.

Floret	Thinet IOC	800–1000 gm/ha	
Chloropyriphos 20% EC	Dersvon	1 ml/lit water	

8. *Redspider mite:* March-October. Adult nymph 0.33 mm. Nov-February. Microcopts light brown. Adult and nymph net a web on lower surface of leaf, plant gets weak and leaves fall.

Muskdana

Dimethoate	0.05	Rogor 30 EC	1.5 ml/lit water

Phosphemedon	0.05	Dimacron	2.0 ml/lit water.
Lemda cyhalothin	0.05	Karat	0.75 ml/lit water.
Prophenophos	0.01	Karina 50 EC	1 ml/lit water.
Dicophal	0.05	Calthene	500–1000 gm/ha

9. *Chinchbug:* May-June. Adult is black with white stripes on wing. Adult plants get affected by excessive sap sucking, also small plants get affected and the parts get yellow or red.

Mint

Dimethoate	0.05	Rogor EC	1.5 ml/lit water
Phosphemedon	0.05	Dimacrron	2.0 ml/lit water
Lemda cyhalothrin	0.05	Karat	0.75 ml/lit water
Monocrotophos	0.05	Cadet 36 LL	0.5 ml/lit water

10. *Scale insect:* Oval-shaped borer or black insect, 0.2–0.4 cm long. Crawls in young stage, later becomes still. Flowers and leaves get affected due to excessive sap sucking, the insect excrete, smut.

Vetiver

Demethoate	0.05	Rogor 30 EC	1.5 ml/lit water
Phosphemedon	0.05	Dimacrron	2.0 ml/lit water
Lemda cyhalothin	0.05	Karat	0.75 ml/lit water
Moncrotophos	0.05	Cadit 36 LL	0.5 ml/lit water

11. *Licebug:* Identification time: April-October, November-March. Adult bug is light yellow and black lower portion. On the body there is lace work insect length 0.3 m; (approx). Adult nymph suck leaf sap, resulting in yellow spots, the leaves appear decayed.

Dimethoate	0.05	Rogor 30 EC	1.5 ml/lit water
Phosphemedon	0.05%	Dimacrron	2.0 ml/lit water
Lemda cyhalothin	0.05	Karat	0.75 ml/lit

12. *White fly:* Throughout the year. Light white-coloured adult. Length 1–1.5 mm, found in lower part of the leaf. The fly sucks the sap so that the plant gets weak and dries up. It spreads many diseaes.

Mint, Citronella, Palmarosa, Lemon grass

Dimethoate	0.05	Rogor 30 EC	1.5 ml/lit water
Phosphemedon	0.05	Dimacsron	2.0 ml/lit water
Lemda cyhalothin	0.05	Karat	0.75 ml/lit water
Prophenophos	0.01	Karina 50 EC	1 ml/lit water

13. *Maho:* January-April. Short green, yellow or red, wingless and winged insect. Sucks the sap and infect plant with several diseases.

Mint, Muskdana and Citronella

Dimethoate	0.05	Rogor 30 EC	1.5 ml/lit water
Phosphemdon	0.05	Dimacrron	2.0 ml/lit water
Lemda cyhalothin	0.05	Karat	0.75 ml/lit water
Ethion	0.05	Phosmite 50 EC	0.75 ml/lit water
Prophenopos	0.01	Karina	1.05 ml/lit water

14. *Thrips:* February to April, July-August, Insect is thin, 1 to 1.5 mm long and in colour, form yellow to brown. Puts saliva on the stem, leaf and fruit, the dead cells sucked and cells turn brown. Plant gets virus diseases. Leaves get curl with white spots. It does not suck sap like Maho.

Crops, Rose, Citronella, Muskdana

Dimethoate	0.05	Rogor 30 EC	1.5 ml/lit water
Phosphemedon	0.05	Dimacron	2.0 ml/lit water

| Lemda cyhalothin | 0.05 | Karat | 0.75 ml/lit water |
| Endosulphan | 0.65 | Theodon | 2 lit/ha spray |

15. *Leafweaver:* Sept-December. Larva 2.5 cm long and green like pea with black spots on all over body. Rolls leaves and stays, get in night and eat leaves.

   Patchouli

Hymonocrotophos	0.05	Cadat	0.5 ml/lit water
Quinlphos	0.05	Eklaks	20 ml/lit water
Pyrethrin + Malathion	0.05	Karat	1 ml/lit water

16. *Sundi with hairs:* July-Nov and March-April. Larva damages the plant and is yellow brown and with hairs. It lives in the leaves and eats green part, leaves look like a membrane and adult eats the crop.

   Mint

| Lemda cyhalothin 5% EC | 0.05 | Karat | 2 ml/lit water |
| Kwinalphos | 0.05 | Ekalos | 2 ml/lit water |

17. *Semilooper:* April-May. Larva yellow-green and damage leaves. Excepts the main nerve every part is eaten.

   Mint, Fennell and Dill seed

| Melathion 0.05 | Hilmala 50 EC | Hilmala 50 EC | 2ml/lit water |
| Kwinalphos | 0.05 | Ek laks | 2 ml/lit water |

18. *Netter:* Mentha leaf August-October. Young stage larva inhabits lower leaf surface, later rolls leaf. It falls down with touch and hangs with thread-like formation. Damages leaves.

   Mint

Hemonocroptophos	0.05	Cedet	0.5 ml/lit water
Kiwnalphos	0.05	Eklas	2 ml/lit water
Pyrethrin + Melathion	0.05	Vrat	1 ml/lit water

Cinchona calycia / oficinalis—Quinine, Cinchonin, Cinconidin.

Hypocyamus niger, H. muticus—Hyocin, Hyocamin.

Colchicum luteum, Gloriosa superba—Colchicin.

Taxus baccata / wallichiana—Taxol.

Artimisia annua—Artimicin.

Rauwolfia serpentina—Resperin.

Commiphora wightii—Gugulipid.

Coleus forskolii—Forskolin.

Bacopa moneri—Bacocide.

Acorus calamus—Acoron.

Andrographis paniculata—Androgracolide and Neoandrogracolide.

Asparagus racemosus—Sapogenin, Glycocide.

*Azadirachta indica—Azadiritin oil.

Carica papaya—Cymopapan.

Centella asiatica—Asiaticacide.

Cissampelos—prariera—Cissampelin.

Cosinium fenestrum—Alkaloid (Berbery).

*Curcuma longa—Curcumine.

Gloriosa superba—Colchicine, Thiocolchinine.

Gossypium—Gossiol.

Mappia foetida—Kemtothesin.

Muluna purpura—Eldopa.

Picrorhiza kurroa—Glycocide.

Phyllanthes emblica (Emblica officinalis)—Lignans.

Cilvium marinum—Cilmarin.

Taxus bacata / wallichiana—Taxol.

Tephrosia purpurea—Flavidoids, alkaloids.

Urginea indica—Cilarin A and B.

Valeriana wallichi—Valeopotratus.

*Patent on *Azadirachta indica* and *Cucurma longa* has recently been vacated by USA.

Distillation, Purification, Quality Control and Storage of Essential Oils

1. Hydrodistillation (Water distillation).
2. Field distillation (Stem and water distillation).
3. Steam distillation.
4. Hydrodiffusion.

In hydrodistillation: The oil may develop knot due to heat that chars the bottom. It takes a lot of time for distillation. It may partly be hydrolysed due to long heating for heat sensitive materials.

Field distillation: It is a modified way to distil essential oils. The vapours of oil and water are condensed in a condensor and are collected in a receiver. The oil is available at the top in the receiver due to different specific gravity. Palmarossa, Citronella, lemon grass, menthol and mint take 7–8 hrs for distillation; however, the unit developed by CIMPO takes about 4–5, hrs thus, a saving of 20–30 percent in fuel consumption.

Steam distillation: The plant is tightly packed for better oil recovery. This method is suitable for hard roots and woody material. The cost of such a plant varies depending upon the capacity which varies from 250 to 2000 kg/per unit.

Hydrodiffusion: In this system steam enters the top of the plant charged and diffuses it through charge by gravity. The process follows the principle of osmotic pressure to diffuse oil from oil glands. Oil and water are collected below the condenser in a typical oil separator.

International price of essential oil undergo large fluctuations, according to the quality of essential oil. The farmers, traders and industrialists store them until return of remunerative prices. However, improper storage may result in deterioration of the quality of oil. General reactions like polymerisation, oxidation, hydrolysis of esters and interaction between functional groups of essential oil constituents, and are responsible for change in oil quality on storage. Exposure to light, oxygen (air), moisture and metals are known to adversely affect the quality. Generally, essential oils are kept in GI containers with epoxy coated stainless steel/Hedp/glass/aluminium/copper containers. To exclude the air, these are completely filled otherwise it may develop rancidity. The essential oils if distilled or rectified give better perfumery and oil factor value than the raw essential oils.

Impact of Aromatic and Drug Plants on Run off and Soil Loss in Hilly Areas

Studies conducted at Ottacamund on certain aromatic and medicinal plants when grown in sloppy areas, have been conducted by Singh et al (2005). Some commonly grown plants like Geranium (Pelargonium). Rosemary (*Rosmarinus officinalis*), Cineraria (*C. maritima*), Thyme (*Thymus vulgaris*), Mentha (*M. Piperita*) and Digitalis (*S. purpura*) were conducted to find out their relative oils and water conservation efficiency. Except were grown on plats of 15 m × 15 m with a spacing of 40 cm × 30 cm); Mentha was grown at a spacing of 40 cm × 20 cm.

It was reported that Geranium had the maximum canopy followed by Digitalis. Cinneraria and Rosemary. Canopy cover in case of Thyme was poor. Metha could survive well only for the first two years. The plant height was maximum in case of

Geranium while it was minimum in the same of Thyme.

Five years data on mean runoff and soil loss indicated that the soil loss was below 3 t ha yr under cultivation of Digitalis followed by Rosemary (1.4 t ha yr., Geranium (2.6 t ha yr and Mentha 2.9 t ha yr). However in case of Thyme cineraria it was 10.9 t ha yr and 6.6 t ha yr respectively. The canopy in case of both the species was poor. Higher runoff was reported, 80.3 mm 9.1% under Cineraria followed by Thyme (73.1 mm). The lowest runoff was recorded under Digitalis (10.3 mm) followed by Geranium (15.5 mm), Rosemary (34.4 mm) and Metha (59.9 mm). Relative soil and water conservation efficiency (mean runoff) of these plants is the highest for Thyme, followed by Digitalis, gerabium and Rosemary. The nutrient losses of N, P and K varied from 2.1–8.1, 1.7–4.8 and 1. p–6 kg/ha yr, which are lower as compared to the cultivation of main crops. Thus it is recommended that the aromatic and medicinal plants should be preferred for cultivation in hilly areas as compared to traditional crops which are economically better than the tradition crops. Since these are more efficient both for soil conservation and economic viewpoint. Moreover it is necessary to take soil and water conservation measures from the hilly slopes in order to maintain the soil fertility on a sustainable basis.

References

Abdul Hameed Khan. 1957. Studies on growth and cultivation of medicinal and other economic plants under semi-temperate conditions. *Pakistan J. Forestry.* **7**: 319–369.

Airi S, Rawat RS, Dhar U and Purohit AN. 2000. Assessment of availability and habitat preference of Jatamansi. A critically endangéred medicinal plant of West Himalaya. *Mountain Res. and Development.* **24(2)**: 141–148.

ANSAB (Asia Network of small scale Bio-sresources) 1997. Nepal NTFD *Entrepreneurs Directory.* Kathmandu, Nepal.

Aryal M. 1993. Diverted Wealth: The trade in Himalayan Herbs. *HIMAL* 1993 **(1)**: 9–18.

Balasubramanian AV, Vijaylakshmi K, Sridhar Sabhasim and Arumug, aswamy 2000. Vrikshayurveda Expts linking ancient text and farmers practices *LEISA. India.* 16(3): Sept 31–34.

Barua AK, Chakravarty D. and Chatterji D. 1955. Steroid hormones from Indian *Dioscorea* plants. *Bull. Inst. Sc. India,* 1955. **4**: 15–20.

Bhaduri DN and Biswas PK. 1965. Increasing root yield of *Rauwolfia serpentina* by Cochiploidy. *Sc. and Cult.* **31**: 197–200.

Biswas K. 1956. Cultivation of *Rauwolfia serpentina*, in West Bengal. *Indian J. Pharm.* **18(6)**: 227–230.

Ceci (Canadian Centre for International Studies and Co-operation) 1997. Inventory of 4-high value non-timber Forest Product in Jumla (*Nardostachys grandiflora, Picrorhiza scrophuliflora, Rheum australe* and *Valeriana jatamansi*). Kathmandu Nepal.

Chandra V. 1956. Studies on Rauwolfia. pp. 11. Preliminary studies on Floral biology of *Rauwolfia serpentina* Benth. *Indian J. Pharm.* **1(4)**: 136–140.

Chandra V. 1967. Studies on cultivation of *Plantago ovata*. Forsk. *Indian J. Phar.* **29**: 331–332.

Chopra IC and Handa KL. 1961. Review of Research on Indian Medicinal and Allied Plants. *ICAR, Review Series.* No **34**. pp. 1–60.

Chopra IC, Sobti SN and Handa KL. 1957. Cultivation of Medicinal Plants in Jammu and Kashmir. *ICAR Res. Ser.* **13**. pp. 1–75.

Chatterjee SK. 1980. Cultivation of Medicinal Plants in India. with reference to Eastern Himalayan region. *Asian Symp. on Medicinal Plants species.* pp. 20–29.

CIMPO 1975. *National Symposium on Recent Advances in the Development, Production and Utilization of Medicinal and Aromatic Plants*. Abstr. pp. 1–46.

Dhar ML, Dhawan BN, Mehrotra BN and Ray C. 1968. Screening of Indian Plants for Biological Activity. Pt. 1. Indian *Journ. Exptl. Biol.* 6(4): 232–247, Pt, 11. *Ibdid*: pp. 250–262.

Dhar R. (Miss) 1965. Variation in the alkaloid content and morphology of four geographical races of *R. serpentina* Benth. *Proc. Indian Acad. Sci.* **62**: 242–244.

Ilieva S. 1967. Improving the Santonin content of *Artemisia maritima* var. *salina* Koch. through selection. *Rasten Nauki*. **4(8)**: 13–17.

Janaki Ammal EK and Sobti SN. 1962. The origin of the Jammu Mint. *Current Sc.* **31**: 38.

Janaki Ammal EK and Gupta BK. 1966. Oil content in relation to Polyploidy in Cymbopogon. *Proc. Indian Acad. Sc.* Dec. B. **64**: 334–350.

Kapoor LD. Some problems facing the commercial cultivation of Medicinal Plants in India. *Indian Drugs*.

Kanitkar UK and Pande GK. 1967. Experimental cultivation of *Plantago ovata* in Maharashtra, India. *Indian J. Pharm.* **29(3)**: 97–98.

Larsen HO and Smith PD. 2004. Stakeholders Perspective on commercial medicinal plants collection in Nepal. *Mountain Res. and Development.* **24(2)**: 141–148.

Larsen HO. 2002. Commercial medicinal plants extraction in the hills of Nepal. Factors, Objective and Power. *Forest Policy* and *Economics*. **17(4)**: 363–374.

Lutkov AN. 1962. Polyploidy and its significance in essential oil crops. Tr. *Moscova Isypt. Periody Ord. Biol*. **3**: 260–273.

Martin FW and Cabinallas E. 1960. The F. Hybrids of some sapogenin bearing Dioscorea species. *American Journ. Bot.* 53(4): 350–58.

Michalsko T. 1968. Radiation induced mutation in *Digitalis lanata*. *Ehrh. Pl. Br. Abstr.* 1968. p. 553. Polish.

Nautiyal, MC. 1995. Agro-techniques of some High Altitude. Medicinal Herbs. In Cultivation of Medicinal Plants in Sikkim Himalaya. pp. 53–64.

Olsen CS and Larsen HO. 2003. Alpine medicinal plant trade and mountain livelihood strategies. *Geographical Journ.* **169(3)** 243–264.

Ozola SM and Eizenberger VT. 1966. Effect of Y-rays on productivity of Peppermint. *Latv. Psr Zinat Akad*. Vestis. 1966: 120–122 (Russia)

Puri HS. 1969. Cultivation of some medicinal plants of United States. *Indian Drugs*. pp. 7–8.

Seale CC. 1955. Agronomic studies on *Dioscorea*, in South Florida. *Ila Agric. Expt. Station*. pp. 273–274.

Shah NC. 1968. The alpine Medicinal Herbs. *Indian Drugs* No. pp 13–16.

Silori CS and Badola. R. 2000. Medicinal Plants cultivation and sustainable develop-ment. A case study in Buffer zone of Nandprayag Biosphere reserve, West Himalaya. *Mountain Res. and Development*. **20(3)**: 272–279.

Singh KN. 1972. Cultivation of Medicinal Plants in Uttar Pradesh. *Bull. No 4. Forest Utilisation* Div. UP. Mimeo. pp. 1–226.

Singh MP. Malia SB. Rajbhandari A. and Mannadha, 1979. Medicinal Plants of Nepal. Restrospects and Prospects. *Econ. Bot.* **33**: 185–198.

Singh DV, Selvi V, Madhu M and Subhash Chand. 2005. Soil and Water Conservation Research and Training Institute, Dehradun, Annual report 2004–2005, p. 12–13.

Sundriyal RC and Sharma. (Eds) 1995. Cultivation of Medicinal Plants and Orchids in Sikkim Himalaya. HimaVikas occasional Pulication No 1. pp. 1–139.

Sobti SN. 1966. *Variation in Progeny of Jammu Mint*. Symp. Recent Advances in the Development, Production and Utilisation of Medicinal Plants in India. CSIR, April 18–20. pp 23.

Wilson AP. 1966–67. *Dioscorea*, medicinal Yam. 1966–67. *ARLONGATON Agric. Hort. Research Stn.* 1966–67. 280–283.

□□□

Agrotechniques for the Cultivation of Drug Plants

1. Artemisia annua Linn
2. Achyranthes aspera
3. Acorus calamus
4. Alpinia galanga Willd.
5. Anethum graveolens
6. Angelica glauca Edgew.
7. Atropa acuminata Royle ex Lindl.
8. Atropa belladona L.
9. Cassia angustifolia Vahl.
10. Carum copticum Benth.
11. Carum carvi L.
12. Catharanthus roseus
13. Chenopodium ambrosioides var.
14. Cinnamomum verum J.S.Presl.
15. Chrysanthemum cinerarifolium Vis.
16. Citrus aurantium
17. Citrus medica var.
18. Coriandrum sativum L.
19. Chlorophytum arundinaceum,
20. Costus speciosus Sims.
21. Crocus sativus L.
22. Curcuma longa L.
23. Cymbopogon species.
24. Datura species
25. Digitalis lanata Ehrh.
26. Digitalis purpurea
27. Dioscorea spp.
28. Elettaria cardamomum Maton.
29. Emblica officinalis Gaertn.
30. Ephedra gerardiana Wall. ex Staff.
31. Eugenia caryophyllus (Spreng.)
32. Foeniculum vulgare Gaertn.
33. Fagopyrum esculentum Moench.
34. Glycyrrhiza glabra L.
35. Humulus lupulus L.
36. Hyoscyamus niger L.
37. Jasminum species
38. Lavendula officinalis Chaix.
39. Matricaria chamomilla L.
40. Mentha arvensis L.
41. Mentha piperita L.
42. Myristica fragrans Houtt.
43. Nardostachys jatamansi DC.
44. Ocimum species
45. Ocimum kilamandcharicum Guerke
46. Ocimum basilicum L.
47. Ocimum sanctum L.
48. Papaver somniferum L.
49. Pelargonium L'Herit (Geranium).
50. Piper species Piper nigrum L.
51. Plantago ovata Forsk.
52. Podophyllum hexandrum Royle.
53. Pogostemon cablin
54. Rauwolfia serpentina Benth. ex Kurz.
55. Rosa damascana Mill.
56. Ricinus communis L.
57. Saussurea costus
58. Tagetes minuta L.
59. Vanila planifolia Andrews.
60. Valeriana officinalis L.
61. Solanum species
62. Salvia moorcroftiana Wall. ex Benth.
63. Vetiveria zizanioides (L.) Nash.
64. Withania somnifera Dunal.
65. Zingiber officinale Roscoe.

10

1. *Artemisia annua* Linn

Family: Compositae nom alt. Asteraceae.

Local names: An Indo-China herb; however, the Indian varieties are called by different names. *A. absinthicum* L. called Vilayati-asfasanthin (Hindi), Kirmala (Hindi) for *A. maritima*; Pardasi-dawana (Mumbai) for *A. persica*; Dona, Jhan for *A. scoparia* in Punjab; Downa (Mumbai) for *A. siversiana*; Nag Douana (Sansk, Hindi) for *A. vulgaris* L.

Description: *Artemisia annua* is a herb considered as a good stomachic, diuretic and prescribed in jaundice and skin diseases. It contains an essential oil (0.3%) and is composed principally of Artemisia ketones, Pinene, cineole, l-camphor, etc. It has also been reported from Peshawar (Pakistan) to Waziristan, ascending to 1700 m. Along with the other species *A. martima* L. is reported to give *Santonin* and a bitter substance *Artemisinin*. The santonin content varies from 1–2% in Kashmir and in Kurram valley. The Indian plant, besides, santonin, contains two more constituents, b-santonin and pseudosantonin devoid of anthelmintic properties. This plant is a native from Kashmir to Kumaon and samples from different regions have been analysed for santonin by the RRL, Jammu. Considering the progressive demand for *artemisinin* (a sesquiterpene lactone endoperoxide) culti-

vation of domesticated *Artemisia annua* has been recommended in order to meet the demands from the leaves and the inflorescences of this plant (S. Kumar, 2005).

Distribution: The enterprises based on *A. annua* are expected to grow and sustain for a long time. The production of artemisinin antimalarial from dried leaves or leaves and inflorescences involves two different kinds of chemical technologies. Firstly artemisinin is solvent extracted and crystallised and subsequently used to synthesise arteethermartem ether and artesunate.

Artemisias are primarily used as vermifuge in traditional system of medicine of India. The main species used are *A. absinthium*, *A. maritima, A. valgaris* and *A. siversiana. A. absinthium* is also used as a tonic in intermittent fevers, and so are the species used like *A. carulifolia, A. maritima, A. persica, A. siversiana* and *A. vulgaris*.

Uses: The content of *artemisinin* in *A. annua* from leaves and inflorescences vary from 0.1 to 0.7%. The cultivars bred in China, Vietnam, Japan, Australia, Netherlands, Switzerland and Brazil have much higher content, in the range of 0.8 to 1.4% (Liersch et al 1986). *A. annua* is a highly cross-pollinated plant; the cultivar quality in terms of 'Artemisinin' content goes down over generation due to segregation, unless

selection for its dry content is exercised continuously to main genes for high arte-misinin content in the population carried forward over generations. Agriculture var. *Arogya* has been bred from segregated material of the cross as for *Jeevanraksha* and *Jwarhati* has been used for breeding this cultivar.

Malarial chemotherapy is based on three classes of antimalarial compounds: anti-foliate, quinolenes and artemisinins. The high artemisinin content variety has been bred from a cross between 2 accessions in which the content of artemisinin was low. Breeded varieties are Jeewanraksha and Jiwanati and has been used for breeding the cultivar *Arogya* (with leaves having upto 1.2% *artemisinin*.

Agrotechniques for the Cultivation

A. annua var. *Jeewanraksha* plants take about one year to complete their life cycle. Plants flower in later September and early October and the achenes comes to maturity in November/December. When the seeds are sown in mid-December seedlings for trans-plantation are available in late January. These plants mature in mid-December of the year of transplanting. Seedlings from Dece-mber sowings can be used in the field transplanting in as late as in May. Seeds can be sown to raise the nursery from mid-December to late April. Nursery raised in April can be used for May transplanting. For transfer the plant density is 70 to 90,000 saplings per ha. This also helps to reduce weed invasion, and allow full growth potential with limited resources input (Gupta et al 1997).

The plantations can be made in the Indo-Gengetic plains. Application of fertilizers and irrigation are much like the food crops. NPK, 80 : 40 : 40 kg per ha under irrigation is given which is as per the *kharif* crops

schedule (Legumes in summer and wheat in winter).

Young plants of *Artemisia annua* have low artemisinin in content during the winter season which rises with the onset of summer in April and remain high until flowering. With the flowering drop, the artemisinin concentration is accompanied by increase in essential oil. Whereas vegetative parts of plants have 0.2% concentration, young leaves are richer in artemisinin content than the older leaves. The crop, thus, may be harvested before the flowering takes place and that harvest is rich in young leaves. Ratoon cropping (multiharvest) can be adopted two to three four times, respec-tively, which gives 175, 207 and 300% higher yield than that harvested once only. Maxi-mum yield of artemisinin is in Brazil (26 kg per ha) against 77.5 kg yield realised in four harvests of annual crops of *A. annua* raised in the Indo-Gangetic plains which has three cropping seasons: July-October/November; during monsoon rain/kharif when rice crop is taken; October, the monsoon rainy/kharif season when rice crop is taken; October/November to March as per the convenience of the farmer. The winter season during which potato, wheat, rape (mustard) and chickpea/lentil and pea crops are taken and March/April-June, the summer/zaid season when *Moongbean* is the principal crop of *Artemisia annua*, in 90–120 days (of *Artemisia annua*) gave 35–45 kg per ha of artemisinin. The crop can also be taken in summer, winter and the monsoon season. Cultivar *Jeewanraksha* 2 can be taken in summer/zaid season without affecting the planting of principal food crops like potato, wheat, mustard, chickpea and paddy.

Post-Harvest Technology

The harvested shoots are shade dried in the summer season or dried by heat circulation

10

from heated air, in drying chambers in the rainy season. The 'artemisinin' obtained is crystallised to obtain the drug.

Among the byproducts is the biomass of the stem and roots, (about 15 tonnes/ha). Vermi-compost from the spent herbage @ seven tons/ha and the essential oil @ 10–100 kg/ha can be obtained. The yield of the essential oil is high when the herbage is rich in the flowers. Herbage is rich in essential oil antioxidants and is suitable for food processing and as pain-reliever, used in dental paste industry. Oil-coated urea can be used as fertilizer and can retard the process of nitrification.

It is felt that *Artemisia annua* shall remain the source of 'artemisinin' for at least more than a decade. No other organism has been found to synthesise artemisinin. Artemisinin and its derivatives are highly potent antimalarials because of their fast action in control of parasitaemia, good toleration activity against multidrug resistant plasmodia and potential to reduce transmission because of action on sexual stages of plasmodia.

References

Sushil Kumar 2005. Prospects for sustainable agricultural production of the antimalarial molecule 'artemisinin' in India. *National Acad. Sciences* Science Letters. **28** (9 and 10) pp. 326–338.

Lierseh R, Soicke H, Stehr C, and Tulhner HU. 1986. *Planta Med*. **52**: 387.

Ram M. Gupta, MM, Dwivedi S and Kumar S. 1997. *Planta Med*. **63**: 372.

Jain DC, Tandon S, Bhakuni RS, Siddiqi MS, Kahol AP, Sharma RP. 1999. US Patent 5: 955, 084.

Jain DC, Bhakuni RS, Saxena S, Kumar S and Viswkarma RA. 2002. Us patent 6: 346, 631.

WHO. Access to antimalarial medicines improving the affordability and financing of 'arte- misinin' based combination therapies. WHO Document (2003).

2. *Achyranthes aspera* L.

Family: Amaranthaceae.

Local names: Apámárga (Sansk), Latjirá (Hindi), Chirchitá, Kutro (Punj), Prickly Chaff-flower (Eng.).

Description: A much-branched more less hairy, straggling or erect undershrub. Stem usually dividing into long rambling branches. Leaves opposite, shortly stalked. Flowers shining, dull green with purple, soon deflected, crowded in long terminal spikes, elongating as the fruit form; bracts and bracteoles spiniscent. Perianth stiff, scarious, 5-parted, segments lanceolate, acute. Stamens 5, alternate with 5-oblong staminodes united with them at the base. Ovary oblong, style long. Stigma capitate. Fruit an oblong utricle, enclosed by the perianth and containing a single seed.

Distribution: Throughout India as a weed, ascending in the Himalaya region upto an altitude of 2000 m.

Three varieties of this plant have been recorded:

1. var. *prophyristachya*, in the Northwest Himalaya.
2. var. *indica* in the Eastern Himalayan region, and
3. Var. *rubro fusca* in peninsular India.

Uses: The plant is valued on account of the ashes, which contains a large quantity of potassium. It is used as an alkali in dyeing, in preparation of the alkaline medicines and caustic potash.

The plant is laxative and used in dropsy, piles, boils and eruptions of the skin. Dried plant is given to children in colic. Water in which the crushed plant is boiled is given to cure pneumonia. It is

also used as an application to wounds caused by *Babool* (*Acacia nilotica* ssp *indica* Thorn).

In large doses, the leaf juice produces abortion or labour pains. Paste of fresh leaves with water is used to allay pain from the bite of wasps and other insects.

Infusion of the root is light astringent and prescribed in bowl complaints and in chest pains. Roots are also given for the night blindness and cutaneous diseases. Seeds are used to cure hydrophobia, also taken in dysentery. Fruit spike is at times used as an expectorant. Seasmum oil and the plant ash is used in the treatment of ear diseases, being poured into meatus.

Extent of utilization: The plant is extensively used in Ayurvedic medicines and also in the Unani system of medicine. Locally, the stem is used as a toothbrush. Petroleum ether extract of the seeds yielded a hydrocarbon an alcoholic extract, a saponin (yield 2%) which is characterised as oleonolic acid-oligosaccharide. The sugar of the saponin is composed of glucose *Galactose* and *Rhancnose*. Oleanolic acid is obtained from the glycosidic fraction of the plant roots.

Other species: *Achyranthes bidentata* Blume, can be distinguished from *A.aspra* by its thin membranous leaves, narrowed in a long slender point. The spikes are slender and are rarely more than 10–12 cm long, as compared to 40 cm in *A.aspera*.

Agrotechniques: In nature, the plant is a weed of wastelands and field bunds, occurs naturally and fairly common which can be collected in large quantities throughout the country. However, it has surprisingly been included by the Uttaranchal Govt as one of the plants to be cultivated. The seeds of the plant are available in plenty which can be collected and stored.

The land can be prepared for sowing after ploughing and the seedlings sown as for any other crop. N, P and K (20 : 20 : 10) can be applied at the time of sowing. Since, the crop is annual, it can be uprooted and stored after drying in a shed and bundled.

References

Gopalachari R, Dhar ML. 1959. *Journ. Sci. and Industr. Res.* 11B: 209.

Gopalachari R. Dhar, ML. 1958. *Journ. Sci. and Industr. Res.* 10B. 276.

Khastagir, HN. and Sengupta P. 1958. *J. Indian Chem. Soc. 35.* 529.

3. *Acorus calamus* L.

Family: Araceae.

Local names: Bach (Hindi), Sweet flag (English), Yekand (Tamil).

Description: A semi-aquatic perennial, aromatic herb with branched rhizomes, creeping underground, horizontal, jointed, somewhat vertically compressed (5–15 cm in length and 1–2 cm thick). These are covered with brownish epidermis, with deep longitudinal wrinkles. Upper surface is covered with triangular leaf scars that encircles the rhizomes, springing from each side alternately. Lower surface bears irregular zigzag line of small raised roots scarcely. Leaves grass-like or sword-shaped, long and slender. Flowers small, yellow green and arranged in a spadix inflorescence. Fruits are green berries. Seeds oblong in shape and 1–2 in each fruit.

A polymorphic species, the Jammu var. shows a tetraploid 2n 184 × 36 chromosomes and 2n = 56, the Kashmir 2n = 54 and hexaploid the lowest is 2n = 18. The tetraploid race has high oil content and *asarone* and has maximum capacity for development.

10

Distribution: In nature it is found on wetlands and freshwater swamps (Mathronwala swamp near Dehradun) and on sandy loam soils. It also occurs in Europe and USA. *Acorus cochinchinensis* occurs in Japan, *A. triqueter* in Northeast Asia and *A. americana* in USA. In India, Himachal Pradesh, Uttaranchal, MP, Bihar near river and rivulets, it has been widely reported. It is also cultivated abundantly in MP and other states, like Karnataka. Renewed interest in the herbal medicines has initiated its cultivation in many regions. Though it is a perennial in nature, yet treated as annual, since planting and harvesting is done on year after year basis.

Uses and utilization in trade: Dried rhizome powder with honey is given to babies as tonic, as a protection against vermin. Infusion in water is used as to wash newborn calves. Rhizome is also used as cure for headache, cough, asthma and strengthens teeth. Root powder is used as aphrodisiac and carminative, also as an antidote to *Croton* poisoning. Dried rhizome is prescribed as antispasmodic, carminative, anthelmintic, used in diarrhoea, dysentery, vertigo, bronchial catarrh, intermittent fevers, flatulence, epilepsy and mental ailments, kidney and liver trouble. The oil possesses anti-allergic and cryoprotective properties.

Asarone, a constituent of oil is a mild sedative, potent tranquilizer, mild-hypotensive and hypothermic. The oil is used for aromatic, cordial and liquors for flavouring beverages and perfumes.

The calamus oil: The period of growth of rhizome, location and the season of collection vary the oil content. Fresh rhizome gives 1.8, dried rhozomes 1.5–3.5, leaves 0.2 and fresh aerial parts 0.12% of oil. Dried and unpeeled rhizomes give the highest yield. It is also reported that the amount of essential oil

increases with chromosomal numbers. Diploid varieties have on an average 2.17%, triploid plants 3.1% and tetraploid have 6.82% oil. Unpeeled dried European roots yield 1.8–4.5% oil, whereas the Japanese material yielded upto 5% oil. The oil possesses anti-tubercular action *in vitro* and has insecticidal and pesticidal, toxic, antigonadal properties against bedbugs, moths, lice, housefly, mosquito, potato-betel and stored grain pests (agricultural pests).

Agrotechniques for the Cultivation

Plantations of rhizomes are done in humid marshy soils, while on sandy loam soils it is planted under irrigation. Soils, suitable for paddy cultivation, are considered to be suitable for this crop also. When it germinates, new sprouts are planted at a distance of 30 cm × 30 cm and about 4 cm deep into the soil during July and August. It can hardly grow on a soil from clayey loam to light alluvial. The tops of previous years crop are planted at 30 cm apart, leaving leafy portion well above the ground. After planting, the field is allowed to go dry for 10–15 days during which the roots emerge. Thereafter, the field is irrigated regularly avoiding drying of the field, while the crop is growing. Since, dry conditions affect adversely the thickness of rhizomes, several roots develop or suckers come up. The crop needs plenty of water, as rainfall, or from irrigation and ploughed before plantation is done. For a uniform planting material several micropropagation protocals for *A. calamus* using different kinds of explants have been reported. About 8000 tops are needed for an ha. Tops from 1 ha of *A. calamus* field provide planting material for 2–2.5 ha. For the land preparation 2–3 ploughings are done before the onset of monsoon so

as to prepare the land as for the paddy cultivation. About 12 tonnes of FYM/ha is applied and mixed in the soil, before planting the rhizomes. Marsh and swampy lands near *nalas* and rivers with standing water are the ideal places for its cultivation. Weeding and hoeing operations are required for a healthy crop.

The crop is generally free from pests and diseases apart from stray attack of leaf spot and mosaic streak diseases. It does not stand extremes of temperature. Temperature between 10°–30 °C seems to be the optimum range. With rise in temperatures the crop may suffer and to avoid such a situation, planting is recommended during March which can be harvested during January-February; though it can also be planted all-round the year.

Though, harvesting can be done after 6–8 months of planting, when the leaves start getting yellow, yet one year is the maximum when the crop develops signs of maturity which is indicated by the leaf fall.

The plant is dug out, rhizomes removed and leaftops kept covered with dry leaves. The rhizomes are taken out by pulling up or digging the entire plant. The rhizomes can also be taken out by ploughing the field. These can be stored in gunny bags in dry places for 2 years, without any deterioration. Rhizomes are also cut into pieces and dried in open shade, so that there is no loss of oil.

Studies on the production and quality of oil obtained from plants raised in *tarai* region showed that it is possible to cultivate this species in areas submerged during monsoon. Physicochemical properties of essential oil were similar to that of plants produced at Bhimtal (at 1330 cm) in Kumaon hills and other parts of the country.

Production of the rhizomes per ha of about 42 qtls/ha can be obtained. The cost of cultivation at the present rate is estimated at Rs 32,000 per unit of ha. Gross profit is estimated at Rs 84,000/per have with a net profit of Rs 52, 000/ha at the existing rate.

It can also be cultivated in combination with paddy cultivation and alternated between paddy cultivation.

Uses: The rhizome is used in gastric, bronchitis, indigestion and hysteria complaints. Some of the brand names of medical preparations are as follows.

HB Pharma, Rajkot	*Ceumina syrup*: Intestinal worms. *Spasmotier pin syrup*: Antispasmodic. *Antospray powder*: softening of skin. *Dantdookh powder*: Dental problems. *Autoar drips*: Infection of middle ear. *Spasmocaronic drips*: antis pasmodic and carminative.
Silpachem, Indore	*Livovel drops*: Diseases of liver. *Livolvel syrup*-Liver diseases. *Saivil tablet*: Tranquilizer. *Silpzyme syrup*- cure for indigestion.
Charak Pharmaceutical, Bombay	*Driconil syrup*-Cold, cough and asthma. *Bed tablet:* epilepsy *Rimenil tab:* Bone swelling, rheumtism and backache. *Cepra tab*: Mental tension. *Tranquinil:* Tranquiliser.
Atsin, Bombay	Tonic for sex. *Suktin:* Gastrocardiac syndrome, flatulence, dyspepsia, gasteritis, peptic ulcer. *Sildin*: Psychosomatic diseases and tranquiliser. *Smyle-syrup*: for cough, throat trouble candy-cough, etc.

10

References

Dandiya PC et al 1958. Studies on *Acorus calamus*. I. Phytochemical investigation. *Canadian Pharm. Journ.* **91**(10): 607–610. 11. Investigation of Volatile oil. *Journ. Pharm. London* **11**: 163–168, 1959. Investigation on mechanism of action in mice, Journ. *Pharmacol. Exptl. Therapy*, **126**(4): 334–336, 1959; V. Pharmacological action of Asarone and beta Asarone on Central nervous system. *Indian Journ. Medical Res.* **50**(1): 46–60.

Kapil VB, Duhans PS. Sinha GK. 1975. Production and quality of Bach (*Acorus calamus*) raised in Tarai region of Nainital District. *Proc. Nat. Symp*, pp. 10.

Sharma JD et al 1962. Studies on the *Acorus calamus*. Pharmacological actions of Asarone and beta asarone on cardiovascular system and smooth muscles. *Indian Journ. Med. Res*. **50**(1): 46–60.

Srivastava JL. 1997. Daldali Kheto–ke–liy Sarvagenic upyukhe, Bach ki-Kheti. *Uddhamita* (Hindi) pp. 31–32.

4. *Alpinia galanga* Willd.

Family: Zingiberaceae.

Local names: Bará-Kulanjan, Motha-Kulinjan (Hindi); Sugandhá- Vach (Sansk); Bari-pán-ki-jar (Mumbai); Péra rattai and Peddadumparashtrá (Tel), Greater or Java Galangal; .

Description: A stout perennial herb; root stock tuberous, slightly aromatic. Leaves oblong-lanceolate, acute, glabrous, green above, paler beneath, with slightly callus white margins, sheaths long, glabrous; ligule short and rounded. Fls. greenish white, in dense fld. panicles; bracts ovate-lanceolate. Calyx tubular, irregularly 3-toothed. Corolla lobes oblong, claw green, blade white, striated with red, rather more than 1 cm long, broadly elliptic, shortly 2-lobed at the apex, with a pair of subulate glands at the base of the apex, with a pair of subulate glands at the base of the claw. Fruit the size of a small cherry, orange red.

Distribution: Throughout India, often cultivated in Konkan, North Kanara. It is mentioned by Marcopolo (AD 1290) to be grown in Bengal and by Varthema (1510) as found in Cambay. Garcia de Orta (1513) and Linschotan (1598) mention that there are two sorts; one Chinese, called *Lavandon* and the other (*Javan*) called *Languas*. The latter they say was sown in Indian gardens. Rhizomes of both the types are mentioned and figured by clusius (Hist. Ext. Plants pp 1–211, 1605).

The rhizomes and the fruits are sold in the market and both are said to be used for calico-printing in UP.

It seems that the greater *Galangal* is only used in medicine as a substitute for lesser Gangal. It has weak antibacterial and antiprotozoal activities (like those of *Pistacia integerrima, Piper betel* and *Nardostachys jatamansi*. In moderate doses it has antispasmodic action on involuntary muscles and tissues, inhibiting excessive peristaltic movement of the intestines.

The cardiovascular and respiratory systems are not much affected. The essential oil has depressant action on the central nervous system of mammals.

Other species: *Alpinia allughas* Roxb. (vern. Taro) a native of Bengal and South Konkan. A stout perennial herb with tuberous aromatic, roots. Leaves oblong lanceolate. Flowers inodorous, pink, in erect decompound lax or dense fld. panicles. Rachis pubescent. Calyx lip more than 2 cm long. Fruit black, globose. Properties are similar as of *A. galanga*. The rhizomes are used in rheumatism, fever and catarrhal affection, especially in bronchial catarrh, stomachic, used as a flavouring agent. The

essential oil is used in respiratory troubles, especially of children.

Alpinia officinarum Hance (Vern Kulanjan-Hindi; Sugandha Bach (Sansk), Kalijana, Choti-pan-ki Jar, Shilla ratti, (Khorudaru). The rhizomes are stimulant and carminative. The essential oil is *galangin*. It is native of China. The oil used to be imported. Principal demand of this plant is in Russia, though it is still an ingredient of old-fashioned English medicine. In India, it is considered as a nervine tonic.

Alpinia speciosa Schum. (*A. nutans* Rosc.) Vern. name: Light galangal (Eng.), Punnag champá (Mumbai), is a native of Eastern Archipilago in place of the Greater or Java Galangal and is, sometimes, mixed with it or with the ginger. It contains essential oils also.

Other species like *A. calcarata* Rosc. is used as substitute for *A. galanga* and is cultivated in Konkan area. *A. malaccensis* Rosc. (–Saliyeridumpa-Tam) in the hills of Assam and Bengal, eastern and western-ghats. The plant also contains an essential oil.

Agrotechniques for the Cultivation

The agrotechniques are similar to that of tumeric and ginger.

5. *Anethum graveolens* L. (A.sowa Roxb.ex Flem, *Peucedanum graveolens* L. in part).

Family: Umbelliferae or Apiaceae-nom alt.

Local names: Indian Dill. (Eng.); Sulphá, (Hindi), Satapushpa, (Sansk).

Description: An annual or biennial herb, aromatic, tall, 1.0–1.2 m high. Leaves compound, light green. Flowers yellow, in compound umbels. The dry and grooved fruit consists of two brown, broadly-oval, compre-ssed, indehiscent, one-seeded carpel (meri-carp), 3–4 cm long, 2–3 mm broad and 1 mm thick. They have 5 ridges, the three dorsal being inconspicuous and brown, the two lateral yellowish and wing-like. It should not be confused with *A. sowa* Roxb. ex Fleming, also the Indian Dill known as *A. graveolens* var *sowa* containing 3.0–3.5% of the essential oil, which is cultivated as a winter crop and used as carminative. The oil distilled from *A. sowa*, however contains appreciable amounts of *Dill Apiol* which is toxic. The flowers are small, light yellow and one borne in umbels. The fruit is broader than that of *A. sowa*. The umbels are also bigger than that of *A. graveolens*. In other respects the two plant species are alike. The oil is borne in the oil canals, known as *vitae* in the fruit.

As Dill is a cross-pollinated plant and visited by the bees, the two varieties should never be cultivated in the same vicinity. The dry and grooved fruits are brown, broadly oval, compressed carpels (mericarps), 3–4 cm long, 2–3 cm broad and 1 mm thick. They have 5-ridges, the 3-dorsal being in conspicuous and brown but the other two lateral, yellowish and wing-like.

Distribution: Dill is native of Europe but wild in Iran, Egypt, Abyssinia and Northern Russia. Cultivation on the commercial scale was in USA, Hungary, Germany, Holland, S. Russia and England. The fruits and its products in India were imported from Oxford upto 1951 but are now successfully being cultivated since 1962 at RRL, Jammu (Chopra and Sobti, 1962). The Indian Dill is also grown as a winter crop. However, the fruits resemble in odour and have a sharp taste akin to it. Calcicole habit of the plant has been indicated by Gupta et al (1962), while studying the effects of salts and hormones.

The fruits resemble Carawy in odour and have a sharp taste similar to *A. graveolens,* is used for flavouring pickles, soups, sauces, stews and salads, etc. The essential oil is an excellent carminative and used in medicine in the form of dill water which is a remedy for flatulence of infants. The oil distilled from *sowa* is used extensively in gastrointestinal disorders of infants.

Dill oil is also used for flavouring various food stuffs and in soap industry. Now Dill herb oil is used in place of oils from fruits. Leafy stem is also used as an aromatic flavour for pickles and soups.

It is recorded to grow well on sandy loam soils, though it can be grown on different soils as a winter crop and in much cooler climate yet not at altitudes where growing season is very short (less than 60 days). It does not do well on poorly drained soils. Trials conducted at Dehradun indicated a seed yield of 154 kg/ha in the first year, while in the second year it was between 16–34 kg per ha under rainfed conditions (Gupta, 1974–75 (Annual Progress report APR CS and WCRTI. DDn).

Propagation

The plants are raised from the seeds and bear primary, secondary and tertiary branches, each terminating in an umbel. *A. graveolens* has been raised as an experimental crop by RRL, Jammu in winters, while in Kashmir as summer crop. The seeds are sown directly in the field, early in spring and the plant matures in the autumn season (Sept-Oct). Heavy rains or late snowfall in spring adversely affect the seed germination. In plains where the summer is very hot, the seeds are sown during the second fortnight of September when here also, heavy rain or snowfall adversely affects the seed germination. The plant ripens in March/April, when the crop is harvested. The depth of

seed sowing is 0.5–0.75 cm on loose soils @ 3 to 3.5 kg/ha.

In the plains the land is prepared after heavy rains, at least a week before sowing and is ploughed 6–8 times, till the soil is completely pulversied. Later, the field is disced and harrowed, FYM @ 20–25 cartloads per acre is added and mixed well with the soil. Before sowing the seeds, all weeds and grasses are removed from the field.

Dill can be grown on different soils but it grows best on fertile loam in mild cold climate. Studies of Gupta et al 1962, on effect of salts and hormones, indicated a calcicol soil habit.

Removal of weeds is necessary for successful cultivation; of course with quality seeds. When the seed is sown in rows, the average seed rate is 5 kg/acre but when broadcast the rate is 7.5 kg per acre. Seed germination starts in 7–10 days after sowing. The first thinning is done a month after the sowing, when the plants are about 2–5 cm high and the other when in bloom. 4–5 hoeings are done to keep the soil open and keep weeds under control.

Application of chemical fertilizers does not affect the oil quality and nitrogenous fertilizers @ 20–25 kg N/acre as ammonium sulphate should be applied, one after the second thinning when the plants are just in bloom; however, 4–5 hoeings are necessary to keep the soil open as stated earlier. 8–9 irrigations are required, till the plants mature. However, larvae of two different kinds of Lepidoptera cause a damage to leaves at this stage.

Harvesting

The seeds do not mature at a time, while the old umbels mature, the new ones are still flowering. The process goes on for about a month. The quality of oil depends upon the time of harvesting. When grown for the herb oil, the crop should be harvested when the fruit is turning brown, while for the fruity-oil it should be harvested when the earliest fruit ripens. Mature seeds are easily shed while plucking the fruits; consequently the yield gets lowered. However, the best time for seed collection is while the umbels are still greenish, yet fully mature.

The umbels should be dried in the shade, which helps to maintain oil percentage. The oil quality too depends upon the process of harvesting.

Yield

Fruit yield per acre is 560–784 kg in USA but under the Indian conditions it is about 448 kg/acre. The yield of green Dill in Hungary is recorded to be 1000–2000 kg per Hungarian "Kasaster Joch", i.e. 1.4222 acres. In some of the areas, the plant flowered but did not form the seeds and are self-incompatible and would form seeds when fertilized by pollen of plants having different genetic constitution.

Though, the entire plant is aromatic, the fruits are the richest in oil (2–4% in the fruit and 0.29–1.5% in the herb). According to the British Pharmacopoeia (BP), Dill oil from fruits contains less than 43% W/W and not more than 63.6% of W/W (wet weight) Carvone. The entire plant on distillation yields 15–16 tones/ha with 100 kg of oil. Average oil content is 2.85% (dry weight). Maximum amount of oil is in the inflorescence (6.24% dw) followed by the leaves (1.14%), stem (0.6%) and root nil. The oil content increases with the plant growth, reaching the optimum at milk stage of fruits.

The average seed yield varies upto 420 kg/acre, from this 15–20 kg of oil can be distilled. Sometimes, the entire green herb is distilled and the herb oil is marketed. An acre yields about 200 maunds of green herb containing 40% of moisture. This, on distillation, yields about 90–100 lbs of herb oil. The physicochemical content of dill herb and herb oil compares favourably with the oil distilled abroad (Kapoor et al 1953, Guenther, 1949).

The oil obtained from plants raised in Kashmir was found to contain 58% carvone, together with d-limonene, d-phillandrene terpenene, and a waxy product (m.p. 63 °C) (Chaudhury et al 1957). Leaf oil contains phellendrine, while ripe fruits contain 6% carvone together with d-limone, d-phillandrene turpentene, and a waxy product.

In case the seed sowing is delayed, the plant remains stunted. The oil content increases with the plant growth reaching an optimum at the milk ripening stage of the fruits. Phellandrine is the principal constituent of the herb, while oil of ripe fruits contains 46% of Carvone. The oil is used in food preparations, perfumery, composition and formulation of certain brands of toothpastes.

References

Chopra IC and Sobti SN. 1962. Cultivation of medicinal plants in Jammu and Kashmir. *ICAR Publ. Res. Series* No **12**, pp. 6–10.

Chaudhary SS, Handa KL. 1956. *Indian Journ. Pharm.* **18**: 421.

Guenther CB. 1949. *The Essential Oils.*

Gupta US, Kaul BK and Kapoor LD. 1962. Individual and Combined Effects of some Salts and Hormones on the Germination and early growth of *Anethum graveolens. Proc. Seminar on sea salt and plants.* pp. 137–142.

Kapoor LD. Handa KL and Chopra IC. 1953. *Journ. Sci. Industr. Res.* **12**: 311.

Kaul BK and Kapoor LD. 1962. Preliminary study on the effect of GA on Dill seeds. *Planta Medica.* **91**: 57.

Happrc 2004. Cultivation of High altitude Medicinal and Aromatic Plants. *Pub. Bull.* No **1**–20.

10

6. *Angelica glauca* Edgew.

Family: Umbelliferae or Apiaceae nom. alt.

Local names: Chorá, Churá, Chorak (Hindi, Himachal), Gandarayan.

Description: A glabrous herb with erect, hollow, finely grooved stem. Leave usually large, 1–3 pinnate; leaflets often in 3's, sometimes, reduced to one, ovate or lanceolate, undivided or lobed, irregularly or sharply toothed, upper surface dark green, lower glaucous. Umbels compound, long stalked; bracts several, linear, bracteoles linear and many. Flowers white or purple, many in an umbel. Calyx-teeth none. Fruit white or purple, oblong, dorsal and intermediate ridges not winged, lateral ridges expanded on into membranous broad, free wings so that the fruit is surrounded by a double or two-leaves border.

Distribution: In temperate regions of the Himalaya, from Kashmir to Shimla in Himachal Pradesh, between altitudes of 2500–3300 m. It requires a cool climate and deep rich porous soils. Heavy input of organic matter (FYM @ 60 qtls per acre) is required. The crop also needs irrigation. Crop harvesting is done after three years of plantation and 4–5 years if sown by seeds.

Extent of utilization and trade: Small quantities of the plant are exported from the hills, when the traders bring the plant (leaves and roots) and sell it or exchange with other products.

Its cultivation has been tried in Garhwal hills through root segments or seeds with success. The roots are harvested in the month of October, after 3 years of sowing; however, a harvest after two years of sowing is reported to given maximum yield.

Uses: The dried roots contain an active principle which is an essential oil (1.2%).

Roots used as condiment in cookery while rootstock, roots and fruits are stimulant, expectorant and diaphoretic. Dry roots are also reported to contain 1.3% of the essential oil.

Agrotechniques for the Cultivation

The technique for the cultivation of this plant is almost similar to that of *Anethum graveolens*. The plant can be put under cultivation at high altitudes of Kashmir (Pirpanjal, Sonmarg, Kagan valley), Himanchal Pradesh (Shimla, Narkanda, Huttoo, Kunawar) and Garhwal (Gangotri region, Nila valley and Lambatch area) where the production is about 400 kg/acre. The herb is useful in dyspepsia and constipation.

Angelica archangelica L. (*A. officinalis* Hoffm.)

Oil from the roots, growing in wild, contains mallic acid, pinene, phellandrene, osthole, 15 hydroxy pentadecnic acid, a solid (m.p. 30–39 °C) and appreciable quantity of terpenes and sesquiterpenes (Chaudhry et al 1958). The roots are stimulant, expectorant and diaphoretic. Dried roots contain 0.35% of essential oil of which phellendrine is the main constituent. The fruit is reported to contain several furo-coumarin as angelicin, bergapeten, xanthotoxin in addition to umbelliprenin and some phenols.

References

* Chaudhary S.S, Singh G. and Handa KL. 1958. *Indian Pharm.* **20**: 107.

7. *Atropa acuminata* Royle ex Lindl.

Family: Solanaceae.

Local names: Angurshafá, Sagangur (Hindi), Mait-brand (Kash), Bantamaku (Punj.), Yebrui, Girbuti, Jharka (Mumbai).

Description: An erect, tall, perennial, bushy herb, 25–125 cm high and dichotomously branched. Leaves oblong elliptical, tapering gradually at both the apex and the base; paired on the flowering branches, green or olive green in colour and each from 12 cm in length and 4–6 cm in breadth. The flowers are axillary, dirty yellow or olive grey, drooping, campanulate, solitary or in groups of two or four. It flowers from June to August. The fruit is globular berry, and looks like cherry, when ripe in October. The rootstock is hard and woody, 2–5 cm in width and on drying, the roots get wrinkled longitudinally.

Distribution: The plant grows in Kashmir, Chamba in Himachal at altitudes between 2000–3500 m. In Kashmir, recorded from Chitral, PirPanjual, Trgol and from Chenab and Kagan valley. The plant grows in shady places on heavy soil particularly where deforestation has taken place. A porous and loamy soil, containing decomposed humus, is suitable for its normal growth. It has been found to flourish as an undergrowth on well-drained places in Pine or Fir forests.

Atropa acuminata was reported as *Atropa belladona* in Hooker's Flora of British India, but now named as *A. acuminata* Royle ex Lindl. Mehra and Sobti (1954) suggested that it may be named as *A belladona* var. *acumintata* due to the morphological, anatomical and cytological affinities with *A.belladona* L.

Uses and extent of utilization in trade: Medicinal properties of this plant were given recognition in 5th Pharmacopoeia in 1932 and the standards were laid down for leaves and roots having not less than 0.15% and not less than 0.25% of alkaloid respectively. These specifications have been modified since then and, brought to the standard as those of *A. belladona*, i.e.

for leaves not less than 0.3% and roots not less than 0.4% of the alkaloids. However, Indian specimens on analysis from the natural habitat varied from 025 to 0.48% and for roots 0.3–0.8% of the alkaloid. The seeds from different countries when analysed contained the alkaloid as Turkey: 0.41–0.52, USA: 0.39–0.4, Oxford: 0.36–0.46, Kew in UK: 0.4–0.62% in the first and second year of plantations, while the corresponding figures for *A. acuminata* are 0.5–0.72%. It has also been recorded that the alkaloid percentage was maximum in the leaves at the time the plant had started flowering (July-August). As the season advanced, the alkaloidal contents of the leaf decrease and are minimum at the stage of fruiting which takes place during the months in November.

The dried leaves and aerial parts constituting the drug are narcotic, sedative, diuretic and used as anodyne, bring a decrease in secretion of sweat and of salivary and gastric glands. It acts as a strong antispasmodic in intestinal colic and other spasmodic indications. It is also useful in asthma and whooping cough.

The drug obtained from the roots of the plant has similar properties, as the one obtained from the leaves and twigs. The roots are, however, believed to have contained certain poisonous substances and are employed chiefly in the preparations prescribed for external applications in rheumatism, neuralgia and inflammation, etc.

The drug is collected at the time when the plant is in flower; roots are, also included. The excessive collection of this drug from the forests of Kashmir and Himachal has enormously decreased the yield of the drug resulting in shortage in the availability for the Indian pharmaceutical industry. Thus, in order to ensure a steady supply, cultivation of this plant was taken up simultaneously by the RRL Jammu at

various places in the temperate climate of Kashmir, where the experiment succeeded with an alkaloid content of 0.64 at Yarkah (2300 m alt.) and 0.52 at Srinagar (about 2000 m alt.) from *A. belladona* and 0.34–0.77% of alkaloid from *A. acuminata* leaves when grown at Katra (300 m), and 0.77% from Srinagar (2300 m).

Agrotechniques for the Cultivation

A. acuminata can be propagated by seeds sown in early March which takes 4–5 weeks for germination. The time can be considerably shortened by pretreating the seeds for 2 minutes with concentrated sulphuric acid. Untreated seeds can be sown in June–July and take 10–15 days to germinate. The seedlings are transplanted in the following spring. Propagation by roots can be done by planting roots of 3–4 years old plant, cut transversely in small pieces, 2 cm thick with a growing bud in each piece, placed well in prepared soil. Propagation by shoot cutting and transplanting is successful. The new buds and leaves emerge in about a month's time. The plants do not stand water-logging. Lime application is reported to increase total alkaloid between 0.15–0.7% in leaves and 0.3–0.66% in roots as reported from plantations in Himachal hills.

The three different methods of propagation through seeds, root divisions and shoot cuttings have been studied by the RRL Jammu at different nurseries at Srinagar (2700 m) and Yarikah (2300 m); however, the nurseries required proper weeding, frequent irrigation and were shaded in order to avoid direct sunlight till the seedlings attained some size (Kapoor et al 1952). Dutta and Mukherji (1950), however, reported that belladona seeds germinate in 14–21 days, when the day temperature is about 70 °F (20–21 °C) and the low night temperature.

The seedlings are transplanted in well-prepared beds after they have developed roots, i.e. in early autumn. The vegetative growth of the autumn transplanting is stunted and the frost adversely affected the young roots of plants; thus, it is considered better to transplant the seedling in the following spring, when they have better and quicker vegetative growth than that of the plants sown in autumn. There is also an advantage, that the cutworms (*Agrotis flammetra*), active in June-July, can be avoided and the damage to young plants can be prevented.

Seedlings transplant is done 5–7 cm deep with roots laid upright with soil around pressed a little. June-July plantations need some shade or cover with pine needles to minimise transpiration from young seedlings, before the roots are well established in the soil; since young seedlings are very sensitive. The roots of 3–4 months or older plants are transversely cut into small pieces, 2–3 cm thick, with a ground bud in each piece which could be planted in the soil for further propagation since from each vigorous plant grows in a short time as the parent plant.

The shoot cutting method consists in cutting a growing shoot and transplanting it in well-prepared beds and irrigating it. Propagation is better, if done in early spring but a large number of shoots of the parent plant may not be available in the early spring. Shoots of 30–45 cm long are cut at their bases, 30–45 cm long and planted about 15–20 cm deep into the soil and large leaves removed from the base of the plants. When planted in spring and irrigated frequently, new buds and leaves appear in about 30 days.

Atropa crop requires a fair amount of irrigation for proper development of the leaves and roots, though water logging of

the field has a fatal effect on the roots. Therefore, the soil should be well-drained and irrigated manually. In dry months of June, irrigation is particularly advantageous. Application of manures and fertilizers (lime, ammonium sulphate, FYM and forest humus) increases the total alkaloids; however, only ammonium sulphate application is not significant. Depth of tilling and the plant spacing also affect the percentage of total alkaloids; the percentage being maximum when the soil is tilled 20 cm deep and the plants spaced 45–70 cm apart (Chopra and Sobti, 1957). The weeds should be rooted out before flowering begins; the common weeds are *Ageratum* species, *Verbascum thapsus*, *Senecio* species.

When the plant starts flowering like, in first week of August, maximum alkaloid has been reported in the leaves. The plants are cut 5 cm above the ground and the field irrigated soon after. New shoots start coming up profusely and this flush is harvested. The harvested plants are chopped into pieces and dried by spreading it into thin layers in the sun and turning them upside down. The first crop of aerial parts averages (leaves and stems) green weight of 1800 kg per acre, in the second year when the area is fully stocked; ratio of leaf to stem being 50:50. On an average to an acre is about 250 kg of dry matter with 44% stem and 56% leaves. Percentage of moisture is more in the stem than in the leaves; average moisture content being about 88% and that of leaves about 80%.

Common Pests and Diseases

Agrostis flamentra, the cut worm or grub is fatal, which cuts the succulent stem, above the ground that causes death of the plant. It is most virulent during dry months, while the incidence decreases during the rainy season. Transplanting seedlings should, therefore, be done and completed during April-May. The larvae of *Gonocephalum* sp. also infest plant-root in addition to a number of caterpillars and beetles which attack leaves, flowers and fruits. Attack of fungi causing downy mildew has also been reported.

Data from experiments conducted by RRL, Jammu have also shown that the leaves raised from seedlings or root cuttings contain low alkaloid percentage in the first year and show remarkable increase in the second year.

8. *Atropa belladona* L.

The generic name *Atropa* takes its origin after the Greek goddess of death "Atropos", while the species *belladona* is from the Italian language where 'bella' means a woman and 'donna' stands for beautiful. The plant occurs wild in Britain, Greece, Italy and most countries of Central and Eastern Europe, former Yugoslavia, USSR, Rumania and the UK providing the drug to the international markets. Its root called *Belladona radix* and leaves *Belladone folium* are of medicinal importance. The plant is a perennial herb, which grows upto 1.5 m in height, branches freely and possesses a large tapering root. The leaves are alternately arranged but adnation exists in the upper portion of the stem, giving them a paired appearance; it is 20–32 cm long and 10–12 cm broad with a very short stalk. The leaves near group of 2–4 flower in their axil during June-September. These are large, bell-shaped and purple with violet brown nerves. The fruit is purplish-black, globose, many-seeded berry.

***Distribution*:** Belladona is one of the few vegetable drugs that has withstood the test of time and its demand rose manifold, mainly for the use by armed forces during the second world war. The

demand has continued to grow and presently it is reported that there is scarcity of this raw material in the world market. It has been cultivated in Kashmir and Himachal at Chamba and in Pakistan (Khan, 1957). In Balkan mountains two varieties are recorded var. lutea and var. nigra. Improved var. atropa, 1029, an early blooming and high herbage yield is an improved variety.

***Uses and extent of utilization and trade*:** The pharmacological drug consists of leaves and roots, which contain three important alkaloids in varying quantities. The roots are used for external applications to relieve neuralgic and other pains, while leaves and their alkaloids are employed for internal administration. The alkaloids atropine and its 1-racemer hypocyamine are used essentially as preanesthetic agents to check secretion in throat and respiratory passages; it also relieves spasm and soothes muscles for which it is given in gastritis, pancreatitis, chronic ureteritis and biliary colic and similar other conditions. Hyocine, which forms only 5–11% of the total alkaloid, is used as a truth confessor in criminological investigations. It also accords protection against violent shocks like bomb blasts.

During middle ages, the 'belladona' herb used to be employed in poisoning since it produces slow, ambiguous and unidentifiable symptoms. It is known that Italian women used its berries in cosmetics, its juice causes dilation of pupil, making the eyes large, lustrous and dreamy. Its use in pharmacy is a chance discovery. It is said that in 1776, Daries, a clerk in a Hamburg Drug Store, inadvertently rubbed the plant into his eyes which led to the finding of its mydriatic property. Its analgesic and antispasmodic properties were discovered later. The isolation of pure atropine in 1931 increased its wider application and popularity in medicine.

The plant was introduced under cultivation for the first time in 1907 in the Kumaon hills and since then, it has been successfully cultivated at a number of experimental stations in North-west Himalaya and the Nilgiris in south. However, its cultivation for a long time has been confined to CIMPO farms and RRL, Jammu. According to a report, the total production in 1969 amounted to 5 tons only of leaves and 1.3 tons of roots, while the demand of the industry at that time was estimated to be at 150 tons. The demand is met mainly by import of the alkaloids, concentrates and their end products.

Agrotechniques for the Cultivation

Soil and climate: Experimental cultivation of the plant was started from the seeds obtained from the Kew Gardens and from other sources like in USA and Turkey. A well-drained silt-loam soil, rich in humus and slightly acidic in reaction is considered suitable for this crop. It is sensitive to water logging, while in nature it has been reported to grow on locations both to direct sun and under light partial shade. The crop required clear dry weather during picking for high alkaloid content in leaves and its rapid drying; continuous dampness and high humidity during winter months favour root rot.

Land preparation is done upto 30–35 cm depth of soil, while deeper digging is considered better. It is grown on deep fertile, slightly acidic soil, as a perennial rainfed crop. Organic manure at the rate of 40–45 tons/ha is applied to the soil at the time of land preparation. 70 kg N, 80–85 kg P_2O_5 and 30 kg K_2O dose of fertilizer, is recommended during the first year with 50 kg N and 40 kg P_2O_5 in second year.

Propagation

The plant could be propagated by seeds, and root-shoot cuttings. However, raising of seedling in nursery or direct broadcasting of seeds in the field are recommended for raising a commercial cultivation. The seeds are very small, brownish-black in colour with hard seed-coat and weigh about 700 seeds per gram. Its germination is slow and erratic; varies from 15–40% in about 30 days time. It requires prechilling or stratification of seeds before sowing to improve the germination. Bhat and Dhar (*Indian J. Agric Sc.* 41(9): 761–4) have shown the presence of a water-soluble inhibitor in the seed-coat. Seed treatment with ethyl alcohol for 3 hrs or with petroleum either for 6 hrs has been recommended which improves the percentage of germination. It could also be improved by treating the seeds with concentrated sulphuric acid for 30 seconds followed by washing in running water for 15 minutes. Such treated seeds are reported to give stunted plants (Lloydia, 1964, 24: 211–216). 4 kg of seeds are required when sown in nursery for 1 ha plantation, while direct sowing requires 20 kg seeds per ha. Hussain and Janardhan (1966) recommended general treatment of the seeds with Captan or Agrosan, before sowing, to protect the seedling from seed-borne diseases. Mixing treatment of the seeds with KOH solution (1.5%) for 24 hrs also enhances seed germination upto 70%. Seeds are sown at 0.5 cm in the soil in late February, when soil temperature is 5–8 °C, at a spacing of 70 cm × 30 cm. Four interculture operations are given.

A good soil levelling with a slight gradient is helpful in maintaining a good crop stand. The seedlings having 1–3 leaves are ready for transplanting in the field during August. It is given spacing of 60 cm × 60 cm and the planting may be done on the ridges in local ties receiving heavy monsoon rains.

Manuring and Interculture

The crop responds favourably to the application of FYM and inorganic fertilizers. In USSR 40 t of FYM, before planting the crop in the field, gave 63% more yield than control. Being highly responsive, 80 t/ha of FYM increased yield by 71% and NPK combined (60 kg/ha each) gave an increase of 38%. Effect of nitrogenous fertilizers also increases the quality and quantity of belladona herbage, while according to some authors there is an increase of alkaloid content with low doses of nitrogen application but the overall total alkaloid and herbage production per ha is increased. Gupta and Shah (1975) recorded that higher levels of nitrogen (calcium ammonium nitrate) (CAN @ 250 kg/ha) and phosphorus (single Super-phosphate @ 150 kg/ha), increased yields of aerial parts from belladona at Govt. Farm, Almora. Higher levels on nitrogen (250 kg/ha CAN) and P_2O_5 (SSP 125 kg/ha) and K_2O (muriate of potash 125 kg/ha) increased alkaloid content significantly as compared to control. Gupta (1973) recorded that the crop on an average is given a basal dose of 100 kg of diammonium phosphate; this is supplemented with 20 kg of N each, when the plants begin to give out branches and immediately after each leaf crop is taken. A basal application of K_2O is found helpful in soils which show potash deficiency.

Belladona fields require 2–3 weeding-cum-hoeing, before receiving the next leaf crop. The crop requires irrigation to a depth of 3–5 cm, after every 10 to 15 days, during dry summer months depending on soil texture and rainfall at a locality. Once the seedlings are established irrigation is given after 8–10 days also.

10

Crop Improvement

Belladona crop is a cross pollinated one and is reported to exhibit a great in-breeding depression on selfing. It has a very wide variation in growth habit and alkaloid contents have been found to range between 0.10 and 0.85% with average being 0.35%. In an effort to double the total yield of active constituent, RRL, Srinagar, have selected plants containing over 0.6% alkaloids. Dettscheff, in Greece in 1957, has also made similar selection from amongst wild populations of a yellowish brown to violet-flowered plant type (containing over 0.89% of total alkaloids). Breeding for improvement has been tried in many European countries and Albros Yoldi (Farmacognosia 1958, 18) produced a triple hybrid with 1% alkaloid content in leaves by crossing *A. boetica* with an existing hybrid of *Atropa belladona, A. lutea* and selfing the progeny. It gave 50 plants of varied form and content, plants containing 1.6% alkaloids were isolated and multiplied vegetatively. However, production from most of these hybrids began to decline, after the third year under irrigation. At Bombay (BARC) work on developing a mutant capable of growing as an annual crop in South Indian plains is reported under progress.

Chemical races in Bulgaria have been isolated in wild growing populations and a mention may be made of 'atropine' race from Samokov region and of a 'scopola-mine' race from Verna region; the latter being reported to be resistent to verticillium attack.

Harvesting, Drying and Yield Levels

The leaf-crop possesses maximum amount of alkaloid during early flowering stage; and is progressively reduced till formation and ripening of the fruits. The leaves contain more active principle when collected after the crop has received clear sunny days, than under cloudy weather. The plants are cut by sickle at 20–25 cm above the ground, except the autumn harvest which is cut at 3 cm above the ground. The alkaloids are synthesised in the roots and are translocated to leaves through the stem; this process continues in the harvested crop so that the stalks are detached only after the crop has been dried completely. The alkaloid content also varies at different stages of plant growth and is at peak during the flowering stage and minimum at the time of fruit formation. The harvested crop is spread in a thin layer and dried rapidly in sun for 2–3 days, turning every now and then, to avoid decomposition of the alkaloids by enzymatic action. Thick roots are sliced into thin pieces of 3–4 cm length and similarly dried. Drying at 40 °C for 24 hrs ensures complete driage. Hot blast of air is used in some countries for quick and uniform drying. A well-dried crop retains its green colour and loses between 70–80% of its weight during drying. High temperature driers (40 °C for 24 hrs) are also suitable and effective. The dry herbage should be stored in a cool, dry and dark room as belladona leaves are hygroscopic; sometimes lumps of calcium chloride are kept to absorb excess of moisture.

Belladona yields 2 crops during the first year of growth, i.e. in August and Oct-Nov months. However, it gives 3–4 leaf crops in the second and subsequent years. The roots are dug out after the third or fourth year of planting. The average reported yield from USSR is 10 qtls/ha (with a range of 35 quintals). In UK 12 qtls. from two cuttings is the average yield. The total leaf crop yield from a uniform crop in India has been reported to vary from 500–600 kg in the first year and about 750 kg/ha in second year and onwards. At CIMPO drug farms, maximum yield of 10 qtls from 4–5 cuttings has been recorded. Five years old plantations may be

uprooted and average yield of 2.5 or 3.75 qtls/ha of dry roots may be obtained.

The dried crop may be pressed hydraulically into large hessian cloth bags and transported to processing units or stored in cool dry place away from light. Cultivation of this drug can also be taken as an intercrop in newly planted irrigated orchards, till they come to full bearing. At present the extension Kutch Act does not permit private farmers for planting this crop but it has potentialities, and if taken would enable the country to become self-sufficient in this drug, much before the scheduled time.

Studies at Abbotabad in Pakistan (Khan, 1957) recorded that when seeds were sown (12 lbs/acre) in November, the yield was 100 lbs per acre. Harvested seeds weighed 20 lbs per acre. Crop can be obtained from the first year of plantation and 2–4 crops can be taken in second year. The leaves are collected, when the plants are just going into flower. In preparation of extracts, the plants which have almost stopped blooming and are full of ripe berries are used. Roots are collected early in summer, before flowering. Leaves are collected in the first and the second year and roots in the third year. The leaves contain most of the alkaloid between flowering and fruiting. Analysis of the leaves, however, showed 0.35% alkaloids.

Pests and Diseases

Cut worms (*Agrostis flammentra*) attack tender seedlings during early summer months and application of 5% aldrin dust at 20–25 g per sq m of nursery beds, before sowing, protects the crop. If the damage occurs, the beds are drenched with 1:19 wettable solution of chlorodan insecticides for 2–3 times after every 10 days during the incidence of the pest. Damping off of the seedlings is caused by *pythium* sp. and is controlled by application of chloropicrine. There is no effective control known for the root, except to pull the plants affected, with adhering soil and burn them separately.

Economics

Average cultivation cost of belladona crop is estimated to be Rs 1800/ha in the first year and about Rs 1500/ in subsequent year (Gupta in 1973). The average market price of leaf and root crop is Rs 5/kg. This provides an approximate saving of about Rs 1000 in the first year and over Rs 200 in succeeding years. Pharmaceutical industry accepts leaves and roots containing not less than 0.30 and 0.04% of alkaloid, respectively. Marketing arrangements for the crop are made in advance to avoid storage risk.

References

Berezinsckaya VV, and Zemlinscki EE. Kushka VF. , Muraveva and FA Satspiperov. 1953. "Belladona". Govt.Press for Medicinal Literature. *mediz*, Moscow. pp. 19–56.

Chopra IC and Sobti SN. 1962. Cultivation of Medicinal Plants in Jammu and Kashmir. *ICAR Res. Series* No 13. p. 18.

Dutta SC and Mukerji B. 1950 *Bull. Pharmacognosy Lab*. Calcutta 1–104.

Gupta, Rajendra. 1973. Belladona. A New Cash Crop for Temperate Regions. *Indian Farming*. December Issue.

Gupta LK and Shah SC. 1975. Response of NPK fertilizers on the yield of Alkaloid content in *Atropa belladona. Proc. Nat. Symp*. pp. 7.

Hussain Akhtar and Janardhanan KK. 1966. Effect of seed dressing. Fungiciods on germination of Belladona. *Labdev*. 4.

Kapoor LD, Handa KL and Chopra IC. 1952. *J. Sci. industr. Res* 11. A 534.

Mehra PN and Sobti SN. 1954. *Indian J. Pharm* 17–230.

Sandhu AS. 1968. Cultivation of Belladona with special reference to Kashmir. CSIR. Symposium.

10

9. *Cassia angustifolia* Vahl. and Cassia acutifolia Delile (Colour Plate 10.1)

Family: Caesalpiniaceae.

Varn. name: Séena-the dried leaflets and pods of both the species constitute the well-known purgative.

Description: Both the plants are shrubs, small, about 1 m height and with peri-pinnate compound leaves. Of dry region (Arabia and Sudan) but cultivated as an annual crop. Both the species provide leaves and pods used in the Indian system of medicine. Its cultivation is beneficial in more than one ways. Apart from the medicinal importance of leaves and pods, the plants (being leguminous) are good green manure and increase fertility of the soil and enhance yield of subsequent crop. *C. acutifolia*, commercially known as the "Alexandrian Senna", is indigenous to tropical Africa and cultivated in upper Nile territories (Kordofan, Sennar). *Cassia angustifolia*, commercially known as *Tiru-nevelli senna*, is indigenous to Somalia and Arabia. It also grows on the wasteland in Punjab and Sindh areas of Pakistan. It is cultivated in South India in Tirunveli, Madurai and Tiruchirapalli. In Ramnad district of Tamil Nadu state, it grows as a self-sown crop on the dry gravelly and red loamy soils, its cultivation has now been extended to other states of the Indian Union, particularly on gravelly and heavy soils. It contains senacide (glucoside) in all parts, leaves and pods contain the maximum quantity; pods contain 3.5% and leaves 2.5 to 4.0% glucoside. Dried leaflets and pods of *C. angustifolia* constitute well known purgative "senna".

Uses and extent of utilization: In Western and Central Europe it is used as herbal tea, also in bakery products and other industrial use, while in UK, the pods are preferred. In India, it is mostly used as purgative agent and about 1,000 tons of leaves are used annually; while all over the world more than 10,000 tones of seena leaves are used, Indians share being more than 5,000 tones valuing more than 20 crores of rupees is exported.

Senna is particularly useful as a purgative, especially in cases of habitual constipation; it lacks the stringent after effects of rhubarb. The tendency to grip caused by seena may be obviated by combining it with aromatics or with a saline laxative. The pods have the same therapeutic effect as the leaves but produce less gripping, probably due to their lesser resin content. Both the pods and the leaves of *C. angustifolia* are official medicine in the British Pharmacopoeia and the Indian Pharmacopoeial list but in USA Pharmacopeia only the leaves are official.

The purgative action of senna leaves raised locally was compared with the official market sample, by RRL Jammu, on albino mice and a fresh infusion of the powdered drug was prepared in water according to Collier et al (1948) and orally administered to the mice. A dose of 200 mg per 100 g body weight of the powdered drug produced purgative effect in 75% of the animals. Thus, the purgative action of Senna is well proved and the clinical action on humans gave similar results.

It is estimated that about 5,000 to 10,000 tonnes of senna is used all over the world from India and in terms of money *Senna* worth more than 50 crore rupees is exported each year.

Agrotechniques for the Cultivation

Sandy loam and lateritic soils with pH 7.0–8.5 are supposed to be best suited for the production of this drug. Dry hot areas with medium temperatures are required; low

temperature and water stagnation are not conducive for this crop. In North India, 130–150 hot dry days and in the South winter months are suitable. Apart from being medicinal it increases the fertility of the soil and enhances yield of subsequent crops. Soil and climatic condition of North India are also suitable, and it can be grown in Kandi areas of Jammu, Haryana and Punjab during rainy season as subsidiary crop in paddy fields immediately after its harvest and raised as an irrigated crop more profitably. Tirunvely senna is indigenous to Somliland and Arab countries, also on wastelands of Sindh and Punjab in Pakistan. It is widely cultivated in Madurai and Tirunelvelly districts of T. Nadu. It also grows as self-sown in Ramnad district on gravelly and loamy soils; also introduced in Karnataka state. Alluvium of rich clay rice fields are also suitable (Yagna Narayan, 1950).

The land is roughly prepared; fine clean tilling is not necessary. The field is ploughed twice, harrowed and levelled. It is also manured with organic manure. The seeds are sown broadcast, however, sowing in rows, 30–40 cm apart, is advantageous and ensures uniform distribution, promotes fair germination and easier weeding. Plants raised by direct broadcast exhibit poor growth and lower yield of pods as well as the leaves.

Planting seedlings is done in February-March or in October and November, in the south, in lines 30–40 cm apart, when the soil moisture is adequate. Since the seed coat is tough, seed abrasion is required to induce rapid germination and, hence, the seeds are mixed with sand and pounded in a mortar. The seeds germinate within 3–4 days to 5–8 days and are normally sown in spring (March-April) or in August. Hot months are avoided, as germination is poor and so is the

survival. At planting time, fungicide application to seed, right time and appropriate depth of sowing should be taken care of. 12–15 kg seeds are required per ha. The crop gets established within a week. After 30–35 days, thinning is done keeping a distance of 30 cm between plant to plant. Hoeing and weeding is necessary and it may be sown in paddy fields, after the crop harvest. `

Irrigation

Though *senna* is principally sown as a dry crop, it grows entirely as a crop in paddy fields on the moisture, still present in the soil or light irrigation may be provided. When raised as a semi-irrigated crop, there is considerable increase in the yield and also fetches better market value, than that from the crop entirely under dry conditions. Heavy irrigation is injurious to the crop. However, hoeing and weeding is necessary when the plants are 8–10 cm tall. Bright sunshine and occasional drizzling promote good growth though continuous rain during the growth period is not conducive to good quality of leaves. The first flowering shoot on emergence induce lateral branching which is also believed to increase potency of the leaves.

Fertilization

For an industrial crop, application of NPK at 80 : 40 : 20 kg per ha is applied; however nitrogen is provided as split dose at the time of thinning and at two pluckings. In North India, irrigation is also provided to the crop, 5–6 times, as required. FYM also improved total leaf and pod yield (as reported by Pareek et al).

Harvesting

Matured leaves, thick and bluish in colour, are plucked by hand leaving the young

leaves on the plant. From spring-sown plants, first flush is obtained at three months after the sowing and the second after another three months. Later, these plants are allowed to bear seeds. Spring-sown plants flower abundantly and pod yield is high. During the rainy season, the plant continues to grow and third flush of leaves is obtained. Plants sown in monsoon produce two flushes and the number of flowers in plants is less with low yield of pods and seeds many of which may be sterile. Normally 1st plucking can be done 70–90 days after sowing when *Cennocide* is average, 2nd plucking in another 90–110 days and the last 3rd plucking after 135–150 days. Monsoon-grown plants provide only two pluckings with less flowers and low pod yield. Annual production is estimated at 480 tons of leaves and 8 tons of pods per ha.

Curing of leaves

Harvested leaves are spread on the floor without overlapping and dried under uniform shade by stirring with a rake, since they contain about 75% moisture. The drying is complete within 7–10 days under normal conditions of weather. Dried yellowish green leaves are packed into bales under hydraulic compression. The pods are also dried and seeds separated (Collier et al 1948). The whole leaves are generally sold, while the half-leaves are used for making 'galenical'.

Diseases of senna

1. *Léaf spot*: Dark brown leaf spots caused by *Alternaria* and *Glomerella*. Spots later on enlarge and leaves fall down.
2. *Damping off*: Caused by *Rhizoctonia solani* affecting germinating seeds or young seedlings. Seedlings either fail to emerge or get killed after the emergence.

Control: Treat seeds with Carbendazim (2 g/kg) before sowing.

Yield

A dry land crop of senna yields 350 kg of cured leaves and 30–60 kg of pods per acre. A wet land crop yields 700 kg of cured leaves and 75 kg of pods. From semi-irrigated land the yield reported was 1100 kg of cured leaves and 80 kg of pods under Jammu conditions (Chopra and Sobti, 1962). Singh (2001) reported 480 tons of leaves and 80 tons of pods, while plants grown in monsoon season give two flushes with less flowers and low pod yield.

Diseases

Symptoms of die-back with leaf shedding and stem turning black, due to fungus attack, have been reported both from South and North India.

Cost benefit

Singh SP (2001) recorded Rs 11,000 per have as the cost of raising this plant, while the income from leaf produced and at the rate of Rs 30 per kg has been estimated at Rs 36,000/ha having an income of Rs 25,000 per/ha of 'senna' cultivation.

References

Chopra IC, Sobti SN and Handa KL. 1962. Cultivation of Medicinal Plants in Jammu and Kashmir. ICAR Research series No 13. pp. 19–22.

Collier et al 1948. (Quoted from the above reference).

Gupta R. 1974. Wild occurring Senna (Cassia angustifolia Vahl.) from Gujarat, Kutch. *Curr. Sci* **43(3)** 89.

Singh SP. 2001, pp. 51–53. Mimeo.

Yagna Narayan 1950 (Quoted from the above reference).

10. *Carum copticum* Benth.

Family: Umbelliferae (Apiaceae).

Local names: Ajawain (Hindi), The Bishop's weed, Lovage (English),

Description: A tall herb, about a metre high; aromatic.

Distribution: The plant is cultivated throughout India but native of Egypt, Persia, Afghanistan and Europe. In Pakistan cultivated as a side crop on field bunds.

Uses and extent of utilization: The plant is astringent, aphrodisiac vermifuge and diuretic; also antiseptic. The oil, distilled with steam, is used in curing cholera, colic and as, carminative, spasmodic and stimulant, also used as a home remedy in rural areas. Besides the oil, the fruits contain thymol (45–55%) which is antiseptic and is a constituent of the famous medicine called *Amritdhara*, along with peppermint and camphor.

The percentage of the oil is 3–4% of the seeds. The spent fruit after distillation contain 25–30% fat which is an excellent supplement in fodder. The distilled water of the seeds called 'Arak' is used as a tonic in dyspepsia and diarrhoea, also used in Gripe water for the children. A crystalline substance separated from the oil during distillation, for which alone the fruits are distilled in Europe. It is identical with thymol, which is mainly antiseptic, besides thymol certain hydrocarbons called 'thymene' are separated from oil and used as soap perfume. The oil is also given in cholera, colic, etc. being considered like the fruits as antispasmodic, stimulant, tonic and carminative.

Cultivation: It is cultivated throughout the country, especially in Madhya Pradesh on heavy soils. Soil is worked thoroughly and FYM at the rate of 23.5 tonnes/acre is applied to the soil. Seeds are sown during the month of March, which germinate quickly as compared to those sown during winter months. The saplings are finally thinned in rows, 45 cm apart, and the distance kept between plants is generally 25–30 cm. After 45 days of germination, the plants produce flowers; fruits are formed in about 15 days after flowering. The crop is harvested before the onset of monsoons. Average seed yield is about 450 kg per acre. The percentage of volatile oil in the seeds is 3–15%.

References

Jame Verghee, Gulati IC and Joshi ML. 1949. Production of Thymol from Ajowain seeds. *Curr. Sci.* **18(1)**.

11. *Carum carvi* L.

Family: Umbelliferae (Apiaceae).

Local names: Kálá-zira, Shingu-zira (Hindi); Jila-Kara (Telgu). Most of the black caraway of Indian origin is the fruit of *Carum bulbocastanum*.

Description: A glabrous, annual to biennial herb. Stem branched, erect or diffuse. Leaves bipinnate, finely dissected, ultimate segments of the lower leaves lanceolate. Flowers white, in groups of 3–8 bracts small, 1–3 linear, or nil, rarely divided. Rays 3–8, unequal; bracteoles small, linear or round. Fruit yellowish brown, almost viscid, oblong, laterally compressed, light to dark brown. Pericarps are usually separate, curved, tapered at each end, 3–7 mm long, 1.5–2.5 mm in diameter and nearly equilaterally pentagonal in cross-section. They have characteristic and agreeable odour and an aromatic pleasant, warm, somewhat sharp taste. The separate mericarps clean and free from stalk are known as under a

separate name in the spice trade. Capsules are terete, narrowed upwards, primary ridges thin but very distinct, vittae solitary, rather large. Carpophore entire or shortly bifid.

Distribution: The plant is rare but found in wild, from Kashmir to Kumaon, in temperate and subalpine regions between 2700–3650 m. It is cultivated in several countries of Asia, Europe and in USA. As met within the Indian market, the fruits (commonly but erroneously referred seeds) are mainly imported. Those grown in Holland are usually preferred.

Uses and extent of utilization: The fruits, its essential oil and aqueous extract are extremely used as mild stomachic and carminative in flatulance, colic and to impart flavour to culinary (flavouring curries, pickles, biscuits, cakes and cheese, in pharmaceutical products as carminative, and stomachic. It is also lactagogue. Oil and carvone are separated from it and used in medicine as mild stomachic and carminative and to some extent in perfumery, mostly in scents and soaps. They are also used in the manufacture of liquor. Caraway-water is a useful remedy in flatulance and colic of children, particularly the infants. An excellent vehicle for children medicine. The chaff oil is used as low grade perfume. The fruits after extraction of the oil are usually employed as an adulterant or substitute for genuine spice. These are also used as cattle food.

Small quantity of the plant is exported to the plains. The plant is a produce of the Lahul forests and estimated yield is about more than 40 qtls per annum, as reported by Sarin in 1967. The agroclimatic conditions of Lahul are considered suitable for the drug. Essential oil from seeds contain ketone, carvone and carvacrol, etc.

Agrotechniques for the Cultivation

The plant flourishes in cool, temperate climate, on fertile, deeply worked soils preferably clay; it usually requires a cover crop in the initial stages for which generally beans, peas, mustard, flax or clover are chosen. Thorough weeding is essential. Fruits are harvested after 15 months of sowing, when it becomes brown. Harvesting is done in early hours of the day. The plants are dried in the field for 7–10 days and thrashed for the fruits. The normal yield is 1000–1600 kg per/ha. Harvesting time reported is 2 yrs after vegetative planting and 3 yrs for seeded crop. Seed production is about 260 kg/acre and total expenditure reported is Rs 18600/per acre, while the net profit is Rs 33,977/acre (HAPPRC. Bull no 11 2004.)

It has been recorded that the Dutch Caraway fruits contain 3–6% volatile oil, while the 'carpathanin' variety contains 4.0–8.6% oil. The oil is isolated by water and steam distillation. Recently (2004) HAPPRC. Bull No. 11 reported 5.8–8.1% essential oil. Estimated rate of oil being Rs 200/kg.

Besides the fruits, the herb, as a whole contains essential oil, having cadinene as the chief component which is absent from the fruit oil. As the plant matures more of limonene is formed.

The fruit chaff also contains a volatile oil which is inferior in odour and flavour to fruit oil.

Other species: Fruits of *Bunonium bulbocastanum* (*Carum bulbocstanum*) called black caraway, Shia, Shahzerah, Guniyun (Kash), Umbhu (Ladakh) are also collected from Kashmir, Lahul and Chamba region and sold as *kalazira*, which is however a distinct plant. It is mainly a weed of the cultivated fields, liable to prove

dangerous owing to the fondness of the pigs to the roots. It also exists truly wild on grazing slopes between 2000–3300 m, when the shepherds collect it, as a valuable source of income; it is nowhere cultivated. The uses are identical with *C. carvi*.

Carum roxburghianum (Vern. Ajmod) is cultivated throughout the country for seeds, which are used in curries and to some extent for leaves, used as a substitute for parsley. As a drug it is carminative and stimulant, useful in dyspepsia and vomiting, etc.

12. *Catharanthus roseus* (L.) G.Don (– Vinca roseo L.)

Family: Apocynaceae.

Local names: Sadábahár (Hindi); Nitya-Kalyani or Kasrali (Tam.), Periwinkle (Beng), Pattipoo (Tam), Nayanthara (Bengal).

Description: An erect perennial herb, woody at the base, 40–80 cm high. Leaves opposite, oblong-elliptic, petiolate, having acute base but rounded tips. Flowers axillary, 2–3 or solitary in a cymose cluster, rose-coloured flowers (Fig. 10.1, Colour Plate 10.2). Two types occur, var. *albus* with white flowers and var. *ocellatus* though with white flowers yet purpling at orifice. It was first described and classified by Linnaeus.

Distribution: Native of West Indies, perhaps Jamaica, but naturalised all over the tropics. It is found growing wild along the sandy coastal tracts of Tamil Nadu and Kerala states. It is a hardy plant, also grown as potherb for its rosy and white flowers, which bloom almost throughout the year; many of the wild growing plants have been exploited. Therefore, now it is introduced in cultivation, extends from 1200–1500 m at Ramnathpuram, Tirunelveli and Madurai districts in south.

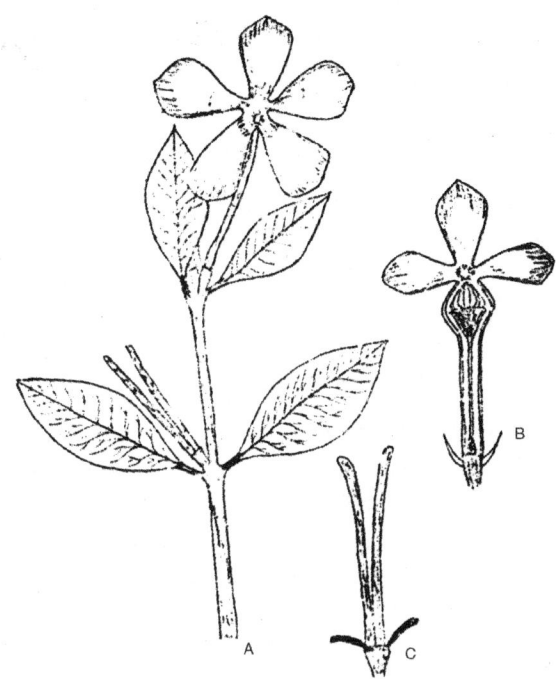

Fig. 10.1: *Vinca rosea* A. flowering twig; B. IS flower; C. fruit.

The 'Madagaskar Periwinkle', also known as the burial ground flower in Britain, is quite a familiar weed, also adoring the wastelands. It is also considered to have come along with the European merch-antmen of the middle ages. From the ubiquitous and wild habitat, one can infer that the plant should be unexacting in its demands and it should be possible to cultivate it in arid wastelands and be successfully exploited by the pharmaceutical industry.

Uses and extent of utilization: The Jamaicans were using the decoction of this plant as a cure for diabetes. Scientists at Eli Lily Coy of the USA and in Canada tested the biological effects of its extract. An alkaloid, 'Vinblastine', isolated from it, and tissue culture studies demonstrated its extract and found that it had no effect on

10

the blood sugar but has profound influence on the blood-forming organs. 'Vinblastine', isolated from it, and tissue culture studies, however, demonstrated its ability to arrest cell division. This property suggested its usefulness in the treatment of cancer. Clinical trials proved that it had good effect against Hodgkin's disease (lymph gland tumour) and acute leukaemia (blood cancer). Another alkaloid isolated from It has established itself as the most powerful and valuable remedy we have at present for securing remission in acute leukaemia. It has also been reported to be of some use in other types of cancer; notably lung cancer.

In the *bazar* the entire dry herb is sold as a raw material but now the roots and thick basal stem which contain heavy percentage of the alkaloid only find market. Annual export of this drug is approximately of more than Rs 10 million.

All the parts of this plant contain almost 66 alkaloids, six of which have a marked potency against various neoplastic diseases, viz. *vinblastine* and *vincristine*, have come to be used in medicine. These drugs reduce arrest or cure of malignant tumours. It is also reported to lessen pain, improve appetite, increase weight and in the process enhance patient's life. The content of the two alkaloids is between 0.000002 and 0.00025%; their dose-levels are equally low. A few firms market the product under the trade names Velbe (Vinblastine sulphate) and Oncorin (Vincristine salt).

As stated earlier, the antihypoglycemic properties have made the Jamaican natives believe that a beverage prepared from its dry leaves provides cure for diabetes.

Agrotechniques for the Cultivation

The plant is easy to propagate and found wild. It can be propagated through seeds on large scale and without much care on the wastelands in the country. However, it has been uprooted near the sea coasts on a large scale and sold by natives in local market. It could be one of the plants to reclaim the large chunks of wasteland in the country. It is not a season-bound crop but can be in either southwest or north-east monsoon season. The crop remains in field from 150–240 days, while the rainfed crop is kept for longer duration. Old crop has higher root yield than the young crop.

The plants are adapted to a wide variety of soil and climatic conditions. It prefers light well-drained, sandy loam soils which facilitates root growth. Mild tropical climate is ideal for high crop yields. Though, raised on black cotton soils yet root yields are low.

The fields are laid into small beds for convenience of irrigation. FYM about 10–25 t/ha is applied during the land preparation; however in some areas tank silt at 50 t/ha has also been profitably used. The land is given 3–4 country ploughings and levelled. Low lying areas are not suitable for this crop.

The usual practice is to raise seedlings in the nursery. The seeds are sown in May-September on soils worked as for vegetable cultivation. 3-cartload of rotted compost is added to per hectare of nursery land. The seeds are sown in small bed, in rows at 10–15 cm distance, and the beds are given immediately a light irrigation. Germination starts in about 10 days which is 85–100%. About 1 kg seeds are sown in 10% of the land which gives enough seedlings for 1 ha. The seedlings grow fast and in 30–45 days attain 10–15 cm height. Seedlings with 2–3 pairs of leaves are ready for transplanting. Some prefer direct sowing in 30 cm rows spacing followed by thinning, to maintain plant distance of 20, 30 and 45 cm, depending upon soil-fertility and the number of irrigations.

Little is known about its nutrient require-ments and effect of the root yield and alkaloid content. Nitrogen fertilizers (50 kg Nitrogen) as CAN (*calcium ammonium nitrate*) or urea/ha is generally applied as split doses at 15 days and 45 days after planting. Some people advocate use of 50 kg of NPK mix (17 kg each of NP and K/ha) whereas, others use 350–440 kg per ha of ground nut cake at the time of land pre-paration.

Mostly, it is raised as a rainfed crop during the monsoon season. As a garden crop it is given 5–6 light irrigations. Irrigated crop induces early branching, forms thick stem and gives high root yield. 25–50% soil moisture deficit is recorded to give maximum alkaloid content, though the total yield is less, since SMD (Soil Moisture Deficit) condition disturbs metabolic path-ways and activity is directed towards alkaloid production.

The crop is given 2–3 inter-culture ope-rations at 20–30 days interval after plan-ting. The plants smoother weed growth and it is difficult to enter the fields.

Information regarding relationship of yield with age class is lacking. In about 150 days, roots go to 15–25 cm depth and give lateral roots. At this stage, it is harvested which is a labour intensive operation. With last irrigation, the plant is pulled, foliage striped off, aerial parts cut at 30 cm above root level and separated thick basal stem and roots from each other before drying them under the sun. Sometimes, the stem is divided into grades, the lower thicker half fetches higher price. Roots are also cut into pieces and dried. Stem and root lose 80–70% of their weight respectively. The yield varies from 2–2.5 t/ha from irrigated crop to 1.5 t/ha

from rainfed crop. In general, 30% of produce is made up of roots, they sell of Rs 9–14 and dry stem Rs 0.75 to 2 per kg. Cost of cultivation is Rs 2500–3000/ha and the return from Rs 7000/- to Rs 10,000/ha.

Local demand for this drug is insignificant and is based entirely on foreign buyers, the produce is exported to USA, Germany and Central Europe.

Diseases

The plant is attacked by thrips and aphids and can be controlled by spraying 10% BHC two to three times. A virus disease 'green rosette' and a 'mosaic' has been reported also to attack the plant. Spray of 0.5% nicotine sulphate is recommended. Virus infected roots do not contain any significant Vinblas-tine, while on the contrary some authors report no difference in contents of healthy and virus-infected plants.

References

Gupta, Rajendra. 1977. Periwinkle produces anticancer drug. *Indian Farming*. July 27(4): 11–13.

Joshi and Rauchaudhri, *Indian J. Hort.* 1964. 21 (3–4) (357–61) and Gard, *Curr. Sc.* 1958, 27. 12: 493–494.

Krishnan KR 1970. Periwinkle– A possible Can-cer Cure. *The Hindu*, June 21. 1970.

13. *Chenopodium ambrosioides* var. anthelminticum A. Grahm.

Family: Chenopodiaceae.

Description: The plant has a branched stem, 60–120 cm high. Leaves lanceolate, lower ones 5 cm long, upper ones smaller. Flowers green (March late to August-Sept.) depending on the time of seedlings transplant, bore in closed spikes mixed with leaves. Fruit small, seeds black, 1 cm in diameter and light, each pound lbs

10

containing two million seeds. The plant has a strong disagreeable smell due to the presence of a volatile oil in glandular hair of seed, stem and leaves.

Several species of Chenopodium occur in India but none yields therapeutically active oil. The plant has been introduced by the RRL, Jammu. The fruits are small, green and roundish, each containing a small seed. The seeds are very small.

Distribution: Experimental cultivation was long started at Bangalore and in Darjeeling, where fruits yielded only 0.48% oil with a very low percentage of ascaridole (46%) as against 65% found in the American samples. No attempt was made to cultivate this plant till strains of seeds, and improved cultivation practices were adopted (Kapoor et al 1954). Subsequently, acute shortage of anthelmintic santonins was felt after 1940 in India. The necessity of the cultivation of Artemisia and other effective anthelmintic drugs was deeply felt in the country. After 1948, seeds from Turkey were tried at various situations in JK state; such as at Jammu (9300 m), Katra (1800 m), Srinagar (1800 m) and Yarikah (2300 m). Fortunately, the seeds germinated within 10–15 days and produced flowers and fruits at Jammu and Katra vigorously, while at higher altitudes the plant growth was stunted and could not survive frost during winter. The seeds yielded a golden yellow oil, with unpleasant odour and burning taste (analytical results of Chopra and Sobti, ICAR Res. Sr. no 13).

Agrotechniques for the Cultivation

The plant grows well on sandy loam soils which are preferred. The land is thoroughly prepared, i.e. ploughed 3–4 times to a fine tilth and kept free of weeds by shallow cultivation. Weeding and hoeing is done before and after harvesting the first flush of crop. In bushy stage, weeding is difficult.

Frequent irrigation during summer season is needed; while water logging should be avoided.

Plants are raised from seeds in early spring (March-April), which germinate in 7–10 days. When the seedlings are ready for plantation (when 10–12 cm tall) at 1 m × 0.5 m distance and require liberal irrigation for establishment. Soon it becomes wild and grows itself in the field, through natural seed dispersal. Seedlings are transplanted in Sept-October months or in the rainy season, also October-November under favourable weather.

The seeds on maturity turn black and are harvested with a sharp sickle, 10 cm above the ground. Since the *ascardole* content of the oil varies with plant maturity, harvesting time is important. Plants transplanted in April start development in May, flower in June and set fruits in July under climatic conditions of Jammu. First harvest of August had erect plants with little branching (Handa et al 1946). First harvested herbage yield was 25–27 maunds per acre; ratio of stem and fruit harvest being 60–40. Analytical results of the sample showed 27% of flowers, 46.8% unmature fruits, 54.8% ripe fruite (in first harvest) and 69% fruit (in second harvest). The quantity of oil in the first and second harvest is 1.2 and 1.1% of oil containing 69.9 and 75.5% ascaridole.

Under irrigation the crop put up fresh shoots on the stumps and attained full maturity in three months. First crop can be taken in the first week of December, the plants are full of seeds with drooping mature branches. The leaves are proportionately smaller than those at the time of first flush. It contained stem and fruits in 60–40 ratio and the yield of the fruit is nearly 300–500 lbs. The second harvest was approximately double of the first harvest.

10

After second harvest the roots are woody and sprout in spring, though the plants are stunted and bear few shoots. It is thus not economical to continue this foliage for harvest. The quality of oil, from the first and the second harvest, is upto BP Standard yielding 1.2 and 1.1% of oil containing 69.9 and 75.5% ascaridolc respectively.

While in USA oil percentage has been recorded maximum at the time of pollination (6.10%). Studies under Indian conditions show that the optimum time for harvest in relation to maximum oil content and ascaridole, is constant at budding and flowering stage (1.12 and 1.1%) but dropped at fruit set. The percentage improved to 1.2% when fruit ripened from first harvest, while 1.15 at seed harvest. Percentage of ascaridole content showed progressive increase from budding to flower stage and fruiting stage. Ripe fruit showed 69.9% ascaridole in first flush, when collected at maturity. Thus, the plants should be harvested only when seeds are fully mature as at this stage ascaridole content is maximum in the oil.

Manuring

The plant required liberal supply of manure as fertilizer. Application of K_2O increases ascaridole content of the oil and well balanced P_2O_5 and nitrogen ratio ensure proper plant maturity. Delayed plant maturity lowers ascaridole content of the oil. Recommended doses of NPK being 3 : 12 : 6 for potash-rich soils, 2 : 9 : 5 in poor soils. The best ratio under Indian soils could be 300–500 lbs per acre.

Distillation

Production of high grade *Chenopodium oil* depends on the stage of maturity of the plant and the method of distillation. Ascaridole, the anthelmintic principle of the oil, is a turpentine oxide. It is very labile and gets decomposed, if the distillation is slow, and carried over a longer period. The oil is brown in colour and possesses a low ascaridole content.

A field distillation still could be used for distillation. The fruits are placed on the false bottom and the water at the bottom is not allowed to come in contact with the drug during distillation. The still is directly heated and the distillate is allowed to flow water. The distillation is completed in about two and a half hours and the oil floating on the surface of the water is removed in the oil separator. It is freed from the suspended impurities, dehydrated over sodium sulphate and filtered. A pale-yellow oil, with characteristic odour, is obtained and is confirmed to BP specification. The oil consisted of ascaridole, and a p. cyamine, while safrole and limone are present in very small quantity.

The oil is also tested for its anthelmintic properties and is comparable to that obtained from USA.

Diseases

Powdery mildew (*Erysiphe cichoracearum*) was reported in the months of February and March at Jammu, and the attack is severe, mainly involving the upper leaf surface. The infected leaves turn yellow to reddish brown with deep red patches, shrivel up and dry. Numerous ascocarps develop ultimately on the undersurface of the leaves.

14. *Cinnamomum verum* J.S.Presl. (Cinnamomum zeylanicum Blume).

***Family*:** Lauraceae.

***Local names*:** Dálchini (Hindi), Darushila (Sansk.), Cinnamomum or Cinnamon (English, Trade name).

The scientific name zeylanicum refers to SriLanka (Cyelon) where the tree is found growing naturally.

Description: An evergreen, small tree 9–12 m, which is extensively cultivated for its aromatic bark. Leaves are large, (12.5–17.5 cm), ovate, thick, leathery, pointed at the tip, shining gossy green above, lighter dull green-coloured beneath, main nerves on the leaves 3–5, running from base of the leaf to middle or tip. Flowers minute, yellow and inconspicuous in large hairy clusters. Fruit oblong or ovate, about 1.5–2 cm long, dark purple or black; seed one.

Distribution: The tree is native of south, though it is planted as an ornamental and a useful bush in Maharashtra, Tamil Nadu and Bengal. As a wild tree, however, it is plentiful in the West and South India from Konkan southwards, rising on the slopes of ghats upto 2000 m in altitude. It is more common on lower altitudes, even below 2000 m in altitude. There is, however, not much known on the cultivation of the tree, though it has been seen under cultivation in some parts of Orissa. It is grown as 'coppied' 'cutback' bushy under cultivation.

Extent of utilization and trade: All parts of plant are aromatic but the bark and the leaves are exploited commercially. The drug is the inner bark of the branches, commercially called the Cinnamon. It is used in diarrhoea, nausea and vomiting. It is also used commonly as a condiment. Oil obtained from the bark is used as a stomachic and as carminative. It cures gastric debility and flatulence, and also has the property of destroying certain germs and fungi. It is most flavouring ingredient, widely used in all kinds of foodstuffs. In small quantities it is used in perfumes of the oriental type. It will give blood circulation a boost and sipping cinnamon tea can help settle an upset stomach. Those suffering from cold can take a tablespoon cinnamon powder which cures the most chronic cough, cold and clear the senses.

It can be also used for the treatment of pimples and the paste is applied directly.

Oil from the leaves is also used as a flavouring agent and preservative for sweets, soaps, etc. and for local application on certain rheumatic pains. It is a convenient source for the extraction of *eugenol*, which is extensively used in perfumery, medicine and for the production of vanillin. It is also a serious competitor of clove oil.

The annual demand for the bark as been estimated at more than 40,000 qtls. South India, Bengal, Assam and Andamans are the most suitable sites for its cultivation. In USA much of the cinnamon comes from different species; *C. loureeifii* Nees, Chinese cassia (*C. cassia* Nees ex Blume) and Indonesian *Cassia* (*C. burmanii*).

Agrotechniques for the Cultivation

The climatic and soil conditions have a profound effect on the bark quality. It flourishes best, when growing on white sand overlying a good rich subsoil along the coast, in warm sunny climate with an average temperature of 27–29 °C and annual precipitation of 2100–2540 mm, without long spells of dry weather. 2 years old plants attained a height of 1.25 m in Orissa (Bhubneshwar).

It is usually propagated from seedlings raised in nursery bed; seeds may also be directly sown in the field. Saplings above 3 yrs old are cut down to a few cm above ground level. Four to six shoots are allowed to grow from the stump and kept straight by pruning. These reach exploitable size about 3 years later. Only young and tender shoots are harvested (1.25–6.5 cm in diam), bark from older shoots is coarse and inferior in quality. The best time for cutting shoots is the rainy season. Outer bark is scrapped off and the soft inner bark is very carefully and deftly peeled off in 15–20 cm long hollow tubes. These are dried in the sun and then in shade for three days each. During this period these are rolled by hand and slightly pressed every day or they will swell and split. During drying the original white colour turns to yellowish brown. These are graded into different qualities according to the aroma and colour.

The quality of the bark improves with the age of bush and with leaf pruning. Best quality bark is of a uniform light brownish yellow colour. Average yield is about 170–225 kg of cinnamon chip and about 1900–2500 kg of fresh leaves.

The chips are employed for oil and an average yield of 0.2% is reported from SriLanka with crude methods of distillation. Madagascar and other places recorded a yield of 0.75 to 2.35%. The chief constituent of the oil are Cinnal-dehyde (65–76%) and eugenol (4–10%). Leaves yield 0.5–2.0% oil which is produced mostly in Sri Lanka, Karnataka and Kerala states of India. However, the yield varies from place to place. Leaf oil content of 2.5 yrs old, that start giving oil increases with age and height during February and May in Bhubneshwar grown plants.

Other Species of Cinnamomum

Cinnamomum tamala Nees et Eberm.

Local names: Patrá (Sans); Gur-andra (Jaunsar), Dalchini (Garh.), Tejpatá (Hindi), Indian cassia (English).

Description: A moderate-sized evergreen; tree, bark wrinkled, thin and brown. Leaves sub-sessile, 7–15 cm long, lanceolate, 3-nerved. Flowers often unisexual, numerous, white in panicles. Perianth small, milky, 6-parted. Fruit ellipsoid, succulent, black when ripe.

Distribution: From Indus to Bhutan, rare west of Yamuna, 1000–2300 m in damp ravines it is sporadic. In Jaunsar and Tehri region it is found upto 2000 m in warm valleys, in association with *Pyrus pashia, Pyracantha crenulata, Rhododendron arboreum* and *Quercus leucotrichophora* formation.

Uses and extent of utilization and trade: The leaves are carminative and stimulant, used in diarrhoea, rheumatism and colic and have a sharp taste. Often it is used as a tonic to brain, good for liver and spleen complaints. They are much used in the manufacture of vinegar and for flavouring food, particularly rice and tea. In decoction it is used for suppression of lochia after childbirth.

The bark is aromatic, when powdered and along with other drugs is used in cough and cold. It is also used as aromatic adjunct in compound preparations. The bark gives an oil which is used in soap manufacture. The flowers from February to May and fruits from June to October are collected, these remain longer on the tree. New leaves appear from April to May and dried ones are largely used in preparation of *Pulao* (a rice preparation).

Leaves and bark are collected during the months of April to May. Annually from Tehri Forests (Kanatal) more than 10 qtls were collected (Uniyal and Issar, 1967).

The leaves also yield an essential oil, resembling cinnamon leaf oil containing phelladrena and 78% eugenol. Oil from the bark is pale-yellow containing 70–85% cinnamic aldehyde.

Cinnamomum camphora (L.) Sieb. is cultivated in India and known as the camphor plant. The oil is distilled from its

10

wood and the leaves. Camphor is chiefly used for local application on sprains, inflammations and rheumatic pains, also given internally in certain types of diarrhoea and as cardiac stimulant. It used to be imported to the extent of 5000 qtls annually; however, camphor is now being produced from other sources.

15. *Chrysanthemum cinerarifolium* Vis.

Family: Compositae (Asteraceae).

Local name: Pyrethrum.

Description: A perennial plant, if properly tended it produces flowers of good quality for 5–7 years. The flowers are toxic and insecticidal (Fig. 10.2, Colour Plate 10.3).

Distribution: After the first world war, Japan became the principal exporter of Pyrethrum to the world market and in 1935 its maximum output was 12,500 tons. Kenya later started production of this drug and was the second largest producer and so Brazil and other countries became significant in the field later on. Since 1941 India began cultivation of this plant which was started in Kashmir and gradually the annual

Fig. 10.2: Cultivation of *Chrysanthemum cinerarifolium* for flowers at manasbal farm of RRL Jammu (Photo R.K. Gupta)

requirement of this plant (flowers) increased as a source of vegetable insecticide to 6000 tonnes by the end of the fifties, (Casida, 1973). Later on, it was introduced successfully to other parts of the country, viz. Himachal, Andhra, Tamil Nadu, etc.

Uses and the extent of utilization and trade: It is one of the most powerful insecticides required for destroying the vector of diseases and plant pests. On account of the less toxic properties, to both plant and animal life, it finds a place in agriculture, horticulture and household, in spite of the availability of synthetic substances in the market.

RRL, Jammu, started experimental cultivation and incidentally Puntatambekar outlined the possibilities of *Pyrethrum* cultivation in Northern India with the result that the area under cultivation increased manifold, even as early as in sixties Kashmir produced 2,000 maunds of Pyrethrum flowers annually. Now, Pyrethrum is cultivated on large areas on a commercial scale.

The flowers are toxic and insecticidal. They contain pyrethrin I and II, chrysanthin, palmatic and linoleic acids and small amount of volatile oils. Pyrethrum is toxic to cold blooded animals and is absolutely harmless to hot blooded animals. Dalzell and Gibson state that 'the flowers are a substitute for Chamomile'. According to a book published by CIMPO, 'people of Deccan administer the plant in conjunction with black-pepper in gonorrhoea'. *Pyrethrum* or "insect flowers" are obtained from several species of Chrysanthemum but there are only two commercial varieties, viz. Dalmation and Persian. The flowers are obtained from var. Dalmation which is cultivated in India. Pyrethrum is efficient against numerous kinds of chewing and

sucking insects. Pyrethrum ointment is effective against scabies. Pyrethrum extracts are used as insect sprays such as mosquitoes and flies. Pyrethrum is a source of vegetable insecticide required for destroying vectors of diseases and plant pests, is less toxic to both vegetation and animal life than the synthetic ones.

Agrotechniques for the Cultivation

It can be propagated by transplanting from nursery stock and by cuttings, the latter method gives poor survival. Seeds are sown in raised nursery-beds during the months of September-October, after pretreatment with cold water for 6–12 hrs. The seedlings are transplanted at 60 cm spacing on 15 cm high ridges at the time of premonsoon showers. Since, it is a perennial, if properly tended, gives flowers of good quality for 5–7 years.

In actual practice 2–3 kg of seeds are required for one-eighth of an acre land and sown during August-September or March-April. The beds are manured with forest humus or FYM. Seeds are sown in rows or broadcast. The seeds germinate in 3–5 weeks time, depending upon the humidity and day temperature. The seedlings raised in spring are transplanted in autumn and the seedlings raised in autumn are planted in spring season. Little showers are beneficial during transplanting in spring season. Seeding is done during spring and autumn seasons. The plant flourishes well on a well-drained and weed-free soil. The plants flower from the end of May till the end of June. These are harvested when three-fourths of the disc-florets are open and are generally hand-picked and dried in sun for 2–3 days, afterwards they are removed to a shady place for rest of the drying process.

Though the plant can be cultivated at an altitude of 1700–2700 m, it is reported to give optimum yields upto altitude of 2000 m. The total pyrethrin content is highest in flowers dried in sun for three days and rest of the drying in complete shade. A gradual fall in the active principle content from year to year has been recorded in the plantations of pyrethrum. No appreciable change in the pyrethrin content has been reported in plants raised at different places and also in JK state. The pyrethrin content in general ranges from 0.98 to 1.10%. However, Pyrethrum plants showed great variations in the number of flowers and the maximum number of flowers recorded per plant is 714. At present the average yield of the flowers per acre in Kashmir reported is 170–4000 lbs against 1000–1500 lbs recorded in Kenya and other countries. The plant flowers only once in its life time, while in some places it has been reported to flower continuously throughout the year. Suitable strains, with a second flowering, if selected, would greatly enhance the yield of flowers. Average flower yield recorded at Chakrata is 100 lbs/acre in 1st yr, 1000 lbs in 2nd year and 1400 lbs in 3rd year and 1300 lb in 4th yr.

The flowers are collected only when the three-fourths of the disc florets are open as suggested earlier, however, there is a gradual increase in the percentage of pyrethrin in flowers from close to open stage, till the development of the seed starts.

The pyrethrin content varies with the time of collection of flowers, being maximum when the flowers have 3–4 rows of disc florets open as stated earlier. Plucking of flowers is done at intervals of 7–10 days from second year onwards. Peak production is reached on 3rd or 4th year; thereafter it falls. In Nilgiris hills the crop rotation is six years, out of which three and a half to four years will be under pyrethrum and the rest under cereals like wheat, barley and rye, etc. The flowers are dried in sun or in specially constructed driers

below 50 °C temperature. At Chakrata pyrethrin content recorded is 0.91–1.9%.

There is a gradual fall in the active principle content from year to year which may be due to soil exhaustion. Pyrethrum trials at different altitudes also show that the percentage of pyrethrin is maximum (1.02%) at 1000 m alt., while at 300 m it is 0.98%, at 600 m 0.90%, at 1500 m it is 1.02%, 2300 m is 1.06% and 2000 m it is 1.105).

Damage by insect pests and animals

The rats (*Microtus rollei*) cause maximum damage at high altitudes, where roots are eaten during winters, by burrowing into ridges. Insect Nelothinie is one of the beetles that feed on cut roots of the plant.

Some diseases of pyrethrum can be controlled as follows.

1. *Root rot*: Caused by *Fusarium solani* during rainy season. Symptoms—yellowing, stunting and finally wilting. Affect plant root show blackening and rotting.
   Control: Pathogen-free nursery. Seeds treated with Carbendazim (2 g per kg) before sowing. In high-risk areas nursery beds should also be drenched with Carbendazim (0.1%).
2. *Leaf spot*: Dark brown spots on leaves caused by *Alternaria*.
   Control: Application of Maneozeb (0.3%) at 10 to 15 days interval.

References

Casida, John E. (ed). 1973. Pyrethrum-*The Natural Insecticide*. Acadenic Press Inc. pp 336.

Fotidar MR. 1940. *Curr. sci.* 9: 360.

Gnadinger, CB. 1955. *Pyrethrum Flowers,* 2nd edn. Mclaughlin Gromley King and Co. Minneapolis. Minnesota.

Singh, Pratap. 1975. Cultivation of Pyrethrum (Chrysanthemum cineari folium) in Kashmir. CSIR. Symp.

Chopra IC, Sobti SN and Handa KL. 1962. Cultivation of medicinal Plants in Jammu and Kashmir. pp. 27–31. *ICAR Research Series* No 12.

16. *Citrus aurantium* L. sub. sp. bergamia (Risso et Puit) Wight and Arn.

Family: Rutaceae.

Local name: Bergamot orange.

Description: A small, evergreen tree of about 4 m height. Flowers fragrant, white and little smaller than that of the orange. Fruits pear-shaped about 7 cm in diam. , very acidic and not edible. Fruit rind is yellow when ripe. Numerous oil glands are embedded in the outer peel, from which the bergamot oil of commerce is obtained by expression.

Distribution: The cultivation of this plant is restricted only on a narrow coastal strip in Rcggio Calabria, the extreme toe of Italy. The plant is very delicate of all citrus species and requires a mild equitable climate with a temperature ranging from 2– 37 °C. Excessive heat, frost, snow, drought and salt winds are all inimical factors; well irrigated alluvial soils are the best for the plant growth.

Uses and extent of utilization and trade: The Bergamot oil is an important perfumery material with a sweet refreshing odour and is indispensable for the manufacture of many perfumes and bouquets like Eau de Cologne and Lavender water. It is also widely used in lotions, creams, powders and soap industry.

The oil is recorded to have strong germicidal properties, just like phenol and is suitable for destroying insects, particularly flies, with particular advantage of a pleasing smell.

India imports Bergamot oil worth a few million rupees and attempts to cultivate this plant within the country had not been very successful and encouraging, for which sincere efforts are still needed and are being made.

The average yield of expressed oil from the pleant in Calabria is 0.48% but new automatic machine "Avena" is reported to give higher yields. The colour of the oil varies from golden brown to greenish brown and has a pleasant sweet odour. Steam distillation of wind fallen and damaged fruits gives inferior oil in small quantities.

The leaves also yield an essential oil known as "petit grain bergamot" which is less valuable than the oil from the peel.

Agrotechniques for the Cultivation

The plant is propagated by building on sour orange stocks. The tree starts to bear fruits in about three years after budding and attains full maturity after 12 years. All necessary precautions are taken as required for a fruit-bearing tree having oil. Annual fruit production per tree is about 300 in numbers. The average life of the plant being 70–80 years. The fruits are harvested when yellow in colour.

17. *Citrus medica* var. limonum Hook. f. (Citrus limonum Risso)

Local names: Jamir (Kumaon), Baráni-mbu, Pahári-nimbu (Hindi), Lemon tree (Eng.).

Description: A small tree or straggling bush, 3–4.5 m high with thorny branches and scented white flowers tinged with pink. The fruit is elliptical or oblong in shape, somewhat collarded with a nipple-shaped extermity. It has leathery rind and abundant acrid pulp, yellowish when ripe.

Distribution: It is cultivated in many countries, principally Sicily and Calabria (Italy) and California (in USA), are the main centres of lemon oil production. In India, it grows wild in the foothills of Northwestern Himalaya and cultivated in the orchards throughout India for the fruits.

Uses and extent of utilization and trade: The root and the fruit juice is given to infants in Kumaon region during stomachache, locally called 'Juka' which have symptoms like loose motions, fever attended with cough.

Fruit is used in various culinary purposes like pickles, cooling beverages, etc. also as a stain remover and bleaching agent Citric acid, pectin and lemon oil are obtained from the fruits. The lemon oil is used for flavouring beverages and soft drinks and other food products, also in perfumery and soap manufacture. Lemon extract which is made by dissolving 5 parts of oil in 95 parts of alcohol, is a well-known flavouring material, next in importance to Vanilla. Hand-pressed Italian oil is also used for the preparation of "turpineless oils" and concentrated oils", which are more stable on storage and have a better solubility in dilute alcohol.

Agrotechniques for the Cultivation

As compared to ordinary orange, it requires a drier climate and is susceptible to rots and fungal diseases. It is, however, adapted to a variety of soils but requires a moderate rainfall. It does not tolerate water-logging.

It is best propagated when budded on a suitable stock or by layering through cutting or the seeds also propagate the plants. Budded plants are planted out in the field, 6–18 months after bud insertion.

Irrigation, especially during dry season, is necessary; however, too liberal a supply of water should be avoided. The crop responds favourably to the application of fertilisers.

Budded and layered trees start fruiting within one and a half year of planting. The fruiting season is mainly late spring or summer and winter, though fruits are borne practically throughout the year. The tree is a prolific bearer and may yield 1000 or more fruits every year.

The outer peel of the fruit contains an oil which is recovered by expression either by hand pressing or by machine, since steam distillation seriously affects some of the constituents and the quality of oil. The oil in the peel is about 0.45–0.7% on the weight of the whole fruit, the recovery being only 50–80%. One thousand lemons yields nearly 400–450 g of the oil.

Properties of the oil, like the yield, depend on a number of factors like locality, climate, weather, seasons of harvest and method of extraction. The main constituent of the oil is D-limonene and aldehydes. On storage, the quality of oil changes, unless it is properly stored in tightly stoppered bottles, filled upto neck and kept at low temperatures.

Several varieties of lemon like *Galgal* and *Eureka, Lisbon, Villafranca, Italian seedless, Nepali oblong* and *Nepal round* are cultivated in India but indigenous production of the oil is negligible and almost the requirements are met by import.

18. *Coriandrum sativum* L.

Family: Umbelliferae (Apiaceae).

Local names: Dhaniyá (Hindi), Kashniz (Urdu), Dháni, Dhana (Raj), Kotumir (Tam).

Description: An annual herb with dissected leaves when young. Flowers small in umbels, white in colour. Mostly used as green *dhania* except on place where commercial cultivation is done for the fruits used as condiment. It has also been brought out that the Indian coriander plant is 30–40 cm high, poorly branched, less leafy and poor yielder (750 kg/ha max.) containing less essential oil (0.1–0.3%). The Bulgarian variety is gigantic (120–150 cm high), very leafy, prolonged growth period with 1% essential oil (Dimri et al 1975).

Distribution: A native of the Mediterranean countries but extensively grown in various parts of the world. It is cultivated in many states of the country as a spice, while MP-Guna district, is famous for a green, small-seeded variety. A Bulgarian strain was introduced by CIMPO, now named as CIMPO-S-33, was tried at Bangalore in 1969. The coriander is grown in India as a commercial crop, besides MP, in Andhra, Tamil Nadu, Maharashtra and Rajasthan. Sizable quantity is exported; the trade being to the tune of more than 30–40 hundred thousand rupees per year. CIMPO introduced varieties from Bulgaria and USSR since 1969 at its Bangalore Farm, however, plants raised from Bulgarian seeds adapted very well to the agroclimatic conditions in Bangalore. Intensive selections have been tried for evolving improved strains in India.

Uses and the extent of trade and utilization: Coriander is used as spice in almost every household in the kitchen. Besides, it is used in digestive troubles. The oil is used as flavouring agent in beverages. The water extract is a better flavouring agent than the fruits or oil. The fruits are used in candies and rich in aromatic essential oil required for flavouring liquors and varieties of foodstuffs.

The improved strain CIMPO S–33 has the capacity to yield more per unit area and has high essential oil content. Table 10.1 gives an insight to the difference between the local

Table 10.1: Comparative hails of ordinary and improved coriander seeds

Character	Local strain	Improved strain S–33
1. Habit	Small, erect, semi-compact less branching, scanty flower and branching	Tall, erect, compact basal branching, profuse flowering
(Average height in cm)	40	120
2. No of days to flower	40	80
3. No of days to maturity	90	145
4. Seeds yield plant (g)	10	38.7
5. Total yield/ha (kg)	750	2117
6. Oil yield/ha (kg)	1.35	27.52
7. Increase in yield over local (%)	...	282
8. Essential oil content (average) (%)	0.18	1.3
9. Average seed weight	1.38	0.712
10. Seed size	bold	small

cultivar and CIMPO-S-33. (after CIMPO note, mimeographed 1976)

Agrotechniques for the Cultivation

Well-drained soil, loamy in texture, is the best. FYM at 23 t/acre with a seed rate of 9 and half lbs per acre was usually recommended. Mericarps are separated by gently rubbing and sown in slightly moist soil at 0.75 cm depth. However, the land is brought to a fine tilth, as for any field crop, and laid in beds of 3 m × 3 m size. Before sowing 15 tons of FYM, 375 kg of Super phosphate and 66 kg of Muriate of potash per ha are incorporated into the soil. Since sowing whole seeds results in 'delayed germination' it is necessary to split the seeds in halves before sowing.

The best sowing time is second week of September. Seed rate is 12 kg/ha and the sowing is done in rows separated 60 cm apart. The seeds take about 7 days to germinate. Seedlings are thinned out after a month of sowing, maintaining a distance of 45 cm between the plants. 204 kg of urea per ha is given in two equal split doses, the first dose is given after thining and the second after two months. Light irrigation is given twice a week,

from germination stage, till the crop comes into full bloom. The frequency of irrigation is cut down when the seed setting takes place and totally stopped two weeks before the crop is ready for harvest. Harvesting is done when the fruits just start browning.

There is no synchronous seed ripening in coriander. The plant is harvested when about 60% seeds mature. The cut plants are left in the field to wither for 2–3 days and then dried in partial shade. Seeds are separated by thrashing.

Pests and diseases

At flowering time the crop is prone to the attack of aphide, which can be checked by spraying 0.1% of Follidol. Powdery mildew, caused by *Erysiphae heraclei*, also attacks the crop during the seed setting time. This can be controlled by spraying Karathens WD at the rate of 1 kg of powder in 1,000 ltrs of water per hacatre.

Yield

CIMPO Strain S-33 is capable of giving about 3 times higher yield of seeds. 2117 kg/ha against 750 kg/ha in local variety. Pod yield per ha in kg is 27.52 against 1.32

kg. Average yield of seeds per plant is 38.7 kg of plant against 10 kg. Days of maturity is 80 against 45 and average height is 120 cm against 40 cm. Average essential oil content is 1.3% against 0.18% for local variety. 100 seed weight is 1.38 g for the local against 0.712 g for improved and seed size is bold.

References

Dimri BP, Khan MNA and Narayana MR. 1975. Introduction and improvement of Bulgarian Coriander (*Coriandrum sativum*) by selection for higher yield and essential oil content. *Proc. Nat. Symp.* pp. 9.

CIMPO 1976. The New Coriander Strain, CIMPO S-33. *Mimeo pp.* 2.

19. *Chlorophytum arundinaceum,*
(=C. boriwillianum, C. tuberosum)

Family: Liliaceae.

Local names: Saféd musli (sometimes confused with 'Satáwari' (Asparagus species) roots.

Description: The plant has succulent roots and can be seen above the ground only for 3–5 months in a year. A pretty herb, suberect with lanceolate, many nerved leaves and erect dense fld. racemes of white flowers.

Distribution: It is naturally found in the forests of Central India but also occurs in Gujarat, Maharashtra and Uttar Pradesh. Out of the 175 species about 13 are under cultivation in India. Found in abundance at Sagar, Panna, Sehore, Dhar, Khandwa, Ratlam and Jabalpore districts in Madhya Pradesh.

Uses and extent of utilization and trade: The roots are used for the preparation of health tonic, used in general and sexual weakness disorders. The roots contain spermatogenic profialis. Decoction is used for curing impotency, as they are rich in glycosides. Planting material for this plant can be obtained from Tropical Forest Research Institute, Jabalpur.

Agrotechniques for the Cultivation
Soil and climate

As it grows well in Central India, the central Indian climate is most appropriate. Sandy-loam well-drained soils with normal pH are suitable. It also grows well on light red or sandy loam soils with slight slopes having pH between 5.5–7.0. On marginal lands with good drainage an ample of organic matter is applied. Rainfall between 900–1000 mm per annum is suited for a good growth.

Land preparation

For land to be ready for sowing two ploughings, in April and May, are applied. FYM at the rate of 75–100 t/ha is also added, while ploughing; if possible green manure could also be added before taking the crop. For the outlet of water, bunds of 40–45 cm distance are created or a raised platform of 3.5 × 1 ft (90 × 45 cm) can also be used. Land ploughed once in the month of May is allowed to weather. On virgin forestland care is taken to root out stumps and the area is cleared of other growth on the ground before planting the material (*Musli discs*) by the end of June. Second ploughing is done, followed by two cross harrowing, to break the soil clods. Finally, the land is dressed by planking, so that the water does not stagnate in the field. It is a *kharif* crop.

Planting

Planting of healthy roots is done on the field-bunds at 2.5–3 cm depth, at a distance of 15 cm. If raised platform is used, 5–6 rows can be planted in each bed. Crown or disc part of the root should be ensured. Each root should

be of 8–10 g and with a population of 1.5 lakh plants per ha, a total of 11–20 qtls of roots are needed. For sprouting of the roots, there should be soil-moisture availability in the ground. Sprouting starts within a weak of planting. End of June is the proper planting time for the sowing of *musli*.

Within 30–40 days of planting, 2–3 weeding and hoeing operations are needed along with putting the soil near the plant. 3–5 t/ha of FYM is added to the field at the time of ploughing, while another 55–60 t of FYM is added after 15–20 days of planting which is mixed in pits and earthing done on bunds. Otherwise 30: 100;60 kg (NPK) per/ha is given to the crop; half of the nitrogen is given in two split doses after planting.

Digging of the roots

After 2–3 months of planting, in early winter season, upper part of the plant gets yellow and the leaves fall down, root starts ripening when it turns dark brown, after 7–8 months the roots are taken out. These are thoroughly washed. Healthy and strong roots are separated, as the produce, and kept in heaps for 3–4 days and then peeled and dried in sun.

For the next year crop, the roots with crowns are kept at 28–30 °C temperature in sand packed plastic bags at 50–65% humidity. Average produce per ha is 50–55 qtls and after drying and peeling it weighs 12–14 qtls. Sometimes, the roots gets affected by the fungus *Fusarium*.

Economics of cultivation

It is reported that about 7 qtls of dry roots cost about Rs 3,50,000/- ha @ Rs 50,000 per/ qtl. The cost of material, planting and processing is estimated at Rs 1,00,000/ha with a net gain of Rs 2,50,000/ha.

References

Singh AK, and Singh Kailash. 2001. Cultivation of Saféd-musli. CIMAP (Mimeo) pp. 65–66.

Bisen SS. 1997. Cultivation of Saféd musli. *Udhamita*. pp. 42.

20. *Costus speciosus* Sims.

Family: Zingiberaceae.

Local names: Keokand (Hindi), Kémka (Mahar); Kia (Bengal); Kiu (Assam).

Description: A perennial, rhizomatous plant.

Distribution: Occurs wild under moist deciduous forests in hilly regions of India. It is widely distributed and large quantities are collected under mixed forests from Assam, Meghalaya, Bihar, Khasi and Jaintia Uttaranchal hills, and Orissa, MP, North Bengal in humid areas. The sub-Himalayan tracts of Himachal Pradesh, Uttaranchal and western ghat are the ideal places for its collection.

Uses and extent of utilization and trade: The rhizomes contain a diosgenin content, ranging from 0.3–2.68% (dwb), and majority of specimens are with more than 1.5% diosgenin. In some places, the plant is grown as an ornamental. Seasonal plant corm sprouts in rainy season and can be seen above the ground for seven months. It flowers in September-October and dries by the month of December, providing seeds. It provides *diosgenin* which gives steroidal hormones used by sport persons and in hormone deficiency.

Agrotechniques for the Cultivation

The plant grows generally on loamy soils but can also be grown on alkaline soils. Humid climate, with 35–45 °C temperature and rainfall ranging from 1200–1600 mm is suitable; however, if the rainfall is deficient additional irrigation is provided. Rhizomes are planted in March, sprout in May; plants attain 1 m height towards the end of July when profuse flowering occurs. Aerial portion withers away in December and the rhizome remains dorment till May.

Planting is dome both through seed or from rhizomes. Seeds in the nursery are sown in July, percentage of seed sprout is 60–65%. The seedlings are sown at a distance of 50 cm × 50 cm. The rhizomes are cut into small pieces and planted in April-May with 2 buds, at about 10 cm in the soil. Sprout percentage is about 90 in this case.

Phosphatic fertilizers are needed for the development of underground parts. NPK at 2 : 1 : 1 ratio are suitable; 60 kg nitrogen, 30 kg P_2O_5 and 30 kg K_2O are generally applied. Under both types of planting, irrigation is given essentially during the months of October-November. Weeding and hoeing operations are done twice a week, the soil is put round the sapling to facilitate formation of rhizomes.

Underground rhizomes after 9 months of planting are extracted in January, and are recorded to weigh 1570 g on an average (weight 35 g at the time of planting). Next harvest is taken in July, when average weight of the rhizome is 1700 g. Rhizome samples extracted in January showed decline in diosgenin content from 2.3% (of parent stock) to 1.6%. The corms contain 85% moisture and if the corms are kept in soil for 2–3 months, diosgenin content increases.

Economics of cultivation

Per hac cost of cultivation is calculated at about Rs 30,000/ha. With an yield of 400–450 qtls of rhizome from an ha (on drying, weight 60–70 qtls), the total income comes to 60 × 1200, 72,000–84000 rupees with a net income of Rs 40,000–54000 per ha.

References

Shukla, S. 1997. Cultivation of *Cosptus speciosus*. CIMAP sym. pp. 45–47.

Sarin YK and Kapahi 1975. Cultivation of *Costus speciosus*.

21. *Crocus sativus* L.

Family: Iridaceae.

Local names: Késar (Hindi); Jafrán (Kashmiri), Saffron (Eng.).

Description: The word *saffron* is derived from 'azaferan", the Arabic name for this spice, while the word Crocus is of Greek origin. Saffron enjoyed great popularity as a drug and several therapeutic properties were ascribed to it (Folch 1957). It is a low lying perennial plant. The corm is somewhat globular, 3–5 cm in diam, produces 6–9 radical, grass-like, channelled leaves surrounded in the lower region by 4 or 5 membranous whitish scales. The plant bears underground corm. The terminal flowers appears in autumn (mid October to early November), are borne enclosed in a spathe, singly or in two or threes. These are lily-like and about 25 cm across. The perianth is light violet, reddish purple, or mauve-coloured, forming a cylindrical tube, 7–8 cm long. The androecium is of 3 stamens, while the gynoecium is 3-carpellary and 3-locular. Placentation is axile with inferior ovary (Colour Plate 10.4). The single elongated pale style branches above into three brilliant orange stigmas, rolled lengthwise to form a funnel-shaped structure. The 3 stigmas, along with an approximately 50 mm portion of style, constitute, in the dry state, the pure saffron of commerce.

Distribution: The valley of Kashmir is famous for its saffron fields (total area nearly 3,350 acres), located on both sides of National Highway on Karéwas (elevated dry table lands of alluvial origin). Pampore, famous for its saffron is situated at 1,700 m altitude, however, saffron has now been successfully tried in other parts like Kistwar region of Jammu, Himachal Pradesh and in Uttaranchal. Mid October to November, saffron fields are in full bloom. Saffron growers of Kashmir have superstitions, notions and methods of their own, where saffron in being grown since ancient times.

Saffron enjoyed special attention from the Egyptians, Greeks, Jews, Romans, Hindus and Muslims and has always been important for worship. Arabs introduced its cultivation in Spain about AD 1921 and has the Arabic name "azafran". It originated in Greece, Asia minor and Persia spreading eastwards to Kashmir. Substantial quantities are grown in UK (since 1728), Italy, Germany, France, Austria but most of it comes from Spain, where it is grown extensively.

Uses and extent of utilization and trade: Saffron is an important ingredient of the prescriptions of Vaghbhatta and Sushruta (about 500 BC). Saffron contains a yellowish red pigment, *crocin*, which is a mixture of glycosides. Saffron yields β-crocetin in reddish crystals and γ-crocetin also in reddish crystals and by acidifying the mother liquor, crocitin is in bluish red crystals. There is a colourless bitter glycoside, *pierocrocin*, the sugar component of which is d-glucose; its aglycone is the volatile oil *safranal*, which possesses the characteristic aroma of saffron. The stigmas are a rich source of riboflavin or its precursors (Karrer and Solomon, 1927, Wealth of India, 1950). Saffron is widely used to colour and flavour foods and is an important ingredient in *Ayurvedic* and *Unani* systems of medicine.

Spain is the main market for saffron and high grade saffron is exported throughout the world at a premium price. French and Italian saffron do not compete with the Spanish product. In India, it is still imported from Spain, France and other sources.

Saffron standards as per the BP Codex provides the following guidelines: (i) Not more than 8% it should consist of the style and filaments of the anthers and not more than 2%, foreign organic matter, (ii) should not yield more than 7.5% of total ash or more than 1% of acid insoluble ash, (iii) imparts a yellow colour to water, alcohol, methanol, ether and chloroform but not to xylene, benzene or carbon tetrachloride. Filtration through charcoal will, however, remove this coloration, (iv) loss on drying should not be more than 14%, (v) when pressed between white blotting paper, no translucent oily spot should appear, (vi) floral parts like ligulate or tubular florets, spinose pollengrains, etc. should be absent, (vii) 0.02 g saffron should impart a colour intensity to 100 lit water not less than that of a 0.1% solution of potassium dichromate, (viii) in sulphuric acid, the saffron stigmas immediately become blue, the colour gradually changing to deep violet or purple and finally to a purplish red (deep wine colour), (ix) the nitrogen content of saffron is remarkably constant at about 2.22 to 2.43% and a *kheldahl* estimation is, therefore, a good test for the purity.

Samples of genuine Kashmir saffron are reported to contain 8.5–10.2% moisture, 5.9–13.3% ash and 1.0–2.4% nitrogen and to have a tinctorial powder, 20–30 times that of potassium dichromate solution (Budhraja, 1942).

Saffron is usually adulterated with various substances like ligulate florets of *Calandula* *officinal*is, dyed with methylorange or a red dye, ligule corollas of 'marigold', tubular florets of *Carthamus acanthium*. Other adulterants stigmas of maize (corn silk) and florets of *Onopordon acanthium* dyed with tetrazine and mixed with ammonium or potassium nitrate, borax and glycerine to increase weight.

Commercially saffron is graded as follows:
1. Very select: Having an average stigma length of 53 mm and, a small portion of the style. Has a lively orange-red colour and a strong penetrating odour.
2. Select: Having an average length of 53 mm of stigmas and style together as a bright red colour and a good odour.
3. Superior: Having an average length of 50 mm of stigmas and style together. Has a dark colour, a good appearance, and a good odour.
4. Medium: Having an average length of 46.50 mm of stigmas and style together. Has a good colour, appearance and odour.

Agrotechniques for the Cultivation

Climate

The plant grows in climatically diverse regions, varying altitude ranges of temperature and humidity. When many other plants are in flower during spring, saffron begins to go in dormancy and is not affected by high temperatures in summer or low temperatures in winter; however, very cold and humid weather is not conducive for its growth. For quality saffron weather at flowering time should be dry and not very humid. Frost and rains, both are harmful for flowering. At high altitudes flowering is delayed, the best situations being of easterly and north easterly aspect. Corms obtained from HP, flowered at Dehradun experimentally which has been tried by the author,

however, it does not indicate, that saffron cultivation could be a profitable adventure.

Soil

Saffron thrives best in warm subtropical climate, requiring a well-drained, sandy or loamy soil, free from clay and decaying humus. Soils of middle structure, more or less loose, with permeable soil or where water flows easily are best for the development of the corms. Saffron develops well in silicious-ferruginous, chalky soils, because of their dryness. Saffron does not grow well on humid soils, especially where water stagnates, and might rot.

Where there is chance for water stagnation raised plots are made. Since calcium-containing soils have the ability to decompose organic manures easily, these are considered to be the best. Very fertile soils contribute mostly to vegetative growth and are, thus, good for the production of quality saffron. The field should have adequate water drainage. The soils of Pampore in Kashmir, locally called 'Gurut', are peculiar in their physical texture for a good growth of saffron. A study of the soils showed a leaching of calcium carbonate, $CaCO_3$ with an accumulation in depper layers, with no difference in micro-nutrient composition of the poor and good soils (Biswas et al 1957). Normally, saffron is not planted in the same soil from which it has been uprooted until after a number of years have gone by (Jalali, 1962).

Planting

Onion-like young corms from the plant, developed in March-April on mother corm are planted, that are solid, somewhat compressed, white and fleshy, covered by a series of fibrous sheaths. The corms reproduce annually and give rise of young cormlets annually. The corm already flowered once withers and dies, producing cormlet and the cycle continues; newer corms replacing the old ones.

Soil is thoroughly prepared, hoed and cleaned of weeds, stones, etc. and ploughed from mid March till the end of August; upto eight ploughings are given. Final soil-preparation is done manually and plots of size 1.5 to 2 m sq with 15–20 cm wide drains are prepared on all the four sides of the plot.

The corms are sown in July and August; the rains during this period are considered beneficial for a good harvest under rainfed conditions. With irrigation, saffron production can be enhanced; which also acclerates blooming. The corms are planted at a depth of 12–18 cm and about 15–20 cm apart in straight rows, with usually a distance of 10 cm between rows, spacing of about 10 cm on either side of row of saffron crop improves yield, than rows planted with different spacings. After every 4–6 rows a deep wide furrow is made and beyond that another wide bed is planted with corms, and so on. Newly formed corms are about 2 cm higher in the soil than those of the preceding year, and after a number of years, the corms lie very close to the surface. In Kashmir, saffron is rotated with crops like wheat and mustard (Jalali 1962). The saffron plant does not demand too much fertiliser because very fertile soils are not conducive to the quality of the final product. Saffron growers in Pampore do not apply any manure to their fields for eight years. Once the plant is established, neither water nor any manure is applied by Pampore farmers. However, experiments conducted at RRL Jammu, have shown (Madan et al 1966) that by the application of sulphate of ammonia, super-phosphate or even potassium nitrate and ordinary FYM, it is possible to obtain higher

yields, contrary to the local belief. In other countries like Spain, manure is applied at the last stage of soil preparation, as a blanket, so that it can mix in the soil and decomposes slowly. The manures used are not too dry stable refuse, sweepings from public roads and sufficiently decomposed refuse from cesspools (Arjona 1945). However, it is considered necessary that the general health of the plant is an important factor and the plant should be provided with proper conditions, like organic matter, aeration and water supply with adequate drainage. A liberal supply of nitrogen increases vegetative growth and phosphate, which encourages vegetative growth, particularly flowering, assumes greater importance in the nutrition of saffron plants. In fact flowering is affected indirectly by the availability of ions that make phosphate available from the soil; trace elements are considered to assume a role in improving flower production, are now proved to be of no primary importance for this process. Photoperiod seems to be an important factor and an optimum 11 hrs of illumination appears to be conducive to floral initiation. Temperature has an effect also on the flowering but unusually low temperatures coupled with high humidity, during the short period of flowering, dictates the success or failure of the saffron crop in a particular season. For a good flower production, therefore, the atmospheric humidity should be low. Thus drainage, optimum moisture, fertilisers and the temperature and photoperiod conducive to flowering appear to be important physiological factors for successful saffron production.

Picking of flowers and production of saffron

The saffron plant is in bloom for only 15 days and the flowering period starts from the middle or late October and lasts only till the first or second week of November. The flowers are picked early morning, before it gets hot and the style and stigmas are separated from the flowers. The operation is labour-intensive and high cost is due to the fact that the yield per ha is very small as compared to other drugs. The flowers left on the plant wilt away once, the duration of picking depends upon the time of blooming; which is dependent upon the climatic regime like the temperatures and the rainfall. A warm spring and long autumn are conducive to early flowering.

The process of separation of stigma is done each day otherwise the flower wilts and the operation becomes difficult to be carried out on wilted flowers. The pistil is removed carefully by hand and immediately deposited in a container.

The method of drying the stigmas is an important work, since value of saffron much depends on the methods the stigmas are dried. In Spain, the method is called *toasting*, the stigmas are placed in sieves, in layers of 2–3 cm thick, over an almost spent fire. The sieves are placed 15 cm above the fire, and by stacking them and changing their order and position, the product is dried very carefully. The process can also be carried out in special stoves for the purpose. The saffron should be kept protected from dampness and light, since light bleaches it to dull yellow (Ramstad 1959).

The stigmas, picked from the flowers and dried are orange-red in colour, constituting the first grade saffron called "Shahi". Flowers are dried in the sun for 3–5 days, then lightly beaten with sticks and passed through coarse sieves. The material which passes is thrown into water. Those parts of the flowers which float are discarded, and the parts which sink, to the bottom, are collected and further dried, constituting the second grade "Mogra" saffron. The discarded parts of the flower are again subjected to the beating procedure, and the process of throwing the entire pounded mass in water is repeated again, the product which sinks is collected, it is very much inferior in value and constitutes the "Lachha" saffron (Jalali 1962).

Yield

One grain of good saffron contains the stigmas and style of 9–10 flowers and an ounce may require 4300–4400 flowers. The average yield of saffron at Pampore in Kashmir valley is recorded at "5–6 lbs" per acre and 1 lb of saffron is obtained from about 75,000 flowers. In Kashmir, 140 lbs of fresh flowers correspond to 5 lbs of dried saffron (Fotidar DN, 1935). In Kishtwar the yield is even lower, while in Spain 1 kg of saffron is obtained from 5 kg of raw stigmas, which are in turn yielded by 80 kg of flowers (Arjona, 1945). In Spain, France and other countries, yields of saffron amounting to 8–11 lbs per acre are obtained normally; proper irrigation, use of suitable fertilizers, etc. account for the higher yield (Wealth of India, 1950). In Kashmir, a corm once planted remains in the field for 10–12 years, while in Italy it is an annual crop, in France, it is uprooted every three years. In Spain, they are uprooted after every 4th harvest and is done in May when the corm is well developed. The corm is then cleaned of the external covering and of the adhering remnants of the old corm.

Diseases and pests

Saffron corms are attacked by the fungus *Rhizoctania corcorum*, causing the death blight and is dominant in Spain and France. Another Fungus *Phomacordophila* is also known to attack the corms, transforming the flesh from white to yellow and turning it

black. Therefore, it is advisable to remove the outermost one or two protective peels (sheaths) and dipping the corms in a 5% solution of copper sulphate before planting.

A mole (*Arvicola arvalis*) is also reported to attack the corm in Spain for which fumigation of the burrows with tobacco smoke, red chillies or sulphur is done to compel the moles to come out of the burrow and get trapped in at the entrance.

Marketing of saffron

Saffron is classified as Asiatic, French, Italian, Spanish, etc. according to its origin. In India, Spanish saffron is quoted at prices higher than that from Kashmir; French and Italian saffron does not compete with the Spanish product; competition to Spain is mainly from saffron growers of Algeria and UAR whose product is in no way inferior to that of Spain.

References

Arjona EM. 1945. Cultivation of saffron and its uses. Infor. Bull. No 16. Dept. Agriculture, Madrid.

Biswas NR, Datta SP, Raychaudhri A and Dkashinamurti C. 1957. Soil conditions for the growth of saffron at Pampore (Kashmir). *Indian J. Agric. Sci.* **27(4)**: 413–418.

Bowles EA, 1952. A hand book of Crocus and Colchicum for gardeners. London. pp 222.

British Pharmacopoeial Codex. 1949. Pharmaceutical Press, London. 276 pp.

Budhiraja KL. 1942. Kashmir saffron with methods of resting its purity. *J. Indian Chem. Soc, Industr.* ed 5, 135–138.

Feinburn KL. 1957. The genus Crocus in Israel and neighbouring countries. *Kew Bull.* 1957 **(2)**: 269–285.

Feinburn KL. 1957. Chromosome numbers in Crocus. *Genetica* 29 (314): 172–192.

Foleh RA. 1957. A drug which is gradually disappearing from the medical armamentarium: *Saffron Framacognosia* **17 (44)**: 145–224; Biol Abstr. K958. 32: 19805.

Fotidar DN. 1935. Saffron cultivation in Kashmir. *Agriculture and Livestock in India.* 4: 242.

Jalali AK. 1962. Saffron in Kashmir. *Prajna-Banaras Hindu Univ. Journ.* **7**: 205–211.

Karasawa, K. 1956. Karyologie studies in Crocus IV. *Genetica* **28**: 31–34.

Karrer, P and Solomon H. Colouring matter of saffron. *Helv. Chim Acta* 10: 397.

Kawatani T and Sanaenosuke F. 1961. Comparison between farm and garden cultures of saffron (*Crocus sativus* L.) In Japanese. *Bull. Natl. Inst. Hyg. Sci.* **79**: 137–145; *Biol. Abstr*, 1962 39: 7766.

Lawrence GHM. 1945 Keys to cultivated plants, Autumn Crocus. *Baileya* **2(3)**: 77–85.

Madan CL, Kapur BM, and Gupta US. 1966. Saffron. *Economic Botany,* **30(4)**: 378–385.

Maw G. 1886. *A monograph of the genus Crocus with an appendix on the etymology of the words Crocus and saffron,* London. pp 326.

Martindale 1943. *The extra pharmacopoeia.* The pharmaceutical press. pp. 262.

Raj Gopalan R, Baliga BR and Bhat JV. 1960. Saffron as a possible source of riboflavin. *Proc. nat. Inst. Sci* (India) Pt. A. Physical Sciences **26** (Suppl) 128–134.

Rafizade N. 1956. Several problems in saffron production. *Sots. Kh Azerbaydzhana* 6: 33–35; *Biol. Abstr.* 1958: 35. 32997.

Shastry LVL, Srinivasan M and Subramanyan V. 1955. Saffron. (*Crocus sativus* L.) *Journ. Sci. and Industr. Res (India)* **14A**: 178–184.

Srinivasamurthey V and Krishnamurty K. 195 Place of spices and aromatics in Indian diet. *Food Sci.* 288.

Warburg EF. 1957. Crocuses. *Endeavour* **16(64)**: 209–216.

Wehmer C. 1929–31. 1936. *Die Pflanzenstoffe,* 2 Vols.

Wealth of India 1950. Raw materials 2. CSIR, New Delhi. pp 370–372.

Youngken HW. 1936. *Text Book of Pharmacognosy.* 4th edn. P. Balkinston's Son and Co. Inc. Philadelphia. 175 pp.

22. Curcuma longa L.

Family: Zingiberaceae.

Local names: Haldi, Haldá, Hálud, Haridrá (Hindi), Turmeric (English).

Description: Tall herb with thick, tuberous rhizomes, transversely ring marked by leaf scars.

Distribution: Largely cultivated for the sake of tuberous rhizomes which yield a yellow dye, also used as condiment. Its cultivated varieties show considerable variation in size and colour of rhizomes (Wealth of India, 1950). Long duration turmeric (CLL 324, 325, 326 and 327) were screened at Kasargod (Nambiar 1972); medium duration (CLL 317) and short duration (CA 73). Var. 24 was released (Thomas Vergheese and Thanakamma, 1975). The existing varieties have low yield potential. A number of varieties were tested from Doon valley (Gupta et al 1982). The varieties tested at Dehradun are Kasturi Tanaka, No 24, CLL 323, Avanigadda, CLL 328, Sugandham, CA 68, Dahgi, CA 66 and GL Puram received from the Central Plantation Crop Research Institute, Kasar-god, against the local variety called Kasturi Tanaka.

Uses and extent of utilisation: The rhizomes are universally used in culinary and curries. A native of India, but best varieties come from Indochina. It is used as a chemical indicator for alkalies and acids. It is put to many uses in native medicine, in diseases of liver, toothache, improvement of eyesight, dissolution of the boils and other congestions. It is also prescribed as a carminative. The uses in ancient literature have been well recognized to the fact that the patent which was recognized in USA has now been vacated in favour of this country.

Large volumes of this crop are not only consumed in the country but turmeric is also exported to many countries. Locally 'laddus' of *haldi* are made during winters and are consumed by elders as a cure for arthritis.

Agrotechniques for the Cultivation

It requires rich, thoroughly ploughed soil. The plant requires plenty of water and can endure shade. For this reason, it has been recommended as an important under-crop during the rainy season under fruit trees like peaches, plums and apple, etc. The crop can also be mixed with colcasia, castor, etc. It is usually sown before the onset of monsoons. Cut rhizomes are sown, 25–30 cm apart, and 30–45 cm distance between the rows. On an average, it takes about 6–8 months to mature and gives an yield of 10–12 qtls per acre.

A 5 years study on this crop at Dehradun showed that var. No 24 gave an yield of 12104 kg/ha followed by CA 6–8 Dahgi 11725 kg/ha followed by the var. Avanigadda (10550 kg). The local variety gave only 8061 kg/ha.

Critical analysis showed that var. 24 was a good yielder and confirms the findings at Kasargod (Thomas Verghese and Thankamma 1975). The variation in yields may be attributed due to climatic factors in Karnataka as compared to Dehradun and the fact that at Dehradun the crop was cultivated under dry land conditions without any supplemental irrigation.

Six varieties of turmeric from Central Plantation Crop Research Institute, Kasargod and one local variety were evaluated at Dehradun in a randomised block design (Gupta and Shukla, 1979). The plot size was 5 m × 2 m and the planting was done at 50 cm row to row and 30 cm from plant to plant. FYM @ 10 tonnes/ha was applied. Average of 5 years showed that var. No. 24 the maximum yield (12104 kg/ha) followed by the variety CA 68- Dahgi (11725 kg/ha).

The yields are higher than that given by the local variety at Dehradun. These varieties are recommended for cultivation in Doon valley region under rainfed conditions (Gupta RK and Shukla, D 1979).

However, a manurial-cum-fertilizer trial on local turmeric provided interesting results. The experiment was laid down with plot size 5.8 m × 1.7 m and the planting was done at 50 cm from row to row and 30 cm plant to plant. An average of five years results showed that under green manuring maximum rhizome yield was obtained from Dhaincha green leaves manuring at the rate of 10 tonnes/ha (7693 kg/ha), followed by Sunhemp (7649 tonnes/ha), followed by Cowpea (7333 kg/ha) and least by Guar (6358 kg/ha). Pooled analysis for five years showed that $N_{50} P_{50} K_{50}$ gave an average yield of 9806 kg/ha followed by $N_{25} P_{25} K_{50}$. Yield under control (untreated) was 7401 kg/ha and was almost equal to that obtained under Dhaincha green manure treatment (7.693 kg/ha) Gupta RK and Shukla D. 1979).

Performance of turmeric varieties in a randomised block design was studied (Gupta and Nambiar, 1977) with 4 replications as intercrops in a three years old plantation of Peach var. FloridaSun and as pure crop. The plots were at 5 m × 2 m size with spacing 50 cm between rows and 30 cm between plants (Tables 10.2 to 10.4).

Table 10.2: Fresh rhizome yield (kg/ha) of different turmeric varieties (after APR of CS and WCR and T Inst Dehradun p. 78).

Varieties	Yield (average of 5 years)
1. Kasturi tanaka	10223
2. No 24.	12104
3. CLL–323 Avanigadda	10550
4. CLL–328 Sagandham	9624
5. CA–68 Dahgi	11725
6. CA–66 GL Puram	10484
7. Local	8061

Table 10.3: Fresh rhizome yield (kg/ha) of local turmeric under different manure and fertilizer treatments. (After APR. 1979. CS and WCR and T. Inst Dehradun pp 80).

Treatment		Average yield of 5 yrs kg/ha
1. Sunhemp10 tonnes/ha		7649
2. Cowpea	do	7333
3. Guar	do	6358
4. Dhaincha	do	7693
5. $N_{25} P_{25} K_{50}$		9762
6. $N_{50} P_{50} K_{25}$		8621
7. N50P50K$_{50}$		9806
8. N50P50K$_{100}$		8803
9. $N_{75} P_{75} K_{150}$		8878
10. $N_{100} P_{100} K_{200}$		9227
11. Control (untreated)		7401

Table 10.4: Fresh rhizome yield (kg/ha) of different varieties of turmeric (after APR, 1977, Central Soil and Water Cons. Res. Trg. Insst. Dehradun).

Variety	Pure crop	Inter crop	Variety	Pure	Intercrop
Kasturi tanaka	9562	5610	CA-68 Daghi.	12650	5140
No 24	11425	5880	CA-66 GL Puram	9975	4280
CLL-323 Avanigadda	11500	4970	Local	7675	3960
CLL-328 Sugandham	9925	4050	S.Em C.D. 5%	1476	375

10

The yield potential of turmeric was 4–5 tonnes per ha of fresh rhizomes when grown as an intercrop in a 3 years old peach orchard. Differences in yield due to various turmeric varieties, however, not statistically significant. Cultivation of turmeric not only get the bonus crop but also helps the main crop since the soil get worked and the leaves of haldi provide manure, and improve the peach crop.

References

Wealth of India (CSIR). 1950. Raw materials pp. 402–403.

ICAR 1969. *Hand book of Agriculture*. 3rd edn. pp. 282–285.

Nambiar MC. 1972. Advances in spices Research. *Indian Farming*. XXII **(8)**: 9–12.

Thomas VP and Thankamma PK. 1975. NPK fertilizer experiment on turmeric. APR. 1974. Central Plantation crop Res. Inst. Kasargpd. 1976. Annual report.

Gupta RK, Shukla D, Nambiar TN and Ghosh SP. 1982. Evaluation of Turmeric varieties for Doon valley. *Indian J. Soil Conservation*. **10**: 101–03.

23. *Cymbopogon* species.

Family: Poaceae (Gramineae nom alt).

The genus Cymbopogon contains about 40 species, dispensed widely in tropics and rare in temperate regions. Several species grow in India and the species of chief economic value are aromatic. Though medicinal properties have been attributed to some of these species, none of them is of very much importance. There are essentially 4 oils, viz. palmarosa oil (rusa), citronella oil, lemongrass oil and ginger grass oil, which are obtained from those grasses.

Description: Grasses of the genus Cymbopogon are usually tall herbs and are aromatic. The species may be distinguished as follows:

Citronella grass belongs to the genus Cymbopogon to which many aromatic grasses belong, such as *Lemongrass*, *Palmarosa grass* and *Ginger grass*. Although botanically there are several varieties from cultivation point of view *C. nardus* Rendle, and *C. winterianus* Jowitt, yields the so called Java Citronella oil. *Cymbopogon nardus* is more extensive in cultivation, since it grows on poor soils.

Cymbopogon martinii (Roxb.) Watson (Rosa, Rohisa, Mirchágandh and Gandhvél) occurs in South India, Uttar Pradesh, Rajasthan and Sikkim, etc. The plant was collected by Mr G Martin at Balaghat in Mysore state (Karnataka) in 1791–92, and cultivated by him afterwards at Lucknow. It occurs in all the drier regions of India and has been recorded to appear in ranges at Pali district (Rajasthan) with appropriate soil conservation practices. The industry based on this plant is rather a modern one. The oil is obtained by distillation of the plant. About 250 kg of dry grass is recorded to yield about 11 kg of oil. The oil is useful in lumbago, baldness and skin diseases. It is internally taken in biliousness complaints. The oil is also used as an ingredient of mosquito repellant ointments. It is reported that plants found of Hyderabad contain low geraniol content (20–50%) and through the selection it was possible to obtain plants giving 92% geraniol.

Cymbopogon flexuous (Nees ex Steud.) Watson, occurs in South India, Uttar Pradesh and Sikkim. It yields lemon grass oil. This oil is a source of citral, starting point for vitamin A. There are plantations of this grass in Kerala state. It is also used in perfumery.

Cymbopogon nardus (L.) Rrendle, grows in several parts of India, chiefly South India. It yields the well-known oil of citronella, which, like inosha oil, is used as insect repellent ointment. The annual demand of this oil in India is estimated at more than 200 qtls.

In the sub-Himalayan tracts of *Bhabar* and the Upper Gangetic Plains, in Uttaranchal, Cymbopogon is abundantly found. A number of physiographic forms of *C. confirtiflorus* Stapf. , *C. martinii* Stapf. and *C. nardus* Rendle are met with. None of these forms is known oil of commercial value and oil of citronella (both Ceylon and Java types). Approximately to the tune of 150 tons are imported annually. The annual requirement of the oil is recorded to be more than 150–200 tons, which may require an area of about 2000 ha. Oil of citronella (Java type) is not produced on any appreciable scale, so far known, in India, except in the cinchona plantations of Ooty in the Nilgiris.

Of the two citronella oils, oil of Java (*C. winterianus* Jowitt) is an important raw material and commands larger market, because of its high citronellol and geraniol contents, the oil of citronella (ceylon) is used in perfumery and mosquito repellant formulations. The Java type of oil, mainly used as perfume bases, is also starting material for the manufacture of various aromatic chemicals. Citronenell can also be used as a raw material for the synthesis of Menthol. Geraniol esters of cinnamic, phenylacetic or lauric acid hold promise of replacing many costly perfume bases (Wealth of India 1956). There is, however, little morphological differences between the Ceylon type (*C. nardus*) and Java type citronella grasses except in their growth and form habit. The former is a narrow-leaved harder plant, growing 1.5–1.7 m in height that lives for longer duration (15–25 years) under cultivation. It is also reported to grow well on less fertile soils and is comparatively drought-resistant. In Java type, Jowitt first gave a

separate specific status (Bor 1965) and as described in detail (Bor, 1954) has broader leaves 1–1.2 m high and bend outwards until their ends almost touch the ground. It requires good fertile soils, need better care for growing and has an active lifespan of only a few years (Guenther, 1950).

Plantation of *C. nardus* may be raised on 10 yrs rotation, however, *C. winterianus*, may require a shorter rotation cycle. 1–2 weedings are necessary. It should be cut ordinarily when 8 months old. The oil content is greater in young leaves. When first leaf is just unfold oil content is the highest. It is harvested about 20 cm above the ground and 3–4 harvests are possible in a year. Under Dehradun conditions three harvests, viz. June, Sept and December are optimum (Gupta, 1978). Sunny days supported with moderate rains is the ideal weather conditions for *C. nardus, C. winterianus*, however, requires a richer soil and more rainfall and is less hardy than *C. nardus*.

Java citronella (*C. winterianus* Jowitt) has been raised in the *Tarai* region of Uttaranchal, and is recorded to require well-ploughed field having a good tilth. It is propagated vegetatively by planting slips, live and rooted, in the month of March and late August. The slips are obtained from a well-grown clump. A clump generally gives 60–80 slips of which nearly 40% have old wiry roots, another 20% have new and developing tender roots and the balance of 40% have no roots at all. Narayana et al (1975) recorded that root type of slips did not cause significant differences in the rooting and tillering percentages, where the leaf area on the slips caused significant differences. This improved rooting and tillering about 15% as compared to conventional practice of retaining 5–10 cm length of leaf blade on slips at the time of planting. Planting of live rooted slips is done in March and late August and for a good establishment, August plantations have been recorded better, otherwise the plantations require irrigation. There is a marked increase in vegetative activity during rainy season and thereafter till the close of growing period in December. Plants began to sprout in February-March and form new leaves. The crop need four weedings; once the crop is established weed growth is subsequently suppressed.

Performance of Cymbopogon species was studied on denuded wastelands at Dehradun under dry land conditions (Table 10.5). The plants were spaced at 75 cm × 75 cm and did not receive any fertilizers. Plot size was 6 m × 4.5 m. Mortality percentage was maximum in C. *martinii* (40.5%) followed by C. *nardus* (33.8%) in the first year. C. *citratus* (23.4%) and C. *winterianus* (5.2%). Casual ties were, however, later replaced. In the first year of the plantation only 1 cut was taken from all the varieties planted (5 months after planting). In the second year 3 cuts were taken, in August, September and November, while in the third year of plantation only 2 cuts could be taken. The yield of various species of Cymbopogon and the oil percentage recovered from all the 4 species is given in Table.

Though the yield differences were not significant in the first year, the year of establishment, during the second and the 3rd year yield differences were significant. It appeared that out of the four species tried C. *citratus* (Colour Plates 10.5 and 10.6) was the most promising on degraded lands under rainfed conditions.

In another study on the spacing trial it was found that out of the four spacings tried in a randomised block design yield differences were significant during the year of establishment. Subsequently, the yield differences were not significant indicating that 60 cm × 60 cm spacing is the optimum followed by 45 cm × 45 cm as shown in Table 10.6.

Table 10.5: Performance of different Cymbopogon species on denuated wast land of D.Dun

Species	Survival (%)	Total	Fresh leaf	Yield	Oil (%)	(Fresh weight)
C. citratus	100	955	31603	31566	0.22	0.20
C. martinii	98.7	262	11553	1645	0.10	0.80
C. winterianus	80.3	1164	16591	9663	0.78	0.80
C. nardus	77.7	1123	14672	16144	0.24	0.13

Table 10.6: Fresh leaf yield of C. citratus at different spacings

Spacing	Total fresh yield (kg/ha)					
	1st year 1 cut	2nd year 2 cut	3rd year 1 cut	4th year	5th year	Total leaf yield kg/ha)
45 cm × 45 cm	2242	18025	43630	24824	11241	99962(6)
60 cm × 60 cm	1588	20771	43006	22240	13694	101299 (6)
75 cm × 75 cm	759	18132	41765	19284	11769	91708 (6)
90 cm × 90 cm	657	13966	30111	15160	10111	70005 (6)

Quoted from APR, pp. 83–84, Annual progress Report, CS and WCR and T lust, Dehradun (RK Gupta and D Shukla) 1976.

Watering and fertilizer application

4 cm depth of water, at a regular interval of 3 weeks to a fortnight from April to June is desirable. Basal dose of Superphosphate at 50 kg/ha, of P_2O_5 and K_2O each in one application before planting and nitrogen at 80 kg/ha in four equal doses as top-dressing with irrigation, soon after harvesting, has proved beneficial. Narayana et al (1975) recommended application of urea @ 450 kg N split in six equal doses, 100 kg P_2O_5 as Superphosphate in two equal doses and 250 kg of K_2O, as Muriate of Potash/ha/year; which improved the yields significantly (50 tons of fresh herbage/ha) and oil yield 392 kg/ha @ 0.79% recovery. Periodical oil analysis showed consistently good quality.

However, in the *Tarai* region the tillering is highest during the months of August and September and continues in November, till it ceases by the close of growing period. On shallow soils, tillering, is, however, poor. Plants raised, under partial shade, show poor growth and less oil content. Where it is continuously wet, during monsoon season, large scale casualty is recorded. In *Tarai* area, four harvests are possible. During rainy season, though the vegetative growth is the highest (2nd clipping) the oil content is highest in October and December harvests. Younger leaves have higher oil content than the mature leaves. With the increasing age, after the monsoon months, when more tillering activity is recorded, more young leaves are given out resulting in higher oil yields. Sample evaluation of the oil from younger and older leaves suggest that the younger leaves give oil with higher specific gravity and optical rotation and exhibit a more fuller aroma. Variation in total geraniol and citronellal contents exists in the different harvests. The same planting stock, when raised at different localities produces oil having wide variation in

Citronellal and Geraniol contents. Since there is variation in yield and quality in different clumps of the same field, selection of material for improvement of type is much to be desired. This should be supplemented with more experiments for nutrient and water requirements, coppicing intervals, etc.

Palmerosa oil grass (rosha grass)

The aromatic oil from this grass is a good source of natural Geraniol and Geraniol acetate. It is extracted from var. *motia* of *C. martinii*, after separating Geraniol and Geraniol acetate; the oils are used in cosmetics and edible product industries. There is a great demand in the national and international markets. Palmarosa oil is light yellow with Geraniol 75–85% and Geraniol acetate 4–15%. The grass occurs naturally in the forests of Maharashtra and Madhya Pradesh and exploited industrially. New varieties of this plant have been evolved to meet the increasing demand of the oil. The oil obtained from *C. martinii* var. *sofia* is commercially known as Ginger grass oil or Lemon grass oil. These oils are used as a base for fine perfumery and are valued because of their geraniol content and alcohol, also found in Rose oil. Palmarosa oil has antiseptic properties and also has wound-healing effect.

Rosha grass is native to mountain slops, near *nallas*, and on well-drained soils but the content of oil differs in different habitats.

Soil and climate

The plant is well distributed in regions having an annual rainfall that is well distributed throughout the year, a temperature between 30°–40 °C and a relative humidity of 80%, are the ideal conditions for the growth of this grass. Temperatures below 20 °C, during winter season, retards

10

the growth of this plant. Sandy-loam soils having good drainage are most suitable. It can, however, be grown on saline soils upto 9.0 cm deep. It does not stand standing water and even 3–4 days of submergence kills the plants. The farming of this grass is only profitable, if proper attention is given to the site. The crop is estimated to transpire about 300 lbs of water for each lb of dry matter produced and so is the Rosha grass which yields 20–30 tons per ha, required 6000–9000 tons of water per ha per year or nearly 15 irrigations with 400 tons of water per irrigation.

Planting

It can be planted by (i) saplings (ii) seeding and (iii) rooted slips. 3–5 kg of seeds are mixed in the soil and are sown in small beds (3 × 1 m, 4 × 4 mm, 4 × 1 m) which should be well above the ground level (10 to 15 cm). The seeds are mixed in the soil after a light hoeing. The beds are watered for 3–4 days and 40–44 seedlings are planted at 60 × 30 or 45 × 45 cm distance and irrigated. For direct seeding, 8–10 kg of seeds in 15 to 20 times of sand are mixed and are sown in lines formed at 60 cm or 45 cm distance at a depth of 2–3 cm followed by light irrigation.

For slip plantation, the rooted slips are taken from 1–2 years old plant. Plantations are made during the rainy season (July-August). Transplanting is carried out in May-June, which requires heavy weeding. In October-November plantations, weeding required is minimum, the only drawback being that the crop is available after a year. When seedlings are 35 cm high, transplanting should be done in prepared plots. Plots should be prepared by ploughing 4–5 times with mould-board plough and then harrowing 5–6 times. Plots should be free

from clods and grass roots and should be done when plants are 40–45 days old. The soils should be weed-free. The seedlings should be planted on ridges, 90 cm apart after giving one irrigation. Ammonium sulphate and muriate of potash is placed below the ridges. Two hand-weedings at 20 days interval are required. The grass has a tendency to form clumps and attains its full yield of 20–30 tons per ha annually. By the end of September, budding of rosha grass plants starts and distillation can be taken up within 45 days of this stage (Singh 1969).

Irrigation

As natural during the rainy season, no irrigation is needed in the North Indian plains (November-February). During the hot season three irrigations are required. It can also be grown under unirrigated condition, however, the production gets reduced by 40–60%. The crop also needs drainage of excess water.

Fertilization

Since it is a perennial crop it requires fertilizer application and recommended dose is 150 kg N, 60 kg P_2O_5 and 60 kg K_2O. The entire doses of phosphorus and potassium are mixed in the field before planting in the first year and after cutting (June-July) with hoeing. Nitrogen is put according to cutting schedule, in 3–4 parts, by broadcasting, when there is moisture in the soil, and it should not to be given after flowering.

Cutting

Under irrigated condition, 3 cuttings can be taken. Sept-Oct, February-March and May-June, while under dry land only 2 cuttings are feasible. When the flowering is light the plant is cut 25–30 cm above the ground and stacked for oil distillation.

Yield

Through water distillation palmarosa oil comes out in 3–4 hrs and the yield is about 150 kg/ha from 3 cuttings (from irrigated land) and two cuttings from irrigated areas having the yield of 75 kg/ha.

Economics

It is estimated that net profit of Rs 44,000/ per ha can be obtained. In plains cost of land preparation and planting being Rs 22,250 in the first year, Rs 36,000 per ha in second year– 4th year; total expenses being Rs 67,750 income of Rs 2,47,000 for total oil production of 550 kg (in three distillations) at Rs 450 per kg of oil (Figures based on CIMAP 2001).

Gupta et al (1975) recorded that palmarosa oil crop raised directly from the seeds gave healthier crop with large number of tillers than transplanted crop. No significant difference in oil composition was recorded under two treatments. Similarly, the ratoon crop (2nd year) gave higher yield than directly sown annual crop without any adverse effect on soil content or composition. The increase being mainly due to progressive increase in tillers per clump throughout the growing period. Further, age of tillers or fertilizer application has little effect on initiation and duration of flowering. May month sown crop showed low Geraniol content but high percentage of Geraniol acetate oil contained in buds and open flowers showed no significant difference in yield or composition. Contrary to the general belief, the leaves are reported to contain high oil content similar in composition with the oil of the flowering tops.

Cymbopogon martinii (Roxb) Watson var. Sofia, *C. flexuous* (Nees ex Sted) Watson. (The Lemongrass oil plant).

In the sub-Himalayan tract of the *Bhabar* region and the upper Gangetic plain of Uttar Pradesh and Uttaranchal Cymbopogon is found abundantly and a number of physiological forms of *C. confertiflorous* Stapf. *C. martinii* Stapf. and *C. nardus* Rendle, also occur. However, studies on *C. martinii* (Gupta et al 1975) show that the age of tillers or fertilizer application has little effect on the initiation of flowering and its duration. May month crop showed low Geraniol content but high percentage of geranyl acetate which is acceptable to industry. Oil in buds or open flowers show no significant difference in yield or composition. The leaves contain appreciably high oil content, almost similar in composition with oil of the flowering tops.

The lemon grass oil is extracted from the distillation of leaves. The main constituent of the oil is citral (80.9%), and has lemon smell, from citrolionon and bionon. Bionon gives vit A and ionin is used for making perfumes. Majority areas, where this grass is cultivated, are Tamil Nadu, Karnataka, Madhya Pradesh, Assam, West Bengal Uttar Pradesh.

In South India, the plantations are done by seeding, while in other parts rooted cuttings or planting of rooted slips are done to propagate this plant. However, hot and humid climate is good for the growth of this plant, particularly where the rainfall varies from 2000–3000 mm and the lands are sloppy. It can also be grown as an unirrigated crop. Rooted slips are planted at 5–8 cm depth and the soil is pressed. Planting is mostly done during the rainy season; if irrigation is available it can the sown during February-March, where production is higher and so is the oil content per ha.

Important varieties are *Pragati, Praman, Kanery* and *Krishna*. The fertilizer application is done at the rate of 150 kg N, 40 kg

P_2O_5 and 40 kg K_2O per ha per year on ordinary soils. Application of potassium and phosphorus is done fully at the time of planting while nitrogen is applied one-third at the time of planting and the rest at the time of cutting the grass (twice usually).

The weed management of the crop is very much needed and hand weeding is done during the first 45 days of the crop. Java grass and remnants of other crops can be spread to control weeds. Chemical weedicide such as Diuron (1.5 kg) and Oxiflorophen (0.5 kg) as insecticide is applied. Weeding is done after every cutting, which improves the grass yield.

Since the crop does not require much water, once the crop is established it does not require irrigation, however, if irrigated after cutting each time it gives good yield. First cutting is taken after 90–100 days of sowing and the second cutting after 60–75 days interval. The crop is cut 10–15 cm above the ground.

Vapour distillation or water distillation is done. The stem is cut in small pieces while distillation is done. Total time taken for distillation is between two and a half and three hrs. The oil production is low during the first year of the crop and later on increases, average being 175–225 kg per ha, reaching from 100 kg to 200 kg in the 4th year. It is an ideal crop for the unirrigated areas. The total oil production during the four years period could be about 750 kg. The cost benefit economics as per the data (based on CIMAP figures) is estimated. As the cost of oil at Rs 350/kg at Rs 262,500 per year and for a period of years Rs 1,85,000, while the cost is estimated at Rs 77,400 total 1st year being Rs 23,400 and for subsequent years Rs 18,000 per year for three years. Net profit is calculated at Rs 16,000/ha per year.

Being an ideal crop for the unirrigated areas, the area is planted during the monsoon period at 45 × 45 cm or 45 cm × 30 cm. Farmyard manure is applied at 10–15 tons per ha. Net profit expected comes to Rs 16,000 per ha per year. The slips can be multiplied to 30 times in one year.

Citronella-Diseases and their Cure

1. *Leafspot/blight*: Fungal pathogen, *Curvularia*. Symptoms are formation of elongated purple red spots on leaves. Later leaf spots coalesce leading to drying of infected leaves. It is severe during monsoon.

2. *Lethal yellowing*: During rainy season it is affected by nematodes, insects and a fungus *Pythium aphani*. Plants suffer from chlorosis. Roots show disintegration of cortical region. Rotting later spreads to stem, causing death of the infected plant.
   Control: avoid water stagnation. Usually diseases spread from slips, and treat it with Ridomit (0.1% solution) before planting. Spray or drench the slips with Redoml (0.1%). Grow water-tolerant varieties like Jal-Pallavi (CIMAP).

3. *Sheath rot or leaf blight*: *Rhizoctinia* in 'Tarai' areas is the fungus. The plant remains stunted. Initially, leaves develop dark brown spots near soil surface and spread inwards causing necrosis of sheath tissue. Severe infection is in areas with stagnant water.
   Control: Spray Propicanazole 0.1% and Mancozeb (0.8%).

Palmarosa

Leaf spot: Caused by *Curvularia* and *Ellisiela*. Small dark brown spots appear on leaf, which enlarge, cause defoliation and reduce oil yield.

10

Control: Spray Mancozeb (0.3%) or Chlorothalonil (0.3%), if required repeat after fifteen days.

Lemon grass-Rust: *Puccinia nakanishki* small orange-coloured spots on leaves. When severe plant growth checked with reduction in total herb and oil yield.

Control: Spray Propicanazole (0.1%) and Mancozeb (0.8%).

Leaf spot: Caused by Curvularia and colletotrichum sp. in rainy season. In severe form brownish elongated spots appear which coalesce to form large spots causing drying of leaves.

Control: Spray Mancozeb (0.3%), Chlorothalonil (0.3%).

References

Ahmad CD and Shind AS. 1948. Cultivation of Rosha in the Punjab *Indian Farming* **9(5)**: 184.

Bor NL, 1954. *J.Bombay Nat Hist Soc*, 52.

Bor NL. 1960. Grasses of Burma, Ceylong, India and Pakistan Pergamon Press. Oxford, London, NewYork. Paris pp. 767.

Dutta SC and Nigam MC 1975. Chemotaxonomic studies on genus Cymbopogon. *Proc. Nat Symp. on Recent advances in Development, production and utilisation of medicinal plants* (Abstr.).

Dutta SC and Nigam MC. 1975. Development of an improved hac. National symp. (Abstr).

Gupta Rajendra, Gulati BC. and Duhan SPS. 1969. Observations on the growth and quality of *Cymbopogon winterianus* (Java Citronella) criop raised in *Tarai of UP. CSIR. Symposium*.

Gupta BK. 1969. Studies in the genus cymbopogon. II Chemocyto taxonomic studies in Indian Cymbopogons. *Plant Science* 3: 120–121 *Proc Indian Acad Sci.* **70**: 241–247.

Gupta RK. 1978. *Cymbopogon martinii* va motia. Mimco pp. 2. Circulated to officer-Trainees of Regular Course on Soil and water

Conservation Des and WCR and T Inst. Dehradun.

Gupta BK. 1971. A note on occurrence of natural hybrids in Indian Cymbopogons. Pp Sc. 3: 120–121.

Gupta R, Maheshwari ML, Jitendra Mohan and Gupta GK. 1975. Effect of fertilisers on yield of oil content and oil composition of Palmarosa oil grass as influenced by seasonal duration *Ibid* pp. 2.

Janaki Ammal EK and Gupta BK. 1966. Oil content in relation to polyploidy in Cymbopogon. Proc. *Indian Acad Sci.* Dec. B. Section. 64: 334–335.

Naryana MR, Ganesha Rao RS, Khan MNA, and aromatic Dimri BP. 1975. Response of Java Citronella to fertilizer application-Ihid.

Narayana MR, Ganesha Rao RS, Khan MNA, 1970. Propagation of Java citronella (*C. wintereanus jowitt*). observation on rooting of slips lbed.

Ravindra Reddy M. 1975. Successful cultivation of genuine Motia strain of *Cymbopogon martinii* of high geraniol content in Hyderabad. National symposium, (Abstract) pp. 1.

Singh Raminder. 1969. Roshagrass farming in Dehradun area CSIR Symposum Abstract.

24. *Datura* species

Family: Solanaceae.

Local names: Dhaturá (Hindi), Sádádhaturá. Apple of Peru, Devil's Apple (English).

The vernacular names given to the species of this genus can hardly be said to distinguish the various forms that exist. The Sanskrit name *Dhustura* or *Dhatura* and *Umatta* mean 'insane' and hence might have been given to the introduced plant (based on its properties being recognised); in fact might be upheld as not necessarily involving an ancient knowledge. The better known vernacular names are either derived from the above or have meanings in the languages to which they belong, they denote the properties of the drug. As met within India, the species of Dhatura have the appearance of introduced plants. They frequent wastelands near human dwellings or invade the border of fields, or cover abandoned cultivation. They do not exist as individual plants that take their own positions in a blended vegetation, appear as invading and here and there become so abundant as to exterminate all other plants.

Specific name is supposed to be a corruption from Greek, on account of the madness produced by its use. According to Stephenson and Churchil the name is said to be derived from Dhatura. Writers like Forskal, described the generic name from the arabic appellation Tatorab, while others made it classical from Do, Dare, Datura, since it is given as a narcotic.

There are about 15 species of Datura; *Datura innoxia*, *D. metel* and *D. stramonium* (Colour Plate 10.7) are important medicinal plants. Dried leaves and seeds of *D. stramonium* are listed in British Pharmacopoeia and also of United states, as antispasmodic in conditions of asthma, whooping cough, etc. The leaves and seeds of *D. innoxia* are used in India for the same purpose as those of *D. stramonium*. It is a possible source of the alkaloid 'scopolamine' used as a pre-anaesthetic in surgery and childbirth. *Datura metel* is recognised in the BPC.

D. stramonium provides hyoscyamine, while *D. inoxia* and *D. metel* provide an alkaloid. The alkaloid Skipalmin is used as anaesthetic during childbirth and anti-diarrhoea. Dried leaves of *D. stramonium* are used as antispasmodic, in conditions of asthma, whooping cough, etc. It is a source of alkaloid 'Scopolamine' as pre-anesthetic in surgery, in ophthalmology and for prevention of motion sickness.

Datura innoxia Mill. (–D. metel auct. non Linn.)

The species is a coarse, bushy annual, attaining a height of 90–120 cm and is native of Mexico, but found in North-western Himalaya, hilly regions of western parts, of the Deccan peninsula and on wastelands near the dwellings in some other parts of India. The plant closely resembles *D. metel* L. and has wrongly been mentioned under the name in some Indian floras. It is distinguished from *D. metel* by its dense pubescence, 10-toothed corolla and long, weak spines on the fruits. The leaves are also dark green, ovate, often somewhat cordate, about 12 cm long and 7–8 cm wide. The flowers are white and fragrant, about 8 cm long. The fruits are ovate-conical, nodding, about 5 cm long and 3 cm in diam, opening at the top into 4 valve-like forms, exposing a central column bearing numerous light-brown seeds. Like other species, *D. innoxia* emits a heavy narcotic odour, common on dust heaps and fallow lands after the rainy season; a potential source of alkaloid 'scopolamine' .

Among the alkaloids, this species has received wide attention. There is also a large collection of the genus like *D. godronii*, *D. fastuca*, *D. metaloides* at various centres. A new variety Mika (2n – 24) has been evolved in Bulgaria. It has been shown that autotetraploids through irradiation are rich sources of Scopolamine and other Tropane alkaloids. It can be directly grown from seeds or transplanted seedlings. Germination is slow and irregular but can be hastened with alternate exposure to freezing and thawing to weaken seed coat. Well prepared land, ploughed 3–4 times is good. Seeds are drilled in spring season 1 m apart. Application of FYM assures healthy growth. Much of the cultivation technology of *Dhatura* is known. However, some salient features are: closer spacing in a row, viz. 70 × 30 cm (-30) gives higher crop yield. The crop is grown by seedling at a depth of 0.5–1.0 cm in row, germination takes about 12–15 days but extends to 35–40 days at low temperatures. Use of herbicide 'belan' is useful for clear cultivation. Plant is sensitive to frost.

The entire herb is harvested at flowering stage, when fruits are mature but leaves are green. 30 cm above the ground and the aerial parts are rich in scopolamines. The leaves are striped and dried separately. Seeds are shaken off from the capsule when fruits beg into bust. Seed placenta contain highest alkaloid concentration (0.4–0.8%), while stem contains the least. In leaves it varies from 0.1–0.6%, pedicel and veins being richer in alkaloid. Fruits in the third (dichotomous) branch when in wax stage. The crop is richer in alkaloid. Under irrigation, two harvests are possible. Yield of the fresh herbage is 30–40 tonnes/ha, while on drying it is reduced to 6–7 tonnes/ha. Average alkaloid content (total) in dry herbage is 0.25%. After the first capsule, the alkaloid contents fall down rapidly to 0.01 to 1.0%.

Datura metel L.

A sub-glabrous, spreading herb which sometimes becomes shrubby. It occurs throughout India and is grown occasionally in gardens. Its leaves are triangular-ovate in outline and unequal at the base. The flowers are 18 cm long, with often double or triple corolla, white, violaceous, reddish purple or purple exterior and white from within. The fruits are globose, tuberculate or muriate, borne on a short, thick peduncle which unlike that of *D. stramonium*, is never erect. Capsule dehisces irregularly exposing a mass of closely packed, light brown, flat seeds which nearly fills the interior. The variety green is called RRL green. Cut plant

is green with light green leaves and is a late variety.

This species can be grown from seeds on any type of soil but for a good crop clayloam soils are the best. The plant for good growth needs good sunlight. The land is prepared well by ploughing it 2–3 times. About 10 tons of FYM per ha is applied along with NPK 25 : 50 : 25 which is mixed well.

Seeds (7–8 kg/ha) are sown directly or in the nursery and transplanted when it is 7–8 cm high, while *D. innoxia* is sown during October-January and *D. metel* is sown during February-March as a summer season crop. For a good germination, the seeds are soaked in water overnight and planted in lines at 1 m interval. It takes about 15 days for the seeds to germinate and within a month the germination is over. Transplanting of the seedlings is done when 10–12 cm long with 4-leaf stage. In general, the seeds are planted in June on the hills and in July in the plains.

It is necessary to do hoeing and weeding. The distance should be 70–90 cm from plant to plant; however, weak saplings should be removed. One week after the plantation the first irrigation is given, in case when there are no rains. If required 10–15 days interval may be kept in irrigations. The plants flower during February and fruits in March till April end. The fruits are plucked when hard and when the leaves are green. Plucking is done 2–3 times and in the end, by May end, the entire plant is cut. Planting of February month should be cut in June. The fruits are dried in sun (in an open place). The yield of leaves and the alkaloid content are influenced by pruning and debudding.

Pruning has an adverse effect on the height, leaf numbers, dry weight and alkaloid content; defoliation enhances these values. It is reported by studies from RRL,

Jammu (1962), that early defloration raised the alkaloid content in a four and half months old plant from 0.2026% to 0.3824% and the fertilizer treatment with ammonium sulphate (at 6,000 g per plot with 42 plants) further increased the alkaloidal content to 0.4025% in 4–5 months plants and to 0.3805% in five and half months old plant. With ripening of the fruit, the alkaloidal content migrates from pericarp to the seed.

Production of leaves and the seeds as reported by CIMAP studies ranges from 1200–1700 kg/ha. In dry seeds, alkaloid content reported varies from 0.2–0.35%. *D. metel* plants are cut for leaves and branches; the leaves are dried in shade. The first cutting is taken in July, while the second in October. The maximum production of leaf and the alkaloid content is reported during July when the first cutting is done.

RRL Neel Rohit (Purple) is an improved variety selected by RRL, Jammu, the entire plant is purple and gives 240–290 qtls of green leaves per ha and 18–24 qtls of seeds/ha. The alkaloid content in the seeds is reported to be 0.2–0.36% and in leaves 0.12–0.19%.

Another var. RRL green is entirely green with light green leaves and is a late variety. Per ha production of green leaves is 210 qtls with alkaloid content 0.24–0.28% and seeds 15–20 qtls per ha having alkaloid content of 0.09–0.128%.

Datura stramonium L. (D. tatula. L.).

A glabrous and farinose erect annual, usually about 1 m high on rich soils. Stem is erect with spreading branches. Leaves pale green, ovate to triangular, 10–15 cm long, irregularly toothed. Flowers large, 10–20 cm long, on short usually axillary-stalks, white or violet. Calyx tubular, 5 toothed, 5-ribbed. Corolla funnel-shaped, 5-lobed, limb spreading, folding at the angle lobes ending in long points. Capsule erect, ovoid and covered thickly with sharp spines, dehiscing into 4 valves. Seeds reniform, numerous (Colour Plate 10.7).

The plant is distributed on the hills, throughout the country upto 1500 m in North-west Himalaya and is found abundantly along road sides and villages but rare in forests or uncultivated land. Though, it grows as common weed in many parts of the world, it is also cultivated for obtaining drug of uniform potency. Near village, it is found on nitrogen-rich soils, also prefers rich calcareous soils. It is considered indigenous to the shores of Caspian sea and considered to have spread throughout Europe, Asia, America and S. Africa. It is widely used in Indian medicine and can also be collected in large quantity. An alkaloid principle similar to atropine has been extracted called *Daturine*. Leaves are the source for Solanaceous alkaloids which vary from 0.30–0.45%.

The plant has a strong disagreeable smell, so powerful that intermittent fever has been ascribed to it. Cow, horses, sheep and goats refuse this plant. When administered in large doses it produces intoxication, nausea, loss of senses, drowsiness, loss of memory, a sort of madness and somewhat permanent convulsions, suffocation, paralysis of limb, excessive thirst, dilatation of the pupil, trembling of body and ultimately death.

Cataplasm of the fresh leaves, when bruised, has successfully been applied to inflammatory tumours. An ointment made with powdered leaves allays the pain of haemorrhoids. Inhalation of smoke from burning leaves is recommended for relieving attacks of asthma. It is supposed to be a better cough remedy than opium. The flowers locally are dried and roughly

10

powdered with leaves and rolled into cigarettes for relief from asthma. The seeds are also smoked like tobacco to cure pyorrhoea in Kumaon.

D. stramonium prefers rich calcareous soil. The seeds are drill sown in spring season about 60 cm apart in rows. The plant is sensitive to frost. When fruits are mature but green, the entire plant is cut down and dried in sunshade. The leaves are stripped and separately dried in sunshade. Seeds are shaken off from capsule when the fruit begins to burst. Per acre yield is recorded to be 1000–1500 lbs of leaves and 2001 lbs of seeds. Nitrogen manuring favours plant growth and ultimately the alkaloid formation.

Var. *inermis* obtained from UK on trial at RRL, Jammu showed seed germination in about 2 weeks time with slow vegetative growth, attaining a height of 45–60 cm and bore fruits when the plant is only 15 cm high. Capsule is spineless. The seeds are germinated well, however, the alkaloid content is almost similar, and is, therefore, not very much advantageous to propagate. Total alkaloid content varies from 0.18 to 0.29%.

Datura fastuosa (Vern. Kálá-Dhaturá): This plant has the capsule in nodding attitude and open irregularly at the apex. It possesses the same properties as *D. stramonium* and regarded as a more deadly poison.

Among the Indian Dhaturas, this plant, because of higher percentage of alkaloids, was considered worth cultivation, though no definite relation of hyoscine and hyoscyamine has been established. Whether the variation is genetic or there are chemical races or the environment has any influence, is still unknown. It would be interesting to study the influence of chemicals (putrescine and methyl putrescine) on the biosynthesis of hyoscine and their percentage. Systematic cultivations of Dhatura species would solve the problem of production of hyoscine and hyoscyamine in the country.

However, the studies of Gupta and Madan (1969) on the physiology of these species is interesting. In order to promote germination a number of physical (abrasion, mechanical injury, prechilling), chemical (acid scarification, acetone treatment, 10% $ZnCl_2$ soln, napthoxy acetic acid) were applied to *Datura metel* seeds. Of all the treatments, immersing seeds in 0.2% ethyl alcohol was found to be most effective. Although there was an increase in germination percentage by other treatments like 10% $CaCl_2$ and gibberellic acid the alkaloid content of the fruit and leaves at different stages of plant growth and maturity showed that in case of purple variety of *D. metel*, percentage alkaloid content of the leaves exhibited a maximum upto bud formation. However, there is a steady decline in the alkaloid content of the leaves during post-flowering; it attains a minimum when the fruits are at the ripening stage. At the start of fruit setting, maximum accumulation of alkaloids was found in the very young fruits at first stage and there was a successive fall during the successive stages of development, until a minimum was reached in fully ripe fruits at dehiscence stage. The stage of maturity of the fruits, thus, is directly related to its alkaloid content. Considerable fluctuations were recorded when leaf materials from a single plant were analysed month-wise to study seasonal variations of the alkaloidal content. It was recorded that the alkaloid content was maximum during March-April, decreased slightly in May and June which could be correlated with the growth stage. During July to August, alkaloid content was low and increased again in September-October, in winter (November-December) it remained constant and after January-

February there is a leaf-fall, shoots started wilting, and consequently the alkaloid content was negligible. Similar observations were recorded in *D. stramonium* (Khan and Hussain, 1962).

References

Gupta US. 1966. Relative alkaloid content in different organs and regions of some *Datura* sp. and varieties. *Curr. Sc.* **35**: 311–312.

Madan CL, Prabhakar VS and Kapur BM. 1966. The location of alkaloids in the capsules and seed-portions of some Datura species and varieties. *Die Naturwissenschaften*. 4:109.

Gupta SC (Mrs) and Madan CL, 1969. *Some physiological observations of interest to cultivation of Datura innoxia and D. metel.* CSIR. Symp.

Khan AK. and Hussain SM. 1962. *Datura stramonium* L. and its cultivation in Abbotabad (Hazara). *Pakistan Journ. For.* **12**; 90–92.

Singh P. and Kaul BL. Induced polyploidy and variations on alkaloid in *Datura innoxia* Mill. *Indian Journ. Exptl. Bio.* 5: 128–129.

25. *Digitalis lanata* Ehrh.

Family: Scrophulariaceae.

Local names: Woolly foxglove.

Description: A biennial or sometimes perennial herb, attaining a height of 60–90 cm. The plant produces a rossette of linear-lanceolate leaves in the first year and a flowering shoot in the second year.

Distribution: It is indigenous to Central and Southern Europe but cultivated in UK, USA and Canada; introduced long back in Kashmir region at an altitude of 2300 m and now cultivated at many places both in North and South India (Palney hills and Udhaya-mangalam).

Uses: The plant produces rosette of linear-lanceolate leaves in the first year and a flowering shoot in the second year. The leaves produce the physiological action of digitalis, with considerable stronger and less cumulative effects. This is used as a source of *digoxin*, an active cardiac glycoside; Digoxin is official in some Pharmacopoeias. *D. campanulata* and *D. alba* show therapeutic action equal to *D.purpurea*, while *D. lanata* is most potent therapeutically.

Agrotechniques for Cultivation

The plant thrives on light loamy soils of reasonable fertility. It is propagated by seeds in spring or in the months of June-July, at higher elevations (1500 m) and October-November at lower elevations. The seeds are sown on well-prepared beds. Some seeds are very small, these are mixed with sand to facilitate uniform distribution in sowing. During excessive hot days the beds are practically shaded and frequently irrigated.

The seeds take 10–15 days for germination and nearly after a month, when the seedlings develop 5–6 leaves, these are transplanted at a distance of 30 cm in rows and 60 cm apart. Regular and frequent irrigations are required during the summer months. The plants put better vegetative growth in shady places.

Lime application increases both the foliage and potency of leaves. FYM and ammonium sulphate application increase the yield but the effect is less marked than that of lime. Tilling the soil 10 cm deep does not lower the yield of the crop, while 30 cm tilling is the maximum spacing required for best yields.

The leaves are collected from April-November from the first year plants at lower altitudes like that of Jammu and during July-November at high altitudes like at Srinagar in Kashmir. The plant starts

flowering during the second year when only two flushes of leaves are taken, one before flowering and the other during flowering after which the plant dies. In the first year, however, several flushes can be taken at regular intervals. In Jammu region, the plant flowers in April-May, while in Kashmir region it flowers during June-July. Lower basal leaves and upper smaller leaves are normally rejected. Three-fourths of the total leaves from each plant are taken, both young and old are mixed during the collection. In the second year, only 2 flushes are taken, one before and other after flowers; after which the plant dies.

The plants are howed and irrigated after every flush. The leaves are plucked by twisting without removing the shoot. The lower basal leaves of poor colour as well are rejected as stated above; practically three-fourths of the total leaves are taken, and thereafter the plants are howed and irrigated. Output of leaves per plant is nearly 240 lbs per acre.

The leaves should be dried rapidly by spreading them in the open on a sunny day and after complete drying stored in dark and airtight containers. While fresh leaves contain 51.9 IU/g potency after drying in sun, it is reduced to 39.9 IU/g in shade for 4 days, 30.21 U/g drying at 100 °C for 2 hrs 23.4 IU/g, 23.0 at 70 °C and drying for 4 hrs, 35.2 IU/g at 50 °C for 14 hrs. The fresh leaves contain 3 natural glucosides, viz. lantosides A, B and C which, on hydrolysis, give digitoxin, gitoxin and digoxin. Digoxin produces safe cardiac effect as digitalis and is 300 times potent than digitalis. Digitalis increases force of systolic contraction and efficiency of decompensated heart. It slows heartbeat and reduces cardiac oedema. The result of biological as says for leaves cultivated at different places showed that samples from J and K and Switzerland show max. potency in plants 43.1 IU/g. Plants grown at Jammu (300 m) show 22.3 and against 31.4 for plants grown in Switzerland and 43.1 at Katra (100 m alt) in J and K state.

26. *Digitalis purpurea*

Commonly known as Foxglove so. *Digitalis purpurea* is a biennial or sometimes perennial herb, 30–60 cm high growing at 1600–2000 m. It is a native of Western Europe, also occurs in south and central areas but cultivated all over Europe in the gardens as a border plant. In India, it has been grown as an ornamental, in different hill stations. In 1880, attempts were made to grow it at Saharanpur and at Mussorie for a regular crop of leaves for medicinal purposes, but the cultivation was not successful. The plant yielded very few leaves. Cultivation of this plant was also started in the eastern part of Darjeeling and in the south at Nilgiris in cinchona plantations but practically abandoned now and the attempts to grow therapeutically active plant in the plains met with little success though again the efforts were made by RRL at Jammu and Srinagar.

Agrotechniques for the Cultivation

It is propagated by seeds and is a calcifuge species, growing on well-drained soils of open texture, reasonable fertility containing traces of manganese. It prefers light over-head shade which gives best results. The soil should be well broken and liberally manured with leaf mould, 4–8 ounces of seeds per acre are needed for sowing which is done on well-prepared nursery beds during March-April at altitudes, while during October-November in plains like those at Jammu.

The seeds are even smaller than those of *D. lanata* and are mixed with fines and to ensure even distribution during sowing.

Spring sowing takes 10–15 days to germinate, when 80–90 cm high the saplings are transplanted in the field at a distance of 30 cm in rows and 60–75 cm apart from plant to plant. The crop is kept clear of weeds and the field is howed 2–3 times. Addition of FYM at the rate of 25 tons per acre, while preparing the field is recommended. The seeds germinate better under slight shade. If direct sowing is done, thinning should be done at the above distance. Average yield of seeds/plant is 22 g (15–34.5 g).

The plant flowers during the next year in April and May. Other operations like collection, drying and storing are done in the same way as in *D. lanata*. Dried leaf yield is 3–4 maunds/acre. Against an international standard of 12.5 units IU/g, the potency obtained from Kashmir samples ranged from 9.17 to 14.21 IU/g, the maximum being at Katra (1000 m alt).

On biological activity it was shown that the activity is equal to one International Unit (IU) present in 96.42 mg of dried leaves. Argentine digitalis leaves yield 0.34% digitoxin, the most potent of digitalis glycosides. Variation in digitoxin content (0.15–0.79 g/kg) and geloxin content (0.0.78 g/kg) is noted in dried leaves. Many other substances such as tannius, luteolin acids and fatty matter are also present. A cardiotonic has also been isolated.

Reference

Michalski T. 1968. Radiation induced mutation in *Digitalis lanata* Ehrh. *Pl, Br,. Abstract.* 1968. 5053. Polish.

27. **Dioscorea** spp.

Dioscorea, mostly cultivated as garden crop, thrives best on deep sandy loam soils with adequate moisture and good drainage. On heavy soils under water-logged conditions tuber development is poor. Tubers deteriorate rapidly or develop insipid taste. It is an exhaustive crop, requiring deep ploughing and liberal manuring. FYM, sheep penning and green manure application to the crop gives good results. Both tuber crop and aerial tubers (bulbils) borne on the stem are used for propagation. Propagation is generally not adopted by aerial tubers. It takes normally about two years to produce edible roots. Tuber tops containing 2–3 healthy eyebuds are planted in pits of size 40 × 40 × 60 cm, in rows 60 to 90 cm apart from April to July after the onset of monsoon season, depending on the local conditions. Vines are allowed to trail on ground, but good results are obtained by staking. The crop comes into bearing in 5–8 months after planting. During this period the field is hoed, weeded and earthened up round the stem when tubers are fully formed or developed the leaves began to drop, vines are then cut and tubers dug out.

Fifty species of this genus are reported from India. Some of the plants such as *Dioscorea deltoidea* have yielded as much as 8% and *D. prazeri* 2–4% diosgenin, which is the most favoured starting material for the synthesis of steroid hormones search of possible sources for cortisone started in 1950 and sapogenin in Agave and Dioscorea species proved as alternative source of Diosgenin in use is recorded to the extent of more than 2.5 million annually with possibilities of export to ones 5 million rupees, about 50 species are reported from this country. Mexico is reported having 63 species out of which *D. cinoisuta* and *D. spiculifera* yielded maximum (13.2 %) Diosgenin.

a. **Dioscorea deltoidea** Wall.

Family: Dioscoreaceae.

Local names: Shingley, Mingly (Himachal), Khis (T. Garhwal).

Description: An extensive climbing herb with stems twining to the left, glabrous. Leaves usually alternate, lower sometimes opposite, 5–15 × 4–14 cm, stalked, sub-opposite, simple, ovate, usually acuminate, membranous, often cordate; the basal lobes rounded or sometimes dilated outwards, 7–9-nerved at the base, glabrous above, with a few minute hairs on the nerves beneath, petiole about as long as the blade, slender. Male spike solitary, unisexual in axillary, simple or sometimes branched. Flower bracteate, in small, distant clusters. Stamens 6, all anther-bearing. Female spikes solitary, flowers distant. Capsule winged all round, often unequally so. Fruit a winged capsule, locular, 2–25 cm. Seeds winged, all around, 2/locule, seed size 24 × 1.5 cm.

The plant came to promineme in 1960–61 and cultivation started in 1995 in H.P. (Barshrey nursery). There can be a little doubt that at least 300 yrs ago various species were cultivated in India .

Distribution: A perennial climbing herb, with geophytic habit. The plant perennates by rhizome (5 cm diam) lying parallel to the soil surface at a depth of 15–40 cm. The plant grows as forest weed in temperate regions of the Himalaya, associated with lower oak (*Quercus incana*)-conifer forests. It is increasingly met within the area on drained loamy soils cleared of forests, on sites of abandoned shifting cultivation in burnt areas, flood deposit along the *nala* and rivers. The plant generally prefers northern aspects and restrict to gentler slopes, excessively grazed or lopped, in oak-conifer forests (1700–2600 m) in temperate regions. In the east of river Beas it has sparse distribution. Other species (*D. glabra, D. Sativa*) occur at lower elevation. In northeast region *D. floribunda* is cultivated.

Sarin (1969) studied in detail the ecological factors affecting distribution of the species. The plant does not thrive well in complete shade. Direct sunlight for a part of the day is preferred more than a continuous exposure to sun. It prefers alluvial soils; best on loamy to clay loamy soils. It has good reproduction capacity through seed. Life cycle of the plant above ground is completed during April-October. Fresh stems sprout every year during the end of March or April and flower during May-June. Fruits ripen by the end of October, when the stem starts drying up. Ripe fruits dehisce during November-December, when still they hang on dried stem. The seeds lie dormant for 3–4 months and germinate during April or May, if the temperature of the substratum is favourable.

Growth of rhizome is longitudinal for the first few years with aerial branches sprouting only at nodes. As the plant grows old sprouting takes place at a number of spots. Food reserves start accumulating at the base of each aerial shoot and this tissue gradually grows into an offshoot of mother rhizome. Repeated branching of rhizome in this manner results in vegetative reproduction. These rhizomes increase in size, resulting in direct growth.

Hooker reported 25 species growing in India; of these Chakravarti et al P (1954) investigated 19 species and found that tubers of *D. deltoidia* and *D. prazeri* contain exceptionally high percentage of sapogenin, i.e. 3.53 and 2.1% respectively.

Uses and extent of utilization: The tubers are employed as a raw material for the manufacture of cortisones (contain 3–4% sapogenin). Diosgenin is the main sapogenin of glycoside (saponin) obtained.

Diosgenin is the raw material for cortisone; effective in rheumatoid arthritis. It was originally isolated from the cortex of the adrenal gland. Since the quantity is small in the cortex it could not be isolated on a large scale from the glands and the cost was exorbitant, besides it occurs in association with a large number of compounds having similar chemical structure which make its isolation in pure-form very difficult. It was commercially prepared by partial synthesis of bile acids (deoxycholic acid) from oxbile which used to be its principal source. The supply from the source could also not be sustained as it depended on the cattle slaughter. It could also be prepared by partial synthesis from naturally occurring steroids which are available in sufficient quantity and can be used as starting material for building up the molecule of cortisone.

Sarmentogenin, the cardiac aglycone from the seeds of *Strophanthus sarmentosus,* an African climber, attracted attention of the scientists as a starting material, but could also not be sustained since the quantity present in it was small and the plant was difficult to propagate. Hecogenin from 'Sisal' (*Agave* species) plants was also tried for the production of cortisone but without much success. *Ergostrol* and *Stigmosterol* have also been used as starting point for this purpose.

Amongst the various raw materials steroid sapogenins have been considered as the best useful starting materials. Diosgenin from various *Dioscorea* species has also come into use, which appeared to be the best material for preparation of cortisone and other steroid hormones. It is the principal sapogenin of the glycoside (saponin) of yamplants.

The tubers are large but not eaten, though other species are used as subsidiary and famine foods. At Kulu, tubers called 'Shingali' are used for washing, while in Kashmir it is called 'Krits(z) used in medicine and for washing wool.

Agrotechniques for the Cultivation

The seed tubers are sown in 2 m × 1 m and 4 m × 1 m, raised beds, manured with FYM. The seed tubers are sown @ of 350 g/ 2 m × 1 m sized bed. The collected tubers are cut into small pieces of 2.5–3 cm each weighing 20–25 g at a spacing of 20–40 cm during the month of February. Dibling is done upto 4–5 cm depth during February. FYM at the rate of 4–5 kg/sq m of bed is given to the field. During dry season watering is done. The tubers sprout in April and about 4–5 stems arise from each tuber. Young shoots are provided with support system by putting up 12–20 cm bamboo sticks when germination is complete. The tuber in nursery bed germinates after 60 days and the percentage germination is about 86. Flowering occurs during the months of August-September and during October-November aerial portion starts drying. In the second year soil is ridged around the tuberline so as to provide more space for development of tuber clump. The plantation is irrigated fortnightly. The aerial part must grow upto 3 m, while the average height is about 1 m.

Digging of tubers is done during the month of February after an interval of 1–2 years. *Dioscorea* tubers collected from nursery are planted, cut into small pieces. The pieces are sown in patches of 45 × 45 cm of 10 cm 3 cube size at 2.5–3 m spacing in forest areas during February-March. Patches are prepared near bushes. Roots in this environment are few, because of lack of aerial support. During the second year tubers of 80–90 g weight are available, though average size of 5-tubers collected after 1 and 2 years of harvesting is about

119 and 168 g respectively. Parent seed tuber bears a few roots. After 1–2 years, of a seed tuber in nursery of 350 g in 2–1 m sized bed is approximately 1.760 kg and 4.100 kg. On well-drained soils upto 4.5 kg of tuber per 2 m × 1 m bed can be obtained. The total yield per ha after 2 years is 200 to 250 qtls and after 3 years 320–370 qtls. Sometimes, tubers interweave and produce mass of tubular branches attaining large sizes.

Processing

1 kg of tuber (2 yrs old) after drying weighs 0.250 kg, and a tuber of 3 yrs may dry upto one-third of its fresh weight. Estimated income from one ha of *Dioscorea* plantation is estimated upto Rs 3.5 lakhs/ha. Marketing of tubers can be done through Village Forest Development Committees or through cooperative forest produce societies, as done in Himachal Pradesh.

Costisone is also used for treating skin diseases, eye infections, blood diseases, endocardiac disorders, and rheumatoid diseases. The starting materials for the preparation of steroid hormone like cortisone are mostly diosgenin obtained from Dioscorea although hecogens or most advanced material from fibrous plants like agave or Stimastrol from soyabeans, cholic acid, Smilagenin and bile are the alternative sources.

Tuber yield ranges from 4.5–6 tons/acre and tubers can be stored for 6 months in cool sheds under dry earth or sand. Where climatic conditions permit and the soil is dry they are best left in the field and dug out when required.

Three species, namely *D. esculenta, D. prazeri* and *D. deltoidea* contain diosgenin in significant quantity. *D.esculenta* contains very low content (0.15–0.2%), *D. praxeri*

yield 4–4.5% though tubers are soft and fleshy, while *D. deltoidea* is the richest Indian source (0.8%).

In Kulu district, (H. Pradesh) the right holders are permitted to collect roots from forest and sell outside the state by paying a royalty to the concerned Panchayat. In 1964 a scheme for rotational extraction of Dioscorea was implemented where excessive extraction caused extinction of Dioscorea from natural sites. Plants other than Dioscorea containing diosgeum are *Trillium govanianum, Paris polyphylla* while *D. prazeri* and *D. dettoidial* contain almost the same quantity of diosgenin advantage that their cultivation is rather easy and practical.

Chemical analysis

Crude diosgenin is obtained which is a yellowish gummy mass. When washed with acetone solvent it gives white gummy pure Diosgenin (MP 203/04 °C). The yield is 2.2% of tuber weight. TLC behaviour of Diosgenin in benzene and ethyl alcohol shows 95% of purity of Diosgenin with RF value of 0.45.

National requirement of Diosgenin is about 150 mt/annum while the total production is about 60 m tonnes and is being imported from China and Mexico, etc.

Hydrolysis of Dioscorea tuber is the main extracting process. Glucoside Diosgenin is attached to glucose molecules through hydroxyl group. In acid hydrolysis, the —C—O bond between glucose and Diosgenin is cleaned and Diosgenin comes into free state which is isolated by extraction with inhexane solvent.

References

Barua AK, Chakravarty D. and Chatterji D. 1955. Steroid hormones from Indian Dioscorea plants. *Bull. Inst. Sci. India,* 1955. **4**: 15–20.

Chander, Ishwar, Handa KL and Kapoor LD. 1955. *Indian Journal of Pharmacy* **17**:142.

Chauhan PL. 2001. Herbs in Himachal. *Dioscorea, shingly-Mingly. Time News Letter.* vol I. pp. 21–23.

Chopra IC and Sobti SN and Handa KL. 1962. Cultivation of Medicinal Plants in Jammu and Kashmir. ICAR Research Series 12. pp. 40–42.

Chopra K and Kapoor LD. 1967. Standardization of steroid sapogenin bearing plants in India. XIth Indian standard convension, Chandigarh DOC. S-1/26.

Martin FW and Cabani Ilas E. The F1 hybrid of some saponin bearing Dioscorea species. *American. Journ. Bot.* **53(4)**: 350–358.

Sarin UK and Gupta S. 1968. A note on the distribution pattern of *Dioscorea deltoidea* Wall. *Proc. Rec. Adv. Trop. Ecology*: 245–247.

Sarin YK. 1968. Some ecological observations of *Dioscorea deltoidea* Wall. of interest to its cultivation. *CSIR. Symposium.*

Seale CC. Agronomic studies of Dioscorea in S.Florida. AR IPA. *Agic Exp.* Stn. 273–274.

Kapoor LD, Datt AK and Sarin YK. 1963. Experimental cultivation of *Dioscorea deltoidea*, Wall. in Kashmir. *Indian For.* **89(9)**: 43.

b. Dioscorea prazeri

This plant grows wild in Bengal, Assam and Sikkim and also in Bhutan while *D. deltoidea* grows wild in the Northwest Himalaya at the foothills from Jammu and Kashmir (Katra Kistwar) to Uttaranchal. It is locally used for washing wool, hairs and for killing lice. Diosgenin content from various places has been analysed by Ishwar Chandra et al (1955) in Kashmir and H. Pradesh: Katra (1000 m) 4.8%, Barot (1400 m): 4.5%, Sanasar (2300 m) 8.9%, Pahalgam (2300 m) 4.6%, Gulmarg (2300 m) 3.8%, Bhadarwah (1600 m) 7.9% and Mahali-HP (2300 m) 6.2%. The tubers contain 60–70% moisture. Tubers collected in different stages of development on analysis showed the results given in Table 10.7 (after Chopra et al 1962).

Table 10.7: Root collection stage of development and content of diosgenin

Root collection time	Stage of development	Diosgenin content (%)
May	Dormant tuber	4.8
July	Sprouting start	3.8
October	Flowering	4.4
December	Mature seed formation	3.8
March	Dormant tubers	4.6

Good quality of *D. deltoidea* tubers contained upto 4.8% diosgenin as the best time of collection is when tubers are dormant and aerial portion of the plant has withered.

Tubers weighing nearly 20 kg from one plant with 8% diosgenin content have also been reported to have been collected (Sobti and Chopra 1960) from Manali and Bhadrawah velley (having 100 cm and 200 cm rainfall) with temperature at freezing point during winters and 30 °C during summer season. Soil analysis showed moisture 12.9%, pH 6.2, N-1.48 w/w, Hydrochloric soluble extract. CaO 1.6% v/v, MgO 0.4, P_2O_5 0.32, K_2O 0.62, AlO_2 3.60, organic matter 48.2%, clay 17.9, silt 6.9 and sand 18.0% v/v.

Propagation

The young tuberous rhizome is cut into small pieces, each segment placed in the soil during spring (March-April) or in rainy season (July-August), Sandy loam soil is preferred, rich in organic manure and natural pH. Before planting the soil is manured (leaf-mould rotten) and tuber planted at 30 cm apart in rows, 60 cm apart. Vines are provided with support or sticks. At low elevations many plants die during summers. Growth is vigorous in March-October and is dormant during winters

(Nov-February). After 18 months, the rhizomes weigh 175–250 g and are collected. Diosgenin content in one-year-old tuber is reported to vary from 3.6–2.8%.

Dioscorea bulbifera L. is shown in Fig. 10.3 and *Dioscorea pentaphylla* is shown in Fig. 10.4.

Distinguishing Features of the Tubers (rhizomes), Vines and Fruits of some Indian Species of Dioscores in India (Fig. 10.5).

1. *Dioscorea alata*: Tubers single or many, polymorphous cylindrical or clavate, globose stout and short, pyriform, lobed variously, fingered and fascicled, sking

Fig. 10.4: *Dioscorea pentaphylla:* Herbarium sheet from BSI, Poone

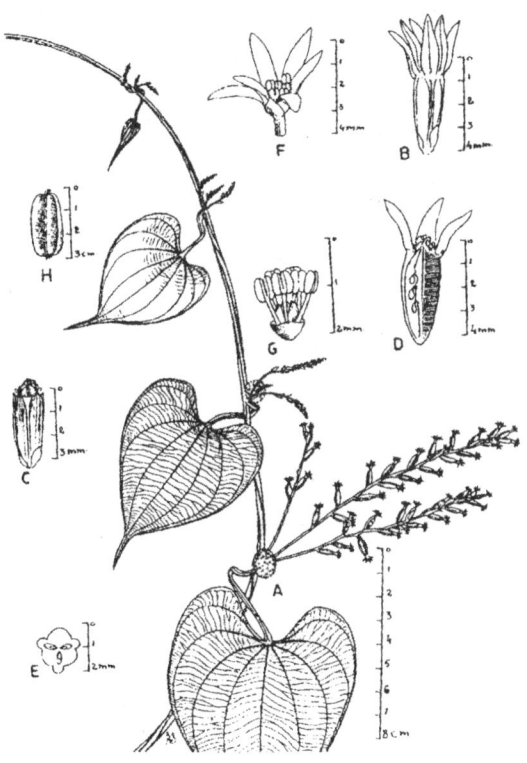

Fig. 10.3: *Dioscorea bulbifera* L. A. flowering twig; B. female flower; C. same with perianth removed; D. longitudinal section of B; E. transverse section of gynoecium; F. male flower; G. perianth of F removed to show androecium and pistillode; H. capsule

brown to black. Flesh white or creamy white or with magenta sap. Vine with quadrangular winged stems, turning to bright; leaves opposite rarely alternate, simple. Bulbils present abundantly or sparingly. Fruit capsules, 2 cm long, wings broader than semicircular; 2 cm long, 1.3 cm wide, reddish brown when dry, apex slightly retuse.

2. *Dioscorea bulbifera*: Tubers solitary, globose to pyriform, usually small and round, large under cultivation, weighing upto one kg and then slightly loberd, skin purplish, black or earth-coloured. Flesh white to lemon yellow, sometimes marked with purple flacks. The vine has unarmed stems twining to the left. Leaves alternate, simple. Bulbils present.

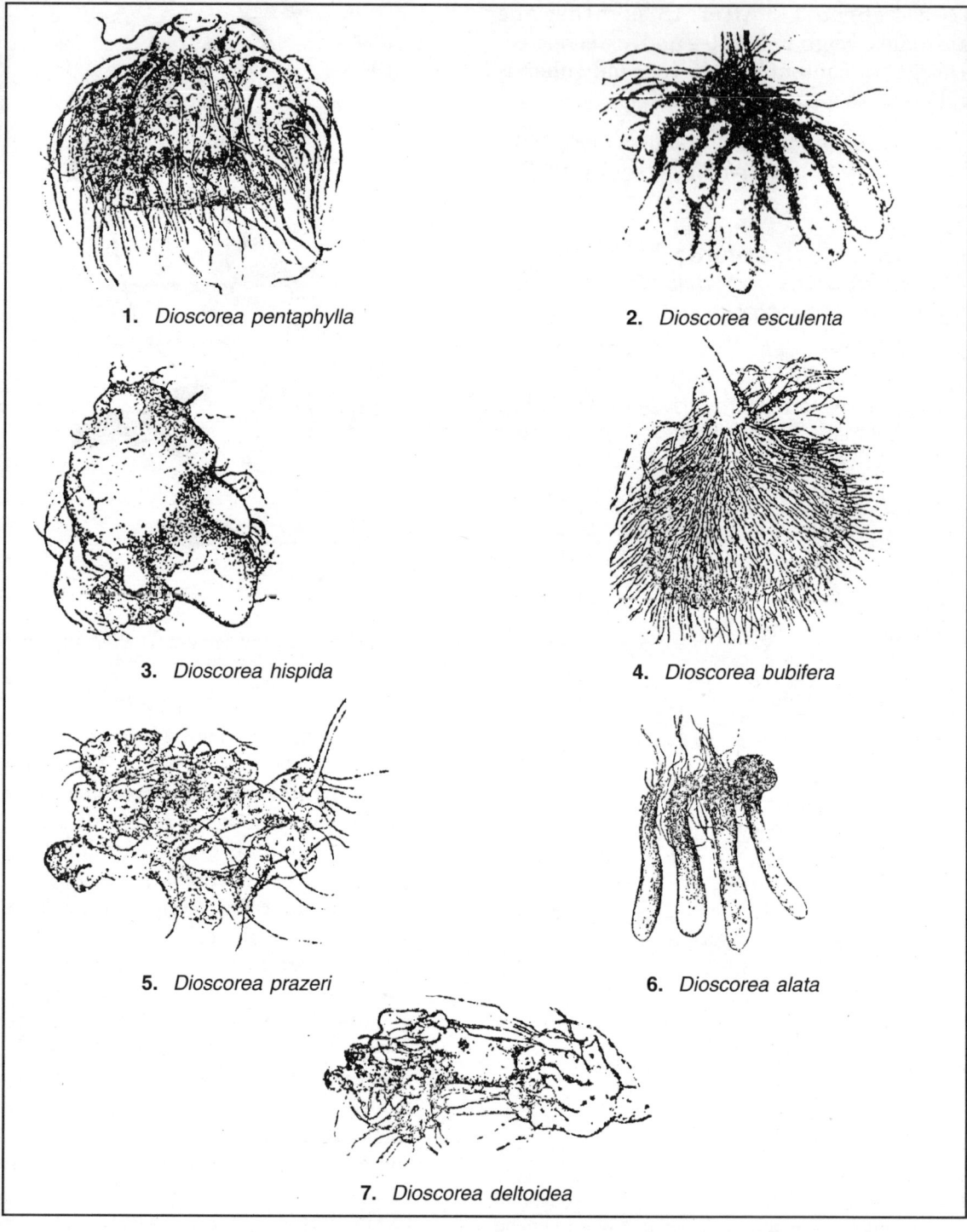

1. *Dioscorea pentaphylla*

2. *Dioscorea esculenta*

3. *Dioscorea hispida*

4. *Dioscorea bubifera*

5. *Dioscorea prazeri*

6. *Dioscorea alata*

7. *Dioscorea deltoidea*

Fig. 10.5: Distinguishing features of some Dioscorea tubers, rhizomes, vines or fruits

The capsules almost straw-coloured when ripe with purple flecks or bronzed, wings rounded at both ends, 2.0 × 2.2 cm –0.9 cm. (Figs 10.3 and 10.4).

3. *Dioscorea deltoidea*: wall ex grisbeck: Tubers horizontal, close to soil surface, 20 cm or more long with scattered roots, skin light chestnut brown with slight rather rectangular cracks.

   Vines unarmed, stems twining to left. Leaves alternate, simple, blade commonly 8 × 6 cm, shortly hispid. Bulbils present.

   Fruits a capsule reddish in colour, straw-coloured when ripe, green with abundant purple flecks when immature, wings slightly broader than semi-circular, 1.5–2.2 cm × 1.4 cm.

4. *Dioscorea esculenta Burkill*: Tubers 4 to many, stalked, in bunches, sausage shaped, globose or flattened and lobed, usually not longer than 12 cm, sometimes reaching upto 50 cm and weight 3 kg, rarely thorny, sometimes reddish.

   Vines are prickly with puberulus or pubescent, villous stems sometimes, twining to the left. Leaves alternate, simple. Bulbils absent.

   Fruits capsular retuse at apex, wing semi-cordate, below almost truncate. 3.0 × 1.2 cm.

5. *Dioscorea glabra* roxb. Tubers 1 and 2 yrs plants unarmed. Later prickly towards base and unarmed and smooth above stems twining to the right. Bulbils absent. Leaves simple.

6. *Dioscorea hamiltonii* Hook. f.: Tubers long stalked or gently smelling from the point of origin; skin dark or black, sometimes unveven and rough, flesh white.

   The stems are angled, glabrous, twining to the right. Leave opposite on thicker stems, often alternating towards the ends. Bulbils present.

   Capsules upto 3.5 cm long, wings 3.0 to 2 cm broader than semicircular, copper coloured.

7. *Dioscorea hispida* Dannst.: Tubers more or less depressed, globose, often lobed, occasionally of considerable size (weight upto 13.5 kg), skin straw coloured grey; flesh white to lemon yellow.

   The stems are prickly, twining to the left. Leaves alternate, trifoliate, rarely simple, in association with the flowers. Bulbils present.

   The capsules are truncate or obtuse at the apex, obtuse or rounded at the base, upto 4.5 cm long, wings semi-elliptic or semi-obovate with stout margins.

8. *Dioscorea oppositifolia* L. Tubers usually single, cylindrical with a few root lets. Skin reddish earth colour, flesh white, soft.

   Stems glabrous or finely pubescent, twining to the right. Leaves alternate, sometimes opposite, simple. Bulbils absent.

   Capsules with wings having evenly rounded to truncate apex, 1.6–2.4 cm 1.2–0.7 cm.

9. *Disoscorea pentaphylla* L. The tubers are almost invariably single, skin brown, yellow or purplish, flesh firm in ovoid tubers that are covered with bristly roots and soft in elongated tubers with few roots, pale cream or lemon yellow with purplish flecks.

   Prickly climber is with stems turning to the left. The leaves are alternate, 3–5 foliate, sometimes simple. Bulbils are present.

10. *Dioscorea prazeri* Prain et burkill. Rhizomes are short, rather stout, grey-brown or nearly black, branching free-

ly; branches about 10 cm long and 1.5–2.0 cm thick; flesh white cream coloured.

Stems smooth or slightly ridged, unarmed, twining to the left. Leaves alternate or rarely opposite, simple, blade glabrous, commonly 12 × 9 cm. Bulbils present though rare.

The capsules form flaucous, green turning as they ripen to a chestnut brown; wings broadly hald obcordate-obovate of subrhomboidal, 2.5–1.5 cm in size.

11. *Dioscorea alata* L.: The tubers are single or in several groups, globose to pyriform, lobed variously fingered and with fasciated skin that is brown to black, flesh white or creamy white or with magenta-coloured sap.

The stem of the vine is quadrangular, winged, turning to the right. Leaves opposite rarely alternate, simple, bulbils present abundantly or sparingly.

The capsules are 2 cm long, wings broader than semicircular, 2 cm long, 1.3 cm wide, reddish brown when dry; apex slightly reduced.

Other sources of diosgenin

Other than dioscoreas which have been examined for diosgenin are *Trillium govanianum* and *Paris polyphylla* both of them belonging to the family liliaceae. The Paris is common in Kashmir in Western Himalaya while the former (*Trillium govanoianum*) grows wild at altitudes of 2700–3700 m from Kashmir to Sikkim. The plant has been collected from other places like in Southeast-Africa (Guinnea), and other hills of Himalaya.

Agave species like promontroi Trel, A. *vilmoriniana* Weber, A. *roseana* Trel, A. soberia, *A. mirabilis* Trel, *A. mapisaga* Trel and A. *aurea, A. atovirens, A. mirabilis* yielded 2.5% of hecogenin; the highest yet

found in Agave. Among the Indian sources *Agave sisalana, A. cantala. A. vera-cruz* and *A. americana* occur in India and have been extensively used to reclaim degraded wastelands such as old riverbed, field boundaries, wasteland on hillu slopes subjected to soil degradation and slope failures and check soil erosion. Propagation on such sites is done by planting bulbils or suckers. Agave flower only once during the lifetime and flowering stem arise from the plant as a thick pole, which rapidly grows and die after fruiting. *A. cantala* has a short ascending root stock and long narrow leaves (more than 1 m and 7.5 cm at the middle), commonly met in Mumbai, Deccan, North Tamilnadu, UP, parts of Punjab and Uttaranchal where it is useful to check soil erosion. *A. sisalana* has short leaves and a short stem; the leaves are dark green fleshy, 1.3–1.9 m in length and has been grown in Assam, Bengal and Karnataka states. The oldest leaves are cut close to the trunk, each plant yield about 250–300 leaves during a period of 7–8 years. Agave have mostly been used for the extraction of fibre and the pulp which is waste product and could be used for the isolation of hecogenin. It is reported that *A. cantala* and *Agave sisalana* contain 0.15% and 0.1% hecogenin respectively in Sept, which may go up to 0.065%.

A. americana var. marginata contain hecogenin (0.15%) and chlorigenin 0.1–0.17%.

Gloriosa superba is being extensively cultivated as an alternative source of diosgenin (Colour Plate 10.8).

28. *Elettaria cardamomum* Maton.

Family: Zingiberaceae.

Local names: Ilayachi, Choti-Ilayachi, Ilachi (Hindi), Yellakai (Tam), Yelakhayalu (Telugu).

10

Description: A perennial herb with thick fleshy rhizomes, having erect, leafy stems, 120–240 cm high. Much branched inflorescence, arise from the ground and bear fruits which is the "Ilyachi" of commerce.

Distribution: Indigenous to South Indian region, growing on a rich, moist forest soil on the hilly tracts of Kanara, Mysore, (Karnataka), Coorg, Wynad, Travancore (Kerala) and Madurai (Tamil Nadu). It is fairly extensively cultivated at altitudes from 200–1700 m. The Malabar Cardamom var. *minor* is with linear-lanceolate leaves, the undersurface with a more or less complete coating of short, white, silky hairs. Inflorescence arising from the very base of stems and creeping on the surface of the ground clumps. Bracts shorter than the spikelets, acute. Fruit white, sub-globular, angled, somewhat coarsely veined. In var. *major* (Badi-Ilayachi), the leaves are oblong-lanceolate, usually quite destitute of silky hairs, Inflorescence erect; bracts longer than the spikelets, obtuse or apiculate. Fruits oblong, fusiform, minutely veined. 2–3 times the length of the minor variety.

Uses and extent of utilization in trade: The plant is extensively used in house as a spice, particularly the major variety. The minor variety is used as mouth-freshener and in medicines.

The agricultural production has very little satisfactory results since the crop is grown in a cool, very shady garden, with soil kept continuously wet, sheltered from sun and rain.

The trade in the fruits of both major and minor varieties is on a very large scale, not as a drug plant but as a condiment.

Agrotechniques for the Cultivation

It requires a fairly deep, rich soil, resting on rock, seems indispensable, and if possible this should be within undulating land, not too remote from the running water, shade and a humid atmosphere, which are essential. It luxuriates in mist and fogs and cooling sea-breezes. Northern and Western exposures, in south peninsular. India, are considered to be the best places. Since all these conditions are not often conveniently combined, the locations for successful production are somewhat restricted.

Small plots of land, within the forest, where wild or acclimatized plant is known to exist are cleared during the months of February and March. The brushwood is cut down and burnt. The roots of the powerful weeds are torn up so as to free the soil. The shade, if too dense, is regulated by a ceratin percentage of trees being felled, thus admitting diffuse day light, and curiously enough, the disturbance or shaking of soil, by the fall of heavy timber on it, is in itself deemed beneficial. At all events, soon after clearing, cardamom plantations sprung up all over the fields, but if any reason exists for doubting the abundance of future crop, young plants are deposited in the soil at the required distances apart. The plots are then left alone, for a couple of years, and by that time cardamom plants will have 8–10 leaves and be 30 cm in height. In the third year, they may be 120 cm high. In the following May-June the ground is weeded and by September-November a light crop may be obtained. In the fourth year a weeding is again given and if the cardamoms are found growing near than 180 cm apart, a few are transplanted to new positions. A full crop is now obtained and the plants may bear for 3–4 years, i.e. the life of the plant is 8–9 years.

Plants may be raised by seeds or rhizomes. In about one year, the seedlings are about 30 cm in length and ready for transplanting.

When raised by division of rhizomes, it should be seen that each cutting contains at least three perfect shoots. Holes, 30 cm deep and 180–200 cm apart each, may be dug all over the specially cleared plot or garden and the seedlings or seeds are deposited within these, but these must not be buried too deeply, as they are liable to rot off.

The flowering season is April-May. It is good to bankup all leaves and rubbish available around the clumps, since this helps to support and encourage the growth of creeping racemes. If at this stage the flowering branches get submerged, the soil is washed away, the roots exposed and the flowers and fruits get ruined with the water and the mud. The fruit should not be allowed to open fully as the capsules will then burst. They should be collected, just as they begin to turn yellow from green. During the months of August to September they swell, and by first half of October have usually attained the desired degree of ripening. The crop is gathered during the months of October and November and in extremely moist weather in December. A dry day is, however, the best for harvest. Scapes bearing clusters of fruits are broken off close to the stems and placed in the baskets lined with fresh leaves. The fruits are spread out on carefully prepared floors and exposed to sun. 4–5 days careful drying is enough; artificial heat may be necessary in rainy season. Capsules are then rubbed by hand or shaken within mats in order to brush off the pedicels, calyces, particles of dust, then are winnowed, hand-picked and assorted to size and degree of ripeness.

Crop may be grown in a cool, very shady garden with the soil kept continuously moist. In Kanara, the crop may be raised from seed and sheltered from sun and rain. If seed germinates too thickly, they should be thinned out and the seedling transplanted into seed buds, shaded by a temporary protection of palm leaves. When 120 cm high and 15–18 months old, they may be carried to their permanent positions and finally transplanted from March to June or again September to October. Pits, 45 cm deep, are dug in the same lines and intermediate between two trees. Into these pits cardamom are deposited and supplied in early March and April with leaf manures. They come into bearing but do not yield much during the first year after transplanting. Flowers appear early in April and May and fruit from June to July. Each should be severed from the scape and plucked. If plucked, the pressure of finger may burst the capsule. After being dried in sun for 2–3 days, the fruits are hand-rubbed to remove the attached stalk and calyx. A too hot sun or too long exposure to sun may dry the fruits quicker. In a fairly stocked betel nut garden, there can be 300–400 cardamom plants to an acre. A well-grown plant may yield upto 250 g of capsules. Light showers in April and May are favourable. The fruits are dried and bleached in a particular way.

29. *Emblica officinalis* Gaertn.

Family: Euphorbiaceae

Local names: Áonla, Amlá, Aolá (Hindi), Avul, Anvalá (Kum.).

Description: A moderate-sized tree, deciduous, bark grey, exfoliating in irregular patches, red inside, branchlets finely pubescent. Leaves subsessile, linear oblong, acute or mucronate, dichotomously close set on deciduous branchlets, together having appearance of pinnate leaves; stipules minute. Flowers apetalous, monoecious, greenish-yellow, in axillary clusters. Male flowers

10

numerous and shortly pedicelled. Stamens 3, joined in a short column. Disc of distinct glands alternating with calyx segments, rarely zero. Female flowers with sepals as in the male flowers. Ovary 3-celled, with 2 ovules in each cell; styles connate at the base, twice bifid. Fruit of three, 2-valved cocci, fleshy, dehiscing when dry (Fig. 10.6).

Fruits on trees grow naturally in the forest and are smaller than the cultivated varieties, however, plants growing in the Panna Divison forest of MP are reported bigger. Important varieties are NA-7, NA-10, Chakia, Francis, Krishna, Kanchan and Pratapgarh. District Pratapgarh in UP, has excelled in the cultivation of this species and has vast area under its cultivation, where it has been promoted on a large scale commercially.

Distribution: Throughout India, upto 1500 m, in mixed deciduous forests of tropical regions. Not common west of the river Ravi and not found in the arid regions. It is found in dry deciduous forests of the Deccan, in westernghats upto 1300 m. Suitable for stony, low fertility lands where it germinates but plants are weak with slow rate of growth (6 m–10 m) with DBH 35–40 cm as compared to 10–15 m height and 60 cm DBH.

Uses and extent of utilization and trade: The fruits are widely used in the Indian medicine and are a source of vitamin C. The famous Ayurvedic medicine 'Chayavan Prash' is made from the juice of the fruits. Dried fruits are sold in the market as *amaika* powder and are one of the constituent of *triphala* (comprising, *Harad, Bahéra* and *Anwla*. The fruits are also canned and sold as *Murabba Anwla*. Now the juice is being sold in the market and the fruit is put to many uses, like sweats (*Laddus*) are made from the fruits in Pratapgarh.

Fig. 10.6: *Emblica officinalis*

Amla is also used in hair oil industry and also in shampoo (Dabur Amla shampoo, Vatika Trade name), also used in chutneys, pickles, etc. The wood is also used.

The rootbark of this plant and tender shoots used as a remedy to cure indigestion and diarrhoea. It is also effective in aphthous stomatitis and a fermentation preparation from the roots is used to cure jaundice, dyspepsia and cough, etc.

The leaves as infusion are used as bitter tonic and a remedy for chronic dysentery. Leaf decoction is used in mouthwash, in aphthae and also as eyewash for curing eyesores.

Juice of the fresh bark with honey and turmeric is useful in curing gonorrhoea and the stem bark is used in diarrhoea.

Fruit is the most useful part and is the richest source of vitamin C. Fresh fruit is refrigerant, tonic, antiscorbutic, diuretic and laxative. A *Sherbat* from fruit is antibilious, diuretic and cooling, used in fevers, hiccup, vomiting, indigestion and habitual constipation, including other complaints connected with the digestive system. The fruits are purgative and used in summers to quench the thirst. Fresh fruit is also a vermifuge. The dried fruit is an excellent intestinal astringent, cooling, stomachic, antiscorbutic and blood purifier, given in diarrhoea, dysentery and haemorrhage. In Ayurvedic system, it is prescribed with iron in anaemia, jaundice, dyspepsia. The liquid that exudes from the cut on the fresh fruit, while in the tree, is used as eyewash and as an external applicant on inflamed eyes. A paste of dried powdered fruit is applied over pubes in case of irritability of bladder and retention of urine. Decoction of dried fruit is injected with benefit in gonorrhoea. The infusion prepared by steeping dried fruit overnight in water, in a new earthen vessel, is effective as an eyewash in ophthalmia and in hair diseases.

The seeds in decoction are used as gargle for loss of taste after fever and an ointment of burnt seeds is useful for itch and fevers.

Agrotechniques for the Cultivation

Anwla seeds can be directly seeded, 2–3 seeds per pit of the size 45 cm × 45 cm × 45 cm. Nursery raised seedlings can also be planted and raising the nursery in a 10 m × 1 m plot 2–3 kg seeds are sufficient. The seeds, after sowing in lines, are covered with 1.5–2 cm soil and much placed on the top. Weeding and sowing is done as a normal practice. 10–12 tons of FYM and NPK is applied. The seedlings are transferred in the polypots from the nursery.

Planting of seedlings is done in pits of size 45 cm³. FYM plus Ammonium sulphate @ 10 g, plus BHC/aldrin is added as insecticide. Pits are placed at 4 m × 4 m distance accommodating 625 plants/ha; at 6 m × 6 m distance only 277 plants can be accommodated in good soil. However, in garden plantation, the distance is kept at 10 m × 10 m.

Weeding and hoeing operations are undertaken as usual farming practice and if needed irrigation is also done. Under irrigation the plant starts fruiting after 7 years and in the garden after 4–5 years. For the first five years the production is estimated at 15 qtls/ha and for the rest 15 years at 60–100 qtls/ha (Table 10.8).

Table 10.8: Production of anwla varieties (in kg/ha)

Variety	4th	5th	6th	7th year after plantation
Banarsi	10.2	13.8	17.5	19.0	
Chakia	20.5	24.2	42.3	55.8	
Francis	10.0	25.2	35.7	43.2	data as per
Kanchan	31.8	48.5	60.3	73.8	Sharma DG
Krishna	14.0	24.2	32.0	41.0	1997. cultiva-
NA-6	20.8	30.3	48.1	56.7	tion of Amla
NA-7	29.2	42.8	38.5	62.7	and its use.
NA-8	6.2	8.3	14.8	16.8	*Udhamita.* Sept
NA-9	11.0	19.5	28.2	37.2	35–37.

An excellent collection of amla plants is available at the Kédia nursery 83, Chilbila, Pratapgarh. Phone (05347) 21361, 321211 and 20377.

The cost of cultivation is estimated at Rs 5000–6000/ha and for the first three years cost of irrigation, hoeing and watch and ward is Rs 3000/ha. For first five years income estimates are @ Rs 500/quintal (15 × 500) Rs 7500/- for a year, while with a production of 60–100 qtls/year the revenue earned varies from Rs 15000–36,000/ha.

Reference

Sharma DG. 1997. Cultivation of *Amla* and advantage. *Udhamita* Sept. issue. pp. 35–37.

30. *Ephedra gerardiana* Wall. ex Staff.

Family: Gnetaceae.

Local names: Tse, Raché (Lahul), Somlata, Jorgantha (T. Garhwal), Tutġantha (Jaunsar).

Description: A small, perennial herb/ shrub, with green branches, striate, opposite or whorled, bark on old stems grey. Leaves scale-like, scale connate into a 2-lobed sheath; yellow or brown. Spikes often in whorled clusters. Fruiting spikes red, often succulent. Fruit ovoid, red and sweet when ripe. Seeds 2 (Colour Plate 10.9).

Distribution: It is common on the dry southern exposures in the Himalayan region, 2500–3000 m, scattered, occurring in Kashmir; Kishtwar, plains of Ladakh; Himachal: Spiti. Lahul Kutti valley in Byans; frequent in Pangi, Lahul, Spiti and in Bashahar region Uttaranchal; Chakrata. It also occurs in China and in Tibet on Kalba Range.

Uses and extent of utilization: The plant flowers in April-June and fruits in August-October and September-October is the best time for its collection in the field. The plant is sold in the market as *Somkalpalata* and collected locally in small quantities. The dried stem is powdered locally and employed as snuff for headache.

E. gerardiana and *E. major* Host. resemble each other and occur side by side and the two can be distinguished by subscandent growth of the former and upright, rarely ascending *E. major*. The young stem, collected from Lahul and Jespa is recorded (Sarin, 1967) to yield 1.7% alkaloid, calculated as ephedrine, however, alkaloid content upto 1.93 and has been reported (Wealth of India, 1948–60), in similar samples previously. The plant has not commercially been exploited during the past; however, a beginning was made where an annual exploitation of more than 375 qtls per annum was allowed with the permission of Himachal Forest Department only from the Lahul valley on royalty basis. The major availability of the drug is confined mostly on the north-eastern and central mass mountains. It appears that at least double the quantity can be obtained, if all the areas are properly explored.

Agrotechniques for the Cultivation

Ephedra can be propagated by seeds or from the root stock cuttings. Seeds are sown 2–3 cm deep into the soil during early spring at about 5 cm spacing in rows 75 cm apart. Percentage of ephedrine in plants grown at Chakrata (Uttaranchal) was about 1.5 which is satisfactory with BP standard. The alkaloid content increases with the age of the plant and the best time to collect the twigs is when the plant blossoms. Stems should be collected from four years old plants. Since the female plants contain little ephedrine male plant need to be collected. The material should not be dried at high temperatures and should be stored at dry place, otherwise alkaloid percentage is seriously reduced.

The total alkaloid content in Indian Ephedrás varies from 0.28–2.79% and the Chakrata collection showed 0.28 and 0.4% of Alkaloid and Ephedrine contents respectively.

Ephedrine is used as useful remedy for asthma. Tincture of ephedra is an excellent cardiac stimulant in toxic conditions. Ephedrine in pharmacological action is similar to adrenaline. The liquid extract of ephedra is useful for controlling asthmatic paraxysms. Tincture of ephedra is cardiac and circulatory stimulant. Decoction of the stems and roots is considered in Russia as a remedy for syphilis. The juice of the berries is useful in the affections of respiratory passages.

Ephedra foliata (Vern. Jangli, Phog), a plant of arid region is locally used on a limited scale. The roots when boiled with milk are considered tonic in seminal debility.

31. *Eugenia caryophyllus* (Spreng.) Bullock and Harrison (–E. caryophyllata Thunb. Syzygium aromaticum (L.) Merr. et Perry Jambosa caryophyllata (Spreng.) Niedenz.

Family: Myrtaceae.

Local names: Clove tree (English), Lavangá (Bengal); Laung (Hindi).

Description: A small bushy or cone-shaped evergreen tree, 4.5–6.0 m high. Leaves lanceolate, dotted with oil glands. Flowers crimson, borne at the end of branches. Flower buds nail-like in shape, consisting of a cylindrical ovary, surmounted by the unopened, ball-like corolla, surrounded by 4-toothed calyx, greenish pink when fresh, becoming brown and brittle on drying.

Distribution: Native of Mollucan islands, but cultivated in many tropical countries like Zanzibar and Pemba. To a limited extent the tree is cultivated in the gardens in Kerala and the Nilgiris.

Uses and extent of utilization: Cloves are widely used as spices and are aromatic, having a fine flavour. In China, it had been used with betel leaves. In Indonesia, shreded cloves used as early as in 250 BC. One of the important uses of cloves along with tobacco in the manufacture of a special type of cigarette.

Clove oil is used for flavouring food products and also oral preparation and chewing gums. Medically, it is antispasmodic and carminative, externally used as rubefacient, having antiseptic and bactericidal properties. It is particularly used in toothache. Oil is also used in perfumes and toilet waters and in scenting soaps. In the laboratory, it is used for microscopy.

Inferior oil from leaf and stem and the budoil are used for the extraction of 'eugenol' for manufacturing Vanilin, and also in toothpaste.

Domestic production of the cloves is negligible, considering its requirement, cultivation of this plant needs to be extended.

Agrotechniques for the Cultivation

Clove requires a warm, moist and equitable climate with well-distributed rainfall. It flourishes best on coastal regions. Propagation of this plant is by seedlings raised in nursery beds, pots or baskets. Young transplants require care and attention. The tree comes into bearing at 4–8 years of planting, the average span of productivity being about 60 years.

Clove buds are ready for harvesting when the calyx assumes a delicate pink colour. They are picked by hand, dried in the sun or kiln after removing flowering stalks.

Cloves yield 15–18% of oil, which darkens with aging, having a strong aroma, typical of the spice. The clove "steams" (flower stalks), which is the by-product, and the leaves also yield essential oil but of inferior quality. The oil is obtained by water and steam or direct steam distillation. The chief content of the oil is "eugenol".

32. *Foeniculum vulgare* Gaertn.

Family: Apiaceae (Umbelliferae)

Local names: Saunf (Hindi); Bádian (Urdu); Fennel (English).

Description: Herbaceous perennial plant with yellow flowers and finely divided leaves.

Distribution: The plant is cultivated for culinary and medicinal properties. There are mainly three varieties. *Saxon*: the most

esteemed of cultivated varieties; *Romanian*: variety smaller in size than the Saxon. It is not camphoraceous in taste. *Indian fennel*: the fruits are darker than of the Saxon variety and the taste is not camphoraceous. Russian variety is almost similar to that of Romania.

It is mostly cultivated as a garden crop throughout India upto 2000 m. Ssp *capillaceum* have var. *vulgare* and var. *dulce,* yielding oils. It requires a mild-climate and is cultivated as a winter weather crop in parts of North India. It does not succeed in South India so well except at high elevations. Fruits are more elongated in Romanian variety than in the variety Saxon.

Uses and extent of utilization: All parts of the plant are utilized. The seeds are used in medicine, perfumery, soap manufacture; confectionery and in liquors. Thickened leaf stalks are used as vegetable.

The seeds of the plant are traded in the market and used extensively in Indian medicine, particularly for distillation and sold as *ark sonf*.

Agrotechniques for the Cultivation

It thrives best on rich, well-drained loam or black and sandy soils, containing sufficient lime. The cultural operations include occasional weeding and irrigation once a week. The soil is ploughed twice or as required to make it loose. FYM at the rate of 23 tonnes per acre is applied at the time of ploughing. The seeds are sown in rows, 45 cm apart, at a depth of 10 cm. Watering is done according to the requirement. About 90% of the seeds germinate successfully after 15 days of sowing. Since the plant is perennial, it spreads to 30 cm or more; hence a distance of 25–30 cm is kept between the plants. The flowers appear during May and the fruit in June. Pruning is done in August after the plant has dried up. Sprouting starts in

October and flowers till December, and ripe in January forming seeds.

An approximate yield of seeds of about 106 kg (212 lbs) per acre has been recorded.

The fruits contain 1.75% volatile oil; normally volatile oil should not be less than 1.4% V/W, however, recently 3% V/W was obtained from plantations grown in Madhya Pradesh.

Reference

Shimada, T. 1959. Induction of flowering in *Foeniculum vulgare* under short days. *Agric. Bull*. Sagar Univ.

33. *Fagopyrum esculentum* Moench.

Family: Polygonaceae.

Local names: Kottu, Ogal, Phápdá (Hindi); Kalátrumba, Kaspat Kathu, Darau (Lahul); Buckwheat (English).

Description: A small herb, scandent sometimes but erect in late stage. Flowers small, white. Seeds trigonous, brown to brown-black in colour. The leaves are cut for use as vegetable in the hilly areas and sold in local market.

Distribution: Extensively cultivated in temperate Himalayan regions at lower hills of India in Nigiris, Himachal Pradesh, Garhwal, Kashmir, Sikkim, Khasi hills and Manipur in the northwest southern and eastern regions of India (700–3000 m). In Doon valley this crop was grown successfully (SewaRam, Gupta RK and Khybri ML, 1979), during the *rabi* season as a short duration quick growth crop (3 months) and valued for its spreading habit and vigorous growth. It forms a good ground cover, thereby checking weed growth. It also prevents erosion in the field during heavy rainfall. Normally, it is sown in July, between 1300–3300 m in western regions and reaped in October. However, in higher

regions of Garhwal it is sown in August as a relay crop in potato fields, after the leaf-fall, and harvested in November. In some areas, it is usually raised more as a vegetable than as a grain crop. In mountains of Assam and Manipur, it is grown between 600–1000 m, below its normal height. It is also utilised, as a first crop, on new clearances and in most alpine tracts in sheltered portions of grassy slopes in giving a crop for one year and left fallow, for succeeding years. It may also be grown on soils too poor for wheat and also succeeds well on rocky soils containing fairly high percentage of granite detritus but not clay. In lower valley irrigation is sometimes necessary and yields a return after only 10 weeks of its sowing.

Agrotechniques for the Cultivation

The crop is sensitive to high temperatures and dry climatic conditions, especially at the blooming stage. It is also succeptible to frost. A further hazard to this crop is the tendency of the matured plants to lodge, as a result of wind or heavy rainfall. Buckwheat does not possess the power to recover from lodging as do most other crops. It is grown under a wide range of soil conditions than any other grain crop. However, it is best suited to slightly acidic, well-drained sandy loams and does not do well on heavy, poorly drained land or sandy soil subject to drought. Buckwheat is often grown on poorly prepared land but best results are obtained on well-prepared soils. At Dehradun one deep ploughing followed by 2–3 harrowings and planking is sufficient for sowing of this crop. In Doon valley seeds were sown during first fortnight of October. The time of seeding for any locality is determined fairly accurately by allowing a period of 10–12 weeks before the first killing frost is expected. Traditionally, the seeds are broadcast but better crop stand is achieved, if seeding is done in rows across the slope by means of bullock-drawn implements or seed drills. It is necessary to sow the seed into moist zone but shallow seeding aids in obtaining rapid and uniform germination; The depth of sowing should be kept about 5 cm, with 20–25 cm spacing between rows and 10 cm between plants. 40–60 kg/ha seed rate is used for grain and 10 kg/ha when grown as cover crop.

Numerous varieties of buckwheat have been distinguished based on the size, shape and colour of the seeds. There are many improved varieties in vogue; common gray, is the most widely known in Canada, Maccan, Tokyo and Tempest are commecially grown. In India no improved varieties are available, but two types, 'Darjeeling' and 'Chinese type' have been tried; Chinese type matures earlier. 30 kg N and 30 kg P_2O_5 is placed below the seed in furrows before sowing. It suffices the need of this short duration crop. Ultimately, the fertilization depends on the soil fertility and for specific recommendations, soils have to be analysed.

One weeding is normally done at 20–25 days after sowing. In case of broadcast crop, one thinning is also necessary to obtain proper spacing between plant to plant.

Irrigation

Normally, it does not require any irrigation. If there are no rains, during the growth period, one irrigation at the time of flowering is done to obtain good yield, but water should not be allowed to stand in the crop.

The crop is generally free from insets, pests and diseases at least in Doon valley.

Harvesting

When the seeds attain brown colour, the crop is ready for harvest. Plants are cut close to the ground and kept for sundrying for 3–4 days. Threshing is done by beating. The produce is thrashed and winnowed like any other cereal. Alternately, buckwheat is swathed when the maximum number of seeds are ripe. Since ripening is rarely uniform, some green seeds are usually present at harvest. Occasionally, blooms are evident when the crop is swathed. Sometimes, buckwheat is harvested premature to avoid frost damage and ripening continues in swath. If the crop matures to a point where swathing begins, the swathing should be done on damp weather or in early morning, when dew is present. The production recorded at Dehradun is about 700–800 kg/ha while in the inner Himalaya of Garhwal production has been recorded from 40–100 kg/nali (20 nalis make an ha).

Uses and extent of utilization: It is used both as vegetable and grain for making *atta* which is extensively used during fasts. The estimated output of the seeds, only from Lahul, has been put to 190–200 m tonnes per annum (Sarin, 1967). However, the plant is a commercial source of rutin (yield 3–5%), a flavinol glycoside, having the properties of reducing increased capillary fragility (Wealth of India). The leaves and the blossoms contain most of the rutin. It reaches at its maximum in early blossoming stage, when the raw material should be harvested. However, In *F. esculentum* the rutin content is low and a techno-economic survey of the rutin resources to find out the feasibility of raising exclusive buckwheat crops for rutin in the Himalaya is very much desired.

F.cymosum Meissn. is an erect, pubescent herb with long-stalked, broadly triangular, acute and pointed leaves, uppermost narrower. Flowers are white, in racemes, forming stalked panicles, flower jointed near the middle. Nut ovate, 3-cornered acutely, more than twice as long as the perianth, enclosing its base.

A native of Eurasia and also found in temperate regions of Kashmir to Sikkim. The grains are recommended as diet in colic, cholera, diarrhoea, flexus of all kinds and abdominal obstructions. *F. tataricum* Gaertn is another species which is found in Himalaya, from Kashmir to Tehri Garhwal, in Bhagirathi and Yamuna valley.

References

Cough JF, Naghksi J and Krewson CF. 1946. Buckwheat as a source of Rutin. *Science*, **103**(2668): 197–198.

Sewa Ram, Gupta RK and Khybri ML. 1979. Buckwheat for Doon valley *Indian Farming*. February Issue. pp. 1–4.

Naghski J, Krewson CF, Porter WL and Cough JF. 1950. Factors affecting the Rutin content of dried Buckwheat meal. *J. American Pharm. Association.* **39(12)**: 696–698.

Singh Ajit and Atal CK. 1969. Cultivation of Buckwheat as a raw material for the manufacture of Rutin. *CSIR Symp.*

Singh HB 1961. Grain Amaranths, Buckwheat and Chenopods. *ICAR Cereal crop Series*. No. 1 pp. 46.

34. *Glycyrrhiza glabra* L.

Family: Papilionaceae.

Local names: Muléthi, Muláthi, Vástu Madhu, Jashtinadh (Hindi); Madhuyashi (Sans.), JasthiMandhu (Bengal); Liqyuorice (Eng.), Atim (Tamil), Atimadhiram (Tel.). Liquorice root of commerce consists of rhizome and root. The name Liquorice is also used for *Abrus precatorius*.

Description: A tall, perennial creeper herb or undershrub upto 1.5 m high reaching upto 2 m. Leaves compound, leaflets 4–7 pairs, oval with pointed tip. Flowers liliac, of light violet to purple, small, in slender axillary spikes which are equal to or slightly longer than the leaves. Fruits pod, 1–3 cm long, flat, densely covered all over with small spinous outgrowth, 2–2.5 cm long. The rootstock throws numerous additional roots or in some other forms, there are several branched stems near the rootstock remaining underground. Seeds are 2–5 cm long. A self-pollinated crop (2 n – 16).

Distribution: Wild in Pakistan and other equatorial countries in northern parts of Kashmir. Cultivated to a limited extent in South Europe, Syria, Iraq, Turkey, Afghanistan, Spain, China, Russia. Recently introduced and cultivated in parts of UP, Haryana, Uttaranchal, Jammu region and MP (Indore). It can be cultivated in northern and central regions of India. It thrives best in warm regions with long growing season.

Uses and extent of utilization and trade: In Europe, the root and underground stem are used for alcohol production, insecticide and for yeast; also used as sweetening agent, such as in chocolate, chewing gum, bear, candy industries, etc. It is adulterated with Manchuria liquorice (*Abrus precatorius*). The root is yellow from inside, fibrous, sweet with some aroma. As a drug, it is used in cold, cough medicines as root powder; peptic ulcer, gastric and duodenal ulcers. Preparations of liquorice are used as laxative, in treatment of cough, such as lozenges for throat and for concealing taste in many nauseating medicines and in tobacco projects. Dried extract of root called 'Rub-us-soos' is commercially available. Finally ground and sifted liquorice is applied to the eyes to remove yellowness of the eyes.

For sore throat and cough, a small piece of raw liquorice is just chewed or sucked. It is reported to promote urination and is diuretic, antibacterial and antibiotic. It is highly used in making syrups. Powder of the drug mixed with butter and honey is applied to cuts and wounds. The leaves when applied as poultice are believed to be useful in scalds of the head. Root is locally used to expel phlegm, relieve constipation, asthma, hoarseness, rattling in throat. It relieves scalding of urine, is diuretic and relieves thirst. In case of cough 40 g of this plant root, 10 g of long pepper finely grounded and mixed with honey, 6 g of it is taken.

Industrial use of liquorice is in the production of it *firefoam* used in fire extinguisher. Spent root is being used in the production of insulated boards of great structural strength. Also used as compost in mushroom culture. Its antioxidant properties help to keep chewing gum in a fresh and pliable shape.

It is also being used as a mild tonic and as regulator in animal feed. It has also been used as a dispersing agent for sulphur as well as for improving wetting and adhesive properties of sprays.

The annual requirement of this drug was estimated to be more than 4,000 qtls in India. Cultivation of this plant has since been encouraged. The active principle *Glycirhizie* is 50 times as sweet as canesugar.

Root powder is also used with betel leaves. It is a common weed in Spain and fresh pieces of roots are sucked by young boys.

Fresh liquorice contains 50% moisture which on drying remains to 10%. *Glycirhiza*, the main constituent varies from 5–20% and 10–14% glycosides in various varieties. It is not present in above ground part. Other constituents of liquorice are cumarin, flovinon, liquiritin, isoliquiritin; livritinizin, apart from glycosides useful for healing wounds.

Agrotechniques for the Cultivation
Climate and soil

Deep fertile, well-drained, sandy loam soils are ideal; heavy soils are unfit for the

rhizomes to penetrate easily. Earlier, it was believed that it can grow under cold climate. It can be grown in the plains of North and Central India having cold winter.

Cultivation

Tractor-drawn harrow cultivator is used for fine tilth and before final land preparation 20–30 t/ha of FYM is used along with 100 kg Urea, 250 kg Superphosphate and 100 kg of muriate of potash and uniformly mixed with the top soil. Since the crops do not tolerate any standing water, ridges are made at 90 cm interval and root cuttings (15–25 cm long) are planted at 60 cm interval on ridges. When grown from seeds, it does not give satisfactory results and the usual practice is to propagate the plants from cuttings of younger parts of rhizomes, suckers or by crown divisions. These are planted in spring season in rows. The plants do not require much attention, except for weeding. However, the growth is slow and marketable rhizomes take 3–4 years to form. Light irrigation is required, immediately after planting. February/March is the ideal time for planting in irrigated areas; in dry areas it can be planted in June and July. 100 kg urea is applied in two equal split doses in the first year. During the second year another 100 kg of urea is applied in two equal splits before the onset of rains and 45 days thereafter. Good soil moisture is maintained; till the establishment of the cuttings and thereafter the crop is irrigated at interval of 20–25 days (excluding rainy season and dormant period). Two weedings are done manually during the establishment phase, in the second year, two hoeings-cum-weedings are done during February and March and June-July. The crop undergoes in dormancy during the winter season and new growth initiation occurs during the months of February and March. There is no severe incidence of pests and disease. However, leaf spot is the common disease; for brown leaf spot 0.1% Malathion or Thiodon of 0.2% solution is applied as spray. The roots of 3–4 years old plant are taken out from the ground using tractor-drawn cultivator, after attaining maturity (two and a half or three years). A light irrigation is given before harvesting, as it helps in loosening the soil. Soil is turned over to considerable depth and the rhizomes are pulled out by hand also. The crowns are carefully separated from the rhizomes and divided for further planting. The rhizomes are dried in shade for about for 30 days, till the moisture content comes to 10% from 50%. The rhizomes are likely to mould or decay under storage. Dried roots contain 5–20% Glycyrrhizin, and 10–14% glycosides. It is best for use upto two years after extracting from the soil.

Leafspot is the common disease and the leaf turns yellow. Diathene M45 (0.2%) is sprayed on the leaves. Aphid sometimes attack and take much leaf sap. Malathion or Thiodon (0.2%) solution is sprayed.

Gross return from the crop is about Rs 3.75–4 lakhs/ha, while the cost of cultivation is estimated at Rs 100,000/ha with a net return of 2.75 to 3.0 hundred thousand or Rs 1,00 thousand per ha/year.

Var. *glandulifera* and var. *typica* have been tried long back at the IARI. The former variety forms long rhizomes while var. *typica* develops strong tap root system, deep into the soil and bears comparatively longer rhizomes. Aerial portion in both the varieties is glandular and hairy. *Mulhati* N-1 or HN-1 is the improved variety recommended. Another var. EC-2195 is suitable for glycyrrhizin.

Young roots that remain in the soil also give rise to new crop, hence the land selection for this crop should be done where it can be permanently established and left for a regular crop. Kapoor et al (1956) analysed results of the produce from J and K state-grown plants and recorded that the drug can be obtained of good quality from lower regions of J and K state.

References

Kothari SK and Singh Kailash. 2001. Cultivation of Liquorice pp. 180–181. (Mimeographed)

Gupta SN. 1997. Cultivation of Liquorice, *Udhamita*. Sept. issue pp. 39–41.

Anonymous. *The storey of Liquorice*. Mac Andrews Forbs Co. USA N. York.

Walker WW. 1952. *Liquorica. Dark Mystry of Industry*. Atlantic Monthly. (Nov).

35. *Humulus lupulus* L.

Family: Canabinaceae.

Local names: Hops (English).

Description: A perennial, herbaceous vine native to Europe and Asia. Pistillate inflorescence are spike-like in appearance and borne in leaf axils of either main stem or lateral branches called strobiles or burrs or comes constituting the commercial Hop.

Distribution: The plant was introduced in India by Peruvian Missionaries in 1862, which has, however, now been exterminated but has been reintroduced in Kashmir after about a century.

Uses and extent of utilization: The female inflorescence, called as Hope, is used in beer manufacture, for giving the beverage its characteristic aroma, taste and sparkle. The flowers are also medicinal. Oil of hops is used for certain perfumes of the Chypre or Fugrere types. Also used as sedative.

There is a good market for this plant and, therefore, efforts for its cultivation were made in a big way so as to be self-sufficient regarding the supply of this drug. India imported all the requirements to the tune of about one million rupees annually. The results of experimental cultivation and exploitation at Ranikhet (in Uttaranchal) were encouraging and since then, arid tracts of Kashmir, Himachal Pradesh and certain districts of Uttaranchal, which escape the violence of summer monsoon, present ideal condition for its cultivation. The quality and yield could further be greatly improved by planting better cultivars from Europe, particularly Germany.

Agrotechniques for the Cultivation

Earlier trials for Hops, appear to have been made in 1840 and met with failure due to the climate being inimical at the flowering time. Later on efforts made in Kashmir and Chamba (HP) in 1880–85 met with some success, however, later experiments at Ranikhet (in Kumaon) at 1500 m with average rainfall of 1270 mm per annum, on clayey loam soil, proved successful.

The cultivation techniques resemble that of polebeans and are expensive, demanding constant care and a great deal of hard labour. Though, the plant can be cultivated on a wide range of soils yet deep, well drained soils or gravelly loams are most successful. It is propagated by rhizome cuttings and not from the seeds as the seedlings vary considerably in the type of Hops produced and their time of ripening. Seedlings also take more than one season to give satisfactory results.

Rhizome cuttings of the plant bearing 2–4 eyebuds, 15–18 cm long and about 1–2 cm thick are planted in rows at a spacing 1.25 × 1.25 m on well-prepared soil mixed with FYM. These are watered. Sprouting is recorded after about 20 days and within a month all the plants sprout. Watering is

done once a fortnight during summer months, with alternate hoeings, to promote growth of the plants. As the plant is a climber and climbs clockwise around the support, long poles, 4–5 m tall and 10–15 cm thick, are placed upright in rows and in between the plants for stretching wire to help trailing of the plant. On set of monsoon, towards the end of June, favoured the growth of plant shoots which are 5–10 per plant (those coming later at the base of the plant were cut and removed). The shoots continued to grow well during monsoon (July-August) and are 5–8 m high, climbing/trailing on the support provided.

The plants flower towards the end of August which continue for about a fortnight. Since, only the cones are desired in the brewing industry, these develop without fertilization and a harvest of fully ripened cones can be handpicked in October.

The yield for mature, sun-dried cone of 3–4 years plant, has been recorded at about 90–100 g plant which works out approximately to 650 kg per ha. This yield is, however, low as compared to the yield reported from the European countries. The yield can be increased by introducing high yielding strains.

The Hop plantations, if cared well, should last for many years but is said to be profitable to renew after 10 years.

The Hops may be collected on clear and dry days when these are fully mature, bright yellowing-green in colour, sticky, crisp or papery to tough and noticeably resilient. The drying temperature should not be high and the hops should be dried in shade, if temperatures are nearly above 10 °C for 18–24 hrs. These may be piled and allowed to stand for about a fortnight to undergo curing or sweating process, before being baled to the breweries.

The quality of the hop should contain sufficient quantity of *lupulin*, though the size might be smaller, the cones should confirm to the local industry. When the cones are subjected to steam distillation they give 0.88% of a thin, light-coloured essential oil, having a pleasant aroma. The yield reported is though encouraging as compared to the highest obtained from various hop-growing European countries (0.86%).

Reference

Sharma BK. 1969. Cultivation and exploitation of *Humulus patens* L. for its utilization in medicine and Brewery in India. *CSIR* symp. *Raising of medicinal Herbs*, 12–15 March.

36. *Hyoscyamus niger* L.

Family: Solanaceae.

Local names: Khursáni ajwain (Hindi); Khurasani-ajwain (Maharashtra); Parasikaya (San); Damtura, Dandura, Bazarban; Indian Henbane (English).

The trade name *Hyoscyamus* is based on the scientific name and is a well-known sedative and having antispasmodic properties.

Description: An annual or biennial herb, bad smelling and densely covered with glandular hairs. The plant has 2 forms, annual and biennial, the latter is usually cultivated. Stem upto 1 m high. Lower basal leaves 15–20 cm, margins toothed; upper leaves smaller and divided into many segments. Flowers 2–3 cm in diameter, pale green, streaked with purple, some borne solitary at the place of branching of stems, others in long terminal spikes. Fruit 1.3 cm in diam, globose (Colour Plate 10.10).

Distribution: It grows wild in the forests of Kashmir valley extending upto Garhwal at an latitude of 1500–3000 m. Common in wasteplaces, forest blanks, on rubbish heaps, dry drains and outer part of habitations. Dry inner valleys of the Northwest Himalayan region are the main places for its collection, and has been collected from Pahalgam, Aru, Gulmerg and Baramula in the Kashmir region, where it is much more common. In several regions of the country, this plant has been cultivated. The biennial form is more common. During the first year of its growth it develops large fleshy branched root, crowned by rosette of large radical leaves, 30 cm long, with long-petiole and coarsely toothed, lobed or cut blades. No aerial stem is produced. In the second year the plant grows erect; thick, coarsely or widely branched and aerial stem is produced bearing flowers in long leafy spikes. The flowers have prominent vein marking of the corolla. Fruit is found within calyx and the plant dies after maturity. Roots become spongy and then hollow.

Uses and extent of utilisation: The dried leaves and the flowering tops are collected, soon after flowering, which constitute the drug. It is useful in relieving certain painful spasmodic conditions of muscles and in nervous irritations, like in hysteria, cough, etc. The properties are similar to that of *Atropa belladona*. Seeds are also used and are locally applied as a paste in pains. The alkaloid content of leaves at low altitudes or in plains is much below the official standard (Handa et al 1947). The leaves are collected from both the annual and biennial forms and have no marked difference in their alkaloidal contents which is much below the official standard. The roots, leaves and the flowering tops contain alkaloid, a bitter glucoside Hyoscpyierin, choline muscilage and albumin. On destructive distillation, leaves yield a poisonous oil. Total alkaloid varies in different parts, viz. root 0.16%, leaves 0.045–0.08%, flowering tops 0.07–0.10% and seeds 0.06–0.10%.

Leaves and flowertops of plants raised in Kashmir contained alkaloid percentage of 0.05 to 0.06 which is well upto BP or USP standard.

Annual requirement of the drug is estimated to be more than 40 qtls and the plant is cultivated on an extensive scale. Malik et al (1964) recorded that with 0.5% colchicine maximum polyploids were produced and there is an increase in total alkaloids in polyploids over diploids.

Agrotechniques for the Cultivation

According to Watt (1890) cultivation of this plant started in India in 1839 by Royle, when the crop produce was entirely consumed. Later, it was imported with dwindling of its cultivation, so much so that the seeds were imported under the name of *Ajwain-khorsani* from Iran and Afghanistan. A well drained, fertile, sandy loam soil, free from weeds, is the best; sloppy areas do not form uniform crop; it is very thick at bottom as most of seeds are washed to lower areas.

The seeds are broadcast at the rate of 1–1.5 kg per 0.5 ha, mixed up with fine dry sand. If sown by drills rows should be 60 cm apart and seed should not be covered more than 0.75 cm of soil. It germinates in 3–4 weeks in spring, and 2–3 weeks in mid-summer. Plants should be spaced at 10 cm in each row. When transplanted at 0.75 cm spacing in field in rows of 60 cm apart when 5–7 cm high are the best for planting. Field should be attended for weeding and hoeing when the small seedlings put up a rosette of small leaves. The seeds sown in spring or mid-summer in nurseries are well prepared with organic manuring. Artificial shade is provided to young seedlings in summer, to avoid transpiration. It takes 1–2 weeks for the seeds to germinate in mid summer which are transplanted when 5–7 cm high. The

nursery is normally watered before seedlings are rooted out. May or Sept-Oct are the best months for transplanting the seedlings. The transplanted seedlings are sometimes shaded with material like *chirpine* needles. These are planted in rows, 60 cm apart, at a distance of 10–15 cm. Though transplanting had not been successful and showed stunted growth, due to tap root injury, better results are obtained if plants are allowed to attain a good size before transplant. Biennial and annual seeds should also not be mixed, since it adversely affects the yield per ha.

The annual forms of the plant grow erect and bear few branches but the biennial bears a rosette of large leaves in the first year. Weeding is done with a curved small blade. With notching of tips, before flowering in early May, plants develop a better vegetative growth. Harvest is done by hand-picking in June-July, when in flower. The annuals die after the seeds mature but the biennials remain dormant in winter season under the snow, after giving first year crop. In spring, a tall branched stem grows with several leaves, which are harvested in June-July. Seeds mature in Sept-Oct. and harvested by cutting the entire plant or rooting it out by hand.

The plant is also sown as winter crop in plain areas of Northern India (like Saharanpur, Jammu, etc.) and the seeds broadcast by drilling in lines, 60 cm apart, in well-prepared soil. Weeding, hoeing and other operations follow. No marked difference in alkaloid content has been recorded from crops raised by transplanting and broadcast. However, frequent irrigation increases the yield and foliage. Water logging is fatal to its growth and surely weeding influence the crop yield. Since, the crop remains stunted, 2–3 weedings are needed during the growing season (June-Sept). Ploughing the land, 2–3 times, before sowing, reduces weed infestation and growth of species like *Stachys sericea, Achillea millefolium, Verbascus thapsus, Chenopodium alba, Plantago tibetica* and grasses.

Larvae of red-cotton bug is one of the main pests attacking the leaves and the capsule; virus attack is also possible.

Drying and collection of crop

Leaves from annual crop are collected at the flowering time, while in biennials first year leaves are collected in late summer and second year leaves at flowering time. Green leaves are spread in thin layers in sun which takes 2–3 days. Shaded places can also be used in inclement weather, where thin layers are placed in trays, which are aerated regularly and might take 7–10 days in shade. Moisture percentage ranges as high as 90% but in Sept-October it falls to 78–80%.

Yield

Depending on the nature of soil, climate, irrigation, weeding and other factors, yield of the crop varies largely. Forest soils rich in organic matter produce good crop. If the field is fully stalked 2–3 mds (0.7–1.5 qts) of dry leaves per acre is normal and may go upto 6 mds (3 qts) per acre, if soil is well worked 3–4 times, and other agricultural operations are well attended.

Hyoscyamus muticus L. has wrongly been reported to grow in India, though stray plants are met; it is indigenous to Egypt, Sudan, Baluchistan and Sindh and hold monopoly for the supply of this (Kahi or Mountain Hemp) drug. The alkaloid content of this plant is more than that of the Indian species, also the narcotic properties are more intense. The leaves are smoked for intoxication. Though, it is a potent and promising species, as a source of alkaloid, it needs to be worked out. Chemical analysis of leaves showed considerable increase of 27.7% in total alkaloid content of polyploid as compared to diploids.

References

Malik NA et al 1963. Effect of different fertilizers on growth and alkaloid contents in *H. muticus*. *Pakistan J. Biol.* and *Agr. Sc.* Dacca; **6(1)**.

Malik NA, Sadiqi AR and Ahmad Arif J 1964 Pt. II *Hyoscyamus muticus*. *Pakistan J. for.* **14**: 130–135.

Malik NA, Siddiqa AR (Miss), Ahmad Arif J and Jahngir M. 1964. Induction of Polyploidy and its effect on alkaloid content in Solanaceous plants. *Pakistan J. For.* **14**: 124–129.

37. *Jasminum* species

Several species of Jasminum are cultivated for their fragrance and ornamental flowers. In France, Italy, Algiers, Egypt and Morocco, the species cultivated for the extraction of Jasminum perfumes is mainly *Jasminum officinalis* var, *grandiflorum*. The other species commonly cultivated in India are: *J. angustifolium* Vahl (Vern. Banmailika), *J. arborescence* (Saptála), *J. auriculatum* (Jai, Juni), *J. humile* (Pili-Jai, Pitmalti); *J. amultilorum* (Burm. f.) Andr. (*J. pubescence* wild. Kund-Phul), *J. officinale* var. *grandiflorum* Bailey (Spanish Jasmine, Chhaméli, Jati-Jai) and *J. sambac* Ait. (Arabian Jasmine; Motia, Mográ, Béla). Of these *J. offocinale* var. *grandiflorum*, *J. sambac* and *J. auriculatum* are cultivated for the perfumery.

Uses and extent of utilization and trade: Jasmine flowers have, since times immemorial, been used in India for ceremonial purposes, decoration of ladies hairs and preparation of cosmetics. The flowers are used for making *attar*, perfumed hair oils and creams. Jasmine absolute is the most important perfume used today. The absolute blends with any floral scent. For use in soaps and creams, Jasmine flower perfume should be compounded to alkali for resistance, otherwise it discolours the products.

In India, Jasmine flowers are extracted by placing them in contact with layers with dehusked sesame seeds for 12 to 24 hours. The exhausted flowers are replaced by fresh flowers, until the seeds get saturated with the odoriferous principle of the flowers. The oil is then expressed from the seeds. The absolute of Jasmine is generally not prepared in India.

The usual practice in this country is to gather flower buds in the evening, although it has been well established that the best yield and quality of perfume is obtained from flowers collected early in the morning. The harvested flowers must be processed for extraction of the perfume, as soon as possible after the collection, otherwise, they loose their perfume and yield perfume of low quality.

The odoriferous principle of Jasmine flowers are extracted either by enfleurage with fats or by extraction with volatile solvents. With petroleum ether, as the solvent, the concentrate is usually obtained; the yield varying from 0.28 to 0.34%; though yields as high as 0.41% have also been reported. The concrete is the reddish brown, waxy mass, possessing a pleasant odour. The so-called volatile oils of Jasmine are derived from these concretes or absolutes by subsequent distillation, but they are not produced commercially.

The leaves of wild Jasmine are also indigenously used to cure sore mouth having some ulcers.

Agrotechniques for the Cultivation

The soil and climatic conditions for the cultivation of Jasmines are practically the

10

same as found in the subtropical forests like *Sal* forests of Dehradun. They flourish well in sunny situations with irrigation facilities. Propagation by cuttings, rooted in manured beds or through layering is resorted. The bushes are trimmed for increased production of the flowers and facilitate flower collection.

Jasminum officinale is a twining or half erect, dark green shrub with opposite leaves; having 3–5 leaflets, the terminal one being the longest. The flowers are white, fragrant in few-flowered corymbose terminal spikes. The calyx tube is linear, more than half the length of corolla. This plant is found in open forests as a secondary growth in temperate regions from Indus eastwards, between 1000–3000 m, such as at Kunawar, Simla, Kulu, Bashahar, Binsar, Bhagirathi velley, Chamba, Kedarnath and in the Yamuna valley. This plant is not used as a trade community but used locally as a household remedy in diseases of the heart, diabetes, thirst, diseases of blood and skin, like the burning sensation. The roots are used to cure ringworm.

Among other species *Jasminum grandiflorum* L. (Chaméli, Jai Vern. Spanish Jai) is a large shrub, suberect, with opposite leaves, having 7–11 leaflets and white flowers with faint pinked streaks outside. Delightfully fragrant. The fresh juice of this plant (leaves) is applied to soften corns between toes. The ulcerations or eruptions in the mucous membrane of the mouth is cured by chewing the leaves. Oil prepared from the leaves is poured to cure otorrhoea, used in toothache. Flowers used to cure skin diseases, headaches and weak eyes. Benzyl acetate is the chief constituent of the oil from flowers which also contains methyl anthranilate and α-linalol.

38. *Lavendula officinalis* Chaix.

Family: Labiatae (Lamiaceae nom. alt.).

Local names: Lavender (English).

Description: Small shrub with linear leaves and terminal spikes of small, fragrant, bluish flowers.

Distribution: The plant is indigenous to S. Europe and grows in the Mediterranean regions on the shores of Mediterranean sea. It grows wild in high hills of S. France and adjacent Italy, but cultivated in several European countries like in UK, and in USA. Introduced successfully in experimental scale in Kashmir region, between altitudes of 2000–2800 m, have cultivation possibilities in Himachal, Uttaranchal and Nilgiris (*L. intermedia*).

Uses and extent of utilization and trade: Lavender oil is used in a variety of pharmaceutical preparations; especially perfumes. Lavender oil is the most popular scent. Its delightfully refreshing odour blends well with many other essential oils. Lavender water, a mixture of oil in water and alcohol, is a highly popular toilet article. Best quality of oil (50% esters), is used in the preparation of high graded perfumes, while the next quality (38–45% esters) for lavender water, toilet waters, eau de Cologne, etc. and a low grade (30–35% esters) in the scenting of soaps.

In medicine, clarysage oil is used as a flavouring agent and sometimes carminative. Lavender flowers apart from being distilled for the oil are also used for making sachets, potporri, etc. The flower heads give oil (2.54%) from fresh flowers which is in considerable demand and is primarily imported. The essential oil is considered to be the finest of the commercially essential oil, derived from Lavendula, because it possesses sweet, refreshing and delightly

fragrance of the true lavender plant. English lavender oils is obtained from a particular strain of the hybrid *L. intermedia* (*L. officinalis* Chaix, (*L. latifolia* Vill) and is considered superior to other lavendula oils. Concrete of lavender is produced by extraction of lavender blossom with petroleum ether and has high fixative properties.

Lavender oil is obtained by water or steam distillation and sometimes by solvent extraction of flowers. The yield per ha is about 30 kg depending upon the cultivated strain.

Agrotechniques for the Cultivation

The plant thrives best on light sandy, dry soils having a pH of 7.0–8.5 but can also grow on well-drained loams, not on lowlands, since well-drained calcareous, rich and fertile soils and sunny situations are the best. It is usually propagated by slips taken from the older, carefully selected plants in October. 10–15 cm long cuttings are prepared from healthy plants of 2–3 years age and planted in 40 to 50 cm plots at 4–5 cm depth, with line to line distance being 1 m in the month of October and November. It required deep ploughing, usually done on May-June and soil is finally prepared by repeated cultivation of the land. Nursery plants are irrigated between 10–11 AM and 3–4 PM with sprinkler for about 15 days. Though, it can also be grown directly from the seeds, mostly propagation is done from cuttings or by root division. Temperatures between 20°–30° C

are optimal for its growth. However, rains during flowering time are not conducive.

Nitrogen and phosphorus each at the rate of 100 kg/ha is also applied at the time of land preparation, with total nitrogen in the succeeding spring (April). On acidic soils (pH 4.5), 5–6 tons of lime/ha is also applied during the ploughing. Deep furrows of 5 cm width are laid out a spacing of 1 m and the lavender propagules are mechanically planted, 15 cm deep into the soil in a row so that 10 cm of leaves are above the ground level. In the first year the plants grow to a height of 5–6 cm and produce 2–3 branches, thus, each plant having a spread of about 20 cm. In the second year, the plant grows upto 22–25 cm tall and covers the rows. During the 3rd, 4th and 5th year the plant is 30, 40 and 50 cm in height respectively and the spread is 40–45 cm in the inter-row spaces.

However, frequent hoeing and irrigation is needed to obtain a good flower yield. 80 kg nitrogen through urea in three split doses of 20, 30 and 30 kg is given in March/April, May/June and August/Sept in the 40 : 40 : 40 in second year and 50 : 50 : 50 in the third year is given as NPK. Use of FYM (800–100 cft) and 80 kg P_2O_5 and 60 kg K_2O is recommended for application at the time of field preparation.

After third year of planting a sizeable harvest of the flowers is expected and 75% of the bloom is expected in the month of November, i.e. 60–70 tons of flowers/ha.

Table 10.9: Flower yield and oil content in Lavendula sp. and varieties

1.	Kazanlak	Flower-yield	47770 kg	Oil 1.18%	50 kg /ha
2.	Karlova	Flower-yield	6000 kg	Oil 1.18%	78 kg/ha
3.	Chemus	Flower-yield	5600 kg	Oil 1.7%	101 kg/ha
4.	Arma	Flower-yield	4000 kg	Oil 1.4%	
5.	Svejest	Flower-yield	4120 kg	Oil 1.2%	49 kg/ha

Flowers are plucked only on sunny days. Since the oil content in the flowers is only 1.2% of the flowers weight, the oil recovered is 72–96 kg per ha of the crop, i.e. 30 kg in the 5th, 6th and the 7th year and 40–50 kg/ha in the 10th year. During the first year average oil yield is nil, while in second year it is about 7 kg, 3rd year 15 kg and 4th year 70 kg.

Flower cutting is done in July and August, when the sky is clear, when 70–80% flowering has taken place. The spike is cut with a sharp-edged knife. However, crop protection measures are also required since fungal infection is common. 0.5% Thiodon 35 EC spray is done twice at 15 days interval; also 1.5% Bavastin or 0.1% Benlet is sprayed.

Only the lower tops are harvested when in full bloom and distilled fresh for oil production. The colour of the oil changes from blue lavender to greyish blue indicating the right stage for harvesting flowers.

The plantations are also rejuvenated (tuft cutting) by cutting all the aerial parts in the 10th year. The plants give new leaves and the oil yield in the rejuvenated crop at 11th and 12th year is 12 kg (60–70% of previous year), 12th year 20 kg, 13th year 40 kg and 15th year 40 kg/ha, when it starts decline again and in the 20th year it remains just at 20 kg/ha. However, there is no difference in the linalyl acetate content in the oil from new and regenerated forms.

Other Varieties of Lavendula

1. *Lavendula latifolia* Vill. It resembles the true lavender and is somewhat taller. The oil is obtained in the same manner as described above but the colour is brownish yellow containing only 2–3% esters, with a harsh camphoraceous odour. This oil is largely used in soap industry and as a substitute for true lavender oil.

2. *Lavendula hybrida* Reverchon. It is a small shrub, 40–90 cm high, harder and more vigorous than true lavender plant. The flower colour ranges from deep blue to grayer shades of spike lavender. It is mainly cultivated in France and Italy. The oil is a substitute for lavender oil whenever the costs are to be brought down and is finding use for disinfecting houses, classrooms and public halls on account of its microbicidal action. The plant is propagated by slips, taken from older carefully selected stock. Flowers are harvested, when in full bloom. It produces extraordinary growth producing 50–60 kg of oil per ha. The oil is obtained in the same way as from true lavender oil. However, the odour is somewhat a combination of delicate fragrance of lavender and camphoraceous harshness of the spikes.

3. *Lavendula vera* is indigenous to France having two varieties, viz. *L. vera* (syn. *L. officinalis*, *L. angustifolia*) which is narrow leaved and *L. apica* (syn *L. latifolia*), broad leaved. In addition, a natural hybrid (sterile) between the two species called Lanendin is also grown in France. *L. vera* (2 n = 24) yields oil of better perfumery grade. Odour of *L. apica* is not as delicate and used mainly in soap perfumery. Oil yield in hybrid lavender is double than that of lavender, but oil is of poor odour value. In temperate climate it is grown on poor soils as a rainfed crop. Since it is a cross-pollinated crop, vegetative propagation is recommended for uniformly high yielding plant. The flower shoot normally contains 0.8% fresh oil but some

selections range between 2–2.5% of oil. Oil of available samples is recorded to contain 35% linalyl acetate, while some selections contain 70% of linalyl acetate (Table 10.9).

The flowers of lavender are protandrous, i.e. anthers dehisce 2–3 days earlier than the stigmas become receptive. For controlled crossing, corolla along with stamens are removed at the bud stage and covered after 2–3 days when the stigma could be seen above the calyx, fresh bloom from male donor flowers is brought near it and bagged, the time of anthesis being between 7–8 AM. It developers mature seeds in 40–45 days. Though no new varieties are reported, mutation breeding work has now been undertaken in countries like Bulgaria.

Cost and benefit analysis

Table 10.10: Economics of lavender cultivation Singhal (2001)

Operations	1st year	2nd year	3rd year
Land pre- paration	Rs 2000	...	...
Planting material at 0.50/per cutting	Rs 1000	...	...
Transplanting	Rs 1000	...	
Fertilizers and FYM	Rs 2000	Rs 1500	Rs 1500
Interculture operations	Rs 2000	Rs 1000	Rs 1000
Irrigation	Rs 2000	...	...
Harvesting	Rs 1000	Rs 1000	Rs 1000
Transport	Rs 500	Rs 500	Rs 500
Landrent	Rs 1000	Rs 1000	Rs 1000
Interest	Rs 1000	Rs 1000	Rs 1000
Unforseen	Rs 500	Rs 500	Rs 500
	Rs 21, 500	Rs 6500	Rs 6500

Table 10.11: Yield of oil 1.2% of flowers weight

Distillation	Rs 1000	Rs 1000	Rs 1000
Yield of oil (per ha)	20 kg	120 kg	120 kg
Rs 1200/kg of oil	Rs 24,000	Rs 1,44400	Rs 1,44400
Net profit	Rs 2500	Rs 1,37500	Rs 1,37,400

References

Clayton DA. 1937. A Taxonomic Study of the genus Lavendula. *J.Linn. Soc.* **51**.

Singhal, A.K. 2001. CIMAP Proc. of Seminar. *Mimco* pp. 36.

39. *Matricaria chamomilla* L.

Family: Compositae (Asteraceae).

Local names: Persian Chamomille (English); Hingarium or small Chamomille (German); Babuna, Sateigul (Urdu).

Description: An annual herb, attaining a height of 40–60 cm, and in some cases about 80 cm. Stem glabrous, branched. Leaves very green, smooth, 2–3 pinnatisect with segments, short and very linear, giving the leaf a finely dissected appearance. Flower head solitary, long peduncled, about 2 cm in diam with white gray and yellow disc flowers.

Distribution: A native of Europe but cultivated in about 40 countries extensively in Hungary, Germany and Yugoslavia. In India, it is reported to grow in the plains. The wild materials exhibit wide variations in plant type, flower size, oil colour (brown to dark blue) and azulene content in the oil. It is possible to induce polyploidy in the species. Countries like Bulgaria, Hungry, Argentina have also started cultivation of this plant.

Uses, extent of utilization and trade: The crop was introduced in India in the fifties with seeds obtained from outside the country, on an experimental basis successfully by the RRL at Jammu (300 m) and Srinagar (1620 m). It is a *rabi* season crop in the plains, while in hills it can be grown from April to September. The plant was brought to India by the Moghuls about 300 years back and the herbal name of the plant is *Babuna* or *Gul é babuna*. In Germany the flowers are used maximum and the world production is 10,000–80,000 tons flowers and 9–10 tons of oil. The flowers are stimulant attenuant, carminative, used in constitutional debility in hysteria, dyspepsia and intermittent fevers.

Blue oil is also used externally in rheumatism, flatulence, colic and in herbal medicine like antiseptic ointment, cream for wounds and antiseptic creams, weakness of body, hysteria, colic fever. Extract of flowers is used in toothpaste, mouthwash, lipstick, medicated soap, baby oil and shampoo. Flowers and the plant are used in ulcer of the stomach, as tonic, in cold and making dye for hairs and for dandruff.

50–60 ha of land is being cultivated in India with this crop producing about 3000–4000 qtls of flowers. It has great export value. Camajuline, Bisebocal and its oxide are the principle used in medicine. International market price is Rs 22,000–36,000 and in India it is Rs 14,000–15,000 per kg.

Agrotechniques for the Cultivation

Subtropical and temperate climates are considered to be the best for this plant. In Hungary, it grows on clayey lime soils which are considered to be too poor for other crops, while in Yugoslavia about 2,30,000 ha of saline soda steppes are reported to support its growth. In the north Indian plains and hilly ranges of Himalaya, it is grown as *rabi* crop, while in the south it is grown both in *rabi* and *kharif* seasons. Temperatures between 18–20 °C are suitable for the seed germination, while 22–25 °C is the optimum for good crop growth and oil production. Regarding the soils, clay loam soils with good water retention capacity are suitable, while *Usar lands* with a pH of 9.5 could also be used for this crop.

Since direct seeding in the field results in poor growth and patchy germination, nursery is prepared for sowing the seeds; as done for other transplanted crops. The seeds are small (1000 seeds weigh 0.09–0.15 g), and kept in nursery for 35–50 days. About 2 kg of seeds are needed to provide enough seedlings for stocking an acre of land. In North Indian plains, first week of October is the suitable month for sowing and the planting is done in mid November. In the hills, the seeds are sown during last week of May and the planting is done during June/July. The size of the plot in the nursery is about 1.0 × 1.5 m. Direct seeding is done at 20 cm distance in close lines. It can also be broadcast. The seeds take about 6 weeks to be ready for transplanting. 500–800 g of seeds are reported to be required for 1 ha of land (CIMAP 2001). 30 cm × 30 cm space for the seedlings is considered to give the highest yield of flowers and the oil content is also high. The crop after being transplanted can either be hand-watered or flood irrigated. Percentage mortality in the transplanted crop is almost negligible.

FYM @ 50–100 qtls/ha is mixed with the soil during land preparation and deep sowing is avoided. In the nursery, mulching can be done to conserve water. Spray watering can also be done for irrigating early stage crop. Repeated irrigation to the crop is necessary; at least 3–4 times it should be done. 50–60 kg of nitrogen, 50 kg phosphorus (P_2O_5) and 50 kg potash (K_2O) is needed. On sandy loam soils 150–180 kg N is given. One-fourth nitrogen, full amount of phosphorus and potash is given at the

preparatory stage of land and the rest is given in split doses.

Weeding and hoeing operations are followed, the first weeding at 20–25 days while third is also done about the same time after the second one. Chemical weeding with 2, 4–D Sodium salt and Linuran at one pound per acre is recommended but should be used at minimum since the flowers are used for tea and as decoction.

There is not much damage from pests and insects however, 'Naisam minor' infection can be controlled with 20% Lindane spray. Leaf curling has also been recorded in the crop.

The crop begins to flower by second week of February/March and April; plucking usually gets delayed so that there are sufficient flowers to pluck. In the higher hills, with June/July plantation, flowers may be plucked in October/November. The flowers are small and the plants have flowers at all stages of growth, viz. bud stage to full bloom stage, hence the plucking has to be selective, as flowers at the bloom stage are plucked. Flower plucking is time-consuming and done at night; the optimum time being when the flowers are deep yellow and the grayflorets droop below. The plucking is done 4–5 times again at 10–15 days interval and during the second and fourth plucking the flowers are usually along with stalk and leaves which should be minimum. 3–4 kg of flowers are usually plucked by a labour in a day, which could also be upto 10 kg, while with a "flower comb" 60–100 kg of flowers can be collected. The flowers are dried in shade carefully, in thin 1–2 cm, layers, spread on the floor and should not be turned during drying but are shaken. Dried flowers are kept in tight containers, since these are liable to fungal attack. Electrical drier can also be made use of.

Oil extraction

The production of fresh flowers ranges between 4.5 to 6 tonnes/ha and on marginal lands the yields are 2.5–4.5 tones/ha. On shade drying 8–10 qtls of flowers are obtained from one ha which may go upto 15 qtls. Air-dried flowers on distillation yield an oil of deep colour with strong smell. The deep blue colour is due to Chamazulene, which is an important constituent of the oil. It decreases during storage of the flowering material. Average flower yield at RRL Jammu is recorded to have been obtained which ranges from 3500–4000 kg of fls/ha which on drying gives an yield of 35 to 37 kg. Oil yield @ of 0.5–0.88% oil from dry flowers, while CIMAP recorded 0.3–1.3% with an average of 0.5% of oil.

After taking 3–4 pickings, the crop is allowed to stand, the remaining flowers mature and set seeds. About 200 kg of seeds can easily be obtained from one ha of land. The cultivation of Matricaria is profitable, if the flowers are collected by technical means, since the concentration of oil is for 10–15 days from the end of March to May —a period which is coincident with *rabi* harvest.

Cost benefit

The cost of producing fresh flowers has been calculated roughly at Re one per kg by RRL, Jammu; while CIMAP has calculated a net profit of Rs 21,000/ha, out of Rs 24000/ cost per ha, Rs 13000/- only on plucking of flowers. Dry flowers give only 0.5% oil. CIMAP var. Valery gives 20–30% more crop. Other varieties are *Camijalm* and *Prashant*.

References

Dutta, PK and Singh Ajit; 1964. Effect of different spacings on fresh flowers and oil yield of

Matricaria chamomilla. Indian Journ. Agron. **9(1)** : 11–12.

Kapoor LD, Dutta PK and Singh Ajit. Cultivation of *Matricaria chamomilla. RRL Bull.* **1(2)** : 104–106.

Kapoor LD and Kaul BK 1964. Preliminary studies on the effect of Gibberellic acid on *Matricaria chamomilla* L. *Indian. J. Agronomy* **9(3)** : 225–228.

Singh, Ajit. 1969. Cultivation of *Matricaria chamomilla*. CSIR Symp.

40. *Mentha arvensis* L.

Family: Lamiaceae (Labiatae).

Local names: Pudiná (Hindi), Mint, Corn Mint (English).

Description: An erect herb, branches upto 60 cm high. Leaves upto 5 cm long; leaf stalk small or none; margins toothed. Flowers small, liliac, in small bunches, borne in the axil of leaves.

Distribution: A strongly scented perennial growing plant abundantly in the Northwest Himalaya from Kashmir to Kumaon, 1500–3000 m. New grown extensive in Sultanpuri district in UP. In its natural habitat, it fairs well on good sandy or loamy soil, rich in humus. Well-drained fertile soil with little rain during harvesting period are considered ideal conditions for its cultivation. It planted on a good sandy soil, with rains in spring and sample sunshine in summer, the plant is recorded to develop high menthol content. Study on its physiology has shown that mint will not grow on saline-alkali soils. The Japanese variety is *M. arvensis* L. ssp. *haplocalyx* Briq. var. *pubescence* Holmes, which has been cultivated for the essential oil, used as a substitute of peppermint oil (obtained from *M.piperata*). It yields bulk supply of menthol and peppermint oil.

Information on the main species, their chemical constituents, cost of oil and average net profit is given in Table 10.12.

Table 10.12: The main species and their chemical constituents

Main species	Chemical constituent	Cost of oil (Rs/kg)	Av.net profit Rs/ha
Japanese mint	Menthol	Rs 300/-	Rs 27,600/-
Spear mint	Carbone	Rs 225/-	Rs 16,975/-
Bergamont mint	Linol (Lillel-acetate)	Rs 250/-	Rs 20,100/-
Peppermint	Menthol	Rs 400/-	Rs 22,600/-

after SIBI, CIMAP 2001.

Several species of Mentha were tried at Dehradun and maximum yield was obtained from *M. arvensis* (5643 kg/ha), followed by *M. piperata* (2699 kg/ha), *Mentha spicata* (1155 kg/ha) and *Mentha citrata* (933 kg/ha) under rainfed conditions.

Uses, extent of utilization and trade: Peppermint oil is one of the most commonly used essential oils used for medicinal and flavouring purposes, such as toothpaste, dental creams, mouthwashes, cough drops, chewing gums, tobacco, confectioneries, alcoholic liquors and medicinal preparations. The chief constituent is *menthol* which is isolated from it and sold separately which was earlier imported from USA, China and Japan.

The oil of peppermint is one of the most commonly used essential oils and is an excellent carminative, gastric stimulant, antiseptic and preservative. Import of this oil in sixties exceeded worth more than six million rupees per annum. Extensive cultivation of M. arvensis in Japan, China and Brazil provided bulk supply of Menthol and Peppermint oil (*M. arvensis* ssp. *haplocalyx* var. *piperascens* Holmes). However, the plant growing naturally to a limited extent yielded oil which did not come to the official standards. Hence, cultivation of this plant was undertaken by the RRL Jammu, at their regional farms, such as at Jammu (300 m), Katra (100 m) and at Srinagar (1700 m), where the analysis of oil obtained compared well with the Japanese and Brazilian natural oils. The plants grown at Srinagar in Kashmir showed stunted growth and low oil percentage. Commercial cultivation of this plant has now been undertaken at a number of places in Uttaranchal (*Tarai* regions) and in UP, where the climate has been found suitable for the growth of this plant. However, water-logged areas do not suit this crop at all.

Agrotechniques for the Cultivation

The plant can grow under a wide range of soil and climatic conditions, provided irrigation facilities are available. Sandy or clay soils do not support a good growth. The land should be open, of an even surface, fertile and rich in organic matter and the pH should be towards the acidic side.

The land should be well prepared before sowing, so that it is free from weeds and stubble. Six ploughings are given before planting the suckers. Root suckers are taken from one-year-old plants lying dormant in the soil. These are planted end to end, in rows of 45 cm apart in fissures, 45–60 cm deep, and followed by liberal irrigation, which is required for profuse vegetative growth. About 150–180 kg of suckers are required to plant an acre of land. The first inter-culture operation should start after the new sprout are clear and is done with hand. After every inter-culture operation, some time should be allowed before it is irrigated. Thus, the crop comes up in a row called rowmint. The field should be irrigated since keeping the soil moist is conducive for a good and uniform germination. In early stages of crop growth, weeding is carried out by hand,

but considerable economy can be made by using wheel hoe; the common country plough. The drawback in using the mechanical means is that the weeds get concentrated within the row of the crop and becomes troublesome later on in the season.

The *rowmint* throws out runners and by July/August these runners cover the inter-spaces and a thick mat of crop is obtained so that it becomes a *meadow mint*. The crop needs to be weeded 4–5 times and if *motha* (Cyperus), *Baru* and *Dub* (Cynodon) grasses predominate, weeding becomes difficult and time-consuming. For a profuse vegetative growth the plant uses good deal of nitrogenous fertilizers and on the alluvial plains of UP, 40 kg of Urea, 50 kg DPAP and 25 kg Muriate of Potash per ha has been recommended where the yield of mint per acre was recorded at 50 kg with an oil content of 0.5%. Preliminary trial on Mentha was conducted at Dehradun and it was felt that Mentha under rainfed conditions is not economical.

Mint crop requires irrigation 5–7 times before the monsoon starts, and after the monsoon crop needs irrigation at least 2–3 times. During the winter, the crop remains dormant in the Northern part of the country, while it is not dormant in the Southern part and irrigated as and when required. Delay in irrigating is harmful and even for regeneration it should be irrigated within two days of harvesting. Persistent water-logging is harmful, though it can withstand water for a couple of days, and during germination it is uneven and poor.

Normally, the crop grown after the land is green manured with Dhaincha (*Crotalaria juncea*), gives good higher yield. Mentha responds also to liberal doses of nitrogenous fertilizers and the increase in herb yield due to application of phosphorus and potassium is not marked as that with nitrogen. 25–30 tons of well rotted manure can also be applied at the time of final ploughing. 120–150 kg nitrogen, 50–60 kg of phosphorus and 40 kg of Potassium per ha is sufficient. Full dose of phosphorus and potassium is given at the time of planting, along with one-third of nitrogen; other two-thirds are given to the standing crop (CIMPO, 2001). RRL Jammu has advocated a uniform dose of ammonium sulphate to each bed at the rate of 20 kg/ acre when the plants are 120–180 cm high. Once the crop gets established, it suppressed any weed to grow and no weeding is required. However, liberal dose of nitrogenous fertilizers increases herb yield. Though it is better to apply potash and phosphate fertilizers as basal dose at the time of planting.

Crop-harvesting requires special care, since improper harvesting seriously affects the oil yield. Harvesting is done when the plants are in full bloom. Crop may be harvested in June/July, or Sept/October when the plant is in blossom. It is advisable to harvest with sickles in the morning on a bright sunny day after the dew has disappeared. Fresh herb is distilled, if weather does not permit drying. Best and most economical method to dry the crop is to separate the leaves for distillation. The harvested crop may be tied in small bundles and hung in open air under the shed. The number of days, required for drying, vary with the season. Care is taken to prevent the leaves from falling. Sometimes, the plant is dried in sun but loss of oil due to resinification and evaporation is reported. The plant gives good crop during the second and the third year, after which the oil yield decreases. It is economical to uproot the crop during the fourth year and replant after rotation. Crop rotation suggested for hilly areas is Fagopyrum-Green manuring-Fagopyrum-Mentha is for the leaves that are edible.

Several diseases have been reported on mentha plantations, of which *Puccinea menthae* is the most common. The healthy green leaves turn yellow and reduces the oil content. A reduction of 13.2% of oil has been reported due to rust infection. The incidence can be checked by taking an early harvest. The other diseases and their control reported are as below.

1. *For termite control*, adequate irrigation is needed. BHC 10% powder, 25 kg/ha is mixed in the soil. At the time of irrigation aldrin is used apart from chloropyrophos.

2. *Leaf cutter control*: 1.25 lit Thiodon 35 EC or Malathion 20 EC in 1000 ltrs of water per/ha is sprayed. For leaf sucking insects 1 lit of Metasystox in 1000 litres of water/ha is sprayed.

3. *Stem cutter*: 5% Alderin in 20–25 kg or Heptachlor 20–25 kg/ha is applied to the field before plantation.

4. *Root decay*: Plants need uprooting. Suckers are dripped in 0.25% solution of Peptan.

5. *Aphid control*: On spear mint, attack is in February. One lit. Metacystox in 1000 ltr. of water is sprayed for one ha. When the intensity of sunshine is low, heavy infestation of white flies rest on the dorsal surface of the leaves, while in mid-day noon, they rest on undersurface. The growth of plant is more or less arrested and the oil content declines. An interesting phenomenon of biological control has been reported in which the grubs of *Coccinella* beetles were found to predate on the nymphs of this fly. Although this is limited yet goes a long way in controlling the infestation.

6. *Rust*: *Puccinia menthae* in Tarai region is the cause. Small reddish brown pustules appear on lower surface of leaves.

10

There is heavy defoliation with reduction in oil yield and mentol content.

Control: Spray Propacinazole (0.1%), Mancozeb (0.3%). Grow rust resistant variety like *Kosi*.

7. *Powdery mildew*: Caused by *Erysiphe cichoraceanum* during April and May. There is deposition of white powdery mass on leaves.

   Control: Spray wettable sulphur (0.2%) or karathane (0.1%)

8. *Leaf spot*: Caused by Coryinespora, Curvularia and Alternaria spp. Out of the three Alternaria blight is common and serious. Infected leaves show dark brown spots which enlarge later and coalesce during the rainy season causing complete necrosis of leaves and heavy leaf fall. At later stage the plant gets affected reducing total yield and oil yield.

   Control: Spray Chlorothalonil (0.3%) or Mancrozeb (0.3%).

9. *Stolon and rootrot*: Caused by Thielavia. Initially the infection appears as small dark lesions on stolons, which increases and turns into soft rot. Later the rotting spreads to roots causing infections and rotting. Infection causes marked reduction in yield of suckers, which should not be used for further propagation of plant.

   Control: Avoid excessive irrigation at later stages. Treat suckers with Carbendazim (0.1%) before planting. Use healthy and disease free suckers.

10. *Root knot nematode*: Glomerella and Cercosporia sp. Irregular to circular dark brown spots appear on leaves. Leaves fall causing reduction in the yield.

    Control: Spray Chlorothalonit (0.2%)

Crop harvest, drying extraction and marketing

The stage and time at which the crop is harvested in *Mentha* is important, since the maximum content of oil in the leaves is around the mid-day, at that time there is very strong characteristic smell of plant all around and within the plantation. During the later half of May month the lower leaves of the Mentha plant start turning yellow and this a very practical index for starting harvesting operations. The yield of green herb per ha in the first flush is around 7500 kg/ha. After the first flush the stumps give out runners, just right and left, and cover the intervening spaces between the rows as stated earlier. There is an increase of about 30–40% in the yield of green herb as compared to first flush. The indications for the 2nd and 3rd harvests of the crop are that large number of Mentha plants are in bloom during Oct-Nov. The oil percentage is also high, i.e. 3.5–4%. In the third flush which is in late November, oil percentage varies between 2–3%. The average proportion of stem to leaf of all the three crops is 1 : 2. The percentage of moisture of all the flushes when harvested in the afternoon, ranges from 50–65% and as high as 45 kg of oil has been obtained from an acre in all the flushes on distillation. Average yield in *Tarai* area is recorded between 27–32 qtls/acre from all the three cuttings. In Japan and Brazil, where plantations cover vast stretches, crop once planted is allowed to be continued for 3–4 years. This practice cannot be followed in smaller holdings because mint cultivation needs fresh planting every year on fresh cultivated land where crop rotation is followed.

Mixed Farming of Sugarcane and Mentha

Sugarcane and mint mixed farming ensure better utilization of space both above ground

and underground water and fertilizers. Sugarcane takes about 35–40 days to establish with slow growth in the beginning. In the interspace weeds cover the ground, thus requiring heavy doses of fertilization and watering. *Bergámont mint, Peppermint* and *Spearmint* can be planted in between sugarcane during spring season.

Sugarcane planting in between the winter mint plantation is the ideal crop (var. Co-1148, Co-1158), in last week of October at 45 cm apart. There should be a difference of 2 weeks in sowing because after planting mint irrigation is required. Sowing in rows is essential for availability of proper light. Spring sowing can also be done. 300 kg nitrogen and 100 kg of phosphorus and Potash/ha is given. One-third of N is applied after sprouting and one-third after last cutting.

Since mint requires repeated irrigation (March-May), it is given at 7–10 days. After monsoon season 75% moisture is maintained in the soil (i.e. 10–12 days). Weeding is done 3–4 times but care is taken not to damage the roots. Winter-sown mint can give 2 cuttings (March and May), while spring-sown one (May), after that sugarcane grows. Mint is cut during sunny days, since oil is extracted from the plant that grows above the ground. The sugarcane production is reduced by 18–38%, while net income from both the crops increase.

Crop drying

Like all other essential oil-bearing plants, drying of mint has important bearing on its oil content. Crop should preferably be dried under shade, since there is 24% loss of oil under sundrying. The best method recommended is to dry the herb in shade, by spreading it in thin layers. The dried material is thrashed to separate the leaves which are collected and distilled. It is good to have big drying sheds with provision for free circulation of air. Sometimes, one has to spread Mentha in thick layer for want of space, so care has to be taken to turn over the herb at frequent intervals to prevent fermentation and ensure proper drying. The plant has 75% moisture and apart from minimizing the loss due to evaporation, it will make the material less bulky and yield oil more rapidly on distillation with less consumption of fuel and time.

Extraction of the Oil

It is done in big stills worked by steam or in smaller stills put in farmers field. The bulky product has to be transported for large stills. The crop should be allowed to wither for 12–24 hrs before taking to stills, for maximum oil yield. If rain is expected it is taken to the still on priority. The oil after distillation contains sometimes dust and other mucilaginous matter which is separated by filtration. A golden oil containing approximately 70–80% Menthol is obtained. On cooling the oil to low temperature, Menthol crystal separates by repeated chilling and centrifuging nearly 51% Menthol is obtained.

Best results were obtained in the *Tarai* region of Uttaranchal, where the yield of dry leaves was 240 kg/acre and the oil content of dry leaves being 2.2% on zero moisture basis. Through good cultural practices it could be further increased.

Oil yield in some other places in given in Table 10.13.

Table 10.13: Oil yield of mint

	Whole herb green	Dry herb	Dry leaves	Av/acre
Jammu	0.4–0.5%	1.02–1.7%	2.8–4%	25
Japan	NA	1.28–1.69	3.37–4.2	30

10

The oil contains menthyl acetate 24.4%, free menthol 44.6%, mentone 24.6% and hydrocarbon 6.2%. Among hydrocarbons alpha pinene, limonene and cadomone are present. Natural oil yields are 40–45% of Menthol and 50–60% dementholised oil which can be used for confectionery and in medicine in place of peppermint oil. This contains total menthol and menthone 88–90%, menthol 82.53%, menthone 1.04–2%, menthol ester 2.6–6.12%. The mint oils are valued for their menthol content and chemical methods of converting methyl esters to menthol by alcoholic hydrolysis have been attempted.

Presently, Richards sons and Bhavan chemicals are the two well-known concerns purchasing oil and paying on the basis of crystal percentage in the oil. Nigam et al (1975) reported bigger crystals directly from the oil by slow cooling and dilution method. Recrystallization with non-polar solvents based upon controlled rates of evaporation and supersaturation has been tried and parameters for getting bigger crystals established. If Mentha has to come on its own it will have to compete with sugarcane and when both of which bring high price due to high yielding substances and if proper incentives are given the crop may come upto meet requirements of the industry.

References

Gupta US. Studies on the physiology of Mints: effect on *M. arvensis* some sodium salts on the sprouting and early growth of suckers of *M. arvensis*. var. purpurescens. Proc. Seminar on sea, salt and plants pp. 121–127.

Gupta US. Effect of some growth regulating substances on growth and essential oil content of *Mentha arvensis* var. piperascens. Holmes. Proc. Indian Symp. on plant growth substanses pp. 453–458.

41. *Mentha piperita* L.

Family: Lamiaceae.

Local names: Pudina (Hindi), Viláyati Pudina (Punjab), The True Peppermint (English). Most of the Indian local names of Mentha species are based on the word Pudina.

Description: A perennial herb, spreading by root stock erect stems 30–90 cm high, branched bearing clusters of bilabiate fls. in axils of the leaves that are lanecolate with triangular blent teeth seeds produced freely. Mentha is a large genus *M. sylocstis* (The Horse Mint) *M. Viris* (Spear mint), *M. arvensis* (Spear mint) and *M. piperita* (Pepper mint) met in garden.

Distribution: The plant is cultivated in several parts of India, e.g. Kashmir, Uttaranchal, UP, Punjab, Himachal Pradesh, Karnataka, Haryana, Bihar, Madhya Pradesh, Tamil Nadu and Maharashtra. In Europe, it is grown as an annual crop and planted in autumn. First in 1881, the plant was raised in the Nilgiri hills and in eastwhile. Mysore state and later at FRI and RRL Jammu when repeated trials showed 40% menthol yield from local plants. *M. piperata* var. *vulgaris* was cultivated in Jammu (Ajit Singh and Balyan, 1975) under irrigated conditions on poor soils and it was recorded that last week of December till the end of January, is the proper time for planting this species.

Uses and extent of utilization and trade: The dried leaves and the flowering tops constitute the drug peppermint, which is used in the treatment of flatulence, vomiting, diarrhoea and nausea. Bruised leaves are applied to headache and other pains. Though, a number of species grow wild and cultivated in India, none of them yield peppermint oil of medicinal value.

It is the source of an essential oil, called the peppermint oil which is used largely in medicine industry. In medicine, the chief uses are for stomach disorders, in ointment for headaches, rheumatism, mouthwashes, etc. The oil is also antiseptic.

More than 20 tonnes of peppermint oil worth over 2 million rupees used to be imported every year for being used in pharmaceuticals, dental creams, chewing gums and confectionery. In medicine it is preferred to the Japanese mint (*M. arvensis*) oil. Recent attempts at CIMPO drug farms have shown that it can be cultivated on large scale economically, though attempts made earlier proved that cultivation of the plant is much costlier than the imported.

Product; Both the oil content and the quantity of oil has been found to be of BP standard. In Pakistan Malik and Khan (1964) reported that composite samples of leaves and twigs gave 1.02% of essential oil with one hour distillation and which contained approximately 56.0% Menthol.

It is estimated that about 20,000 tonnes of Mentha oil is produced in India (SIDBI, 2001, Singh K.).

Agrotechniques for the Cultivation

Except peppermint all the varieties are grown on sandy loam to clayloam soils (pH 6.5–7.5). Cold climate is necessary in places where irrigation and water disposal facilities are available.

The plant thrives well on humus-rich and fertile soils showing good response to organic fertilizers which do not affect the physico-chemical properties of the oil. Nitrogen application increases the yield of the green herb as well as the oil considerably. Application of 30 kg of N/ha gave an increase of 22.7%, whereas 60 kg of N produced an increase of 47.7%; while phosphate and potash did not show any response. At Jammu var. *vulgaris* gave very economic returns and were reported with P_2O_5 and K_2O at the rate of 50 kg/ha each as basal dose and nitrogen at 100 kg/ha applied in 3 split doses. The second and third year crop showed decline in yield and oil content. Three cuttings were possible at Jammu (300 m); average yield being 418.37 qtls, 307.58 qtls and 232.4 qtls and the oil recovery being 0.22% of oil yield being 0.92 qtls 0.68 qtls and 0.58 qtls respectively.

Earlier works indicated that *M. piperita* var. *Black milcham* gave maximum oil content of good quality and Menthol content of at least 45% at the full bloom stage. Studies of Duban et al (1975) in *Tarai* region, at the Haldwani station in tarai region of Kumaon, showed that the herb in the first harvest in June yielded maximum (81.06 qtls/ha) corresponding to 18.00 kg of oil, followed by harvest which gave 77.5 qtls/ha of herb responding to 15.25 qtls of oil. In the first and second case the age of the crop was 160 and 145 days respectively. The oil quality is not affected in the second harvest. In Assam, it is estimated that the yield of two harvests with application of N at the rate of 120 kg/ha can reach upto 12.5 tonnes of green herb and the oil content may vary from 0.4–0.5% on fresh weight basis (Singh et al 1975).

Cost-benefit analysis of mint cultivation is given in Table 10.14.

In Europe, *M. piperita* is planted in autumn, which sprouts in April/May, thus the first cutting is received by the end of July or early August and the second in October. The first is distilled for oil while the second is dried as one of the ingredient of "Bulgarian tea" and exported. Dry mint leaves are also used in food preparation, confectionery and home folk medicine. Now there are many

improved varieties available, evolved through clonal selection and polyploid breeding, viz.

- K-63, a triploid early maturing with 0.3% oil containing 43–44% menthol.
- Sofia, a polyploid, blooming 15–20 days earlier than Milcham variety, rust resistant, 0.3% oil containing 43–44% Menthol.
- Marilsa, a mutant of Soviet var. *Prelenlaskya*, but suspectable to verticillium disease.
- K-64 evolved by inducing mutation in K-63 with higher yields, both in herbage and oil content, but with low Menthol content by 3% to var. *Sofia*. K-63.

A fertile tetraploid form (2 n = 134) has recently been isolated from amongst Colchiploids of *M. piperita* which breeds true when grown in isolation. It is now used in making crosses with *M. arvensis* and hybrid produced yield 4.5% of oil (dried leaves), containing about 70% Menthol. The oil shows presence of Menthafuran. Another hybrid is obtained by crossing *M. arvensis M. longifolia*, which is an early blooming type with high percentage of oil content in the leaves and 60% menthol content.

It has been shown that all the varieties (except Spearmint) cuttings are taken when upper leaves get smaller and lower leaves start yellowing. Japanese mint-first cutting is taken when of 100–120 days and during clear days and plant cut at 4–5 cm from ground. It should be immediately collected and if distillation is to be delayed, the harvest is kept is shade.

Spearmint-at 50% flowering harvest is taken during clear days and cut at 4–5 cm from ground.

The harvest is immediately collected and if distillation is to be delayed the harvest is kept in shade.

The following rotations can generally be adopted for higher yields

1. Maize–Lahi–Mentha.
2. Maize–Potato-Mentha.
3. Pegionpea–Mentha (medium age variety).
4. Paddy (early)– Brassica–Mint (Japanese).
5. Paddy–Potato–Mint.

Table 10.14: Cost and benefit analysis of mint cultivation (after CIMAP. 2001).

Cost	
Land preparation	Rs 2000/ha
Uprooting of plants	Rs 2000/ha
Fertilizers	Rs 1800/ha
Sowing/planting	Rs 1000/ha
Irrigation	Rs 2000/ha
Protection	Rs 1000/ha
Weeding	Rs 1800/ha
Plant cutting	Rs 2000/-
Distillation	Rs 1800/-
Others. miscellaneous expenses	Rs 1500/-
Total income	Rs 17,400

	Oil/ha	Price/kg	Total income Rs	Net gain Rs
Japanese mint	15 kg	Rs 300/-	45,000	27,600
Spear mint	25 kg	Rs 275/-	34,375	16,975
Peppermint	100 kg	Rs 400/-	40,000	22,600
Bergamot	150 kg	Rs 250/-	37,000	20,100

References

Bradu Bl. 1968. Some observations of growing *Mentha arvensis* in Jammu. *CSIR Symp.*

Chaudhury S, Singh S and Handa KL. 1957. Chemical composition of *Mentha piperita* oil from plants raised in Jammu and Kashmir. *Indian J. Pharm.* **19** : 74–75.

Chopra IC, Handa KL and Kapoor LD. 1947. Preliminary notes on essential oil bearing plants raised in Jammu and Kashmir. *Indian J. Agric Sc*. **14**.

CIMAP. 2001. Pudine ka Tel ke utpadan Hetu Unnat Pryodhki. Hindi' pp 1–9.

Dulian SPS, Garg SN and Gulati BC. 1975. Effect of period of harvest on the yield and quality of oil of *Mentha piperita* L. *Proc. Nat. Symp*. p. 4.

Dutta PK, Bradu, BL Choudhry SB and Rao PR. 1964. Effect of soil types on yield of *Mentha arvensis* L. *Indian J. Agron*. **9(4)**.

Ellis et al 1941. A study of some factors affecting the yield and market value of Peppermint oil. *Perdue Univ. Agric. Expt. Stn*. Lafyette. *Indiana. Bull*. 461.

Green RG. 1963. Mint farling. *Agric. Inf. Bull* No **212**. Purdue Univ. Agric. Expt. Station.

Gupta US. Effect of some growth regulating substances on growth and essential oil content of *Mentha arvensis* L. var. *piperiscens* Holmes. *Proc. Intern. symp. Plant Growth substances*. pp. 453–458.

Gupta US. Studies in the physiology of mint s1. Effect of some Sodium salts on the sprouting and early growth of sucker of *Mentha arvensis* L. var. *Pipurescens* (Japanese mint) *Proc. Seminar of Seasalts and Plants*. pp. 121–127.

Janaki Ammal EK and Sobti SN. 1962. The origin of the Jammu Mint. *Curr. Sci*. **31**: 38–78

Kapoor LD, Handa KL, Chopra IC and Abrol BK. 1955. Cultivation of *Mentha arvensis* in Jammu and Kashmir. *J. Sci. and Industrial Res*. **14(8)** A 374–378.

Malik MN and Khan FW. 1964. Essential oil and Menthol content of Mentha species. *Pakistan J. For* **14** : 81–84.

Malik MN and Khan FW. 1964. Essential oil and Menthol content in Mentha spp. *Pakistan Journ. Forestry* **14** : 81–84.

Nigam MC, Sen T, Siddiqui MS and Datta SC. 1975. Production of bigger crystals of Menthol. *Proc. Nat. symp*. p. 43.

Ozola SM and Eizenberger VT. 1966. Effect of γ rays on productivity of Peppermint. Lalv. PSR Zinat Akad. Vestis 1966. 12 : 120–122. (Russiay).

Singh, Ajit. 1969. Commercial cultivation of *Mentha arvensis* in *Tarai* area. *CSIR Symp*.

Singh Ajit and Balyan SS. 1975. Cultivation of *Mentha piperita* in Jammu. *CSIR Symp*. pp. 5–6.

Singh Anup Sandhu, Khan MA, and Mahmood SM. 1969. Cultivation of *Mentha piperata*. in Jammu and Kashmir. *CSIR. Symp*.

Singh AK, Tomar VKS and Kumar S. 1997. Pudina (Menthol Mint) ke-tél ka-utpadan. *Udhyamita* Sept. **6** : 9–13.

Singh, Kailash 2001. Pudina ke tel-ka-utpadan hetu unnat Prayodhki. *Udhyamita*.

Singh KK, Upadhayay DN, Gupta NK and Ganguly D. 1975. Cultivation of Black peppermint (*M. piperita*) in Assam. *Problems and Possibilities. CSIR Symp*. p. 5.

Singh, Lal. 1969. Commercial cultivation of *Mentha arvensis* in *Tarai* area. *CSIR Symp*.

Sobti. SN. 1966. Variation in progeny of Jammu Mint. *Sym. Recent Adv*. in *Development, Production and Utilization of medicinal and Aromatic plants* CSIR in India pp. 23. April 18–20.

42. *Myristica fragrans* Houtt.

Family: Myristicaceae.

Local names: Jaiphal (Hindi), Nutmeg (English).

Description: A handsome evergreen tree 9–18 m high. Leaves dark. Normally unisexual having either male or female flowers that are pale yellow and aromatic orange or golden yellow fruit, when ripe rearable an apricot or peach but variable in form and size. When ripe fully, husk splits along the grove show is shiny brown seed covered with bright red branching aril. Inside the seed is aromatic Kernel which is seed as dried Nutmeg of commerce.

Distribution: Native of Mollucas but cultivated in India not to a very great extent. It has succeeded best in Barliyar in Conoor valley.

10

Uses and extent of utilization and trade: Nutmeg fruit is an aromatic stimulant, carminative and narcotic in large doses. Widespread use of both Nutmag and Mace is made in European cookery. In rural medicine it is given to children during winters in cold affections.

It gives an essential oil and a fixed oil, from the fixed oil nutmeg butter is obtained by expression, the powdered nut being steamed and pressed, while hot. It occurs in blocks of yellow colour. The essential oil is obtained by distillation, and is white in colour with the odour of nutmeg. Largely it is used in perfumery.

Agrotechniques for the Cultivation

Deep-rich loam soils with good drainage are suitable for the cultivation of this plant. It does not thrive on sandy soils and stagnant water above the roots soon kills it. The climate best suited for it should be hot and moist with an annual rainfall of 150–180 cm.

It can be raised from fresh seeds sown in nursery beds, sheltered from the sun and wind. In dry weather they require watering daily. When 60–90 cm high the saplings are transplanted at a distance of 8–10 m apart in pits of 45 cm³, dug in advance. The young trees must be shaded and well watered with constant weeding. Should dry weather commence, the ground around the stem is mulched with straw, leaves or stable litter. All parasitic or epiphytic plants on stem or branches should be removed.

When the trees flower the sexes must be determined and about one male left to every 8–10 females. The male should be on the windward side, so that pollen may be carried to the female plants.

The tree starts bearing after about seventh year of plantation and the produce increases till about 15th year of the age.

The fruits are picked up every morning, after it has fallen from the tree. 1500–2000 nuts must be obtained from each tree in full bearing. The Mace is stripped off and the nuts dried in trays raised above the smouldering fires. When dry, the shells are broken with mallets and the nuts rubbed with lime to prevent attack from the worms and packed in light cases. The Mace after being stripped off is spread on mats or trays to dry and, when it turns yellow brown it becomes the Mace of commerce.

43. *Nardostachys jatamansi* DC.
(-Valeriana jatamansi DC.)

Family: Valerianaceae.

Local names: Jatámansi, Bhutkéshi, Bálchard (Hindi), Jatámánsi (Bombay, Gujarat), Bhatijatt (Kashmir), Balchar (Dehradun).

The trade name of this plant, Jatamansi, is derived from the Hindi name and refers to the beared appearance of the rhizome.

Description: A perennial herb, about 60 cm tall; rhizomes woody, covered with the fibres. The lower basal leaves are about 20 cm long, narrowed into the petiole; upper much smaller, almost ovate in shape. Flowers small, several in small bunches, covered with minute hairs.

Distribution: The herb is mainly found in the alpine reaches of the Himalaya, on dry stony aspects ranging from 3300 to 4200 m. east of Garhwal; in Sikkim it is found above 1700 m. Found in Pangachuli, Pindar Nilavalley, Damdar valley, Bhillangna valley and other places.

Uses, extent of utilization and trade: The drug constitutes a portion of the rhizome which is as thick as the little finger, surrounded by a bundle of reddish flowers brown in colour, the remains of the radical

10

leaves. Rhizomes used as *dhoop*; also as a tonic.

The rhizome is cardiac and respiratory stimulant, nervine tonic, aromatic, carminative, stomachic, hypertensive, laxative, antispasmodic, diuretic, emmenagogue and deobstruent. It is used in various disporders of the digestive system, respiratory organs and in hysteria, in which it acts as a palliative, not as a cure. In mild cases of mental derangement, nervous and convulsive disorders and certain disturbances due to menopause, it is one of the efficacious palliatives. In milder doses it is used for cholera, flatulence, palpitation of heart, jaundice, disorders of the digestive system, bronchitis, etc. Also used as a vermifuge, dysmenorrhoea, and polysuria. Oil from rhizome promotes growth of hairs and impart black colour.

The roots paste in Kumaon hills is used as a topical application in haemorrhoids. Dried roots of *V. jatamansi* from the drug and are laxative, stimulant, sedative, nervine tonic, cardiac and used to cure cholera, epileptic fits, eye diseases, hysteria and blood diseases. All the parts are collected in April-May. Rootstock is separated and dried in partial shade. In local market thus products are sold, one as *Tagar ganth* (nodular tagar) and the other *Tagar-shass* (leafy portion). Annual requirement of 20 qtls is estimated. Drug is collected during August-Sept.

Another species *V. hardwickii* Wall. (differs from Nardostachys jatamansi *V. jatamansi*) in having ovate radical leaves that soon disappear. The flowers are in axillary compound corymbs forming a long terminal panicle. The properties are same as of *V. jatamansi*.

The drug is also used for improving urination, menstruation and digestion. Locally, it is used as a tonic for hairs by the ladies and is an ingredient of many hair washes. An oil given out from distillation of rhizomes is aromatic and used in making medicinal oils.

Requirement of this plant in trade was estimated to be more than 20 qtls per annum, however, small quantities used to be collected by locals and sold in the market by traders who used to visit plain area during the winter season.

Attempts have now been made to grow this plant on a commercial scale (in Tons valley) by the farmers in their fields along with other alpine plants like Atis (*Aconitum hetrophyllum*), Pharan (*Allium carollinianum*), Jambu (*A. wallichii*), Gandrayan (*Angelica glauca*), Kalajira (*Carum carvi*), Hattajari (*Dactylorhiza*), Katuki (*Picrorhiza kurroa*), Chora (*Pleurospermum angelicoides*), Bankakri (*Podophyllum hexandrum*), Dolu (*Rheum australe*) and Kuth (*Saussurea costus*).

The plant has largely been used also as an aromatic adjunct in the preparation of medicinal oils and is considered to increase the growth of black hairs. As early as in 1790, the perfume was first identified by Sir Charkes W. Jones (Asiatic Research. 1790, 2 : 405–417).

It is multiplied both through rhizome cuttings and seeds. Moist and partially sunny area with sandy loam and acidic soils, rich in organic carbon and nitrogen are suited for good growth at, above 2200 m altitude. About 240 g of seeds or 44,000 seedlings are required per acre if sown at 30 × 30 cm distance. Seed sowing is done during winters, inside green house during March. April, soil depth being 0.5 cm. Transplantation is done during March-April. About 60 qtls of FYM/acre is required. It is harvested after 3 years when grown vegetatively and after 5 years for seeded plants, in Sept-Oct. Production is 45 kg/acre after 3 years. Expenditure including infra-

10

structural development for 3 years recorded at Rs 50,700 and the net profit at Rs 63,542/—(as per HARPRC. Bull No 5).

44. *Ocimum* species

Family: Lamiaceae (Labiatae).

Local names: Surabhi, Sursa, Báhupatri, Bahumanjari, Ramya, Bhooteshta, Bhoot-priya, Haripriya, Krishnapriya, Surejya, Bhutaghni, Papagni, Pootpatri; Kayastha, Deodundbhi, Gramyasulbha (Hindi); Shaparam (Persian), Sgaha–am–farm (Arabic), Rehan (Urdu-Quaran).

- *Ocimum sanctum*. Holybasil, (Eng.) Tulsi (Hindi).
- *Ocimum americanum*. Kali-tulsi (Hindi), Kukka-Tulsi (Telgu).
- *Ocimum gratissimum*. Vridhatulsi (Sans); Ramtulsi (Telgu).
- *Ocimum basilicum*. Sweet-basil, Munjariki (Sans.), Indian basil Babil tulsi (Hindi), Karpur tulsi (Tamil), Sweet Basil (Eng.)
- *O. basicilicum* var. *minimum* Bush-basil (Eng.)
- *O. basilicum* var. *odoratum*. Scented basil (Eng.)
- *O. basilicum* var. *purpurascens*. Purple basil (Eng.)
- *O. Kilmandcharicum*. Kapur tulsi (H), Camphor basil (Eng.).

The spiritual name 'Ocimum' comes from the Greek. meaning 'to be fragrant'. The common name, Basil, comes from Greek *basileus*, a king, because its smell is so excellent that it is for a king's house. According to others, it is derived from basilik, a creature that could kill with a look, based on a belief that basil could "drive a person insane". It came to be associated with hate, poverty and misfortune and was known as a *devil's plant* in the west.

In Northern India, the two most common varieties are *Ram Tulsi* and *Shyam Tulsi*. The first one is the garden basil with bright green leaves (pale green), and vivid purple bloom, found in most Indian houses. In Vribda Kunj, there are attempts to have mega Tulsi plants and some of them about 3 m tall with a trunk 5–7 cm thick in diam.

45. *Ocimum kilamandcharicum*
Güerke, Kapur tulsi (H) Camphor Basil (Eng.)

Description: An aromatic, much branched, perennial herb. Leaves ovate to broadly elliptic, dentate, prominently veined, hairy, upto 6 cm long. Flowers pale-purple or pinkish-white in racemes, at the end of branches. The leaves when bruised emit a strong camphor odour.

Distribution: Native of East Africa, but introduced in India in 1942 and now cultivated at a number of a places, like Uttar Pradesh, Maharashtra, Bengal and Karnataka states. It was planted at Dehradun. A three years old plantation gave herb yield at 8800 kg/ha in July, 2200 kg/ha in October. The oil content was 0.19%; the plantation dried up in the 4th year.

Uses and extent of utilization and trade: The volatile oil obtained from the leaves is a valuable source of camphor, which is used in religious ceremonies, in the manufacture of celluloid, explosives and in various pharmaceutical preparations such as disinfectants, toothpaste, powder and ointments, etc. It is also used as plasticizer. The leaves do not lose camphor, even if stored for a year, nor the content gets affected by drying or rotting of leaves. The residual oil left after separation of camphor is likely to be suitable as a cheap substitute for perfumes and also as disinfectant; also

used in the application of liquid gold on glass and ceramic wares.

Though a considerable quantity of camphor is obtained synthetically, no extensive plantations of *Cinnamomum camphora* Nees and Ebrem, have been undertaken and this plant has considerable scope of cultivation as a source for natural camphor. This could be grown as a cover crop in forest plantations, like teak or an intercrop in Orchards so as to reduce its cost.

Annual requirement of more than 400,000 qtls is partly met by indigenous resources.

Agrotechniques for the Cultivation

The plant can grow on a variety of soils and climates. Clayey soils with impeded drainage appear to be the most suitable. Areas with more than 125 cm of rainfall per year, with even distribution, can be taken up, upto an altitude of 1000 m.

It is propagated by direct sowing, through planting seedlings raised in nursery, and transplanting in the field. The seed rate recommended is about 70 g per ha. The best time of transplanting is at the break of monsoon. Self-grown plants can also be transplanted. The seeds are light and give high percentage of germination. The seeds are sown in March, on flat-beds with pulverised soil. The seeds are mixed with 20 parts of sand for uniform sowing and watered to keep the soil surface moist. It takes 5–10 days for germination, which is completed in 15–20 days and is ready for transplanting during the monsoon season when 5–7 weeks old, at a distance of 60 cm in rows at about 60 cm apart.

The crop is ready for harvesting within a few months of seed sowing (4–5 months) and two to three harvests can be taken for a number of years. Irrigation and weeding must be done as and when necessary.

The plants are cut by sickle at about 10–15 cm from the ground, bundled and stacked for 'heating' and drying up for 4–10 days, depending upon the weather conditions. The leaves are separated from the stem by thrashing and thoroughly dried in sun when plants become bushy they give better leaf yield.

The residuary oil left after separation of camphor is a cheap substitute for soap perfumes and also as disinfectant.

Plants grown by RRL at Jammu, yielded 43% of oil but no camphor could be obtained on chilling. Camphor content in the oil was 70.5% and camphor oil content from the plant is recorded in Table 10.15 (after Chopra, Sobti and Handa, ICAR Res. Sr. No.13).

Table 10.15: Camphor oil content in Tulsi plant

	Entire plant %	Leaves %	Flower %	Stem %
Fresh herb	–	34.10	31.70	34.1
Air-dried herb	–	26.00	35.40	38.6
Moisture	0.70	76.40	64.40	65.0
Camphor and Camphor oil (Fresh herb)	0.43	0.54	0.53	0.054
Camphor and Camphor oil (in air-dried herb)	1.38	2.27	1.30	0.056
Camphor content of oil	70.5	–	–	–
MP of camphor	–	174.75°	–	–

Cost benefit

Cost of cultivation has been calculated at Rs 6400/ha (land preparation Rs 1000/, seeds Rs 200/, fertilizers Rs 1200/, plant nursery Rs 500/, planting Rs 500/, weeding and hoeing Rs 1000/, cutting Rs 500/, distillation Rs 1500/-). Total Rs 6400/-.

Oil produced 85 kg. Cost of oil Rs 200/kg, i.e. Rs 17,000/ha. Net profit (17,000–6400)

10

is Rs 10,600/ha (after Singh Kailash, CIMPO).

46. *Ocimum basilicum* L.

Indian Basil, Sweat Basil. French Basil. Occurs naturally in NW India. Useful in fever, cough, worms, stomach complaints and gout.

Commercial cultivation of *Ocimum basicicum* was started in the *Tarai* of Udhamsingh Nagar in Uttaranchal state, but low yield per ha, susceptibility to diseases and particularly uneconomic production level, had eliminated chances of its future large-scale cultivation. It has been again revived in Barieley, Moradabad, Sitapur districts of Uttar Pradesh. Preliminary trials at Dehradun gave an yield of 1377 kg/ha in July with 0.14% oil from degraded area.

It can be cultivated on well-drained sandy loam soils. The land is well-prepared. The soil is ploughed to a depth of 15–20 cm and 15 t/ha of FYM is mixed in the soil. Plots of 1 m × 1 m size are prepared, 20 kg of P_2O_5 and K_2O each are mixed in the soil. Direct seeding is not done and the seedlings are prepared for planting. 750–1000 g of seeds/ha are required.

Seeds are mixed in ratio of 1 : 10 with soil, and sown in lines 8–10 cm apart, but not very deep. The seeds require 15–20 days for germination. Nitrogen at 20 kg/ha is useful; 5–6 weeks old seedlings are planted. After planting, irrigation is done. Transplanting is done when it is cloudy or on light rainy day. For planting seedlings line to line distance of 60 cm and plant to plant 30 cm is maintained. While no irrigation is required during the rainy season, till September and after light irrigation is given. Weeding and hoeing is done one month after planting and the second weeding is done after 3–4 weeks of the first weeding. 15 tonnes/ha of FYM, plus

75–80 kg N, 40 kg P_2O_5 and K_2O is given. One-third of nitrogen, P_2O and K_2O is applied before planting and balance of nitrogen is given in two split doses.

Cutting of the herb is done after full bloom 10–12 weeks after planting. Plants should be left in the field for 4–5 hours which helps in distillation. Basil oil is obtained with distillation both through water or vapour; however, the latter is useful. Average production is about 20–25 t/ha and the oil recovery is 80–100 kg/ha.

47. *Ocimum sanctum* L.

This is a much-branched herb with stem upto 75 cm, all hairy. Leaves about 5 cm long, margins entire or toothed, dotted with minute glands. Flowers purplish or reddish, in small clusters on slender spikes. Seeds yellowish or reddish. The plant is mostly cultivated in the houses and near temples.

The leaves and seeds are medicinal. The oil obtained from the leaves, has the property of destroying bacteria and insects. Juice or infusion of leaves is useful in bronchitis, catarrh, digestive complaints and is locally applied on ringworm and other skin diseases like earache (dropped in ears). Decoction of leaves is used as a home remedy to cure common colds. Seeds are useful in complaints of urinary system, while root decoction is given in malarial fever to bring about sweating. Though, used extensively, seldom exploited on a commercial scale and is more an article of worship. The spiritual name given to this plant by *Mother* (Sri Aurobindo Ashram) is *Devotion*. Mention of this plant is made in Devi Bhagwat 9.25 41–43; Padampurana, Sant Kumar Samhita, Kashkand, Skand Puran, Garur Puran, where it is regarded as Divinity, and incarnation of Goddess Lakshmi or Vrinda. Since Lord Vishnu likes garland of Tulsi leaves it has great demand at Badrinath temple in Uttaranchal. Offering of Tulsi flowers are made to Radhe-Govinda. There is mythology about its origin and worship, and use in rituals.

Since ancient times *Tulsi* plant is regarded as very efficacious in giving strength to the body and has been used in traditional home remedies, like acidity, flatulence, soar-throat, hoarseness, eye trouble, sinus congestion and pneumonia, blood pressure, heart ailments, epilepsy, hysteria, insomnia, paralysis, jaundice, kidney ailments, leucoderma, tuberculosis and sluggish fever. USDA has listed Basil's actions as "antispasmodic, aphrodisiac, appetizer, carminative, demulcent, diaphoretic, diuretic, emmenagogue, stimulant and stomachic. Modern research has proved that regular intake of this 'plant-leaves' helps in relieving heart ailments. It has antibacterial additivity against *Staphylococcus aureus* and *Mycoplasma tuberculosis in vitro* as well as other pathogens including fungi.

During the 19th century many Englishmen used to wear bead necklaces made of wood of 'tulsi' plant like the one worn by holymen because of its curative properties. In Europe, physicians were unable to agree as to Basil's medicinal value, while Pliny recommended Basil tea as remedy for nerves and headache. Greeks used Basil in aromatic baths to strengthen nerves and ancient Egyptians used Basil for snake bites, scorpion stings, eye troubles and rheumatism. In aroma therapy, the essential oil is used for clarifying mind from mental fatigue, while in cosmetics the seeds ground into a paste and mixed with honey make a very beneficial face pack, since the content of mercury in seeds lightens unwanted blemishes. Paste of 'tulsi' leaves with lime-juice removes black spots on the face.

The volatile oil consists of Linalool and Methyl chavicol and small quantities of Methyl cinnalate, cincole and other terpenes and its medical effects are mostly due to thymol, eugenol and camphor. It has both pro and anticaner properties, since *estragol* in the herb has been found to produce tumours in rats.

48. *Papaver somniferum* L.

Family: Papaveraceae

Local names: Posta, Dodá, Kaknár (Hindi), Gule-é-lala (Urdu); Opium Poppy (Eng), Ahiphena (Sans).

Description: A stout herb, 60 cm to 120 cm. Leaves glaucous with oblong, cimplexical lobed, toothed and serrated leaves. Flowers usually white, sepals glabrous. capsule large, 2 cm in diam. Seeds white (or black).

Distribution: A well-known drug plant of West Asia, from where it was introduced into Greece. Arabs brought it to India and China. Now cultivated in Turkey, Afghanistan, Pakistan, Iran and Europe. Cultivation of this plant in some places is banned and licences are issued for its cultivation such as in India (mostly MP and East UP). The wild variety which is a poor yielder has been improved by cultivation and selection. The three varieties like *nigrum* has purple, red flowers with roundish oblong capsules. Seeds are dull greenish-black, there are pores under the stigma. Var. *album* has white flowers. Capsule roundish-ovate. Seeds white. No pores under the stigma. Var. *abnormal* is an unimportant variety. Kaicker et al (1975) studied the varietal yield of opium under different fertilizer conditions. In Europe var. *turcicum* is grown in some countries since it is characterised by winter hardiness, abundant latex during capsule ripening

10

and high morphine content, say about 16%. New selections S-188, S-230 and P-360, are available from Bulgaria where opium is extracted from ripe capsules cut at harvest stage. The reported yield is about 300 kg of capsule shells, 100 kg of seeds per ha and 3.1 kg of morphine per ha. Selections S-188 and S-230, have white blossoms, while P-300 has blue flowers. Morphine content in S-188 and P 300 is between 0.48 and 0.58% in dry capsules which is equivalent to some best varieties like *Novina* from USSR (0.61), *Katvanskii* (0.62) and *morphine poppy* (0.65) from Hungary. Some of the other cultivated varieties in vogue are as below:

1. *Telia*: Most popular and productive. Capsule oblong-ovate, pale, olive-green, without any powdery coat. Grown for early crop.

2. *Sufaid-danthi*: Late cropping and less productive than the above variety.

3. *Kutila*: Or Katila can withstand hail and high wind. Leaves much divided and blight-resistant. Capsules oblong-ovate and glaucous. Latex red when fresh. Occasionally, the stalk is bristly. Thrives on sandy loam soils.

4. *Kali-danthi*: Flower's stalk of bluish black colour after the petals are shed. Gives better yield as an early crop.

5. *Kali danthi-baunia*. A dwarf form. Produces more opium than the other standard form.

6. *Monoria*: Requires heavy clay-loam soil and good manure for the best growth. Capsule large, roundish, glaucous. Produces large amount of opium.

7. *Dheri-danthi*: Somewhat light resistant, produces less opium but contains high morphine. Comparatively a new race.

8. *Saféd-danthi-monoria*: Robust hybrid with large roundish capsules.

9. *Galania*: About 10 cm tall, white petals with pink border. Opium yield more than that of Danthia.

10. *Ghotia*: Flowers white with pink border.

11. *Chaglia*: Petals red or pink with white dots at the bottom. Average yield is higher than *Ghotia*, *Galania* and *Ramzatak*.

The following varieties are of lesser importance than the above but also cultivated.

12. *Subza-kala-dhanti*: Sensitive to heat and excessive moisture. Early crop on light soils.

13. *Choura-kutla*: Grows well on clay loam soils, in moist localities.

14. *Variegated poppy*: New race of saféd-Dheri. Little affected by blight.

15. *Monoria teyleash*: A hybrid.

16. *Sandhapa-dhodhua*: A tall variety but little of opium.

17. *Sahbania*: It yields little opium.

18. *Bhatphoria* or *dhaturia*: Poor opium yielder.

Agrotechniques for the Cultivation

It grows best on clay and loamy soils but weather conditions are important in determining the ultimate yield, since excessive rains, hailstorms and wind adversely affect the opium yield. The crop is sown in the first half of October and is of 220–230 days duration. The rossette phase takes 90–100 days. Water requirement is estimated at approximately 350–400 lit per sqm of the cultivated area. FYM at the rate of 23–24 t/ha at the time of field preparation is beneficial. Application of superphosphate at 100–120 kg/ha is given as a basal application with additional 200 kg/ha is given as ammonium sulphate, when applied at rossette stage, immediately after thinning as top dressing gives optimum yield. Seed rate recommended is about 6 kg per/acre and the seeds are sown by drill in rows, 30 cm apart, about 15 to 20 cm deep in the soil. This seeds germination normally takes place after 12 days of sowing. Final spacing is 60 × 15 cm. Capsules are formed in April and harden. Lancing is done in morning and exuded juice collected in the evening. In plains, it is done during afternoon till evening. Maximum opium yield is obtained when 75% of the central capsules are cut horizontally, somewhat above their bulging section by a skilled man. Thinning in Feb-March gives good opimum yield and delay of a 15 days reduces 10% opium yield, 15% seed and 25% poppy capsules. Care is taken that the epidermis is not torn. In the plains it is done usefully during the afternoon, till the evening and the product collected the following morning. Crude opium is collected in boxes attached to the belt of the collector, needed together in metal trays later and made into cakes of 200–300 g which are dried in air. FYM, if applied at a rate of 10 t/ha reduce inorganic requirement to two-third. Generally an yield of 4 kg of opium per acre is recorded with a morphine content of 8.7–10.2%.

Aphis and Centrorhynchus nucula alba is recorded to deform the capsule, also the blight attack is reported. Spraying of Bordeaux mixture and dry treatment with mercuric preparations are effective.

Normally, an yield of about 20 kg of opium per ha is recorded having a Morphine content of 8.7–10.2%.

Kaicker et al (1975) tried six cultivars, viz. *Ranjratak, Dhuturia, Kantia white, Kantia pink* and a *F2 line* of *Bhakua X Haryana* with three different spacings (45 ×22.5, 30 × 30 and 37.5 × 375) and three fertilizer doses, NPK (124 : 80 : 80) (100 : 50 : 50) (45 : 30 : 30). The results at Mandsaur (MP) showed maximum yield of opium followed in Cv Dhaturia with NPK 100 : 50 : 50 kg/ha, followed by Hybrid Bhakna and NPK (124 : 80 : 80) as the best dose for over-all averages of opium. 45 × 22 cm spacing between the plants in the same cultivar gave max. yield, i.e. 49.55 kg/ha. Av. per plant yield was maximum in Haryana, followed by Dhauria. Maximum percentage of mor-phine (11.28%) was in Cv. Dhaturia followed by B × H and with application of NPK (124 ×80 × 80). For Mandsaur in Rajasthan recommended is the variety with a *B × H* (F$_2$) fertilizer dose of: 120 × 80 × 80. Spacing is 45 × 22.5 cm; Cv. Dhaturia. Fert. 124 : 80 : 80, spacing 45 × 22.5 cm. At IARI maximum yield recorded was in Cv Dhaturia (with NP 100 : 50 kg/ha) at spacing of 30 cm with 20 cm × 30 m row to row.

Reference

Kaicker US, Singh BS, Turkhede BB. Singh SP, Chaudhury AR. 1975. Varietal yield of opi-um at different fertilizer levels.

49. *Pelargonium* L'Herit (Geranium).

Family: Geraniaceae.

Local name: Geranium (English).

Description: This is a large genus of perennial/annual, highly ornamental plants (Colour Plate 10.11). The Geraniums differ from Pelargonium in having regular flowers, while the flowers in Pelargoniums are highly irregular like that of Pansy and have a spur-like appendage. Many species of Pelar-gonium are widely cultivated throughout the world, though many species are native of South Africa, Mediterranean countries and Australia. The highly aromatic species are *P. graveolens* L' Herit, *P. odoratissimum* (L.) Ait. , *P. capitatum* (L.) Ait. and *P. radula* (Cav.) L; Herit, which are cultivated for leaf oil in Algeria, Morocco, South France, Spain,

Re-Union, Italy, Corsica, Russia and Africa. All these are perennial herbs, 1 to 1.2 m high. They have great opportunity and popularity as pot-plants because of their ornamental scented foliage. The great ease with which the species crossed among themselves and quick method of preparation by stem-cutting makes it a favourable plant for plant breeders and floriculturists, and gave birth to a number of horticultural varieties and cultivars, many of which have runwild in the neighbouring countryside. Earnest Sen and J. Prioris, introduced geranium to Yercaud on Shevroy hills, in early part of the last century (Gulati 1963). The various oil-producing forms are considered by Guenther (1949) to have been descendents from hybrids created in Europe and reexported, particularly from South France, to colonies in Africa for a large-scale cultivation.

In India, *P. graveolens* and *capitatum* are cultivated in S. India. *P. graveolens* L' Herit ex Ait, is the parent plant for all commercial varieties, now grown on plantation scale and is recognised by its characteristic 5-lobed leaves and rose to pink flowers. It has very pleasing strong rose-like odour, for which it is called Rose Geranium. *P. capitatum* and *P. odoratissimum* are the other species having rose-like scented leaves and whitish flowers. *P. citriodorum* and *P. crismum*, have lemon-scented leaves, while the foliage in *P. tomentosum* smell like of peppermint. The leaves of *P. quercifolium* and *P. radens* are credited to possess pungent scent. Some interesting horticultural species are the hybrids in which they resemble aroma of apricot (P. × 'M'), *pine* (*P. denticulatum*, nutmeg (*P. fragrans*) and that of strawberry (*P. crispum* var. *minimum*). Surprisingly, nurserymen, planters and horticulturists list the plant yielding perfumery grade oil as *Pelagonium roseum,* which has added confusion to the identity of *P. roseum*, although the hybrid nature of the plant has been universally recognised. Rose geranium has been identified as a hybrid between *P. graveolensx* and *P. radens* and closely resembles *P. graveolens* except in having leaves with narrower, more deeply cut segments, covered with stiff hairs—the character it is most likely to have been inherited from *P. radens*. The vast differences in the soil and climatic conditions in different countries, where it is planted, show certain amount of variation in chemotaxonomical build-up of plant stock which accounts for the variation in odour value.

Uses and extent of utilization and trade: The oil is used in perfumery which is obtained through distillation. About half a century ago, Geranium was introduced from Africa (Cape provine). The world production of oil is more than 300 tonnes, while in India itself its consumption exceeds 65 t/year. In North India, it was introduced by CIMPO for increasing oil production. The oil is colourless to yellowish green with a strong rose-like odour, occasionally slightly harsh and minty. In general Spanish oils are found to contain 70% total alcohols of which 65% is geraniol and those grown in Israel 45% citronellol oils from plant grown in India contain between 60–80% total alcohol which is accepted by the industry.

Rose Geranium grows upto a height of 1 m under favourable conditions of growth and is a branched perennial, bushy herb with large cordate leaves, covered with hairs. The leaves have 5–7 lobes, each divided further into lobes. The plant bears small flowers, rose to pink-rose with purple veins. The characteristic spur is a long-extended appendage of the upper petal which is adnate to flower stalk and contains slightly swollen nectary at its base.

Agrotechniques for the Cultivation

Soil and climate: Rose geranium grows luxuriantly in the subtropical climate in the North Indian plains, well on deep, permeable light soils-sandy loam, gravelly to clay loam in texture with a pH between 7–8.5, while at Shevroy hills it is grown as a perennial crop on soil having pH 5–6. Soils having pH more than 8.5 in Indian plain adversely affect the growth and oil production. Borbon and Algerian variety are suitable for North India; the former is suitable for industrial production. In South Indian hills it prefers moderate rainfall receiving areas in sheltered, warm locations with well-defined seasons between 1000–2000 m. Well-drained red soils with poor organic matter could also be managed to support a good crop (Gupta R, 1972). It can also be grown in the mild-climate like that of Bangalore but is sensitive to frost and prolonged winter rains, similarly excessive summer heat and low winter temperatures are not suitable. In France, it is cultivated as an annual crop, because of severe winters. Though it withstands drought to some extent, yet the crop yield is low.

Preparation and planting of cuttings

Normally new growth of 30–50 days of plant having 3–4 buds provided rooting in nursery. Nov.–Dec. is the safest period for planting the cuttings which is done in 50 cm lines (CAMAP) plant to plant distance is 40 cm. Rooted plant could also be obtained from CIMPO, Bangalore and Govt. Cinchona Deptt, Udham-anagalm (T. Nadu). The crop can also be raised from stem-cuttings, planted in February, so that those are ready for planting in May-June, with the first monsoon showers. Normal size of cuttings is 8–15 cm with 2–4 nodes. The foliage dries out and falls off slowly, if not removed before planting. The cut-end of the stem is dipped in solutions of root-promoting hormones, like Seradix and embedded 6–8 deep in rows, at a spacing of 10 cm × 10 cm (R. Gupta 1972). Nursery is provided with partial shade, in the initial stage, since low night temperatures adversely affect rooting. It has been observed that during the rainy season the plant gets infected normally and so availability of the planting material for next crop is a problem. About 80% of cuttings sprout in three weeks period and the plants put up new leaves and ready for planting in the field in May-June. Root cuttings have also been tried in bamboo baskets or punctured polythene bags, facilitating their transport to the plantation site. Since the crop remains in the field for about three years, the land is ploughed deeply and soil is well worked. FYM at 10 t/ha (25–50 t/ha as per Singh 2001) is applied to the soil, while preparing the land, 60 kg Superphasphate and 40 kg K_2O is also added. 150–180 kg nitrogen/ha is given in split doses. In S. India 150 : 80 : 60 kg of NPK is usually applied. The crop in general is given one weeding and one to two hoeings, before a harvest of leaves is taken, it depends upon the growth of plant and the weeds, including soil moisture conditions. Pruning of branches after onset of first monsoon rain also hastens bushy growth.

8–10 irrigations are needed with normally 3 weedings and hoeings are required. Heavily moist soils or too dry conditions lower the yield. At Nilgiris, two types, viz. *Algerian type* with slender plants and dark pink flowers and *Bourbon type* with light pink flowers have been separated at Govt. Cinchona Plantation, Nilgiris: Algerian type being less suitable for humid situations. The average biomass yield recorded is 32 t/ha with a price of Rs 2/kg of biomass.

10

Crop harvesting and distillation

After 100–120 days, when the leaves get *kukura,* (lemon scented turns rose scented), is the right time for harvest. The leaves contain 0.12 to 0.2% volatile oil (freshy in comparison to those harvested after monsoon or in winter season). Oil yield in the second year and onwards varies between 8–18 kg per ha, while in other countries it is reported to be 30–35 kg/ha. After sunrise, 15–20 cm upper portion of the plant is cut; small branches are left which can be cut after 40–45 days. Antifungal treatment is given after first cutting. Yield of leaves (green) after first cutting is 250–300 qtls per ha, giving oil as above. Second crop can be obtained with good management.

The crop can be planted, as catch crop, in young orchards which could bring back the expenses on laying out of an orchard, before it comes into bearing. However, it can also be fitted into the crop-rotation such as Paddy-Potato-Geranium, Maize-Potato-Gernium, Pigeonpea-Geranium, Geranium-Garlic, Geranium-Onion, Geranium-Capsicum and Geranium-Tagetes (wild).

The crop is distilled by hydro-distillation for which 10 gallons of water is filled in a still, in which 250–300 kg of material is packed for distillation. Small boilers can also be used for distillation, which is complete in about an hour and a clear, light yellow to brownish coloured oil is collected. The oil has a strong, pronounced rose-like smell, which improves upon storage.

Diseases

Termite attack is common which hollows out the stem. 21 lit/ha Chlorooyriphos, is used with water. Fusarium attacks the crop and take a high toll. Misty conditions and soil moisture aggravated pathogen attack. Spraying 5% Bordeaux mixture, controls the disease due to fungal attack, coupled with pruning of diseased part. With root attack the plant dried fully and should be pulled out along with the root. Application of Bavastin (0.1%) in 600 litres of water to soil through spray is also recommended in general.

Diseases of pelargonium and their control

1. *Stem rot: Rhizoctonia* affects the terminal stem cuttings planted in the nursery for rooting. Can cause 70–80% mortality in cuttings. Initial symptoms appear as yellowing and drying of cuttings, due to rottings of basal portion. Rottings extend upwards and cause death of cuttings.
   Control: Raise nursery in disease-free area. Drench the stem in solution with Mancozeb (0.3%) before planting.
2. *Root rot and wilt*: *Rhizoctonia solanii.* Visible symptom is drooping of leaves and finally wilting of the entire plant. When uprooted, such plants show severe root rotting.

Control: Use healthy and disease-free rooted cuttings.

References

Singh AK. 2001 *CIMPO Symposium (Mimeo)* pp 30–31.

Gupta, R. 1972. Grow Scented-leaved Geranium. *Indian Farming*. December issue. pp 3.

Singh. Puran. 1916. Constituents of Indian Geranium oils. *Indian For. Rec.* **5(7)**.

50. *Piper* species Piper nigrum L.

Family: Piperaceae.

Local Names: Golmirch, Káli or Saféd mirch, Dakhni-mirch (Hindi) Kaph-virodhi (Sans.), Kalluvalli (Tamil), Black or white Pepper (English).

Description: A robust woody climber, usually dioecious. Leaves large, broad-ovate or round leaves. Flowers borne in long

spikes. Fruits small, globose, 6–7 mm in diam.

Distribution: Wild in the forests of Kérala and can be cultivated in hot and damp places of South India. On the west coast it was cultivated from very early times, also in Java, Sumatra, Malaysian peninsula, while Malabar region has always been considered to produce the best pepper. Today, it is mostly, in India, cultivated extensively in Tamil Nadu, Maharashtra, Andhra Pradesh and Kerala states. In Bengal, its cultivation is to a limited extent, while in Assam it seems not favourable with the farmers.

Uses and extent of utilization and trade: Dried unripe fruits are stimulant, carminative and stomachic, if taken in larger quantity it gives a feeling of warmth and causes sweating. It is also diuretic, but can cause irritation in urinary tract. Antibiotic activities of the seeds have also been recorded. Some attribute the property of curing diabetes to seeds, if taken 4–5 daily.

It is traded in the market on a regular basics as a spice and a condiment.

Agrotechniques for the Cultivation

It is propagated either by suckers, which spring from the underground roots or from shoots of the stem. When shoots are used they are bent down to the ground to strike root before they are severed from the mother plant. Young plants are uprooted at he commencement of the rains, and planted at the base of the trees on which they grow—as a rule one plant only is placed along the side of each tree and at first it has to be tied carefully to its support.

Like betelvine it requires liberal manure as FYM. Cowdung and household refuge are also used at the end of rains. This is simply heaped around the base of the tree on which the vine climbs. To keep the moisture in the manure heap, pieces of leaf sheaths of the plantain tree are laid over the top and reviewed from time to time.

The plantation must be hoed and cleaned once a year, at the close of the monsoon period. In May, the manure heaps are levelled down and spread over the ground.

The vines begin to bear in 3–5 years after planting which continues for almost 20 years. It flowers in the month of May and the berries are plucked in December, when just beginning to ripe. If intended for cultivators use, the berries are boiled in water for a few minutes to stiffen the husk which is removed by rubbing over a bamboo basket. If intended for the market as black pepper they are simply dried in the sun after boiling and allow to retain the husk, which assumes a black colour. Highest output from a single vine is about 3 kg of dried pepper, while average yield is about 1 kg per vine.

a. *Piper longum* L.

Local names: Piplamul (Hindi), Pippali, Magadhi (Sans.), Piplu (Assam), Pipili (Tam. Tel.), Jatya, Pipul (Bengali), Shawppa (Jalpaiguri); Longpepper (English). The trade name is based, because of the long spikes which distinguish it from *P. nigrum*. The Sanskrit name Magadhi indicates, that the plant is indigenous to Magadh region of Bihar.

Description: A smaller aromatic plant, trailing on the ground; also climbing. Lower leaves 6–10 cm long, broadly ovate, deeply cordate, with big lobes at the base; upper leaves oblong-ovate, cordate, dark green and shining above, pale on lower surface. Stipules are conspicuous, 1–3 cm long but falling off soon. Spikes of flower solitary; bract of male spike narrow and of female

circular. Fruit ovoid, sunk in fleshy spike, which is 2.5–4 cm, ovoid, blackish green and shining.

Distribution: Native of hotter parts of India from Nepal eastwards and westward to Maharashtra, south to Kerala state, all in the warm region.

Uses and extent of utilization and trade: Dried fruits and the roots form the drug, which is used as tonic and in making irritating snuffs, in liniments for rheumatic pains and paralysis. Decoction from the immature fruit (dried) is useful in curing bronchitis. Ripe fruits are aromatic, stomachic and carminative but used extensively in *ayurvedic* medicine for various *Churna* (powders).

The compound 'piperine' extracted from the fruits, helps in absorption of antibiotics such as rifampicin, when taken in combination with piperine.

The drug is collected on an extensive scale and sold in the market. It is also cultivated to a limited extent.

Agrotechniques for the Cultivation

It requires a hot and moist climate and flourishes on a rich and well-drained loamy soil on shady situations. It is cultivated in South India and in Bengal.

The field is properly mowed and manured with FYM before planting. The propagation is done by suckers or cuttings which are planted in the field after the rainy season at a distance of 1.5 m from each other. Application of fertilizers is necessary on poor soils and for increasing the yield potential; however, no information on fertilizer requirements is available.

The fruits are picked annually in the month of January and plucked when still green and preserved by drying in the sun.

After three years of cultivation the roots are dug up. The yield is about 370 kg per ha. Roots are in much greater demand than the fruits, which are exported also to West Asia and Africa. It is recorded that each *Bigha* gives 2 mds in 1st year, 4 in 2nd and 6 mds in the 3rd year. Though no irrigation is needed yet in hot season roots are covered with straw.

b. *Piper cubeba* L.f.

Local names: Cubeb, *Shitalchini,* Sugandh-Mariha, Kababchini.

Description: A climbing perennial, with a smooth, zigzag stem. Flowers unisexual, arranged in spikes. Fruit a globose drupe, sessile when young. The stalk distinguishes the dried cubed berries from black pepper which is otherwise more or less similar.

Distribution: A native of Java but cultivated to a small extent in India (Karnataka state) for the fruits-the *Chaba* of Indian medicine. Mainly the fruits are imported.

Uses and extent of utilization and trade: The fruits yield a thick colourless oil with an aromatic odour and flavour of camphor and peppermint. The annual demand of this drug in India is estimated to be more than 200 qtls. The southern and eastern parts of India are suitable places for its cultivation.

Cubeb is used in medicine, a urinary antiseptic, a kidney stimulant and in the treatment of catarrh, gleet and internal inflammations. In Indian medicine it is used as an expellent of gravel from the kidneys and bladder. It is also believed to clear the throat.

Though Cubeb oil is seldom used in medicine yet used in the form of *lozenzes,* for flavouring butters and spicy table sauces.

All the requirements are met generally through import, though the plant can be cultivated extensively in pepper growing regions of Karnataka and Kerala states of the country.

Agrotechniques for the Cultivation

It thrives under the same conditions of soil and climate as the pepper, requiring artificial support to climb over. It requires a fairly rich soil and a hot humid climate but does not tolerate direct sun. In Indonesia, it is grown in coffee and coco plantations.

It is propagated by cuttings taken from the top or fruiting shoots. For distillation, crushed berries are used and an yield of 10–18% essential oil is obtained. The oil is somewhat viscous with a spicy odour and a camphoraceous acid taste. Major constituent of the oil is *Dordl-sabinene* forming 33%. The colour of the oil is light green to bluish green. *Piper chaba* or chaba is a native of Mollucas, cultivated for the fruits.

51. *Plantago ovata* Forsk.

Family: Plantaginaceae.

Local names: Isabgol, Isabgolá (Hindi).

In Persian language Isapaghula means ear of horse. The name reveals the origin of the drug in West-Asia, such as Persia, Arabia. It represents the characteristic boat shaped appearance of seeds.

Description: An annual herb, 30–45 cm high under cultivation. Nearly stemless, hairy. Leaves narrowly linear, 7.5–20 cm long and scarlet 0.6 cm broad. Large number of flowering shoots arise from the base. The spikes are cylindrical to ovoid, in shape and 1.2–4 cm long. Flowers protogynous, favouring cross pollination. Fruits mature progressively from bottom upwards, a capsule of ellipsoid shape, 8 mm long, each containing two seeds. Seed 3 mm long, boat-shaped, smooth, rosy-white coating provides the husk on mechanical milling.

Distribution: An important cash crop in Northwest India, cultivated in more than 16,000 ha in North Gujarat under irrigation as a *rabi* crop. Small areas are in Rajasthan (Sirohi district) Haryana-Rewari and Bihar (Sasaram), but the produce is of inferior quality. Experimental cultivation was successful at Jodhpur by the present author. It flourishes in a cool and dry climate; showers at maturity are detrimental to the crop even a light shower or dew at maturity spoils seeds.

Use and extent of utilization and trade: Isabgol husk is a rosy plant white outer part of seed, a light membranous covering on the seed which constitutes the commercial drug and has the capability to absorb moisture forming a tasteless, mucilaginous substance, which is given as a safe laxative. It is beneficial in cases of habitual constipation, chronic diarrhoea and dysentery. It does not irritate the intestine and is specific in a condition when mucous membrane is disturbed by inflammatory affections. The drug appears to be a recent introduction, since no mention is found in ancient medical books from 'Charka' to 'Vaghbata'. The earliest therapeutic reference appears to have been made by Dey in 1896. Now, it is widely used in USA and Western European countries as a household medicine and included in the Indian, British and many other country's national Pharmacopoeias. It is demulcent, soothing for gastrointestinal tract, chronic amoebic and bacillary dysentery, chronic constipation; diarrhoea of children are also reported to be cured. Pharmacological studies reported the effect of lowering blood pressure in anaesthesised cats and dogs. The extract showed cholinergic activity. Also used in printing, food, ice-cream industries. Swelling property of the seed in water is the simplest assay method for its acceptance by Pharmacopoeias. The husk is the dry seed coat and has the same property as the seeds but having an advantage there is no risk of mechanical obstruction or irritation in alimentary canal and hence it is always taken without presoaking, as such with water or hot milk. The muscilage is supposed to bind and increase the mass of stool and smoothen its passing out. Combination of liquid extract of (Holarrhena antidysentria *Kurchi* with Plantago seeds are beneficial to patients suffering from prolonged diarrhoea and dysentery.

In all, about 75% of the total *Isabgol* production is exported and India continues to hold a monopoly in the production and trade throughout the world. The plant has wide scope and more than 500 tons of seed and 300 tons of husk are exported. The husk is removed by a machine, grinding by the process called dehusking. This gives 30% husk by weight and about 69% residue as the by-product, which is used as cattle feed or manure.

Agrotechniques for the Cultivation

Soil and climate

Marginal lands with well-drained soils, having pH 7.0– 8.0, are suitable for this crop. Presently it is an irrigated *rabi* crop, which grows extremely well in light well-drained sandyloam to loamy soil rich in organic matter (organic carbon 0.8) and available potash content (280 kg/ha) but low in phosphate. It can also be taken after fallow in the rainy season, where the yields are as high as 12–15 qtls/ha as against 6.0 to 7.5 qtls when taken as *rabi* crop after harvest of *kharif* crop like *bajra*. It requires clear sunny days and dry weather during maturity. A cloudy weather, mild dew or light showers may cause heavy shedding of seeds and inflict large losses to the yield.

Preparatory tillage

After the harvest of kharif crops, in Oct-Nov. like *bajra*, *jowar* or ground nut, the land is ploughed crosswise two to four times, in order to remove the stubbles of the previous crop. If necessary, a soaking irrigation is given to facilitate the preparatory tillage operations. The land which is left fallow for *kharif*, the stubbles are collected and removed, the land is harrowed 2–3 times to pulverise the soil thoroughly. After the last ploughing, the land is planked to get levelled and firm seed bed of convenient dimensions, viz. 1 m × 3 m or 2.5 m × 2.5, 2.5 × 5 m or even longer strips of 6 m × 3 m, depending upon the source of irrigation. However, application of FYM, 15 cartloads per ha, is beneficial.

Seed sowing

The seed is broadcast or sown in the second half of October, in lines (25–30 cm apart); 5–7 kg of seeds per ha are needed. After broadcast by hand on dry beds, these are covered lightly by raking. The seeds retain their viability for several years but it is advisable to use fresh seeds for a 100% germination. Sowings can be done till middle of January, in general the practice being after the sowing of 'cumin' is over; *Isabgol* sowing is taken up the farmers. The late-sown crop in December or early January, however, is caught by heat at maturity resulting in hastening of maturity and low yields. The seed drilled upto a depth of 7 cm fail to germinate; hence, if drilling is not practised, seeds should be dropped very shallow.

The beds are given light irrigation after sowing of seed and germination starts in

8–10 days period. A second irrigation, is given three weeks after the sowing, and thereafter at an interval of 15 days, till all the spikes are out. Irrigation demand of *Isabgol* crop is medium; in all, it needs 6–7 irrigations, till the spaces are filled with seeds.

Manuring and interculture operations

If *Isabgol* crop is taken as a *rabi* crop on the same land, where *bajra, jowar* and ground-nut have been taken during *kahrif,* no additional FYM is applied. Some farmers find a top dressing of 25 kg N/ha as useful, in split doses, half at the time of first irrigation and the second half at flowering. Some authors reported application of 50 kg , nitrogen, 40 kg P_2O_5 and 24 kg K_2O for good yields; however, Gupta (1974) recorded that higher application of nitrogen has not given any consistent increase in the yield, since it is a non-exhaustive crop and its nutrient requirements are low.

The crop is given one to two weedings, as and when required, during the entire growing period. In Gujarat no sort of any after tillage operations are followed, except that at places where plants are too close, thinning to 20–30 cm is done one month after the sowing. The crop starts forming flowering shoots in about 60 days of sowing, each plant gives about 2–100 tillers. This being an important period, the crop is given an irrigation at this stage. The crop is ready for harvest in about 120 days, when the plants turn yellow. Proper maturity is tested by pressing the ear between fingers; when ripe full grown seeds come out. A prolonged cold-weather favours good harvest, which is done during March when the max. temp. is between 32–41 °C. In order to avoid losses due to seed shedding, the plant is cut close to the ground in early morning hours and the material is stored on a thrashing yard for 1–2 days. On the previous evening of thrashing, a small quantity of water is sprinkled on the heap to facilitate easy separation of seeds from the spikes. Next day early morning, the harvest is spread over the thrashing yard and trempled under the feet of bullocks. Seeds are winnowed and separated for storage. The straw is used as fodder for cattle. Average seed yield comes to 400–500 kg/ha, but higher seed yields are common (Gupta, 1974). CIMAP (2001) recorded yield of 4–5 qtls of seeds/ha. The husk which is 20% of the seed weight is the drug, about 5 qtls/ha is the produce.

The factors which enhance the yield include 1. Prolonged season in the early stage of growth, 2. at least eight irrigations during the growth period and three in absence of rains during the maturity period. Since, Gujarat farmers are not using fertilizers adequately, the yield potential can easily be enhanced by the judicious use of fertilizers.

The main varieties of *isabgol* are; Guja-rat-2, Trombay Selection (1–10) and EC-124 or 345. Gujarat-2 is a commercially accepted variety and is presently available at 30–40 rupees for a kg. Though there is enough scope for genetic improvement and selection of suitable strains, a fair proportion of plants in all commercial crop are male steriles, the sterility being controlled by factors in the cytoplasm. Such plants can be differentiated from fertile plants by shrivelled appearance of their anthers. Cytoplasmic male sterlity is a valuable tool in the hands of plant breeders and should be effectively utilized for this crop improvement. A study of flowers and pollination behaviour show that the male fertile plants produce abundant seeds by both self and cross pollination.

Cross pollination takes place when the papillose hairs on the stigma and style are moist and fully expanded. Tetraploid plants produced by Colchicine treatments are reported to have better seed size, quantity of muscilage and plant vigour. A study of the diploid and tetraploid seeds showed that the long photoperiod hastens the development of flowering, while short photoperiods delay the flowering. Low night temperatures result in good vegetative growth and overall production of healthy seeds, while high night temperatures reduce plant growth and number of flowers. Exposure of young plants to low temperatures for 3–4 days has no harmful effect on the growth, but flowering is slightly delayed. Sandy soils give healthy and better plants as compared to plants grown on clay soils as shown by studies of the author at Jodhpur.

The crop is subjected to heavy infestation of *Sclerotium*, the symptoms being appearance of a white moist mycellium on the peduncle, followed by wilting and appearance of black sclerotis. The disease is confined to beds receiving heavy irrigation.

Downy mildew, blight, powdery mildew are the common diseases. Diathane M-453 g per lit, Captane or Captafall 2 g/lit in water or Bawastin 1 g/lit for white grub and aphid is sprayed till the plant is drenched.

Marketing and processing

The main marketing places for the drug are Sidhpur, Unjha and Patan in Mehana, district, Palanpur and Banaskanta (in Gujarat state). The crop is sold through open auction to wholesale dealers and factory owners during and after the harvesting season. Market is organised through commission agents and commercial banks advance loans to farmers and traders against their stock. The market rate usually depends upon the buyers demand from abroad, including the crop forecasts and its weekly turnover in the *mandies*. A bold seed crop fetches a better price and the Indian buyers now prefer for a shining white seed crop which sells at a premium.

There are factories for the processing of *isabgol* seeds having capacity from 35–150 qtls a day and are mainly located at Sidhpur Patan, Unjha and Palanpur towns. The seed crop as well as the husk needs storage in cool, dry place to prevent absorption of atmospheric humidity.

The seeds are not graded and no standards, with well-defined quality clauses are in vogue. The colour of seeds from an old crop is generally brownish, while the shade found in the new crop are preferred in the trade. *Isabgol*-based main drugs use husk and powder. 'Nature Cure' is the brand named by Dabur, which is presented in a variety of tastes and packings. Softovac is another brand name by Lupin. Zilax, Cilax, Sona, Telephone and Hiran are the other brand names.

The seeds are processed through a series of grinding mills to separate the husk and each grinding mill is provided with an exhaust fan to do the job and an elevator to lift the balance produce to the successive mill. It is a simple mechanical milling process, whereby husk's material separates out at each grinding, till the produce passes through last mill. The general recovery of husk at various grinding stages is: 1st grinding-40%, 2nd grinding-25%, 3rd grinding-15%, 4th grinding-10% and 5th and 6th grindings comparatively smaller in size having a reddish tinge. The husk of 2nd and 3rd grinding is regarded to be of superior quality. The husked grain is separated into three marketable grades, known as *Lali, Gola* and *Khaka*. The *Gola* is nothing but husked grains, while *khaka* contains small husk powder and sand is collected as the sieved material when each grinding is

passed through a 70 mesh sieve. The *Gola, Lali* and *Khaka* are used as cattle feed. The percentage recovery of the husk and the three by-products is Husk 30%, Gola 65%, Lali 3% and Khaka 2%. In India Gujarat, Punjab and UP are main centres. It has been introduced in MP (Mandser, Malhargarh, Neemuch, Mander) where about 5000 ha are under cultivation. Experimental cultivation of Isabgol at many places has proved successful such as at Poona, Ranikhet, Jodhpur, etc. The time of sowing affects the yield which gets reduced during winter sowing. The yield from plants sown in December is the highest in plain areas. Lack of response to fertilizer application may be due to basal dose of FYM. There is no ill effect of thick stand and it might be of interest to know the extent to which the plant stand can be increased without affecting the performance of the individual plant, so that additional yields can be obtained due to plant number alone.

In Gujarat and Unjha Isabgol is sold at Rs 1100–1200 per quintal while medicine based on it sells at Rs 10–15 per 100 g.

Results of a study conducted at Jodhpur showed as below: Date of sowing 5th Sept; germination 12th Sept. Flowering- *P. ovata* 5th Jan; Seed setting 18th Feb, Maturity 4th March, *P. psyllium*: Seed setting 28th Feb. flowering 18 Feb and maturity 11th March.

Other Species of Plantago

Plantago major

Plantago major is indigenous to Maharashtra. *P. psyllium* occurs in Punjab, *P. lanceolata* and *P. brachyphylla* in the Himalaya from Kashmir to Kumaon. All these species contain mucilage, however, very little attention has been given to these species from medicinal or commercial viewpoint.

Apparently *P. ovata* was introduced from the Unani system of medicine and was first introduced in Punjab.

Description: A stemless herb. Leaves distinctly stalked, broadly ovate, entire. Scape cylindric, erect, furrowed. Flowers numerous, green. Stamens protruding. Capsule containing 8–16 minute seeds.

Uses: Expressed juice of the plant has curative value in tubercular consumption with spitting of blood. Roots and leaves are used against intermittent fevers. The leaves are considered cooling and fresh leaves are rubbed on parts of the body stung by insect, nettles, etc. and stay bleeding of minor wounds. The seeds are considered stimulant, tonic and are an efficient remedy for dysentery and are used as good substitute for *P. ovata*. Watery extract of the seeds is sometimes given to cure whooping cough.

Plantago psyllium L. (Psyllium, Fleaseeds, Fleawort).

The drug has come out of obscurity into much greater demand in recent times. It is chiefly cultivated in France and Spain, where two crops are raised annually. It is used as a mild mechanical laxative. Experiments on the cultivation of this plant have shown that it is resistant to hailstorms, to some extent, while *P. ovata* is susceptible, and is of longer duration.

References

Atal CK. 1969. Cultivation of Isabgol in North Gujarat. *CSIR Symposium.*

Atal CK and Kapoor KK. 1963 *Indian J. Pharmacy* **25**: 326.

Atal, CK and Kapoor KK and Siddiqui HH. 1964. *Ibid, 26*: 163.

Chandler Clyde. 1954. Improvement of Plantago for muscilage production and growth in US. *Center Boyce Thompson Inst. 17*: 249–259.

10

Chandra V. 1967. Studies on cultivation of *Plantago ovata* Forsk. *Indian J. Pharm.* **29**: 331–332.

Dube, Sudershan; 1977. *Isabgol.* Export oriented medicinal crop. *Uddhamita* Sept. pp 73–74.

Gupta RK and others 1972. Plantago cultivation at Jodhpur, in Rajasthan: Effect of soil types of the yield and growth of Plantago. Ovata IC 7739 and Plantago psyllium EC 32125.

Joshi P, Tiwari SC, Vadava BBl. 1968. *Indian For.* **94(5).**

Kanitkar UK and Pendse GS. 1969. Experimental cultivation of Isapaghula in Maharashtra. CSIR Symp. *Indian J. Pharm.* **29(3):** 97–98.

Modi, J.M, Mehta KG and Gupta Rajendra. 1974. Isabgol. A Dollar earner of N. Gujarat. *Indian Farms.* January.

Siddiqui HH, Kapoor KK and Atal CK. 1964 *Indian Journ. Pharmacy* **26:** 266.

Sirohi SS, Chauhan JS and Chaurasia BD. 1970. Plantago cultivation in Punjab. *Progressive Farming* **6(1):** 11–13.

52. *Podophyllum hexandrum* Royle. (syn. P. emodi Wall. var. hexandrum (Royle) Chatterji and Mukherji).

Family: Berberidaceae.

Local names: Bankáru, Banbaigan, Pápri, Bakrachimaka, BanKakri, (Hindi). Rikpat (TG), Venivel (Gujarat), Banwagan (Kashmir), Padwal (Maharashtra). Bankakri, Gulkakri (Punjab).

The trade name, Indian Podophyllum, is based on the scientific name of the plant.

Description: A succulent, glabrous herb; scapigerous; rootstock creeping, scape 12–15 cm high with palmately divided leaves and white or pink flowers (Colour Plate 10.12). Sepals 3, petals like, soon falling off. Petals 6. Stamens 6, anthers opening by lateral slits. Ovules many. Style short, stigma crest-like, ridged. Berry ovoid, scarlet. Seeds many, enveloped in the pulp.

Distribution: The plant is met in the interior Himalayan ranges lying westwards beyond river Satluj, including JK state, Himachal Pradesh and some adjoining hills of Uttaranchal. It loves moist situations, between 2500–4000 m and flourishes well as undergrowth with humus and decayed organic matter. Generally, associated with *Rhododendron, Salix, Juniperus* and *Viburnum* species. Some plants are also met in open alpine meadows but the frequency is much less. On exposed slopes, particularly under high altitude conifers, and the tract lying between the alpine meadows and upper extremity of the timberline the frequency is well distributed. The scare resource has made it obligatory to grow this plant in the region. HAPPRC (2004) provided cultivation Techniques Bull. No 14. for this plant in hills. About 44000 plants/acre are planted at 30 cm × 30 cm in March-June at an alt. 2000–2500 m. Seedling of 60 days or more are planted on forest litter soil. Harvesting is done after 3 yrs for vegetatively grown crops and 4–6 yrs for seeded crop. Harvesting time is Sept/Oct. Production is 1600 kg/acre. Net profit estimated is about Rs 46,000 for an acre.

Uses and extent of utilization and trade: The dried rhizomes constitute the drug. The rhizomes are locally used as purgative while the fruits are edible. Attempts have been made successfully to grow this plant in Tons Division of Tehri district on farmer and field.

The resin *podophyllin*, from the rhizome is medicinal, its action is slow but severe; in large doses it causes irritation and gripping. It is usually administered in mixture with belladona or aloes, etc. It has been reported to be useful in many skin diseases and tumorous growth, also used in curing cancerous tissues.

Small quantity of this plant earlier used to be collected by the natives and sold in the market; though total annual requirement of this plant is more than 40 qtls.

Podophyllum peltatum L. An American species which is also cultivated. The rhizomes on an average contain 10% Podophyllin, a resin with *podophyllotoxin* as active principle (Kapoor and Sarin, 1962). A glucoside with marked anticancerous properties has been isolated from the roots and rhizomes. The demand of the drug has increased within recent years manifold and still there is a vast scope for its use and exploitation in medicine. In 1889, the Indian Podophyllum was analysed and regarded to contain higher percentage of the 'Podophylloresin' than in the American Podophyllum. In 1884, the price of this drug was about Rs 300 per maund which increased between Rs 2200–2500 in fifties. On account of the specific anticancerous properties, the drug has a bright future. It has been estimated that from the hilly regions of JK, HP and Uttaranchal 1,0,000 kg could be collected (Kapoor et al 1962).

Time of collection and phenology : The plant flowers during August-September and the rhizomes are harvested in autumn, dried in sun, cleaned and sold in the market. The quality and quantity of the resin varies with the locality and altitude. Maximum percentages of resin occurs in May when the plant is in flower, while the percentage of *Podophyllotoxin* resin increases from 35.35% in May to 49.77% in autumn.

The rhizomes can be harvested three or four years after transplanting. The estimated cost of roots is Rs 60/kg. The approximate quantity available from Western Himalaya is estimated to about 80,000 kg, a substantial quantity can be exploited by systematic extraction, without damaging the natural crop. However, commercial cultivation in forest areas needs to be encouraged and the rhizomes can be harvested 3–4 yrs after plantations. The drug has a very bright future.

Reference

Kapoor LD and Sarin YK. 1962. Indian Podophyllum; its distribution and availability in *Northwest Himalaya*. RRL Bull, 1(1): 38–39.

53. *Pogostemon cablin* (Blanco) Benth.(Syn. P. patchouli Pellet var. suavis Hook.f.).

Family: Pogostemonaceae.

Local names: Patchouli, Panri, Patchá (Hindi).

Description: Perennial herb, 1–1.2 m high. Leaves aromatic, slightly lobed, toothed, hairy, about 5–10 cm long and 3–9 cm broad. The plant differs from the Indian species *P. heyanus* Benth. (Syn. *P. patchouli* Hook.f; non Pellet), in having thicker and pubescent leaves. It is said to flower only in Philippines, where it is indigenous.

Distribution: Cultivated extensively in Malaysia, Indonesia, the Seychelles, China, etc. including other states like Madagascar, Reunion, Paraguay, Brazil, etc. In India successful cultivation on an experimental scale was achieved in Shevoary hills, Nilgiris and Annamalai hills, Karnataka and Dehradun.

The plant thrives well in a warm and damp climate with evenly distributed rainfall. It can also be cultivated as intercrop in mango orchards. The plant does not tolerate direct sun heat. It requires fertile soils. Undulatings lands are most suitable though it flourishes on lowlying, well-drained lands. Temperature in most of sites ranges from 20°–28 °C with relative humidity at 75–80%.

10

Uses and extent of utilization and trade: Considerable amount of oil is imported in the country. The dried leaves are kept in clothes to repel insects. The oil is obtained by steam distillation of the dried leaves. Industrial production is mostly in Indonesia, Malaysia, Brazil, China, and also in S. India. The oil is used in perfumery, cosmetic and flavouring industry. Medicinally, it has been reported to stop blood in urine. In perfumery oil is the best fixative for heavy perfumes, used in soaps, cosmetics and for incense. It is also used for flavouring tobacco.

Agrotechniques for the Cultivation

The plant thrives well in warm and damp climate with an evenly distributed rainfall. Hot and humid climate is suitable. Rainfall between 150–500 cm, during the rainy season, and average annual temperature between 24–28 °C and RH (Relative Humidity) of about 75% are suitable climatic conditions for a successful cultivation.

It requires fertile soils, which is well drained, loamy or sandy loam in nature having a pH of 5.5–6.2. High pH soils with stony or hard layer are not suitable.

Propagation is by cuttings which may be first planted in the nursery beds in shade, 2–3 months before planting in the field (in February and March) or may be directly planted in the field during monsoon season. Layering can also be practised. Young plants should be protected from direct sunrays by covering them with palm leaves. Careful weeding, irrigation and manuring are essential, throughout the growing period. The cuttings should be 10–12 cm to 15–20 cm long with 4–5 nodes or with 4–6 upper leaves planted at 2–3 cm distance in the nursery. The cuttings root in 4–5 days and 8–10 weeks old cuttings can be planted. On the lower part of the cutting root-promoting hormone, like Ceradex-B2, should be applied and planted in shady places. The cutting should be taken from the upper part of the motherplant.

In Northern India, it can be cultivated as an intercrop in mango orchards, since the plants cannot tolerate direct sun heat.

Many commercial varieties are available such as *Johar, Singapur* and *Indonesia*. Var. *Johar* has been recorded to give higher yields than the other varieties.

Planting

30–35 days cutting is suitable for transplanting, which can be done from July to Sept at 60 cm × 60 cm from plant to plant and row to row, 10–20 t/ha of FYM is incorporated along with NPK 150 : 50 : 50 kg/ha. 20 kg of N along with full dose of other fertilizers is given, while preparing the field. Long and wide beds with provision for water-outlet are made. Furadon at 20 kg/ha is also mixed and before planting initial irrigation is required. Planting is usually done in the evening at a spacing of 30–45 cm in lines at, 60 cm apart. It needs frequent irrigation. After 2 months of planting second dose of N at 25 kg/ha is applied, followed by 50 kg N after first cutting and 50 kg N at second cutting, thus making a total of 150 kg nitrogen per ha.

Cutting

The first cutting is taken 4–5 months after planting, when the leaves start turning yellow. It is irrigated after 3–4 months interval. Crop can be taken for 3–4 years. Upper part of the plant is cut in the morning, so as to avoid loss of oil. Plant at 25–50 cm with leaves should be cut and 1–2 branches are left on the plant.

Leaves after cutting are spread under shade on a floor and should be turned at

10

least twice to avoid decay. Total production of the leaves is about 100 qtls/ha of green matter, in three cuttings; which remains 16–20 qtls after drying. Availability of oil is about 2.5%. In dry leaves, on an average, availability of 2 tons of dry leaves and 50–60 kg of oil can be obtained per year and per ha.

The oil contains more than hundred compounds.

Diseases

Insects do not attack the crop; only root knot nematode and wilt attack is reported for which 20 kg Furadon and Diathane M-45 at 3 g per lit of water is sprayed, for the leaves are affected by *Zhulsa* disease. Application of crop rotation, nematicides or *neemcake* is useful, since 50% of damage to the crop can be expected. Other common diseases and their control are as follows.

1. *Leaf blight*: *Colletotrichanil gloeosporoidea* appears as lesions on leaf margins. Under high humidity these lesions extend inwards causing complete necrosis of leaves. In advanced stage of infection, axillary and terminal buds are affected reducing herb and oil yield.

Control: Spray Chlorothanil (0.3%) and repeat after 5 days. Remove infected plant material and debris from the field and destroy.

2. *Root knot*: Caused by nematodes. Infected plant remains stunted, with root infected plants bear big galls.

Control: Apply Crabofuron (1–1.5 kg /ha) before planting. Use disease-free cuttings, if infection is serious, treat beds with Carbofuran.

The oil is obtained by steam distillation of dried leaves which blends easily with geranium oil, sandal oil, clove oil and is a good fixative.

Cost benefit: *Pogostemon heyneanus* Benth (*Indian Patchouli*) is found in western and southern parts of the country and the oil is prepared by passing steam through the leaves heated in large copper cylinders and condensing the distillate, which is used largely in perfumery.

The plant which affords the greater part of the patchouli perfume of European commerce is *P. suave* (considered as a var. by Hooker). Patchouli of Assam is obtained from *Microtaena cymosa,* in which various adulterants are used such as *Ocimum basilicum* var. *pilosa,* and *Urena lobata.* Locally, about 6 kg of material yields about 24–30 ounces of oil. The crop is gathered by cutting down all but one stalk on each root and placed to dry in sun during day and under cover at night. The dried stems are then made into bales and sold.

Net profit has been estimated at Rs 26,900/year. Cost of cultivation including land preparation (Rs 1000 /ha), planting material (Rs 800), fertilizers (Rs 1600), planting (Rs 800), irrigation (Rs 2000), weeding (Rs1500), cutting (Rs 1050), Drying (Rs 400), pest-control (Rs 2800), distillation (Rs 2000) and other expenses (Rs 1150) makes a total investment of Rs 13,100/ha. Income from 40 kg oil/ha at Rs 1000/ kg comes to Rs 41,000/ ha. The net profit has been estimated at Rs 26,900/year. Many commercial varieties are available like Johar, Singapore, Indonesia Var. Johar gives higher yield (data based on CIMAP, Lucknow).

54. *Rauwolfia serpentina* Benth. ex Kurz.

Family: Apocynaceae.

Local names: Chotachánd (Hindi), Sarpgandhà, Chandriká (Sans.), Chandrá (Beng.), Harkáyà (Maharashtra), Dhanbaruá (Urdu), Chavanda-avalpori (Tam.) Patala-gandhi (Tel.).

10

The trade name Rauwolfia is based on the scientific name of the plant. The word Rauwolfia refers to Leonard Rauwolf German botanist and physician.

Description: An erect shrub, glabrous, 30–75 cm high. Leaves whorled, 8–20 cm long; gradually tapering into a short petiole. Flowers white or pinkish, 1.5 cm long, in small clusters. Fruit dark purple or black when ripe (Colour Plate 10.13).

Distribution: Chandra (1955) discussed the distribution of this species and in detail cultivation practices, diseases and pests for both *R. serpentina* and *R. cane-scens*. The plant prefers clayey soils and grow well under semi-shady conditions in nature. *R. canescence* has naturalised in North and Central Bengal and not mentioned in Hooker's classic book *F1. Brit India*. It also grows side by side with *R. serpentina* inhabiting most tropical regions of India. *R. serpentina. R. canescence* show, however, long period of flowering and fruiting from April to next February.

Uses and extent of utilization and trade: In the Ayurvedic and Unani systems of medicine, the roots are used in diseases of the brain, epilepsy, intestinal desiccation, menstruation, etc. In modern medicine the roots are used in the preparation of medicines for blood pressure (BP) control. Dried roots with bark intact, constitute the drug. The roots contain several alkaloids, used as sedative and hypnotic and for reducing BP. Now the drug is largely used in insanity. Sedative action of the drug is slow and hence in acute cases it is not useful. It is considered to be more suitable for mild anxiety cases or on patients of chronic mental illness. The drug has a tranquilizing effect. Roots are also useful in disease of bowels and in fever, care being taken that the patients do not suffer from bronchitis, asthma or gastric ulcers.

The drug is believed to have been known in Indian medicine for the last 4000 years and a mention is found in Charak's work. The annual requirements are estimated at more than 12,000 qtls and its cultivation is gradually expanding. On an average the yield varies from 10–12 qtls per ha. The roots in naturally growing plants are collected, with bark intact, during the autumn from 3–4 years old plants.

Agrotechniques for the Cultivation

Soil and climate

Sandy loam and clay loam soils are suitable and the plant grows well under semi-shady conditions. Net and highly humid areas are not suitable. pH of the soil should not exceed 8.5 and can be cultivated under light shade. Temperatures ranging from 10°–30 °C are suitable fot its cultivation.

Land preparation

Deep ploughing is done during the summer months and left. If the site is near the forest, roots of sprouts are taken out. After the first shower, 10–12 tonnes of FYM is mixed to the soil and ploughed again. Before sowing, two ploughings and then levelling is done, provided with irrigation channels in the farm.

Plant saplings

Seeds, root cuttings or rooted stumps are used as the planting material. Seed germination is stated to be low (7–30%) and direct sowing gives poor results, but transplanting of nursery stock is satisfactory. Black fruits formed in May-June, produce high percentage of viable seeds. High germination of viable seeds is obtained by *in vitro* culture of embryos in a nutrient medium under controlled light and temperature condition (Mitra, 1975). *R.*

serpentina can, however, be best propagated from root cuttings after treatment with IAA (indol acetic acid). Treated cuttings, however, root after 15 days.

5 kg of seeds/ha are generally used for the nursery. Bhadwar et al recommended 4 ft (120 cm²) beds for sowing seeds. Pre-sowing seed treatment with 5% sodium chloride solution in water is done, so that seeds are removed from the surface, it also protects against damping off and improves germination. Seeds are put overnight in water before sowing. In North India the seeds are sown during May in the nursery and in the first week of July the seedings are ready. In South India, the nursery is raised after the rains have started and after 6 weeks the transplants are ready. Under Dehradun conditions, seeds are sown in May and the transplants are ready for planting in July.

In North India plantations above 2100 m have failed. Studies from Pakistan show that in Abbotabad, plants can tolerate semi-temperate conditions and even some snow (1300 m). The aerial parts dry up during the winter and the seeds reach maturity before cold sets in. Aerial parts sprout after winters are over.

R. serpentina like *R. canescence* have long period of flowering and fruiting from April to next February but the percentage germination is high and can be sown by dibbling in holed fields. It can also be raised by transplanting nursery-raised seedlings from April-October and the survival percentage is high. Vegetative reproduction is difficult, though there is possibility of planting root-shoot cuttings with application of Ciradix-B3. *R. canescens* show long period of flowering and fruiting from April to Sept-February. Alkaloid in stem is 0.50% while in air dry leaves it ranges from 1.72–1.8%.

Root cuttings

The roots have more alkaloid content and are used so that plants yield better alkaloid. 7.5–10 cm long root cuttings are used and put 5 cm in drains, about 100 kg of root (fresh) are required for one ha of land. The rainy season is the best time for planting. The roots should not be damaged, while taking out from the nursery; the roots are covered with the green leaves or damp gunny bags. While planting, line to line distance is kept at 60 cm and plant to plant distance 30 cm apart.

Irrigation is provided every day after transplanting. While preparing the field 8–10 t/ha well rotten FYM is added and NPK each 30 kg/ha is mixed with the soil before planting. During the first, second and third year 25 kg of nitrogen is applied each time in two split doses during August-Sept and February-March.

In *R. serpentina* the seeds take about 21 days to germinate, establishing in 40 days. Bhadwar et al recommended 120 sq cm beds for seed sowing. 1/20 acre nursery from 5 ounces of seeds being enough for one acre of land. The seedlings are transplanted at 60 cm × 60 cm. Weeding and hoeing is done in both the species regularly to ensure a good crop after 15–20 days and two times in a year. Flowers are plucked is the morning which is also done for roots, so that there is no seeding. Root yield increase with number of stems per unit area upto 25–26 plants/m² of bed. Optimum spacing is 80 × 80 cm having air dry stem yield of 3050 kg/ha.

Under irrigated condition, the roots are ready for harvest within 2–3 years. These are dug in December, since the plant is dormant and have few leaves. The roots are dried in shade after washing. Roots weigh 100–400 g per plant. If the distance is 60–30 cm, root production expected is

1175 kg of dry root or on an average 10–12 qtls/ha. In *R. serpentina* the roots are collected in the 3rd year, when the alkaloid content is at the maximum. Total alkaloids vary from 0.8–1.3%. If this species is propagated by cuttings, the alkaloid content remains one-third. In case, the plant is grown under the existing forest, the alkaloid content in various localities in UP has been recorded from 1.87–2.25%. The root yield increases with the number of stems per unit area (upto 25–26 plants/sqm) with optimum spacing of 80 cm × 80 cm. Experiments conducted on the effect of fertilizers, in Assam, did not show significant results. The alkaloid percentage in the stem is 0.50, while in air dry leaves it ranges from 1.72–1.8 and in the rootbark it is 0.1. The alkaloid *rauwolscine* is a cardiovascular depressant. Alkaloid content of plants grown in various localities of Uttar Pradesh varies from 1.87–2.25%.

The average root yield is higher in plants raised from root cuttings while those raised from seeds and stem cuttings showed the lowest and intermediate yields respectively. Proper time for the root harvesting is at the end of 2nd and 3rd year, in case of diploid and tetraploid strains respectively. Successful propagation of root cuttings, however, depends more on the relative diameter of the propagating roots. Both root and shoot primordia initiated better on thinner roots than on thicker roots. IBA 500 pm is more suitable auxin than NAA/IAA for both root and shoot formations. Tomar (2001) recorded 100–400 g root per plant, if distance is 60 × 30 cm per ha. Yield of root is 1175 kg (dry) weigh 10–12 qtls on an average.

Cost-benefit analysis is given in Table 10.16.

Table 10.16: Cost benefit analysis (after data from CIMAP, 2001)

Expenditure	1st year/ ha	Expenditure	2nd year/ ha
Nursery	Rs 2000/-	Hoeing and weeding	Rs 1000/-
Land preparation	Rs 1500/-	Irrigation	Rs 500/-
Fertilizer	Rs 1500/-	Digging	Rs 4000/-
Irrigation	Rs 1500/-		
Others	Rs 1000/-		
	Total 7500		Rs 5500/-

Total expenditure in 2 years Rs 13,000/-
Root production in 2 years 1000 kg and at a rate of Rs 60/kg
The total cost 60,000, thus giving a net profit of 47000 in 2 years, i.e. Rs 23500/year.

References

Bhadwar RL, Karira GV and Ramaswami F. 1955. *Rauwolfia serpentina,* the wonder drug of India, Sarpagandha. *India For.* **81(4).**

Bhaduri, RN and Biswas PK. 1965. Increasing root yield of *Rauwolfia serpentina* by Cochiploidy. *Sci. and Cull.* **31:** 197–200.

Biswas K. 1956. Cultivation of Rauwolfia in W. Bengal. *Indian J. Pharm.* **18(6):** 227–230.

Biswas PK and Bhaduri PN. 1975. Induced tetraploid. *Rauwolfia serpentina* Benth. and its propagation. *Proc. Nat. Symp.* p. 11.

Chandra V. 1956. Studies on Rauwolfia. *Journ. Sc. and Industrial Res. India,* **15** (2–3) Pt II Preliminary studies on floral biology of *R. serpentina* Benth. *Indian. J. Pharm.* **18(4):** 136–140

Dhar, R. (Miss). 1965. Variation in alkaloid content and morphology of four geographical races of *R. serpentina* Benth. *Proc. Indian Sci. Acad.* **62:** 242–244.

Gupta, R. 1968. Commercial cultivation of *Rauwolfia serpentina.* Need for quality seed. *ISI Bull.* **20(9):** 363–365.

Khan AK. 1958. Preliminary cultivation trials of *R. serpentina* in W. Pakistan. *Pakistan J. For.* **8:** 128–129.

Khan AH and Hussain SM. 1955. Some observations on *R. serpentina;* its occurrence and scope of cultivation in east *Pakistan J. For.* **5(4).**

Mitra GC. 1975. Studies in the formation of viable and non-viable seeds in *Rauwolfia serpentina. Proc. Nat.*

Sahu BN. 1970. Studies on *Rauwolfia serpentina* Benth. Effect of FYM, Amonium sulphate and Superphosphate on growth and yield of roots. *Indian Far.* **96(9):** 680–689.

Tomar VKS. 2001. Cultivation of Rauwolfia–agrotechniques. CIMAP. Symposium.

Yonkman FM, Mohr FL. 1954. Reserpine (Serpasil) and other alkaloids of *R. serpentina.* Chemistry, Pharmacy and clinical applications. *New York Acad. Sci.* **59**: 1–140.

55. *Rosa damascana* Mill.

Family: Rosaceae.

Local names: Guláb (Hindi), Scented Rose, Damask Rose, Persian Rose (English).

Description: An erect shrub, armed with curved, unequal-sized prickles. Flowers rosy pink, sweet scented in bunches of 5–10. The sepals are bent backwards after flowering. Flowers from February to April.

Distribution: It is a cultivated variety, not known in wild state, grown in many countries. The best and most extensive cultivation, being in Bulgaria, France, Mexico. Though rose is cultivated practically everywhere in India, most important essence producing centres are Kannauj, Ghazipur, Aligarh, Etah, Kanpur, etc. in UP.

It has now been grown in Himachal Pradesh, J and K, Rajasthan, Bihar in approximately 2500–3000 ha of land.

Uses and extent of utilization and trade: For production of rose oil, Turkey and Bulgaria are the foremost countries, Mexico produces only rose water, while countries like India, Egypt, China, France and Russia, produce rose oil, rose water, concrete and absolute. The yearly production is approximately 15 tonnes of all products. Production in India for rose oil is about 150 lit per annum (Table 10.17).

Scented rose is being cultivated since the Moghul times. Cultivation of this rose started from the Mediterranean areas. There are two varieties, viz. *trignipetala* and *bifera.* Since the plantations are mixed, the flower production is reduced. Two forms *Jwala* and *Himrose,* have been selected for the North Indian conditions; *Jwala* is suitable for the plains and low-hills, while *Himrose,* for the hilly areas having altitudes from 1200–2500 m, which can tolerate cool climate of the region.

Agrotechniques for the Cultivation

Deep sandy loam soils, with pH 6.0–7.5, well drained are suitable for its cultivation, while to some extent it can be cultivated on saline and *usar* soils. Sunlight and good relative humidity in the atmosphere are essential for proper growth and flowering. It has also been recommended that for quality rose oil,

Table 10.17: Oil content in Rosa damascana

Type	No of fl./plant		Oil content	Duration of days	Flg. time
	3 yrs old	*5 yrs old*			
K	900–950	1200–1400	0.05	20	29/30 May to 19–20 June.
A	850	11,00–1200	0.043–0.045	20	Early by 2–3 days. Completes early than K rose

(Himalaya Sampada Poyodhiki Sansthan symp. p 23 2001.)

it should be cultivated in valleys of the hills in Himalaya ranges and Siwalik hills and plains of North India, where irrigation is available. Coastal areas of South India are not suitable and no cultivation can be done under the crop. For a good growth, temperature between 0–5 °C and at flowering temperatures between 25–30 °C, are required, along with a humidity of 60% for good flowering. Severe frost, cold climate with cold northerly or hot westerly winds are all inimical to good growth. Water-looged soils are avoided.

Since the roots get deeper into the soil, deep ploughing, 3–4 times, are done before levelling. Commercial plantations of rose is done by planting cuttings, while for ornamental purposes roses are grown by budding or grafting. Cuttings of about 50 cm circumference from at least one-year-old plant, 15–20 cm long and 1–1.5 cm thick are transplanted. In each cutting, there should be at least 3–4 buds. These cuttings are tied in bundles for sprouting and kept for 20–25 days in 10–15 cm deep soils (or in 45 cm deep pits having at least 3 kg of rotten manure and 100 g of Superphosphate, Malathion powder kept for 2–3 weeks). The sprouted cuttings are planted in the nursery at a spacing 10 × 45 cm with at least half the portion under the soil to protect the plants from termites. The nursery-raised plants are planted in July, thus 6–7 months are required to prepare the plants for the field.

Cuttings if obtained from outside are immersed in water for 7–10 days before planting in the nursery beds. At the beginning of November and by February the leaves appear or it can be planted in pits as described above. Early and excessive drought is harmful and so is the excessive rain. By the following March the plantation is ready for flowers collection.

8–10 t/ha of FYM is mixed in soil at the time of land preparation. 100–200 kg of N, 80 kg P_2O_5 and 60 kg K_2O, is also applied with half dose of nitrogen and full phosphorus and potash is applied after cutting and pruning and half nitrogen as second dose after cutting and bud sprouting. It has been recorded that spray of 1% urea solution improves bud formation, which ultimately increases the flower production. Irrigation is provided at 20 days interval in November-April/June, depending upon the rainfall. Since weed-free crop gives more buds, weeding is done 3–4 times, to maintain good flowering. Spray of Kinetin, 20 mg/lit, after cutting in the first year and second spray during bud formation and third after bud formation improves bud formation and the flowering, (30 days–15 days and 15 days respectively). The following pruning methods are adopted.

1. Intensive planting (1.0–1.25 m × 0.5–0.75 m) Intensive pruning to ground level in August.
2. Intensive planting (1.0–1.5 m × 0.3–0.75 m) pruning in Nov-December, intense pruning.
3. Planting in rows (1.5–1.75 m × 0.5–0.75 m), cyclic pruning, every year respectively, above a height of 10–12 cm yearly.
4. Wide plantation (1.5–2.0 m × 1.5–2.5 m), pruning once a year in Nov. and December, keeping height of the plant.

Dried and infected plants are removed regularly. Important insects that curl the leaves are aphids, thrips, chaffer beetles, red spider mite, caterpillars, leaf hopers, etc. (controlled by spray of Malathion 1 ml/lit). For die back disease, 0.1% of Bawastin powder is sprayed, 3–4 times at 15 days interval, after cutting and pruning. At the bud-formation no treatment is given.

Flower harvesting and oil distillation

Harvesting of flowers is done during the early hours of the day and an yield of about 560–670 kg/ha can be obtained in Northern India, though in Bulgaria an yield of 4000 kg has been reported to have been obtained.

The roses are distilled, soon after the harvest with water distillation. Distilled water should be collected as rose oil which is soluble in water. With improved methods, an yield of 0.01–0.3% is recorded. In Bulgaria l kg of oil can be obtained from 4,000 kg of flowers. A small distillation plant of 10 kg flowers (0.25–0. 4 ha) can give about 1000–1200 liters of rose water. For 1.2 to 2.0 ha of land, unit of medium size can serve the purpose, while for a unit of 3 ha of land, with capacity of 400 kg flowers are also available.

Singh et al estimated direct and indirect selection of flower methods. The selection gain for flower yield was 19.97% while among component trait it was minimum for plant spread (27.85%), followed by flower plant (35.69) and branch per plant (13.80%). The correlated response in flower yield was the highest through flower per plant. Relative selection efficiency (RSF) was more than unity and in proportion to correlated response (CRF) for these three traits indicating the role of these traits in multiple-selection index to enchance flower yield.

Characteristics of rose oil also depend on many factors. It is colourless at first, turning to yellowish or greenish in colour gradually. Being one of the most expensive items, it is frequently administrated with natural isolates such as feraniol from palmrosa oil, L-citronellol from geranium oil, etc.

Concretes and absolutes in addition to oil are also prepared with suitable solvents like petroleum ether. The yield of Concrete on the weight of the flowers also ranges from 0.22–0.25%, while the absolute (alcohol-oil soluble portion) is 50–60% yield on the weight of the concrete.

Since pruning is an important operation to increase the production of flowers, pruning at different dates, at different height have remarkable effect on the yield of flowers. Pruning at a height of 50 cm from the ground level is reported (Srivastava and Chandra, 1975) to have given maximum flowers yield.

Plant improvement

Some of the crosses made between Kazanlak rose and *R. gallica, R.aliba; R. rugosa, R. centipetala* and *R. demascana* (Cremean redrove) show that the hybrids give as much as 15,000 kg of flowers per harvest, but the oil percentage is either low or the odour is not comparable to Kaunlak rose oil. Also mutagenic studies by irrigating the pollens with doses of 50 r to 10,000 r (with interval of 200 r) at 50 r per minute have been attempted in various species, but no conclusive results worth reporting are available. Since rose is a naturally cross-pollinated crop, the seed formation is poor, by the use of stimulated pollens more seeds were found. Chemical mutagination has also been tried at different centres. Physiological and biochemical studies in rose, have shown that it contains more sugar than starch, especially oligosaccharides which provide winter hardiness. Sugar quality, similarly, changes during the annual growth cycle, i.e. oligosaccharides appear during November and disappear in spring season. Flower production increases with age, till 7–8 years, when it declines; plants are rejuvinated by cutting aerial parts. Biochemical differences are reported to occur in normal and rejuvinated bushes. Free amino acids are known for their frost resistance and are less susceptible to mist. In *R. damascana*

(Kazanlak rose), white-flowered rose (*R. alba*) and a hybrid 77/5 (*R. gallica X R. damascana*), record 17 free amino acids and amides, asparagine and glutamine. The root and shoots at the beginning of vegetative growth possess largest number and highest content of free amino acid (f.a.a.), whereas at reproductive phase and during period of physiological rest plants have fewer number and less content of amino acids. However, there is no correlation between frost resistance and highest content of f.a.a.

Since the oil contains more than 120 compounds, the odour is dependent upon a complex combination of different fractions of oil. However, quality is estimated on percentage composition of alcohol, viz. citronellol, geraniol, linalol and furnasol. The alcohol varies with site, such as Kaunlake rose contains 72–75% of the alcohol of citronellol fractions being always more than geraniol in ratio of 1.2–1.2 : 1.0. Oil of rose *from Morocco* and *Iran* possesses the two alcohols in equal quantity. In the oils of *Crimean* rose (of USSR), *Rocen-tipetala* and some other forms citronellol ratio varies from 0.1 to 1.0, 0.5 to 1.0. The *Indian* rose oil (Aligarh type) flowers once in a year, compares favourably with Bulgarian rose, while the twice flowering form contains 21.3% of citronellol, 24–36% geraniol and 19.4% of nerol and is poor on odour.

***Plant improvement*:** A plant type blooming for a longer duration, bearing more bloom, containing higher percentage of oil and of good odour value is desired. Various methods viz. selection of individual plant with desired characteristics followed by vegetative propagation, reported screening of progeny and selection till new type material is evolved, are practised. In Bulgaria, two var. are reported, viz. *Iskara* and *Svejen*. These types give higher yield of flowers and better oil content. A 4–5 years old *Iskara* bush is estimated to give 11,560 kg of blossom and a *Svejen* bush 10,860 kg of blossom per ha. Oil percentage varies from 0.055 to 0.54, but the plant is susceptible to rust and black-spot diseases and cannot withstand long freezing temperatures.

Rosa damascana Mill; though known to be grown throughout the world, bears profuse flowers under the Mediterranean climate, yet comes up in J and K state along with Gazipur-Kannauj-Sikandrabad region, where it is also currently grown. Var. *trigintripetala* Dieck (2 n = 14) or thirty petalled rose, is common in Bulgaria which is also considered to be the home for 'rose oil'. This bush requires a long winter and can tolerate temperatures upto 20 °C. The bush blossoms during the later part of May and early June. A var. *bifera* bears a second bloom during Sept-Oct. and grows in India and China. Since such biennial flowering is not considered very healthy for the bush, it sets in early decline. In general, rose plant requires clear and sunny days during flowering with high temperature (25–28 °C), high humidity (80–85%) night temperatures ranging from 5–6 °C, it can be tried on a large scale in the state of Uttaranchal, Himachal and other places. In India, other oil-bearing species is *R. centipetala* having large-sized flowers with 100 petals, each weighing about 11 g, against 2–2.5 g; it is used for making *Gulkand* and rose-water. It is inferior in oil content, as compared to Bulgaria. The average oil content in the blossoms from Aligarh type material and that of Kazanlak rose of the same age is as below.

A rose called 'Ranisahiba' has been developed from the oil producing rose varieties found in Kannauj, UP, by the CIMAP and is a superior genotype of *damascana,* which has been patented in

US. This genotype has a light green stem and red purple flowers with a distinct 'globular canopy'. It is also reported to have higher oil content than other roses, has a low wax content, meaning that the oil does not freeze at room temperature during winter season. The type has revealed its higher stability and adaptation at Pantnagar (in subtropical plains) and Purara (in temperate hills).

Diseases and their treatment in *Damascana* rose are as follows:

1. *Dieback:* Diploidea-blight of young twigs. Nectrotic lesions on twigs extending downwards causing dry twigs. Adversely affect flower yield.

Control: Copper oxide (Blitox 0.3%) at cutends after pruning, if required apply fungicide at 15 days interval.

2. *Leaf spot*: Black spot on leaves (Alternaria, Phoma sp.) under high humidity. Spots enlarge, causing necrosis of leaves and heavy defoliation.

Control: Spray Chlorothalonil (0.3%).

References

Narayanswamy V and Biswas K. 1955. *Survey of Rose growing centres and rose industry in India*. CSIR, New Delhi.

Nigam MC, Singh KN, Gupta GM and Nigam CI. 1959. Studies on the cultivation of *Rosa damascana* in UP. *Indian Perfumer,* **8(2)** 76–80.

Singh LB and Deolia SK. 1963. Cultivation, extraction and economics of essential oil of rose. (*R. damascana* Mill. *R. bourboniana* Desf.) *Indian Perfumer.* **7(2).**

Singh LB and Sharma ML. 1969. Rose oil from flowers of *Rosa damascana* Mill. , raised on saline-alkali soil. Perfumerie and Kosmetik.

Singh LB. 1970. Utilization of saline-alkali soils without prior reclamation-*Rosa damascana*, its botany, cultivation and utilization. *Economic Botany*, **24(2)**: 175–179.

Srivastava HP and Chandra V. 1975. Effect of pruning at different heights on flower yield of *Rosa damascana* Mill. National Seminar. (Abstr). p. 13.

Himalaya Jaivsampada Pryodhiki Sansthan 2001. Damascus gulab (*Rosa damascana*) ki-Sugandh Utpadon Ke lie khéti; Mimeo pp. 1–6.

56. *Ricinus communis* L.

Family: Euphorbiaceae.

Local names: Arandi (Hindi), Eri, Bherénda (Beng.); Divéli (Guj); Eranda (Sansk). Amanakkam (Tam.). Erandamu (Tel), Avanakka (Malya.); The Castor oil plant (English).

The trade name is based on the common English name. In early literature, the tree has been referred to as *Chitravija, Panc-hangula* and *Vatri,* these refer to the shape and properties of the plant and its parts. *Chitravija* refers to o-mentations on the seeds. *Panchangula* to five nerved leaves, *Vatari* as the enemy of rheumatism (*Vata*).

Various varieties of this plant have been produced, which are of short duration and are suitable for drought-affected areas, these are called as *Bhatrendi* for the variety having white seeds, *Joggia*-rendi, for varieties having pale seeds, *Maru* suitable for desert and *aruna,* etc.

Description: A tall shrub, sometimes tree-like. Leaves very large, broad, roundish in outline but partly divided into 7, sometimes 9 lobes, margin toothed. Flowers large, in big terminal bunches. Fruit a prickly capsule, rather marked into six parts. Seeds oblong, seedcoat crustaceous. One of the forms is perennial with large red seeds; while other forms are annual.

Distribution: It is probably indigenous to Africa, but occurs throughout India, either cultivated or wild as escape. Seldom it is grown as a pure crop. It suffers from frost. Largely cultivated on borders of fields,

gardens and habitations in lower regions of Himalaya, found as weed on wastebland rubbish heaps. In Andhra cultivated as single crop.

Uses and extent of utilization and trade: The seeds and the oil from the seeds are used medicinally. Oil is used as purgative, when administered with milk or fruit juice. Grounded Kernel of seeds, when boiled with milk, can be taken as breakfast to cure rheumatic conditions and has been used and profitably tested by the author. The oil is also used in the ointments as a soothing agent and as an oil base in eyedrops. The castor oil is used in making contraceptive jellies and creams. The tribals in Bastar are reported to rub the leaves on joints to get relieve from pain. They also use young leaves as purgative. The leaves when rubbed with oil, relieve inflammation and pain of joints, when applied locally and tied with a cloth. It is also the remedy in local and rural households well known to people. The seeds contain 20–40% oil. The oil cake is poisonous but a good organic manure.

The drug is used extensively in trade and also for oil used in lubrication and illumination.

Agrotechniques for the Cultivation

It takes about 15 days for the seeds to germinate. Distance between the plant kept is about 150 cm and in case of annual variety 75 cm. The flowering starts from July to September. Flowering is irregular and the fruit ripens at different periods. The crop takes about 5–8 months to flower, after it has been sown in the field. Capsules are collected before they open out.

It is a long duration crop in India, of 240–270 days, and much cultivated on sandy soils with poor water-holding capacity, particularly in dry areas of Andhra.

Var. *aruna* has been recorded to flower in 35 days and the crop matures in 120 days. Different orders of the spike emerge in quick succession. Pollination and fertilization are effective, because staminate flowers are interspersed and not restricted to lowermost portion. It produces small and fewer leaves under scarce rainfall, thus reducing hazards of low transpiration. The yield potential is about 420 kg of seeds per ha.

57. *Saussurea costus* (Falc.) Lipsch. (Saussurea Iappa C.B. Clarke).

Family: Asteraceae (Compositae).

Local names: Kuth (Hindi), Kur (Beng.), Agada, Kushta (Sans.), Koshtam (Tam. and Tel.), Sepuddy (Mal.); The costus roots (English). The trade name 'costus roots' is based on the scientific and Hindi names. Sanskrit name *Kushta* has given it the present name Kuth.

Description: A tall perennial herb. Stem about 2–3 m long. Leaves large, basal ones upto 1.2 m long, on winged stalks; upper leaves smaller, sometimes with stalk, 2 small lobes at the base of these leaves, almost clasping the stem. Flowers bluish-purple or almost black, borne on round flower heads; few flower heads clustered together in the axil of leaves or at the top of stems. Pappus feathery, giving a curious fluffy appearance to the fruiting flower heads (Colour Plate 10.14). Fls in July-August. Attains maturity after 5 yrs. Stem dies back each year after rains and new shoots come up in following spring, though new shoots are produced even after 5 years, roots delay and turn hollow, hence it is best to dig roots after 5 year from 3rd year and onwards.

Distribution: Found at the subalpine levels, from Kashmir eastwards between 2500–4000 m altitude. The plant was introduced in Tehri Garhwal hills by the Maharaja of J&K at a very early stage which was a very successful introduction, but could not be sustained for various reasons. In Kashmir, it is naturally found on the upper reaches of Kishen Ganga and Kishtwar valleys and is sporadic in adjoining Bhandal, Bharmour in Chamba district of HP. Subsequently, it was domesticated and grown commercially in arid-temperate valleys of Pangi, Lahul and Spiti tehsils. A widely talked related species is *S. sacra* (Yogishpada-Sansk.), which is found on stony alpine regions, above 4500 m.

Uses and extent of utilization and trade: The Lamas call it, "Pachak" (thrice scared) and offer it to "Tri-Budha". *Sadhus* in Chamba district of HP drink extract of powdered roots, boiled in water, as a warm cordial drink during winters. Its main use is as an incense. Resinoids essential oil, alkaloid, insulin, a fixed oil and minor constituents like tannins and sugars are reported in roots. The roots contain an alkaloid *saussurine* used for medicine requirements. An essential oil (1.0–1.5%) is a disinfectant and antiseptic which possesses deep penetrating, characteristic fragrance; used as fixative in blending perfumes. Kuth roots are also used in rheumatism, scabies, itching and incurable sores. Forest areas, near man-made lakes and areas primarily used for grazing during summer, locally called *Kharaks*, are the suitable sites for the cultivation of this species. The roots in addition to alkaloid and essential oil, also contain alkaloid *Saussurine* (1.5%), a bitter resin (6%), tannin, inulin (18%), and a fixed oil.

The essential oil is antiseptic and disinfectant. The oil fetches high price, but its use as fixative and prohibitive cost of production, limit its extensive use in perfumery. Area under *Kuth,* in Lahul accounts for more than 600–700 acres of irrigated land; commercial plantations have been made in the south and south-eastern slopes of Ain Pukri (3480 m) in Chamba district and an experimental farm on northern slops of Kalatope near Dalhousie, was developed in early seventies.

A few other species of *Saussurea* are found in the Himalaya, but they have some restricted medicinal value.

1. *Saussurea obovalata* Wall. (Sthalkamala (Sansk), Sourajkamal (Tehri-Garhwal). The flowers are sacred by local people. The roots are applied on bruises and cuts.
2. *Saussurea heteromella* (D.Don) Hand-Mazz. (–S. scandens C.B. Clarke). An erect herb with cottony stem, branching near the top. In subtropical and subtemperate regions of Northwest India ascending 2500 m. The seeds are carminative and considered as a cure for horsebite.

Medicinal uses of S. costus

The dried roots are reported to be aphrodisiac, tonic, antiseptic, astringent, sedative and carminative. It is commonly prescribed in spasmodic diseases like bronchial asthma, cough and persistent hiccup, as a stomachic and tonic in advanced stages of typhous fever, as a stimulant in cholera, alterative in chronic diseases and rhmatism. Internally, it is given as infusion and externally the roots are powdered for treatment of ulcer and skin diseases. Powdered roots mixed with mustard oil are used for scalp in prurigo and as hairwash, are said to have property of turning grey hairs to black. Fumigatory plastills are made out of the root and also smoked in place of opium. It is a cardiac stimulant. It promotes urination and has the property of relaxing involuntary muscles. Drug is also

reported to be used in skin diseases. Tincture prepared from the defatted roots is beneficial in bronchitis.

Agrotechniques for the Cultivation

Soil and climate

It requires deep and porous, moist, well drained clay loam to sandy loam soils, that are rich in organic matter. Since roots are used, which are fleshy and large, depth of the soil is of great importance. In nature, it prefers open canopy from the trees and does not tolerate overhead shade of broad-leaved associates, found in nature. It avoids high rainfall regions but cherishes more snow than the rains. Deep rich forest soils at lower elevations (2600 m) with warmer summer days and longer growing period induces early flowering and faster vegetative growth, including formation of large thick roots.

Nursery techniques

It is a plant that bears profuse seeds during autumn season which are kept in cool dry places and retain viability for a year, after which the viability is lost. The seeds are small and light in weight; about 400 seeds weigh 10 g. Seeds are sown in the nursery, where the soil has been dug to a depth of about 50 cm and heavily manured with rotted leaf litter or FYM. Small sunken beds are made along the contour. About 1 kg of seeds are sown in 0.1 ha of land and can provide saplings for planting an ha. Seeds collected during the month of Sept. are sown in beds during November or April. Spring sowing gives better results when snow has melted completely. The seeds are either boradcast or sown in lines at a spacing of 30 cm × 6 cm, which starts germination in about 10 days and is more than 75%. The seedlings are ready for transplanting in the next 12 to 15 months. Nursery requires regular weeding and hoeing, with occasional irrigation when season continues as dry for long. Kuth plantation can also be raised from roots and root-shoot cutting. Healthy roots are cut into pieces of 3 to 4 cm length and 2 cm thick; thicker roots are longitudinally cut leaving peripheral portion intact, since it puts out new shoots. The cuttings are planted during the autumn season by putting lower side of cut root downwards. It takes about 10–30 days for the roots to sprout. The growing root stumps require more moisture and survival percentage is poor.

Land preparation for planting

The field is worked out to a depth of 30 cm and manured with well-rotted FYM or leaves. Pits are dug at a distance of 1 m to 1.5 m, of 30–45 cm^3 size. Seedlings or sprouted stumps are transplanted during July to August. Stumps (root-shoot cuttings), generally prepared in the nursery, consist of 25–45 cm of root portion and 3–2 cm of stem. Three weedings are done in the first and second years of plantation. In the third year two weedings are done, followed by one in the fourth and none in the fifth year. The transplants normally take time to start fresh growth and may remain dormant till the next growing season. The shoots die back during winters annually and the plants put forth new growth again after snow melts in the following year. The roots are taken out, when it is mature and fully grown.

Most suitable time for the root extraction is October/November. After being dug, the roots are washed and cut into short pieces, 8–10 cm long, and dried in the sun or on fire. Instead of wood, burning charcoal is recommended. If necessary, the roots are spread on perforated tin rakes, placed at a

distance, one over the other, and the firewood/charcoal is burnt. The dry roots are graded and packed for marketing. Freshly dug out roots contain 65–75% moisture. It could easily be stored for a few years in dry cool places.

Kuth plantation yields between 1 to 1.3 t/ha of dry roots, which is higher than from the plants grown under forestry conditions. Gupta (1973) estimated the plantation cost to be Rs 2400/ha giving a net return of Rs 1800/ (1973 level). However, on poor soils the yield may be poorer than on well-cared fertilized soil conditions. HAPPRC (2004) reported the root production at 1417 kg after 3 years of planting in the Himalayan region. Net profit is estimated at Rs 9312 per acre.

Reference

Gupta R. 1973. Kuth has many uses. *Indian Farming*. Sept issue. pp. 1–2.

Varma BS. 1951. Isolation of Costus oil from Costus roots. *Indian For.* **77**: 513.

HAPPRC. 2004. Cultivation of High Altitude Medicinal and Aromatic Plants. Bull. No. 12. Srinagar.

58. *Tagetes minuta* L.

Family: Asteraceae (Compositae).

Local names: Marigold (English); Gainde-ka phool, (Hindi).

Description: A common flowering plant having leaves resembling hemp and are diuretic. Leaves and flowers are medicinal (Colour Plate 10.15).

Distribution: The plant is extensively cultivated in northern parts of the country for the flowers. The small variety is normally grown in the hills of Kumaon and Garhwal. Cultivated mostly in South Africa, Brazil and Australia. Wild in India for a long time and recently being cultivated in Himachal Pradesh, J and K state and Uttaranchal industrially. In India, the oil production is about 10 tones per ha. CIMAP released and improved a variety called *Vanphool*.

Uses and extent of utilisation and trade: Leaves ground in water, strained and drunk expel along with urine, stone of bladder and kidneys, stop haemorrhage from piles. Also an antidote against poison of wasp sting, both externally and internally. It resolves swellings of female breasts in initial stages, when applied as a paste if suppuration has set in poultice of cooked leaves is applied. Ringworm and eczema is cured by application of juice of marigold flowers, earache is relieved by dropping 2–4 drops of lukewarm juice of leaves in ears.

Marigold flower is extensively used in northern part of the country and offered to the deity. However, it contains an oil which is largely used in perfumery and other industries.

Agrotechniques for the Cultivation

Climate and soil

Well-drained, sandy loam, sandy clay soils with pH 4.5–7.5; having good quantity of calcium in the soil are suitable of this crop. The subtropical and temperate climate in the plains and lower hills are suitable for the germination of the seeds and long summer days are required for plant growth. Low humidity and dry season help during harvesting.

Nursery practices and germination

The seeds are sown during the month of October in North Indian plains, while on the hills nursery is done during March-April. 10–15 cm long seedlings are sown in the soil. Well-rotten FYM is mixed with the soil for

nursery. After seeding, plants are irrigated from time to time. 2 kg of seeds per ha are sufficient for sowing which are mixed with the soil and put in lines or broadcast. For the plain areas 750 g seeds/ha are needed for planting. It is sown in lines 45 cm apart with a 30 cm distance from plant to plant is adopted. After 45 days weak saplings are uprooted. It is also essential to irrigate the field after transplanting the saplings. In the plains 6–7 and in hills 3–4 irrigations are given.

Field preparation

As the field is ploughed, 10–12 t/ha of FYM is mixed in the soil. 100 kg N, 60 kg P_2O_5 and 40 kg K_2O per hac. is also applied at the time of planting. Nitrogen is given in two equal split doses at the first weeding (after 30–40 days of planting) and the other a month after.

Harvesting is done, when the crop is fully mature, which is in the month of October in plains and March-April in hills. Plants are cut 30 cm above the ground. Ratoon crop can also be taken after the first cutting in the *tarai* areas.

Improved variety of marigold flowers gives about 200–300 qtls of herb which give about 50–55 kg of oil/ha. Distillation is done immediately after the crop cutting. Singh and Singh (1975) reported that combination of NPK significantly improved the herb and the oil yield. Maximum yield of herb (324.5 qtls) and essential oil (50.637 kg/ha) can be obtained by the application of 120–120–120 kg/ha of NPK fertilizers, which enhanced the herb and oil production to the extent of 73.6 and 55.7% as compared to the control.

Cost-benefit analysis of cultivation is given in Table 10.18.

Table 10.18: Cost-benefit ratio of cultivation (after Krishna and Singh, CIMAP 2001)

Expenditure / ha		Income / ha
1. Seeds	Rs 2000/	Oil/ha 50 kg (on an average).
2. Nursery	Rs 500/-	Cost of oil = 800 × 50 =Rs 40,000/-
3. Land preparation	Rs 1200/-	Net profit Rs 40,000/– Rs 12,500/-
4. Fert./ manure	Rs 2000/-	=Rs 27,500/ha.
5. Planting	Rs 1500/-	
6. Weed mgt.	Rs 1500/-	
7. Irrigation	Rs 300/-	
8. Cutting/ harvest.	Rs 1000/-	
9. Distillation	Rs 1500/-	
10. Other expenses	Rs 1000/-	
Total	12,500/ha	

References

Krishna Alok and Singh BR. 2001 *CIMAP symposum*. pp 22–23. (Mimeo)

Singh RS and Singh KK. 1975. Effect of combination of NPK on *Tagetes minuta* L. *National Symposium on recent advances* (abstr) pp. 3.

59. *Vanila planifolia* Andrews.

Family: Orchidaceae.

Local names: Vanila (English).

Description: A large, climbing orchid, bearing pod-like fruits, 1.0–2.5 cm in length and 8–14 mm in diameter. The 'vanila beans' of commerce are fully grown but unripe dried fruit are taken.

Distribution: Indigenous in southeast Mexico, Central and South America, but now cultivated in several countries of the old world. The main producing regions are Mexico, Madagascar, Mauritius, Seychelles Reunion, Comoron islands, Tahiti, Dominica, Puerto-Rica and Guadelope. The plant has

10

been successfully cultivated on an experimental scale in many parts of the country like Bengal, Bihar, Assam, Nigiri hills, Tamil Nadu and Karnataka states.

Uses and extent of utilization in trade: Inspite of the synthetic product available in the market, natural *vanila* is one of the world's favorite flavouring materials. It is used chiefly as flavouring agent in chocolates, ice-creams, confectionery, cakes, puddings, etc. One bean is considered sufficient to flavour about 0.7 kg of chocolate. It is also employed in scenting of soaps, blending of perfumes and in sachet powders. It has, of late, been used as a poison bate for pests like fruit flies, grasshoppers, etc.

Considering the importance of vanila in trade, the plant is still not extensively cultivated. It can safely be grown in most of the pepper-growing districts of Southern India and also in Bengal, Bihar Assam and Orissa states.

The characteristic fragrance of the plant is due to a substance called *Vanilin*. There are other substances known as *Balsam*, responsible for the distinct aroma of cured beans. Vanila tinctures are used in medicine and are obtained by a cold percolation of the beans and sucrose with dilute alcohol under specified conditions.

Agrotechniques for the Cultivation

Vanila flourishes in hot, moist tropical climate with well-distributed rainfall of about 2500 cm per year. Dry spells and strong winds are harmful and so are the extensive rains. It is most successful on islands of the coastal regions with typical island climate. The soils needed are light, porous, not subjected to excessive drainage, friable, rich in humus. Heavy clay soils with impeded drainage are not suitable. Gently sloping hillsides covered with trees are the suitable sites for vanila plantations. The support provided could be the living trees which also provide the shade.

The plantations are raised from the cuttings of about 1 m or more length. The vines have to be frequently pruned to induce flowering and to train them at convenient heights. The diseased plants and parasites are removed promptly; care has to be taken to keep a mulch on the soil, especially over the roots of the vines which are surface feeders. The plant comes into bearing in about two years time and remains productive for 7–8 years, giving maximum yields.

Vanila beans are picked by hand, as and when they mature, but while they are still unripe, i.e. when they are yellow in colour, develop hard black tip. With proper curing the beans develop a proper aroma. The process consists of heating the beans in sun or by hot water and sweating them under the blankets during the day. These are then packed in airtight blanket-lined boxes, in which they sweat all the night. The process of curing may last for several weaks; the quality depending on the efficiency of the process. Well-cured beans are oily and smooth, very pliable, black or dark brown in colour and aromatic in flavour. When the beans are slightly over-ripe at the time of harvesting, they split and lose some of their seeds during the process of curing. These are called splits in trade.

60. *Valeriana officinalis* L.

Family: Valerianaceae.

Local names: Billilotan, Jalakan, Badrangboya (Hindi), Kalavala (Mah).

Description: An erect perennial herb, root stock horizontal, thicker than the stem; producing suckers. Flowers white or dull white, in small clusters at the top of branches, often unisexual. Female and male on different plants. Calyx tube adnate to the ovary. Corolla funnel-shaped. Stamens 3, inserted on corolla tube. Ovary inferior. Fruit crowned with pappus-like calyx limb. (*V. jatamansi*).

Distribution: The plant occurs in the Western Himalayan region, particularly on high altitudes (2500 m) of Kashmir. It is also indigenous to North and Central Europe and North Asia. Cultivated on a commercial scale in Europe-Belgium, Holland, France, etc. Occurs wild in Europe. A new variety "Shipka" was released by Rose Res. Inst. Kazanlak (Bulgaria), having 20–25% more iso-valeric acid than the wild parents. Two other varieties, *Samokov* 39 and *Samokov* 40, were also released by the Plant Breeding and Research Institute at Sofia.

Though the Valerian has been grown as a biennial crop, new varieties yield well as annual crop.

Uses and extent of utilization and trade: The drug has depressant action on central nervous system and is used in the treatment of hysterical fits, flatulence and other nervous disorders. The fresh juice is used as narcotic in insomnia and in certain cardiac preparations. The properties of the drug are lost on drying. *Valerian oil* is also used as tonic and as stimulant; also as an adjunct in the flavouring of certain brands of tobacco.

Collection of the drug in the autumn months and slow drying of the raw drug are recommended.

Agrotechniques for the Cultivation

The plant requires a cool temperate climate and can be grown on a variety of soils; but thrives best on moist heavy loamy soils.

It is propagated from the rhizomes but seedings can also be done.

Rhizome cuttings are sown before the snowfall. The seeds germinate during the month of February and the crop is in the field for about 10 months. The fields are well prepared in early spring, before planting the seedlings and kept weed-free. FYM application at 30–35 t/ha is recommended. Also NPK at the rate of 40 : 45 : 60 kg per ha as supplement is also given, potash and phosphorus are applied before sowing, while nitrogen is applied after germination/sprouting, in two equal installments. First split dose is applied with the first irrigation and the second dose of nitrogen is given when the new sprouts come out. Spacing for sprouts is kept at 70 cm between rows. Shorter interspace is reported to yield more roots. Three intercultural operations are required, to keep plantations free of weeds. Liberal application of FYM, supplemented by fertilizer application naturally promotes growth of the rootstock.

It is advisable to harvest the plant after two years of planting, which is done during the month of September, when the leaves began to turn yellow. Rhizomes and roots are pulled during the dry weather. The rootstock goes to a depth of 15–18 cm. The rootstock is dried at a temperature of 40 °C. The harvested material is cut into small pieces after cleaning and drying. An yield of 2250 kg/ha has been reported. Var. *Shipka* gives about 10–12 tones of roots per ha, containing 0.35–0.44% of essential oil on dry matter basis.

The oil is collected by direct steam or water distillation but the distillation may be cohobated, as it affects the oil quality. The yield varies from 0.1 to 1.0%, but an yield of 0.4–0.6% is considered quite satisfactory.

No serious pests and diseases have so far been reported in this crop.

61. *Solanum* species

Family: Solanaceae.

Potentialities of the utilization of glyco-alkaloids found in the fruits of *S. khasianum,* on a commercial basis have been explained by Saini (1969) and it was recorded that in the first year of experimental cultivation, the material showed up its heterozygous nature. Based on growth habits, three distinct types were recorded. Yield of the glycoalkaloids content in fruits of *S. khasianum* reached a maximum value at certain stage of development. The maximum concentration occurred when the fruits were 50–55 days old. This stage of fruit growth coincided with its colour changing from green to yellow. Value of glycoalkaloid content (on dry weight basis) observed in various experiments is a much higher figure than any other Solanum species, so far recorded. The fruits of *Solanum khasianum* are a high yielding material and can form an alternative source for supply of this alkaloid, as a starting material for the synthesis of cortisone and steroid hormones on a commercial basis. A high value, 15% glyco-alkaloid content, reported from the fruits collected from the Nilgiris appear to show the possibility of obtaining higher yield. Since the alkaloid is situated in a particular layer, which can easily be exposed to the action of the solvent, it is not necessary to dry and powder the fruit material. The fresh fruits after opening under pressure are directly extracted repeatedly with hot 3% acetic acid through musclin cloth and crude glycoalkaloid is precipitated on basification. In view of the large scale availability of this plant, the method may be applied at site and the precipitate may be partially dried in sun, after concentration and bottled for onward processing. On cultivation it has been shown that comparatively in warmer conditions the plant put up fairly large amount of vegetative growth and flower number, fruiting is greatly hampered. Maximum temp. 30–31 °C is detrimental though overall number of flowers are reduced due to poor branching. On an average, dry matter of the fruit is made up of 80–90% and is made up of seeds, and 10–12% the pericarp, etc. Difference in dry weight of fruit development under warmer conditions are large due to poor seed setting.

Other species of Solanum used in medicine are as below.

a. *Solanum surattense* Burm.f. (S. xanthocarpum Schrad. et Wendl.)

Local names: Kateri, Ringni (Hindi), Bhoyaringani (Gujarat), Brihati-bangani (Orissa); Kantakári (Bengali, Canarese, Sanskrit), Kandiyari, Mokryan (Punj.), Laghukant Karwi (Tehri Garhwal); Kandanattari (Tam.), Challamulaga, Nelamulaka (Tel), Kantakari (Trade name). The trade name is based on the Indian name of the plant.

Description: A low diffuse herb, rough with scattered hairs. Stem prickly, pro-cumbent, branching. Leaves oblong, pinna-tifid, nerves armed with long straight yellow prickles. Flowers purple, solitary or in cymes, opposite the leaves.

Distribution: Throughout India, ascending to 2000 m in the hills. Commonly found on the wastelands and in open scrublands on dry sandy places.

Uses and extent of utilization and trade: The dried roots constitute the drug and are expectorant, used in cough, asthma, catarrhal fever and pain in the chest. Roots beaten up and mixed in wine, is given to check vomiting and is one of the commonest

ingredients of the famous drug *Dashmul*, an ayurvedic drug. It is diuretic and considered useful in concretions and stones in the bladder. The fruits are bitter and carminative. The fruits are useful in a number of diseases, such as sorethroat, bronchitis, muscular pains and fever. Bastar tribals, put the juice of crushed fruits into aching ears. Stem and leaves have been tested for any antifertility properties and did not show any promise as contraceptives.

It is widely used in Indian medicine and exploited on a commercial scale. The fruits yield *Solanine* which finds use as a starting material for the synthesis of cortisone and sex hormones.

The plant is found just as a weed and exploited from natural occurrence but not grown so far commercially as a crop.

b. *Solanum nigrum* L.

Local names: Black Night shade (English); Gewain, Nan-Gewain (Kumaon), Kayankothi (Jammu); Ghamai, Makoi (Hindi); Pilak (Punjab), Kachigida (Kanarese).

Description: An erect leafy annual herb, without any prickles. Leaves stalked, thin, ovate, waved or bluntly lobed with a few broad teeth. Flowers white, in axillary or lateral, drooping, umbellate cymes. Berries round, green, later turning to red or black or yellow, often found wild in nature.

Distribution: A weed of wastelands during the rainy season.

Uses and extent of utilization and trade: Infusion of the plant is useful in dysentery, stomach complaints and fevers. It promotes urination. Juice of the plant is useful in ulcers and other skin diseases. The fruits are tonic, oleaginous, bitter, laxative, diuretic and improve appetite and taste.

Useful in heart and eye diseases, leucoderma and itch. The juice is very effective in enlarged liver. Decoction of the plant is taken in jaundice, enlargement of liver and hepatic dropsy. A decoction of *Makoi, Kasaundi* and *Kasni* leaves, when boiled with a bit of naushadar, is sieved and taken to cure jaundice, enlargement of liver and other problems. The product has been tried on the author and was found very useful. Syrup from the herb is expectorant, diaphoretic and a cooling drink in fever. Not only internally, but externally the plant juice is used for local dressing as an anodyne, in nephrotic colic, severe burn, rheumatic and gouty joints. In decoction, it is used for washing inflamed, enema and painful parts. An enema of the infusion is given to patients having abdominal pains.

Rootbark is used in diseases of the ear, eye and nose and is good for ulcers, throat burning, inflammation of the liver and in chronic fever. The most used part is the leaves, which are externally used as poultice in ringworm, gout and earache. The juice mixed with vinegar is dropped in the eye. Plant made into vegetable, but without salt, is taken as a remedy in the swelling of body.

Dried berries are sold commercially in the market as it preserves honey and is reported to be useful in pulmonary tuberculosis fever, diarrhoea and hydrophobia. Juice is effective in enlarged liver. The berries contain solanine in combination with malic acid. No commercial cultivation has so far been tried. The requirements met from natural existence mainly during rainy season.

c. *Solanum dulcamara* L.

Local names: Bitter Night shade (English), Kakmachi (Sanskrit).

Uses and extent of utilization and trade: The stalks, fresh or dried, have bitter taste followed by remarkable sweetness, somewhat resembling liquorice. It produces symptoms of narcotic poison and is diuretic and diaphoretic. In large doses, it produces nausea and diarrhoea. Patients suffering from chronic rheumatism, gout, incipient phthisis and jaundice are benefitted by the use of this drug.

d. *Solanum aviculare* Forst. f.

A source of solasodine, forming the starting material for the synthesis of cortisone. In USSR, it showed encouraging results, when cultivated on an experimental basis in Kashmir (Dutt et al 1963). It is reported that an average plant in the second year yields 1 kg of dry leaves in 4 pluckings and an estimated yield of 4000 kg of dry leaves can be harvested per acre. Leaf sample shows 0.3% of alkaloid calculated as Solasodine.

The plant has been found susceptible to wilt and mosaic disease on cultivation. The alkaloid content is reduced in infected leaves.

Other species worthy of consideration, are *Solanum incanum* L. (Gaglibhata, Asind-Hindi); *Solanum indicum* L. (Tita bhekuri-Beng, Brihatika-Hindi). This plant is used as an expectorant and the juice of the leaves with fresh ginger is given locally to stop vomiting. Leaves and fruits rubbed with sugar are used as an external application to cure itch.

References

Bordoloi DN, Rabha LC and Ganguly D. 1975. Some observations on the growth and yield of *Solanum paciniatum* Ait. at conditions of Jorhat. *Proc. Nat. Sympocum.* (Abst).

Dutt AK, Kapoor LD and Sarin YK. 1963. Introduction of *Solanum aviculare* Forst. f. in Kashmir. *Indian J. Pharm.* **25(5)**: 160–61.

Saini AD. 1969. Physiological consideration of *S. khastanum* Clarke, as source of Solsodine. National Symp. (Abstr)

Shastry KSM, Thakur RN, Gupta JS and Pandotra VP. Expts on cultivation of *S. aviculare* Forst. under Jammu conditions. *CSIR symposium.*

Singh HS, Hazarika JN and Ganguli D. 1975. Studies on *Solanum khasianum* Clarke, as a source for Solasodine. *Proc. Nat. Symp* (Abstr) p2.

62. *Salvia moorcroftiana* Wall. ex Benth.

Family: Lamiaceae (Labiatae).

Local names: Clerey sage (English).

Description: An erect herb clothed with woolly hairs. Leaves thick, long-stalked, crenate or sharply toothed, upper surface cottony tomentose, lower white tomentose. Flowers pale-blue in distant whorls, bracts large.

Distribution: In temperate regions from Kashmir to Kumaon, between 2000–3000 m. An aromatic plant which was imported in 1978, from Bulgaria and farmed in Himachal and Uttaranchal hills.

Uses and extent of utilization: The roots are given in cough, while the leaves used a medicine for the 'guinea worms' and in form of poultice applied to wounds. The seeds are emetic and given in colic and dysentery, also applied to the boils. An aromatic plant and the oil obtained from flowers by distillation is used in perfumes.

The plant is exploited as drug on a very "limited scale", however, the oil is used extensively.

Agrotechniques for the Cultivation

Soils and climate

Sandy loam soils with pH ranging from 5.0–7.0, well-drained, are suitable for the cultivation. Temperate regions having day temperature ranging from 20°–30 °C and humidity from 60–70% are suitable.

Planting material and field preparation

The field is harrowed, 2–3 times, and levelled. Plots of 20 m × 1 m are demarcated having well laid out drainage. 1 cft of FYM, 300 g of DPAP and 200 g of potash, is well-mixed in the soil of the plot. During August-September, 500 g of seed per plot are sown by broadcast and covered with the mulch. The plots are irrigated till the seeds sprout and the mulch is removed. The saplings are ready after 40–50 days. These are transplanted in the month of October and November, after dipping in a solution of Bavastin, at a distance of 60 cm × 60 cm.

Two to three hoeings and weedings are done during the period March to May. The weeds can also be controlled chemically (Flavimetason 2 kg/ha before planting). About 5–6 irrigations are required. The fertilizer schedule during different periods (Table 10.19).

Table 10.19: Fertiliser schedule during different periods

Season	1st year			2nd to 4th year (in kg/ha)		
	N	P	K	N	P	K
Oct-November	20	80	80	20	60	40
March-April	30	...	...	30	...	...
May	40	...	...	40	...	...
August (After 1st cutting)	40	...	...	40	...	...
Total	130	80	80	130	60	40

The crop is of 3–4 yrs duration, fertilizer is applied at 5–6 cm depth and 10 cm away from the plant.

Cutting the crop

The flowers gets brown (catechu-coloured), when the first cut is taken during July and the second cut is taken in the month of September. Total flowers yield ranges from 6000–7000 kg per ha of flowers, when fresh and reported to yield from 18–21 kg oil on distillation. On cloudy days, oil percentage decreases. Distillation for the oil is done immediately after plucking the flowers.

Cost benefit and net income

Per ha cost for two years is given in Table 10.20 (based on data from Ashok K. Singh 2001, pp 36–39 (CIMAP).

Table 10.20: Cost and net income over two years

Items	1st year (in Rs/ha)	2nd year
Land preparation	2200/-	...
Planting material	800/-	...
Cost of planting	850/-	...
FYM and fertilizer	2660/-	2240/-
Irrigation	300/-	100/-
Hoeing and weeding	2500/-	2500/-
Plant protection	1000/-	1000/-
Cutting and storing	750/-	750/-
Distillation	2000/-	2000/-
Land on rent	2500/-	2500/-
Total	15,560	10,990
Weight of flowers (kg/ha) (0.25% of oil)	6000/- 15 kg	6000 15 kg
Total income (Rs) 3000 × 15	45,000	45,000
Expenses (Rs)	15,560	15,560
Net profit	Rs 29,440	Rs 34,010/ha

63. *Vetiveria zizanioides* (L.) Nash.

Family: Poaceae (Gramineae).

Local names: Khaskhás, Khas (Hindi), Vetiver grass (English).

Description: A densely tufted perennial grass, growing in stout clumps, about 2 m high. The rootstock is branched into a number of thin, spongy and aromatic roots (Colour Plate 10.16).

Distribution: The grass is found over a major part of the country, along river banks and marshy soils, particularly in Rajasthan, Haryana, Punjab, Gujarat, Maharashtra,

10

Bihar, Assam, Andhra Pradesh and Kerala states. Sometimes, it is not only cultivated as a crop but planted on the bund as graded bunds for soil and water conservation, has recently been promoted by the World Bank officials. It is also found in Africa, Myanmar, Sri Lanka and Southeast Asia. The grass being hardy, extends to tropical regions, tolerant to different soils, climate and geological conditions from 500 mm to 2000 m rainfall regions. Near Dehradun, it is found on forest edges and clearing, in tufts on sandy loam soils, water-logged areas and on alkaline soils.

***Uses and extent of utilization and trade*:** The grass is aromatic and gives an oil called the 'oil of vetiver' (*Khaskhas oil*). The roots are fragrant and used for making mats and hand fans. It is unique in having an outstanding feature of deep strong fibrous root system. It also acts to form a closing hedge, along contour and can act to filter both soil and water. The roots are also medicinal, used as tonic, refrigerant, stomachic, stimulant, antispasmodic, diaphoretic and diuretic, also used as emmenagogue.

Oil is used extensively in perfumery and cosmetics. It imparts long-lasting notes to perfumes, acting at the same time as natural fixative. It blends with the sandalwood, patchouli and rose oils.

More than 50% of the roots produced in India are used in the preparation of mats, etc. Principal producer of vetiver oil is Indonesia, Reunion and West Indies. It has also been estimated that more than 2900 metric tones of roots can be made available from different parts of the country. Studies of Nair and Chinnamma (1975) have shown that the difference from provenance is marked and the types collected from Nilambur in Kerala gave maximum root and oil yield. It is estimated that 100,000–1,10,000 tons of aromatic oil is produced in the world, India being on the 3rd position. The quality of the produced oil is also good.

Agrotechniques for the Cultivation

It can be cultivated on a variety of soils but flourishes best on rich sandy loams, though clay loams, saline sloppy and watery soils can also be used for its cultivation (pH ranging from 4.0–7.5) yet neutral to slightly alkaline soils should be preferred. It is usually propagated by roots, which are planted just before the rainy season and harvested 18–24 months after planting during dry season. 1–2 ploughings are sufficient and the field is prepared. Sowing is done in blocks called *kyaris* Area should be free of weeds. The fields are ploughed 2–3 times, deep enough for root penetration. Though the planting could be done from February to October, best time is during the rainy season from July to August on drylands and sloppy areas. The rooted slips are planted in lines, 60–75 cm apart. Plantations are made during the monsoon season. On *Usar* and sandy soils NPK fertilizers at 40 : 40 : 30 kg/ha are applied, before planting and next year on a rainy day 40 kg of N/ha is given as top dressing in the standing crop. FYM at 20 trollies/ha, if applied before planting is useful.

The plant growth is poor upto 60–70 days, thereafter 3–4 weedings, hoeing is done or Ultragen-10 at 500 kg AI/ha is used, which kills 60–75% of the weeds. Weeding can also be done by cutting the crop before flowering. Third cutting is done before digging the roots.

Root digging is done 15 months after planting, between November to February,

if done after February the oil quality is reduced.

Oil distillation takes about 12–14 hours. If the tanks are made of stainless steel it takes more time. Roots are dipped in water for 12–14 hrs and then cut, 2.5–5.0 cm long pieces, before distillation is done.

Improved varieties: Different varieties are available such as *Pusa Hybrid*-8, *Kanpur Hybrid*-40, *Sugandha and KS*–1.

CIMAP var Dharni-scented oil and fit for soil conservation.

Var. Gulabi is rose scented

Var. Kesri is saffron scented.

On an average, 10–15 qtls of root/ha are produced which on distillation give 15–30 kg of oil. Hybrid varieties give 8–10% more oil. The quantity of roots and oil is reduced from plants grown on different type of soils and are given in Table 10.21:

Table 10.21: Oil content on different soil from Vetiveria

Fertile loamy soil	45–50 qtls/ha	Oil 20–30 kg
Medium type of soils	35–40 qtls/ha	Oil 20–25 kg
Sandy soils	10–15 qtls/ha	Oil 8–10 kg
Submerged soils	15 qtls/ha	Oil 10–12 kg
Saline and problematic Soil	10–15 qtls/ha	Oil 8–10 kg

Pests

White Sundi/termites affect the crop adversely. Lindane at 125 kg/ha, is applied before the last ploughing in the field. Leafspot disease also adversely affects the crop, which can be controlled with the application of the fungicide. Roots also get affected by the rodents for which zinc phosphate may be used.

Table 10.22 gives economics of cultivating vetiver.

Table 10.22: Economics for cultivating Vetiver (after Lal and Singh 2001)

1. Cost of cultivation		
Field preparation at	Rs	1000/ha
Ploughing 4 h		400/-
Harrowing 2 × 2 ha		400/-
Levelling 2 ha		200/-
Cost-benefit for cultivation of Vertivaria zisanoides.		
2. Planting		
Cost of slips at 50 p/slip for 55,600.	Rs	27,800/-
Cost of planting 20 man days @ 70/day	Rs	4,00/-
3. Fertilizers (80 kg N, 40 kg P_2O_5, 40 kg K_2O		
Dap 86.96 kg at 320/50 kg	Rs	731/-
Urea 139.98 kg at Rs 420/20	Rs	574/-
Muriate of Potash 66-67 kg @ 185/50kg		247/-
Application cost. 4 man days		280/-
4. Intercultural operations		
5. Weedings-4 (30 × 2 man days) and gap filling.	Rs	8,400/-
6. Irrigation 4, at Rs 200/ha	Rs	800/-
7. Chopping/cutting 3 at 10 man days	Rs	2100/-
8. Plant protection: 1 spray; broadcasting	Rs	375/-
9. Roots digging-100 man days	Rs	7000/-
10. Distillation (fuel, chopping, etc.)	Rs	50,000/-
11. Interest on working capital at 15% for 15 months.	Rs	10,632/-
Total	Rs	67,339/-

Return

Available oil yield/ha	15 kg/ha
Oil price/kg. Rs 7000/-	Rs 1,05,000
Other dried biomass	5,000/-
Total cost	1,10,000/-
Total cost	67,339/-
Net income 1,10,000–67,339	Rs 42,660/-

References

Menon AK and Ihychan CT. 1945. Survey of Indian Vetiver (*Khas*) and its oil. CSIR. Essential oil Advisory Committee. N. Delhi.

Nair, FVG and Chinnamma NP. 1975. Studies on Vetiver. *Proc. Nat. Symp.* pp 10.

Raj Kishore Lal and Sandan Singh 2001. Khas-Kétél-ka-Utpadan hetu Unnat Pryodhki CIMAP. pp 18–19.

Rikshit JN and Dutt KB. 1947. *Khaskhas* (Vetiver) oil. *Indian Soap, Journ.* **15.**

64. *Withania somnifera* Dunal.

Family: Solanaceae.

Local names: Ashvagandhá (Hindi), Asan, Godha-Asor, Santhiana-popda (Guj). Hiren-addinegida (Kash), Amukkiram (Mal).

The trade name *Ashwagandha* is based on the local name Nagauri Asgandh.

Description: A small or middle-sized undershrub, upto 1.5 m high. Stem and branches covered with minute, star-shaped hairs. Leaves ovate, hairy-like the, branches, upto 10 cm long. Flowers pale green, small about 1 cm long, few flowers borne together in short axillary clusters. Fruit about 6 mm in diam., globose, smooth, enclosed in the inflated and membranous calyx.

Distribution: Throughout India, particularly in Maharashtra, Gujarat, Madhya Pradesh, Bengal, Rajasthan, Haryana, West UP, Karnataka, Kerala and in the Himalaya upto a height of 1500 m.

Common on wastelands and fallow lands but also cultivated. Also in Spain, Morocco, Jordon, Egypt, East Africa, SriLanka and Pakistan. In MP, being cultivated at Mandsor district, Bharatpur Mansa and Neemush tehsils in about 3000 ha. In Rajasthan at Kota, Jodhpur, Jaipur and Mehsana in Gujarat. About 4000 ha in Punjab and Haryana. Annual demand is about 7000 t. Dry wasteland soils can be used for the cultivation of this plant. It can be grown in pot near the houses.

Uses and extent of utilization and trade: Roots, leaves and seeds of this plant are used as drug, however, the dried roots constitute the main part used which are tonic and used for arthritis, skin diseases, problem of lungs, ulcers and many other complaints. Roots are given in fever. It is diuretic, acts as narcotic and removes functional obstruction of body. Useful in consumption, sexual and general weakness. Antibacterial and antibiotic properties have been experimentally confirmed.

Sometimes *Ashwagandha* of commerce is a tuber of *Convolvulus* which though much smaller and different in habit, does not appear to differ botanically from *Ipomoea digitata*. In Bombay, seeds of this species are employed to coagulate milk in the same way as *Withania coagulans*, which is a small herb (Akri-Hindi, Khamjira) which is used in liver complaints. The fruit has the property of coagulating milk.

Agrotechniques for the Cultivation

Late *kharif* is the season for the cultivation of this plant. Dry climate and moist soils are essential. Light black, sandy loam or red soils (pH 7–8), well-drained are needed. Salty, flooded soils are not suitable. It can be sown as a dryland crop in regions having annual rainfall of 650–700 mm per year and with 1–2 rainy days during the *rabi* season it helps to get good crop. Humid and shady soils are not suitable. Withania seeds can be purchased from Mandsor Agriculture College or Mansa Tehsil farms in MP at Rs 30–40/per kg.

Normally, it can be cultivated on unirrigated areas where other crops are not possible to be taken up and are not economic to grow. A single ploughing is done to turn the soil, during the hot season. It is followed by 2–3 ploughings, followed by levelling. 10–12 kg of seeds per ha are needed. Sowing can be done, either as broadcast or line-sowing and is covered with soil not more than 1 cm deep. The seeds sprout in

8–10 days. Thining is done, keeping 60 plants per m² or sqm 600,000 plants per ha. The line to line distance kept is generally 20–25 cm, and plant to plant distance is 8–10 cm.

The seedlings can also be transplanted, if nursery is available. Nursery can be raised in July, where seeds raised in 500 m² with 5 kg of seeds. 5-8 weeks old seeds gives plantable seedlings. Planting could be done at a spacing of 60 cm × 60 cm in the afternoon, followed by irrigation which could be done in late monsoon season (August). The seeds can also be treated with Diathane M-45, Therum Dalton 3 g/kg of seeds. In the normal course fertilizers are not used. However, 15 kg N and 15 kg P_2O_5 are recommended. Spacing of seedlings is done after 20–25 days of sowing. Second operation of hoeing and weeding is done, two months after the sowing.

In the normal course no irrigation is needed during the rainy season and if needed it is done as per the requirements.

Common diseases are the decay of seeds, leaf damping off, blight (seedlings), dye back and leaf blight. Virus attack can also be seen in the crop. The seeds are treated with Diathane M-45 at 3 g/kg of seeds. 30-days old crop can be sprayed with Diathane 3 g/lit of water, at an interval of 7–10 days. For caterpillar attack Rogor or Nuwan 0.6% is sprinkled, 2–3 times or 0.5% Melathion plus 0.1–0.3% Kalthene is sprinkled at 10–15 days interval.

Digging of roots, their drying and storage

Flowering time for the crop is December, when leaves get yellow, the crop is dug during January to March. The crop is ready in 100–150 days. The entire plant is taken out, along with the roots when the soil is damp. Roots can also be collected after top cutting and ploughing the field. The roots are cut into pieces and dried. About 7–10 cm cuttings can be made and sun shade dried, till the water content is 10–12%, and then graded as below.

A grade: Solid white and shining root with length 7 cm dia. 1 to 1.5 cm.

B grade: Root solid, shining, length 5 cm, dia. 1 cm.

C grade: Roots 3–4 cm long, dia upto 1 cm.

Low grade: Small thin roots of yellow colour protected from mould and fungi.

Production: Normally, 3–4 qtls of root per ha is the produce, which may go upto 6–7 qtls per ha. Apart from roots, seed production is about 50 kg/ha. The roots of 6–15 mm diam, and length 7–10 cm contain 0.13–0.31% alkaloid.

Net profit with a production of 5 qtls/ha roots at the rate of Rs 5000/ qtl (Rs 25,000) plus 50 kg seeds at Rs 50/kg (total Rs 2500) is Rs 27,500/ha. The cost of production being Rs 9000/ha (as per the details): Land preparation Rs 1600. ha, seed sowing and treatment Rs 1500/ha, Irrigation Rs 500/ha; Thining, hoeing and weeding Rs 1500/, Protection Rs 500/ha; Root digging Rs 2500/ha and others Rs 500/-). The net profit is Rs 14,000/ha (after Hari Om and Tomar VKS. 2001 CIMAP. pp. 54–57).

Reference

Ray SS, Sahoo, S and Das R. 2002. *In vitro* shoot proliferation and propagation of Aswagandha (*Withania somnifera*) from germinated seedlings. Nat. Acad. Sci. (Biology section) 72 and session. p.27. Abstract of papers.

65. *Zingiber officinale* Roscoe.

Family: Zingiberaceae.

Local names: Adrak, Ginsen (Hindi), Ginger (English).

10

Description: An important spice crop, considered to be native of Southeast Asia. Cultivated from ancient times in Asia. Arabs disseminated the knowledge about this plant in Europe.

Distribution: India is the largest producer of dry ginger in the world, accounting for more than half of the total world production. Other important ginger producing countries are Taiwan, Jamaica Sierra Leone, Australia, Mauritius, Nigeria, etc. Though ginger is cultivated in all the states of the country, Kerala accounts for more than 60% of the total production. Other states where ginger is cultivated are Himachal, West Bengal, Orissa, Uttaranchal, Karnataka, TamilNadu, Andhra and Gujarat, etc.

Uses and extent of utilization and trade : Though the rhizomes are used as spice, dried rhizomes called 'Sonth'are also used medicinally as powder for indigestion, cold and in arthritis. Decoction of ginger with tea is taken usually during winter months to cure nose-running and for curing sore throat. Produce locally are sold in the market as dry or fresh ginger rhizomes and there are big *mandies*, where produce from hills is exported throughout the region. Fibreless ginger is usually preferred and is sold at a premium price.

Agrotechniques for the Cultivation

Ginger grows best in humid and warm climates and prefers light soil, rich in organic matter. The crop needs plenty of manure (70–100 t/ha). It is cultivated from almost sea level to an altitude of 1500 m. The crop requires heavy and well-distributed rainfall. In low or moderate rainfall areas, it needs good irrigation and for a successful crop, moderate rainfall at the time of sowing till sprouting of rhizomes, fairly heavy showers during growing period, and dry weather before harvesting are necessary. It thrives well on wide range of soils, like sandy or clay loam, red loam or lateritic loam soils. A feriable loam rich in humus is considered to be the ideal.

Planting material

The rhizomes are used for planting and the seed-rhizomes are treated with 0.25%. Ceresan wet solution for 30 minutes, drained and stored in a cool dry place or in earthen pits plastered with mud. A layer of paddy husk may be provided before storing the rhizome. A number of varieties are available.

For dry ginger-Maran, Wynad, Mananthody, lppampady Valluvanad, Ernad, and Kuru.

For green ginger-Rio de Janerio (preferable for the extraction of oleoresin), China, Wynad local or Tafengiya. Varieties, like Nadia, Poona, Wynad, Kunnamangalam, Burdwan, etc. have been tried at Kasargod and at Dehradun and gave good results.

Land preparation

The land is ploughed thoroughly (5–8 times) to bring the soil to fine tilth, and beds of 1 m wide, 15 cm high and convenient length are prepared at an interspace of 10 cm in between beds. In case of irrigated crop, ridges are formed 40 cm apart. Early sowing with receipt of good summer showers ensures good yield. In Northern India, the earliest ginger in the market is in August and is sown in wheat fields. Seed rhizomes are cut into pieces, each having one or two buds and weighing about 15 g. Seed pieces are planted in rows, 20–25 cm apart, at a distance of 15–20 cm within the row and covered with soil. Seed rate per ha is 1,800 kg. After sowing the rhizomes are covered with dry leaves of *Sal* in areas where it is available.

10

Manuring and mulching

Well decomposed compost at 20–25 t/ha is applied and is broadcast over the beds or applied in pits or seed rhizomes pits before covering. Fertilizers at 75 kg N, 50 kg P_2O_5 and 50 kg K_2O/ha are applied; as usual half potash and phosphate applied at the time of sowing and half of N at 40 days after planting and the balance N and K_2O three months after planting. The beds are earthened up after each top dressing, with the fertilizers. Two weedings are generally given, the first weeding just before the second mulching and repeated, depending upon intensity of weed growth. If necessary, weeding is repeated third time also. The crop is ready in about 6–8 months. Rhizome yield depends upon the locality and crop condition, and taken out from soil when the leaves turn yellow.

Rotation and mixed cropping

Ginger is commonly rotated with other crops, such as tapioca, chillies, dry paddy and gingelly in Kerala in rainfed areas and with *ragi,* ground nut, maize and vegetables in irrigated areas. In Karnataka, ginger is cultivated mixed with ragi, red gram and castor. It is also grown as an intercrop in coconut, arecanut, coffee and orange plantations.

Harvesting

The clumps are carefully lifted with a spade or digging fork and the adhering soil removed. Average yield per ha ranges from 15 to 30 tonnes. For green ginger it is harvested earlier and the rhizomes are washed thoroughly so as to remove soil and dirt. For preparing dry ginger, the skin is peeled off from the rhizome and the produce, after cutting into small pieces, is kept soaked in water overnight. The rhizomes are then rubbed well between the palms of hand

and after cleaning removed from water and dried in sun uniformly for a week. Last bit of the skin is removed by rubbing dried rhizomes together. Yield of dry ginger is 16–25% of the green ginger. For seed, the rhizomes are stored in pits, under shade and rhizomes harvested in December are preserved for about 4 months till April-May. The rhizomes are treated with 0.25% Cerasan (250 g in 100 lit water) for 30 minutes and placed under shade till the solution has dried. If the seed rhizomes are infected, these are treated with 0.05% malathion/dimethoate, and stored in pits of convenient size (10–15 cm high), covered with a wooden plank and leaving a gap of 10 cm between the seed and the plant. In some areas rhizomes are loosely heaped over a layer of sand or paddy husk and covered with dry leaves.

Plant protection measures

Pests: Shoot borer causes damage to rhizome and pseudostems. Controlled by spraying 0.05% Dimethoate/Phosphamidon or soil application of granular insecticide.

Leaf roller

Caterpillar can be controlled by spraying 0.10% Carbaryl or 0.05% Dimetholate. Phosphamidon. *Scales* suck up rhizome and can be controlled by dipping seed rhizome in 0.05% Malathion Dimetholated at the time of planting.

Soft rot is the most destructive disease and can be controlled by dipping rhizomes 0.25% Cerasen for 30 minutes. *Leafspot* disease can be controlled by spraying 1% Bordeaux mixture of 0.2% Thiram.

Processing of ginger

Since India is the largest producer of ginger in the world, the bulk of it is marketed in

the whole raw condition or in dried (usually sundried) form. A little part of the production is processed for the extraction of oil oleoresin. Ginger is also preserved in the form of "murabbas" in many places. As discussed, dry ginger is prepared from the green underground shoots or rhizomes and only dry ginger is exported, which is prepared by peeling off the outer skin and then drying in sun for about a week. The dry ginger so prepared is called *unbleached ginger*. Another type the *bleached ginger* is prepared by peeling off the outer skin, dipping the peeled ginger in a solution of milk of lime and then drying it in sun. There are wide fluctuations in the price pattern of commercial dry ginger, depending upon the season, place of origin and method of processing.

1. *Raw material* selection is important, since the time of harvest of ginger is very important not only from yield point of view but also from quality considerations as influenced by oil and fibre contents. In case of dry ginger, the optimum maturity at harvest will depend upon its one use, whether for powdering or for oil extraction. For powdering, the fibre content should be low, hence to be harvested fairly early. Since tender ginger has a very thin skin it will possess a lighter shade of colour.

2. *Peeling*: The skin being the almost impervious barrier for the transmission of moisture, partial peeling of rhizomes is an essential prerequisite for satisfactory dehydration of ginger. Complete peeling can be done only by hand, since the size and shape of the ginger rhizome is of odd size. In some countries like Jamaica, ginger is peeled by immersing for one minute in boiling water, when the skin comes off easily.

However, on Indian varieties this method has negative effect and it is not possible to peel by this method. Partial peeling can be done through an abrasive peeler, such as is used for potato, beetroot, carot, etc. Both the loss of volatile oil and the time of drying, fully hand scrapped ginger approximate closely the sample abrasive peeled for 60 seconds corresponding to a peeling loss of 10–20%.

In certain countries, like Australia, rotary drier is the preferred type of dehydrator, for ginger compensates for the high labour cost in some countries. However, in countries where labour is not so costly, the tray and tunnel type of drier should be satisfactory. The critical temperature of dehydration of ginger is 60 °C. If higher temperatures are attained during dehydration, darker colour is unattractive for use. Since higher dry bulb temperatures upto 80 °C has no adverse effect on the volatile content of the dried ginger, higher temperatures can be used where the material is meant for oil extraction. It may not also be necessary to peel ginger and it can be more advantageously sliced.

The drying time of ginger (medium size) abrasive peeled for 60 seconds at a tray load of 1.0 kg. sq. ft. and dry bulb temperature of 60 °C is 24 hrs in cross flow drier and 14 hrs in through flow drier. The lower drying time in the through flow drier is premium drier. The over drying time in the through flow drier is presumably due to more intimate contact of material and the drying air.

Packaging studies with sliced and ground ginger in different types of multi-wall paper bags over a 5-month-period have shown that there is a 50% loss in volatile oil in case of

ground ginger, whereas there is practically no change in the oleoresin content in the case of sliced samples. It is, therefore, preferable to store dried ginger in the form of slices than powder, until use.

Details regarding the equipment, capital investment and production cost are available with the Central Food Technological Research Institute, Mysore, Department of Industrial development, and consultancy services.

References

Central Plantation Crops. Research. Inst. 1976. *Package of Practices for Ginger,* Pamphlet No 9 E. (2nd Review edn.) pp.17.

CFTRI. Process information sheet. Mimeo pp. 14 Process: Manufacture of dehydrated Ginger.

Hira Lal and Gupta RK 1988. Gingar-a rainfed cash crop for hill farmers. *Indian Farming Digest* **21(5)**: 17–20.

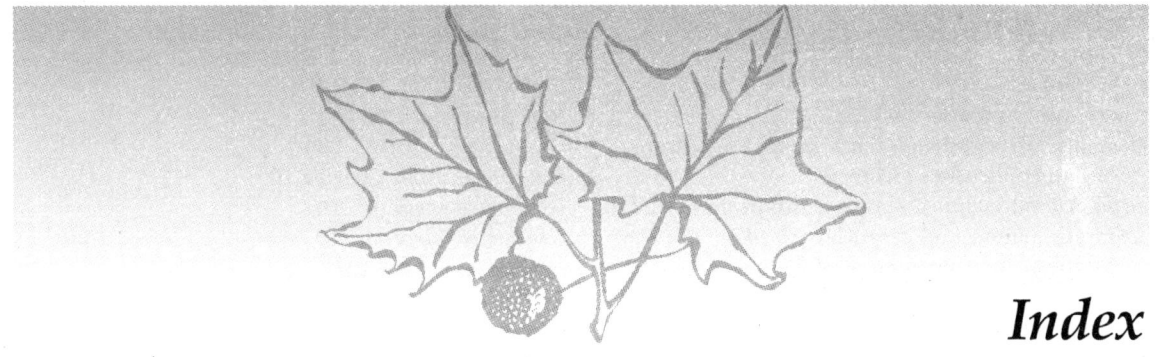

Index

❑❑❑